Tobias **Neckel**
Florian **Rupp**

Random Differential Equations in Scientific Computing

VERSITA

Versita Discipline: Mathematics

Published by Versita, Versita Ltd, 78 York Street, London W1H 1DP, Great Britain.

ISBN (paperback): 978-83-7656-024-3

ISBN (hardcover): 978-83-7656-025-0

ISBN (for electronic copy): 978-83-7656-026-7

Managing Editor: Aleksandra Nowacka-Leverton

Language Editor: Nick Rogers

Cover illustration: © Florian Rupp and Tobias Neckel

www.versita.com

To our beloved parents

Rosemarie and Hans-Georg

&

Erika and Heinrich

Preface

It is interesting thus to follow the intellectual truths of analysis in the phenomena of nature. This correspondence, of which the system of the world will offer us numerous examples, makes one of the greatest charms attached to mathematical speculations.

PIERRE-SIMON LAPLACE
(1749-1827)

Mathematical discoveries, small or great are never born of spontaneous generation. They always presuppose a soil seeded with preliminary knowledge and well prepared by labour, both conscious and subconscious.

HENRI POINCARE
(1854-1912)

This book on the theory and simulation of random differential equations came into being as the result of our lecture "Dynamical Systems & Scientific Computing – Introduction to the Theory & Simulation of Random Differential Equations" during the summer term 2012. This novel interdisciplinary way to cover dynamical systems and scientific computing brought Master students into contact with cutting edge research and was awarded the Ernst-Otto Fischer prize for innovative and trend-setting teaching paradigms by the department of Computer Science of the Technische Universität München.

Figure 1 provides a short overview of the pieces of this lecture and their fit. We will discuss this course and in particular the workshop on random differential equations and their application to ground motion excited multi-storey buildings in chapters 17 and 18. These hands-on lecture notes serve as the theoretical foundation for our lecture and the workshop. MATLAB commands[1] blend theory with application, and provide a solid foundation of the principles of deterministic ordinary differential equations and their numerics.

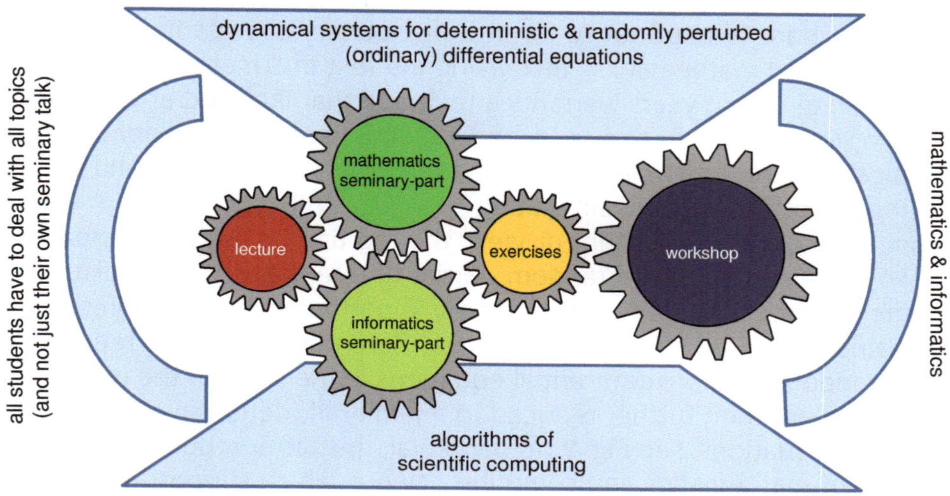

Figure 1. Design of the lecture "Dynamical Systems & Scientific Computing – Introduction to the Theory & Simulation of Random Differential Equations" combining the workshop with seminar contributions, exercises, and few lectures to increase student-centred learning effects.

Concerning random differential equations there is just a limited amount of literature discussing them on a graduate level, and despite their significance in scientific and engineering applications there is only one book in German by Helga Bunke [45] from 1972 and another one in Polish by Dobieslaw Bobrowski [30] from 1987 totally dedicated to the theory of random differential equations. T. T. Soong's monograph [233] from 1973 dedicates about half of his work to these equations. Arnold Kistner's PhD thesis [157] from 1978 (in German, too) covers essential aspects of linear random (ordinary) differential equations. There are a couple of recent papers and the one or other book chapter on the numerics of random differential equations. However, a holistic approach is missing in particular taking into account recent results of Ludwig Arnold on random dynamical systems, cf. [11].

This book is a holistic and self-contained treatment of the analysis and numerics of random differential equations from a problem-centred point of

[1] Note that the MATLAB examples presented throughout this book are mainly meant to illustrate certain individual aspects in a compact manner; hence, these examples do not represent a "nice" implementation from a software engineering point of view (mostly, we skipped documenting comments for the sake of a compact representation, e.g.).

view. We take an interdisciplinary approach by considering state-of-the-art concepts of both dynamical systems and scientific computing. Our intended audiences are those of beginning graduate/ master level courses on theory and numerics of stochastically perturbed differential equations. The areas covered here are of importance for interdisciplinary courses in informatics, engineering and mathematics. Increasing interest in "Uncertainty Quantification" during recent years warrants a textbook that is aimed at a new generation of researchers in this field, and which is rooted in the principles of dynamical systems and scientific computing. This will foster a solid understanding of both theory and simulation.

From a methodological point of view, the red line pervading this book is the two-fold reduction of a random partial differential equation disturbed by some external force as present in many important applications in science and engineering. First, the random partial differential equation is reduced to a set of random ordinary differential equations in the spirit of the method of lines. These are then further reduced to a family of (deterministic) ordinary differential equations, see Fig. 2. In particular, this latter reduction step and the fields of mathematics and computer science which support it, form the basis of this exposition.

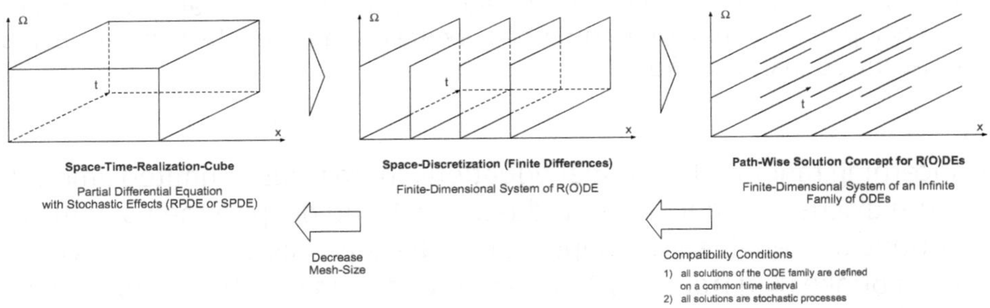

Figure 2. Reduction from a given continuum mechanical random partial differential equation to a family of deterministic ordinary differential equations.

Hereby, our main example is the motion of multi-storey buildings subject to stochastic ground motion excitations. The (simplified) buildings are either modeled as solids by standard assumptions of continuum mechanics (and their corresponding partial differential equation description) or wireframe-structures based on deterministic oscillators. The external forcing is due to a linear filtered white noise that describes the earth's surface as a layer between the ground surface and the nearest bedrock where the source of an earthquake is located and treats the wave propagation in this layer as being one-dimensional and vertical. The corresponding stochastic models are known as the Kanai-Tajimi filter or the Clough-Penzien filter.

From a didactical point of view, we put much effort into providing the deterministic foundation of the theory and simulation of ordinary differential equations as well as that of random variables and stochastic processes in order to give a self-contained presentation. Every chapter begins with a list of key concepts and key questions the reader should keep in mind while studying the contents of the respective chapter. Moreover, quite uniquely for a mathematics text book, every sub-chapter ends with a set of quizzes in the type of oral exam questions, allowing the knowledge obtained to be consolidated quickly, and to enable a successful self-study of the materials covered.

Outline of the Chapters

Figure 3 sketches the rough outline of this book and focuses on randomly perturbed phenomena in science and engineering, their mathematical analysis and effective as well as efficient numerical simulation. In contrast to the "classical" bottom-up textbook approach, we follow an application oriented top-down procedure, and proceed from discussions of complex applications to simpler known concepts for the following reason: This allows us to start with the complete picture and introduce the reader to applications, numerics and general theory quickly. Thus, in part I, we proceed from random partial differential equations (RPDEs), to random ordinary differential equations (RODEs) and then finally to ordinary differential equations (ODEs). During the lecture on which the book is based, we saw that the students struggled with the new concept of randomized ODEs at first and actually required knowledge on stochastic ordinary differential equations (SODEs) and other solution concepts to fully place RODEs into their body of knowledge and to fully appreciate the stochastic concepts. After the discussion of the "complete picture" we continue, in parts II and III, with a recap in the classical way, because of the interdisciplinary background of the intended readership, we believe that this is necessary in order to give a self contained representation. In particular, the chapters are such that they may be skipped by those readers familiar with the corresponding concepts. The main part of the exposition proceeds in part IV with a discussion of those RODEs that can be treated more or less easily: the linear ones. Here, the existence and uniqueness results are based on the general theorems provided in part I. Finally, RODEs and simulations together with their evaluation are joined in the workshop part V.

In particular, the single chapters of this book contain the following specific information:

Part I serves as an introduction to the modelling of randomly perturbed phenomena in science and engineering by random partial differential equations and their reduction to random ordinary differential equations. Here, we discuss the following aspects:

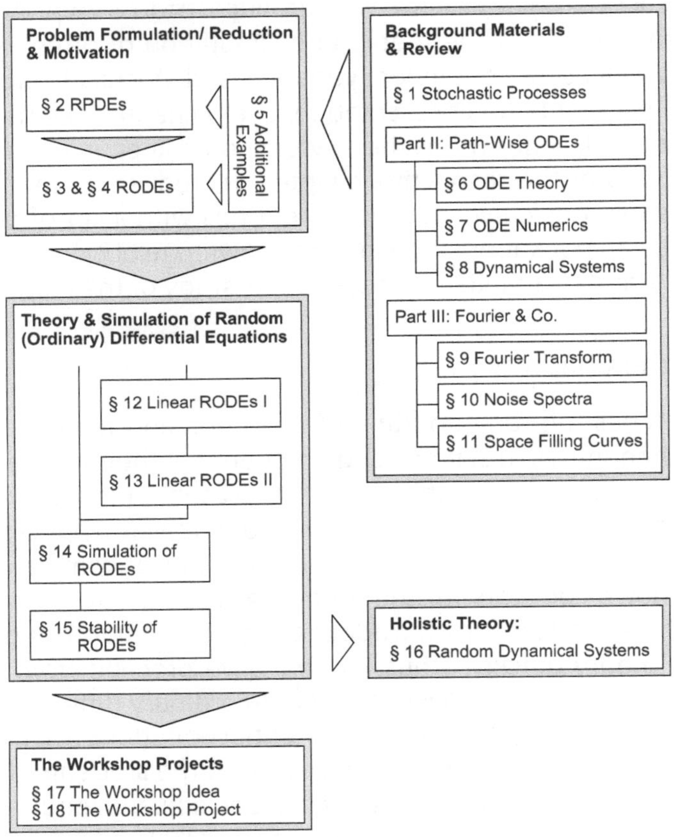

Figure 3. Outline of the book from the point of view of theory and simulation of random (partial/ ordinary) differential equations.

Chapter 1 provides a friendly review of the central concepts of probability theory focusing on random variables and their properties that eventually lead to the notion of a stochastic process. Our aim is to recall the basic definitions and equip them with tailored illustrations and MATLAB commands rather than emphasize the most general and abstract mathematical concepts.

Chapter 2 discusses how specific Random Partial Differential Equations are transformed to Random Ordinary Differential Equations by applying classical spatial discretisations. Variants from a variety of applications, are discussed, leaving the time discretisation for Chap. 7. The derivation of the underlying system of (deterministic) partial differential equations and corresponding boundary conditions is shown for the example of elastic body motion. We discuss different types of meshes with an emphasis on regular Cartesian grids. The three main spatial discretisations—finite differences, finite volumes, and finite elements—are briefly explained before delving deeper into finite difference schemes (FD). We derive the corresponding FD approximations for the fundamental equations of elastic body motion and simulate steady-state scenarios of buildings which are bent.

Chapter 3 motivates and mathematically rigorously discusses existence and uniqueness of path-wise solutions of random (ordinary) differential equations. We start by modelling external and ground motion excitations by means of stochastic processes which motivates the study of random (ordinary) differential equations. Their solution, existence and uniqueness concepts are then discussed together with the correspondence between stochastic and random (ordinary) differential equations. In particular, we study the conditions that lead to the existence of path-wise unique solutions. Solutions in the extended sense are analysed as well as the dependence of solutions on parameters and initial conditions. As an excursion we finally give the equations of motion for single- and multi-storey (wireframe) buildings. Our main source for the set-up and discussion of random differential equations is Helga Bunke's book [45].

Chapter 4 adds the notions of $\mathbb{P}$- and mean-square solutions to our discussion. The special nature of random (ordinary) differential equations often requires additional refined solution concepts going beyond that of a path-wise solution. Taking, for instance, into account that a solution may fulfill the given random differential equation with probability one, or that the solution is a square integrable stochastic process leads to the notion of a $\mathbb{P}$-solution or a mean-square solution, respectively. Their properties and interconnections, in particular with respect to path-wise solutions, are studied here.

Chapter 5 widens the scope to additional categories of applications for random differential equations. In particular, flow problems are discussed in more detail. These problems represent an important class of applications in computational science and engineering. The various possible random effects in the model, the geometry, the boundary conditions, and the parameters may be generalised to other flow scenarios, involving coupled scenarios such as fluid-structure interaction or biofilm growth.

Part II is a mini-course on the interplay between dynamical systems and scientific computing in itself. Here, we cover the analytical and numerical foundations of deterministic ordinary differential equations. A special emphasis is given to dynamical systems theory including essential phase space structures (equilibria, periodic orbits, invariant sets) as well as fundamental tools (Lyapunov exponents and Lyapunov functions).

Chapter 6 serves as a holistic introduction to the theory of ordinary differential equations (without singularities). After some preliminaries, integral curves in vector fields are discussed, i.e., ordinary differential equations $\dot{x} = F(t, x)$. Hereby, we start with continuous right hand sides F and their ε-approximate solutions as well as the Peano-Cauchy existence theorem and its implications. We continue our discussion for Lipschitz-continuous functions F and the existence and uniqueness theorem of Picard-Lindelöf. In particular, we analyse maximal integral curves, give the three types of maximal integral curves that can occur in autonomous systems and show the transformation of a d-th order equation into a first order system. Next, we deal with the existence of solutions in the extended sense where the right hand side function may be continuous except for a set of Lebesgue-measure zero. Caratheodory's existence theorem and its implications are studied together with maximum and minimum solutions. Then, we study the broad class of linear ordinary differential equations by discussing the unique existence of their solutions and their explicit construction. Applications of the theory focus on first integrals and oscillations for the deterministic pendulum and the Volterra-Lotka system. Finally, we provide a first glance into the existence, uniqueness and extension of solutions of ordinary differential equations on infinite-dimensional Banach spaces.

Chapter 7 contains the relevant aspects of the numerical simulation of ordinary differential equations. Classical explicit one-step methods such as the explicit Euler or Runge-Kutta schemes are presented before motivating implicit approaches for stiff ODEs. A variety of example implementations show the behaviour of the different schemes applied to different initial value problems. The brief discussion of the Newmark family of schemes and of symplectic methods widens the scope of this chapter to approaches that are typically neglected but that provide useful features worth being on the radar in the context of RODE simulations.

Chapter 8 provides a brief review on deterministic dynamical systems. Fundamental notions and concepts are introduced, like that of (continuous) dynamical systems, long-time behavior, invariance and attraction. This paves the way to analyze stability in the sense of Lyapunov by utilizing Lyapunov-functions for proving (asymptotic) stability in non-linear systems. Next, we analyze the correspondence between the stability properties of non-linear systems and their linearisation. Here, we give the famous theorem of Hartman and Grobman, a classification of equilibria in planar systems with respect to their stability properties as well as the techniques for the determination of the position of Lyapunov exponents of a linear system, like the Routh-Hurwitz criterion or the Lozinskii-measure method.

Part III covers important concepts and algorithms in Scientific Computing: the discrete Fourier transform and its variants, the frequency domain method for response analysis, as well as space-filling curves as paradigms for effective and efficient data storage.

Chapter 9 discusses the basic aspects of the continuous and the discrete Fourier transform, with the focus on the latter including various MATLAB examples. The famous Fast Fourier Transform is derived. We briefly present the trigonometric variants of the discrete Fourier transform related to symmetry properties of the underlying input data. These trigonometric transforms allow us to realise fast Poisson solvers on Cartesian grids which are needed in the workshop problem (cf. Chap. 18). Frequency domain aspects and the Fourier transform are essential to understand key characteristics of stochastic processes (spectrum, power spectrum) and the propagation of excitations through mechanical structures.

Chapter 10 starts with the basic definitions and implications related to the spectral representation of stationary and periodic stochastic processes. Based on these, we study the notions of energy, power and spectral density. We give several examples for colored noise processes, the frequency domain method for response analysis, and linear filters. In particular, we apply this method to our problem of multi-storey excitation due to seismic impacts and their propagation through wireframe structures.

Chapter 11 introduces the fundamental concepts, definitions, and properties of space-filling curves such as the Hilbert and Peano curves. We briefly present three different categories of possible applications motivating the usage of these special curves in the context of computational simulations. Two variants for the construction of (discrete iterations of) the curves are explained in detail such that the reader is in the position to use space-filling curves for a tangible tasks like ordering Cartesian mesh cells. Here, strong connections to spacial discretisation (cf. Chap. 2) and its efficient implementation are provided.

Part IV is devoted to a more in depth study of the theory and simulation of random (ordinary) differential equations. It analyses the theory of linear random differential equations. Numerical schemes for (non-linear) random differential equations, like the the averaged Euler and Heun method are discussed. Stability of the null-solution is considered and Lyapunov-type methods are applied to the various concepts of stochastic stability. Finally, the recent theory of random dynamical systems and its impacts on the study of random (ordinary) differential equations is presented.

Chapter 12 treats linear inhomogeneous ordinary random differential equations of the type $\dot{X}_t = A(t)X_t + Z_t$ where the randomness is located just in the inhomogeneous driving process Z_t. These types of equations can be analysed in ways analogous to their deterministic counterparts already exhibiting a wealth of interesting phenomena. Of importance are the stochastic characteristics of the solutions process as well as periodic and stationary solution types. In particular, we give first stability conditions with respect to which solutions converge towards periodic or stationary ones.

Chapter 13 extends this body of knowledge on linear random ordinary differential equations by also allowing stochastic effects in the coefficients. We give the general solution formulas for these types of equations together with equivalence result for path-wise and mean-square solutions. Moreover, on the one hand, we analyse the asymptotic properties of path-wise solutions focusing on (exponential) decay towards the null-solution as well as on upper bounds for path-wise solutions. On the other hand, we also study the properties of the moments of path-wise solutions with respect to the (exponential) decay as well as the existence of asymptotically θ-periodic solutions. As an excursion, the general solution formula of linear non-commutative path-wise continuous noise systems is constructed.

Chapter 14 discusses all relevant aspects for simulation of path-wise RODE problems. We present lower-order explicit RODE schemes (Euler and Heun) as well as higher-order K-RODE Taylor schemes. Detailed information on the corresponding MATLAB implementation for the wireframe model are given and numerical results show the validity of the approach.

Chapter 15 studies the various notions of stability of the null solution of a random (ordinary) differential equation with a focus on path-wise equi-stability, h-, $\mathbb{P}$-, and W-stability. In particular, the relations/ implications and inter-connections between these concepts are discussed and the results of Chap. 13 on the path-wise stability of linear random differential equations with stochastic coefficients are re-framed in the context of these concepts. Moreover, we extend the deterministic Lyapunov method to random differential equations. Based on suitable Lyapunov-functions, necessary conditions for h-stability and path-wise equi-stability are given. Finally, the stability of deterministic systems subject to different classes of continuously acting random perturbations is analysed.

Chapter 16 provides a glimpse into the very recent theory of random dynamical systems. We give the fundamental definitions of metric, measurable and random dynamical systems together with some illustrative examples. Moreover, we study the notions of forward and backwards stability and their implications.

Part V gives the problem set of the workshop associated to the course we gave in the summer term 2012 together with some key results and lessons learnt from this experiment in higher education.

Chapter 17 focuses on the didactic aspects of the workshop. We discuss the integration of workshop as a central part of the complete course. Details on the design of the workshop are presented covering in particular the concept of a virtual software company, the choice of the environment, and the team role descriptions.

Chapter 18 contains the project specification used in the workshop. We present a selection of example results which our students produced at the end of the project. Finally, we summarise the lessons learnt—both from the point of view of the participants and the supervisors—providing interesting hints for future or similar projects.

Acknowledgments

A number of chapters of this book have been read and criticized in manuscript. In alphabetical order, we would like to acknowledge the support and comments of Michael Bader, Peter Gamnitzer, Miriam Mehl, Philipp Neumann, Horst Osberger, Alfredo Parra, Benjamin Peherstorfer Christoph Riesinger, Konrad Waldherr, and last but not least Jonathan Zinsl.

We thank our student assistants Andreas Hauptmann, Veronika Ostler, and Alexander Wietek for their support in preparing the course. Furthermore, we thank Thorsten Knott, who thoroughly designed the tasks in the context of solving the Poisson equation via continuous Fourier transform in Sec. 9.6.1, as well as Michael Bader for providing basic course material w.r.t. the Fourier transform and, in particular, the space-filling curves easing the development of the corresponding chapters. Peter Gamnitzer contributed via fruitful discussions and valuable hints in the context of structural dynamics in Chap. 2, a help that is gratefully acknowledged. Special thanks go to Alfredo Parra Hinojosa for various contributions, in particular concerning the simulation routines for random ordinary differential equations in Chap. 14.

We are very grateful to Versita's publishing team and our editors Aleksandra Nowacka-Leverton, Marcin Marciniak, and Grzegorz Pastuszak for realising this book in the open access format.

Garching bei München,
June 26, 2013

Tobias Neckel and Florian Rupp

Contents

Part I

Motivation and Decomposition of Multi-Storey Building Excitation Problems

He who seeks for methods without having a definite problem in mind seeks for the most part in vein.

DAVID HILBERT *(1864 - 1943)*

Chapter 1

Recap: Random Variables & Stochastic Processes

This chapter provides a friendly review of the central concepts of probability theory focusing on random variables and their properties that eventually lead to the notion of a stochastic process. Our aim is to recall the basic definitions and equip them with tailored illustrations and MATLAB commands rather than emphasize the most general and abstract mathematical concepts.

1.1 Key Concepts

This chapter sums up the material on *continuous random variables* and *stochastic processes* suitable for an undergraduate/ beginning graduate lecture, see [62, 25, 61, 122, 14]. The blending of MATLAB commands into the text is motivated by [79], [153], and [177].

As illustrative introductory examples, we motivate the concepts of (i) convergence of random variables that will be essential to set-up stochastic stability as well as (ii) of ergodicity of stochastic processes.

Example 1.1 (Convergence of Random Variables, cf. [225], p. 148). Let $\{X_i\}_{i=1}^n$ be a sequence of normally distributed random variables with vanishing mean and variance i^{-1}, i.e. $X_i \sim \mathcal{N}(0, i^{-1})$. Figure 1.1 displays the (cumulative) distribution functions of the first elements of this sequence. Based on this figure, it seems as if $\lim_{i\to\infty} X_i = X$ with the limiting random variable $X \sim \text{PointMass}(0)$.

Though, $\mathbb{P}(X_i = X) = 0$ for any i, since $X \sim \text{PointMass}(0)$ is a discrete random variable with exactly one outcome and $X_i \sim \mathcal{N}(0, i^{-1})$ is a continuous random variable for any $i \in \mathbb{N}$. In other words, a continuous random variable, such as X_i, has vanishing probability of realising any single real number in its support.

Thus, we need more sophisticated notions of convergence for sequences of random variables.

Example 1.2 (Ergodic & Non-Ergodic Stochastic Processes). Figure 1.2 shows some sample paths of a parameter-dependent stochastic process X_t, called "geometric Brownian motion", cf. [222]. In particular,

$$X_t = X_0 \exp\left(\left(a - \tfrac{1}{2}b^2\right) t + bW_t\right) ,$$

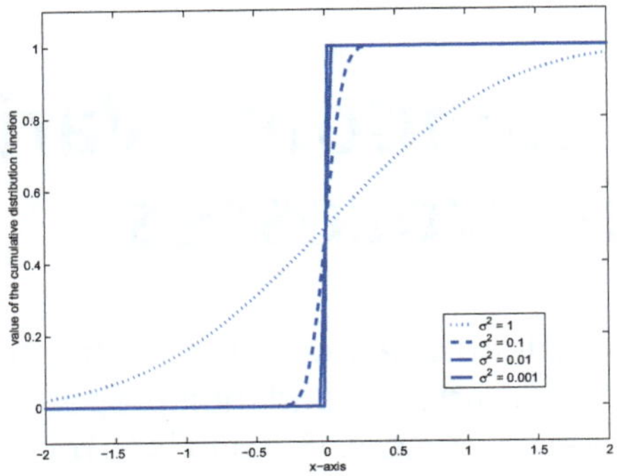

Figure 1.1. Cumulative distribution functions of several normally distributed random variables $\mathcal{N}(\mu, \sigma^2)$ with $\mu = 0$ and $\sigma^2 = 1, 10^{-1}, 10^{-2}, 10^{-3}$.

with initial value X_0 at time $t = 0$, $a, b \in \mathbb{R}$, and a (standard) Wiener process W_t (we will discuss this fundamental stochastic process in Chap. 3.2.1). The geometric Brownian motion is, for instance, used to model stock prices in the famous Black-Scholes model and is the most widely used model of stock price behavior.

With some deeper understanding of the properties of the geometric Brownian motion we see on the one hand that the expected value follows the deterministic exponential function $\mathbb{E}(X_t) = \exp(at)$. On the other hand, though, its path's converge to zero for all $a < \frac{1}{2}b^2$.

I.e., for $a \in (0, \frac{1}{2}b^2)$ we are in the paradoxical situation that the expectation of the process diverges and all its samples converge to zero-solution.

Processes for which statistical properties, like the expected value, can be derived from the sample paths are called "ergodic". In Fig. 1.2 (a) and (b), sample averages will provide an excellent estimator for the expected value. This is not the case in the non-ergodic case displayed in Fig. 1.2 (c) and (d).

When reading this chapter note the answers to the following questions

1. What is a (real valued) stochastic process over a probability space?
2. What does expectation and variance of a random variable or a stochastic process tell us?
3. What does conditional expectation of a random variable or a stochastic process mean?
4. Which concepts describe the convergence of one random variable towards another?
5. What does the Borel-Cantelli lemma state?

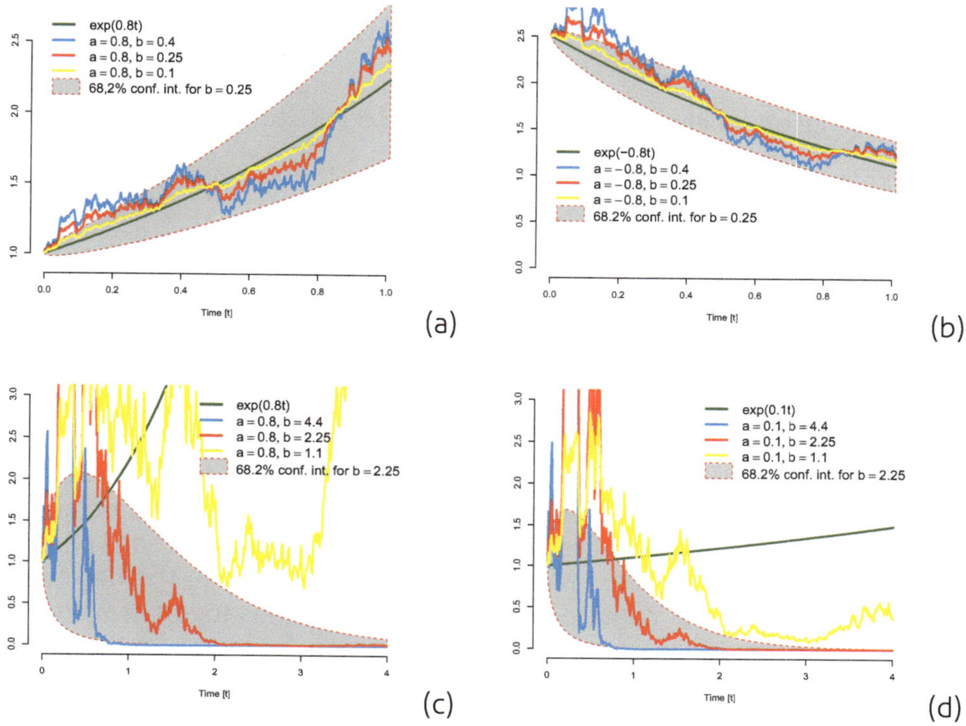

Figure 1.2. Paths of the Geometric Brownian Motion for $a = 0.8$, $X_0 = 1$ (a) $a = -0.8$, $X_0 = 2$ (b), $a = 0.8$, $X_0 = 1$ (c), and $a = 0.1$, $X_0 = 1$. In (a) and (b) the values of b are the same ($b = 0.4, 0.25, 0.1$) and the 68.2% confidence interval for the $b = 0.25$ paths is shown as well as the expectation value $X_0 \exp(at)$. In (c) and (d) the values of b are the same ($b = 4.4, 2.25, 1.1$) and the 68.2% confidence interval for the $b = 2.25$ paths is shown as well as the expectation value $X_0 \exp(at)$.

6. Under what conditions are two stochastic processes indistinguishable?
7. What are the characteristics of Gaussian processes?
8. How can we utilize MATLAB to simulate random variables, stochastic processes and their properties?

as well as the following key concepts

1. σ-algebras, probability measures and probability spaces,
2. Random variables as well as their density, distribution and moment functions,

3. Independence and conditional expectation,
4. Convergence in distribution, in probability, in the r-th mean, almost sure convergence and sure convergence,
5. Stochastic processes, their continuity and indistinguishability,
6. Kolmogorow's fundamental and continuity theory,
7. Martingales, super-martingales, and filtrations, and
8. Gaussian processes.

This chapter is structured as follows: In Sec. 1.2, we start with the fundamental concepts of random variables, generated σ-algebras and density functions. Next, Section 1.3, discusses moments of random variables like the expectation value and variance as well as integration with respect to probability measures. In Sec. 1.4, the essential concepts of independence of random variables and conditional probabilities and conditional expectation are studied. In particular, it is here that we give the various definitions of convergence of random variables. Moreover, in Sec. 1.5, we give the basic definitions and concepts of continuous stochastic processes together with a brief discussion of Gaussian processes. Finally, Section 1.6 wraps up the contents of this chapter.

Prerequisites: Some pre-knowledge on probability theory and stochastic processes are helpful.

Teaching Remarks: Though labeled chapter 1, we certainly do not suggest to start a course for beginning graduate students with the basics presented in this chapter as its contents are heavily loaded with technical definitions that are not very motivating for the student interested in applications. In view of our top-down approach, we assume the concepts of this chapter as prerequisites to be considered in a lecture after chapters 3 or 5 when required. For a lecture class, it seems to be most appropriate to give this chapter as a homework and discuss some relevant exercises together in the classroom.

1.2 Random Variables, Generated σ-Algebras and Density Functions

With respect to the key elements and notations of probability theory we start with the introduction of random variables and especially the σ-algebras they generate. Next, density and distribution functions will be discussed followed by the definition of (central) moments and moment generating functions. Finally, we define what we mean by integration with respect to a probability measure and give some useful inequalities.

1.2.1 Continuity, Measures and Probability Spaces

Let us first recall the definition of Hölder- and Lipschitz-continuity as well as that of $\mathcal{C}^{k,\alpha}$-functions, cf. for instance [3], p. 40:

Definition 1.3 (Hölder and Lipschitz Continuity/ $\mathcal{C}^{k,\alpha}$-functions). Let $(X, \|\cdot\|_X)$, $(Y, \|\cdot\|_Y)$ be normed spaces and $0 < \alpha \leq 1$. A function $f : X \to Y$ is called *globally Hölder continuous* of order α if there is a positive constant C such that

$$\|f(x) - f(y)\|_Y \leq C\|x - y\|_X^\alpha \qquad \forall x, y \in X\,. \tag{1.1}$$

f is called *locally Hölder continuous* of order α if it satisfies the condition (1.1) on every bounded subset of X. f is called *globally (or locally) Lipschitz continuous* if it is globally (or locally) Hölder continuous of order $\alpha = 1$. f is called a $\mathcal{C}^{k,\alpha}$-function if it is k times continuously differentiable and the k-th derivatives are locally Hölder continuous of order α for some $k \in \mathbb{N}$.

The central problem in measure theory is to find a measure/ volume for as many elements of the power set $\mathcal{P}(\mathbb{R}^d)$ as possible, such that this measure/ volume is additive, translation invariant and normalized. As there is no solution to define a measure/ volume for all elements of $\mathcal{P}(\mathbb{R}^d)$, we have to restrict ourselves to special sub-set systems:

Definition 1.4 (σ-Algebra). Let Ω be a nonempty set. A collection of sets $\mathcal{A} \subset \mathcal{P}(\Omega)$ is called *σ-algebra*, if

- $\mathcal{A}$ is a *algebra*, i.e.,
  - $\Omega \in \mathcal{A}$,
  - $A \in \mathcal{A} \Rightarrow A^c \in \mathcal{A}$ and $A, B \in \mathcal{A} \Rightarrow A \cup B \in \mathcal{A}$
- $\forall n \in \mathbb{N} : A_n \in \mathcal{A} \Rightarrow \cup_{n \in \mathbb{N}} A_n \in \mathcal{A}$

Trivial examples for σ-algebras are $\mathcal{A} = \{\emptyset, \Omega\}$ and $\mathcal{A} = \mathcal{P}(\Omega)$, moreover, for any $A \subset \Omega$ the σ-algebra properties of $\mathcal{A} = \{\emptyset, A, A^c, \Omega\}$ are easily verified. In particular, if $\mathcal{E}$ is a collection of subsets of Ω, then the smallest σ-algebra generated by $\mathcal{E}$, and denoted by $\sigma(\mathcal{E})$, is defined as

$$\sigma(\mathcal{E}) := \bigcap \{\mathcal{A} : \mathcal{E} \subset \mathcal{A} \text{ and } \mathcal{A} \text{ is a } \sigma\text{-algebra on } \Omega\}\,.$$

For instance, the smallest σ-algebra containing all open subsets of $\mathbb{R}^d$ is called the *Borel σ-algebra*, denoted by $\mathcal{B}^d$ or simply by $\mathcal{B}$ if the dimension d requires no specific mentioning.

Let Ω be a nonempty set and $\mathcal{E} \subset \mathcal{P}(\Omega)$. The set-system $\mathcal{E}$ is called *intersection-stable*, if

$$\forall E_1, E_2 \in \mathcal{E} \implies E_1 \cap E_2 \in \mathcal{E}\,.$$

Obviously, every σ-algebra is intersection-stable[1].

Let Ω be a nonempty set and $\mathcal{A}$ be a σ-algebra on Ω. The pair $(\Omega, \mathcal{A})$ is called *measurable space* and the elements of $\mathcal{A}$ are called *measurable sets*.

Definition 1.5 (Measurable Function). Let $(A, \mathcal{A})$ and $(B, \mathcal{B})$ be measurable spaces. A function $f : A \to B$ is called *$\mathcal{A}$-$\mathcal{B}$-measurable*, if $f^{-1}(\mathcal{B}) \subset \mathcal{A}$.

For instance, every continuous function $f : X \to Y$ between two metric (or topological) spaces X and Y is measurable.

Definition 1.6 (Measure and Probability Measure). Let Ω be a nonempty set and $\mathcal{A}$ be a σ-algebra on Ω. Then a set-function μ on $\mathcal{A}$ is called a *measure*, if

- $\mu(A) \in [0, \infty]$ for all $A \in \mathcal{A}$,
- $\mu(\emptyset) = 0$,
- μ is σ-additive, i.e., for any disjoint collection of sets $A_1, A_2, \dots \in \mathcal{A}$ with $\cup_{n\in\mathbb{N}} A_n \in \mathcal{A}$ it holds that

$$\mu\left(\bigcup_{n\in\mathbb{N}} A_n\right) = \sum_{n=1}^{\infty} \mu(A_n) .$$

Moreover, a measure μ is called a *probability measure* if it additionally satisfies

- $\mu(\Omega) = 1$.

A measure μ on a measurable space $(\Omega, \mathcal{F})$ is called *σ-finite*, if there exist $E_1, E_2, \dots \in \mathcal{F}$, pairwise disjoint, s. t. $\Omega = \cup_{n\in\mathbb{N}} E_n$ and $\mu(E_n) < \infty$ for all $n \in \mathbb{N}$. Moreover, for two measures μ, ν on a measurable space $(\Omega, \mathcal{F})$, the measure ν is called *absolutely continuous* with respect to μ, if every μ-nullset is a ν-nullset. The notation for this property is $\nu \ll \mu$.

If μ is a measure on the σ-algebra $\mathcal{A}$ of a measurable space $(\Omega, \mathcal{A})$, then the triplet $(\Omega, \mathcal{A}, \mu)$ is called *measure-space*. In particular:

Definition 1.7 (Probability Space). Let Ω be a nonempty set and $\mathcal{A}$ be a σ-algebra on Ω. The triplet $(\Omega, \mathcal{A}, \mathbb{P})$ is called *probability space*, if $\mathbb{P}$ is a probability measure on the measurable space $(\Omega, \mathcal{A})$.

Let $(\Omega, \mathcal{A}, \mathbb{P})$ be a probability space: points $\omega \in \Omega$ are usually addressed as *sample points* and a set $A \in \mathcal{A}$ is called *event*, hereby $\mathbb{P}(A)$ denotes the *probability* of the event A.

[1] In an extended course on measure theory, the property of set-systems to be intersection-stable motivates the discussion of Dynkin systems, see, e.g., [93], pp. 24.

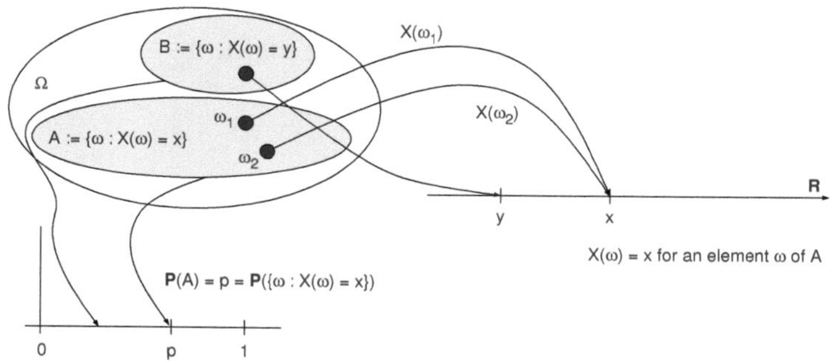

Figure 1.3. Relation between a random variable X and its probability function, following [62], p. 53.

A property which is true except for an event of probability zero is said to hold *almost surely* (abbreviated "a.s.") or *almost everywhere* (abbreviated "a.e.").

1.2.2 Random Variables and the σ-Algebras They Generate

Probabilities are a set-functions that assign a number between 0 and 1 to a set of points of the sample space Ω, cf. [62], pp. 53. Their domain is the *set of events* of a random experiment and their range is contained in the interval $[0, 1]$. A random variable is also a function whose range is a set of real numbers, but whose domain is the set of sample points $\omega \in \Omega$ making up the whole sample space Ω (not subsets of Ω), see Fig. 1.3.

Definition 1.8 (Random Variable). Let $(\Omega, \mathcal{A}, \mathbb{P})$ be a probability space. Then a function $X : \Omega \to \mathbb{R}^d$ is called *random variable*, if for each Borel-set $B \in \mathcal{B} \subset \mathbb{R}^d$

$$X^{-1}(B) \;=\; \{\omega \in \Omega : X(\omega) \in B\} \;\in\; \mathcal{A}\,.$$

I.e., a random variable is a $\mathbb{R}^d$-valued $\mathcal{A}$-measurable function on a probability space $(\Omega, \mathcal{A}, \mathbb{P})$.

We usually write X and not $X(\omega)$. This follows the custom within probability theory of mostly not displaying the dependence of random variables on the sample point $\omega \in \Omega$. We also denote $\mathbb{P}(X^{-1}(B))$ as $\mathbb{P}(X \in B)$, the probability that X is in $B \in \mathcal{B}$.

Example 1.9 (Indicator and Simple Functions are Random Variables). Let $A \in \mathcal{A}$. Then the *indicator function* of A,

$$\mathbb{I}_A(\omega) \;:=\; \begin{cases} 1 & \text{if } \omega \in A \\ 0 & \text{if } \omega \notin A \end{cases}$$

is a random variable.

More generally, if $A_1, A_2, \ldots, A_n \in \mathcal{A}$ are disjoint sets, such that $\Omega = \dot{\bigcup}_{i=1}^n A_i$ and $a_1, a_2, \ldots, a_m \in \mathbb{R}$, then

$$X = \sum_{i=1}^{n} a_i \mathbb{I}_{A_i}$$

is a random variable, called a *simple* or *elementary function*.

Example 1.10. Sums and products of random variables are themselves random variables, too.

MATLAB can generate distributed pseudo-random numbers between 0 and 1 can be generated utilizing the rand command. For instance, the code below generates a one row vector with five column entries the values of which are uniformly distributed on the interval $[1, 10]$:

```
r = 1 + (10-1).*rand(1,5);
r =
   7.8589    5.1082    1.1665    8.3927    5.0023
```

The MATLAB manual says: "MATLAB software initializes the random number generator at startup. The generator creates a sequence of [pseudo-]random numbers called the global stream. The rand function accesses the global stream and draws a set of numbers to create the output. This means that every time rand is called, the state of the global stream is changed and the output is different." Due to this algorithmic procedure, pseudo-random numbers are created that depend on the initial value/ seed of the generation algorithm. This initial value/ seed can be controlled to allow for a repetition of the generated sequence of pseudo-random numbers. The randn command generates normally distributed pseudo-random numbers. For instance,

```
randn('state',100)          % set the initial
r = 1 + 2.*randn(1,5);        % value/ seed of randn
r =
   2.8170   -3.4415    0.5219    1.1375   -3.0404

randn('state',100)          % reset the initial
r = 1 + 2.*randn(1,5);        % value/ seed of randn
r =
   2.8170   -3.4415    0.5219    1.1375   -3.0404
```

generates two identical one row vectors with five column entries the values of which are normally distributed with mean 1 and standard deviation 2. In both cases the same initial value/ seed was applied to MATLAB's random number generator.

Lemma 1.11 (σ-Algebras Generated by Random Variables). *Let $X : \Omega \to \mathbb{R}^d$ be a random variable on the probability space $(\Omega, \mathcal{A}, \mathbb{P})$. Then*

$$\mathcal{A}(X) \ := \ \{X^{-1}(B) : B \in \mathcal{B}\}$$

is a σ-algebra, called the σ-algebra generated by X. *This is the smallest sub-σ-algebra of $\mathcal{A}$ with respect to which X is measurable.*

Proof. It is easy to verify that $\{X^{-1}(B) : B \in \mathcal{B}\}$ is a σ-algebra, and, moreover, that it is indeed the smallest σ-algebra of $\mathcal{A}$ with respect to which X is measurable. □

It is essential to understand that, in probabilistic terms, the σ-algebra $\mathcal{A}(X)$ can be interpreted as "containing all relevant information" about the random variable X.

In particular if a random variable Y is a function of the random variable X, i.e., if $Y = \Psi(X)$, for some reasonable function Ψ, then Y is $\mathcal{A}(X)$-measurable. Conversely, suppose $Y : \Omega \to \mathbb{R}$ is $\mathcal{A}(X)$-measurable. Then, there exists a function Ψ such that $Y = \Psi(X)$. Hence, if Y is $\mathcal{A}(X)$-measurable, Y is in fact a function of X. Consequently, when we know the value $X(\omega)$, we in principle know also $Y(\omega) = \Psi(X(\omega))$, although we may have no practical way to construct Ψ.

1.2.3 Density and Distribution Functions

Our discussion of properties of random variables starts with the one-dimensional setting: A scalar random variable X is associated to the *(probability) density function* $X \mapsto f(X)$ defined by the property

$$\mathbb{P}(x_1 \le X \le x_2) \ = \ \int_{x_1}^{x_2} f(X) \mathrm{d}X$$

where $\mathbb{P}(x_1 \le X \le x_2)$ denotes the probability of the event $x_1 \le X \le x_2$.

The *(cumulative/ probability) distribution function* $F(X)$ is given by

$$x_0 \ \mapsto \ F_X(x_0) \ := \ \mathbb{P}(X \le x_0) \ = \ \int_{-\infty}^{x_0} f(X) \mathrm{d}X$$

I.e., the probability that x lies in the interval $(x_1; x_2]$ is $F_X(x_2) - F_X(x_1)$ with $x_1 < x_2$. Of course $F(\infty) = 1$, see Fig. 1.4.

The following theorem ensures the existence of a density:

Theorem 1.1 (Radon-Nikodym). *Let $(\Omega, \mathcal{F})$ be a measurable space and μ a σ-finite measure on $\mathcal{F}$. Moreover, let $\nu \ll \mu$ be a measure. Then ν has a density*

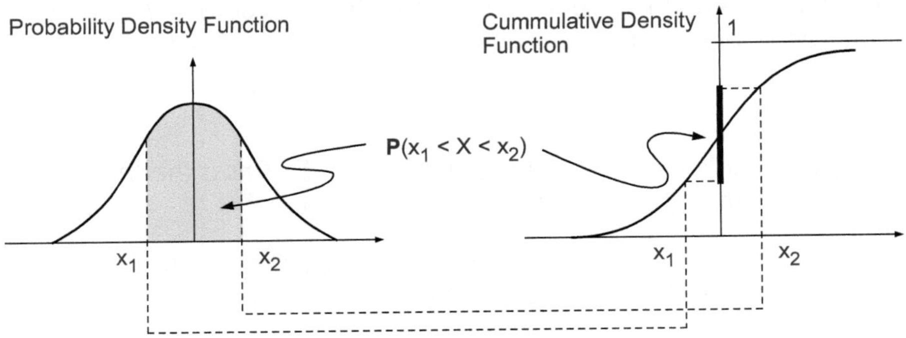

Figure 1.4. Sketch of a density and its corresponding distribution function. Shown is the connection between the probability $\mathbb{P}(x_1 < X < x_2)$ and the graphs of these two functions.

with respect to μ, i.e., there exists a measurable function $f : (\Omega, \mathcal{F}) \to ([0, \infty], \mathcal{B})$ such that

$$\textit{for all} \quad E \in \mathcal{F} \quad \textit{it holds that} \quad \nu(E) \;=\; \int_E f \, d\mu \,.$$

Proof. see [93], pp. 277 □

In applications one often has to deal with data samples of random variables and tries to gather some information about the density and distribution of these random variables. The following MATLAB Example takes, for convenience, 1000 sample points of a normally distributed random variable with mean 3 and variance 25, i.e. of a $\mathcal{N}(3, 25)$-random variable, and graphs its distribution curve using histograms (see Fig. 1.5 for a visualisation of the resulting plots).

MATLAB Example 1.1. plotting.m: Generating the images visualised in Fig. 1.5.

```
num = 1000;  x = 3 + 5.*randn(num,1);
figure(1); plot([1:length(x)], x)

[count bins] = hist(x,sqrt(num));
figure(2); hist(x,sqrt(num))
figure(3); plot(bins,count,'-b','LineWidth',2)

count_sum = cumsum(count);
figure(4); plot(bins,count_sum,'-b','LineWidth',2)
```

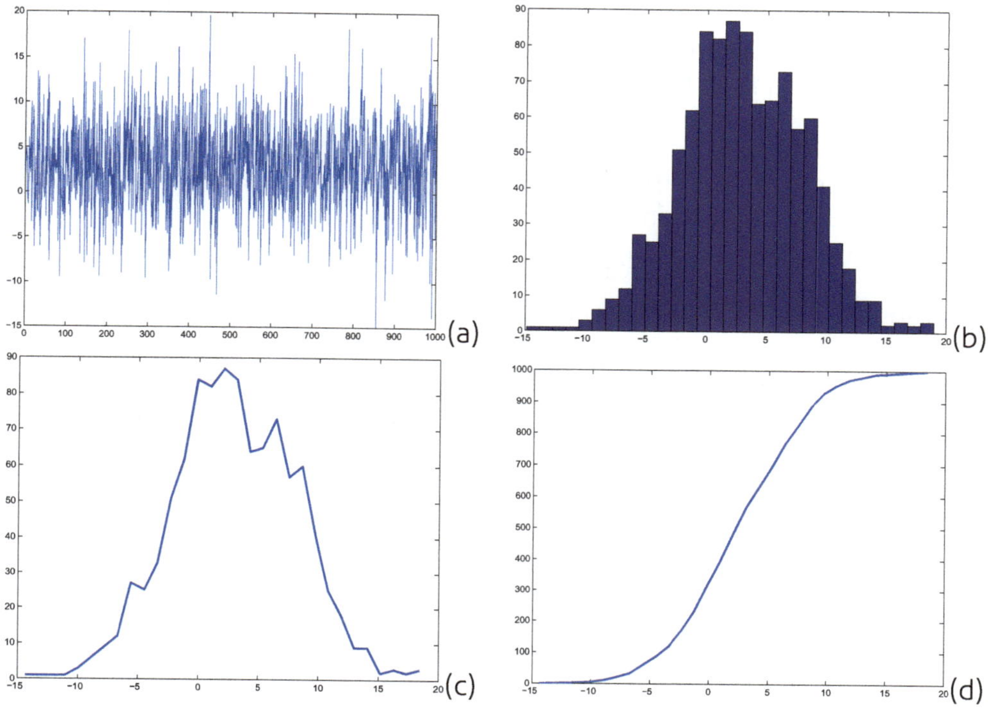

Figure 1.5. (a) 1000 sample points of a $\mathcal{N}(3, 25)$-random variable, (b) histogram, (c) empirical density and (d) empirical distribution of these points.

The density function of a random variable, the *joint (probability) density function* $f_{X,Y}(x, y)$ of two random variables X and Y is given analogously as

$$\mathbb{P}(x_1 \leq X \leq x_2, y_1 \leq Y \leq y_2) = \int_{x_1}^{x_2} \int_{y_1}^{y_2} f_{X,Y}(x, y)\mathrm{d}y\mathrm{d}x$$

with

$$f_{X,Y}(x, y) = f_{Y|X}(y|x)f_X(x) = f_{X|Y}(x|y)f_Y(y)\,,$$

where $f_{Y|X}(y|x)$ and $f_{X|Y}(x|y)$ are the *conditional densities* of Y given $X = x$ and of X given $Y = y$ respectively, and $f_X(x)$ and $f_Y(y)$ are the *marginal densities* for X and Y respectively[2]. In particular we have for the marginal densities:

$$f_X(x) = \int_{-\infty}^{\infty} f_{X,Y}(x, y)\mathrm{d}y \quad \text{and} \quad f_Y(y) = \int_{-\infty}^{\infty} f_{X,Y}(x, y)\mathrm{d}x\,.$$

2 E.g., the marginal density of X simply ignores all information of Y, and vice versa.

Two random variables are *independent* if the conditional probability distribution of either given the observed value of the other is the same as if the other's value had not been observed, e.g.

$$f_{Y|X}(y|x) = f_Y(y)$$

In particular, two random variables are independent, if their joint density is the product of the marginal densities:

$$f_{X,Y}(x,y) = f_X(x) f_Y(y) \,.$$

Moreover, for independent random variables X, Y it holds, that $\mathbb{E}(X \cdot Y) = \mathbb{E}(X) \cdot \mathbb{E}(Y)$, $\mathbb{V}ar(X+Y) = \mathbb{V}ar(X) + \mathbb{V}ar(Y)$ and $\mathbb{C}ov(X,Y) = 0$.

For our two random variables X and Y, their *joint distribution* is the distribution of the intersection of the events X and Y, that is, of both events X and Y occurring together. Consequently, the joint distribution function $F_{X,Y}(x,y)$ is given by

$$F_{X,Y}(x,y) = \mathbb{P}(X \le x, Y \le y) = \int_{-\infty}^{x} \int_{-\infty}^{y} f_{X,Y}(u,v) \mathrm{d}v \mathrm{d}u \,.$$

In the case of only two random variables, this is called a bivariate distribution, but the concept (as well as that of the bivariate densities and independence) generalizes to any number of events or random variables.

Example 1.12 (The Normal-Distribution). For a normally or Gaussian distributed[3] random variable X, the density $f(x)$ is given by

$$f(x) = \frac{1}{\sqrt{2\pi}\sigma} \exp\left(-\frac{(x-\mu)^2}{2\sigma^2} \right) ,$$

where μ and σ^2 denote the mean value and the variance of X, respectively. To denote that a real-valued random variable X is normally distributed with mean μ and variance σ^2, we write $X \sim \mathcal{N}(\mu, \sigma^2)$.

The relation between the joint density of two random variables, the marginal and conditional densities is sketched in Fig. 1.6. Note that in general, the conditional probability of X given Y is not the same as Y given X. The probability of both X and Y together is $\mathbb{P}(XY)$, and if both $\mathbb{P}(X)$ and $\mathbb{P}(Y)$ are non-zero this leads to a statement of *Bayes Theorem*:

$$\mathbb{P}(X|Y) = \frac{\mathbb{P}(Y|X) \cdot \mathbb{P}(X)}{\mathbb{P}(Y)} , \quad \text{and} \quad \mathbb{P}(Y|X) = \frac{\mathbb{P}(X|Y) \cdot \mathbb{P}(Y)}{\mathbb{P}(X)} .$$

Conditional probability is also the basis for statistical dependence and statistical independence as we will see in Sec. 1.4.

---

3 The normal distribution was first introduced by Abraham de Moivre in an article in 1733, which was reprinted in the second edition of his "The Doctrine of Chances", 1738, in the context of approximating certain binomial distributions for large natural numbers.

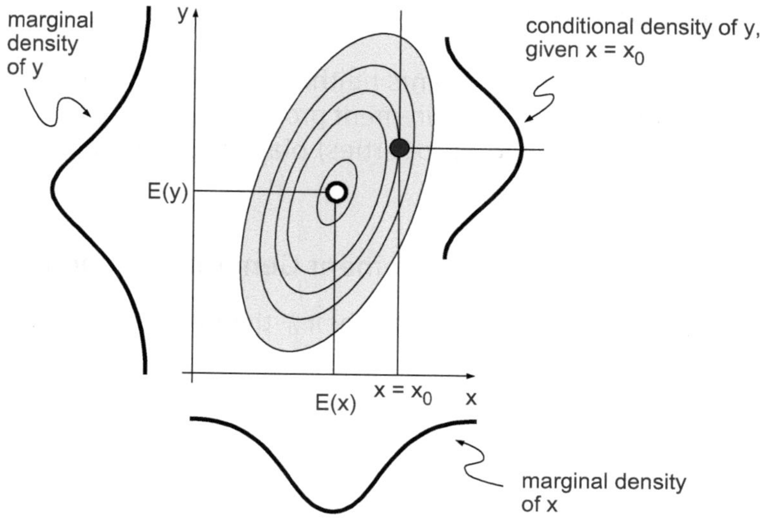

Figure 1.6. Relation between the joint, marginal and conditional densities.

Before you continue, make sure to answer the following questions:

Quiz: Section 1.2

Q1 Give the definitions of Hölder and Lipschitz continuity, and give an example of a Hölder-continuous function that is not Lipschitz-continuous.

Q2 Give the definition of a σ-algebra, a probability measure, and a probability space. Give two examples for probability spaces.

Q3 Give the definition of a random variable. Give two examples for random variables.

Q4 How can σ-algebras be generated utilizing random variables?

Q5 Give the definitions of the density and the distribution function of a random variable.

Q6 What are conditional densities and marginal densities?

Q7 What does the theorem of Radon-Nikodym state and why is it important?

1.3 Moments and Integrals

Particularly in (physical) experiments neither distributions nor densities are available easily throughout measurement processes; though the expectation and moments (as well as their properties) play an important role for those applications.

1.3.1 (Central) Moments and Moment Generating Functions

Random variables can be described by their k-th moments which are defined bas:

$$\mathbb{E}(x^k) := \int_{-\infty}^{\infty} x^k f(x) \mathrm{d}x$$

and their k-th central moments,

$$\mathbb{E}((x - \mathbb{E}(x))^k) := \int_{-\infty}^{\infty} (x - \mathbb{E}(x))^k f(x) \mathrm{d}x \,.$$

The most important moments are the *mean*/ average value/ expected value/ 1st moment

$$\mu := \mathbb{E}(x) = \int_{-\infty}^{\infty} x f(x) \mathrm{d}x$$

and the *variance*/ 2nd central moment

$$\sigma^2 := \mathbb{V}ar(x) := \mathbb{E}((x - \mu)^2) := \int_{-\infty}^{\infty} (x - \mu)^2 f(x) \mathrm{d}x \,,$$

whereby the quantity σ is called the *standard deviation.*

Example 1.13 (Moment Generating Function and First Moments of Normally Distributed Random Variables). For normally distributed random variables all higher moments ($k > 2$) can be expressed by the mean μ and the variance σ^2.

Given a real random variable X, the so-called *moment generating function* is defined as

$$M_X(t) := \mathbb{E}(\exp(tX))$$

Provided the moment generating function exists in an open interval around $t = 0$, the n-th moment is given as

$$\mathbb{E}(X^n) = M_X^{(n)}(0) = \left. \frac{\mathrm{d}^n M_X(t)}{\mathrm{d}t^n} \right|_{t=0} .$$

Thus, for a normally distributed random variable X, the moment generating function is

$$\begin{aligned} M_X(t) &= \mathbb{E}(\exp(tX)) = \int_{-\infty}^{\infty} \frac{1}{\sqrt{2\pi}\sigma} \exp\left(-\frac{(x-\mu)^2}{2\sigma^2}\right) \exp(tx) dx \\ &= \exp\left(\mu t + \tfrac{1}{2}\sigma^2 t^2\right) . \end{aligned}$$

This leads to $\mathbb{E}(X) = \mu$, $\mathbb{E}(X^2) = \mu^2 + \sigma^2$, $\mathbb{E}(X^3) = \mu^3 + 3\mu\sigma^2, \ldots$ and $\mathbb{E}(X - \mu) = 0$, $\mathbb{E}((X-\mu)^2) = \mathbb{E}(X^2 - 2\mu X + \mu^2) = \sigma^2$, $\mathbb{E}((X-\mu)^3) = 0, \ldots$.

Following [146], example 3.4, we apply the moment generating function to derive the geometric distribution: First, via MATALB, we obtain a closed form for the moment generating function by

```
ML = simplify( symsum(exp(t*k)*p^k*q, k, 0, inf) );
pretty(ML)
                q
      - ------------
         exp(t) p - 1
```

The first and second moments are generated by differentiation and substitution for $t = 0$ in the resultant expression. For the first moment this leads to

```
MLP = limit(diff(ML), t, 0)
MLP =
        1/(p-1)^2*q*p
```

We repeat the process for the second moment

```
MLPP = limit(diff(ML,2), t, 0)
MLPP =
        -q*p*(p+1)/(p-1)^3
```

The variance is now computed and simplified by noting $q = 1 - p$ and substitution $\mathbb{V}ar(X) = \mathbb{E}(X^2) - \mathbb{E}(X)^2$:

```
VARL = subs(MLPP, q, 1-p) - subs(MLP, q, 1-p)^2
pretty( simplify(VARL) )

           p
        --------
               2
        (p - 1)
```

The first and second moments of a random vector $x = (x_1, \ldots, x_n)^T$ are defined by

$$\mu := \mathbb{E}(x) := (\mathbb{E}(x_1), \ldots, \mathbb{E}(x_n))^T$$

and by the *covariance matrix* (symmetric and positive definite)

$$P := \mathbb{E}((x-\mu)(x-\mu)^T)$$

$$:= \begin{pmatrix} \mathbb{E}((x_1-\mu_1)(x_1-\mu_1)) & \mathbb{E}((x_1-\mu_1)(x_2-\mu_2)) & \dots & \mathbb{E}((x_1-\mu_1)(x_n-\mu_n)) \\ \mathbb{E}((x_2-\mu_2)(x_1-\mu_1)) & \mathbb{E}((x_2-\mu_2)(x_2-\mu_2)) & \dots & \mathbb{E}((x_2-\mu_2)(x_n-\mu_n)) \\ \vdots & \vdots & \ddots & \vdots \\ \mathbb{E}((x_n-\mu_n)(x_1-\mu_1)) & \mathbb{E}((x_n-\mu_n)(x_2-\mu_2)) & \dots & \mathbb{E}((x_n-\mu_n)(x_n-\mu_n)) \end{pmatrix}.$$

The diagonal elements $\mathbb{E}((x_i-\mu_i)^2)$ of P are the *variances* and the off-diagonal element are the *covariances* of the vector components. The *standard square deviation* is given by the trace of P:

$$tr(P) = \mathbb{E}\left(\sum_{i=1}^{n}(x_i-\mu)^2\right).$$

For instance, with the covariance matrix P, the density of a normally distributed n-vector is

$$f(x_1,\dots,x_n) = \frac{1}{\sqrt{(2\pi)^n}\sqrt{det(P)}}\exp\left(-\tfrac{1}{2}(x-\mu)^T P^{-1}(x-\mu)\right).$$

To generate a realisation of a multivariate Gaussian random variable $X \sim \mathcal{N}(\mu, P)$, $P \in \mathbb{R}^{d\times d}$ we can proceed as follows. cf. [153], pp. 487:

1. Perform a Cholesky decomposition of P to yield the non-singular $d \times d$-matrix G such that $P = GG^T$.
2. Generate a realisation $u \in \mathbb{R}^d$ of a random vector $U \sim \mathcal{N}(0, \mathbb{I})$ in $\mathbb{R}^d$, where $\mathbb{I} = \text{diag}(1,1,\dots,1) \in \mathbb{R}^{d\times d}$ denotes the $d \times d$ unit matrix.
3. Form the realisation of X as $x = Gu + \mu$.

As an example, let us assume $\mu := 0$ and

$$P := \begin{pmatrix} 1 & 2/3 & 1/3 \\ 2/3 & 1 & 1/3 \\ 1/3 & 2/3 & 1 \end{pmatrix} \quad \Rightarrow \quad G = \begin{pmatrix} 1 & 0 & 0 \\ 0.6667 & 0.7454 & 0 \\ 0.3333 & 0.5963 & 0.7303 \end{pmatrix}.$$

In Fig. 1.7, 100 realisations of X are plotted by use of the following MATLAB commands:

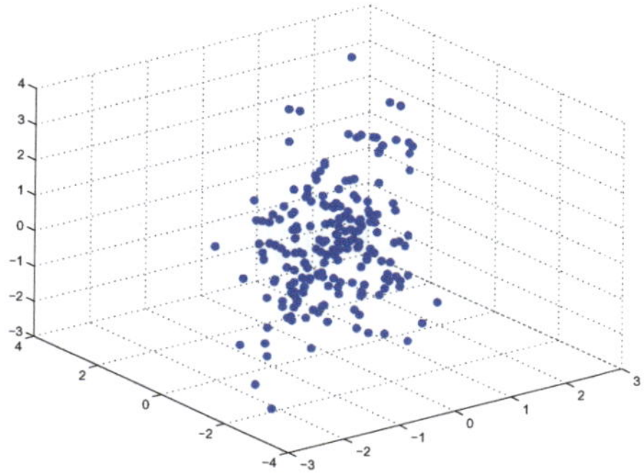

Figure 1.7. Realisations of a 3-dimensional multivariate Gaussian random variable, cf. [153], p. 488.

MATLAB Example 1.2. plotting3D.m: Generating Fig. 1.7.

```
P = [1 2/3 1/3;2/3 1 2/3;1/3 2/3 1];
G = chol(P) ';          % perform Cholesky decomposition
                        % MATLAB produces P=A'*A so G=A'
M = 200;
for m = 1:M             % generate realisations of x
   u = [randn(1,1) randn(1,1) randn(1,1)]';
   x = G*u;
   scatter3(x(1),x(2),x(3),'filled')
   hold on;
end
```

Cross-expectations of products of random variables are generally difficult to obtain, though the following proposition allows us to calculate an arbitrary product of normally distributed random variables[4].

Proposition 1.14 (Expectation of a Product of Normally Distributed Random Variables). *Suppose* $X = (x_1, x_2, \dots, x_n)^T \sim \mathcal{N}(0, P)$, *where* P *is an* $n \times n$-*positive semi-definite matrix. For non-negative integers* $s_1, s_2, \dots, s_n$, *we have*

$$\mathbb{E}\left(\prod_{i=1}^{n} x_i^{s_i}\right) = \begin{cases} 0 & \text{if } s \text{ is odd}, \\ \frac{1}{(\frac{s}{2})!} \sum_{\nu_1=0}^{s_1} \cdots \sum_{\nu_n=0}^{s_n} (-1)^{\sum_{i=1}^{n} \nu_i} \cdot & \\ \cdot \begin{pmatrix} s_1 \\ \nu_1 \end{pmatrix} \cdots \begin{pmatrix} s_n \\ \nu_n \end{pmatrix} \left(\frac{1}{2} h^T P h\right)^{s/2} & \text{if } s \text{ is even}, \end{cases}$$

where $s = s_1 + s_2 + \cdots + s_n$ *and* $h = \left(\frac{1}{2}s_1 - \nu_1, \frac{1}{2}s_2 - \nu_2, \dots, \frac{1}{2}s_n - \nu_n\right)^T$.

[4] See, [138], too. In the physics literature such closed form solutions for the expectation of the product of normally distributed random variables are associated with *Wick's formula*, cf. [150], p.546.

Proof. see [150], p.547 □

Chebysev's inequality gives an estimate how far a random variable deviates from its mean.

Proposition 1.15 (Chebysev's Inequality). *Let X be a random variable with finite mean and variance, then*

$$\mathbb{P}\left(|X - \mathbb{E}(X)| \geq \varepsilon\right) \leq \frac{\mathbb{V}ar(X)}{\varepsilon^2},$$

for every $\varepsilon > 0$.

Proof. cf. [122], p. 121 □

As an extension of our integration concept for random variables, let us discuss the integration with respect to a probability measure.

1.3.2 Integration with Respect to a Probability Measure

Up until now, we have used the concept of a density to set up integrals over random variables (with respect to the Lebesgue measure). Often, these densities are not available easily, hence we briefly define the integration with respect to a random measure itself — of course, all what comes can be played back to our previous discussions by applying the theorem of Radon-Nikodym (Theorem 1.1) in order to gain the appropriate density function.

Integration with respect to a probability measure is commonly defined in a three step process (often called stochastic induction):

1. If $(\Omega, \mathcal{A}, \mathbb{P})$ is a probability space and $X = \sum_{i=1}^n a_i \mathbb{I}_{A_i}$ is a real-valued simple random variable, we define the integral of X with respect to $\mathbb{P}$ by
$$\int_\Omega X \mathrm{d}\mathbb{P} := \sum_{i=1}^n a_i \mathbb{P}(A_i).$$
2. If next X is a non-negative random variable, we define
$$\int_\Omega X \mathrm{d}\mathbb{P} := \sup_{Y \leq X, Y \text{simple}} \int_\Omega Y \mathrm{d}\mathbb{P}.$$
3. Finally, if $X : \Omega \to \mathbb{R}$ is a random variable, we write
$$\int_\Omega X \mathrm{d}\mathbb{P} := \int_\Omega X^+ \mathrm{d}\mathbb{P} - \int_\Omega X^- \mathrm{d}\mathbb{P},$$
provided at least one of the integrals on the right hand side is finite. Here, we used the positive part $X^+ := \max(X, 0)$ and the negative part $X^- := \min(X, 0)$ of X, so that we have $X = X^+ - X^-$.

Next, suppose $X = (X^{(1)}, X^{(2)}, \ldots, X^{(d)})^T : \Omega \to \mathbb{R}^d$ is a vector-valued random variable. Then

$$\int_\Omega X \mathrm{d}\mathbb{P} := \left(\int_\Omega X^{(1)} \mathrm{d}\mathbb{P}, \int_\Omega X^{(2)} \mathrm{d}\mathbb{P}, \ldots, \int_\Omega X^{(d)} \mathrm{d}\mathbb{P} \right) .$$

We will assume the usual rules for these integrals to hold. Moreover, we can write the expected value $\mathbb{E}(X)$ and the variance $\mathbb{V}ar(X)$ of a vector-valued random variable X as

$$\mathbb{E}(X) = \int_\Omega X \mathrm{d}\mathbb{P} \quad \text{and} \quad \mathbb{V}ar(X) = \int_\Omega |X - \mathbb{E}(X)|^2 \, \mathrm{d}\mathbb{P} ,$$

where $|\cdot|$ denotes the Euclidean norm. Observe in particular that

$$\mathbb{V}ar(X) = \mathbb{E}(|X - \mathbb{E}(X)|^2) = \mathbb{E}(|X|^2) - |\mathbb{E}(X)|^2 .$$

The good thing about defining mean and variance with respect to probability measures is that it allows us to use the same symbols and formulas for both continuous and discrete random variables. One may note that, with respect to discrete random variables and their discrete probability measures (counting measures) the above integrals become sums.

Let $X \sim \mathcal{N}(0, 1)$. Utilizing MATLAB, we estimate the mean and variance of X^2:

```
x = randn(1,10000);   y = x.^2;
m = mean(y);          v = var(y);
m = 0.99762           v = 2.0120
```

1.4 Independence and Conditional Expectation

Let $(\Omega, \mathcal{A}, \mathbb{P})$ be a probability space, and $A, B \in \mathcal{A}$ be two events, with $\mathbb{P}(B) > 0$. We want to find a reasonable definition of

$$\mathbb{P}(A|B) = \text{"the probability of } A \text{ given } B\text{"}$$

Suppose some point $\omega \in \Omega$ is selected "at random" and we are told $\omega \in B$. What then is the probability that $\omega \in A$ also?

Since we know $\omega \in B$, we can regard B as being a new probability space. Therefore we define $\tilde{\Omega} := B$ endowed with the trace σ-algebra $\tilde{\mathcal{A}} := \{C \cap B : C \in \mathcal{A}\}$ and the measure $\tilde{\mathbb{P}} := \frac{\mathbb{P}}{\mathbb{P}(B)}$, so that $\tilde{\mathbb{P}}(\tilde{\Omega}) = 1$. Then, the probability that ω lies in A is

$$\mathbb{P}(A|B) = \tilde{\mathbb{P}}(A) = \frac{\mathbb{P}(A \cap B)}{\mathbb{P}(B)} .$$

This motivates the following definition:

Before you continue, make sure to answer the following questions:

Quiz: Section 1.3

Q1 Give the definitions of the expected value $\mathbb{E}(X)$ and the variance $\mathbb{V}ar$ of a real valued random variable X. What do these two concepts illustrate?

Q2 Let the density function $f(x)$ of a random variable x be given as $f(x) := \sin(x)$, $x \in [0, \frac{1}{2}\pi]$. Compute the expected value and variance of x.

Q3 Let the density function $f(x)$ of a random variable x be given as $f(x) := 6x - 6x^2$, $x \in [0, 1]$. Compute the expected value and variance of x.

Q4 Why are moment generating functions useful? Illustrate your answer with the example of normally distributed random variables.

Q5 What does Chebysev's inequality state?

Q6 Let $(\Omega, \mathcal{A}, \mathbb{P})$ be a probability space and X be a real valued random variable. What does the symbol $\int_\Omega X d\mathbb{P}$ mean? How does this relate to $\mathbb{E}(X)$ and $\mathbb{V}ar$.

Definition 1.16 (Conditional Probability). Let $(\Omega, \mathcal{A}, \mathbb{P})$ be a probability space, and $A, B \in \mathcal{A}$ be two events, with $\mathbb{P}(B) > 0$. Then, the *conditional probability* $\mathbb{P}(A|B)$ of A given B is defined as

$$\mathbb{P}(A|B) := \frac{\mathbb{P}(A \cap B)}{\mathbb{P}(B)} .$$

Now, what should it mean to say "A and B are independent"? This should mean $\mathbb{P}(A|B) = \mathbb{P}(A)$, since presumably any information that the event B occurred is irrelevant in determining the probability that A has occurred. Thus

$$\mathbb{P}(A) = \mathbb{P}(A|B) = \frac{\mathbb{P}(A \cap B)}{\mathbb{P}(B)} \implies \mathbb{P}(A \cap B) = \mathbb{P}(A) \cdot \mathbb{P}(B) ,$$

if $\mathbb{P}(B) > 0$. We take this for the definition[5], even if $\mathbb{P}(B) = 0$:

Definition 1.17 (Two Independent Events). Let $(\Omega, \mathcal{A}, \mathbb{P})$ be a probability space. Two events A and B are called *independent*, if

$$\mathbb{P}(A \cap B) = \mathbb{P}(A) \cdot \mathbb{P}(B) .$$

[5] Visually independence of two events A and B means, that the ratio $\mathbb{P}(A)$ to $\mathbb{P}(\Omega) = 1$ is the same as the ratio $\mathbb{P}(A \cap B)$ to $\mathbb{P}(B)$, or more sloppy, the A to Ω is the same as the part of A in B to B.

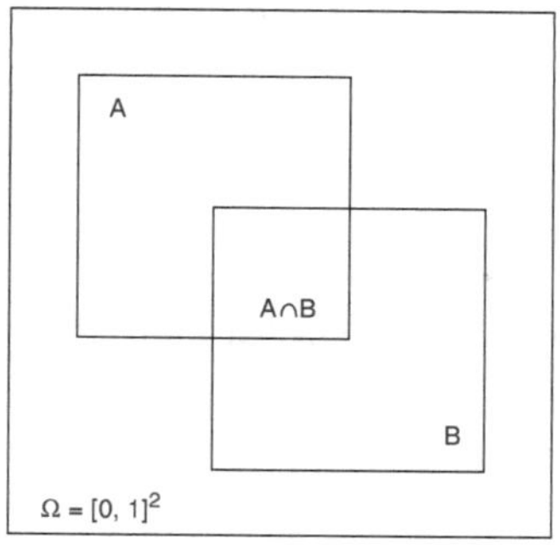

Figure 1.8. An example with "geometric probabilities": $\lambda(\Omega) = 1$, $\lambda(A) = \lambda(B) = 0.25$ s. t., $\lambda(A \cap B) = 0.25^2$. Thus, A and B are independent. Moreover, $\lambda(A^c) = \lambda(B^c) = 1 - 0.2.5 = 0.75$, $\lambda(A^c \cap B) = \lambda(B \setminus (A \cap B)) = 0.25 - 0.25^2 = 0.1875$ and $\lambda(A^c)\lambda(B) = 0.75 \cdot 0.25 = 0.1875$, i.e., A^c and B are independent.

This concept and its ramifications are the hallmarks of probability theory.

It is easy to check, that if A and B are independent, then so are A^c and B, or likewise, A^c and B^c, see Fig. 1.8, e.g.

Two events A and B are called *mutually exclusive*, if $A \cap B = \emptyset$. One can show, that two events cannot be both, independent and mutually exclusive, unless one of the two is a null set.

The following definition extends the concept of independence first to an arbitrary number of events under consideration, then to σ-algebras and last to random variables:

Definition 1.18 (Independence of Events/ σ-Algebras and Random Variables). Let $(\Omega, \mathcal{A}, \mathbb{P})$ be a probability space.

- Let $A_1, A_2, \ldots$ be events in $(\Omega, \mathcal{A}, \mathbb{P})$. These events are *independent* if for all choices of $1 \leq k_1 < k_2 < \cdots < k_n$, it holds that

$$\mathbb{P}(A_{k_1} \cap A_{k_2} \cap \cdots \cap A_{k_n}) = \mathbb{P}(A_{k_1}) \cdot \mathbb{P}(A_{k_2}) \cdot \cdots \cdot \mathbb{P}(A_{k_n}) .$$

- Let $\mathcal{A}_i \subset \mathcal{A}$, $i = 1, 2, \ldots$, be σ-algebras. The $\{\mathcal{A}_i\}_{i=1}^{\infty}$ are *independent* if for all choices of $1 \leq k_1 < k_2 < \cdots < k_n$ and of events $A_i \in \mathcal{A}_i$, it holds that

$$\mathbb{P}(A_{k_1} \cap A_{k_2} \cap \cdots \cap A_{k_n}) = \mathbb{P}(A_{k_1}) \cdot \mathbb{P}(A_{k_2}) \cdot \cdots \cdot \mathbb{P}(A_{k_n}) .$$

- Let $X_i : \Omega \to \mathbb{R}^d$, $i = 1, 2, \ldots$, be random variables. The random variables $X_1, X_2, \ldots$ are *independent* if for all integers $k \geq 2$ and all choices of Borel-sets $B_1, B_2, \ldots, B_k \subset \mathbb{R}^d$, it holds that

$$\mathbb{P}(X_1 \in B_1 \,,\, X_2 \in B_2 \,,\, \ldots \,,\, X_k \in B_k) = \mathbb{P}(X_1 \in B_1) \cdot \mathbb{P}(X_2 \in B_2) \cdot \ldots \cdot \mathbb{P}(X_k \in B_k) .$$

This is equivalent to saying that the σ-algebras $\{\mathcal{A}(X_i)\}_{i=1}^{\infty}$ generated by the X_i are independent.

Next, we give the connection to distribution functions and densities of independent random variables:

Theorem 1.2 (Independence: Criteria for Distribution and Density Function). *Let $(\Omega, \mathcal{A}, \mathbb{P})$ be a probability space. The random variables $X_1, \dots, X_n : \Omega \to \mathbb{R}$ with distribution functions F_{X_i} of X_i, $i = 1, \dots, n$, are independent if and only if*

$$F_{X_1,\dots,X_n}(x_1, \dots, x_n) \;=\; F_{X_1}(x_1) \cdot \ldots \cdot F_{X_n}(x_n)\,, \tag{1.2}$$

for all $x_i \in \mathbb{R}^d$, $i = 1, \dots, n$.

Moreover, if the random variables $X_1, \dots, X_n$ have densities (1.2) is equivalent to

$$f_{X_1,\dots,X_n}(x_1, \dots, x_n) \;=\; f_{X_1}(x_1) \cdot \ldots \cdot f_{X_n}(x_n) \qquad \forall\, x_i \in \mathbb{R}^d, i = 1, \dots, n\,,$$

where the functions f_{X_i} are the densities corresponding to X_i.

Proof. For the assertion w.r.t. the distributions, let us assume first, that the X_i, $i = 1, \dots, n$, are independent. Then

$$\begin{aligned} F_{X_1,\dots,X_n}(x_1, \dots, x_n) &= \mathbb{P}(X_1 \le x_1, \dots, X_n \le x_n) \\ &= \mathbb{P}(X_1 \le x_1) \cdot \ldots \cdot \mathbb{P}(X_n \le x_n) \\ &= F_{X_1}(x_1) \cdot \ldots \cdot F_{X_n}(x_n) \end{aligned}$$

Next, we prove the assertion w.r.t. the densities: Here, we prove the converse statement for the case that all random variables have densities. Select $A_i \in \mathcal{A}(X_i)$, $i = 1, \dots, n$. Then, $A_i = X_i^{-1}(B_i)$ for some $B_i \in \mathcal{B}$. Hence,

$$\begin{aligned} \mathbb{P}(A_1 \cap \dots \cap A_n) &= \mathbb{P}(X_1 \in B_1, \dots, X_n \in B_n) \\ &= \int_{B_1 \times \dots \times B_n} f_{X_1,\dots,X_n}(x_1, \dots, x_n) \mathrm{d}x_1 \dots \mathrm{d}x_n \\ &= \prod_{i=1}^{n} \left(\int_{B_i} f_{X_i}(x_i) \mathrm{d}x_i \right) \;=\; \prod_{i=1}^{n} \mathbb{P}(X_i \in B_i) \;=\; \prod_{i=1}^{n} \mathbb{P}(A_i)\,. \end{aligned}$$

Therefore $\mathcal{A}(X_1), \dots, \mathcal{A}(X_n)$ are independent σ-algebras. □

Two of the most important properties of independent random variables are given in the following theorem.

Theorem 1.3 (Independence: Implications for Mean and Variance). *Let $(\Omega, \mathcal{A}, \mathbb{P})$ be a probability space and $X_1, \ldots, X_n : \Omega \to \mathbb{R}^d$ are independent random variables.*

- *If $\mathbb{E}(|X_i|) < \infty$ for $i = 1, \ldots, n$, then*

$$\mathbb{E}(|X_1 \cdot \ldots \cdot X_n|) < \infty \quad \text{and} \quad \mathbb{E}(X_1 \cdot \ldots \cdot X_n) = \mathbb{E}(X_1) \cdot \ldots \cdot \mathbb{E}(X_n) .$$

- *If $\mathbb{V}ar(X_i) < \infty$ for $i = 1, \ldots, n$, then*

$$\mathbb{V}ar(X_1 + \cdots + X_n) = \mathbb{V}ar(X_1) + \cdots + \mathbb{V}ar(X_n) .$$

Proof. For[6] the first part, let us suppose that each X_i is bounded and has a density. Then

$$\begin{aligned} \mathbb{E}\left(\prod_{i=1}^{n} X_i\right) &= \int_{\mathbb{R}^d} x_1 \cdot \ldots \cdot x_n f_{X_1,\ldots,X_n}(x_1, \ldots, x_n) \mathrm{d}x_1 \ldots \mathrm{d}x_n \\ &= \prod_{i=1}^{n} \left(\int_{\mathbb{R}} x_i f_{X_i}(x_i) \mathrm{d}x_i \right) = \prod_{i=1}^{n} \mathbb{E}(X_i) , \end{aligned}$$

where independence was used to split the integral over the joint densities into a product.

For the second part, let us use induction and consider the case $n = 2$ (the arbitrary case follows by an analogous train of thoughts). Let $\mu_1 := \mathbb{E}(X_1)$ and $\mu_2 := \mathbb{E}(X_2)$. Then $\mathbb{E}(X_1 + X_2) = \mu_1 + \mu_2$ and

$$\begin{aligned} &\mathbb{V}ar(X_1 + X_2) = \int_{\Omega} \left(X_1 + X_2 - (\mu_1 + \mu_2)^2\right) \mathrm{d}\mathbb{P} \\ &= \int_{\Omega} (X_1 - \mu_1)^2 \mathrm{d}\mathbb{P} + \int_{\Omega} (X_2 - \mu_2)^2 \mathrm{d}\mathbb{P} + 2 \int_{\Omega} (X_1 - \mu_1)(X_2 - \mu_2) \mathrm{d}\mathbb{P} \\ &= \mathbb{V}ar(X_1) + \mathbb{V}ar(X_2) + 2 \underbrace{\mathbb{E}(X_1 - \mu_1)}_{=0} \underbrace{\mathbb{E}(X_2 - \mu_2)}_{=0} , \end{aligned}$$

where we used independence in the next last step. □

1.4.1 Conditional Expectation

An elegant approach to conditional expectation is based upon projections onto closed sub-spaces and motivated by the following finite dimensional example.

---

6 We will follow the arguments outlined in [34], pp. 76, and [97], pp. 19.

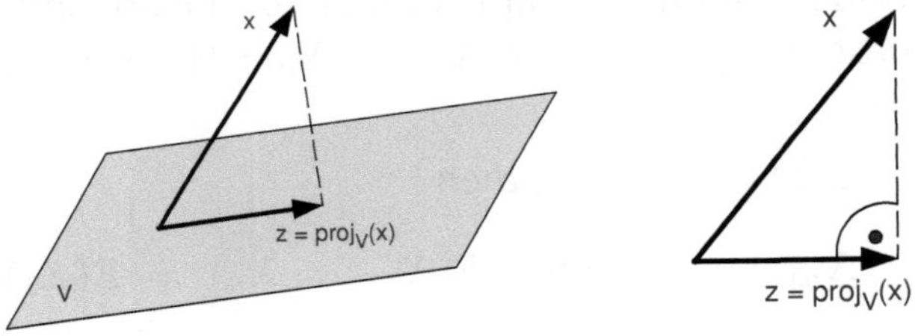

Figure 1.9. Illustration of the finite dimensional least squares approach.

Least squares method: Consider for the moment $\mathbb{R}^d$ and suppose that $V \subset \mathbb{R}^d$ is a proper subspace. Suppose, we are given a vector $x \in \mathbb{R}^d$. The least squares problem asks us to find a vector $z \in V$ such that

$$|z - x| \; = \; \min_{y \in V} |y - x| \, .$$

One can show that, given x, there exists a unique vector $z \in V$ solving this minimisation problem. We call z the projection of x onto V: $z = proj_V(x)$, cf. Fig. 1.9.

Now, we want to find a formula characterising z. For this we take any other vector $w \in V$ and define

$$i(\tau) \; := \; |z + \tau w - x|^2 \, .$$

Since $z + \tau w \in V$ for all τ, we see that the function $i(\cdot)$ has a minimum at $\tau = 0$. Hence, $0 = i'(0) = 2(z - x)w$, i.e.,

$$xw \; = \; zw \qquad \text{for all } w \in V \, .$$

The geometric interpretation is that the "error" $x - z$ is perpendicular to the subspace V.

Projection of random variables: Motivated by the example above, we return now to conditional expectation. Let us take the linear space $L^2(\Omega) = L^2(\Omega, \mathcal{A})$, which consists of all real-valued, $\mathcal{A}$-measurable random variables Y such that

$$||Y|| \; := \; \left(\int_\Omega Y^2 \mathrm{d}\mathbb{P} \right)^{1/2} \; < \infty \, .$$

We call $||Y||$ the norm of Y, and if $X, Y \in L^2(\Omega)$, we define the inner product to be

$$(X, Y) \; := \; \int_\Omega XY \mathrm{d}\mathbb{P} \; = \; \mathbb{E}(XY) \, .$$

Next, take as before $\mathcal{S}$ to be a sub-σ-algebra of $\mathcal{A}$. Consider then

$$V := L^2(\Omega, \mathcal{S}),$$

the space of all square-integrable random variables that are $\mathcal{S}$-measurable. This is a closed sub-space in $L^2(\Omega)$. Consequently, if $X \in L^2(\Omega)$, we can define its projection

$$Z := proj_V(X),$$

by analogy with the finite dimensional case just discussed. Almost exactly as before, we can show

$$(X, W) = (Z, W) \qquad \text{for all } W \in V.$$

Take in particular $W = \mathbb{I}_A$ for any set $A \in \mathcal{S}$. In view of the inner product, it follows that

$$\int_A X d\mathbb{P} = \int_A Z d\mathbb{P} \qquad \text{for all } A \in \mathcal{S}.$$

Since $Z \in V$ is $\mathcal{S}$-measurable, we see that Z in fact is $\mathbb{E}(X|\mathcal{S})$, as defined in our earlier discussion. That is,

$$\mathbb{E}(X|\mathcal{S}) = proj_V(X).$$

We could therefore alternatively take the last identity as a definition of conditional expectation. This point of view also makes it clear that $Z = \mathbb{E}(X|\mathcal{S})$ solves the least squares problem:

$$||Z - X|| = \min_{Y \in V} ||Y - X||,$$

and so $\mathbb{E}(X|\mathcal{S})$ can be interpreted as that $\mathcal{S}$-measurable random variable which is the best least squares approximation of the random variable X.

1.4.2 Further Properties of the Conditional Expectation

Other important properties of the conditional expectation are as follows[7]:

1. $|\mathbb{E}(X|\mathcal{S})| \leq \mathbb{E}(|X||\mathcal{S})$.
2. $X \geq 0$, then $\mathbb{E}(X|\mathcal{S}) \geq 0$.
3. X $\mathcal{S}$-measurable, then $\mathbb{E}(X|\mathcal{S}) = X$.
4. $X = a =$ const., then $\mathbb{E}(X|\mathcal{S}) = a$.

[7] Note, all these equations and inequalities hold almost surely. For the proofs see, e.g., [8], pp. 18, or [122], pp. 471.

5. $X, Y \in L^1$, then $\mathbb{E}(aX + bY|\mathcal{S}) = a\mathbb{E}(X|\mathcal{S}) + b\mathbb{E}(Y|\mathcal{S})$.
6. $X \leq Y$, then $\mathbb{E}(X|\mathcal{S}) \leq \mathbb{E}(Y|\mathcal{S})$.
7. X $\mathcal{S}$-measurable and $Y, XY \in L^1$, then $\mathbb{E}(XY|\mathcal{S}) = X\mathbb{E}(Y|\mathcal{S})$ (in particular $\mathbb{E}(\mathbb{E}(X|\mathcal{S})Y|\mathcal{S}) = \mathbb{E}(X|\mathcal{S})\mathbb{E}(Y|\mathcal{S})$).
8. $X, \mathcal{S}$ independent, then $\mathbb{E}(X|\mathcal{S}) = \mathbb{E}(X)$.
9. $\mathcal{S}_1 \subset \mathcal{S}_2 \subset \mathcal{A}$, then $\mathbb{E}(\mathbb{E}(X|\mathcal{S}_2)|\mathcal{S}_1) = \mathbb{E}(\mathbb{E}(X|\mathcal{S}_1)|\mathcal{S}_2) = \mathbb{E}(X|\mathcal{S}_1)$.

The *conditional probability* $\mathbb{P}(A|\mathcal{S})$ of an event A under the condition $\mathcal{S} \subset \mathcal{A}$ is defined by

$$\mathbb{P}(A|\mathcal{S}) \ := \ \mathbb{E}(\mathbb{I}_A|\mathcal{S}) .$$

Being thus a conditional expectation, the conditional probability is a $\mathcal{S}$-measurable function on Ω. In particular, for a $\mathcal{S}$ generated by at most countably many atoms $\{A_n\}$,

$$\mathbb{P}(A|\mathcal{S})(\omega) \ = \ \frac{\mathbb{P}(A \cap A_n)}{\mathbb{P}(A_n)} \quad \forall \omega \in A_n$$

such that $\mathbb{P}(A_n) > 0$[8]. From the properties of the conditional expectation, it follows that $0 \leq \mathbb{P}(A|\mathcal{S}) \leq 1$, $\mathbb{P}(\emptyset|\mathcal{S}) = 0$, $\mathbb{P}(\Omega|\mathcal{S}) = 1$ and

$$\mathbb{P}\left(\bigcup_{n=1}^{\infty} A_n|\mathcal{S}\right) \ = \ \sum_{n=1}^{\infty} \mathbb{P}(A_n|\mathcal{S}) \qquad \text{with } \{A_n\} \text{ pairwise disjoint in } \mathcal{A},$$

almost surely.

However, since $\mathbb{P}(A|\mathcal{S})$ is defined only up to a set of measure zero depending on A, it does not follow that $\mathbb{P}(\cdot|\mathcal{S})$ is, for fixed $\omega \in \Omega$, a probability on $\mathcal{A}$. On the other hand, for a random variable X consider the conditional probability

$$\mathbb{P}(X \in B|\mathcal{S}) \ = \ \mathbb{P}(\{\omega : X(\omega) \in B\}|\mathcal{S}) , \quad B \in \mathcal{B}^d .$$

There exists a function $p(\omega, B)$ defined on $\Omega \times \mathcal{B}^d$ with the following properties:

- for fixed $\omega \in \Omega$, the function $p(\omega, \cdot)$ is a probability on $\mathcal{B}^d$,
- for fixed B, the function $p(\omega, \cdot)$ is a version of $\mathbb{P}(X \in B|\mathcal{S})$, i.e., $p(\cdot, B)$ is $\mathcal{S}$-measurable and

$$\mathbb{P}(S \cap (X \in B)) \ = \ \int_S p(\omega, B) d\mathbb{P}(\omega) \quad \forall S \in \mathcal{S} .$$

[8] For instance the elementary conditional probability of an event $A \in \mathcal{A}$ under the condition $B \in \mathcal{A}$, with $\mathbb{P}(B) > 0$, is $\mathbb{P}(A|B) \ = \ \frac{\mathbb{P}(A \cap B)}{\mathbb{P}(B)}$.

Such a function p (uniquely defined up to a set of measure zero in $\mathcal{S}$ that is independent of B) is called *conditional (probability) distribution of X for given $\mathcal{S}$*. For $g(X) \in L^1$,

$$\mathbb{E}(g(X)|\mathcal{S}) = \int_{\mathbb{R}^d} g(x)p(\omega, \mathrm{d}x)\,. \tag{1.3}$$

1.4.3 Convergence Concepts for Sequences of Random Variables

Next, we give the most common convergence notations for random variables, see, for example, [220], pp. 83.

Convergence in Distribution: Let $\{F_n\}_{n\geq 0}$ be the sequence of distribution functions or laws[9] corresponding to the random variables $\{X_n\}_{n\geq 0}$ and F be the distribution corresponding to the random variable X. Then, the sequence X_n converges toward X *in distribution* (or *weakly* or *in law*) if

$$\lim_{n\to\infty} F_n(x) = F(x) \qquad \text{for all } x \in \mathbb{R} \text{ at which } F \text{ is continuous}\,.$$

Convergence in distribution is often denoted by $X_n \xrightarrow{\mathcal{D}} X$.

Since $F(x) = \mathbb{P}(X \leq x)$, convergence in distribution means that the value of X is in a given range very similar to the probability that the value of X_n is in that range, provided that n is sufficiently large[10].

In the following definitions, $\{X_n\}_{n\geq 0}$ denotes a sequence of random variables and X is a random variable.

Convergence in Probability: The sequence X_n converges toward X *in probability* if

$$\lim_{n\to\infty} \mathbb{P}(\|X_n - X\| \geq \varepsilon\} = 0 \qquad \text{for all } \varepsilon > 0\,. \tag{1.4}$$

Convergence in probability is often denoted by $X_n \xrightarrow{\mathcal{P}} X$. The defining equality (1.4) is equivalent to $\lim_{n\to\infty} \mathbb{P}(\|X_n - X\| < \varepsilon) = 1$ for all $\varepsilon > 0$.

Convergence Almost Everywhere The sequence X_n converges toward X *almost surely* (or *almost everywhere* or *with probability 1* or *strongly*) if

$$\mathbb{P}\left(\lim_{n\to\infty} X_n = X\right) = 1\,. \tag{1.5}$$

Almost sure convergence is often denoted by $X_n \xrightarrow{\text{a.s.}} X$. The equality (1.5) is equivalent to $\mathbb{P}\left(\{\omega \in \Omega : \lim_{n\to\infty} X_n(\omega) = X(\omega)\}\right) = 1$.

9 Recall, the word "law" is sometimes used as a synonym of "probability distribution".

10 An equivalent definition is $X_n \xrightarrow{\mathcal{D}} X$ if $\mathbb{E}(f(X_n)) \to \mathbb{E}(f(X))$ for $n \to \infty$ and all real-valued, bounded and continuous functions f, see, e.g. [220], p. 84.

Sure Convergence: The sequence X_n converges toward X *surely* (or *everywhere* or *point-wise*) if

$$\lim_{n\to\infty} X_n(\omega) = X(\omega) \qquad \text{for all } \omega \in \Omega\,. \tag{1.6}$$

The defining equality (1.6) is equivalent to $\{\omega \in \Omega : \lim_{n\to\infty} X_n(\omega) = X(\omega)\} = \Omega$.

Convergence in the r-th Mean: The sequence X_n converges toward X *in the r-th mean* (or *in the L^r norm*) if

$$\lim_{n\to\infty} \mathbb{E}\left(\|X_n - X\|^r\right) = 0\,.$$

Convergence in the r-th mean is often denoted by $X_n \overset{L^r}{\to} X$. If $r = 1$, we say X_n converges toward X *in mean* and if $r = 2$, we say X_n converges toward X *in mean square* or *in quadratic mean.*

We will use the so called Borel-Cantelli lemma to prove that $X_n \overset{\mathcal{P}}{\to} X$ implies sure convergence for some subsequence $(X_{n_j})_{j\in\mathbb{N}} \subset (X_n)_{n\in\mathbb{N}}$. In order to state this lemma, we define what we mean by an event occurring infinitely often: Let $A_1, A_2, \ldots$ be events in a probability space. Then, the event

$$\begin{aligned} A &:= \limsup_{n\to\infty} A_n := \bigcap_{n=1}^{\infty}\bigcup_{m=n}^{\infty} A_m = \\ &= \{\omega \in \Omega : \omega \text{ belongs to infinitely many of the } A_m\} \end{aligned}$$

is called A_n *infinitely often*, abbreviated "A_n i.o.".

Lemma 1.19 (Borel-Cantelli Lemma). *If $\sum_n \mathbb{P}(A_n) < \infty$, then $\mathbb{P}(A) = 0$. If the sequence $\{A_n\}$ is independent, then conversely, $\sum_n \mathbb{P}(A_n) = \infty$ implies $\mathbb{P}(A) = 1$.*

Proof. Let us start with the first part[11]: By definition A_n i.o. $= \cap_{n=1}^{\infty} \cup_{m=n}^{\infty} A_m$, and hence for each n

$$\mathbb{P}(A_n \text{ i.o.}) \le \mathbb{P}(\cup_{m=n}^{\infty} A_m) \le \sum_{m=n}^{\infty} \mathbb{P}(A_m)\,.$$

The limit of the right hand side is zero as $n \to \infty$, because $\sum_{n=1}^{\infty} \mathbb{P}(A_n) < \infty$.

For the second part, we obtain, because of independence:

$$\mathbb{P}(A_n \text{ i.o.}) = \mathbb{P}\left(\bigcap_{n=1}^{\infty}\bigcup_{m=n}^{\infty} A_m\right) = 1 - \mathbb{P}\left(\bigcup_{n=1}^{\infty}\bigcap_{m=n}^{\infty} A_m^c\right)$$

[11] see [122], pp. 96

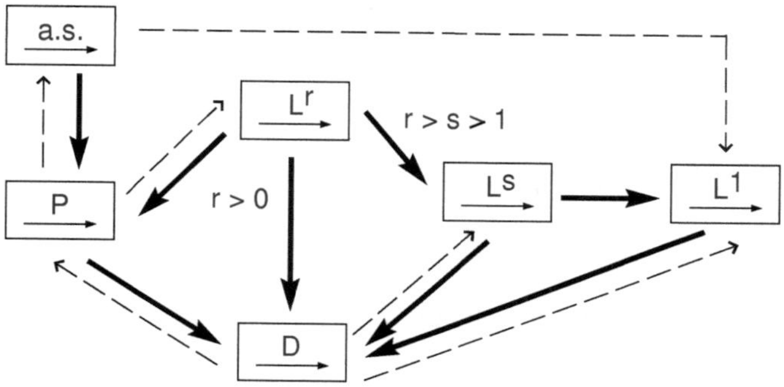

Figure 1.10. Implications for the different convergence concepts (bold arrows). The further shown implications (dashed arrows) require additional assumptions.

$$
\begin{aligned}
&= 1 - \lim_{n\to\infty} \left(\bigcap_{m=n}^{\infty} A_m^c \right) = 1 - \lim_{n\to\infty} \prod_{m=n}^{\infty} \mathbb{P}(A_m^c) \\
&= 1 - \lim_{n\to\infty} \prod_{m=n}^{\infty} (1 - \mathbb{P}(A_m)) = 1 - 0 = 1 ,
\end{aligned}
$$

since $\sum_{n=1}^{\infty} \mathbb{P}(A_n) = \infty$ is equivalent to $\prod_{m=1}^{\infty}(1 - \mathbb{P}(A_m)) = \infty$. □

The following proposition illustrates the application of the Borel-Cantelli lemma:

Proposition 1.20. *If $X_n \xrightarrow{\mathcal{P}} X$, then there exists a subsequence $(X_{n_j})_{j\in\mathbb{N}} \subset (X_n)_{n\in\mathbb{N}}$ such that*

$$X_{n_j}(\omega) \to X(\omega) \qquad \forall \omega \in \Omega .$$

Proof. For each positive integer j we select n_j so large that $\mathbb{P}(|X_{n_j} - X| > j^{-1}) \leq j^{-2}$, and also $\cdots < k_{j-1} < k_j < \ldots, k_j \to \infty$. Let $A_j := \{|X_{n_j} - X| > j^{-1}\}$. Since $\sum_j j^{-2} < \infty$, the Borel-Cantelli lemma implies $\mathbb{P}(A_j \text{ i.o.}) = 0$. Therefore, for almost all sample points ω, we have $|X_{n_j}(\omega) - X(\omega)| \leq j^{-1}$, provided $j \geq J$, for some index J depending on ω. □

Of course, this was just the tip of the iceberg concerning relations between the various convergence notations. For instance, sure convergence implies all of the other kinds of just noted convergence concepts. Moreover, the following implications hold (see Fig. 1.10 for a visualisation):

Proposition 1.21 (Convergence Concept Implications). *Let $\{X_n\}_{n\geq 0}$ be a sequence of random variables and X a random variable. Then, we have*

- $X_n \stackrel{a.s.}{\rightarrow} X \Rightarrow X_n \stackrel{\mathcal{P}}{\rightarrow} X \Rightarrow X_n \stackrel{\mathcal{D}}{\rightarrow} X$,
- *For all* $r > 0$: $X_n \stackrel{L^r}{\rightarrow} X \Rightarrow X_n \stackrel{\mathcal{P}}{\rightarrow} X$,
- *For all* $r > s \geq 1$: $X_n \stackrel{L^r}{\rightarrow} X \Rightarrow X_n \stackrel{L^s}{\rightarrow} X$.

Proof. see, e.g., [109], pp. 8, [122], pp. 218, or [14], pp. 61 □

The reverse implications are in general false, as the following example[12] illustrates:

Example 1.22 (A Sequence of Random Variables that Converges Almost Surely to X but does not Converge in the r-th Mean to X). Consider as sample space the unit interval $(0;1)$ endowed with uniform probability (Lebesgue) measure and define the sequence $\{X_n\}_{n\in\mathbb{N}}$ of random variables as follows:

$$X_n(\omega) \;=\; 2^n \mathbb{I}_{(0,1/n)}(\omega)\,.$$

Then, $X_n \stackrel{\text{a.s.}}{\rightarrow} 0$, as we have

$$\mathbb{P}\left(\lim_{n\to\infty} X_n = 0\right) \;=\; 1\,.$$

Moreover, the complement of the set for which X_n converges not only has probability zero, but is also empty. Thus,

$$\lim_{n\to\infty} \mathbb{P}(X_n \geq \epsilon) \;=\; 0 \qquad \forall \epsilon > 0$$

it additionally follows that $X_n \stackrel{\mathcal{P}}{\rightarrow} 0$. However,

$$\mathbb{E}\left(|X_n - 0|^r\right) \;=\; \mathbb{E}\left(X_n^r\right) \;=\; \int_0^{1/n} (2^n)^r \mathrm{d}\omega \;=\; \frac{2^{nr}}{n}\,,$$

which tends to infinity as $n \to \infty$ for every $r > 0$. Hence, X_n does not converge to zero in r-th mean.

Under further assumptions reverse and additional implications can be shown[13]:

Remark 1.23 (Further Convergence Concept Implications). Let $\{X_n\}_{n\geq 0}$ be a sequence of random variables and X a random variable. Then, we have

---

12 see [220], p. 97.

13 See [220], pp. 83, for a variety of 55 illustrative (counter-)examples on why the implications in proposition 1.21 can not be reverted in general. For the proofs see also, e.g., [109], pp. 8, [122], pp. 218, or [14], pp. 61.

1. Let $c \in \mathbb{R}$ be a constant, then $X_n \overset{\mathcal{D}}{\to} c$ implies $X_n \overset{\mathcal{P}}{\to} c$.

2. Let $X_n \overset{\mathcal{D}}{\to} X$ and $\mathbb{P}(\|X\| \leq b) = 1$ for all n and some constant b, then $X_n \overset{L^r}{\to} X$ for all $r \geq 1$[14].

3. X_n is said to converge toward X *almost completely* (or *fast in probability*) if for all $\varepsilon > 0$ it holds that $\sum_{n=1}^{\infty} \mathbb{P}(\|X_n - X\| \geq \varepsilon) < \infty$.

   If X_n converges almost completely toward X, then it also converges almost surely[15].

4. If $S_n := \sum_{n=1}^{N} X_n$, $N \in \mathbb{N}$, then S_n converges almost surely if and only if S_n converges in probability.

5. If $X_n \overset{\text{a.s.}}{\to} X$ and there is a random variable Y such that $|X_n| < Y$ and $\mathbb{E}(Y) < \infty$, then $X_n \overset{L^1}{\to} X$.

1.5 A Primer on Stochastic Processes

A random variable/ function of time is called a *stochastic process* or time series (in the case of discrete time $t \in \mathbb{Z}$).

Definition 1.24 (Stochastic Processes and Their Sample Paths). Let $I = [t_0, T] \subset \mathbb{R}_0^+$ be a time interval, $(\Omega, \mathcal{A}, \mathbb{P})$ be a probability space and $(E, \mathcal{E})$ a measurable space. In general, a random element of E is a map from Ω into E that is $\mathcal{A}$-$\mathcal{E}$-measurable. A *stochastic process* X_t is a collection $\{X_t : t \in I\}$ of random elements of E.

For each $\omega \in \Omega$ the map $I \ni t \mapsto X_t(\omega)$ is called a *sample path*, trajectory or realisation of X_t.

A stochastic process is said to be *cadlag* if it almost surely has sample paths which are right continuous with left limits. A stochastic process is said to be *caglad* if it almost surely has sample paths which are left continuous with right limits.

Applying the so-called Kolmogorow construction we can derive stochastic processes from a given set of distribution functions, see, e.g., [100], pp. 1. Here, we follow a shortcut by [8], pp. 22.

---

14 In other words, if X_n converges in probability to X and all random variables X_n are almost surely bounded above and below, then X_n converges to X also in any r-th mean, $r \geq 1$

15 In other words if X_n converges in probability to X sufficiently fast (i.e. the above introduced sequence of tail probabilities is summable for all $\varepsilon > 0$), then $X_n \overset{\text{a.s.}}{\to} X$ holds, too.

Before you continue, make sure to answer the following questions:

Quiz: Section 1.4

Q1 Give the definitions of conditional probability and independence of random variables.

Q2 Let $X_1, \dots, X_d$ be d independent random variables. What do we know about $F_{X_1,\dots,X_d}$, $f_{X_1,\dots,X_d}$, $\mathbb{E}(X_1 \cdot \dots \cdot X_d)$ and $\mathbb{V}ar(X_1 + \dots + X_d)$?

Q3 Suppose X and Y are discrete random variables whose values are the non-negative integers and their joint probability function is

$$p_{X,Y}(x,y) = \frac{1}{x!y!}\lambda^x \mu^y \exp(-(\lambda+\mu)), \quad x,y = 0,1,2,\dots .$$

Are X and Y independent? What are their marginal distributions?

Q4 Illustrate the concept of conditional expectation.

Q5 Consider the joint density

$$f_{X,Y}(x,y) = \begin{cases} \lambda^2 \exp -\lambda y & 0 \le x \le y \\ 0 & \text{otherwise} \end{cases},$$

and find the conditional density of X given Y as well as the conditional density of Y given X.

Q6 Give a sequence of random variables $\{X_n\}_{n\in\mathbb{N}}$ such that $X_n \stackrel{\mathcal{P}}{\to} X$ as $n \to \infty$, but for which $X_n \stackrel{\text{a.s.}}{\to} X$ does not hold for $n \to \infty$.

Q7 Give a sequence of random variables $\{X_n\}_{n\in\mathbb{N}}$ such that $X_n \stackrel{L^r}{\to} X$ as $n \to \infty$, but for which $X_n \stackrel{\text{a.s.}}{\to} X$ does not hold for $n \to \infty$.

The *finite-dimensional (joint) distributions* of a stochastic process $\{X_t : t \in I\}$ are given by

$$\begin{aligned} \mathbb{P}(X_t \le x) &= F_t(x) \\ \mathbb{P}(X_{t,1} \le x_1, X_{t,2} \le x_2) &= F_{t_1,t_2}(x_1,x_2) \\ \mathbb{P}(X_{t,1} \le x_1, X_{t,2} \le x_2, X_{t,3} \le x_3) &= F_{t_1,t_2,t_3}(x_1,x_2,x_3) \\ \vdots \quad & \quad \vdots \end{aligned}$$

where $t, t_i \in I$ and $x, x_i \in \mathbb{R}^d$ (the symbol "$\leq$" is to be interpreted for every component) and $n \geq 1$.

This system of distribution functions satisfies the following two conditions:

Condition of Symmetry: If $\{i_1, \dots, i_n\}$ is a permutation of the numbers $1, \dots, n$, then for arbitrary instants and $n \geq 1$,

$$F_{t_{i_1},\dots,t_{in}}(x_{i_1}, \dots, x_{in}) = F_{t_1,\dots,t_n}(x_1, \dots, x_n).$$

Condition of Compatibility: For $m < n$ and arbitrary $t_{m+1}, \dots, t_n \in I$,

$$F_{t_1,\dots,t_m,t_{m+1},\dots,t_n}(x_1, \dots, x_m, \infty, \dots, \infty) = F_{t_1,\dots,t_m}(x_1, \dots, x_m).$$

In many practical cases, we are not given a family of random variables defined on a probability space but rather a family of distributions or their distribution functions $F_{t_1,\dots,t_n}(x_1, \dots, x_n)$ which satisfy the symmetry and compatibility conditions. The equivalence of these concepts is demonstrated by the following theorem.

Theorem 1.4 (Kolmogorow's Fundamental Theorem). *For every family of distribution functions that satisfies the symmetry and compatibility conditions, there exists a probability space $(\Omega, \mathcal{A}, \mathbb{P})$ and a stochastic process $\{X_t : t \in I\}$ defined on it that possesses the given distributions as finite-dimensional distributions.*

Proof. see [100], pp. 3 □

Throughout our considerations on stochastic processes and stochastic differential equations, we will always assume that the state space of the stochastic process is $\mathbb{R}^d$ endowed with the σ-algebra of Borel sets, i.e., $(E, \mathcal{E}) = (\mathbb{R}^d, \mathcal{B}^d)$. Moreover, if not stated otherwise, we choose

$$\Omega = \left(\mathbb{R}^d\right)^{[t_0,T]},$$

i.e., the space of all $\mathbb{R}^d$-valued functions on the interval $I = [t_0, T]$, and

$$\mathcal{A} = \left(\mathcal{B}^d\right)^{[t_0,T]},$$

i.e., the product σ-algebra generated by the Borel sets in $\mathbb{R}^d$, and $X_t = \omega(t)$ for all $\omega \in \Omega$. Then, $\mathbb{P}$ is the probability uniquely defined (according to Kolmogorow's fundamental theorem) by the finite dimensional distributions of the process X_t on $(\Omega, \mathcal{A})$. If we have further information about the analytical properties of the sample functions, we can choose for Ω certain subspaces of $(\mathbb{R}^d)^{[t_0,T]}$, e.g., the space of all continuous functions.

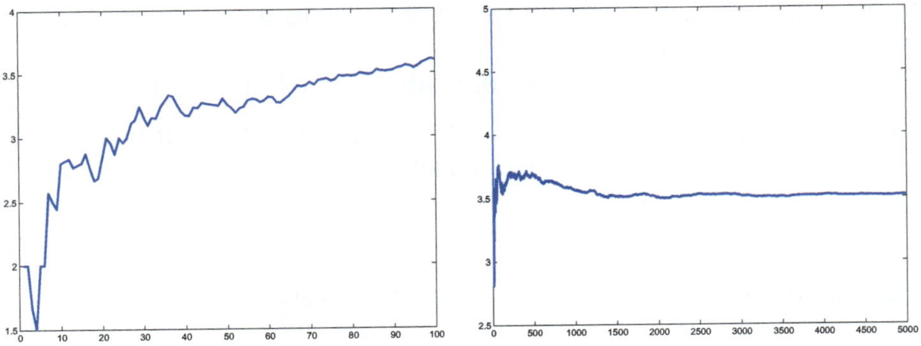

Figure 1.11. Illustration of the law of large numbers with (a) 100 throws of a dice and (b) 5000 throws.

Finally, a stochastic process $\{X_t : t \in I\}$ is said to satisfy the *law of large numbers* if for all $\varepsilon, \delta > 0$ there exists a $T > 0$ such that for all $t > T$

$$\mathbb{P}\left(\left| \frac{1}{t}\int_{t_0}^{t_0+t} X_s \mathrm{d}s - \frac{1}{t}\int_{t_0}^{t_0+t} \mathbb{E}(X_s)\mathrm{d}s \right| > \delta \right) < \varepsilon .$$

A stochastic process $\{X_t : t \in I\}$ is said to satisfy the *strong law of large numbers* if

$$\mathbb{P}\left(\left(\frac{1}{t}\int_{t_0}^{t_0+t} X_s \mathrm{d}s - \frac{1}{t}\int_{t_0}^{t_0+t} \mathbb{E}(X_s)\mathrm{d}s \right) \stackrel{t\to\infty}{\longrightarrow} 0 \right) = 1 .$$

Traditionally, the law of large numbers is illustrated by throwing a dice many times and by observing that the successive means convergence to the expectation. The MATLAB commands read as follows for a sequence of 100 throws:

```
x = unidrnd(6,100,1);
xbar = cumsum(x)./(1:100)';
plot(1:100,xbar,'-b','LineWidth',2)
```

Figure 1.11 shows the results for 100 and 5000 throws of a Laplacian dice.

1.5.1 Continuous Stochastic Processes

If for almost all $\omega \in \Omega$ the sample paths are continuous functions for all $t \in I$, then we say that the stochastic process is *continuous*. Analogously, if for almost all $\omega \in \Omega$ the sample paths are right/ left continuous, the stochastic process itself is called *right/ left continuous*.

The mean, the variance and the covariance matrix of a scalar (stochastic) process $X_t = X(t)$ are

- $\mu(t) = \mathbb{E}(X_t)$,
- $\sigma^2(t) = \mathbb{V}ar(X_t) = \mathbb{E}((X_t - \mu(t))^2)$,
- $\mathbb{C}ov(X_t, X_s) = R_X(t,s) = \mathbb{E}((X_t - \mu(t))(X_s - \mu(s)))$, where R is the so-called *auto-correlation function* of the stochastic processes X_t with itself (compared to the correlation/ covariance function $\mathbb{C}ov(X_t, Y_s)$ of two different processes X_t and Y_t).

Only two of these three quantities are independent, as $\sigma^2(t) = R(t,t)$.

The concepts of auto-correlation and cross-correlation of one or two stochastic processes respectively, are important and will be discussed in detail in Chapt. 10. The autocorrelation function of a random signal describes the general dependence of the values of the samples at one time on the values of the samples at another time. MATLAB provides a function called xcorr which may be used to implement both the auto- and the cross-correlation function, cf. [247]:

We start by plotting the auto-correlation function of a sine-wave with frequency 1 and sampling frequency 200:

MATLAB Example 1.3. autoCorr.m: Generating the auto-correlation function in Fig. 1.12 (a).

```
N = 1024;                  % Number of samples
f1 = 1;                    % Frequency of the sine-wave
FS = 200;                  % Sampling Frequency
n = 0:N-1;                 % Sample index numbers
x = sin(2*pi*f1*n/FS);     % Generate the signal, x(n)
t = [1:N]*(1/FS);          % Prepare a time axis
subplot(2,1,1);            % Prepare the figure
   plot(t,x,'-b','LineWidth',2);  % Plot x(n)
   title('\bf{Sine-Wave of Frequency 1000Hz [FS = 8000Hz]}');
   xlabel('Time [s]');
   ylabel('Amplitude');
   grid;
Rxx = xcorr(x);               % Estimate its autocorrelation
subplot(2,1,2);               % Prepare the figure
   plot(Rxx,'LineWidth',2);   % Plot the autocorrelation
   grid;
   title('\bf{Auto-Correlation Function of the Sine-Wave}');
   xlabel('Lags');
   ylabel('Autocorrelation');
```

The result is shown in Fig. 1.12 (a). Notice that when using the function xcorr, to estimate the autocorrelation sequence, it has double the number of samples as the signal $x(n)$. An important point to remember when using the function xcorr is that the origin is in the middle of the figure (here it is at $lag = 1024$).

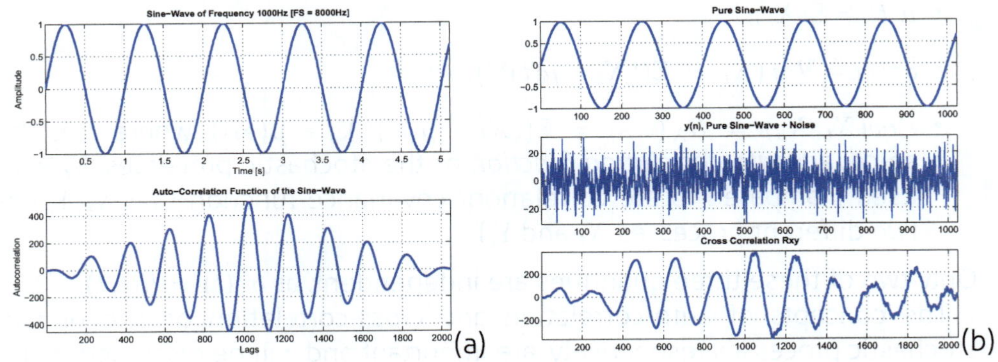

Figure 1.12. (a) auto-correlation of the deterministic process $x(t) := \sin(2\pi t)$ and (b) cross-correlation of $x(t) = \sin(2\pi t)$ and $y(t) := x(t) + \gamma_t$, with a zero mean, unit variance Gaussian stochastic process γ_t.

Next, cf. [247], we plot the cross-correlation of the signals

$$x(t) = \sin(2\pi t)\,, \qquad y(t) = x(t) + \gamma_t\,,$$

where γ_t is a zero mean, unit variance Gaussian stochastic process (see Fig. 1.12 (b) for a visualisation).

MATLAB Example 1.4. crossCorr.m: Generating the cross-correlation function in Fig. 1.12 (b).

```
rng('default')
Y = simulate(model,50,'numPaths',1000);

plot(Y,'Color',[.85,.85,.85])
   title('Simulated AR(2) Process')
   hold on
   h = plot(mean(Y,2),'k','LineWidth',2);
   legend(h, 'Simulation Mean', 'Location', 'NorthWest')
   hold off

plot(var(Y,0,2),'r','LineWidth',2)
   title('Process Variance')
   hold on
   plot(1:50,.83*ones(50,1),'k--','LineWidth',1.5)
   legend('Simulation', 'Theoretical', 'Location', 'SouthEast')
   hold off
```

A stochastic process is said to be *ergodic* if its statistical properties (such as its mean and variance) can be deduced from a single, sufficiently long sample (realisation) of the process.

An important class of ergodic stochastic processes are *stationary (stochastic) processes*, cf. [263]. These processes X_t are characterised by the property that the joint (probability) distribution of any finite subsequence $(X_{t_1}, \ldots, X_{t_k})$ $(k \in N)$ does not change when shifted in time, i.e.,

$$F_{X_{t_1},\ldots,X_{t_k}}(x_1, \ldots, x_k) \;=\; F_{X_{t_1+t},\ldots,X_{t_k+t}}(x_1, \ldots, x_k) \qquad \text{for all times } t\,.$$

In problem 1.43 we will show that for a stationary process the mean is a constant, the variation is finite (if they exist), i.e.,

$$\mathbb{E}(X_t) \;=\; \mu \;=\; \text{const.} \quad \text{and} \quad \mathbb{E}(X_t^2) \;<\; \infty \qquad \text{for all times } t$$

and the correlation function R depends on the difference $\tau := |t_{k_i} - t_{k_j}|$ only, i.e.,

$$R(t_{k_i}, t_{k_j}) = R(t_{k_i} + t, t_{k_j} + t) \;=:\; R(\tau)\,.$$

Coming back to the notion of an ergodic process we see that the properties of a stationary processes X_t allow the statistical sampling of, for instance, the mean by performing a time average:

$$\frac{1}{2T}\int_{-T}^{t} X_t \mathrm{d}t \;=\; \hat{\mu}_T \;\overset{T\to\infty}{\longrightarrow}\; \mu\,,$$

where $\hat{\mu}_T$ stands for the estimation of the mean μ with respect to the observation interval $[-T, T]$. If $\hat{\mu}_T$ converges in mean-squared to μ as $T \to \infty$, then the process X_t is said to be *mean-ergodic* or *mean-square ergodic in the first moment.* Likewise, one can estimate the auto-correlation by performing a time average, and if this expression converges in the mean-squared sense to the true auto-correlation, then the process is said to be *auto-correlation-ergodic* or *mean-square ergodic in the second moment.* Finally, a process which is ergodic in the first and second moments is sometimes called *ergodic in the wide sense.*

Besides the white noise process we will introduce in Sec. 3.2.1, an illustrative example of such a stationary process is the level of the sea on a sunny summers day. Tide-driven seas, on the other hand, change the average level with time – a non-stationary process. A non-stationary process is sometimes called an *evolutionary process.*

Following the MATLAB help pages[16], a special class of stationary stochastic processes are so-called autoregressive (AR) processes of order p that take past values up to time $t - p$ into account. Let an AR(2) process be specified by the equation

$$x_t \;=\; 0.5 + 0.7x_{t-1} + 0.25x_{t-2} + w_t\,,$$

[16] http://www.mathworks.de/help/econ/simulate-stationary-arma-processes.html

where w_t is a Gaussian white noise process with zero-mean and, in this case, variance 0.1. With MATLAB this process is conveniently simulated via the econometrics toolbox function arima incorporated in the MATLAB release 2012a:

model = arima('Constant',0.5,'AR',{0.7,0.25},'Variance',.1);

1000 sample paths with 50 observations each are generated via MATLAB Example 1.5:

MATLAB Example 1.5. process.m: Generating Fig. 1.13 (a).

```
rng('default')
Y = simulate(model,50,'numPaths',1000);

plot(Y,'Color',[.85,.85,.85])
   title('Simulated AR(2) Process')
   hold on
   h = plot(mean(Y,2),'k','LineWidth',2);
   legend(h, 'Simulation Mean', 'Location', 'NorthWest')
   hold off

plot(var(Y,0,2),'r','LineWidth',2)
   title('Process Variance')
   hold on
   plot(1:50,.83*ones(50,1),'k—','LineWidth',1.5)
   legend('Simulation', 'Theoretical', 'Location', 'SouthEast')
   hold off
```

MATLAB Example 1.6. processTrans.m: Generating Fig. 1.13 (b).

```
rng('default')
Y = simulate(model,150,'numPaths',1000);
Y = Y(101:end,:);

plot(Y,'Color',[.85,.85,.85])
   title('Simulated AR(2) Process')
   hold on
   h = plot(mean(Y,2),'k','LineWidth',2);
   legend(h, 'Simulation Mean', 'Location', 'NorthWest')
   hold off

plot(var(Y,0,2),'r','LineWidth',2)
   xlim([0,50])
   title('Process Variance')
   hold on
   plot(1:50,.83*ones(50,1),'k—','LineWidth',1.5)
   legend('Simulation', 'Theoretical', 'Location', 'SouthEast')
   hold off
```

The results are displayed in Fig. 1.13 (a). The simulation mean is constant over time. This is consistent with the definition of a stationary process. The process variance is not constant over time, however. There are transient effects at the beginning of the simulation due to the absence of pre-sample data.

Continuing with the MATLAB help pages[17], we can reduce transient effects by oversampling the process. For instance, to sample 50 observations, we generate paths with more than 50 observations, and discard all but the last 50 observations. Here, we simulate paths of length 150, and discard the first 100 observations.

The realisations now look like draws from a stationary stochastic process. The simulation variance fluctuates (due to Monte Carlo error) around the theoretical variance, cf. Fig. 1.13 (b).

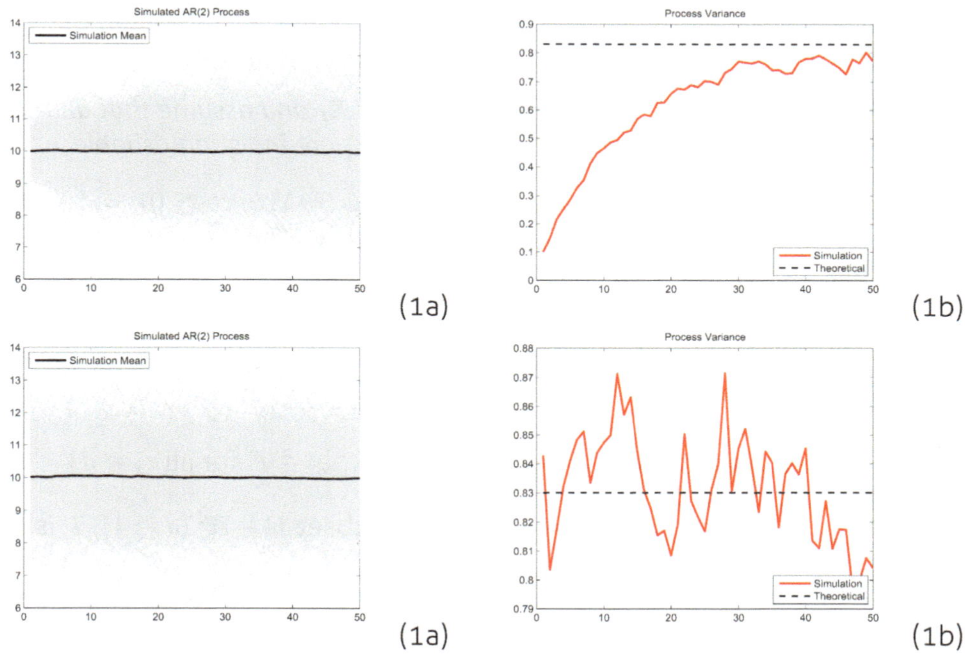

Figure 1.13. Simulation of an AR(2) process: its mean (a) and variance (b). Row (1) displays the results without suppressing transient effects, whereas row (2) discards the first observations of the simulations.

Often, one has to specify in which sense two (stochastic) processes are the same (see [234]). Hereby, the following definitions are used:

[17] http://www.mathworks.de/help/econ/simulate-stationary-arma-processes.html

Definition 1.25 (Indistinguishable Processes and Modifications). Let $I \subset \mathbb{R}_0^+$ be a time interval. Two $\mathbb{R}^d$-valued processes X_t and Y_t are called *indistinguishable* or *stochastically equivalent* if the set $\{X_t = Y_t : \forall t \in I\}$ contains a set of probability one — hence the paths of indistinguishable processes are a.s. equal.

They are called *modifications* of each other if $\mathbb{P}(X_t = Y_t) = 1$ holds for all $t \in I$.

The processes are said to have the *same finite dimensional distributions* if for any n-tuple $(t_1, \dots, t_n)^T$ with $t_i \in I$ for $i = 1, \dots, n$, the probability distributions of the random vectors $(X_{t_1}, \dots, X_{t_n})^T$ and $(Y_{t_1}, \dots, Y_{t_n})^T$ coincide.

Clearly, the first of these three concepts is the strongest, it implies trivially the second, which in turn yields the third. Now, let T be a non-negative real random variable with a continuous distribution. Let $X_t \equiv 0$ and Y_t be defined as $Y_t(\omega) := \mathbb{I}_{\{t=T(\omega)\}}$, $t \in [0, \infty)$. Then, X_t is a modification of Y_t, whereas $\mathbb{P}(X_t = Y_t : \forall t \geq 0) = 0$.

Proposition 1.26. *Let Y_t be a modification of X_t and assume that all paths of X_t and Y_t are right-continuous. Then X_t and Y_t are indistinguishable.*

Proof. Right-continuity allows us to write $\{X_t = Y_t : \forall t \geq 0\} = \{X_t = Y_t : \forall t \in [0, \infty) \cap \mathbb{Q}\}$. Since Y_t is a modification of X_t, the last set (is measurable and) has probability one. □

A d-dimensional stochastic process is called *separable* $\{X_t = X(t), t \in I \subset \mathbb{R}_0^+\}$ if there exists a countable sequence $T = \{t_j\}$ which is a dense subset of I and a $N \subset \Omega$ with $\mathbb{P}(N) = 0$ such that, if $\omega \notin N$,

$$\{X(t, \omega) \in F \text{ for all } t \in J\} \;=\; \{X(t_j, \omega) \in F \text{ for all } t_j \in J\}$$

for any open subset $J \subset I$ and any closed subset $F \in \mathbb{R}^d$ ($d \geq 1$). T is called a *set of separability* or briefly a *separant*.

It is easy to see that a right (or left) continuous stochastic process is separable and any dense sequence in I is a set of separability.

Theorem 1.5. *Any d-dimensional process is indistinguishable to a separable stochastic process.*

Proof. see [87], pp. 57 □

The random variables of the separable process in Theorem 1.5 may actually take the values $\pm\infty$ although the random variables of the original process are real valued.

A stochastic process $\{X_t : t \in I \subset \mathbb{R}_0^+\}$ is said to be *continuous in probability* if, for any $s \in I$, $\varepsilon > 0$,

$$\mathbb{P}(|X_t - X_s| > \varepsilon) \to 0, \qquad \text{if } t \in I \text{ and } t \to s.$$

The following theorem is due to Kolmogorow:

Theorem 1.6 (Kolmogorow's Continuity Theorem). *Let $\{X_t : t \in I \subset \mathbb{R}_0^+\}$ be a d-dimensional stochastic process such that*

$$\mathbb{E}(|X_t - X_s|^\beta) \leq C \cdot |t - s|^{1+\alpha}$$

for some positive constants C, α, β. Then there is a version of X_t that is a continuous process.

Proof. see [100], pp. 7 □

1.5.2 Filtrations, Martingales and Super-Martingales

Let $(\Omega, \mathcal{A}, \mathbb{P})$ be a probability space. A *filtration* $\mathbb{F} = \{\mathcal{A}_t : t \geq 0\}$ is a collection of sub-σ-algebras of $\mathcal{A}$ such that $\mathcal{A}_s \subset \mathcal{A}_t$ for all $s \leq t$. Given a stochastic process X_t we denote by $\mathcal{A}_t^X$ the smallest σ-algebra for which all X_s, $0 \leq s \leq t$, are measurable and $\mathbb{F}^X := \{\mathcal{A}_t^X : t \geq 0\}$.

Given a filtration $\mathbb{F}$ a process X_t is called $\mathbb{F}$-*adapted, adapted to* $\mathbb{F}$, or simply *adapted*, if for all times $t \geq 0$ the random variable X_t is $\mathcal{A}_t$-measurable. Clearly, every process X_t is adapted to $\mathbb{F}^X$.

We write $X \hat{=} Y$ if $X(\omega) = Y(\omega)$ holds for almost all $\omega \in \Omega$. Next, let X_t and Y_t be *equivalent* on I, i.e., $\mathbb{P}(Y_t = X_t) = 1$ for each fixed $t \in I$, such that $X_t \hat{=} Y_t$, $t \in I$, then there are sets $\Omega_t \in \mathcal{A}$, $t \in I$, such that $\mathbb{P}(\Omega_t) = 1$ and $X_t(\omega) = Y_t(\omega)$ holds for $(t, \omega) \in \cup_{t \in I}[\{t\} \times \Omega]$.

Let Y_t, $t \in I := [t_0, \infty)$ be a one-dimensional separable stochastic process which is adapted to the filtration $\mathcal{F}_t$ such that $\mathbb{E}(Y_t|\mathcal{F}_s) \hat{=} Y_s$ holds for all $t_0 \leq s \leq t$. Then, the pair $(Y_t, \mathcal{F}_t)_{t \in I}$ is called non-negative *martingale*. For martingales the (conditional) expectation or estimate of a future state (given the whole history up to the present state) is the present state, i.e. $\mathbb{E}(Y_t|\mathcal{F}_s) \hat{=} Y_s$. The concept of a martingale is essential in financial mathematics in order to forecast prices (cf. [259], [229]), an in particular for Itô's stochastic calculus as we will see in section 4.5.

Let Y_t, $t \in I := [t_0, \infty)$ be a one-dimensional separable stochastic process which is adapted to the filtration $\mathcal{F}_t$ such that $Y_t \hat{\geq} 0$ and $\mathbb{E}(Y_t|\mathcal{F}_s) \hat{\leq} Y_s$ holds for all $t_0 \leq s \leq t$. Then, the pair $(Y_t, \mathcal{F}_t)_{t \in I}$ is called non-negative *super-martingale*. In particular, for arbitrary $\lambda > 0$ the super-martingale inequality

$$\mathbb{P}\left(\sup_{t \in I} Y_t \geq \lambda\right) \leq \frac{1}{\lambda}\mathbb{E}(Y_{t_0})$$

holds. Moreover, there is a one-dimensional random variable Y_∞ such that

$$\lim_{t\to\infty} Y_t \triangleq Y_\infty \quad \text{and} \quad \lim_{t\to\infty} \mathbb{E}(Y_t) = \mathbb{E}(Y_\infty) .$$

1.5.3 Gaussian Processes

If for arbitrary times $0 = t_0 < t_1 < \cdots < t_k$ the corresponding process increments ($X(t_0) = 0$)

$$X(t_1),\ X(t_2) - X(t_1),\ \ldots,\ X(t_k) - X(t_{k-1})$$

are independent, then the process $X_t = X(t)$ is called a process with *independent increments* and it is

$$X(t_k) = \sum_{i=1}^{k} (X(t_i) - X(t_{i-1})) .$$

A process $X_t = X(t)$ is said to have *stationary increments*, if the probability distribution of any increment $X_t - X_s$, $t > s$, depends only on the length $t - s$ of the time interval; increments with equally long time intervals are identically distributed.

Let X_t be a d-dimensional second order stochastic process on the real interval I. The mean $\mathbb{E}(X_t) = m_X(t)$ is a vector-valued function on d. The $d \times d$-matrix function

$$C_X(t_1, t_2) = \mathbb{E}\left((X_{t_1} - m_X(t_1))\,(X_{t_2} - m_X(t_2))^T\right)$$

is the covariance function of X_t.

A *Gaussian process* is a (stochastic) process which generates samples over time X_t such that no matter which finite linear combination of the X_t one takes (or, more generally, any linear functional of the sample function X_t), that linear combination will be normally distributed[18].

Stated differently, the stochastic process X_t is called a Gaussian process if all finite dimensional distributions are normally distributed. This means that for every finite subset $I_m = \{t_1, t_2, \ldots, t_m\}$ of I and every vector $a \in \mathbb{R}^{m \cdot d}$ the scalar product $Y = Z^T a$, with $Z^T = (X_{t_1}^T, X_{t_2}^T, \ldots, X_{t_m}^T)$, is normally distributed, i.e., Y has a probability density of the form $p(y) = c \exp\left(-h^2(y-\eta)^2\right)$. If the covariance matrix

$$C_X(I_m) = \begin{pmatrix} C_X(t_1,t_1) & \ldots & C_X(t_1,t_m) \\ \vdots & & \vdots \\ C_X(t_m,t_1) & \ldots & C_X(t_m,t_m) \end{pmatrix} \tag{1.7}$$

[18] Some authors also assume the random variables X_t to have mean zero.

of the random vector Z is non-singular, then, for $z \in \mathbb{R}^{m \cdot d}$,

$$p(\dot{z}) = \frac{\exp\left(-\frac{1}{2}(z - \mathbb{E})^T C_X^{-1}(I_m)(z - \mathbb{E})\right)}{\sqrt{(2\pi)^{m \cdot d} \cdot \det(C_X(I_m))}} \tag{1.8}$$

is the probability density of $Z \in S^{m \cdot d}$.

The probability distribution of a Gaussian process X_t is completely determined by its mean $A(t) := \mathbb{E}(X_t)$ and covariance $B(t,s) := \mathbb{E}((X_t - A(t))(X_t - A(t)))$. For any function $A(t)$ and any positive-definite function $B(t,s)$ there exists a Gaussian process X_t with expectation $A(t)$ and covariance function $B(t,s)$. Moreover, a d-dimensional stochastic process with vector values $X_t = (X_t^{(1)}, X_t^{(2)}, \ldots, X_t^{(d)})^T$ is called Gaussian if the joint probability distributions of arbitrary variables

$$X_{t_1}^{(i_1)}, \; X_{t_2}^{(i_2)}, \; \ldots, \; X_{t_n}^{(n_1)}$$

are Gaussian.

Lemma 1.27. *Let $m(t) : I \to \mathbb{R}^d$ and $C(t_1, t_2) : I \times I \to \mathbb{R}^{d \times d}$ such that for every subset I_m of I the matrix $C(I_m)$ is symmetric and non-negative definite (i.e., all eigenvalues of $C(I_m)$ are non-negative). Then, there is a Gaussian process with mean $m(t)$ and covariance function $C(t_1, t_2)$.*

Proof. See [87]. □

Let $X_t^{(r)}$, $r = 1, 2, \ldots$, be Gaussian processes on I and

$$\lim_{t \to \infty} X_t^{(r)} \hat{=} X_t \quad \text{or} \quad \text{m.s.-}\lim_{t \to \infty} X_t^{(r)} = X_t$$

hold for a stochastic process X_t for all $t \in I$. Then X_t is a Gaussian process, too. (Here, m.s.-lim denotes the mean-square limit.)

Besides the Wiener process W_t, which is not-stationary, but has stationary increments (and which we will discuss more deeply in Sec. 3.2.1), the *Ornstein-Uhlenbeck process*, see Fig. 1.14, is an example of a stationary Gaussian process. Such stationary Gaussian processes with continuous spectra are important examples of ergodic (stochastic) processes.

The Ornstein-Uhlenbeck process solves the following stochastic initial value problem

$$\mathrm{d}X_t = \lambda(\mu - X_t)\mathrm{d}t + \sigma \mathrm{d}W_t, \qquad X_0 = x_0,$$

with initial value x_0, mean reversion rate/ stiffness λ, mean μ and diffusion/ volatility σ. The symbol $\mathrm{d}W_t$ denotes the symbolic "differential" of the

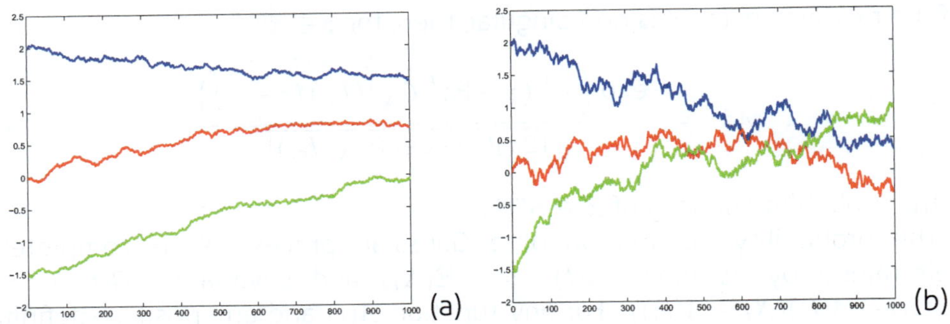

Figure 1.14. Sample paths of an Ornstein-Uhlenbeck process for initial values $x_0 \in \{-1.5, 0, 2\}$, with $\lambda = 1$, $\mu = 1.2$ and $\sigma = 0.3$ in (a) and $\lambda = 1$, $\mu = 0.5$ and $\sigma = 1$ in (b) displayed over the time axis.

Wiener process (though the Wiener process is nowhere differentiable as we will illustrate in Sec. 3.2.1), in particular, dW_t is the mathematically correct way of denoting white noise $w_t = dW_t$, cf. e.g. [97, 100, 159].

The paths of the Ornstein-Uhlenbeck process as displayed in Fig. 1.14 were plotted by applying the following MATLAB Example which represent a very simple explicit scheme for solving stochastic differential equations:

MATLAB Example 1.7. plotOU.m: Generating the plots of the Ornstein-Uhlenbeck process in Fig. 1.14.

```
T = 10;     N = 2^10;         dt = T/N;
randn('state',100);          dW = sqrt(dt)*randn(1,N);
lambda = 1; mu = 1.2;        sigma = 0.3; X0 = 0;
X = zeros(1,N);
for i = 1:N
   X(i) = X0 + lambda*(mu - X0)*dt + sigma*dW(i);
   X0 = X(i);
end
plot([0:dt:T], [X0, X], '-b','LineWidth',2)
```

The Ornstein-Uhlenbeck process will play a crucial role when converting stochastic differential equations into random differential equations, see Sec. 3.3.2. We will discuss this interesting process in more depth in Sec. 10.3.

Gaussian processes that are stationary in the narrow sense may be realised by way of certain dynamical systems (a shift in the space of trajectories [87]). The dynamical systems obtained (which are sometimes denoted as normal, on account of the resemblance to the normal probability distributions) are of interest as examples of dynamical systems with a continuous spectrum which can be exhaustively studied using a decomposition technique (see [140, 141]). The first actual examples of dynamical systems with "non-classical" spectral properties have been constructed in this way.

Nowadays, Gaussian processes are applied in such diverse fields as regression analysis (see [214, 113, 35]), approximation (see [72]), statistics for spatial data (see, [71, 238]), machine learning (see [215, 232]), prediction of deformed and annealed micro-structures (see [21]), or the discovery of bio-markers in micro-array experiments (see [59]). Another nice application is learning to control an octopus arm with Gaussian process temporal difference methods as discussed in [95].

Before you continue, make sure to answer the following questions:

Quiz: Section 1.5

Q1 Give the definition of a stochastic process and its sample paths. Give two examples of stochastic processes.

Q2 What does Kolmogorow's fundamental theorem state?

Q3 Give the definition of a stationary process.

Q4 What does indistinguishability of two stochastic processes mean?

Q5 Give the definition of a separable process.

Q6 Give the definition of a continuous stochastic process.

Q7 What does Kolmogorow's continuity theorem state?

Q8 Give the definition of a filtration and a super-martingale together with the correct formulation of the super-martingale inequality.

Q9 Give the definitions of a stochastic process with a) independent increments and b) stationary increments.

Q10 What is a Gaussian process and which properties does it have with respect to its mean and variance?

Q11 What does ergodicity mean for a stochastic process? Give two examples for ergodic (stochastic) processes.

1.6 Chapter's Summary

In this chapter, we briefly recalled the basic definitions and concepts of random variables, namely probability, expectation, variance, conditional expectation, convergence of random variables, as well as the fundamental definitions and concepts of (time-continuous) stochastic processes. Thereby, a bunch of MATLAB examples rounded the exposition.

The contents of this chapter are well documented in the literature as they form the core of every probability class being worth its name. In case of difficulties or in search for a deeper exposition, we encourage the reader to consult the following works that we considered to be very helpful during our own studies and research (sorted alphabetically): [14], [25], [61], [79], [122], and [153].

We assume that at least the part on random variables is already known to the reader, and that the review on stochastic processes was not too technical for those who have never been exposed to deeper ideas in stochastic analysis. The contents of this chapter are more than sufficient to succeed with the topics of the first part of this book "Motivation and Decomposition of Multi-Storey Building Excitation Problems".

Problems

Classification: ☼ easy, ◷ easy with longer calculations, ✩ a little bit difficult,
challenging.

Exercise 1.28. [☼] **Recalling the Basics of Probability Theory – pt. 1**
Suppose X is a random variable having the probability density function

$$f(x) = \begin{cases} R \cdot x^{R-1} & \text{for } 0 \leq x \leq 1, \\ 0 & \text{elsewhere,} \end{cases}$$

where $R > 0$ is a fixed parameter.

1. Determine the distribution function $F_X(x)$.
2. Determine the mean $\mathbb{E}(X)$.
3. Determine the variance $\mathbb{V}ar(X)$.

Exercise 1.29. [☼] **Recalling the Basics of Probability Theory – pt. 2**
A special piece of control hardware for the ARIANE 5 rocket consists of two

main components: A and B. The operating times until failure of the two components are independent and exponentially distributed random variables with parameter 2 for component A, and 3 for B. The system fails at the first component failure making it impossible to land the ARIANE rocket safely.

1. What is the mean time to failure for component A? For component B?
2. What is the mean time to system failure?
3. What is the probability that it is component A that causes the system failure?
4. Suppose that it is component A that fails first. What is the mean remaining operating life of component B?

Exercise 1.30. [☼] **Cycling Home**
When cycling home at night, I notice that sometimes my rear light is switched off when I arrive home. Presumably the switch is loose and can flip from on to off or back again when I go over bumps. I suppose that the number n of flippings per trip has a Poisson distribution $\exp(\lambda)\lambda^n/n!$. If the probability that the light is still on when I arrive home is p, find λ.

Exercise 1.31. [☼] **MATLAB Exercise: Random Variables and Densities**

1. Generate 100 uniformly distributed random variables $\mathrm{X} = \mathrm{rand}(1,\ 100)$, and plot the histogram with 10 bins, $\mathrm{hist}(\mathrm{X},\ 10)$. This histogram bins the elements of X into 10 equally spaced containers (non-overlapping intervals of length 0.1) and returns the number of elements in each container. Next, generate 10^5 uniformly distributed random variables by $\mathrm{rand}(1,\ 10^5)$, plot the histogram $\mathrm{hist}(\mathrm{X},\ 100)$, comment on the difference. What density function is the histogram approximating?
2. Repeat the first part for a Gaussian random variable generated by $\mathrm{X} = \mathrm{randn}(1,\ 10^5)$, with a plot of histogram $\mathrm{hist}(\mathrm{X},\ 100)$.
3. Let $\mathrm{X} = \mathrm{rand}(1,\ 10^5)$, $\mathrm{Y} = \text{-}\ln(1\text{ - }\mathrm{X})/0.3$. Find the density function of Y and compare with $\mathrm{hist}(\mathrm{Y},\ 100)$.

Exercise 1.32. [☼] **Conditional Expectation and the Normal Distribution**

1. Let $X \sim \mathcal{N}(0,1)$ be a normally distributed random variable with mean 0 and variance 1. Suppose that $x \in \mathbb{R}$, $x > 0$. Find upper and lower bounds for the conditional expectation

$$\mathbb{E}\left(X|X \geq x\right) .$$

2. Now suppose that X has a power law distribution, $\mathbb{P}(X \geq x) = \alpha x^{-\beta}$, for $x \geq x_0 > 0$, and some $\alpha > 0, \beta > 1$. Calculate the conditional expectation

$$\mathbb{E}\left(X|X \geq x\right) , \qquad x \geq x_0 .$$

Exercise 1.33. [☼] **An Example for a Binomial Random Variable**
Let the uniform probability measure on $\{0,1\}^d$, or more generally the d-fold product of a measure which puts mass p on 1 and mass $(1-p)$ on 0 be given. Show that the function $f(x) = \sum_{i=1}^{d} x_i$, $x = (x_1, \ldots, x_d)^T$, is a binomial random variable.

Exercise 1.34. [☼] **Drug Tests and Competitions**

1. A drug trial gives the result that the drug works better than the placebo, with 95% confidence. What exactly does this statement mean? What further assumptions are needed to be able to deduce that the probability of the drug working is actually 95%?

2. A company has a competition to win a car. Each contestant needs to pick a positive integer. If there is at least one unique choice, the person who made the smallest unique choice wins the car. If there are no unique choices, the company keeps the car and there's no repeat of the competition. It turns out that there are only three contestants, and you're one of them. Everyone knows before picking their numbers that there are only three contestants. How should you make your choice?

Exercise 1.35. [☼] **Joint Density of Normally Distributed Random Variables**
Consider two independent normally distributed random variables $X \sim \mathcal{N}(\mu_1, \sigma_1)$ and $Y \sim \mathcal{N}(\mu_2, \sigma_2)$. Determine the joint density function of X and Y.

Exercise 1.36. [☼] **Escape From a Maze**
A mouse in a maze has three possible doors to escape from. If he escapes from door 1, he gets caught in 4 hours and goes back to the maze. If he leaves through door 2 he gets caught in 2 hours and goes back. Finally, if he leaves through door 3, then he is gone forever. Assume that the mouse forgets which door he escaped from each time he is put back in the maze. Find the average number of hours it takes for the mouse to escape forever.

Hint: Let x be the random variable corresponding to the number of hours it takes the mouse to be free. Recall that

$$\mathbb{E}(X) = \int_{-\infty}^{\infty} \mathbb{E}(X|Y=y) f_Y(y) \mathrm{d}y .$$

In our setting, $f_Y(y)$ is a density function consisting of three delta functions. Hence

$$\mathbb{E}(X) = \mathbb{E}(X|D_1)\mathbb{P}(D_1) + \mathbb{E}(X|D_2)\mathbb{P}(D_2) + \mathbb{E}(X|D_3)\mathbb{P}(D_3)\,,$$

where D_j ($j = 1, 2, 3$) is the event that the mouse escapes through door j. Now set $\mu = \mathbb{E}(x)$ and solve for μ.

Exercise 1.37. [☼] **MATLAB Exercise: Properties of Gaussian Random Variables**

1. Let $Z := (N_1, N_2)$, where N_1 and N_2 are independent zero mean unit Gaussian random variables. Let S be an invertible 2×2-matrix, show that $X = S^T Z$ is jointly Gaussian with zero mean, and covariance matrix $S^T S$.

2. Write a program to generate a pair of Gaussian random numbers (X_1, X_2) with zero mean and covariance $E(X_1^2) = 1$, $E(X_2^2) = 1/3$, $E(X_1 X_2) = 1/2$. Generate 1000 pairs of such numbers, evaluate their sample averages and sample covariances.

Exercise 1.38. [☼] **Expectation & Standard Deviation of Discrete Random Variables** Consider a set Ω with N distinct members, and a function $f : \Omega \to \{0, 1\}$, such that $\frac{1}{N}\sum_{x\in\Omega} f(x) = p$. For a subset $S \subset \Omega$ of size n, define the sample proportion

$$p := p(S) = \tfrac{1}{n}\sum_{x\in S} f(x)\,.$$

If each subset of size n is chosen with equal probability, calculate the expectation and standard deviation of the random variable p.

Exercise 1.39. [☆] **The Poincare Limit**
Let the rotation-invariant probability measure on the d-dimensional sphere $\sqrt{d}\mathbb{S}^{d-1}$ be given. Show that the projection $f(x) = x_1$, $x = (x_1, \ldots, x_d)^T$, is a random variable which converges in distribution to the standard Gaussian distribution when $d \to \infty$. (This is sometimes called the "Poincare limit" because it was first observed by Maxwell and first rigorously proved by Borel.)

Exercise 1.40. [☆] **On Convergence in Distribution**
Let the uniform probability measure on the permutation group S_n be given. Show that the number of fixed points of a permutation is a random variable which converges in distribution to a Poisson distribution with mean 1.

Exercise 1.41. [☼] **Expectation & Covariance of a Stochastic Process – pt. 1**
Consider a random variable X uniformly distributed on $[0, \pi]$, and define the stochastic process $(Y_t)_{t\geq 0}$ by $X_t := X\cos(\omega t)$.

1. Determine the mean/ expectation function of the process $(Y_t)_{t\geq 0}$.

2. Determine the covariance function of the process $(Y_t)_{t\geq 0}$.

Exercise 1.42. [☼] **Expectation & Covariance of a Stochastic Process – pt. 2**
Consider two independent random variables X uniformly distributed on $[0,\pi]$ and φ uniformly distributed on $[-\pi,\pi]$. Let the stochastic process $(Y_t)_{t\geq 0}$ be defined by $X_t := X\cos(\omega t+\varphi)$.

1. Determine the mean/ expectation function of the process $(Y_t)_{t\geq 0}$.

2. Determine the covariance function of the process $(Y_t)_{t\geq 0}$.

Exercise 1.43. [☼] **Properties of Stationary Processes**
Let $I\subset\mathbb{R}_0^t$ and $(X_t)_{t\in I}$ be a stationary stochastic process.

1. Show that the mean function is constant.

2. Show that for all s,t the covariance function $R(t,t+s)$ only depends on s.

Exercise 1.44. [☼] **Continuity and Gaussian Processes**
Let $I:=[0,T]\subset\mathbb{R}_0^t$ and suppose that the stochastic process $(X_t)_{t\in I}$ is such that $\mathbb{E}(X_t)=0$ and $\mathbb{E}(X_t^2)=1$ for all $t\in I$.

1. Show that $|\mathbb{E}(X_tX_{t+h})|\leq 1$ for any $h>0$ and $t\in[0,T-h]$ by applying Jensen's inequality.

2. Suppose that for some $\lambda<\infty$, $p>1$, and $h_0>0$, as well as $|\mathbb{E}(X_tX_{t+h})|\geq 1-\lambda h^p$ for all $0<h\leq h_0$. Using Kolmogorow's continuity theorem show that then $(X_t)_{t\in I}$ has a continuous modification.

3. Suppose that $(X_t)_{t\in I}$ is a Gaussian stochastic process such that $\mathbb{E}(X_t)=0$ and $\mathbb{E}(X_t^2)=1$ for all $t\in I$. Show that if $(X_t)_{t\in I}$ satisfies the inequality

$$\mathbb{E}\left((X_{t+h}-X_t)^{2n}\right) = \frac{(2n)!}{2^n n!}\left(\mathbb{E}\left((X_{t+h}-X_t)^2\right)\right)^n \leq \left(\frac{(2n)!}{n!}\right)\lambda^n h^{pn}$$

for some $\lambda<\infty$, $p>1$, and $h_0>0$, then for any $0<\gamma<\frac{1}{2}p$, the process $(X_t)_{t\in I}$ has a modification which is locally Hölder continuous with exponent γ.

Chapter 2

Reduction of RPDEs to RODEs

In this chapter, specific Random Partial Differential Equations (RPDEs) are transformed to Random Ordinary Differential Equations (RODEs) by applying classical spatial discretisations. Variants from a variety of applications, are discussed, leaving the time discretisation for Chap. 7. The derivation of the underlying system of PDEs and corresponding boundary conditions is shown for the example of elastic body motion. We discuss different types of meshes with an emphasis on regular Cartesian grids. The three main spatial discretisations—finite differences, finite volumes, and finite elements—are briefly explained before delving deeper into finite difference schemes (FD). We derive the corresponding FD approximations for the fundamental equations of elastic body motion and simulate steady-state scenarios of buildings which are bent.

2.1 Key Concepts

Random partial differential equations (RPDEs) are partial differential equations that include random effects. A partial differential equation (PDE) is an equation for an unknown quantity (function) u involving derivatives with re-

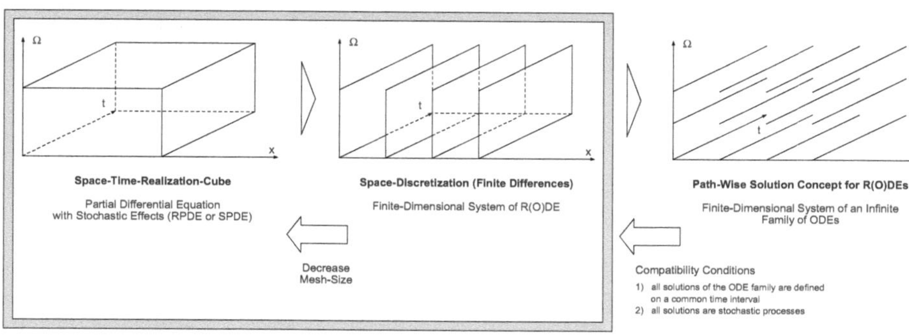

Figure 2.1. Reduction from a given continuum mechanical random partial differential equation to a family of deterministic ordinary differential equations. In this chapter we focus on the first step which involves getting from a random partial differential equation to a finite-system of random (ordinary) differential equations.

spect to more than one independent variable (i.e. partial derivatives). Partial differential equations are necessary when spatially distributed effects play a non-negligible role. The criteria for using a PDE vs. an ODE are not problem-dependent but depend on the desired effects: For instance, ODEs can be used to model population dynamics of individuals of a given species over time (see Chaps. 6 and 7 for examples). When a spatial distribution becomes relevant (where do more rabbits or foxes live in a forest, etc.), partial differential equations are typically the method of choice.

As an example, consider the time-dependent heat equation in d dimensions as a relatively simple example of a partial differential equation[1],

$$\frac{\partial T}{\partial t}(t,x) = \kappa \cdot \Delta T(t,x) + f(t,x) \qquad \text{for} \quad t \in I := [t_0, t_e] \subset \mathbb{R}, x \in \Omega \subset \mathbb{R}^d \tag{2.1}$$

Derivatives with respect to time ($\partial T/\partial t$) and space (Laplacian ΔT) appear and actually make (2.1) a PDE. The coefficient κ represents the thermal diffusivity of the material under consideration. We are interested in values of the temperature $T = T(t,x)$ at various points in time and space. The source term f may contain heat sources or sinks in the domain due to chemical reactions or a heating radiation, for example. Depending on the desired effects to be modeled, there is still a freedom in the choice of the dimensionality: Using $d = 1$, $d = 2$, or $d = 3$ for a heat bar ($T(t,x)$), a axially symmetric setup ($T(t,x,y)$), or a fully unsymmetric problem ($T(t,x,y,z)$), changes the accuracy but also increases the complexity in solving the resulting systems.

Note that a PDE describes physical effects inside the domain of interest (temperature variation in space and time in (2.1), e.g.). The equation determines a number of solutions. To identify a unique solution to the problem, *initial conditions* and *boundary conditions* have to be fixed. Of course, as the underlying equations, the boundary and initial conditions have to be derived in that they respect the underlying physics. In the case of the heat equation, these conditions imply

$$\begin{array}{rcll} T(t_0,x) & = & g(x) & \text{in } \Omega\,, \\ T(t,x) & = & T_0(t,x) & \text{on } \Gamma_0 \subset \partial\Omega\,, \\ \frac{\partial T}{\partial n}(t,x) & = & r(t,x) & \text{on } \Gamma_1 \subset \partial\Omega\,, \end{array} \tag{2.2}$$

i.e. a given temperature field on the whole domain at start time t_0 and temperature data in the form of prescribed values or changes on disjoint parts Γ_0 and Γ_1 of the boundary $\partial\Omega$. The former type of boundary conditions are generally called *Dirichlet conditions*, whereas the latter are *Neumann conditions*.

For RPDEs, the randomness may appear in different forms. For instance, material parameters may be available only in the form of (error-prone) mea-

[1] since it the heat equation is scalar, linear, and only of second order because the highest derivatives are given in ΔT

surements, or boundary conditions stemming from measurements. See also Chap. 5 for a more detailed discussion on various random effects in the context of flow scenarios. Formally, RPDEs are (deterministic) PDEs in a path-wise sense, a concept that will be explained in Chap. 3 and excludes stochastic influences that are too weird.

In the following, we assume random effects that do not have any spatial distribution but vary in time only. One example would be randomly perturbed boundary conditions in time that are identical for all boundary points at a given moment in time. For the case of our earthquake scenario described in Chap. 3 and computed—in a RPDE context—in Chap. 18, such boundary conditions are the earthquake-incited motion of the earth surface representing a given motion for the bottom boundary of a building. With this restriction, we can directly apply a spatial discretisation to the RPDE to reformulate it as a system of RODEs of large dimension (see Fig. 2.1 for a visualisation of the approach). Then, the classical RODE approach and mathematical instruments discussed in Chaps 3, 4, 12, and 13 can be applied and used. The technical steps of discretising the RPDE in space will be discussed in more detail in this chapter.

Typical fields of applications using (R)PDEs comprise many problems and phenomena from physics and continuum mechanics such as:

- **Thermodynamics**: How does the heat distribution in a combustion engine look?
- **Fluid Mechanics**: Where is a tornado going to take place? What are the drag and lift coefficients of a wing or car?
- **Structural Dynamics**: What is the maximum stress or strain in a under time-dependent loads? Will an anti-crash construction in a car do the job it has been designed for? Will a bridge be stable under loads from trucks or wind effects?
- **Chemical Engineering**: Where and how fast do certain reactions take place in a chemical reactor?
- **Electromagnetism**: What is the spatial distribution of the electron density in a transistor? How has an MRI machine to be designed to deliver images of a desired quality?
- **Geology**: Where and/or when will an earthquake take place? What are the far-range effects of a tsunami?

Modern problems may involve even more complex scenarios by coupling different fields such as in fluid-structure interaction scenarios[2]. We cover the

[2] This coupling between different fields of application is typically called *multi physics*.

topic of structural dynamics as the main source of applications in this chapter and briefly consider flow problems in Chap. 5.

Numerical approaches to classical PDEs are numerous and complex. Solving PDEs (numerically) comes at a certain price: resolving the spatial distribution (approximately) requires more computational resources, and correct (in the sense of consistency and convergence) schemes need to be developed.

When reading this chapter note the answers to the following questions

1. How do the governing equations for elastic body motions look?
2. What possibilities for space-time discretisations exist?
3. How does the discretisation of the elastic equations look for finite difference approximations?

as well as the following key concepts

1. Derivation of PDEs via continuum mechanics considerations,
2. Categorisation of various meshes to discretise the spatial domain,
3. Main variants of spatial discretisation schemes,
4. Finite difference approximations and their quality.

This chapter is structured as follows: We derive the governing equations of motions in form of a PDE system for elastic materials in Sec. 2.2. Section 2.3 contains basic information on space-time discretisations. Different types of meshes, in particular Cartesian meshes, are discussed as well as the three main methods of discretising the operators of the underlying PDE. Details on finite difference approximations are presented in general and for the elastic equations in particular in Sec. 2.4. Finally, Section 2.5 wraps up the contents of this chapter.

2.2 Elastic Materials & Material Laws

We start this chapter with a detailed discussion on elastic material behaviour, mostly following [37]. This serves a double purpose. First, the governing equations for elastic body motion are an example how PDEs are typically derived in continuum mechanics. Second, we will get acquainted with the continuum mechanics view for a straightforward modeling approach of moving buildings: The governing equations are used for the discretisations in the following sections and chapters. In the workshop, in particular, a very simple but spatially dependent version of a multi-storey building will be used: The building is modeled as a 2D domain fully covered by material without any rooms. To the discretisation of this type of building, we will apply the concept of space-filling curves (see Chap. 11) and variants of the Fast Fourier transform (cf. Chap. 9).

2.2.1 Basic Aspects of Continuum Mechanics

The hypothesis of a continuous distribution of the material throughout the complete domain of interest is the classical assumption for continuum mechanics. In contrast to investigations on microscopic or even atomic levels, where individual molecules and their corresponding interactions are studied, the macroscopic properties of a large number of particles is of central importance. Hence, the material is assumed to be continuously distributed at each geometric point x in space.

In continuum mechanics, two basic approaches exist which differ with respect to the point of view of the observer: the **Lagrangian** and **Eulerian** description. In the Lagrangian or material-based approach, the point of view of the observer is fixed to a particle of the material and follows its movement. Solid mechanics problems, which often show relatively small movements, are typically described with the Lagrangian approach. In contrast, the Eulerian point of view is restricted to a fixed control volume. The observer watches different material passing through a kind of window, keeping track of the over-arching properties but not those of each particle. Using the Eulerian approach, large changes and movements can easily be described. This fact is particularly advantageous for classical fluid mechanicsand will be used in Sec. 5.2.

We are now going to apply the Lagrangian description to elastic bodies. In the following, $[t_0, t_1] \subset \mathbb{R}$ represents an interval of time and $x \in \mathbb{R}^3$ denotes a coordinate in space. Let an elastic body B be represented in a reference configuration $\overline{\Omega}$ with respect to a fixed coordinate system $\{O_B, e_B\}$. In Fig. 2.2 (a), the body B is sketched. We denote the domain (i.e. the inner part of B) by Ω while Γ represents the boundary of the body. The behaviour of the body under loads is typically described via deformation and displacement mappings.

Definition 2.1. The *deformation* of an elastic body B is a mapping

$$\begin{aligned} \varphi : \overline{\Omega} \times [t_0, t_1] &\rightarrow \mathbb{R}^3 \\ (x, t) &\mapsto \varphi(x, t) \end{aligned}$$

Definition 2.2. The *displacement* of an elastic body B is a mapping

$$\begin{aligned} u : \overline{\Omega} \times [t_0, t_1] &\rightarrow \mathbb{R}^3 \\ (x, t) &\mapsto u(x, t) := \varphi(x, t) - x \,. \end{aligned}$$

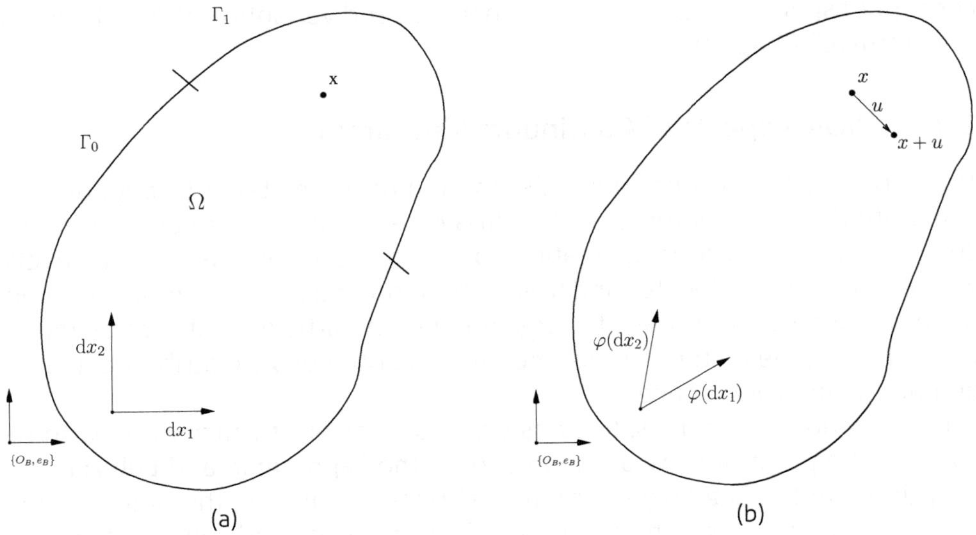

Figure 2.2. Visualisation of an elastic body: reference configuration (a) and deformed configuration showing the displacement u of an example point x and the deformation φ applied to infinitesimal vectors dx_i.

In Fig. 2.2 (b), the displacement and deformation are sketched. The deformation contains, in particular, possible axis-parallel scaling as well as changes of angles. If external conditions of a scenario demand a prescribed displacement of (parts of) the boundary of the body B, the solution (i.e. the mappings u and φ) need to respect this additional information. Such boundary conditions are typically called Dirichlet boundary conditions. They will allow us to incorporate random excitations of the ground into the discrete PDE system which we now derive.

2.2.2 Stress & Strain

In order to derive and describe the underlying equations for elastic bodies, the definitions of the strain and stress tensors are important.

Checking the strain or distortion of a certain control volume of material, we can formulate the tensor E as

$$E_{ij} = \frac{1}{2}\left(\nabla\varphi^T\nabla\varphi - I\right) = \frac{1}{2}\left(\nabla u + \nabla u^T + \nabla u^T \nabla u\right) . \qquad (2.3)$$

The tensor E is typically called the *Green-St. Venant strain tensor*[3]. It is easy to show that this strain tensor is invariant under rigid body motions (see Prob-

[3] also called *Green-Lagrangian strain tensor*

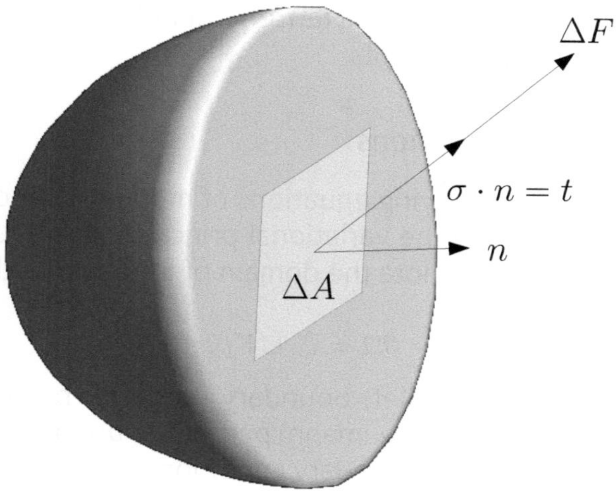

Figure 2.3. Sketch of an axis-aligned slice of an infinitesimal volume element. The stress vector t is equal to the infinitesimal force ΔF divided by the infinitesimal surface element ΔA.

lem 2.7), which is an essential property if we want to describe distortions. In case of the linear elasticity theory, where only first-order terms in the displacement u are considered, the tensor E reduces to the strain tensor ε:

Definition 2.3. The *strain tensor* ε has the form

$$\varepsilon \;=\; \varepsilon(u) \;=\; \left(\nabla u + \nabla u^T\right) . \tag{2.4}$$

The definition of stress relates the surface force to a surface element ΔA. The stress can be interpreted as the action of inner forces of a body on the surface of an infinitesimally small volume element (cf. Fig. 2.3). The three possible slices through the surface point x orthogonal to one Cartesian coordinate axis, respectively, correspond to the three columns of the stress tensor σ. The stress vector t at location x for a surface described by the normal vector n is defined as

$$t(x,n) := \lim_{\Delta A \to 0} \frac{\Delta F\,(\Delta A)}{\Delta A} = \frac{dF(A)}{dA} .$$

and may then be written as

$$t(x,n) = \sigma \cdot n \tag{2.5}$$

The notion of material laws now relates the stress and the strain tensors. The simplest case is Hooke's Law:

$$\sigma \;=\; \lambda tr\{\varepsilon(u)\}I + 2\mu\varepsilon(u) , \tag{2.6}$$

where tr represents the trace of the tensor, I denotes the identity, and λ and μ are the so-called Lamé constants.

2.2.3 Fundamental Equations

In order to derive the underlying equations of motion (a set of PDE) for elastic bodies, we are applying the variational principle to the functional of energy of the system. Let Ω denote the domain of interest (i.e. the elastic body, cf. Fig. 2.2) and

$$\partial\Omega = \Gamma_0 \cup \Gamma_1 \,, \tag{2.7}$$

where Γ_0 denotes the (Dirichlet) boundary part with prescribed displacements and Γ_1 represents the (Neumann) part of the boundary $\partial\Omega$, where surface forces are prescribed as boundary conditions, respectively.

The inner energy (or strain energy) of an elastic body is defined as

$$W := \frac{1}{2}\int_\Omega \sigma : \varepsilon dx \,.$$

Here, $\sigma : \varepsilon = tr\{\sigma^T \varepsilon\}$ represents the double contraction of the two tensors (cf. [226] for a nice representation of tensor calculus).

Together with volume and surface forces, the strain energy W defines the potential energy U of the system,

$$U = \underbrace{\frac{1}{2}\int_\Omega \sigma : \varepsilon dx}_{=W} - \int_\Omega u^T \beta dx - \int_{\Gamma_1} u^T \tau ds \,, \tag{2.8}$$

with the density of volume forces β and the surface stress vector τ.

Finally, we use the kinetic energy

$$T = \frac{1}{2}\int_\Omega \rho \dot{u}^T \dot{u} \tag{2.9}$$

as the integrated square of the velocity $\dot{u}$ weighted with the constant mass density ρ. The variational principle[4] characterises the motion of an elastic body via the requirement that the functional

$$\int_{t_0}^{t_1} (T - U)\, dt \tag{2.10}$$

has to be stationary, i.e. the solution of the motion—the function u—represents an extremal point (a minimum, a maximum, or a saddle point)

4 Hamilton's principle, in this case

for the functional compared to competitor functions. Assuming that we know this solution u, we formally may define

$$\text{competitor functions} \quad : \quad u + \theta v \quad \text{with } \theta \in \mathbb{R}$$
$$\text{and}$$
$$\text{variations} \quad : \quad \delta u := \theta v \,.$$

Furthermore, we assume that v is in a "suitable space" (without going into details) and that

$$v(x, t_0) \;=\; 0 = v(x, t_1) \quad \forall x \in \overline{\Omega}\,, \tag{2.11}$$
$$v(., t) \;=\; 0 \quad \text{on } \Gamma_0 \quad \forall t \in [t_0, t_1] \tag{2.12}$$

holds. The condition (2.11) implies that the trajectories of u and $u + \theta v$ are identical at the temporal start and end points, whereas (2.12) states that the competitor functions have to respect the given Dirichlet boundary conditions of the problem. Using the functional

$$J(\theta) \;:=\; \int_{t_0}^{t_1} \left(T(u + \theta v) - U(u + \theta v)\right) dt\,, \tag{2.13}$$

the stationarity of (2.10) now reads

$$0 \;=\; \frac{d}{d\theta} J(\theta)|_{\theta=0} \tag{2.14}$$

since u is the assumed solution. Plugging in the definitions (2.9) and (2.8) for T and U, (2.14) reads

$$\begin{aligned} 0 \;=\; & \frac{d}{d\theta} \int_{t_0}^{t_1} \left[\frac{1}{2} \int_\Omega \rho \overline{\dot{(u + \theta v)}}^T \overline{\dot{(u + \theta v)}} dx - \frac{1}{2} \int_\Omega \sigma(u + \theta v) : \varepsilon(u + \theta v) dx \right. \\ & \left. \left. + \int_\Omega (u + \theta v)^T \beta dx + \int_{\Gamma_1} (u + \theta v)^T \tau ds \right] dt \right|_{\theta=0} \end{aligned} \tag{2.15}$$

Performing the derivation with respect to θ, we obtain with some calculations the following form:

$$0 = \int_{t_0}^{t_1} \left[\int_\Omega \rho \dot{v}^T \dot{u} dx - \int_\Omega \sigma(u) : \varepsilon(v) dx + \int_\Omega v^T \beta dx + \int_{\Gamma_1} v^T \tau ds \right] dt\,. \tag{2.16}$$

We apply an integration by parts with respect to time on the first term in (2.16) to get

$$\int_{t_0}^{t_1} \left[\int_\Omega \rho \dot{v}^T \dot{u}\, dx \right] dt \;=\; \underbrace{\left. \int_\Omega \rho v^T \dot{u}\, dx \right|_{t_0}^{t_1}}_{=0 \text{ due to (2.11)}} - \int_{t_0}^{t_1} \int_\Omega \rho v^T \ddot{u}\, dx\, dt\,. \tag{2.17}$$

Inserting (2.17) in (2.16) results in

$$0 = \int_{t_0}^{t_1} \left[\int_\Omega \rho v^T \ddot{u} dx + \int_\Omega \sigma(u) : \varepsilon(v) dx - \int_\Omega v^T \beta dx - \int_{\Gamma_1} v^T \tau ds \right] dt \,. \quad (2.18)$$

Applying the divergence theorem to the double contraction in (2.18) we obtain

$$0 = \int_{t_0}^{t_1} \left[\int_\Omega v^T (\rho\ddot{u} - \mathrm{div}(\sigma) - \beta)\, dx + \int_{\Gamma_1} v^T (\sigma \cdot n - \tau)\, ds \right] dt \;. \quad (2.19)$$

Under the condition that v vanishes not only on Γ_0 but also on Γ_1, the boundary integral term in the right-hand side of (2.19) vanishes and the fundamental lemma of the calculus of variations (see [68]) gives

$$0 \;=\; \rho\ddot{u} - \mathrm{div}(\sigma) - \beta\,. \quad (2.20)$$

If $v \neq 0$ on Γ_1, the condition $\sigma \cdot n - \tau = 0$ is necessary to again obtain (2.20) via the fundamental lemma of the calculus of variations. Hence, (2.20) together with the boundary conditions on the Dirichlet and Neumann boundary Γ_0 and Γ_1 forms the set of equations of motion for our elastic body:

$$\begin{aligned} \rho\ddot{u}(x,t) &= \mathrm{div}\sigma(u(x,t)) + \beta(x,t) && \text{in } \Omega && (2.21)\\ u(x,t) &= u_0(x,t) && \text{on } \Gamma_0 && (2.22)\\ \sigma(u(x,t)) \cdot n(x) &= \tau(x,t) && \text{on } \Gamma_1\,. && (2.23) \end{aligned}$$

Applying suitable initial conditions, the continuous problem is now fully described. Note that random effects can be considered in the boundary conditions, i.e. in u_0 and τ. For the case of earth-quake induced random motion of the ground—as we will use it in the Workshop, see Chaps. 17 and 18—the random effects are purely time-dependent. Hence, random effects in the PDE directly translate to random effects in the ODE resulting in an RODE setup.

We are now going to derive a formulation of Eqs. (2.21)–(2.23) that implies derivatives on the degrees of freedom $u = (u^1, \ldots, u^d)^T$ directly. Therefore, we have to replace the σ terms in the formulas. In three dimensions, the divergence of the tensor σ is defined as

$$\mathrm{div} \begin{pmatrix} \sigma_{11} & \sigma_{12} & \sigma_{13} \\ \sigma_{21} & \sigma_{22} & \sigma_{23} \\ \sigma_{31} & \sigma_{32} & \sigma_{33} \end{pmatrix} = \begin{pmatrix} \frac{\partial \sigma_{11}}{\partial x_1} + \frac{\partial \sigma_{12}}{\partial x_2} + \frac{\partial \sigma_{13}}{\partial x_3} \\ \frac{\partial \sigma_{21}}{\partial x_1} + \frac{\partial \sigma_{22}}{\partial x_2} + \frac{\partial \sigma_{23}}{\partial x_3} \\ \frac{\partial \sigma_{31}}{\partial x_1} + \frac{\partial \sigma_{32}}{\partial x_2} + \frac{\partial \sigma_{33}}{\partial x_3} \end{pmatrix} \quad (2.24)$$

Using the special case of two dimensions, we are now going to simplify the underlying equations in two steps. First, we derive the 2D form of (2.21)–(2.23) using the explicit form of the divergence of the stress tensor σ in 2D.

Similar calculations could be carried out for the three-dimensional case, too, so this demonstrates how to derive the corresponding explicit form. Second, we need to distinguish two variants of "2D materials" as limit cases of physical 3D scenarios: the plane stress and the plane strain. This will result in a special condition on our material parameters. Performing the first of the two mentioned steps, we derive the first component of the divergence of the stress tensor as an example case using its definition (2.6):

$$\begin{aligned}(\mathrm{div}(\sigma))_1 &= \frac{\partial}{\partial x_1}\left(\lambda\sum_{k=1}^{2}\frac{\partial u^k}{\partial x_k}+\mu\left(\frac{\partial u^1}{\partial x_1}+\frac{\partial u^1}{\partial x_1}\right)\right)+\frac{\partial}{\partial x_2}\mu\left(\frac{\partial u^1}{\partial x_2}+\frac{\partial u^2}{\partial x_1}\right)\\ &= (2\mu+\lambda)\frac{\partial^2 u^1}{\partial x_1^2}+\mu\frac{\partial^2 u^1}{\partial x_2^2}+(\lambda+\mu)\frac{\partial^2 u^2}{\partial x_1\partial x_2}\end{aligned}$$

In an analogous manner, we compute the second component as

$$(\mathrm{div}(\sigma))_2 = (2\mu+\lambda)\frac{\partial^2 u^2}{\partial x_2^2}+\mu\frac{\partial^2 u^2}{\partial x_1^2}+(\lambda+\mu)\frac{\partial^2 u^1}{\partial x_1\partial x_2}\,.$$

Hence, we obtain the following 2D system of coupled PDEs formulated in the unknowns u:

$$\rho\ddot{u}^1(x,t) = (2\mu+\lambda)\frac{\partial^2 u^1}{\partial x_1^2}+\mu\frac{\partial^2 u^1}{\partial x_2^2}+(\lambda+\mu)\frac{\partial^2 u^2}{\partial x_1\partial x_2}+\beta_1(x,t)\text{in} \quad \Omega \tag{2.25}$$

$$\rho\ddot{u}^2(x,t) = (2\mu+\lambda)\frac{\partial^2 u^2}{\partial x_2^2}+\mu\frac{\partial^2 u^2}{\partial x_1^2}+(\lambda+\mu)\frac{\partial^2 u^1}{\partial x_1\partial x_2}+\beta_2(x,t)\text{in} \quad \Omega \tag{2.26}$$

$$u(x,t) = u_0(x,t) \qquad \text{on} \quad \Gamma_0 \tag{2.27}$$

$$\sigma(u(x,t))\cdot n(x) = \tau(x,t) \qquad \text{on} \quad \Gamma_1 \tag{2.28}$$

For our application of moving buildings, we may assume a Dirichlet boundary (Γ_0) on the bottom and force boundary conditions on the rest of a rectangular domain (Γ_1), for the sake of simplicity. Hence, we have to replace σ in (2.28) by derivatives of u. The normal vector n changes its direction on each boundary. If we formulate the boundary conditions explicitly, we obtain the following equations on the boundary edges:

- Upper boundary: $n(x) = (0,1)^T$

$$\tau(x,t) = \begin{pmatrix}\mu\left(\frac{\partial u^2}{\partial x_1}+\frac{\partial u^1}{\partial x_2}\right)\\ (\lambda+2\mu)\frac{\partial u^2}{\partial x_2}+\lambda\frac{\partial u^1}{\partial x_1}\end{pmatrix}$$

- Right boundary: $n(x) = (1,0)^T$

$$\tau(x,t) = \begin{pmatrix} (\lambda+2\mu)\frac{\partial u^1}{\partial x_1} + \lambda\frac{\partial u^2}{\partial x_2} \\ \mu\left(\frac{\partial u^2}{\partial x_1} + \frac{\partial u^1}{\partial x_2}\right) \end{pmatrix}$$

- Left Boundary: $n(x) = (-1,0)^T$

$$\tau(x,t) = \begin{pmatrix} -(\lambda+2\mu)\frac{\partial u^1}{\partial x_1} - \lambda\frac{\partial u^2}{\partial x_2} \\ -\mu\left(\frac{\partial u^2}{\partial x_1} + \frac{\partial u^1}{\partial x_2}\right) \end{pmatrix}$$

At the corners of the domain, one should use an averaged normal vector using boundary information from both borders:

- Left upper corner: $n(x) = (-1,1)^T$

$$\tau(x,t) = \begin{pmatrix} -(\lambda+2\mu)\frac{\partial u^1}{\partial x_1} - \lambda\frac{\partial u^2}{\partial x_2} + \mu\left(\frac{\partial u^2}{\partial x_1} + \frac{\partial u^1}{\partial x_2}\right) \\ -\mu\left(\frac{\partial u^2}{\partial x_1} + \frac{\partial u^1}{\partial x_2}\right) + (\lambda+2\mu)\frac{\partial u^2}{\partial x_2} + \lambda\frac{\partial u^1}{\partial x_1} \end{pmatrix}$$

- Right upper corner: $n(x) = (1,1)^T$

$$\tau(x,t) = \begin{pmatrix} (\lambda+2\mu)\frac{\partial u^1}{\partial x_1} + \lambda\frac{\partial u^2}{\partial x_2} + \mu\left(\frac{\partial u^2}{\partial x_1} + \frac{\partial u^1}{\partial x_2}\right) \\ \mu\left(\frac{\partial u^2}{\partial x_1} + \frac{\partial u^1}{\partial x_2}\right) + (\lambda+2\mu)\frac{\partial u^2}{\partial x_2} + \lambda\frac{\partial u^1}{\partial x_1} \end{pmatrix}$$

In the equations above, the Lamé constants appear as parameters. These constants are material properties influencing the dynamic behaviour through the stress tensor (see definition 2.6). An alternative set of parameters is also very common: Poisson's ratio ν and Young's modulus E. The corresponding definition and relation to the Lamé constants is as follows:

$$\begin{aligned} \nu &:= \frac{\lambda}{2(\lambda+\mu)}, \\ E &:= \mu\frac{3\lambda+2\mu}{\lambda+\mu}, \\ \mu &= \frac{E}{2(1+\nu)}, \\ \lambda &= \frac{\nu E}{(1+\nu)(1-2\nu)}. \end{aligned}$$

Hence, we may directly use ν and E from standard tables for material properties in our computations.

In the following considerations, we use two simplifications concerning the description of elastic body motion. First, we assume the whole building modelled with one homogeneous material without holes[5]. Hence, we are going to describe "buildings" with no rooms; a more realistic approach would immediately involve a whole bunch of additional challenges such as complex geometries, combinations of different material, realistic material parameters, etc. Since we want to keep things simple to demonstrate mathematical and computational concepts instead of solving a complex engineering problem, this approach is justified. Second, we will implement programs only for two-dimensional scenarios—3D setups would immediately require considerably more computational resources than we are willing to spend for our MATLAB implementations and would not offer additional insights into the characteristics of the problem.

The restriction to two dimensions requires an adaption to the general setting. For a purely two-dimensional setup, Poisson's ratio has to be chosen as zero: $\nu \stackrel{!}{=} 0$. The reason for this is that real materials live in three dimensions. If we restrict the scenario to a thin 2D plate (in the $x - y$ plane, for the sake of simplicity), two variants of effects may appear for real materials:

1. **Plane Stress:**
   For stresses (or forces) purely arising in the $x-y$ plane (i.e. $\sigma_{33} = 0$), strain values arise in the orthogonal z direction: Some stretching or squeezing effects will take place (i.e. $u_3 \neq 0$; imagine a rubber ribbon torn apart that becomes thinner).

2. **Plane Strain:**
   Here, we assume that strains appear only in the $x - y$ plane (i.e. length changes only in the plane: $\varepsilon_{33} = 0, u_3 = 0$). Then, forces or stresses in z direction would be necessary to keep the thickness of the material constant (i.e. $\sigma_{33} \neq 0$).

Only with an artificial "material" value of $\nu = 0$, the two scenarios of plane stress and plane strain do not differ. This translates into special values for the Lamé constants: $\lambda = 0, \mu = E/2$.

Now that the continuous, two-dimensional setting is complete, we are going to discuss aspects of spatial discretisation techniques in the next section.

2.3 Spatial and Temporal Discretisation of PDEs

We do not go into the details of PDEs (their categorisation, analytical aspects, etc.) since this would be far beyond the scope of this book. Instead, we focus

[5] Imagine a rubber block, e.g.

Before you continue, make sure to answer the following questions:

Quiz: Section 2.2

Q1 What is the difference of the Eulerian description compared to the Lagrangian one?

Q2 What are the two categories of boundary conditions for PDEs?

Q3 Where do the Lamé constants for elastic materials come into play?

Q4 Why is $\nu = 0$ necessary for purely 2D scenarios?

in this section on the aspects relevant for a basic understanding of numerical (i.e. approximate) solutions of PDEs, following mainly [161, 160] and [37].

We start with a brief survey on the space-time problem before discussing different types of meshes and the discretisation of spatial operators.

2.3.1 From Space-Time to Space & Time

Let us consider the example of the time-dependent heat equation (2.1),

$$\frac{\partial T}{\partial t}(t,x) = \kappa \cdot \Delta T(t,x) + f(t,x) \tag{2.29}$$

The task for a numerical solution is to find an approximate solution T_h of (2.1) in the sense that

- $T_h \approx T$ at given points (or in a globally measured norm),
- T_h can be represented in a discrete form (at certain discretisation points $(t_h, x_h) \in I_h \times \Omega_h$ or via a discrete (i.e. finite) number of degrees of freedom).

Hence, the infinite-dimensional problem to find a quantity of interest (such as $T(t,x)$) for arbitrary values of x and t is restricted to a finite-dimensional one of finding approximations only at certain points (nodes) or patches (cells). To realise this task, we need to discretise the domain of interest $I \times \Omega$ (see Sec. 2.3.2) and to discretise the differential operators and function evaluations in the PDE (cf. Sec. 2.3.3). Different variants of the discretisation of $I \times \Omega$ exist and are visualised in Fig. 2.4. The red arrows represent the method of horizontal lines (also called Rothe method) which first discretises the time interval I and solves a steady-state problem in space in each time step. This method is often used if a numerical solver for steady-state problems

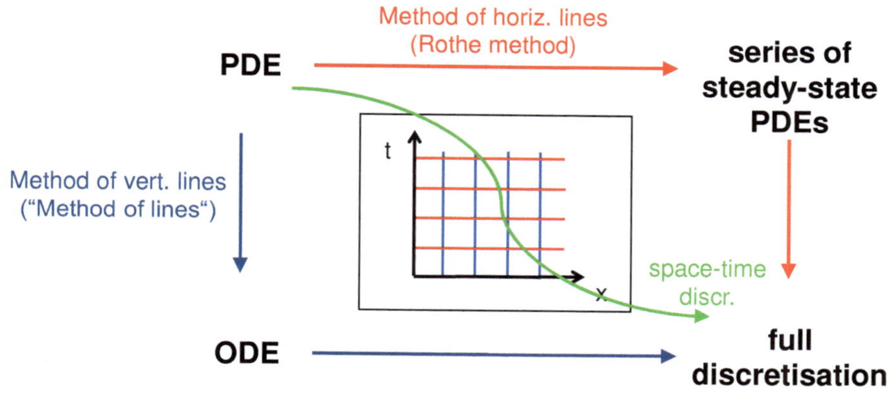

Figure 2.4. Visualisation of the three different discretisation approaches for space-time PDEs: The method of horizontal lines (also called Rothe method; red), the method of vertical lines ("the" method of lines, blue), and the combined space-time approach (green).

is already available (and efficient). In contrast, the method of vertical lines (sometimes simply called "the" method of lines)—visualised by the blue arrows—uses a spatial discretisation of the PDE to formulate a large (and coupled) system of ODEs involving each spatial degree of freedom as a component of the system. The method of lines is frequently used since efficient implementations for ODE systems are available on which developers of software for PDEs can rely. Both the Rothe method and the method of lines use a full spatial mesh in combination with a certain amount of the previous time step information for each of those grid points (i.e. these schemes work with a discrete window instead of a full mesh in time). Finally, the green arrow symbolises a full, combined discretisation in space and time simultaneously. This is the most complex variant (at least from the point of view of mesh generation and handling), since it considers the higher-dimensional space directly. Hence, for the 3D heat equation, a four-dimensional grid has to be used, whereas the former two approaches only need a 3D grid in space plus data from a few time steps[6]. The higher dimensionality of the mesh is non-negligible challenge both from the point of view of visualisation or debugging (how to check whether the mesh is correctly generated) and of the storage requirements throughout the simulation. Therefore, the full discretisation is not (yet) used to a large extent.

6 The amount of time steps under consideration depends on the numerical scheme for ODEs in use (see Chap. 7). For one-step schemes, e.g., two time levels are relevant and have to be stored.

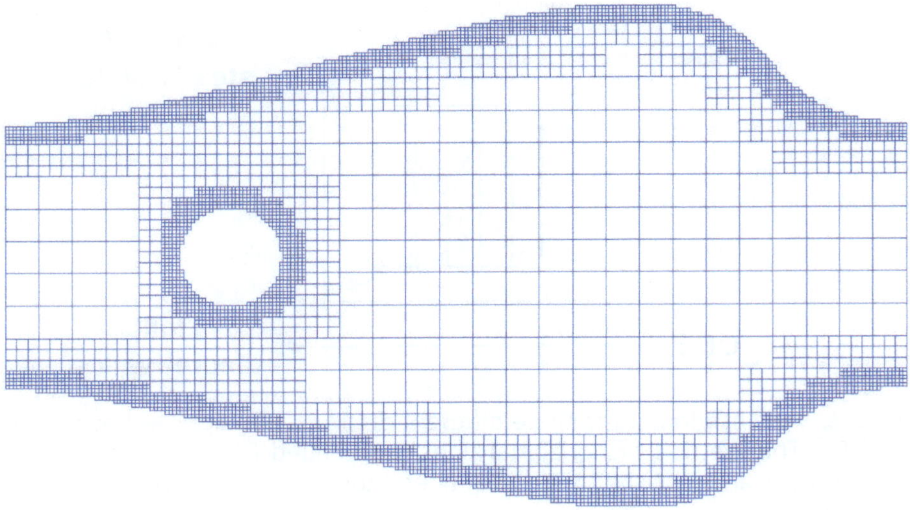

Figure 2.5. Example of an adaptive Cartesian mesh for flow problems in specific micro pumps.

In the following, we concentrate on the classical (and simpler) approach of a separate discretisation in space and time via the method of lines. Hence, a PDE problem is converted to a (large) system of ODEs (see Chap. 7 for the corresponding ODE discretisation). We now discuss the basic ingredients for spatial discretisations: meshing and operator discretisation.

2.3.2 Spatial Discretisation: Meshing

To represent the solution of a PDE at a finite amount of points in the spatial domain, $x_h \in \Omega$, a *mesh* or *grid* is typically used. We denote a mesh for a given domain Ω with Ω_h.

Many different variants of meshes exist. Before categorising meshes, we start with two examples: A two-dimensional Cartesian grid using squared cells is given in Fig. 2.5. Figure 2.6 shows a 2D triangular grid. This latter type of grid is very popular and sometimes, "triangulation" is even used synonymously for meshing or mesh generation. Usually, finer meshes with smaller cells deliver more accurate results both with respect to the approximation of the boundary geometry and with respect to the discretisation error inside the computational domain.

The following categories or properties of meshes are typically important:

- Hexahedral vs. Tetrahedral Meshes:
  The amount of corners one cell has seems to be not so important:

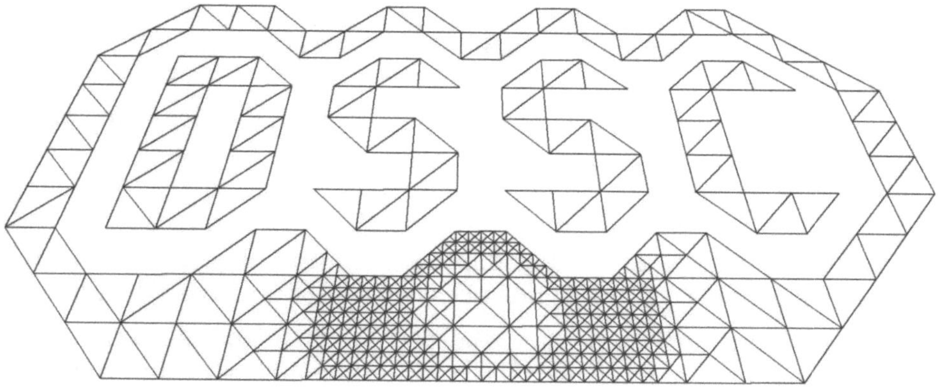

Figure 2.6. Example of an adaptive triangular grid for an academic problem of tsunami simulations (image courtesy of Martin Schreiber, see also http://www5.in.tum.de/sierpinski).

Whether triangular cells with three corners or rectangular cells with four are used in 2D makes no large difference with respect to computational costs. In fact, discussions on which of these categories to prefer sometimes have the flavour of religious battles. However, there are fundamental impacts of the choice of cell type on the discretisation of operators (see Sec. 2.4). Therefore, a lot of implementations prefer to support only one category; hybrid variants exist but are more complex to handle.

- Structured vs. Unstructured Meshes:
  Structured meshes possess an inherent rule of production. Consider the Cartesian mesh of Fig. 2.5: All cells are squared ones, and smaller cells are created by subdividing larger ones into child cells (this is often called a *spacetree* approach). This approach is very simple, but has the drawback of a "LEGO"-like[7] behaviour: In order to obtain good approximations of a curvilinear boundary, more/finer grid cells are necessary.

  In contrast, unstructured meshes have the full "freedom of shape", i.e. the corner points defining corresponding cells do not have to obey an inherent structure but are stored individually. Such meshes have advantages in the representation of arbitrary geometries but come with a considerable overhead in storage.

7 For those who do not know LEGO (http://www.lego.com/): This toy allows to construct 3D objects by putting together mainly brick-shaped items of various sizes. When one constructs a sphere with bricks, the quality or accuracy of the approximation depends on the size of the bricks w.r.t. the radius of the sphere ("smaller parts allow finer spheres").

- Regular vs. Adaptive Meshes:
  For structured meshes, adaptivity is an important concept. The idea is to refine a grid in those areas of the domain where accuracy demands it, but to save resources via coarser refinement in the remaining parts of the domain. Hence, equally accurate results may be obtained as with a regularly refined structured grid, but at a much lower cost. As an example, the geometric approximation of the curvilinear boundary in Fig. 2.5 is visually quite good due to the adaptively refined Cartesian cells. Imagine how many more cells would be necessary if all squares had the same size.

  Unstructured meshes possess the feature of adaptivity inherently in their construction. However, formulating efficient algorithms for the construction of such adaptive unstructured meshes with high quality requirements also is a challenging task.

  The crucial point for adaptivity is the development of criteria that measure the necessity of grid refinement. This involves a considerable amount of knowledge of the operator discretisation (see next section) and the physics of the application one is trying to solve numerically.

- Fixed vs. Dynamic Meshes:
  In certain cases, meshes will stay constant throughout the whole simulation cycle. A classical example are steady-state problems without time-dependency. Time-dependent scenarios may use fixed grids if the changes are not too large (see the example of flow simulations in Chap. 5). If possible, the usage of fixed meshes is favorable because it is computationally less expensive.

  If the geometry or the approximate solution of a problem undergo considerable changes over time, the grid used for the numerical approximation must be changed as well. Certain types of grids can cope with this task rather easily if their construction principle is local (adaptive Cartesian meshes, e.g.), whereas other meshes have a larger overhead due to their full, global reconstruction of the grid. In the latter case, a frequent approach is to restrict remeshing to every 50th time step or some similar choice to reduce the remeshing costs.

  Note that for our structural dynamics simulations, we have elastic objects changing their shape possibly in each time step. The Lagrangian approach is to compute the corresponding displacements with respect to the reference geometry. Hence, only the mesh of the reference geometry needs to really be created, at least if deformations are sufficiently small.

For all types of meshes, the ordering of the geometric items (such as vertices or cells) is not prescribed. Different variants exist on how to realise that

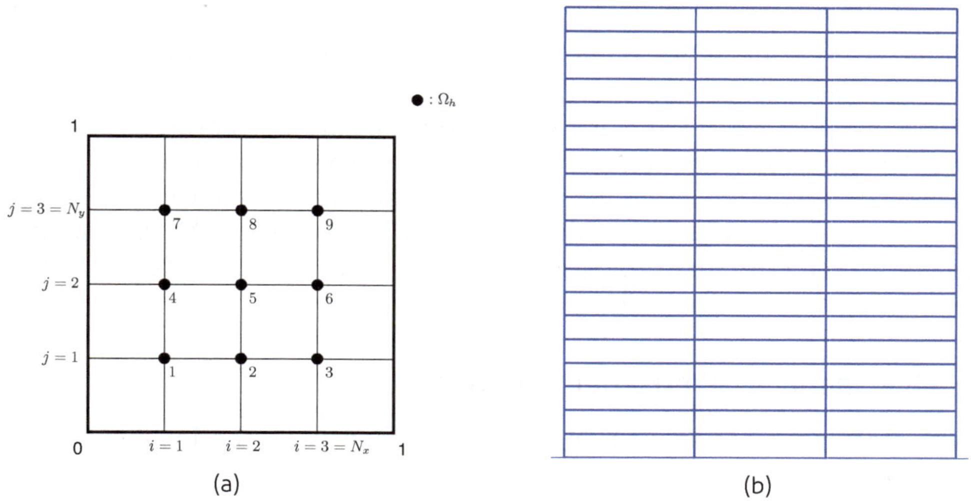

Figure 2.7. Two regular Cartesian meshes: A simple regular Cartesian mesh for the unit square using 4×4 cells to obtain $N = 3$ inner nodes in each direction (a); a regular Cartesian mesh for a 2D domain $\Omega := [0, a] \times [0, b], a \neq b$, using different mesh sizes $h_x \neq h_y$ in each direction (b).

aspect in concrete implementations, and we are going to discuss one clever approach via space-filling curves in Chap. 11.

In the following, we are going to restrict the discussion to regular Cartesian grids for two reasons. First, regular Cartesian meshes are very simple to construct and implement and do not demand huge computational resources for their storage and handling. This fits very well with our hands-on approach using MATLAB. Second, the reference geometry of our problem (a 2D building) is a rectangle which we can exactly represent with a Cartesian mesh of any resolution. Hence, we do not introduce a considerable discretisation error due to our approximate computational domain. In Fig. 2.7 (a), a very simple version of a regular Cartesian grid is visualised for two dimensions. We define the discrete mesh as

$$\begin{aligned} \Omega_h &:= \{(ih, jh) \mid i = 1, \ldots, N, j = 1, \ldots, N\} && (2.30) \\ &= \{(x, y) \in \Omega \mid x = ih, y = jh \text{ with } i, j \in \mathbb{Z}\} && (2.31) \end{aligned}$$

with N inner grid points in each direction on a rectangular 2D domain $\Omega := [0, a] \times [0, b], a, b \in \mathbb{R}^+$, for a given mesh size $h_x = h_y = h := 1/(N + 1)$. The discrete boundary $\partial\Omega_h$ is then defined as

$$\partial\Omega_h := \{(ih, jh) \mid i \in 0, N + 1, j \in 0, \ldots, N + 1 \text{ or } i \in 0, \ldots, N + 1, j \in 0, N + 1\}$$

$$= \{(x, y) \in \partial\Omega \mid x = ih, y = jh \text{ with } i, j \in \mathbb{Z}\} \tag{2.32}$$

Equations (2.30) and (2.32) contain all relevant information to implement a regular Cartesian grid in a straightforward manner.

In general, of course, the amount of grid cells or nodes does not have be equal in both spatial directions; furthermore, the mesh size in each direction may be different ($N_x \neq N_y$). In Fig. 2.7 (b), an example of such a more general Cartesian mesh is visualised; this mesh has been used in the workshop to discretise the reference geometry of a building for structural dynamics simulations under earthquake excitations.

2.3.3 Spatial Discretisation: Operators

With meshing in hands, we turn to the problem of representing the continuous PDE in a discrete context without losing physical properties. The evaluation of certain quantities such as source terms (cf. f in the heat equation (2.1) or β in (2.21), e.g.) is relatively straightforward: we may just evaluate these functions at the discrete points in space and time. A more complex task is the translation of the continuous differential operators in the equation.

Several approaches exist to discretise the differential operators or the equation. The three most prominent ones are finite differences (FD), finite volumes (FV), and finite elements (FE). A vast amount of literature is available for all these methods. For further reading, we recommend [160] for a survey, [171] for FD[8], [98][9] and [172] for FV, and [16, 265] for FE.

The basic idea of FD schemes is to approximate differential quotients by difference quotients. We discuss the finite difference schemes in more detail in Sec. 2.4. Finite volumes apply a volume integration over each discrete cell on the underlying PDE, then those integrands containing a divergence term are converted to surface integrals on borders of the cells by applying the divergence theorem (Gauß' Theorem). This approach assures the conservation of global fluxes and is very popular for flow problems. Concerning the basic formulation of the approach, the finite elements are mathematically the most challenging variant of discretisation. Here, one does not use the strong form of the PDE, which restricts the solution to be smooth enough to fulfil the equations in each continuous point[10], but a so-called weak form that allows for discontinuities such as shock waves. The basic idea is to weaken the problem by averaging it via an integration over the whole domain Ω and

8 accompanied by nice MATLAB examples and exercises: http://faculty.washington.edu/rjl/fdmbook/

9 See http://www.cmi.univ-mrs.fr/~herbin/PUBLI/bookevol.pdf for an updated version which is accessible online.

10 For the heat equation (2.29), $T \in \mathcal{C}^2$ would be necessary to fulfil the PDE in a strong sense.

multiplying it with a test function. The weak form has to be respected by the (weak) solution for all test functions. In the final step, the weak form—which is still continuous—is formulated and solved on a discrete, finite-dimensional subspace. A proper definition of the underlying function spaces for the test functions and weak solutions involves a considerable amount of functional analysis. The convergence features of the approach are, however, strong and relatively easy to obtain. FE methods gained more attention for flow problems in recent years, but they are surely the most common approach to problems of structural dynamics.

All three approaches are different in various aspects, but they have one feature in common: After the discretisation of the continuous problem, a coupled linear or non-linear system of equations remains to be solved. As an alternative to classical solution methods for these linear systems in the context of certain classes of Poisson problems, we are going to discuss an alternative (and very efficient) approach in Chap. 9: the so-called *fast Poisson solvers*. Considerations of Fourier transforms and corresponding fast implementations thereof are highly relevant for another class of schemes, the *spectral methods* (for further details, see [51], e.g.).

Before you continue, make sure to answer the following questions:

Quiz: Section 2.3

Q1 What is the "Method of lines"?

Q2 Of which order is the ODE system resulting from a spatial discretisation of the equations of motion for elastic bodies, when the MoL is used?

Q3 In which sense is the heat equation (2.1) simpler than the governing equations for elastic body motion (2.21)?

Q4 Which categories of different meshes for the discretisation of the spatial domain exist? List at least three!

Q5 What are the three most prominent methods to discretise the underlying equations and, in particular, the differential operators?

2.4 Finite Difference Approximations

Of the three above-mentioned approaches, finite differences are the simplest to understand. However, complexity strikes back when it comes to mathematical proofs of convergence etc. We will stick to our hands-on approach and explain the general concept following [161, 160] before presenting the specific form of the FD discretisation of elastic body motion (2.21)–(2.23).

2.4.1 General Concept

The name "finite difference method" contains the basic idea of this approach. In a differential equation such as the heat equation (2.29), the spatial differential quotients are approximated by difference quotients,

$$\frac{\partial T}{\partial x}(x) = \lim_{h\to 0} \frac{T(x+h)-T(x)}{h} \approx \frac{T(x+h)-T(x)}{h} \quad \text{for a fixed } h \in \mathbb{R}\,, \tag{2.33}$$

ignoring the time-dependency of T on t in these spatial considerations. We assume $x \in \mathbb{R}$ in (2.33), for the sake of simplicity. Since h is a fixed mesh size of a regular Cartesian grid Ω_h (as visualised in Fig. 2.7 (a)), Eq. (2.33) represents a way to compute approximations of the differential operators on a given mesh. This results in a discrete set of linear or non-linear algebraic equations to be solved numerically. Often, the numerical solution is denoted by T_h at given grid points to distinguish it from the analytical solution T of the continuous problem. T_h is now a vector of unknowns with length equal to the number of degrees of freedom in the discrete domain.

The question of how well the finite difference on the right-hand side of (2.33) approximates the real derivative arises naturally. Assuming a sufficient smoothness of T, a Taylor expansion gives the corresponding answer:

$$\begin{aligned} T(x+h) &= T(x) + \frac{\partial T}{\partial x}(x)\cdot h + \frac{1}{2}\frac{\partial^2 T}{\partial x^2}(x)\cdot h^2 + \dots \\ &= T(x) + \frac{\partial T}{\partial x}(x)\cdot h + \mathcal{O}(h^2) \end{aligned} \tag{2.34}$$

Hence, restricting the expansion after the first term delivers an approximation of the derivative $\partial T/\partial x$ which is first-order accurate (i.e. the error is of $\mathcal{O}(h)$). For $x \in \mathbb{R}^d$, one uses a one-dimensional Taylor expansion in one component x_i of x to obtain the corresponding partial derivative w.r.t. x_i.

The approximation (2.33) is called the *forward difference quotient* (often abbreviated as $\partial^+ T(x)$). Other variants such as the *backward* or *central* difference quotient (denoted by $\partial^- T(x)$ and $\partial^0 T(x)$, respectively) exist and are derived in Problem 2.8. If the underlying PDE contains spatial derivatives of higher order such as the Laplacian in the heat equation, corresponding finite difference schemes can be derived either via a successive application of FD schemes for first-order derivatives or via a combination of Taylor expansions using different mesh sizes; Problem 2.8 contains such an example (5-point-stencil for the Laplacian).

Up to now, we considered the $M_0 := N_x \times N_y$ inner points of the domain Ω_h. For boundary conditions, points $(ih, jh) \in \partial\Omega$ have to be considered. Continuous Dirichlet conditions on a part $\Gamma_0 \subset \partial\Omega$ directly translate to discrete counterparts by simply evaluating the corresponding function at the discrete points. In the case of the heat equation, this corresponds to

$$T_h(ih, jh) = T_0(ih, jh) \text{ for } (ih, jh) \in \Gamma_0$$

Such conditions simply modify the right-hand side of the resulting algebraic system of equations. For Neumann boundary conditions (again ignoring the time-dependency),

$$\frac{\partial T}{\partial n}(ih, jh) = r(ih, jh) \text{ for } (ih, jh) \in \Gamma_1 ,$$

we have to use forward or backward difference quotients depending on which data is available (i.e. on the orientation of the outer normal vector n). Since the values of T_h at such points are also unknown, additional M_2 equations enter the linear system for the corresponding degrees of freedom. T_h is frequently called a grid function with $T_h(ih, jh) = T_{ij}$ for $i = 1, \ldots, N_x$, $j = 1, \ldots, N_y$; it is formally different from the vector $\mathbf{T}_h = T_{ij}$ for a specific choice of the order of unknowns at indices (i, j) (see Chap. 11 for more details).

Let

$$A_h \mathbf{T}_h = \mathbf{q}_h \tag{2.35}$$

be the linear system of equations resulting from the discretisation of a linear partial differential equation such as the heat equation with finite differences. The matrix $A_h \in \mathbb{R}^{M_1 \times M_1}$ is created by line-wisely inserting the FD approximations for the total amount of $M_1 := M_0 + M_2$ degrees of freedom (the inner points plus the Neumann boundary points). The right-hand side $\mathbf{q}_h$ has the form

$$\mathbf{q}_h := -\hat{A}_h \mathbf{T}_0 + \mathbf{f} \tag{2.36}$$

and contains the discrete source terms represented by $\mathbf{f} = f(ih, jh)$ as well as the Dirichlet boundary contributions. The values of $\hat{A}_h$ are all zero except for stencil contributions in the case of a grid node i being an orthogonal neighbour of a Dirichlet boundary node j. Of course, all algebraic quantities still depend on the choice of h. We do not go into the details on how to solve this typically sparse linear system of equations since this would be beyond the scope of this book; so readers should just rely on MATLAB's backslash operator to do a decent job.

2.4.2 Quality of FD Approximations

In this section, we consider the question how well a numerical solution T_h approximates the associated continuous one. To measure the difference $T_h - T$ of the two solutions, we need a suitable norm. Possible norms are the *maximum norm*,

$$\|T_h - T\|_\infty := \max_{\substack{i=1,\ldots,N_x \\ j=1,\ldots,N_y}} |(T_h - T)(ih, jh)| \tag{2.37}$$

or the *discrete L^2-Norm*

$$\|T_h - T\|_{0,h} := h \left(\sum_{i=1}^{N_x} \sum_{j=1}^{N_y} [(T_h - T)(ih, jh)]^2 \right)^{\frac{1}{2}} . \tag{2.38}$$

One should note that the above definitions assume a two-dimensional domain but generalise to arbitrary dimensions in a straightforward manner. Furthermore, we skipped the Neumann boundary indices in these definitions for the sake of simplicity.

Denoting with $\|.\|_h$ now one suitable norm, we may define the consistency and convergence of an FD scheme:

Definition 2.4 (Consistency of FD Approximations). Let (2.35) be the linear system of equations resulting from the finite difference discretisation of the continuous steady-state PDE with mesh size h. Let $\mathbf{T}$ be the vector representation of the solution T at the grid points. Furthermore, let $\|.\|_h$ be a norm in the space of grid functions on Ω_h and $|.|_h$ be the corresponding vector norm on $\mathbb{R}^{M_1}$.
An FD approximation is *consistent* w.r.t. $\|.\|_h$ if

$$|A_h \mathbf{T} - \mathbf{q}_h|_h \to 0 \quad \text{for} \quad h \to 0 \tag{2.39}$$

Furthermore, the approximation has the *order of consistency* $p > 0$ if

$$|A_h \mathbf{T} - \mathbf{q}_h|_h \leq C h^p \tag{2.40}$$

holds with a constant $C > 0$ independent of h.

Thus, the *error of consistency* $|A_h \mathbf{T} - \mathbf{q}_h|_h$ measures how well the exact solution satisfies the approximate equations. We emphasise that the error $|\mathbf{T}_h - \mathbf{T}|_h$ does not necessarily behave like the error of consistency, since[11]

$$|\mathbf{T}_h - \mathbf{T}|_h = \left|A_h^{-1} A_h (\mathbf{T}_h - \mathbf{T})\right|_h \leq \left\|A_h^{-1}\right\|_h |A_h (\mathbf{T}_h - \mathbf{T})|_h . \tag{2.41}$$

We are, of course, interested in finite difference schemes that converge.

[11] for a matrix norm $\|.\|_h$ compatible with the vector norm $|.|_h$

Definition 2.5 (Convergence of FD Approximations). An FD approximation *converges* w.r.t. a given norm $\|.\|_h$ if

$$\|T_h - T\|_h \to 0 \quad \text{for} \quad h \to 0 \tag{2.42}$$

and has the *order of convergence* $p > 0$ if

$$\|T_h - T\|_h \leq Ch^p \tag{2.43}$$

holds with a constant $C > 0$ independent of h.

Due to (2.41), consistency alone does not guarantee convergence; we need stability in addition:

Definition 2.6 (Stability of FD Approximations). In the situation of definition 2.4, the approximation is *stable* w.r.t. $\|.\|_h$ if a constant $C > 0$ exists independent of h such that the following holds:

$$\|A_h^{-1}\|_h \leq C\,. \tag{2.44}$$

From the definitions 2.4 and 2.5 together with (2.41), we directly have the

Theorem 2.1. *A finite difference scheme that is consistent and stable converges and its order of convergence is at least equal to its order of consistency.*

The notion of consistency and convergence are analogue to the numerical solutions of ODEs in Chap. 7. Proving stability for a given FD approximation is not straightforward at all (see [161] for the Laplacian of the Poisson equation). This is the mathematical price for the simple construction of the scheme.

2.4.3 FD Approximations for Elastic Body Motion

In this section, we formulate a finite difference scheme for the equations of elastic body motion (2.25)–(2.28). In contrast to our example of the scalar heat equation above, we now need to find a vector-valued grid function $u_h = (u_h^1, u_h^2)$ which approximates the analytical solution u at the degrees of freedom: $u_h^1(ih_x, jh_y) = u_{ij}^1$, $u_h^2(ih_x, jh_y) = u_{ij}^2$.

We would like to use the finite difference quotients of second-order ($\partial^0 u^i(x)$, central difference) for the sake of accuracy. For inner points of the domain, this is possible. At the Neumann boundaries, however, we do not have the necessary degrees of freedom available outside the domain. Therefore, we have to switch to forward or backward quotients $\partial^+ u^i(x)$ and $\partial^- u^i(x)$ depending on which boundary part (top, left, or right) we need to evaluate the Neumann condition. The bottom boundary does not have to be considered here since the corresponding Dirichlet conditions only enter the right-hand side of the linear system of equations.

Discretising the spatial derivatives of the expanded system (2.25) and (2.26) for inner points $(ih_x, jh_y) \in \Omega_h$ results in the following terms for the first component u^1:

$$\underbrace{(2\mu+\lambda)}_{=:\xi}\frac{\partial^2 u^1}{\partial x_1^2} + \mu\frac{\partial^2 u^1}{\partial x_2^2} + (\lambda+\mu)\frac{\partial^2 u^2}{\partial x_1 \partial x_2} \doteq$$

$$\doteq \xi\left[\frac{u^1_{i-1,j} - 2u^1_{ij} + u^1_{i+1,j}}{h_x^2}\right] + \mu\left[\frac{u^1_{i,j-1} - 2u^1_{ij} + u^1_{i,j+1}}{h_y^2}\right] +$$
$$+(\lambda+\mu)\left[\frac{u^2_{i+1,j+1} - u^2_{i-1,j+1} - u^2_{i+1,j-1} + u^2_{i-1,j-1}}{4h_x h_y}\right]. \quad (2.45)$$

For the second component u^2, we obtain

$$\underbrace{(2\mu+\lambda)}_{=:\xi}\frac{\partial^2 u^2}{\partial x_2^2} + \mu\frac{\partial^2 u^2}{\partial x_1^2} + (\lambda+\mu)\frac{\partial^2 u^1}{\partial x_1 \partial x_2} \doteq$$

$$\doteq \xi\left[\frac{u^2_{i,j-1} - 2u^2_{ij} + u^2_{i,j+1}}{h_y^2}\right] + \mu\left[\frac{u^2_{i-1,j} - 2u^2_{ij} + u^2_{i+1,j}}{h_x^2}\right] +$$
$$+(\lambda+\mu)\left[\frac{u^1_{i+1,j+1} - u^1_{i-1,j+1} - u^1_{i+1,j-1} + u^1_{i-1,j-1}}{4h_x h_y}\right]. \quad (2.46)$$

Concerning the Neumann boundary conditions, we have to distinguish different parts of the boundary due to the different orientation of the outer normal vector n.

- Upper boundary: $n(x) = (0,1)^T$

$$\tau(x,t) = \begin{pmatrix} \mu\left(\frac{\partial u^2}{\partial x_1} + \frac{\partial u^1}{\partial x_2}\right) \\ \xi\frac{\partial u^2}{\partial x_2} + \lambda\frac{\partial u^1}{\partial x_1} \end{pmatrix} \doteq \begin{pmatrix} \mu\left[\frac{u^2_{i+1,j} - u^2_{i-1,j}}{2h_x} + \frac{u^1_{ij} - u^1_{i,j-1}}{h_y}\right] \\ \xi\frac{u^2_{ij} - u^2_{i,j-1}}{h_y} + \lambda\frac{u^1_{i+1,j} - u^1_{i-1,j}}{2h_x} \end{pmatrix} \quad (2.47)$$

- Right boundary: $n(x) = (1,0)^T$

$$\tau(x,t) = \begin{pmatrix} \xi\frac{\partial u^1}{\partial x_1} + \lambda\frac{\partial u^2}{\partial x_2} \\ \mu\left(\frac{\partial u^2}{\partial x_1} + \frac{\partial u^1}{\partial x_2}\right) \end{pmatrix} \doteq \begin{pmatrix} \xi\frac{u^1_{ij} - u^1_{i-1,j}}{h_x} + \lambda\frac{u^2_{i,j+1} - u^2_{i,j-1}}{2h_y} \\ \mu\left[\frac{u^2_{ij} - u^2_{i-1,j}}{h_x} + \frac{u^1_{i,j+1} - u^1_{i,j-1}}{2h_y}\right] \end{pmatrix} \quad (2.48)$$

- Left Boundary: $n(x) = (-1, 0)^T$

$$\tau(x,t) = \begin{pmatrix} -\xi\frac{\partial u^1}{\partial x_1} - \lambda\frac{\partial u^2}{\partial x_2} \\ -\mu\left(\frac{\partial u^2}{\partial x_1} + \frac{\partial u^1}{\partial x_2}\right) \end{pmatrix} \doteq \begin{pmatrix} -\xi\frac{u^1_{i+1,j} - u^1_{ij}}{h_x} - \lambda\frac{u^2_{i,j+1} - u^2_{i,j-1}}{2h_y} \\ -\mu\left[\frac{u^2_{i+1,j} - u^2_{ij}}{h_x} + \frac{u^1_{i,j+1} - u^1_{i,j-1}}{2h_y}\right] \end{pmatrix} \tag{2.49}$$

Note that the function τ of surface forces has to be evaluated at the discrete boundary points, and the corresponding contributions have to be considered in the right-hand side $\mathbf{q}_h$ of the resulting linear system of equations. At the corners of the domain, we use the averaged normal vector of both adjacent borders to obtain the following finite difference approximations.

- Upper left corner: $n(x) = (-1, 1)^T$

$$\tau(x,t) = \begin{pmatrix} -\xi\frac{\partial u^1}{\partial x_1} - \lambda\frac{\partial u^2}{\partial x_2} + \mu\left(\frac{\partial u^2}{\partial x_1} + \frac{\partial u^1}{\partial x_2}\right) \\ -\mu\left(\frac{\partial u^2}{\partial x_1} + \frac{\partial u^1}{\partial x_2}\right) + \xi\frac{\partial u^2}{\partial x_2} + \lambda\frac{\partial u^1}{\partial x_1} \end{pmatrix}$$

$$\doteq \begin{pmatrix} -\xi\frac{u^1_{i+1,j} - u^1_{ij}}{h_x} - \lambda\frac{u^2_{ij} - u^2_{i,j-1}}{h_y} + \mu\left[\frac{u^2_{i+1,j} - u^2_{ij}}{h_x} + \frac{u^1_{ij} - u^1_{i,j-1}}{h_y}\right] \\ -\mu\left[\frac{u^2_{i+1,j} - u^2_{ij}}{h_x} + \frac{u^1_{ij} - u^1_{i,j-1}}{h_y}\right] + \xi\frac{u^2_{ij} - u^2_{i,j-1}}{h_y} + \lambda\frac{u^1_{i+1,j} - u^1_{ij}}{h_x} \end{pmatrix} \tag{2.50}$$

- Upper right corner: $n(x) = (1, 1)^T$

$$\tau(x,t) = \begin{pmatrix} \xi\frac{\partial u^1}{\partial x_1} + \lambda\frac{\partial u^2}{\partial x_2} + \mu\left(\frac{\partial u^2}{\partial x_1} + \frac{\partial u^1}{\partial x_2}\right) \\ \mu\left(\frac{\partial u^2}{\partial x_1} + \frac{\partial u^1}{\partial x_2}\right) + \xi\frac{\partial u^2}{\partial x_2} + \lambda\frac{\partial u^1}{\partial x_1} \end{pmatrix}$$

$$\doteq \begin{pmatrix} \xi\frac{u^1_{ij} - u^1_{i-1,j}}{h_x} + \lambda\frac{u^2_{ij} - u^2_{i,j-1}}{h_y} + \mu\left[\frac{u^2_{ij} - u^2_{i-1,j}}{h_x} + \frac{u^1_{ij} - u^1_{i,j-1}}{h_y}\right] \\ +\mu\left[\frac{u^2_{ij} - u^2_{i-1,j}}{h_x} + \frac{u^1_{ij} - u^1_{i,j-1}}{h_y}\right] + \xi\frac{u^2_{ij} - u^2_{i,j-1}}{h_y} + \lambda\frac{u^1_{ij} - u^1_{i-1,j}}{h_x} \end{pmatrix} \tag{2.51}$$

Using all seven approximations (2.45)–(2.51), we can implement a MATLAB function to construct the matrix entries of the corresponding linear system of equations (this is actually part of the workshop tasks, see Sec. 18.2):

$$A_h\mathbf{u}_h = \mathbf{q}_h$$

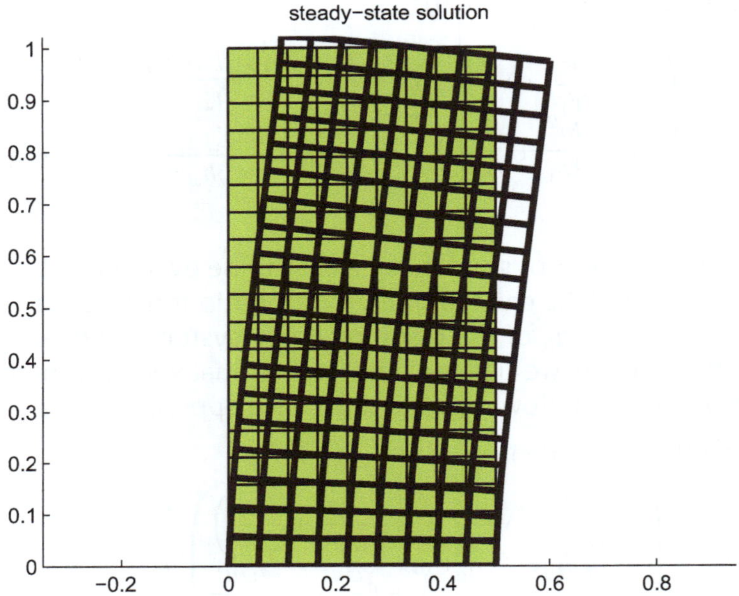

Figure 2.8. Qualitative steady-state solution of a building under windload from left.

For the formulation of the above system, one needs to specify a certain order of the degrees of freedom: The 2D mesh data of the two components, (u^1_{ij}, u^2_{ij}), has to be serialised in a vector (see Sec. 11.3 for different variants of the serialisation of DoF).

To demonstrate the correct derivation of the FD approximations, we simulated a building under surface forces acting on the left side of the domain (imagine wind loads coming from the left). The steady-state solution visualised in Fig. 2.8 shows the corresponding bending of the building for $E = 500$ and $h_x = h_y = 0.05$ for a coarse mesh with 10×19 grid nodes. Of course, results for scenarios with large displacements will not be physically meaningful since we derived the continuous equations under the assumption of linear elasticity theory (cf. Def. 2.3).

In a time-dependent context, the same approach is used in combination with a discretisation of the ODE system (compare Chap. 7); see Sec. 18.3 for corresponding results including also random ground motion using the Kanai-Tajimi model of Sec. 3.2.2.

Before you continue, make sure to answer the following questions:

Quiz: Section 2.4

Q1 Where do finite difference schemes get their name from?

Q2 Which technique can be used to derive both a finite difference scheme and its order of approximation?

Q3 How do you create FD schemes for higher-order derivatives?

Q4 Why do Neumann boundaries have to be treated in a special manner compared to Dirichlet boundaries?

Q5 Which technique may be applied to use central finite differences also at Neumann boundaries?

2.5 Chapter's Summary

Various applications that require random effects to be added to partial differential equations motivated our discussion of the first step in a reduction process of such RPDEs to RODEs via classical deterministic discretisation schemes in space. The resulting coupled system of (R)ODEs can be solved by applying methods discussed in Chap. 7 and 14. For the example of elastic body motion, we saw how the underlying system of PDEs and corresponding boundary conditions are derived. Random effects that appear purely in time may be incorporated in these (Dirichlet) boundary conditions to tackle the resulting RODE numerically as discussed in Chap. 14. We discussed different types of meshes with emphasis on regular Cartesian grids. The three main variants of spatial discretisations—finite differences, finite volumes, and finite elements—have briefly been explained. Certain aspects of finite difference schemes have been presented in more detail, and the corresponding FD approximations for the fundamental equations of elastic body motion have been derived in order to simulate steady-state scenarios of buildings that are bent.

Real-world applications of structural dynamics will typically be tackled with finite elements on highly refined unstructured grids. Hence, the (R)PDE discretisation presented in this chapter represents a simplified approach, but the field is prepared for a full realisation in the workshop (cf. Part V) in interest of the highly restricted amount of time for participants of interdisciplinary background.

Problems

Classification: ☼ easy, ⊕ easy with longer calculations, ☆ a little bit difficult,
challenging.

Exercise 2.7. [☼] Strain Tensor under Rigid Body Motions
Show that the Green St. Venant strain tensor E defined in (2.3) is invariant under rigid body motions.

Exercise 2.8. [☼] Finite Differences in 1D

1. Derive the following 3 finite difference (FD) approximations for the first order derivative of a 1D function $u(x) \in C^3(x-h, x+h)$:

$$\partial^+ u(x) := \frac{u(x+h) - u(x)}{h}, \tag{2.52}$$

$$\partial^- u(x) := \frac{u(x) - u(x-h)}{h}, \tag{2.53}$$

$$\partial^0 u(x) := \frac{u(x+h) - u(x-h)}{2h}. \tag{2.54}$$

   What is the corresponding order of approximation of each variant (2.52)–(2.54)? Why?
   Hint: Use the Taylor expansions of $u(x \pm h)$.

2. A FD approximation for the second order derivative of the 1D function $u(x) \in C^4(x-h, x+h)$ is defined in the following manner:

$$\partial^- \partial^+ u(x) := \frac{u(x+h) - 2u(x) + u(x-h)}{h^2}. \tag{2.55}$$

   - Derive (2.55) via the Taylor expansion. What is the order of approximation?
   - Derive (2.55) in a direct way using (2.52)–(2.54).

Exercise 2.9. [⊕] Finite Differences in 2D
We consider the two-dimensional Poisson Equation

$$\Delta u(x,y) = -\sin(\pi x) \cdot \sin(\pi y) \cdot 2\pi^2, \qquad (x,y) \in \Omega := [0,1]^2, \tag{2.56}$$

$$u(x,y) = 0, \qquad (x,y) \in \Gamma := \partial\Omega, \tag{2.57}$$

where $\Delta u(x,y)$ represents the Laplacian $\Delta u(x,y) = \frac{\partial^2 u}{\partial x^2}(x,y) + \frac{\partial^2 u}{\partial y^2}(x,y)$.

1. Apply the 1D FD approximations of Problem 2.8 to derive a FD approximation for the 2D Laplacian $\Delta u(x,y)$.

2. Let

$$\Omega_h := \{(ih, jh) \mid i = 1, \ldots, N, j = 1, \ldots, N\} \qquad (2.58)$$

be a mesh of N inner grid points in each direction on the unit square $[0,1]^2$ for a given mesh size $h_x = h_y = h := 1/(N+1)$.

- Draw a sketch how the mesh Ω_h looks.
- Which neighbouring points are necessary in order to evaluate the FD approximation of task 1 for a given point (i, j)?
- Sketch the matrix A of the linear system of equations $Au_h = b$ that is created by applying the FD of task 1 to the given problem (2.56)–(2.57) using the mesh (2.58).

3. Write a short MATLAB script which implements the numerical solution of (2.56)–(2.57).
Hint: Use the MATLAB backslash operator ("\") to solve the linear system of equations.
How can you verify that your implementation is correct?

Exercise 2.10. [☼] **Finite Difference Approximation of Advection Terms**
In chemistry, engineering and earth sciences, advection is a transport mechanism of a substance or conserved property by a fluid due to the fluid's bulk motion. An example of advection is the transport of pollutants or silt in a river by bulk water flow downstream.

The advection operator is given as $u \cdot \nabla = \langle u, \nabla \rangle$, where u is a given vector field.

1. In 1D the application of the advection operator to the vector field u itself reads as $u(x) \cdot u_x(x)$. Give its space discretisation.

   Hint: Apply Taylor approximation analogous to the derivation of the forward time discretisation.

2. In 2D the application of the advection operator to the vector field $u = (u_1, u_2)$ itself reads as

$$\left\langle \begin{pmatrix} u_1(x,y) \\ u_2(x,y) \end{pmatrix}, \begin{pmatrix} \partial_x \\ \partial_y \end{pmatrix} \right\rangle \begin{pmatrix} u_1(x,y) \\ u_2(x,y) \end{pmatrix} = \\ = \begin{pmatrix} u_1(x,y)\partial_x u_1(x,y) + u_2(x,y)\partial_y u_1(x,y) \\ u_1(x,y)\partial_x u_2(x,y) + u_2(x,y)\partial_y u_2(x,y) \end{pmatrix}.$$

   Give its space discretisation.

Exercise 2.11. [☆] **9-Point Stencil**
Neglecting boundary conditions, how does the linear system and especially the matrix of discretisation for the 2D Laplace equation look like, if instead of the 5-point stencil the **9-point stencil**

$$\begin{aligned}\Delta u(x,y) \approx\ & \frac{1}{12h^2}(-u(x+2h,y)+16u(x+h,y)-u(x,y+2h) \\ & +16u(x,y+h)-60u(x,y)+16u(x-h,y)-u(x-2h,y) \\ & +16u(x,y-h)-u(x,y-2h))\end{aligned}$$

is used on a regular grid with $h=\Delta x=\Delta y$?

1. Set-up the matrix of discretisation for the 2D Laplace equation with this stencil.
2. What are the advantages and disadvantages of using a 9-point stencil?
   Hint: The 9-point stencil leads to an accuracy of $\mathcal{O}(h^4)$.
3. Derive the 9-point stencil with the help of suitable Taylor expansions.

Chapter 3
Path-Wise Solutions of RODEs

This chapter motivates and rigorously discusses both existence and uniqueness of path-wise solutions of random (ordinary) differential equations. We start by modelling external and ground motion excitations by means of stochastic processes which motivates the study of random (ordinary) differential equations (RODEs). Their solution, existence and uniqueness concepts are then discussed together with the correspondence between stochastic and random (ordinary) differential equations. In particular, we study the conditions that lead to the existence of path-wise unique solutions. Solutions in the extended sense are analysed as well as the dependence of solutions on parameters and initial conditions. As an excursion, we give the equations of motion for single- and multi-storey (wireframe) buildings.

3.1 Key Concepts

Following our program to

This chapter brings us one step closer to solving random partial differential equations by equipping us with the theoretical background knowledge required to convert them back to deterministic (ordinary) differential equations (see Fig. 3.1). Our main source for the set-up and discussion of random differential equations is the fabulous book [45].

Random differential equations occur quite naturally in plenty of real situations, as the following example illustrates:

Example 3.1 (Pollution-Induced Biochemical Oxygen Demand and Dissolved Oxygen in Streams). Following [205], p. 467, it has become increasingly evident that the world's most valuable natural resources – air and water – are being endangered by the activities of civilized man. The water supply is being endangered by the disposal of organic (and other) waste materials into natural bodies of water by municipalities and industries. This pollution has become a major concern of the scientific community and various regulatory agencies have specified minimum levels for dissolved oxygen (DO) in lakes and streams. These minimum levels of DO are extremely important since if DO falls below a certain threshold value, the fish and other living organisms in the body of water may die.

The situation we are modelling is that of a pollutant running into a stream which is transported downstream, see Fig. 3.2. We assume a steady and uniform stream flow and that the conditions at every cross section are unchanged with time. Next, let $x(t)$ denote the biochemical oxygen demand (BOD) of pollution degrading bacteria in the body of water at the distance t downstream from the pollution source, and $y(t)$ the concentration of DO at distance t downstream. Third, let us assume that the actual pollution at a distance t downstream is proportional to BOD.

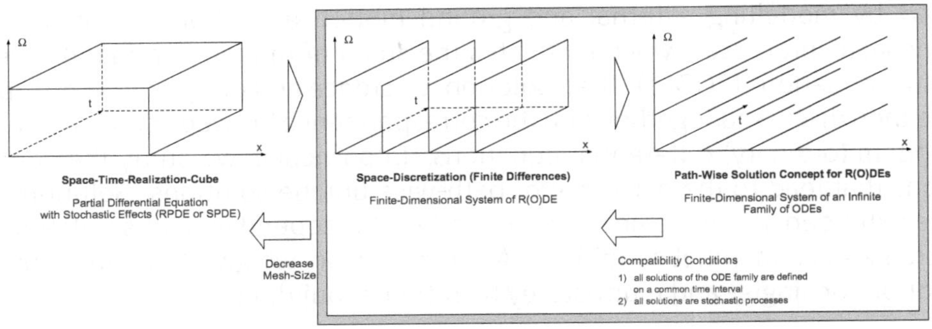

Figure 3.1. Reduction from a given continuum mechanical random partial differential equation to a family of deterministic ordinary differential equations. In this chapter, we focus on the second step, i.e., the correspondence between random differential equations and deterministic ordinary differential equations.

Finally, we assume that there are five major activities in the stream, see [205], p. 468:

1. The pollution (BOD) and DO are decreased by the action of bacteria at a rate related to the amount of pollution present with proportionality constant $k_1 > 0$.

2. The dissolved oxygen is increased due to reaeration at a rate proportional to the dissolved oxygen deficit (i.e. the DO saturation concentration minus the actual DO concentration) with proportionality constant $k_2 > 0$.

3. The pollution is only decreased by sedimentation and adsorption at a rate proportional to the amount of pollution present with proportionality constant $k_3 > 0$.

4. The pollution is increased from small sources along the stretch of stream with rate $x_a > 0$ independent of the amount of pollution present.

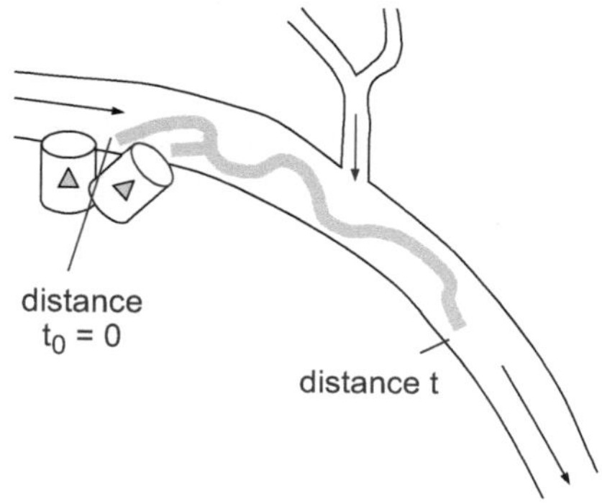

Figure 3.2. Sketch of the situation described in example 3.1: a pollutant is running into a stream and is transported downstream.

5. The dissolved oxygen is decreased at a rate d_O which may have positive or negative values and represents the net change in dissolved oxygen due to the Benthal demand and respiration and photosynthesis of plants.

In [205] the following model for BOD and DO is considered

$$\begin{aligned} 0 &= -u\dot{x}(t) - (k_1 + k_3)\,x(t) + x_a\,, \\ 0 &= -u\dot{y}(t) + k_2\,(y_s - y(t)) - k_1 x(t) - d_O\,, \end{aligned}$$

where $u > 0$ is the average velocity along the stretch, and $y_s > 0$ is the saturation concentration for dissolved oxygen. The initial conditions are denoted by $x(0) = x_0$ and $y(0) = y_0$.

Hereby, the quantities $k_1, k_2, k_3, u, x_a, y_s, d_O, x_0, y_0$ are not modelled as constants but are more realistically considered as random variables or (t-dependent) continuous stochastic processes which have certain probability distributions.

Such randomly perturbed ordinary differential equations go beyond the scope of the theory of deterministic ordinary differential equations, and at a first glance, it is not trivial to determine a notion of a solution for such problems that carefully makes sense for the random perturbations involved. In this chapter, we will show how to convert random (ordinary) differential equations to ordinary differential equations.

In particular, we will give existence and uniqueness proofs for the solutions of random (ordinary) differential equations. These proofs are of constructive/ algorithmic nature and will foreshadow the numerics that can be applied to simulate the solutions of random (ordinary) differential equations.

When reading this chapter note the answers to the following questions

1. What are the commonly accepted differential equation models for earthquake excitations?
2. What is the difference between stochastic (ordinary) differential equations and random (ordinary) differential equations?
3. How can stochastic (ordinary) differential equations with constant diffusion terms be transformed to random (ordinary) differential equations?
4. How can we prove existence and uniqueness of random (ordinary) differential equations, and what do the notions of existence and uniqueness mean when stochastic perturbations are present?
5. How do the solutions of random (ordinary) differential equations depend on parameters and initial conditions?

as well as the following key concepts

1. Gaussian white noise and Wiener processes (incl. their continuity and differentiability properties),
2. The Doss/ Sussmann & Imkeller/ Schmalfuss correspondence,
3. Path-wise existence and uniqueness of solutions of random (ordinary) differential equations,
4. Solutions of random (ordinary) differential equations in the extended sense, and
5. Equations of motion for single- and multi-storey (wireframe) buildings.

This chapter is structured as follows: In Sec. 3.2, we model external and ground motion excitations by means of stochastic processes, mainly white noise driven ones. This motivates the study of random (ordinary) differential equations. Their solution, existence and uniqueness concepts are discussed in Sec. 3.3 together with the correspondence between stochastic and random (ordinary) differential equations. For simplicity of notation we often suppress the term "ordinary" when it is clear which kind of differential equation we mean. Next, in Sec. 3.4 we discuss the conditions that lead to the existence of path-wise unique solutions. Here, solutions in the extended sense are analysed as well as the dependence of solutions on parameters and initial conditions. As an excursion, Section 3.5 models deterministically the equations of motion for single- and multi-storey (wireframe) buildings. Finally, Section 3.6 wraps up the contents of this chapter.

Prerequisites: The contents of Chap. 2 and a basic knowledge of ordinary differential equations are required for this chapter. Some pre-knowledge on stochastic processes would be good. The required MATLAB commands for stochastic processes are given in Chap. 1. In order to understand the proofs in Sec. 3.4 a sound knowledge of basic measure theory, including measurable functions and product measures, is required, cf. [14].

Teaching Remarks: As indicated in the preface our top-down approach intentionally start with the more complex topic of random differential equations before giving a review of deterministic ordinary differential equations in Chap. 7. For a beginning graduate course, we assume this to be the right approach and encourage everyone to start a class explicitly with the Chaps. 2 and 3. If required, the basics of stochastic processes and deterministic ordinary differential equations can be recalled later on. In particular, these chapters would be well motivated in the eyes of the students who appreciate the big picture of the class.

Moreover, Sec. 3.4.1 on solutions of random differential equations in the extended sense can be viewed as an add on. The second theorem on existence and uniqueness of solutions (Theorem 3.2) is a special case of Theorem 3.4 contained in Sec. 3.4.1. Thus, it seems beneficial to encourage students to study Theorem 3.4 and its proof extensively and convert it to the easier setting from Theorem 3.2.

3.2 Stochastic Processes as Models for External and Ground Motion Excitation

The Earth's crust is a system of constantly moving plates[1]. A sudden fracture of these plates can cause an earthquake where a sudden burst of energy is released causing seismic energy waves that damage surrounding areas. One assumes that these waves originate from a focal point which was the original site of the friction, and this focal point is called the epicenter of the earthquake.

Despite seismic activities caused by volcanic eruptions in the interior of a plate we can assume that earthquakes occur at the plate boundaries. There are three essential types of boundaries and two of them can have a high probability of causing seismic activities:

- *Divergent boundaries* occur when two tectonic plates move away from one another, and are most common between oceanic plates and can also form volcanic ridges.
- *Convergent boundaries* (or destructive plate boundaries) cause deformities where two tectonic plates come together and collide; they are common sites for volcanic activity. There are three types of convergent

[1] See http://meighanhenry.wikispaces.com/What+is+an+Earthquake\%3F.

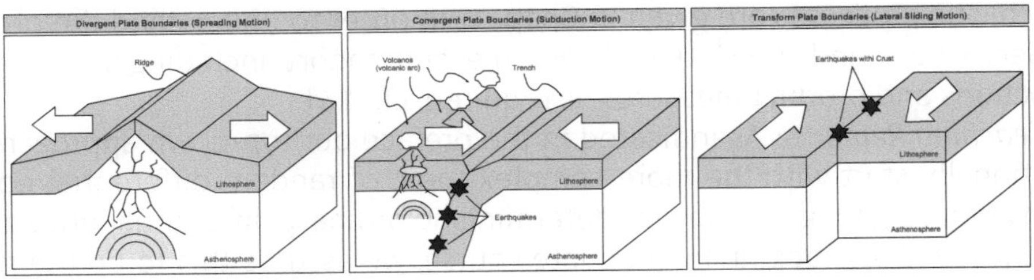

Figure 3.3. Overview of the three types of plate motion: (left) divergent plate leading to spreading motion, (middle) convergent plate boundaries leading to subduction motion, and (right) transform plate boundaries with lateral sliding motion.

boundaries: oceanic-continental, continental-continental and oceanic-oceanic. The friction between the convergent boundaries can cause earthquakes.

- *Transform boundaries* occur when two plates slide past each other. They are mostly found on the ocean floor where they form spreading ridges in a zig-zag pattern. Here the friction between the transform boundaries can cause earthquakes, too.

Figure 3.3 illustrates these boundary types.

As said, in the event of an earthquake energy is transmitted through the crust which leads to a ground motion excitation which effects buildings. Let us study how we can model this ground motion excitation. Before we can start, we require a suitable model for the stochastic effects. Here, we will discuss the most famous of all: white noise. In particular, we follow its discovery under the microscope of Robert Brown.

3.2.1 Brown's Experiment & White Noise

In 1827 the Scottish botanist Robert Brown (21.12.1773 - 10.06.1858) observed that microscopic particles suspended in liquid make very strange and highly irregular movements. These movements result from the collisions of the particle with the liquid molecules. Since the mass of the suspended molecules are small enough, each of these collisions will have a visible effect on the particle's path.

If we assume that the stochastic influence has no memory of the past (i.e., is a Markov process), is not correlated with itself (i.e., the auto-correlation is δ-distributed), and that at each time the probability of obtaining a specific value $x \in \mathbb{R}$ is the same as obtaining $-x$ (i.e., the stochastic influence has zero mean), this leads to the notion of white noise w_t. From the illustrative point of view, white noise w_t is a signal (or process), named in analogy to white light, with equal energy over all frequency bands (we come back to the spectral properties of stochastic processes later in part III). White noise w_t is characterised by the following properties:

Definition 3.2 ((Gaussian) White Noise). A *white noise process* $(w_t)_{t\in\mathbb{R}_0^+}$ has vanishing mean, i.e., $\mathbb{E}(w_t) = 0$, and its auto-correlation is δ-distributed, i.e., $\mathbb{E}(w_t w_s) = \delta(t-s)$.

The following MATLAB Example[2] visualises key properties of the Gaussian white noise process.

MATLAB Example 3.1. visWhite.m: Generating the visualisation of Gaussian white noise in Fig. 3.4.

```
L = 50 ;              % number of samples for auto-correlation
Fs = 1000;            % sampling rate
Fc = 10;              % carrier frequency for the dummy signal
t = 0:1/Fs:2;         % time base
variance = 1;         % variance of white noise
% generate a dummy signal - later on this can be replaced
% by the signal that we are interested in
signal=5*sin(2*pi*Fc*t);
% Generate Gaussian white noise with mean 0 & variance 1
subplot(3,1,1);
   whiteNoise = sqrt(variance)*randn(1, length(signal));
   plot(t,whiteNoise);
   title('\bf{Gaussian White Noise}');
   xlabel('Time (s)'); ylabel('Amplitude');
% Calculate auto-correlation of the white noise, L is the
% number of samples used in auto-correlation calculation
subplot(3,1,2);
   [whiteNoiseCov, lags] = xcov(whiteNoise, L);
   stem(lags, whiteNoiseCov/max(whiteNoiseCov));
   title('\bf{Normalized Auto-Correlation of White Noise}');
   xlabel('Lag [samples]');
% Frequency domain representation of noise
subplot(3,1,3);
   NFFT = 2^nextpow2(length(whiteNoise));
   NoiseSpectrum = fft(whiteNoise,NFFT)/length(whiteNoise);
   f = Fs/2*linspace(0, 1, NFFT/2+1);
   stem(f, 2*abs(NoiseSpectrum(1:NFFT/2+1)))
   title('\bf{Frequency-Domain Representation of White Noise}');
   xlabel('Frequency (Hz)'); ylabel('|Y(f)|') ;
```

These three plots are displayed in Fig. 3.4. We can clearly see the vanishing mean in the time-domain as well as the δ-correlation and the flat spectrum in the frequency-domain. The integral over the frequency function corresponds

2 cf. http://www.gaussianwaves.com/2012/05/colored-noise-generation-in-matlab/

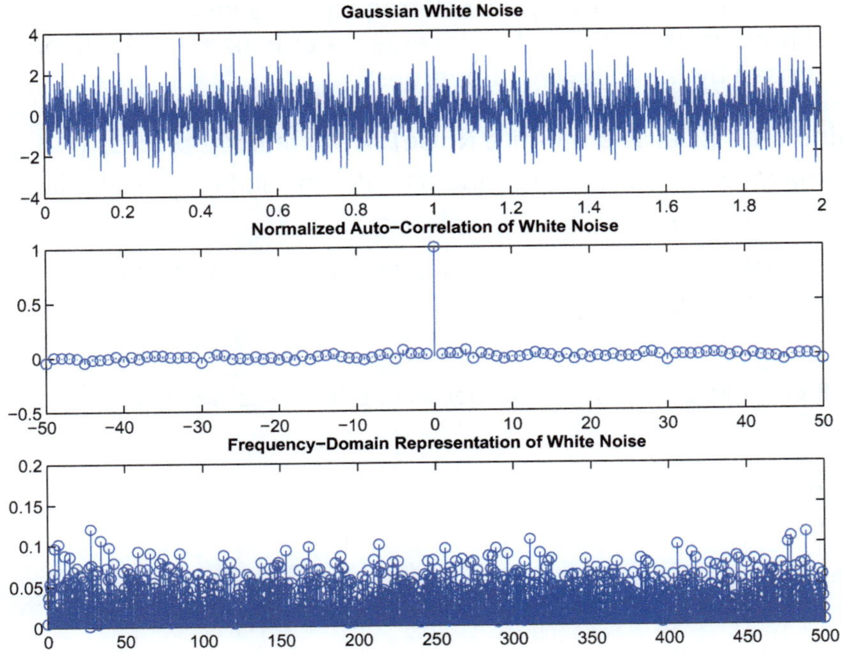

Figure 3.4. Properties of the Gaussian white noise process: (top) sample path in the time-domain, (middle) normalized auto-correlation and (bottom) frequency representation.

to the energy of a process and white noise has infinite energy. No real process has this property, which means that we should consider white noise as a fictitious "completely stochastic" process.

The first mathematical treatment of the Brownian motion dates back to Thorvald Nicolai Thiele (24.12.1838 - 26.09.1910), who effectively created a model of Brownian motion while studying time series in 1880, see [250], [126]. In 1900 Louis Bachelier (11.03.1870 - 26.04.1946) inspected the dynamic behavior of the Paris stock market and thereby created a model of Brownian motion, see [17], [18], [69]. Those models of Thiele and Bachelier had little impact for a long time, cf. [147], the most influential model was that which was proposed in 1905 by Albert Einstein (14.03.1879 - 18.04.1955) in his famous paper [92]. In 1923 Norbert Wiener (26.11.1894 - 18.03.1964)[3] gave the first acknowledged and concise definition of a mathematical model for Brown's experiment, see [258].

3 See, e.g., [88], for a short summary of Wiener's contributes to probability theory.

Formally, the Wiener process W_t is the "integral" of the white noise process w_t[4]:

Definition 3.3 (Wiener Process). The *Wiener process* $(W_t)_{t\in\mathbb{R}_+}$ is characterised by the following three properties:

1. $W_0 = 0$ (with probability 1),
2. W_t has independent increments (with distribution),
3. $W_t - W_s \sim \mathcal{N}(0, t-s)$ for $0 \leq s < t$.

If W_t is a Wiener process every process aW_t, $a =$ const., is a Wiener process, too. Thus, every Wiener process can be expressed by the *standard Wiener process* with $\mathbb{E}(W_t^2) = t$.

One can show that the Wiener process is almost surely continuous and nowhere differentiable in the mean square sense. Both properties follow from $W_{t+h} - W_t \sim \sqrt{h}\mathcal{N}(0,1)$, as on the one hand this implies $\mathbb{E}((W_{t+h} - W_t)^2) = h \to 0$ for $h \to 0$ (continuity in mean square), and, on the other hand the normal distribution of the increments implies $\mathbb{E}(\frac{1}{h^2}(W_{t+h} - W_t)^2) = \frac{1}{h} \to \infty$ for $h \to 0$ (nowhere differentiable in mean square).

MATLAB Example 3.2. wiener1.m: Implementation of the realisation of a standard Wiener process, variant 1 (cf. [129]).

```
randn('state',100)                  % set the state of randn
T = 1; N = 500; dt = T/N;
dW = zeros(1,N);                    % preallocate arrays ...
W = zeros(1,N);                     % for efficiency

dW(1) = sqrt(dt)*randn;             % first approximation outside
                                    % the loop ...
W(1) = dW(1);                       % since W(0) = 0 is not allowed
for j = 2:N
   dW(j) = sqrt(dt)*randn;          % general increment
   W(j) = W(j-1) + dW(j);
end
```

In particular, applying Kolmogorov's continuity theorem (Theorem 1.6), we have that W_t is continuous: For instance by choosing $\beta = 4$ and $\alpha = 1$ we get[5]

$$\mathbb{E}((W_t - W_s)^4) = 3|t-s|^2, \tag{3.1}$$

[4] Recall: Let $(\Omega, \mathcal{A}, \mathbb{P})$ be a probability space and $I \subset \mathbb{R}_0^+$ a real time-interval. A real valued function $X_t(\omega)$ on $I \times \Omega$ is called a stochastic process on I if it is $\mathcal{A}$-measurable for each fixed $t \in I$.

[5] Recall, that for any zero mean Gaussian random variable ξ with $\mathbb{E}(\xi^2) = \sigma^2$ it holds that $\mathbb{E}(\xi^4) = 3\sigma^4$.

i.e., the constant in Theorem 1.6 is $C = 3$. MATLAB Examples 3.2 and 3.3 [129] show two equivalent variants to calculate a realisation of a Wiener process.

MATLAB Example 3.3. wiener2.m: Implementation of the realisation of a standard Wiener process, variant 2 (cf. [129]).

```
randn('state',100)              % set the state of randn
T = 1; N = 500; dt = T/N;

dW = sqrt(dt)*randn(1,N);       % increments
W = cumsum(dW);                 % cumulative sum
```

In particular, if we do not have any insights into the structure of noise, white noise is a suitable model (though its infinite energy spectrum is rather unphysical). More clever models take white noise as an input and transform it by applying a differential equation. This method is called filtering and we are going to use it to model earthquake excitations in the way Kanai & Tajimi and Clough & Penzien suggested it in the next section.

3.2.2 Stochastic Models for Earthquake Excitations

Kanai [151], [152] and Tajimi [248] introduced a now commonly used model for earthquake induced ground excitations which is based on the following observations, see [175], pp. 78.

> For most earthquake engineering purposes, the earth can be considered as a stratified half-plane, with generally lighter material in an upper layer than the one below. If the source of an earthquake is reasonable deep, then as seismic waves propagate to the ground surface, their direction of propagation is almost vertically upward, as can be explained by Snell's law of refraction. As a first approximation, we may take into account only the uppermost layer between the ground surface and the nearest bedrock and treat the wave propagation in this layer as being one-dimensional and vertical. In the Kanai-Tajimi model, this layer is further approximated by a single-degree of freedom linear system, [175], p. 80.

The described situation is illustrated in Fig. 3.5 (a). In view of this picture, the ground excitation can be regarded as the response of a linear filter with zero-mean Gaussian white noise as input.

In [207], pp. 18, the stochastic differential equation formulation for the ground motion excitation $\ddot{u}_g(t)$ in the sense of the Kanai-Tajimi model is given as

$$\ddot{u}_g = \ddot{x}_g + w_t = -2\zeta_g\omega_g\dot{x}_g - \omega_g^2 x_g ,$$

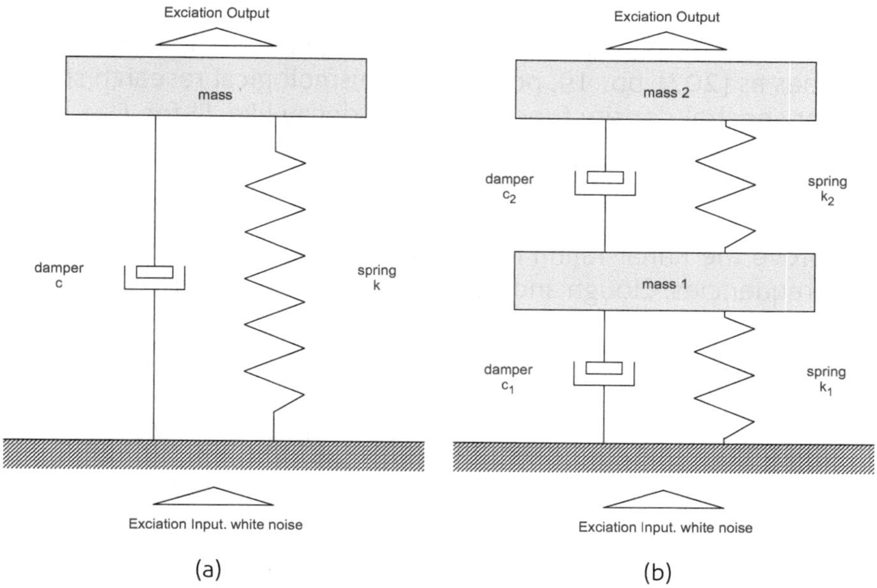

Figure 3.5. Sketch of the Kanai-Tajimi model (a) and its refinement, the Clough-Penzien model (b).

where x_g is the solution of a zero-mean Gaussian white noise w_t driven stochastic oscillator, i.e.

$$\ddot{x}_g + 2\zeta_g\omega_g\dot{x}_g + \omega_g^2 x_g = -w_t\,, \qquad x_g(0) = \dot{x}_g(0) = 0\,. \tag{3.2}$$

Here, ζ_g and ω_g are model parameters reflecting the local geological site conditions. For instance, in [133] the use of $\zeta_g = 0.64$ and $\omega_g = 15.56$ (in rad/s) is recommended for firm soil conditions in a frequency range from $f = 2.1$ (in rad/s) to $f = 21$ (in rad/s).

The power spectral density function[6] of $\ddot{u}_g(t)$ with respect to the Kanai-Tajimi filter (KT) given in equation (3.2) is

$$S_{KT}(f) = \frac{D_0}{2\pi} \cdot \frac{\omega_g^4 + 4\zeta_g^2\omega_g^2 f^2}{(\omega_g^2 - f^2)^2 + 4\zeta_g^2\omega_g^2 f^2}\,,$$

where $S_{w_t} = \frac{D_0}{2\pi}$ is the power spectral density function of the applied white noise (we come back to the spectral properties of stochastic processes later in part III, and particularly in Sec. 10.4.1 where we derive S_{KT}). Figure 3.6 (a) illustrates S_{KT} for different values of the model parameter ζ_g.

6 For details on the power spectral density function and the frequency domain representation of stochastic processes, see Chap. 10.

The Kanai-Tajimi model displays the characteristic properties of real earthquakes for high frequencies quite well, though has its inaccuracies at low frequencies as [207], pp. 19, points out. Seismological research shows that the power spectral density function should decay like f^4 for $f \to 0$, cf. [32]. But, $S_{KT}(f) \to \frac{D_0}{2\pi}$ for $f \to 0$. These low frequencies of an earthquake have strong influence on the dynamics of nonlinear inelastic structures, cf. [207], [4], [142].

To improve the Kanai-Tajimi model with respect to a better resolution of the low frequencies, Clough and Penzien suggested in [63] to take the Kanai-Tajimi response as the input of a second linear filter, see Fig. 3.5 (b). For the corresponding stochastic differential equation we follow [264], [255]: In the Clough-Penzien model the ground motion excitation $\ddot{u}_g(t)$ is given as

$$\ddot{u}_g = \ddot{x}_f = -2\zeta_f\omega_f\dot{x}_f - \omega_f^2 x_f - 2\zeta_g\omega_g\dot{x}_g - \omega_g^2 x_g\,,$$

where x_g and x_f are the solutions of the coupled stochastic oscillators

$$\begin{cases} \ddot{x}_g + 2\zeta_g\omega_g\dot{x}_g + \omega_g^2 x_g = -w_t & , \quad x_g(0) = \dot{x}_g(0) = 0\,, \\ \ddot{x}_f + 2\zeta_f\omega_f\dot{x}_f + \omega_f^2 x_f = -2\zeta_g\omega_g\dot{x}_g - \omega_g^2 x_g & , \quad x_f(0) = \dot{x}_f(0) = 0\,. \end{cases} \tag{3.3}$$

Here, ζ_g, ζ_f and ω_g, ω_f are model parameters and w_t is zero-mean Gaussian white noise. For instance, in [130] the use of $\zeta_g = 0.623$, $\zeta_f = 0.619$ together with $\omega_g = 15.46$ (in rad/s), $\omega_f = 1.636$ (in rad/s) is recommended for firm soil conditions.

The power spectral density function of $\ddot{u}_g(t)$ with respect to the Clough-Penzien filter (CP) given in equation (3.3) is

$$S_{CP}(f) = \frac{D_0}{2\pi}\cdot\frac{\omega_g^4 + 4\zeta_g^2\omega_g^2 f^2}{(\omega_g^2 - f^2)^2 + 4\zeta_g^2\omega_g^2 f^2}\cdot\frac{f^4}{(\omega_f^2 - f^2)^2 + 4\zeta_f^2\omega_f^2 f^2}\,,$$

where $S_{w_t} = \frac{D_0}{2\pi}$ is the power spectral density function of the applied white noise. Note, that now $S_{CP}(f)$ decreases like f^4 for $f \to 0$. Again, see Sec. 10.4.1 for the derivation of S_{CP}.

One should note that the Kanai-Tajimi model and the Clough-Penzien model lead to stationary ground motion excitations, whereas real earthquakes are reported to be non-stationary. In more advanced seismic discussions envelope functions are combined with stationary models or non-stationarity is introduced by time-dependent additional functions in the Kanai-Tajimi or the Clough-Penzien model, cf. [175], [143], [217], [255].

For instance, based on the observation of 1940s El Centro earthquake Fan and Ahmadi extended the Kanai-Tajimi model by introducing a time-dependent (deterministic) envelope function $a(t)$ together with a time-dependent frequency $\omega_g(t)$ such that the ground motion excitation became

$$\ddot{u}_g = -\left(2\zeta_g\omega_g(t)\dot{x}_g(t) + \omega_g^2(t)x_g(t)\right)a(t)\,,$$

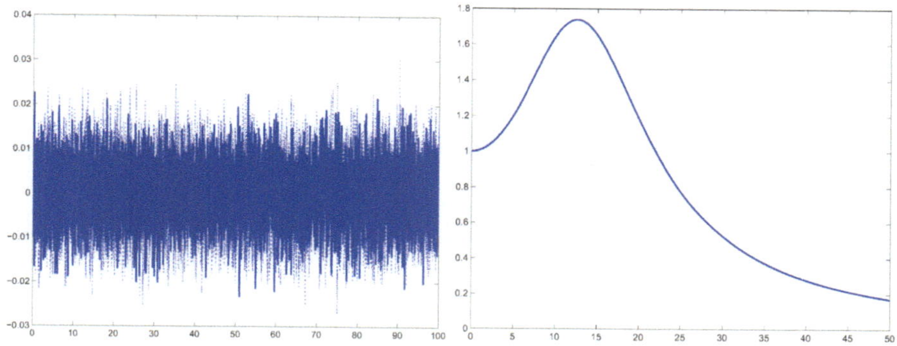

(a) (b)

Figure 3.6. Simulation excitation $\ddot{u}_g$ with respect to the Kanai-Tajimi filter (a) and its power spectral density function (b) for firm soil conditions.

where $x_g(t)$ is the solution of the modified Kanai-Tajimi stochastic differential equation (3.2), i.e.

$$\ddot{x}_g(t) \;+\; 2\zeta_g\omega_g(t)\dot{x}_g(t) \;+\; \omega_g^2(t)x_g(t) \;=\; -w_t\,.$$

In particular for the El Centro earthquake the following choices were applied:

$$\begin{aligned} a(t) &= 9.44t^3\,\mathsf{exp}(-1.1743t) + 3.723\,, \\ \omega_g(t) &= \pi\,(10.01\,(\mathsf{exp}(-0.0625t) - \mathsf{exp}(-0.15t)) + 3.0)\;, \\ \zeta_g &= 0.42\,. \end{aligned}$$

Figure 3.7 illustrates the excitation $\ddot{u}_g$ under the influence of the envelope function for the El Centro earthquake parameters.

Another model to describe in some qualitative manner the nature of an earthquake disturbance was considered by [31], see [233], pp. 164. This model for ground accelerations is given as follows

$$y(t) \;=\; \begin{cases} \sum_{j=1}^{n} ta_j\,\mathsf{exp}(-\alpha_j t)\,\mathsf{cos}(\omega_j t + \Theta_j)\,, & \text{for } t \geq 0 \\ 0\,, & \text{for } t < 0 \end{cases}\,, \tag{3.4}$$

where a_j, α_j and ω_j are given real positive numbers and the parameters Θ_j are independent random variables uniformly distributed over an interval of length 2π. A short calculation (see problem 3.30) shows

$$\begin{aligned} \mathbb{E}(y(t)) &= 0\,, \\ \mathbb{C}ov(y(t), y(s)) &= \sum_{j=1}^{n} \tfrac{1}{2} tsa_j^2\,\mathsf{exp}\left(-\alpha(t+s)\right)\mathsf{cos}\left(\omega_j(t-s)\right)\,, \quad \text{with } t, s \geq 0\,. \end{aligned}$$

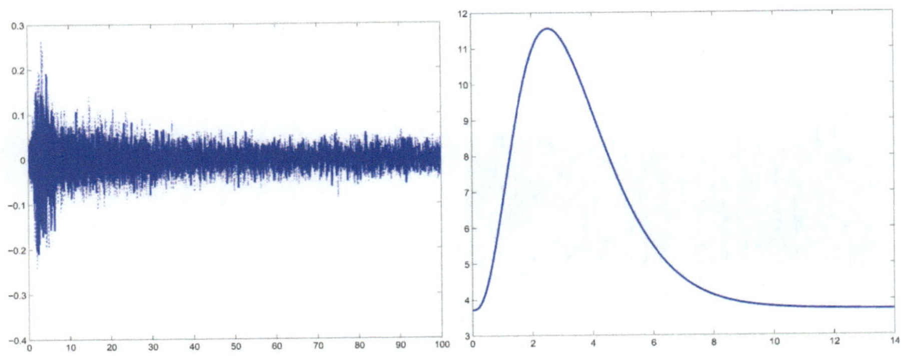

(a) (b)

Figure 3.7. Simulation excitation $\ddot{u}_g$ with respect to the Kanai-Tajimi filter (a) and plot of the envelope function for the El Centro earthquake parameters (b).

Following [233], pp. 164, the plausibility for choosing $y(t)$ that particular way stems mainly from the fact that the sample functions generated by (3.4) possess, like those of the Fan-Ahmadi modification, the usual appearance of earthquake accelograms: they show a general trend of an gradually increasing and then decreasing absolute value in an erratically behaving graph.

Following [144], p.267, the coefficients of (3.4) are given as $\omega_1 = 6$ (in rad/s), $\omega_2 = 8$ (in rad/s), $\omega_3 = 10$ (in rad/s), $\omega_4 = 11.15$ (in rad/s), $\omega_5 = 20,75$ (in rad/s), $\omega_6 = \dots, \omega_{11} = 0$ (in rad/s), $\omega_{12} = 21.50$ (in rad/s), $\omega_{13} = 23.25$ (in rad/s), $\omega_{14} = 25$ (in rad/s), $\omega_{15} = 27$ (in rad/s), $\omega_{16} = 29$ (in rad/s), $\omega_{17} = 30.5$ (in rad/s), $\omega_{18} = 32$ (in rad/s), $\omega_{19} = 34$ (in rad/s), $\omega_{20} = 36$ (in rad/s), as well as $\alpha_1 = \dots = \alpha_{20} = 1/3$, $a_1 = \dots = a_{20} = 0.5$. For the uniformly distributed random variables Θ_j we can use the MATLAB command

```
theta=2*pi*unifrnd(0,2*pi,20,1);
```

by this the resulting function $y(t)$ is plotted in Fig. 3.8. We note that $y(t)$ is highly oscillating, as we would suspect for a suitable model for the ground acceleration of an earthquake.

3.3 Random Differential Equations

Our main source for the set-up and discussion of random differential equations will be [45]. Throughout this text $(\Omega, \mathcal{A}, \mathbb{P})$ always stands for a probability space. We refer to Chap. 1 for an introduction of the basic concepts of stochastic processes that we require.

The entries of a *random-vector* $X = (X_1, \dots, X_d)^T \in \mathbb{R}^d$ are d random variables $X_1, \dots, X_d : \Omega \to \mathbb{R}$. In other words, a random-vector is an $\mathcal{A}$-

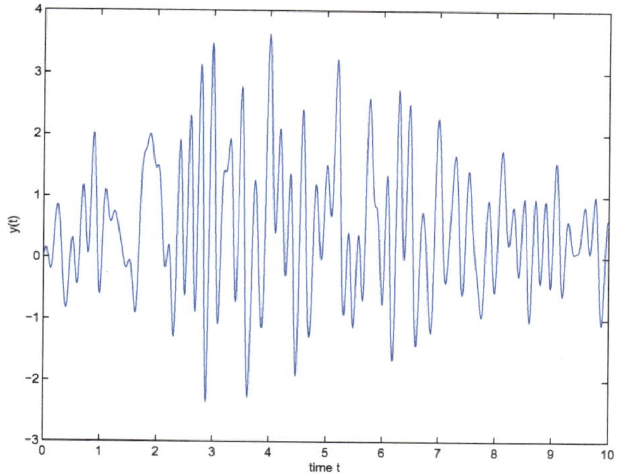

Figure 3.8. The ground acceleration $y(t)$ from (3.4) plotted over the interval $[0, 10]$.

Before you continue, make sure to answer the following questions:

Quiz: Section 3.2

Q1 Give the definitions of the Gaussian white noise process and the (standard) Wiener process.

Q2 What do you know about continuity and differentiability of the Wiener process?

Q3 Why can the white noise be considered as unphysical?

Q4 Describe the Kanai-Tajimi and the Clough-Penzien model for seismic excitations. Give their defining differential equations with respect to white noise.

measurable function that maps from Ω to $\mathbb{R}^d$ and for which the pre-images of Borel-sets are elements of the σ-algebra $\mathcal{A}$, i.e., $X^{-1}(B) \in \mathcal{A}$ for all $B \in \mathcal{B}$.

We denote by S_d the set of all d-dimensional random-vectors. S_d will serve as the set from which the initial values of our random differential equations will be taken (of course $\mathbb{R}^d \subset S_d$ which allows deterministic vales as well).

Let $I \subset \mathbb{R}_0^+$ be a real interval (most often starting at zero, i.e., $I = [0, T]$). A real valued function $X_t(\omega)$ on $I \times \Omega$ is called a *stochastic process* on I if it is $\mathcal{A}$-measurable for each fixed $t \in I$. For notational simplicity, we often suppress the argument ω if no ambiguities result.

We write $X \hat{=} Y$ if $X(\omega) = Y(\omega)$ holds for almost all $\omega \in \Omega$. Next, let X_t and Y_t be *equivalent* on I, i.e., $\mathbb{P}(Y_t = X_t) = 1$ for each fixed $t \in I$, such that $X_t \hat{=} Y_t$, $t \in I$, then there are sets $\Omega_t \in \mathcal{A}$, $t \in I$, such that $\mathbb{P}(\Omega_t) = 1$ and $X_t(\omega) = Y_t(\omega)$ holds for $(t,\omega) \in \cup_{t\in I}[\{t\} \times \Omega]$. If, in addition, one can choose the sets Ω_t independently from t ($\Omega_t = \Omega^*$), then $X_t(\omega) = Y_t(\omega)$ holds for all $(t,\omega) \in I \times \Omega^*$; then we denote such an equivalence between the processes X_t and Y_t by $X_t \stackrel{I}{=} Y_t$. (Later on, we will use this notation for processes that are only defined on subsets of I, too.)

If the realisations of a stochastic process X_t are continuous on I for almost all $\omega \in \Omega$ then the process X_t is called *path-wise continuous* on I. Moreover, a stochastic process X_t is called *path-wise differentiable* on I if there exists a set $\Omega_1 \subset \mathcal{A}$ such that $\mathbb{P}(\Omega_1) = 1$ and that for all $\omega \in \Omega_1$ the derivative

$$\dot{X}_t(\omega) := \frac{\mathrm{d}}{\mathrm{d}t} X_t(\omega) := \lim_{h\to 0} \frac{X_{t+h}(\omega) - X_t(\omega)}{h}, \qquad t \in \mathrm{int}(I),$$

exists. (As in the deterministic setting the "derivative" at the boundary points ∂I is defined accordingly.) The mapping $\dot{X}_\cdot(\cdot) : I \times \Omega_1 \to \mathbb{R}^d$ is called the *path-wise derivative*. Note, there is always a path-wise continuous process Y_t, $t \in I$ such that $Y_t \stackrel{I}{=} \dot{X}_t$. Analogously, the *path-wise integral* is defined: We say that X_t has a path-wise integral on I if the integral $\int_I X_t(\omega)\mathrm{d}t$ exists for almost all $\omega \in \Omega$.

Now, let $X : I \times \Omega \to \mathbb{R}^m$ be an $\mathbb{R}^m$-valued stochastic process with continuous sample paths, and $f : \mathbb{R}^d \times I \times \Omega \to \mathbb{R}^d$ be a continuous function. To underline the dependence of the stochastic process X_t on (arbitrary) elements of Ω, we sometimes write $X_t(\cdot)$.

Definition 3.4 (Random (Ordinary) Differential Equation). A *random (ordinary) differential equation* (short: *random differential equation*) on $\mathbb{R}^d$,

$$\frac{\mathrm{d}X_t}{\mathrm{d}t} = f(X_t(\cdot), t, \omega), \qquad X_t(\cdot) \in \mathbb{R}^d, \tag{3.5}$$

is a non-autonomous ordinary differential equation

$$\dot{x} = \frac{\mathrm{d}x}{\mathrm{d}t} = F_\omega(x,t), \qquad x := X_t(\omega) \in \mathbb{R}^d \tag{3.6}$$

for almost all $\omega \in \Omega$. Again, for notational simplicity, we often suppress the argument ω if no ambiguities result.

A stochastic process X_t defined on the interval I is called *path-wise solution* of the random differential equation (3.5) if almost all realisations of X_t on I are solutions of the non-autonomous deterministic ordinary differential equation (3.6).

Let $t_0 \in I$ and let $X_0 \in S_d$ such that $X_{t_0} \hat{=} X_0$. Then, X_t is called path-wise solution of (3.5) with respect to the initial condition (X_0, t_0).

The path-wise solution of (3.5) with respect to the initial condition (X_0, t_0) is called *unique* on I if for an arbitrary pair X_t and X_t^* of path-wise solutions with respect to the initial condition (X_0, t_0) the following holds true:

$$X_t \stackrel{I}{=} X_t^* ,$$

i.e., $X_t(\omega) = X_t^*(\omega)$ for all $(t, \omega) \in I \times \Omega^*$ with $\Omega^* \subset \Omega$ such that $\mathbb{P}(\Omega^*) = 1$. To be extremely precise here we use $X_t(\omega)$ to denote the dependence on one specific $\omega \in \Omega$ in (3.6) and we use $X_t(\cdot)$ to indicate that a specific choice of $\omega \in \Omega$ is not required as in (3.5).

We associate a deterministic (ordinary) differential equation to any specific realisation $\omega \in \Omega$. The solutions $X_t(\omega)$ of these deterministic differential equations are the realisations of a stochastic process $X_t(\cdot)$ which is the path-wise solution of the random differential equation (3.5).

Thus, in order to show the existence of path-wise solutions one first plays this issue back to the theory of deterministic (ordinary) differential equations and shows that (3.6) has a solution on an ω-independent interval I for almost all $\omega \in \Omega$.

Remark 3.5 (How to Read a Mathematical Definition or Theorem). As this book is intended for both students of engineering/ sciences and mathematics we want to point out that mathematics has the tendency to be rather compact in its notation. In particular when it comes to definitions and theorems, every word counts.

Recall, that most of the definitions you will see and every theorem consists of assumptions/ prerequisites together with one ore more statements. Be very precise when reading a definition or theorem: ensure that you understand the notations (and their subtleties), assumptions and statements. The best way to understand the assumptions is to ask what happens if they are violated. The statement of a theorem is true only under the assumptions (i.e., before you can apply the theorem you must check the validity of the assumptions), so one way to understand a theorem is to apply it to some easy examples.

Before we come to existence and uniqueness of solutions in Sec. 3.4, let us examine the property of being a path-wise solution more carefully. Recall, a path-wise solution of a random differential equation is a stochastic process that is defined on an ω-independent interval. In the following section we attempt to clarify these two points by considering counterexamples for path-wise solutions.

3.3.1 Counterexamples for Path-Wise Solutions

The solutions of $\dot{x} = F_\omega(x,t)$ do in general not have a common (and ω-independent) interval of existence.

Example 3.6 (A RDE Without Path-Wise Solutions, cf. [45], p. 13). Let $a(\omega)$ be an exponentially distributed random variable[7]. The random differential equation

$$\dot{X}_t = a(\omega)X_t^2, \qquad X_0 = 1,$$

has a unique solution for every $\omega \in \Omega$. Though, it has no path-wise solution for this initial condition, as the solutions

$$X(t,\omega) = \frac{1}{1-a(\omega)t}$$

of this random differential equation exist only in the interval $[0,(a(\omega))^{-1}]$. Thus, there is no common interval $J \subset [0,T]$ such that all solutions exist in J for almost all $\omega \in \Omega$.

Likewise, a function $X(t,\omega)$ defined on $I \times \Omega$ that satisfies for every fixed $\omega \in \Omega$ the differential equation (3.5) on I does not necessarily have to be a stochastic process.

Example 3.7 (A RDE Solution That is Not a Path-Wise Solution, cf. [45], p. 14). Let $\Omega^* \subset \Omega$ such that $\Omega^* \notin \mathcal{A}$. For every $\omega \in \Omega$ the function

$$X_t^*(\omega) = \begin{cases} 0, & \text{for } \omega \in \Omega^* \\ \frac{1}{4}t^2, & \text{for } \omega \in \Omega \setminus \Omega^* \end{cases},$$

is a solution on $\mathbb{R}_0^+ = [0,\infty)$ of

$$\dot{x} = \sqrt{|x|}, \qquad \text{with the initial condition } X_0(\omega) = 0.$$

Though, X_t^* is not a stochastic process and hence not a path-wise solution.

In definition 3.4 the $\mathcal{A}$-measurability of a path-wise solution X_t for every fixed $t \in I$ is required. Otherwise the mean $\mathbb{E}(X_t)$ or the probability $\mathbb{P}(X_t > c)$, $c \in \mathbb{R}$, may not be defined.

If a path-wise solution exists and if the solutions of (3.6) are unique for almost all $\omega \in \Omega$ on I then the path-wise solution is unique, too. We will make this more precise in Sec. 3.4.

---

7 The exponential distribution describes the time between events in a Poisson (jump) process. Its probability density function is given as $f(x,\lambda) = \lambda\exp(-\lambda x)$ for $x \geq 0$ and $f(x,\lambda) = 0$ for $x < 0$, where $\lambda > 0$ is the rate parameter of the distribution.

3.3.2 Connections between Random and Stochastic Differential Equations

The most prominent class of randomly perturbed differential equations are *stochastic differential equations* (often abbreviated as SDE or SODE for stochastic ordinary differential equation). These are formally given as

$$\mathrm{d}X_t \;=\; a(X_t,t)\mathrm{d}t + b(X_t,t)\mathrm{d}W_t\,, \qquad X_0 \;=\; X(0)\,,$$

where a and b are suitable functions. As for deterministic ordinary differential equations the above equation is defined in terms of its integral representation

$$X_t \;=\; X_0 + \int_0^T a(X_t,t)\mathrm{d}t + \int_0^T b(X_t,t)\mathrm{d}W_t\,.$$

Here, the first integral is a usual Riemann- or Lebesgue-integral whereas the second integral is a stochastic integral with respect to a Wiener process. The interpretation of this second integral is somewhat tricky. Since, W_t is nowhere differentiable with probability 1, the integral $\int_0^T f(t,\omega)\mathrm{d}W_t$ cannot be defined in the usual Lebesgue-Stieltjes sense. Moreover, the sample paths of W_t are not of finite variation on compact time intervals. Thus, as a consequence of the Banach-Steinhaus theorem the usual Riemann-Stieltjes integral cannot be applied either (cf. [212], pp. 43).

What Itô saw in [139] was that if one restricts the class of potential integrands $b(x,t)$ to those that are adapted to the underlying filtration of σ-algebras generated by the Wiener process, and if one restricts the choice of the evaluation points in a Riemann sum approximation of the integral to the start of the respective time interval, then one can use the independence of the increments of the Wiener process in a clever way to obtain the convergence of the sums to a limit which is a martingale, cf. [213]. Though, new rules of calculus are required. The above stochastic integrals are called Itô stochastic integrals. Another common interpretation goes back to Stratonovich and is compatible with the usual rules of calculus but does not lead to martingales.

For a thorough study of stochastic differential equations we refer to the literature: Excellent publications include [97], [100], [227] which focus on theory and [8], [11], [155], [156], [211] that focus on stability as well as [149] and [159] with a focus on the numerics of stochastic differential equations. Maple and MATLAB toolboxes are, e.g., discussed in [77], [78], [79], [129] and [210] (MATLAB SDE Toolbox).

Random differential equations with Wiener processes can be re-written as stochastic differential equations, so results for one type of these equations can be applied to the other and vice versa, cf. [149], pp. 10.

Example 3.8 (From a RDE to a SDE, cf. [149], pp. 10.). The scalar random differential equation $\dot{X}_t = -X_t + \sin(W_t(\omega))$ can be re-written as the following two-dimensional stochastic differential equation

$$\mathrm{d}\begin{pmatrix} -X_t \\ Y_t \end{pmatrix} = \begin{pmatrix} X_t + \sin(Y_t) \\ 0 \end{pmatrix} \mathrm{d}t + \begin{pmatrix} 0 \\ 1 \end{pmatrix} \mathrm{d}W_t .$$

Following [149], pp. 10, we see that any (finite-dimensional) stochastic differential equation can be transformed to a random differential equation: In the case of commutative noise this is a famous result obtained by Halim Doss and Hector J. Sussmann (see [89], [245]), which was generalized in 1998 to all stochastic differential equations by Peter Imkeller and Björn Schmallfuss (see [136]). This *Doss/ Sussmann & Imkeller/ Schmalfuss correspondence* is easily illustrated for a scalar stochastic differential equation with additive noise.

Example 3.9 (From a SDE to a RDE, cf. [149], pp. 10.). The scalar stochastic differential equation

$$\mathrm{d}X_t = f(X_t)\mathrm{d}t + \mathrm{d}W_t$$

is equivalent to the random differential equation

$$\dot{Z}_t = f(Z_t + O_t) + O_t ,$$

where $Z_t := X_t - O_t$ and O_t is the stationary *Ornstein-Uhlenbeck process* satisfying the stochastic differential equation $\mathrm{d}O_t = -O_t\mathrm{d}t + \mathrm{d}W_t$. Note, the Ornstein-Uhlenbeck process is a Gaussian process, and thus it is already uniquely determined by its first two moments.

Important and meaningful examples of well-recognized models where this correspondence can be applied include the continuous Vasicek model, the extended Vasicek model, the Hull-White model (see problem 3.26), the Black-Karasinski model or the Black-Derman-Toy model (see problem 3.25) for interest rates, cf. [40], or the Kanai-Tajimi model for earthquake induced ground motion excitations discussed earlier in this chapter, just to name a few.

In particular, applying the Doss-Sussmann/ Imkeller-Schmalfuss correspondence to the Kanai-Tajimi earthquake model allows us to transform this stochastic differential equation (3.2), alias

$$\mathrm{d}\begin{pmatrix} x \\ y \end{pmatrix} = \begin{pmatrix} -y \\ -2\zeta_g\omega_g y + \omega_g^2 x \end{pmatrix} \mathrm{d}t + \begin{pmatrix} 0 \\ 1 \end{pmatrix} \mathrm{d}W_t ,$$

into the path-wise equivalent random differential equation

$$\begin{pmatrix} \dot{z}_1 \\ \dot{z}_2 \end{pmatrix} = \begin{pmatrix} -(z_2 + O_t) \\ -2\zeta_g\omega_g(z_2 + O_t) + \omega_g^2 z_1 + O_t \end{pmatrix} . \tag{3.7}$$

From a broader perspective, the Doss-Sussmann/ Imkeller-Schmalfuss correspondence opens the whole field of diffusion processes typically modelled via stochastic differential equations, see Fig. 3.9. For a complete discussion of the indicated correspondences amongst diffusion processes we refer to the literature, e.g. [243], [204] or [222].

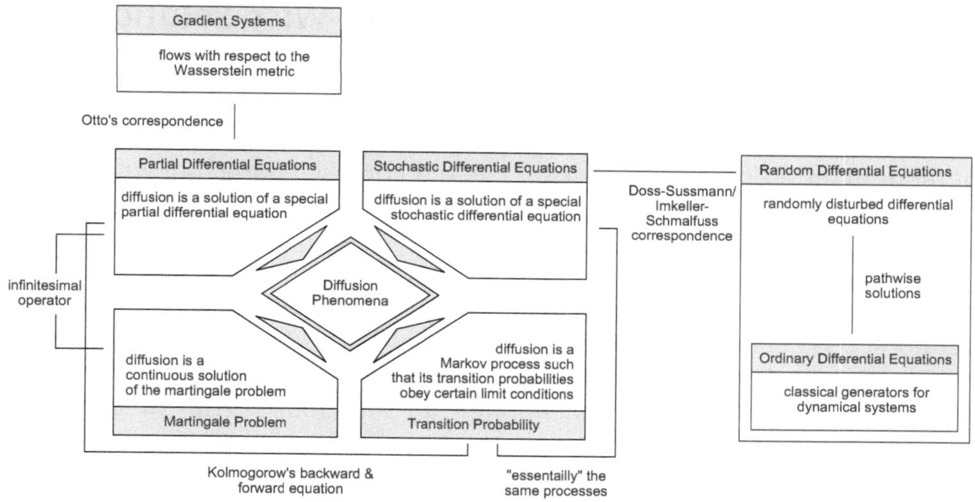

Figure 3.9. Some important correspondences between diffusion processes and random (ordinary) differential equations.

Before you continue, make sure to answer the following questions:

Quiz: Section 3.3

Q1 Give the definition of a random (ordinary) differential equation and its path-wise solutions. What does uniqueness mean in this setting?

Q2 Do random (ordinary) differential equations always have path-wise solutions? Give examples.

Q3 What is a stochastic (ordinary) differential equation and in which sense is it different from a random (ordinary) differential equation?

Q4 Can one convert stochastic (ordinary) differential equations into random (ordinary) differential equations?

Q5 Transform the linear stochastic (ordinary) differential equation $\mathrm{d}X_t = aX_t\mathrm{d}t + b\mathrm{d}W_t$, $a, b \in \mathbb{R}$, into its corresponding random (ordinary) differential equation.

3.4 Existence & Uniqueness of Path-Wise Solutions of Random Differential Equations

Before we state the existence and uniqueness results for path-wise solutions of random differential equations we cite a technical lemma that will allow us to continue the unique solution of the family of deterministic equations (3.6) to a stochastic process.

Lemma 3.10 (Completion to a $\mathcal{B}^d \times \mathcal{A}$-Measurable Function). *Let $\mathcal{B}^d$ the product Borel-σ-algebra on $\mathbb{R}^d$, $\Omega_1 \in \mathcal{A}$ and $f(x, \omega) : \mathbb{R}^d \times \Omega \to \mathbb{R}^m$ be continuous on $\mathbb{R}^d$ for $\omega \in \Omega_1$ as well as $\mathcal{A}$-measurable in ω for all $x \in \mathbb{R}^d$. Then there exists a $\mathcal{B}^d \times \mathcal{A}$-measurable function $g(x, \omega)$ such that $f(x, \omega) = g(x, \omega)$ for $(x, \omega) \in \mathbb{R}^d \times \Omega_1$.*

Proof. see [87], p. 60, Theorem 2.2.5 (cf. problem 3.24) □

As postulated in Definition 3.4, the path-wise solution of a random differential equation does not only have to solve a family of ordinary differential equations (given by fixing $\omega \in \Omega$), but is also required to be a stochastic process that is defined on an ω-independent interval. The following theorem proves existence and uniqueness of path-wise solutions under rather weak conditions[8]:

Theorem 3.1 (Existence & Uniqueness of Path-Wise Solutions). *Let $X_0 \in S_d$, $t_0 \in I$, and $a, b \in \mathbb{R}^+$ such that $[t_0 - a, t_0 + a] \subset I$ as well as*

$$Q(\omega) := \left\{ (x,t) \in \mathbb{R}^d \times I \,:\, \|x - X_0(\omega)\| \leq b \,, |t - t_0| \leq a \right\} .$$

Moreover, let the following four prerequisites be satisfied:

1. *For each fixed $(x, t) \in \mathbb{R}^d \times I$ the function $f(x, t, \omega)$ is $\mathcal{A}$-measurable in ω, and for each fixed $t \in I$ and almost all $\omega \in \Omega$ this function is continuous on $\mathbb{R}^d$.*
2. *There is a set $\Omega^* \in \mathcal{A}$ of full measure, i.e., $\mathbb{P}(\Omega^*) = 1$, and for every $\varepsilon > 0$ there is a $\delta(\varepsilon) > 0$, such that*

$$\|f(x_1, t_1, \omega) - f(x_2, t_2, \omega)\| \leq \varepsilon \,,$$

under the assumptions that $\omega \in \Omega^*$, $(x_i, t_i) \in Q(\omega)$ ($i = 1, 2$), $\|x_1 - x_2\| < \delta(\varepsilon)$ and $|t_1 - t_2| \leq \delta(\varepsilon)$.

---

8 Recall: The path-wise solution of (3.5) with respect to the initial condition (X_0, t_0) is called unique on I if for an arbitrary pair X_t and X_t^* of path-wise solutions with respect to the initial condition (X_0, t_0) if $X_t \stackrel{I}{=} X_t^*$ holds.

3. *For almost all $\omega \in \Omega$ it holds that $\|f(x,t,\omega)\| \le M < \infty$ for $(x,t) \in Q(\omega)$, i.e. $\|f(x,t,\omega)\|$ is bounded on $(x,t) \in Q(\omega)$ by a constant M.*

4. *For almost all $\omega \in \Omega$ there are unique solutions of the deterministic differential equation (3.6) with respect to the initial condition $(X_0(\omega), t_0)$ on the interval $I_\alpha := (t_0 - \alpha, t_0 + \alpha)$, where $\alpha := \min(a, M^{-1}b)$.*

Then on I_α there exists a unique path-wise solution of (3.5) with respect to the initial condition (X_0, t_0).

Proof. Following [45], p. 15, we have: Due to the Cauchy-Peano's existence theorem (Theorem 6.2) for deterministic ordinary differential equations we have for almost all $\omega \in \Omega$ that there exists a solution of the deterministic differential equation (3.6) on I_α with respect to the initial condition $(X_0(\omega), t_0)$. These solutions can be constructed as follows:

- Let $\{\varepsilon_k\}_{k\in\mathbb{N}}$ be a monotonously decreasing sequence of real positive numbers such that $\lim_{k\to\infty} \varepsilon_k = 0$, and for each ε_k let there be a $\delta(\varepsilon_k)$ given according to condition 2. With the division points $t_0 = t_1^k < \cdots < t_m^k = t_0 + \alpha$ we divide the interval $I_\alpha^+ = [t_0, t_0 + \alpha]$ such that

$$\max |t_l^k - t_{l-1}^k| \le \min\left(\delta(\varepsilon_k), \frac{\delta(\varepsilon_k)}{M}\right)$$

  holds, $l = 2, \ldots, m$ (the proof for $I_\alpha^- = [t_0 - \alpha, t_0]$ is analogous).

- Let the functions $X_k(t,\omega)$ be defined in the spirit of Euler polygons as, cf. Fig. 3.10,

$$X_k(t_0, \omega) = X_0(\omega),$$

  and for $t_{l-1}^k < t \le t_l^k$

$$X_k(t,\omega) = X_k(t_{l-1}^k, \omega) + f\left(X_k(t_{l-1}^k, \omega), t_{l-1}^k, \omega\right)\left(t - t_{l-1}^k\right).$$

- In particular (cf. ODE theory), the functions $X_k(t,\omega)$ are equi-continuous[9], uniformly bounded[10] and converge on I_α^+, due to condition

[9] A family of functions is equi-continuous if all the functions are continuous and they have equal variation over a given neighborhood. More precisely, let (X, d_X) and (Y, d_Y) be two metric spaces, and $\{F_k\}_{k\in\mathbb{N}} : X \to Y$ a family of functions. $\{F_k\}_{k\in\mathbb{N}}$ is equi-continuous at a point $x_0 \in X$ if for every $\varepsilon > 0$, there exists a common $\delta > 0$ such that $d_Y(F_k(x_0), F_k(x)) < \varepsilon$ for all x such that $d_X(x_0, x) < \delta$. The family is equi-continuous if it is equi-continuous at each point of X.

[10] For a family of bounded functions $\{F_k\}_{k\in\mathbb{N}}$, the constant c_k that bounds the functions $|F_k| < c_k$ can vary from function to function. If one can find one constant c which bounds all functions of the family, i.e., $|F_k| < c$ for all $k \in \mathbb{N}$, then this family is called uniformly bounded.

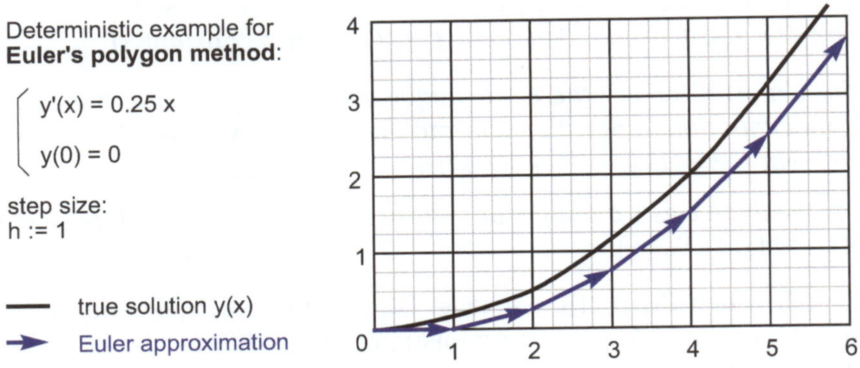

Figure 3.10. Purely deterministic example of the Euler polygon method for the simple equation $y'(x) = \frac{1}{4}x$ with $y_0 = 0$ and step size $h = 0$. In this case, the polygonal approximation can be calculated by pencil and paper.

4. to the unique solution $X(t, \omega)$ of the deterministic equation (3.6) with respect to the initial condition $(X_0(\omega), t_0)$.

Because of condition 1 we can apply Lemma 3.10 to the function $f(x, t_l^k, \omega)$. Together with $X_0 \in S_d$ it follows by induction that there is a stochastic process X_{kt}, $t \in I_\alpha^+$ such that $X_{kt} \stackrel{I_\alpha^+}{=} X_k(t, \omega)$. Consequently, there is a stochastic process X_t, $t \in I_\alpha^+$, such that $X_t \stackrel{I_\alpha^+}{=} \lim_{k\to\infty} X_k(t, \omega)$ which is the unique path-wise solution of (3.5) on I_α^+ with respect to the initial condition (X_0, t_0). □

Remark 3.11 (On the Usefulness of Proofs). Unfortunately, in many textbooks for engineering/ science, proofs are kept to less than a minimum. From a mathematicians' point of view nothing may be believed without a proof (at least in mathematics). In particular, when combining analysis and numerics/ scientific computing constructive proofs like that of Theorem 3.1 serve as a wonderful resource and inspiration for the construction of algorithms: start with a suitably good time discretisation, then use Euler polygons to approximate the function and finally interpret the result in terms of an ω-dependent path-wise solution. (Numerical methods for random (ordinary) differential equations will be discussed more deeply in Chap. 14.)

We apply Theorem 3.1 to an illustrative example.

Example 3.12 (Application of Theorem 3.1, cf. [45], p. 15). Let Y be an m-dimensional random-vector whose realisations lie in a compact set $H \subset \mathbb{R}^m$, $x_0 \in \mathbb{R}^d$, $t_0 \in \mathbb{R}$, and

$$Q = \left\{ (x, t) \in \mathbb{R}^d \times \mathbb{R} : \|x - x_0\| \le b, |t - t_0| \le a \right\},$$

where $a, b \in \mathbb{R}^+$. Moreover, let $f(x, t, \omega)$ be a continuous function on $\mathbb{R}^d \times \mathbb{R} \times H$. If for each fixed $y \in H$ there is a unique solution on I_α of the deterministic differential equation $\dot{x} = f(x, t, y)$ with respect to the initial condition (x_0, t_0), then Theorem 3.1 guarantees the existence of a unique path-wise solution on I_α of the random differential equation $\dot{X}_t = f(X_t, t, Y)$ with respect to the initial condition (x_0, t_0).

Finally, the following theorem gives the unique existence of a path-wise solution in a way which is analogous to the uniqueness and existence theorem of solutions for deterministic ordinary differential equations. In principle this theorem is derived from Theorem 3.4, that we will discuss later on, by taking the empty set as the exceptional set H used in theorem 3.4.

Theorem 3.2 (Existence & Uniqueness of Path-Wise Solutions). *Let the following three prerequisites be satisfied:*

1. *The functions $f(x, t, \omega)$ are $\mathcal{A}$-measurable for all $(x, t) \in \mathbb{R}^d \times I$.*
2. *$f(x, t, \omega)$ is continuous on $\mathbb{R}^d \times I$ for almost all $\omega \in \Omega$.*
3. *For almost all $\omega \in \Omega$ there is a real continuous function $L(t, \omega)$ on I such that*
$$\|f(x_1, t, \omega) - f(x_2, t, \omega)\| \leq L(t, \omega)\|x_1 - x_2\|,$$
where $t \in I$ and $x_1, x_2 \in \mathbb{R}^d$.

Then for any initial condition $(X_0, t_0) \in S_d \times I$ there exists a unique path-wise solution on I of the random differential equation (3.5).

Proof. As stated this theorem is a special case of Theorem 3.4 and follows from Theorem 3.4 by taking the exceptional set H used there as $H = \{\}$ in the case of Theorem 3.2. □

We apply this theorem to an illustrative example.

Example 3.13 (Application of Theorem 3.2, cf. [45], p. 16). Let Z_t, $t \in [0, \infty)$, be a path-wise continuous d-dimensional random-vectorial process. Let A_t be a $d \times d$-matrix whose elements are path-wise continuous stochastic processes on $[0, \infty)$. Under these condition Theorem 3.2 ensures the existence of a path-wise unique solution on $[0, \infty)$ of the random differential equation $\dot{X}_t = A_t X_t + Z_t$ for any initial condition $(X_0, t_0) \in S_d \times [0, \infty)$.

Before you continue, make sure to answer the following questions:

Quiz: Section 3.4 – Part 1 (Existence & Uniqueness of RDEs)

Q1 Under which conditions does a unique path-wise solution of a random (ordinary) differential equation (RDE) exist?

Q2 Sketch the proof of the existence and uniqueness theorem you just stated.

Q3 What do you know about the existence and uniqueness of linear inhomogeneous random (ordinary) differential equations of the form $\dot{x}_t = A_t x_t + z_t$, with $x_t \in \mathbb{R}^d$?

Q4 Let the scalar random (ordinary) differential equation $\dot{x}_t = \arctan(t) - x_t + \beta O_t$ be given, where O_t is an Ornstein-Uhlenbeck process. What can you then say about the behavior of the solution x_t for $t \to \infty$, if we additionally assume $\beta \ll 1$?

3.4.1 Path-Wise Solutions in the Extended Sense

Let $\mathcal{S}_I$ denote the set of all Lebesgue-measurable sub-sets of the interval $I \subset \mathbb{R}$. We say that a relation holds for almost all $t \in I$ if there is a set $I_0 \in \mathcal{S}_I$ of vanishing Lebesgue-measure such that the respective relation holds for all $t \in I \setminus I_0$ (in particular I may be empty).

In the deterministic setting, an absolutely continuous function $x(t)$ that satisfies the deterministic ordinary differential equation $\dot{x} = f(x,t)$ for almost all $t \in I$ is called a solution of this equation in the extended sense. Following [45], pp. 16, we introduce the corresponding concept of solutions in the extended sense for random differential equations.

Definition 3.14 (Solutions of Random Differential Equations in the Extended Sense). A stochastic process X_t is called a *path-wise solution in the extended sense* (e.s.) of the random differential equation (3.5) on the interval I if for almost all $\omega \in \Omega$ the functions $X_t(\omega)$ are absolutely continuous on I and solve the deterministic differential equation (3.6) for almost all $t \in I$.

If additionally $X_{t_0} \hat{=} X_0$ holds for a $t_0 \in I$ and an $X_0 \in S_d$, then X_t is called a path-wise solution of (3.5) in the extended sense with respect to the initial condition (X_0, t_0).

Analogously to Theorem 3.1 the conditions for the unique existence of the path-wise solution of (3.5) in the extended sense with respect to the initial condition (X_0, t_0) can be stated:

Theorem 3.3 (Existence & Uniqueness of Path-Wise Solutions in the Extended Sense). *Let $(X_0, t_0) \in S_d \times I$, and $a, b \in \mathbb{R}^+$ such that $[t_0 - a, t_0 + a] \subset I$ as well as*

$$Q(\omega) := \left\{ (x,t) \in \mathbb{R}^d \times I : \|x - X_0(\omega)\| \leq b, |t - t_0| \leq a \right\}.$$

Moreover, let the following three prerequisites be satisfied:

1. *For almost all $\omega \in \Omega$ let $f(x,t,\omega)$ be continuous on $\mathbb{R}^d$ for fixed $t \in I$, and for fixed $x \in \mathbb{R}^d$ let $f(x,t,\omega)$ be $\mathcal{S}_I \times \mathcal{A}$-measurable.*

2. *For almost all $\omega \in \Omega$ let $\|f(x,t,\omega)\| \leq m(t)$ for $(t,x) \in Q(\omega)$, where $m(t)$ is a Lebesgue-integrable function on the interval $[t_0 - a, t_0 + a]$.*

3. *For almost all $\omega \in \Omega$ the solutions of (3.6) on $I_\alpha = (t_0 - \alpha, t_0 + \alpha)$ in the extended sense with respect to the initial condition $(X_0(\omega), t_0)$ are unique, where $\alpha \leq a$ is such that $\int_{t_0-\alpha}^{t_0+\alpha} m(\tau)d\tau \leq b$.*

Then there exists a unique path-wise solution of (3.5) on I_α in the extended sense with respect to the initial condition (X_0, t_0).

Proof. Following [45], pp. 16, we have: Due to the assumptions we can apply the existence theorem of Caratheodory (cf. Theorem 6.10). Thus, on I_α there are solutions $X(t,\omega)$ in the extended sense of the deterministic ordinary differential equation (3.6) with respect to the initial conditions $(X_0(\omega), t_0)$. Each of these solutions can be interpreted as the limit of a sequence of equicontinuous and uniformly bounded functions $X_j(t,\omega)$ such that

$$X_j(t,\omega) := \begin{cases} X_0(\omega) & t_0 \leq t \leq t_0 + \alpha j^{-1} \\ X_0(\omega) + \displaystyle\int_{t_0}^{t_0 - \alpha j^{-1}} f(X_j(\tau,\omega),\tau,\omega)\mathrm{d}\tau & t \leq t_0 + \alpha j^{-1} < t < t_0 + \alpha \end{cases}.$$

(The following considerations hold analogously for $t_0 - \alpha < t < t_0$.)

For $r = 1, \ldots, j$ let $I_r := [t_0 + r j^{-1} \alpha]$ and assume that there is a stochastic process $X_{j,t}^r$, $t \in I_r$ $(r < j)$, such that $X_{j,t}^r \overset{I_r}{=} X_j(t,\omega)$. The functions $X_j(t,\omega)$ are continuous on I_r for almost all $\omega \in \Omega$. Thus, by Lemma 3.10, there is a $\mathcal{B}_{I_r} \times \mathcal{A}$-measurable function $Y_j(t,\omega)$ such that $Y_j(t\omega) \overset{I_r}{=} X_j(t,\omega)$. Moreover, due to condition 1 together with Lemma 3.10 there is a $\mathcal{B}^d \times \mathcal{B}_{I_r} \times \mathcal{A}$-measurable function $g(x,t,\omega)$ such that $g(x,t,\omega) \overset{\mathbb{R}^d \times I_r}{=} f(x,t,\omega)$.

Hence, $g(Y_j(t,\omega),t,\omega)$ is a $\mathcal{B}_{I_r} \times \mathcal{A}$-measurable function. According to Fubini's theorem the existence of a stochastic process $X_{(j)t}^{r+1}$, $t \in I_{r+1}$ follows such that

$$X^{r+1}_{(j)t} = \begin{cases} X_0(\omega) & t_0 \le t \le t_0 + \alpha j^{-1} \\ X_0(\omega) + \int_{t_0}^{t_0-\alpha j^{-1}} f(X_j(\tau,\omega),\tau,\omega)\mathrm{d}\tau & t \le t_0 + \alpha j^{-1} < t < t_0 + \alpha \end{cases}$$

$$\stackrel{I_{r+1}}{=} X_j(t,\omega)\,.$$

By induction the existence of a stochastic process $X^j_{(j)t}$, $t \in I^+_\alpha = [t_0, t_0 + \alpha)$, follows such that $X^j_{(j)t} \stackrel{I_{\alpha+}}{=} X_j(t,\omega)$. Consequently, there is a stochastic process X_t, $t \in I$ such that $X_t \stackrel{I^+_\alpha}{=} \lim_{j\to\infty} X_j(t,\omega)$, too. X_t is the path-wise unique solution of (3.5) on I_α in the extended sense with respect to the initial condition (X_0, t_0). □

It is not uncommon to be interested in stochastic processes that are known to have just piece-wise continuous realisations, like Poisson jump processes which describe the behavior of queues. For such cases the following theorem can be applied:

Theorem 3.4 (Existence & Uniqueness of Path-Wise Solutions in the Extended Sense with Piecewise Continuous Realisations). *Let the following four prerequisites be satisfied:*

1. *The function $f(x,t,\omega)$ is $\mathcal{A}$-measurable for all $(x,t) \in \mathbb{R}^d \times I$.*
2. *For almost all $\omega \in \Omega$ the function $f(x,t,\omega)$ is continuous on $\mathbb{R}^d \times (I \setminus H(\omega))$, where $H(\omega)$ is an at most countable subset in I the elements of which do not have a finite accumulation point.*
3. *For almost all $\omega \in \Omega$ and all $x \in \mathbb{R}^d$ the function $f(x,t,\omega)$ is bounded on any finite interval $J \subset I$.*
4. *For almost all $\omega \in \Omega$ there is a continuous function $L(t,\omega)$ on I such that*

$$\|f(x_1,t,\omega) - f(x_2,t,\omega)\| \;\le\; L(t,\omega)\|x_1 - x_2\|\,,$$

where $t \in I$ and $x_1, x_2 \in \mathbb{R}^d$.

Then, for any initial condition $(X_0, t_0) \in S_d \times I$ there exists a unique path-wise solution in the extended sense on I of the random differential equation (3.5).

Proof. Following [45], p. 18, we construct the usual sequence of Picard-approximations, cf. Fig. 3.11,

$$\begin{cases} X_0(t,\omega) & := & X_0(\omega) \\ x_k(t,\omega) & := & X_0(\omega) + \int_{t_0}^{t} f(X_{k-1}(s,\omega),s,\omega)\mathrm{d}s \end{cases}, \tag{3.8}$$

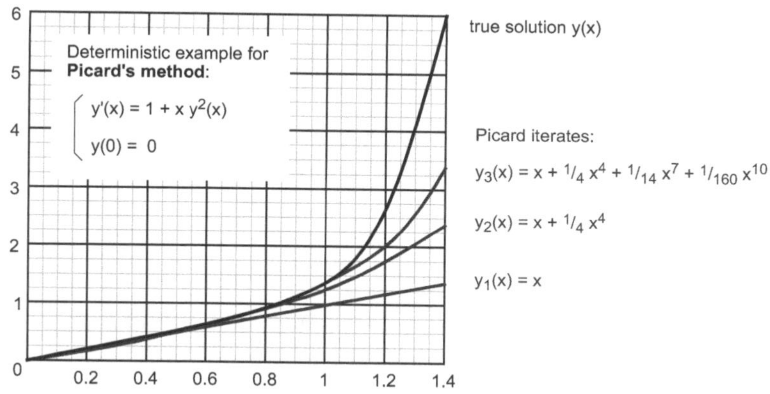

Figure 3.11. Purely deterministic example of the Picard iteration method for the simple equation $y'(x) = 1 + xy^2(x)$ with $y_0 = 0$.

where the continuity of $X_k(t, \omega)$ on I for almost all $\omega \in \Omega$ follows by induction, as well as that the above Riemann-integrals (i) are defined as the integrand functions $f(X_{k-1}(s, \omega), s, \omega)$ are defined on $I \setminus H(\omega)$ due to condition 2, and (ii) are continuous functions of the upper limit t.

Due to condition 4 we obtain for $t_0, t \in J = [A, B] \subset I$:

$$\|X_{k+1}(t,\omega) - X_k(t,\omega)\| \overset{I}{\leq} \left(\sup_{s\in J} f(X_0(\omega), s, \omega)\right) \left(\sup_{s\in J} L(s,\omega)\right)^k \frac{(B-A)^{k+1}}{(k+1)!} .$$

For almost all $\omega \in \Omega$ this implies the uniform convergence on J of $X_k(t, \omega)$ towards a continuous function $X(t, \omega)$ with the properties

$$\frac{\mathrm{d}X(t,\omega)}{\mathrm{d}t} = f(X(t,\omega), t, \omega) , \qquad t \in J \setminus H(\omega) ,$$

and $X(t_0, \omega) = X_0(\omega)$. By choosing J in a suitable way we obtain such a solution of (3.6) in the extended sense for any $t \in I$.

If we are now able to find a stochastic process X_t, $t \in I$, such that $X_t \overset{I}{=} X(t, \omega)$ holds, then we have proven that X_t is a path-wise solution of (3.5) in the extended sense with $X_{t_0} = X_0$.

As (i) $X_0 \in S_d$, (ii) by condition 4 and Lemma 3.10 there is a $\mathcal{B}^d \times \mathcal{A}$-measurable function $g(x, t, \omega)$ for every $t \in I$ such that $f(x, t, \omega) \overset{\mathbb{R}^d \times I}{=} g(x, t, \omega)$, and (iii) that the integral in equation (3.8) is the limit of Riemann sums, it follows by induction that a stochastic process $X_t^{(k)}$, $t \in I$, exists such that $X_t^{(k)} \overset{I}{=} X_k(t, \omega)$.

Thus, there is also a stochastic process X_t, $t \in I$ such that

$$X_t \stackrel{I}{=} \lim_{k\to\infty} X_k(t,\omega) \stackrel{I}{=} X(t,\omega)\,.$$

Finally, let Y_t be an arbitrary path-wise solution of (3.5) in the extended sense with $Y_{t_0} = X_0$, then we have by condition 4

$$\|X_t - Y_t\| \stackrel{I}{\leq} \int_{t_0}^{t} L(s,\omega)\|X_s - Y_s\|\mathrm{d}s\,,$$

which implies together with Gronwall's lemma (see Lemma 3.15 in Sec. 3.4.2) that $\|X_t - Y_t\| \stackrel{I}{=} 0$ holds. This shows the asserted uniqueness of our path-wise solution in the extended sense. □

For instance, the conditions of Theorem 3.4 are satisfied for the random differential equation

$$\frac{\mathrm{d}X_t}{\mathrm{d}t} = f(X_t, t, Z_t)\,, \tag{3.9}$$

if $f(x,t,z)$ is continuous on $\mathbb{R}^d \times I \times \mathbb{R}^m$ and satisfies the Lipschitz-condition

$$\|f(x_1,t,Z_t) - f(x_2,t,Z_t)\| \stackrel{\mathbb{R}^d\times I\times\mathbb{R}^m}{\leq} L_t\|x_1 - x_2\|$$

with a path-wise continuous stochastic process L_t and if Z_t is an m-dimensional stochastic process on I such that almost all of its realisations $Z_t(\omega)$ are piecewise continuous functions with discontinuities of order one, i.e. the discontinuities are either removable or jumps of finite height.

Following [45], p. 19, let us denote by $G_t := \sigma(Z_\tau : \tau \in [t_0, t])$ the smallest σ-algebra with respect to which the vector-valued stochastic processes Z_τ, $\tau \in [t_0, t]$ are measurable.

If we replace $X_0(\omega)$ by $\xi \in \mathbb{R}^d$ in the proof of Theorem 3.4 then we can show for every $k = 1, 2, \ldots$, analogously to the existence of a vector-valued function $X_t^{(k)}(\omega)$ defined on $I \times \Omega$, the existence of a vector-valued function $x^{(k)}(t,\omega,\xi)$ defined on $I \times \Omega \times \mathbb{R}^d$ that has the following properties:

1. For each fixed $t \in I$ the function $x^{(k)}(t,\omega,\xi)$ is $G_t \times \mathcal{B}^d$-measurable.
2. If $X_k(t,\omega,\xi)$ is defined recursively by

$$\begin{aligned} X_0(t,\omega,\xi) &:= \xi\,, \\ X_k(t,\omega,\xi) &:= \xi + \textstyle\int_{t_0}^{t} f\left(X_{k-1}(s,\omega,\xi), s, Z_s(\omega)\right)\mathrm{d}s\,, \end{aligned}$$

then the identity

$$x^{(k)}(t,\omega,\xi) \stackrel{I\times\mathbb{R}^d}{=} X_k(t,\omega,\xi)$$

holds.

Taking limits leads to the existence of a vector-valued function $x(t,\omega,\xi)$ that is $G_t \times \mathcal{B}^d$-measurable for each fixed $t \in I$ and for which

$$x(t,\omega,\xi) \overset{I\times\mathbb{R}^d}{=} X(t,\omega,\xi)$$

holds, where for almost all $\omega \in \Omega$ and every $t \in \mathbb{R}^d$ the function $X(t,\omega,\xi) := \lim_{k\to\infty} X_k(t,\omega,\xi)$ denotes the unique solution in the extended sense of the deterministic ordinary differential equation $x = f(x,t,Z_t(\omega))$ with respect to the initial condition (ξ, t_0).

By modifying $\omega \in \Omega$ in the initial condition $x(t_0,\omega,\xi)$ on a $\mathbb{P}$-zero set we gain that $x(t_0,\omega,\xi) = \xi$, $\xi \in \mathbb{R}$, indeed holds for all $\omega \in \Omega$.

By the way, $X_t = x(t,\omega,X_0)$ is indeed the path-wise solution of (3.9) in the extended sense with respect to the initial condition $X_{t_0} = X_0 \in S_d$.

Note that condition 4 in Theorem 3.4 is restrictive and may thus cause problems when studying stability of solutions (cf. Chap. 15) where the unique existence of a path-wise solution on $I = [t_0, \infty)$ is demanded. In this context it is desirable to replace condition 4 by a local Lipschitz-condition. This is possible if, for instance, the existence of a Lyapunov-function is guaranteed as in the case of most theorems on stability.

Theorem 3.5 (Existence & Uniqueness of Path-Wise Solutions in the Extended Sense Under a Lyapunov-Condition). *Let the conditions 1, 2 and 3 from Theorem 3.4 be satisfied, i.e.:*

1. *The function $f(x,t,\omega)$ is $\mathcal{A}$-measurable for all $(x,t) \in \mathbb{R}^d \times I$.*

2. *For almost all $\omega \in \Omega$ the function $f(x,t,\omega)$ is continuous on $\mathbb{R}^d \times (I \setminus H(\omega))$, where $H(\omega)$ is an at most countable subset in I the elements of which do not have a finite accumulation point.*

3. *For almost all $\omega \in \Omega$ and all $x \in \mathbb{R}^d$ the function $f(x,t,\omega)$ is bounded on any finite interval $J \subset I$.*

Moreover, define

$$R_k^n := \{x \in \mathbb{R}^n : \|x\| < k\},$$

and assume that the additional conditions are satisfied:

1. *For almost all $\omega \in \Omega$ there is a sequence $\{L_k(t,\omega)\}_{k\in\mathbb{N}}$ of continuous functions in I such that for every k, $t \in I$ and $x_1, x_2 \in R_k^n$ it holds that*

$$\|f(x_1,t,\omega) - f(x_2,t,\omega)\| \leq L_k(t,\omega)\|x_1 - x_2\|.$$

2. *There exists a real-valued function v on $\mathbb{R}^d \times I \times \Omega$ such that for almost all $\omega \in \Omega$ there exist continuous partial derivatives of v with respect to t and x and that the following properties hold:*

- *v is radially bounded from below, i.e.,*

$$v(x,t,\omega) \overset{\mathbb{R}^d\times I}{\geq} \varphi(\|x\|) \geq 0,$$

with $\lim_{r\to\infty}\varphi(r)\to\infty$, and

- *the orbital derivative of v is bounded by v from above, i.e.*

$$\partial_t v(x,t,\omega) + (\partial_x v(x,t,\omega))^T f(x,t,\omega) \overset{\mathbb{R}^d\times I}{\leq} c\cdot v(x,t,\omega),$$

with some $c\geq 0$.

Then for any initial condition $(X_0,t_0)\in S_d\times I$ there exists a unique path-wise solution on I in the extended sense of the random differential equation (3.5).

Proof. Following [45], p. 20, we choose $(X_0,t_0)\in S_d\times I$ as a fixed initial condition. We can define functions $f^k(x,t,\omega)$ on $\mathbb{R}^d\times I\times\Omega$ such that for each k it holds that

$$f^k(x,t,\omega) = f(x,t,\omega), \qquad \|x\|\leq k, \quad (t,\omega)\in I\times\Omega,$$

and such that the random differential equation

$$\frac{\mathrm{d}X_t}{\mathrm{d}t} = f^k(X_t,t,\omega)$$

fulfills the conditions of Theorem 3.4 and thus has a unique path-wise solution X_t^k on I in the extended sense.

Next, we consider

$$t_r^k(\omega) := \begin{cases} \infty & \text{if } X_t^k(\omega)\| \leq r,\, t\in I \\ \inf\{t\in I : \|X_t^k(\omega)\| > r\} & \text{else} \end{cases}.$$

Due to the construction of the path-wise solutions X_t^k in (3.8) we have that for almost all $\omega\in\Omega$ the realisations of X_t^k ($k\geq r$) coincide for $t\leq t_r^r(\omega)$. Hence, for $r=1,2,\dots$, we obtain the relations $t_r^k = t_r^r(\omega) = t_r(\omega)$ ($k\geq r$) as well as $t_r(\omega)\leq t_{r+1}(\omega)$.

Finally, we show that $t(\omega) = \lim_{r\to\infty} t_r(\omega) = \infty$ for almost all $\omega\in\Omega$ such that

$$\lim_{k\to\infty} X_t^k(\omega) = X_t(\omega) = X_t^r(\omega), \quad (\text{with } r \text{ such that } t_r^r\geq t,\, t\in I),$$

exists and that thus X_t is the unique path-wise solution on I in the extended sense of the random differential equation (3.5). In order to prove this last assertion, we use condition 2, which says that for almost all $\omega\in\Omega$:

$$\frac{\mathrm{d}}{\mathrm{d}t}\left(v\left(X_t(\omega),t,\omega\right)\exp\left(-c(t-t_0)\right)\right) \leq 0, \qquad \text{for almost all } t\in I,$$

as well as for r such that $t_r(\omega) < \infty$

$$\varphi(r) = \varphi\left(\|X_{t_r(\omega)}\|\right) \leq v\left(X_{t_r(\omega)}(\omega), t_r(\omega), \omega\right)$$
$$\begin{cases} \leq & v\left(X_t(\omega), t, \omega\right) \exp\left(c(t - t_0)\right) & \text{if } t(\omega) < \infty \\ < & \infty & \text{if } t(\omega) = \infty \end{cases}.$$

Due to $\lim_{r\to\infty} \varphi(r) = \infty$ the identity $t(\omega) = \infty$ holds for almost all $\omega \in \Omega$. □

Before you continue, make sure to answer the following questions:

Quiz: Section 3.4 – Part 2 (Path-Wise Solutions in the Extended Sense)

Q1 What is the difference between the concept of a solution of a random (ordinary) differential equation (RDE) and the concept of a solution in the extended sense of a RDE?

Q2 Given a RDE with continuous right-hand side. Do its path-wise solution and its path-wise solution in the extended sense differ from each other (provided they exist)?

Q3 How do things change if you replace the continuity assumption in Q2 by piece-wise continuous (with finitely many jumps)?

Q4 Under which conditions does a unique path-wise solution in the extended sense of a RDE exist?

Q5 Sketch the proof of the existence and uniqueness theorem you just stated.

Q6 As for deterministic stability considerations, unique path-wise solutions on $[t_0, \infty)$ are required to study stability of RDEs. How can such a $[t_0, \infty)$-existence be ensured?

3.4.2 Dependence on Parameters and Initial Conditions

Based on the well known results on deterministic ordinary differential equations that satisfy certain Lipschitz-conditions, we round up our exposition on random (ordinary) differential equations by providing results on the dependence of path-wise solutions on parameters and initial conditions, see [45], pp. 21. Our main tool is Gronwall's lemma:

Lemma 3.15 (Gronwall's Lemma (for Stochastic Processes)). *Let $I = [t_0, b]$ ($b = \infty$ may hold in which case we assume $I = [t_0, \infty)$), and $C : \Omega \to \mathbb{R}_0^+$ be a non-negative real-valued scalar function on Ω. Moreover, let U, V, W be real-valued scalar functions on $I \times \Omega$ with the following properties:*

1. *For almost all $\omega \in \Omega$ the function $U(t, \omega)$ is continuous on I and $U(t, \omega) \overset{I}{\geq} 0$.*

2. *For almost all $\omega \in \Omega$ the functions $V(t, \omega)$ and $W(t, \omega)$ are Lebesgue-integrable on any bounded subinterval of I and $V(t, \omega), W(t, \omega) \overset{I}{\geq} 0$.*

Then, the inequality

$$U(t, \omega) \overset{I}{\leq} C(\omega) + \int_{t_0}^{t} W(\tau, \omega) d\tau + \int_{t_0}^{t} U(\tau, \omega) \cdot V(\tau, \omega) d\tau$$

implies

$$U(t, \omega) \overset{I}{\leq} C(\omega) \exp\left(\int_{t_0}^{t} V(\tau, \omega) d\tau \right) + \int_{t_0}^{t} W(\tau, \omega) \cdot \exp\left(\int_{\tau}^{t} V(s, \omega) ds \right) d\tau \,.$$

Proof. see problem 3.31 □

Utilizing this version of Gronwall's lemma allows us to estimate the dependence of path-wise solutions on initial conditions and parameters which may be random in general.

Theorem 3.6 (Dependence on Initial Conditions & Parameters). *Let $I = [t_0, b)$, Λ be a set of random variables with values in $H \subset \mathbb{R}^d$ (the parameter set), $Y_0 \in \Lambda$ and $X_0 \in \Xi \subset S_d$ (the random initial condition). For any $X \in \Xi$ and $Y \in \Lambda$ let there be a path-wise solution $X_t(\omega; X, Y)$ on I of the random differential equation*

$$\frac{dX_t}{dt} = f(X_t, t, \omega, Y)$$

with respect to the initial condition (X, t_0).

Assume that f is defined on $\mathbb{R}^d \times I \times \Omega \times H$ such that for almost all $\omega \in \Omega$ the function $f(x, t, \omega, y)$, $y \in H$ is continuous on $\mathbb{R}^d \times I$ and that there are functions V and W such that

$$\| f(x_1, t, \omega, Y_0(\omega)) - f(x_2, t, \omega, Y_0(\omega)) \| \overset{\mathbb{R}^d \times \mathbb{R}^d \times I}{\leq} V(t, \omega) \| x_1 - x_2 \| \,,$$

and

$$\| f(x, t, \omega, y_1) - f(x, t, \omega, y_2) \| \overset{\mathbb{R}^d \times I \times H \times H}{\leq} W(t, \omega, y_1, y_2) \,,$$

which satisfy the condition 2 of Gronwall's lemma (Lemma 3.15) for each pair $y, y_2 \in H$. *Then for arbitrary* $X \in \Xi$ *and* $Y \in \Lambda$ *it holds that*

$$\|X_t(\omega; X_0, Y_0) - X_t(\omega; X, Y)\| \overset{T}{\leq} \|X - X_0\| \cdot \exp\left(\int_{t_0}^{t} V(\tau, \omega) d\tau\right) + \int_{t_0}^{t} \exp\left(\int_{\tau}^{t} V(s, \omega) ds\right) \cdot W(\tau, \omega, Y_0, Y) d\tau . \tag{3.10}$$

If, additionally, the following conditions are satisfied

$$V(t, \omega) \overset{I}{\leq} V_0(\omega), \quad V_0 \hat{>} 0, \quad \text{and} \quad W(t, \omega, y_1, y_2) \leq W_0(\omega, y_1, y_2), \tag{3.11}$$

then it holds that

$$\|X_t(\omega; X_0, Y_0) - X_t(\omega; X, Y)\| \overset{T}{\leq} \|X - X_0\| \cdot \exp\left(V_0(\omega)(t - t_0)\right) + \left(\exp\left(V_0(\omega)(t - t_0)\right) - 1\right) \frac{W_0(\omega, Y_0, Y)}{V_0(\omega)} . \tag{3.12}$$

Proof. Following [45], p. 22, we have that inequality (3.10) follows immediately from Gronwall's lemma (Lemma 3.15), because with $U_t := \|X_t(\omega; X_0, Y_0) - X_t(\omega; X, Y)\|$ we gain the estimate

$$\begin{aligned}
U_t &\overset{I}{=} \left\| X_0 - X + \int_{t_0}^{t} \left(f(X_t(\omega; X_0, Y_0), \tau, \omega, Y_0) - f(X_t(\omega; X, Y), \tau, \omega, Y)\right) \mathrm{d}\tau \right\| \\
&\overset{I}{\leq} \|X_0 - X\| + \int_{t_0}^{t} \|f(X_t(\omega; X_0, Y_0), \tau, \omega, Y_0) - f(X_t(\omega; X, Y), \tau, \omega, Y_0)\| \, \mathrm{d}\tau \\
&\quad + \int_{t_0}^{t} \|X_t(\omega; X, Y), \tau, \omega, Y_0) - X_t(\omega; X, Y), \tau, \omega, Y)\| \, \mathrm{d}\tau \\
&\overset{I}{\leq} \|X_0 - X\| + \int_{t_0}^{t} W(\tau, \omega, Y_0, Y) \mathrm{d}\tau + \int_{t_0}^{t} U_\tau V(\tau, \omega) \mathrm{d}\tau .
\end{aligned}$$

Inequality (3.12) is obtained from (3.10) by estimation and integration. □

Theorem 3.6 gives rise to several important continuity implications. For the following corollaries let the conditions of Theorem 3.6 together with (3.11) be satisfied with $V_0 \in S_1$ and $\mathcal{A} \times \mathcal{B}^d \times \mathcal{B}^d$-measurable W_0.

Corollary 3.16. *Let* $\{X_n\}_{n \in \mathbb{N}} \subset \Xi$ *with* $\lim_{n \to \infty} X_n \hat{=} X_0 \in \Xi$ *and* $\{Y_m\}_{m \in \mathbb{N}} \subset \Lambda$ *with* $\lim_{m \to \infty} Y_m \hat{=} Y_0 \in \Lambda$. *Moreover, let* $\omega \in \Omega_1$ *such that* $\mathbb{P}(\Omega_1) = 1$ *as well as*

$$\lim_{y \to \nu} W_0(\omega, \nu, y) = 0, \quad y, \nu \in H$$

then

$$\lim_{n,m\to\infty} X_t(\omega; X_n, Y_m) \stackrel{I}{=} X_t(\omega; X_0, Y_0)$$

holds. The above limit is uniform in t on each bounded subinterval of I.

Corollary 3.17. *Let $W_0(\omega,\nu,y) \overset{H\times H}{\leq} K(\omega)u(\|\nu - y\|)$, where $K \in S_1$ and u is a non-negative function on $\mathbb{R}^+$ such that $\lim_{r\to 0} u(r) = 0$. Then, for any $t \in I$ and given $\varepsilon > 0$ and given $\gamma \in (0,1)$ there exist numbers $\varepsilon^* > 0$ and $\gamma^* \in (0,1)$ such that*

$$\mathbb{P}\left(\|X_t(\omega; X, Y) - X_t(\omega; X_0, Y_0)\| > \varepsilon\right) < \gamma,$$

provided $\mathbb{P}(\|Y - Y_0\| + \|X - X_0\| > \varepsilon^*) < \gamma^*$. On each subinterval of I the constants ε^* and γ^* can be gained independently of t.

Proof. Following [45], p. 23, fix $\tau \in I$ and let k be a positive number such that

$$\mathbb{P}(A(\omega) < k) > 1 - \tfrac{1}{4}\gamma, \quad \text{and} \quad \mathbb{P}(K(\omega)B(\omega) < k) > 1 - \tfrac{1}{4}\gamma,$$

where $A(\omega) := \exp((\tau - t_0)V_0(\omega))$ and $B(\omega) := V_0(\omega)^{-1}(\exp((\tau - t_0)V_0(\omega)) - 1)$. Moreover, let $\delta > 0$ be such that $u(r) \leq (2k)^{-1}\varepsilon = \beta$ for $r \in [0,\delta)$ and $\varepsilon^* = \min(\beta,\delta)$ as well as $\gamma^* = \frac{1}{4}\gamma$.

Because of (3.12) the following estimates for $t \in [t_0, \tau]$ hold

$$\begin{aligned}
&\mathbb{P}\left(\|X_t(\omega; X_0, Y_0) - X_t(\omega; X, Y)\| > \varepsilon\right) \\
&\leq 1 - \mathbb{P}\left(\|X - X_0\|A < \tfrac{1}{2}\varepsilon\,,\, W_0(\omega, Y_0, Y)B < \tfrac{1}{2}\varepsilon\right) \\
&\leq 2 - \mathbb{P}\left(\|X - X_0\| < \beta\,,\, A < k\right) - \mathbb{P}\left(u(\|Y - Y_0\|) < \beta\,,\, KB < k\right) \\
&\leq 2 - \tfrac{1}{2}\gamma - \mathbb{P}\left(\|X - X_0\| < \beta\right) - \mathbb{P}\left(u(\|Y - Y_0\|) < \beta\right) \\
&\leq 2 - \tfrac{1}{2}\gamma - 2\mathbb{P}\left(\|X - X_0\| + \|Y - Y_0\| < \varepsilon^*\right) \leq 2 - \tfrac{1}{2}\gamma - 2(1 - \gamma^*) = \gamma\,.
\end{aligned}$$

This shows the assertion. □

Corollary 3.18. *Let*

$$\mathbb{E}\left(\exp\left(2V_0(t - t_0)\right)\right) < \infty, \quad \textit{and} \quad \mathbb{E}\left(\frac{\left(\exp\left(V_0(t - t_0)\right) - 1\right)^2}{V_0^2}\right) \leq \infty\,. \tag{3.13}$$

Assume that for almost all $\omega \in \Omega$ it holds that $W_0(\omega, y, \nu) \leq C\|y - \nu\|$ for $y, \nu \in H$ and $C > 0$. Let $\{X_n\}_{n\in\mathbb{N}} \subset \Xi$ and $\{Y_m\}_{m\in\mathbb{N}} \subset \Lambda$ be square-integrable sequences in L^2_n and L^2_m, respectively, such that $Z_k := (X_k, X_0, Y_k, Y_0)$ and V_0 are independent for all $k \in \mathbb{N}$. Moreover, assume

$$q.m.\lim_{n\to\infty} X_n = X_0 \in \Xi, \quad \textit{and} \quad q.m.\lim_{m\to\infty} Y_m = Y_0 \in \Lambda,$$

(see Sec. 4.3 for the definition of the mean-square/ quadratic mean limit $q.m.\lim$), then it holds that

$$q.m. \lim_{n,m\to\infty} X_t(\omega; X_n, Y_m) \hat{=} X_t(\omega; X_0 Y_0) \,. \tag{3.14}$$

If (3.13) holds for all t from a bounded subinterval of I then (3.14) is uniformly in t there.

Finally, let us apply our results to specific examples:

Example 3.19 (A Linear Inhomogeneous Random Differential Equation, cf. [45], p. 24.). Let us study the dependence of solutions of the following random differential equation on its initial conditions:

$$\frac{\mathrm{d}X_t}{\mathrm{d}t} = A_t X_t + Z_t \,,$$

with path-wise continuous stochastic processes as the entries of the $d \times d$-matrix A_t and vector Z_t. Moreover, let A_t be stationary with $\mathbb{E}(\|A_t\|) < \infty$ and

$$\lim_{t\to\infty} \tfrac{1}{t} \int_{t_0}^{t} \|A_\tau\| \mathrm{d}\tau \hat{=} \mathbb{E}(\|A_{t_0}\|) = \mathbb{E}(\|A_t\|) \,,$$

where $\|A\| = (\sum_{i,k} a_{i,k}^2)^{1/2}$ is the Frobenius norm. Let X_0 and X be two arbitrary elements of S_d.

According to theorems 3.2 and 3.6 it holds that

$$\|X_t(\omega; X_0) - X_t(\omega; X)\| \overset{I}{\leq} \|X - X_0\| \exp\left(\int_{t_0}^{t} \|A_\tau\| \mathrm{d}\tau \right) \,,$$

and thus, due to the presumed ergodicity of A_t, it holds for every $a > \mathbb{E}(\|A_{t_0}\|)$ that

$$\lim_{t\to\infty} \exp(-at) \|X_t(\omega; X_0) - X_t(\omega; X)\| \hat{=} 0 \,,$$

as the left hand side of this identity is almost surely not bigger than

$$\lim_{t\to\infty} \|X_0 - X\| \exp\left(\left(\tfrac{1}{t} \int_{t_0}^{t} \|A_\tau\| \mathrm{d}\tau - a \right) t \right) \,.$$

Example 3.20 (A Nonlinear Random Differential Equation, cf. [45], p. 24.). Let us consider the random differential equation

$$\frac{\mathrm{d}X_t}{\mathrm{d}t} = A_t(\omega) X_t + Z_t(\omega) + (Y_1(\omega) h(X_t) + Y_2(\omega)) \, g(t) \,, \tag{3.15}$$

where, as in example 3.19, A_t and Z_t again have path-wise continuous stochastic processes as entries, such that $\sup_{t\in I} \|A_t(\omega)\| = \alpha(\omega) \in S_1$.

Let Y_1 be a $d \times d$-matrix whose elements are random variables and Y_2 be a d-dimensional random vector, i.e., $Y_2 \in S_d$. Moreover, let $h(x)$ be a d-dimensional vector-valued function on $\mathbb{R}^d$ such that $\|h(x)\| < h_0$ and $\|h(x_1) - h(x_2)\| \leq L\|x_1 - x_2\|$ for all $x_1, x_2 \in \mathbb{R}^d$. Let g be a continuous real-valued function defined on I with $\|g(t)\| \leq g_0$.

Let us study the dependencies of the solutions of (3.15) on the initial conditions and the parameters Y_1 and Y_2. The requirements of the Theorems 3.2 and 3.6 hold, and equation (3.12) is valid with

$$\begin{array}{rclrclrcl} \Sigma & = & S_d & \Lambda & = & S_{d^2+d} & V_0(\omega) & = & \alpha(\omega) + Lg_0\|Y_{10}(\omega)\| \\ Y & = & (Y_1, Y_2) & Y_0 & = & (Y_{10}, Y_{20}) & & & , \end{array}$$

and

$$W_0(\omega, Y_0, Y) = g_0 \left(h_0\|Y_{10} - Y_1\| + \|Y_{20} - Y_2\| \right) .$$

Before you continue, make sure to answer the following questions:

Quiz: Section 3.4 – Part 3 (Dependence on Parameters and Initial Conditions)

Q1 Give Gronwall's lemma for stochastic processes and compare it to its usual deterministic version.

Q2 How do the solutions of random (ordinary) differential equations depend on initial values and parameters?

Q3 Sketch the proof of this dependence assertion you just used.

Q4 Which continuity implications do these dependencies on initial values and parameters have?

3.5 Excursion: Deterministic Description of the Vibrations of Single & Multi-Storey Buildings

Following [57], pp. 50, the forces acting on the mass on top of the weightless frame of the one-storey building shown in Fig. 3.12 (1a) are the *external force* $p(t)$ as well as the *damping* resisting force $f_D(t)$ and the *elastic* (or *inelastic*) resting force $f_S(t)$ of the structure.

The external force p is, for the moment, taken to be positive in the direction of the x-axis. The displacement $u(t)$, the velocity $\dot{u}(t)$ and the acceleration $\ddot{u}(t)$ are also positive in the direction of the x-axis. The damping (f_D) and elastic/ inelastic (f_S) forces are acting n the opposite direction, because they are internal forces and resist the velocity and deformation respectively.

The resultant force along the x-axis is $p - f_D - f_S$ and finally, Newton's second law of motion gives

$$p - f_S - f_D = m\ddot{u} \qquad \text{or} \qquad m\ddot{u} + f_D + f_S = p. \tag{3.16}$$

Now, we would like to understand the resting forces f_D and f_S. This will be done in the following sections.

3.5.1 Vibrations of a Single-Storey Building

The standard model for mechanic oscillations is that of a mass $m > 0$ being connected to a wall by a spring with stiffness $k > 0$. Hooke's law ($F = -kx$) together with Newton's second law of motion ($F = m\ddot{x}$) lead to

$$m\ddot{x} + \tilde{c}\dot{x} + kx = F(t), \quad \text{with } x(0) = x_0 \text{ and } \dot{x}(0) = x_1,$$

where

- x denotes the displacement of the mass point from the equilibrium coordinate x_0,
- $\tilde{c}\dot{x}$ denotes the (viscous) damping of the mass point, and
- $F(t)$ is a time-dependent external force.

Scaling time t via $t = \sqrt{m/k}\tau$ and the position coordinate x via $x = x_0 y$, we obtain the dimension-free form of the free oscillation (without external force) as

$$\ddot{y}(t) + c\dot{y}(t) + y(t) = 0, \quad \text{with } y(0) = 1 \text{ and } \dot{y}(0) = v = (x_0^{-1} x_1 \sqrt{m/k},$$

where $c := \tilde{c}\left(\sqrt{m/k}\right)^{-1}$.

Next, to the case of earthquake induced oscillations: As we have already seen, in mechanic equilibrium we have

$$f_S + f_D + f_I = 0$$

for

- the deformation force $f_S = ku$,
- the damping force $f_D = c\dot{u}$, and
- the mass inertia $f_I = m\ddot{u}^T$,

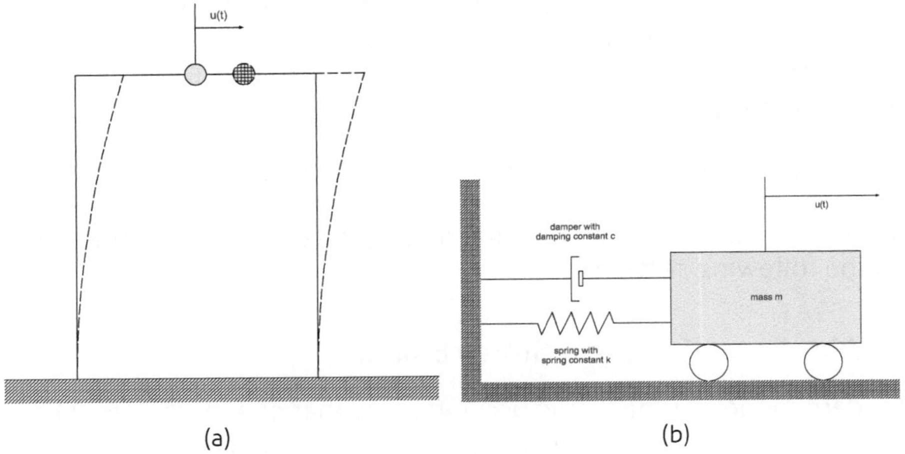

Figure 3.12. Sketch of a one-storey building subject to an external excitation $u(t)$ such that the top floor of the building swings (a), and of the analogous mass-damper system (b).

where $u^T = u + u_g$ denotes the total displacement of the position coordinate u subject to the influence of the earthquake induced soil motion u_g. Here, only the relative displacement of the mass point from the initial position has an influence on the deformation and the damping forces. Hence,

$$m\ddot{u} + c\dot{u} + ku = -m\ddot{u}_g .$$

For instance, with the periodical ground excitation ($A > 0$)

$$\ddot{u}_g = -A \sin\left(\omega\sqrt{k/m}t\right)$$

and neglecting damping, we get

$$m\ddot{u} + ku = A \sin\left(\omega\sqrt{k/m}t\right) .$$

Again, scaling $t = \sqrt{m/k}\tau$ and $u = (A/k)y$ leads to

$$\ddot{y}(t) + y(t) = \sin(\omega t) .$$

We consider a one-storey building that is at rest at $t = 0$ and let $X(t)$, $t \geq 0$, denote the relative horizontal displacement of its roof with respect to the ground. Then, based upon an idealized linear model, the relative displacement $X(t)$ subject to ground accelerations is governed by

$$\ddot{x}(t) + 2\zeta\omega_0\dot{x}(t) + \omega_0^2 x(t) = -y(t) , \quad \text{for } t \geq 0 . \tag{3.17}$$

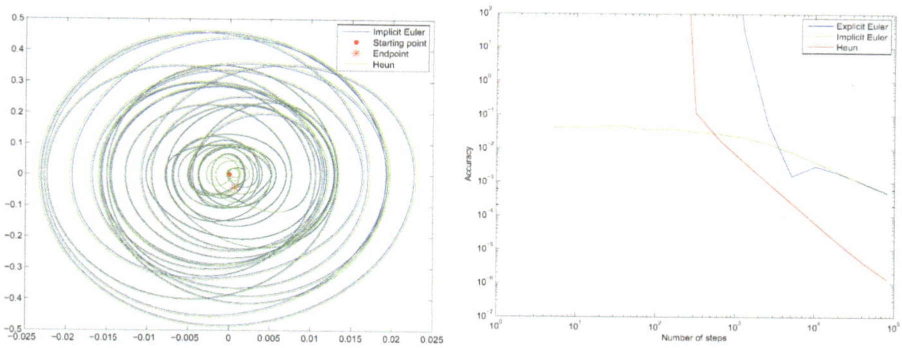

(a) (b)

Figure 3.13. (a) Solution for the one-storey excitation model (3.17) in the x-$\dot{x}$-phase plane, and (b) convergence of the different methods compared to a high order solution computed with MATLAB.

For this equation let the parameters be given according to [144], p. 267, as $\omega_0 = 20$ (in rad/s) and $\zeta = 0.05$.

Let us first display the solution of (3.17) in the x-$\dot{x}$-phase plane with initial conditions $x(0) = 0$ and $\dot{x} = 0$, i.e. the building is at rest at $t = 0$, see Fig. 3.13 (a). We see that the acceleration $\dot{x}$ increases quickly and alternates between positive and negative acceleration. The solutions produced by the deterministic Euler and Heun scheme behave similarly, in Figure 3.13 (b) the convergence of the known methods is illustrated. The convergence rate of the implicit and explicit Euler is $\mathcal{O}(\Delta h)$, but the implicit Euler produces an useful solution even for small step sizes. Heun's method has a convergence rate of $\mathcal{O}(\Delta h^2)$, which is clearly illustrated in the convergence plot. Like the explicit Euler, Heun's method needs a minimal step size to produce an accurate solution.

3.5.2 Vibrations of a Multi-Storey Building

In the case of a d-storey building ($d = 1, 2, \dots$), we have that the forces F_j that act on a floor j can be split into those resulting from a component that belongs to the floor above (F_j^{j+1}) and one that belongs to the floor below it (F_j^{j-1}), i.e.,

$$F_j = F_j^{j-1} + F_j^{j+1},$$

where we set $F_d^{d+1} = 0$ as there is no external force acting on the roof and F_1^0 equal to the forces induced by the earthquake. This leads to the following forms of the deformation and damping forces where displacements are measured relative to the j-th floor:

- Deformation force:

$$f_S^{(j)} = k_j\,(u_j - u_{j-1}) + k_{j+1}\,(u_j - u_{j+1}) \;=\; -k_j u_{j-1} + (k_j + k_{j+1}) u_j - k_{j+1} u_{j+1}\,.$$

- Damping force:

$$f_D^{(j)} = c_j\,(\dot{u}_j - \dot{u}_{j-1}) + c_{j+1}\,(\dot{u}_j - \dot{u}_{j+1}) \;=\; -c_j \dot{u}_{j-1} + (c_j + c_{j+1}) \dot{u}_j - c_{j+1} \dot{u}_{j+1}\,.$$

For $u := (u_1, u_2, \ldots, u_d)^T$ the dimension-free equation in matrix-vector notation reads as

$$\ddot{u} + C\dot{u} + Ku \;=\; F(t)\,, \tag{3.18}$$

with a time-dependent external force F corresponding to the earthquake excitation, and where, e.g.,

$$K \;=\; \begin{pmatrix} k_1 + k_2 & -k_2 & & & \\ -k_2 & k_2 + k_3 & -k_3 & & \\ & -k_3 & k_3 + k_4 & -k_4 & \\ & & \ddots & & \\ & & & -k_i & +k_i \end{pmatrix}$$

and C analogous. (If necessary, masses are included into the model via diagonal matrices.)

Figure 3.14 displays a simulation run of a three-storey building with initial displacement $u_0 = (0.1, 0.02, -0.1)^T$ and constants $k_1 = k_2 = k_3 = 1$ and $c_1 = c_2 = c_3 = 0.5$. The effects of damping are clearly visible.

3.6 Chapter's Summary

In this chapter, we gained insight into the modelling of seismic activities due to the application of linear stochastic differential equations. Important models were the Kanai-Tajimi and the Clough-Penzien filter which both use additive white noise as the essential driving term.

This additive dependence on white noise allowed us, using the Doss-Sussmann/ Imkeller-Schmalfuss correspondence, to rewrite these stochastic (ordinary) differential equations as Ornstein-Uhlenbeck process driven random (ordinary) differential equations. This motivated the analytic study of random (ordinary) differential equations.

We discussed existence and uniqueness of path-wise solutions of random (ordinary) differential equations by converting them to (deterministic) ordinary differential equations in a path-wise sense. Thus, in order to gain unique

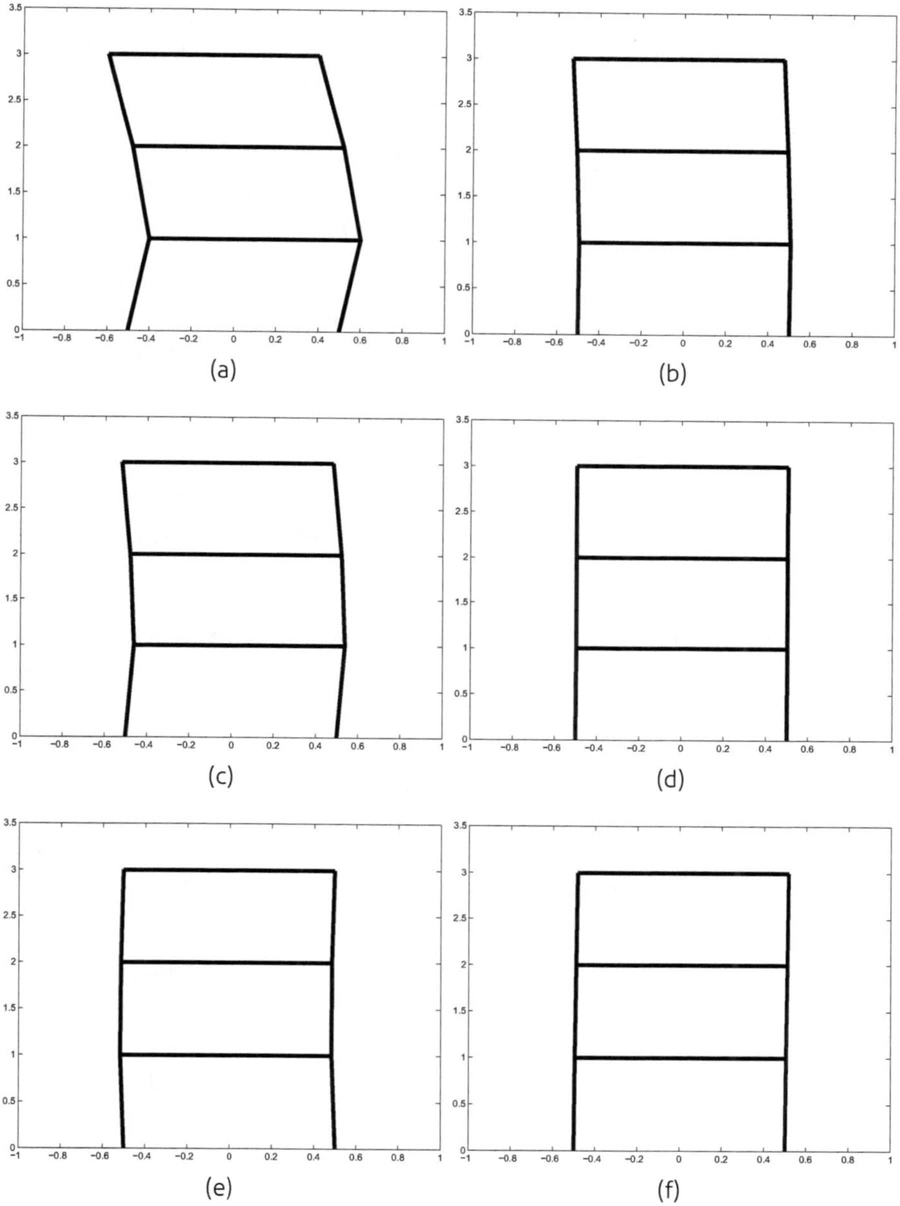

Figure 3.14. Simulation of the oscillatory movement of a structurally damped three-storey building staring with a slight displacement from the equilibrium position.

path-wise existence each element of the ω-dependent family of ordinary differential equations is required to have a unique solution, that these solutions share a common time-interval of existence and that the thus derived ω-dependent solution is a stochastic process, again. In particular, Theorem 3.2 gave easy to apply assumptions for the unique existence of such path-wise solutions.

Let us conclude with the model presented in the introduction of this chapter: a simple model for pollution induced biochemical oxygen demand and dissolved oxygen in streams (see example 3.1).

Example 3.21 (Pollution Induced Biochemical Oxygen Demand and Dissolved Oxygen in Streams: Existence and Uniqueness of Solutions). If we assume all parameters to be stochastic processes this two-dimensional system reads, after some algebraic modifications as

$$\begin{aligned} \dot{x}(t) &= -a_t(\omega)x(t) + \alpha_t(\omega) , \\ \dot{y}(t) &= -b_t(\omega)y(t) - c_t(\omega)x(t) - \beta_t(\omega) , \end{aligned}$$

where $a_t, b_t, c_t, \alpha_t, \beta_t$ are suitable bounded continuous stochastic processes defined for $t \in [t_0, T)$ on the probability space $(\Omega, \mathcal{A}, \mathbb{P})$, where $T = \infty$ may hold.

Applying Theorem 3.2 the following conditions have to be fulfilled in order to guarantee the existence of a path-wise unique solution of this system of random differential equations:

1. The functions $f_1(x, y, t, \omega) = -a_t(\omega)x(t) + \alpha_t(\omega)$ and $f_2(x, y, t, \omega) = -b_t(\omega)y(t) - c_t(\omega)x(t) - \beta_t(\omega)$ must be $\mathcal{A}$-measurable for all $(x, y, t) \in \mathbb{R} \times \mathbb{R} \times [t_0, T)$.

2. $f_1(x, y, t, \omega)$ and $f_2(x, y, t, \omega)$ must be continuous on $\mathbb{R} \times \mathbb{R} \times [t_0, T)$ for almost all $\omega \in \Omega$. For this model this is immediately verified.

3. For almost all $\omega \in \Omega$ there must be a real continuous function $L(t, \omega)$ on $[t_0, T)$ such that

$$\begin{aligned} &\sqrt{(f_1(x_1, y_1, t, \omega) - f_1(x_2, y_2, t, \omega))^2 + (f_2(x_1, y_1, t, \omega) - f_2(x_2, y_2, t, \omega))^2} \\ &= \sqrt{(a_t(x_1 - x_2))^2 + (b_t(y_1 - y_2) + c_t(x_1 - x_2))^2} \\ &\le \sqrt{(a_t^2 + c_t^2)(x_1 - x_2)^2 + b_t^2 (y_1 - y_2)^2} \\ &\le L(t, \omega)\sqrt{(x_1 - x_2)^2 + (y_1 - y_2)^2} , \end{aligned}$$

   where $i \in [t_0, T)$ and $x_1, x_2, y_1, y_2 \in \mathbb{R}$. As our manipulations already show, such a function $L(t, \omega)$ indeed exists by setting $L(t, \omega) \ge \sqrt{\max\{a_t^2 + c_t^2, b_t^2\}}$.

Thus, our pollution stream model has a unique path-wise solution on $[t_0, T)$ for any initial condition $(x_0, y_0, t_0) \in S_1 \times S_1 \times [t_0, T)$.

Finally, solutions in the extended sense are analysed that allowed us to extend the concept of a path-wise solution to systems with (mildly) non-continuous right-hand sides. Moreover, the dependence of solutions on parameters and initial conditions was studied and, as an excursion, we gave the equations of motion for single- and multi-storey (wireframe) buildings.

Before you continue, make sure to answer the following questions:

Quiz: Section 3.5

Q1 Derive the equations of motion of a one-storey building (wireframe structure) subject to ground motion excitations.

Q2 Now, assume a three-storey building. How does the equations of motion from Q1 change?

Q3 Assume a constant wind force acting on each floor of a three-storey building with the same amount. How do the equations of motion from Q2 change?

Q4 Assume that a building consists of three identical building blocks that are arranged such that they form an L-shaped structure with two blocks at the basement and the remaining one directly above one of the basement blocks. How do the deformation and damping matrices of the thus built wireframe structure look like?

Problems

Classification: ☼ easy, ⌚ easy with longer calculations, ☆ a little bit difficult,
⌧ challenging.

Exercise 3.22. [☼] Combination of Wiener Processes
Let W_t and W_t^* be independent standard Wiener processes and a, b positive real constants. Determine the relationship between a and b for which $Z_t := aW_t - \sqrt{b}W_t^*$ is again a Wiener process.

Exercise 3.23. [☼] Scaled Wiener Processes
Consider the process $X_t := \sqrt{a}W_{a^{-1}t}$, where W_s stands for a standard Wiener process and a is a real positive constant. This process is known as *scaled Wiener process*. The time scale of the Wiener process is reduced by a factor a, and the magnitude of the Wiener process are multiplied by a factor $\sqrt{a}$.

This can be interpreted as taking snapshots of the position of a Wiener process with a shutter speed that is a times as fast as that used for recording a standard Wiener process, and magnifying the results by a factor $\sqrt{a}$.

1. Derive the expected value and the variance of X_t.
2. Derive the probability distribution as well as the probability density of X_t.
3. Next, consider the increments: Derive $\mathbb{V}ar(X_{t+s} - X_t)$, for $s \geq 0$.
4. Argue whether X_t is a Wiener process.

Hint: By employing the properties of the distribution of Wiener processes this exercise can be done without elaborate calculation.

Exercise 3.24. [☼] Completion to a $\mathcal{B}^d \times \mathcal{A}$-Measurable Function
Show the assertion of Lemma 3.10: I.e., let $\mathcal{B}^d$ the product Borel-σ-algebra on $\mathbb{R}^d$, $\Omega_1 \in \mathcal{A}$ and $f(x,\omega) : \mathbb{R}^d \times \Omega \to \mathbb{R}^m$ be continuous on $\mathbb{R}^d$ for $\omega \in \Omega_1$ as well as $\mathcal{A}$-measurable in ω for all $x \in \mathbb{R}^d$. Then there exists a $\mathcal{B}^d \times \mathcal{A}$-measurable function $g(x,\omega)$ such that $f(x,\omega) = g(x,\omega)$ for $(x,\omega) \in \mathbb{R}^d \times \Omega_1$.

Exercise 3.25. [☼] The Black-Derman-Toy Model as a Random Differential Equation
Let $0 \leq t \leq T$. Black, Derman and Toy developed a discrete time model for the short-term interest rate r. Its equivalent in continuous time is the stochastic differential equation

$$\mathrm{d}\ln(r_t) = a(t)\mathrm{d}t + b\mathrm{d}W_t \,.$$

Rewrite this equation as a random differential equation by applying the Doss/ Sussmann & Imkeller/ Schmalfuss correspondence. Under which conditions on the parameters has this random differential equation a unique path-wise solution?

Exercise 3.26. [☼] The Hull-White Model as a Random Differential Equation
Let $0 \leq t \leq T$. Hull and White developed a model for the short-term interest rate r in which the long-run mean is specified via the non random time-dependent function $a : [0,T] \to \mathbb{R}$. The corresponding stochastic differential equation for this model reads as

$$\mathrm{d}r_t = (a(t) - \alpha r_t)\,\mathrm{d}t + b\mathrm{d}W_t \,,$$

where $\alpha, b \in \mathbb{R} \setminus \{0\}$ and $r_0 \in \mathbb{R}$ is known. Rewrite this equation as a random differential equation by applying the Doss/ Sussmann & Imkeller/ Schmalfuss correspondence. Under which conditions on the parameters has this random differential equation a unique path-wise solution?

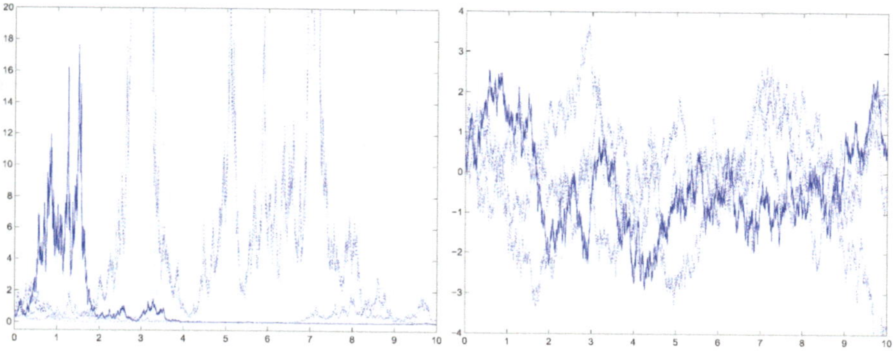

Figure 3.15. 5 simulation runs displaying the solution of the Black-Derman-Toy equation (a) and the Hull-White equation (b) for parameters $a(t) = \exp(-t)$, $\alpha = 1$ and $b = 2$.

Exercise 3.27. [☼] **The Clough-Penzien Model as a Random Differential Equation**
In Sec. 3.2.2, we discussed the Clough-Penzien model for ground motion excitations, see equation (3.3). Rewrite this second order equation as a first order system of stochastic differential equations and then convert it to the corresponding system of random differential equation by applying the Doss/ Sussmann & Imkeller/ Schmalfuss correspondence. Under which conditions on the parameters has this random differential equation system a unique path-wise solution?

Exercise 3.28. [☼] **Functions of Path-Wise Solutions – Part 1**
Let X_t be a path-wise solution of (3.5) on I, $\{t_1, \dots, t_k\} \subset I$ an arbitrary parameter set, and φ an arbitrary Borel-measurable function[11]. Show that the function $\varphi(X_{t_1}, \dots, X_{t_k})$ is $\mathcal{A}$-measurable.

Exercise 3.29. [☼] **Functions of Path-Wise Solutions – Part 2**
Let X_t be a path-wise solution of (3.5) on I.

1. Show that almost all realisations of X_t on I are continuous functions.

2. Let ψ be an arbitrary continuous functions on $\mathbb{R}^d$. Show that the functions $\sup_{t\in I} \psi(X_t)$ and $\inf_{t\in I} \psi(X_t)$ which are defined on Ω are equivalent to $\mathcal{A}$-measurable functions.

[11] A Borel-measurable function φ is a function for which all subsets of the type $\mathbb{E}(x : \varphi(x) \geq c)$, $c \in \mathbb{R}$, in its domain of definition are Borel sets.

Exercise 3.30. [☼] **RDE Properties**
As discussed in Sec. 3.2.2, [31] considered the following model in order to describe in some qualitative manner the nature of an earthquake disturbance:

$$y(t) = \begin{cases} \sum_{j=1}^{n} ta_j \exp(-\alpha_j t)\cos(\omega_j t + \Theta_j)\,, & \text{for } t \geq 0 \\ 0\,, & \text{for } t < 0 \end{cases}\,,$$

where a_j, α_j and ω_j are given real positive numbers and the parameters Θ_j are independent random variables uniformly distributed over an interval of length 2π.

Let us assume a one-storey building that is at rest at $t = 0$ and let $X(t)$, $t \geq 0$, denote the relative horizontal displacement of its roof with respect to the ground. Then, based upon an idealized linear model, the relative displacement $X(t)$ subject to ground accelerations is governed by

$$\ddot{x}(t) + 2\zeta\omega_0\dot{x}(t) + \omega_0^2 x(t) = -y(t)\,, \quad \text{for } t \geq 0\,. \tag{3.19}$$

1. Show that $y(t)$ has vanishing mean and covariance

$$\mathbb{C}ov(y(t), y(s)) = \sum_{j=1}^{n} \tfrac{1}{2} tsa_j^2 \exp\left(-\alpha(t+s)\right)\cos\left(\omega_j(t-s)\right)\,, \text{with } t, s \geq 0\,.$$

2. Show that (3.19) has a path-wise solution.
3. Compute this solution analytically.

Exercise 3.31. [☼] **Gronwall's Lemma (for Stochastic Processes)**
Prove Gronwall's lemma for Stochastic Processes as stated in Lemma 3.15.

Chapter 4

Path-Wise, $\mathbb{P}$- & Mean-Square Solutions of RODEs

The special nature of random (ordinary) differential equations requires additional refined solution concepts beyond that of a path-wise solution. Taking, for instance, into account that a solution may fulfill the given random differential equation with probability one, or that the solution is a square integrable stochastic process leads to the notion of a $\mathbb{P}$-solution or a mean-square solution, respectively. Their properties and inter-connections, in particular with respect to path-wise solutions, are studied in this chapter.

4.1 Key Concepts

In our lectures and, of course in our own experience when learning new concepts, we have seen that boundaries and connections to other points of view are extremely important in order to understand concepts with all of their implications. Therefore, this chapter is dedicated to alternative notions of a "solution" to a given random (ordinary) differential equation as well as their connections to the concept of a path-wise solution introduced in Chap. 3.

These concepts assume a solution either (i) to be a stochastic process that fulfills a given random differential equation with probability one ($\mathbb{P}$-solution) or (ii) to be a second order stochastic process that again fulfills the given random differential equation with probability one such that the right hand side of this random differential equation is equivalent to a square integrable process.

During this chapter note the following key concepts of

1. $\mathbb{P}$-solutions, and
2. Mean-square/ quadratic-mean solutions ($q.m.$-solutions),

as well as the answers to the following questions

1. How are the different notions of solutions connected?
2. In particular, how are they connected to the concept of a path-wise solution?
3. Are there random differential equations that have path-wise solutions but no $\mathbb{P}$- or m.s. solutions?

4. What about the reverse? Are there random differential equations that have $\mathbb{P}$- or m.s. solutions but no path-wise solutions?
5. How does this all relate to Itô solutions of stochastic differential equations?

This chapter is structured as follows: Section 4.2 discusses the notion of $\mathbb{P}$-solutions and subject to which conditions a given random differential equation has a unique $\mathbb{P}$-solution. Next, Section 4.3 recalls the essential concepts of mean square calculus which lay the foundation for the following discussions. In Sec. 4.4, the concept of mean-square solutions is studied together with the conditions which a given random differential equation must possess in order to have a unique mean-square solution. As an excursion, Section 4.5 provides a brief primer on Itô's stochastic integration with respect to white noise which is essentially a mean-square calculus. Finally, Section 4.6 wraps up the contents of this chapter. The main source for this chapter is [45], chapter 2.

4.2 $\mathbb{P}$-Solutions of Random Differential Equations

Engineering and the natural sciences use the concepts of path-wise solutions to describe processes on $\mathbb{R}^d$ as path-wise solutions of

$$\dot{X}_t := \frac{\mathrm{d}X_t}{\mathrm{d}t} = f(X_t(\cdot), t, \omega), \qquad X_t(\cdot) \in \mathbb{R}^d, \tag{4.1}$$

which means that they are solutions of the non-autonomous deterministic ordinary differential equation

$$\dot{x} = \frac{\mathrm{d}x}{\mathrm{d}t} = F_\omega(x, t), \qquad x := X_t(\omega) \in \mathbb{R}^d \tag{4.2}$$

for almost all $\omega \in \Omega$, see Definition 3.4.

The random differential equation (4.1) can be interpreted as an equation on the Banach-space $\mathcal{L}_d^2$ of two-times integrable functions on $\mathbb{R}^d$, or as an equation that holds at each fixed $t \in I$ for almost all $\omega \in \Omega$. Though the concept of path-wise solutions is well suited for for a large vaiety of practical purposes, it is beneficial to study its connection to other solution types.

In contrast to path-wise solutions, the $\mathbb{P}$-solutions of a random differential equation (4.1) satisfy equation (4.2) just at each fixed $t \in I$ for almost all $\omega \in \Omega$. Here, the exceptional set $\Omega_t^* \subset \Omega$ can be different for each $t \in I$ and the union of these exceptional sets may not have vanishing probability, i.e., $\mathbb{P}(\cup_{t\in I}\Omega_t^*) \geq 0$.

Definition 4.1 ($\mathbb{P}$-Solution of a Random Differential Equation). A stochastic process on an interval $I \subset \mathbb{R}_0^+$ is called $\mathbb{P}$-solution (solution with probability one) of the random differential equation (4.1) if it holds that

$$\frac{\mathrm{d}X_t(\omega)}{\mathrm{d}t} \mathrel{\hat{=}} f(X_t(\omega), t, \omega), \qquad X_t(\cdot) \in \mathbb{R}^d, \quad t \in I.$$

Of course, path-wise solutions of (4.1) are also $\mathbb{P}$-solutions, but $\mathbb{P}$-solutions do not need to be path-wise solutions, as the following example shows:

Example 4.2 (A $\mathbb{P}$-Solution That is Not a Path-Wise Solution, cf. [45], pp. 25). Let Z be a uniformly distributed random variable on $(0, 1)$. The stochastic process X_t with the realisations

$$X_t(\omega) = \begin{cases} 0, & t = Z(\omega) \cdot n^{-1}, \quad \text{for } n = 1, 2, \dots \\ Z(\omega) \cdot t^2, & \text{elsewhere} \end{cases}$$

is a $\mathbb{P}$-solution of the random differential equation $\dot{X}_t = 2 \cdot Z \cdot t$ on $[0, \infty)$.

This can be verified as follows: Fix $t > 0$ and let

$$\Omega_t := \left\{\omega \in \Omega : Z(\omega)n^{-1} = t \text{ for an } n \in \mathbb{N}\right\}.$$

Then, $\mathbb{P}(\Omega_t) = 1$, and if $\omega \in \Omega \setminus \Omega_t$ there is an $n^*(t, \omega) \in \mathbb{N}$ such that $t \cdot n^* < Z(\omega) < t \cdot (n^* + 1)$. Hence, there is a real value $H > 0$ such that $(t + h) \cdot n^* < Z(\omega) < (t + h) \cdot (n^* + 1)$ for all $h \in (-H, H)$ holds. Consequently, $\omega \notin \Omega_{t+h}$ for all $h \in (-H, H)$. We finally get

$$\frac{\mathrm{d}X_t(\omega)}{\mathrm{d}t} = \lim_{h \to 0} \frac{X_{t+h}(\omega) - X_t(\omega)}{h} = \lim_{h \to 0} \frac{Z(\omega)(t + h)^2 - Z(\omega)t^2}{h} = 2Z(\omega)t$$

for all $\omega \notin \Omega_t$. For $t = 0$, we trivially have $\frac{\mathrm{d}X_t(\omega)}{\mathrm{d}t} = 0$. Thus, on $I = [0, \infty)$ we have that X_t is a $\mathbb{P}$-solution.

On the other hand, in an arbitrary interval $[0, b)$, $b > 0$, every realisation $X_t(\omega)$ has countably many t-values $t_n = n^{-1}Z(\omega)$ for which it is discontinuous and does not have a derivative. Thus, X_t is not a path-wise solution of the random differential equation $\dot{X}_t = 2 \cdot Z \cdot t$ (not even in the extended sense).

As this example illustrates, realisations of $\mathbb{P}$-solutions can be quite irregular. Though $\mathbb{P}$-solutions with continuously differentiable realisations are path-wise solutions, too.

Theorem 4.1 ($\mathbb{P}$-Solutions are Path-Wise Solutions if Their Realisations are Path-Wise C^1). *For almost all $\omega \in \Omega$ let $f(x, t, \omega)$ be continuous on $\mathbb{R}^d \times I$ and $\mathcal{A}$-measurable for every fixed (x, t). Let X_t be a $\mathbb{P}$-solution of (4.1) on I such that it has a path-wise continuous path-with derivative $\dot{X}_t$ on I. Then, X_t is a path-wise solution of (4.1), too.*

Proof. Following [45], p. 26, we have that according to Lemma 3.10 there is a path-wise continuous stochastic process Z_t such that $Z_t \stackrel{I}{=} f(X_t, t, \omega)$. Moreover, there is a path-wise continuous stochastic process Y_t such that $Y_t \stackrel{I}{=} \dot{X}_t$ and (due to the definition of a $\mathbb{P}$-solution X_t) $Y_t \hat{=} \dot{X}_t \hat{=} f(X_t, t, \omega) \hat{=} Z_t$, for $t \in I$. The path-wise continuity of Y_t and Z_t now imply

$$\dot{X}_t \stackrel{I}{=} Y_t \stackrel{I}{=} Z_t \stackrel{I}{=} f(X_t, t, \omega)\,.$$

This shows the assertion. □

To properly set-up the notion of a mean-square or quadratic mean ($q.m.$) solution of a random differential equation, we first recall some important results of mean-square analysis for second order stochastic processes.

Before you continue, make sure to answer the following questions:

Quiz: Section 4.2

Q1 Give the definition of a $\mathbb{P}$-solution.

Q2 How are $\mathbb{P}$-solutions and path-wise solutions connected?

Q3 How is this connection proven?

Q4 Give an example for a random differential equation that has a $\mathbb{P}$-solution but no path-wise solution.

4.3 Review: Mean-Square Analysis of Second Order Processes

For a sequence $\{X_n\}_{n\in\mathbb{N}} \subset L^2_d$ convergence towards $X \in L^2_d$ in *mean-square* or *quadratic-mean* ($q.m.$) is defined via

$$q.m.\lim_{n\to\infty} X_n \hat{=} X \qquad :\Leftrightarrow \qquad \lim_{n\to\infty} \mathbb{E}\left(\|X_n - X\|^2\right) = 0\,.$$

A stochastic process X_t, $t \in I$, with $X_t \in L^2_d$ for all $t \in I$ is called a *stochastic process of second order* (or short: process of second order). For the remainder of this section, let X_t always denote a process of second order.

The process X_t is called $q.m.$-*continuous* at t if

$$q.m.\lim_{h\to 0} X_{t+h} \hat{=} X_t$$

holds. If there exists a stochastic process Y_t of second order such that

$$q.m.\lim_{h\to 0}\tfrac{1}{h}(X_{t+h}-X_t)\mathrel{\hat{=}}Y_t\,,$$

then Y_t is called *q.m.-derivative* of X_t. A Riemann *q.m.-integral* $\int_a^b X_t g(t)\mathrm{d}t$ with a real function $g: I\to\mathbb{R}$ is defined analogously to the usual integral as the limit

$$q.m.\lim_{m\to\infty}\sum_{j=1}^{m} g(t_j)X_{t_j}(t_{j+1}-t_j)$$

together with a partitioning-sequence $a=t_1<t_2<\cdots<t_{m+1}=b$ and $\max(t_{j+1}-t_j)\to 0$, provided this limit exists and is independent of the partitioning-sequence used. If for every sequence $a_m\to-\infty$ it holds that

$$\lim_{m\to\infty}\int_{a_m}^{b} X_t g(t)\mathrm{d}t \;=\; X\in L^2_d\,,$$

then we call $X=\int_{-\infty}^{b}X_t g(t)\mathrm{d}t$ an *improper q.m.-integral.*

Finally, if $q.m.\lim_{m\to\infty}X_m\hat{=}X$, then $\lim_{m\to\infty}\mathbb{E}(X_m)=\mathbb{E}(X)$ holds. In particular, $q.m.$-differentiation and Riemann-$q.m.$-integration thus commute with taking the expectation, provided the derivative or the integral exists, respectively.

In the remainder of this section, we are only considering (second order) processes $X_t=(X_t^{(1)},\dots,X_t^{(d)})^T\in\mathbb{R}^d$ that have vanishing mean, and for which $\Gamma_X(t,s)$ denotes the vector-valued auto-correlation functions with their components given as $\Gamma_X^{(l)}(t,s)=\mathbb{E}(X_s^{(l)}X_t^{(l)})$, $l=1,2,\dots,d$. (For processes Y_t such that $\mathbb{E}(Y_t)\neq 0$, define $X_t:=Y_t-\mathbb{E}(Y_t)$.)

The following (non comprehensive) list summarizes some important results on second order stochastic processes, cf. [45], pp. 27 (as well as [179], [180], and [70]):

1. X_t is $q.m.$-continuous if and only if $\Gamma_X(t,s)$ is continuous on $I\times I$. If $\Gamma_X(t,s)$ is continuous for every element $(t,s)\in I\times I$ then it is continuous on $I\times I$.

2. Let $X_{m,t}$ be $q.m.$-continuous second order processes such that $q.m.\lim_{m\to\infty}X_{m,t}\hat{=}X_t$ holds uniformly for $t\in[a,b]$. Then, the limit process X_t is $q.m.$-continuous on $[a,b]$.

3. X_t has a $q.m.$-derivative Y_t if and only if the partial derivative $\frac{\partial^2}{\partial t\partial s}\Gamma_X(t,s)$ exists on $I\times I$. In this case $\Gamma_Y(t,s)=\frac{\partial^2}{\partial t\partial s}\Gamma_X(t,s)$. If the derivative exists for every element $(t,s)\in I\times I$ then it exists on $I\times I$, too.

4. If X_t has a $q.m.$-derivative, then it is $q.m.$-continuous.

5. Let $\Gamma_X(t,s)$ be continuous on $I \times I$, $g : I \to \mathbb{R}$ be a real function and assume that the Riemann-integral $V = \int_a^b \int_a^b g(t)g(s)\Gamma_X(t,s)\mathrm{d}s\mathrm{d}t$ exists. Then, the Riemann-$q.m.$-integral $Z := \int_a^b X_t g(t)\mathrm{d}t$ exists and $\mathbb{E}((Z^{(l)})^2) = V^{(l)}$ holds for every component $l = 1, 2, \ldots, d$.

6. If X_t is $q.m.$-continuous for all $t \geq t_0$, then $Z_t := \int_{t_0}^t X_\tau \mathrm{d}\tau$ is $q.m.$-differentiable and it holds that X_t is the $q.m.$-derivative of Z_t, i.e., $\dot{Z}_t \hat{=} X_t$.

7. Let the $q.m.$-derivative $\dot{X}_t$ of X_t be $q.m.$-continuous on I and $[a,b] \subset I$, then $\int_a^b \dot{X}\mathrm{d}t \hat{=} X_b - X_a$.

8. Let $\frac{\partial^2}{\partial t \partial s}\Gamma_X(t,s)$ be continuous at each point $(t,s) \in I \times I$ (according to 1 and 3 this is equivalent with the $q.m.$-derivative of X being continuous on I). Then, for X_t there exists an equivalent stochastic process with continuous realisations on I.

   Proof: This follows from Kolmogorow's Continuity Theorem as both, the $m.s.$-derivative $\dot{X}_t$ of X_t and, due to item 4, X_t itself are $m.s.$-continuous. Because of

$$
\begin{aligned}
&\lim_{h,k\to 0}\left(\mathbb{E}\left(\dot{X}_{t+h}^{(l)} X_{s+k}^{(l)}\right) - \mathbb{E}\left(\dot{X}_t^{(l)} X_s^{(l)}\right)\right)^2 \\
&\leq 2 \lim_{h,k\to 0} \mathbb{E}\left(\left(\dot{X}_{t+h}^{(l)}\right)^2\right)\mathbb{E}\left(\left(X_{s+k}^{(l)} - X_s^{(l)}\right)^2\right) \\
&\quad + 2\lim_{h\to 0}\mathbb{E}\left(\left(X_s^{(l)}\right)^2\right)\mathbb{E}\left(\left(\dot{X}_{t+h}^{(l)} - \dot{X}_t^{(l)}\right)^2\right) \;=\; 0\,,
\end{aligned}
$$

   this implies the continuity of $\frac{\partial}{\partial t}\Gamma_X(t,s) = \mathbb{E}(\dot{X}_t X_s)$ on $I \times I$. Thus, on any bounded interval $J \subset I$ there are positive numbers c and d such that

$$
\begin{aligned}
\mathbb{E}\left(\left(X_{t+h}^{(l)} - X_t^{(l)}\right)^2\right) &= \left(\Gamma_X^{(l)}(t+h,t+h) - \Gamma_X^{(l)}(t,t+h)\right) \\
&\quad + \left(\Gamma_X^{(l)}(t,t+h) - \Gamma_X^{(l)}(t,t)\right)\,,
\end{aligned}
$$

   for $t \in I$ and $|h| < d$.

9. Let the random vectors $X_k \in L_d^2$ be normally distributed and assume their limit $q.m.$ $\lim_{k\to\infty} X_k \hat{=} X$ exists. Then, X is normally distributed, too.

10. Let X_t be a Gaussian process such that $\Gamma_X(t,s)$ fulfills the assumptions in 3 and 5, respectively. Then its $q.m.$-derivative $\dot{X}_t$ and its Riemann-$q.m.$-integral $Z_t = \int_{t_0}^t X_\tau \mathrm{d}\tau$ are Gaussian processes, too such that $\Gamma_{\dot{X}}(t,s) = \frac{\partial^2}{\partial t \partial s}\Gamma_X(t,s)$ and $\Gamma_Z(t,s) = \int_{t_0}^t \int_{t_0}^s \Gamma_X(\tau,\sigma)\mathrm{d}\sigma\mathrm{d}\tau$, respectively.

11. Let $a(t)$ be a differentiable real function on I and assume that X_t has a $q.m.$-derivative $\dot{X}_t$ on I. Then, the $q.m.$-derivative of $a(t) \cdot X_t$ equals $\dot{a}(t)X_t + a(t)\dot{X}_t$. Recall, $\dot{a}(t)$ is the usual derivative with respect to t, whereas $\dot{X}_t$ has to be interpreted in the sense of a $q.m.$-derivative.

12. Let X_t be both $q.m.$-differentiable as well as path-wise differentiable. Then its $q.m.$-derivative and its path-wise derivative are equivalent.

13. Let X_t be both $q.m.$-continuous as well as path-wise continuous. Then its $q.m.$-integral $⨍_{t_0}^{t} X_\tau \mathrm{d}\tau$ and its path-wise integral $\int_{t_0}^{t} X_\tau \mathrm{d}\tau$ exist for $t_0, t \in I$, and are equivalent.

14. Let $\{X_t^k\}_{k\in\mathbb{N}}$ be a sequence of second order stochastic processes. Then there is a limit (stochastic) process X such that $q.m.\lim_{k\to\infty} X_t^k \hat{=} X_t$, $t \in I$, if and only if for each $t \in I$ and each $l \in \mathbb{N}$ the limit $\lim_{i,k} \mathbb{E}\left(X_t^{i(l)} X_t^{k(l)}\right)$ exists. Then, it holds that

$$\lim_{k\to\infty} \mathbb{E}\left(X_{t_1}^k \left(X_{t_2}^k\right)^T\right) = \mathbb{E}\left(X_{t_1} X_{t_2}^T\right) .$$

15. Assume $X_k \in L_d^2$, $X_k \hat{\geq} 0$ and $q.m.\lim_{k\to\infty} X_k = X$ for each $k = 1, 2, \ldots$. Then $X \hat{\geq} 0$ holds, too.

16. Let X_t be a stationary process the $q.m.$-derivative $\dot{X}_t$ of which exists. Then, $\mathbb{E}(X_t^T \dot{X}_t) = 0$.

4.4 Mean-Square Solutions of Random Differential Equations

The random differential equation

$$\dot{X}_t := \frac{\mathrm{d}X_t}{\mathrm{d}t} = f(X_t(\cdot), t, \omega) , \qquad X_\cdot \in L_d^2 , \tag{4.3}$$

is now interpreted as a differential equation in L_d^2 and the derivative $\dot{X}_t$ as $q.m.$-derivative of X_t. This gives rise to the following definition:

Definition 4.3 ($q.m.$-Solutions of Random Differential Equations). A second order stochastic process X_t, $t \in I$ is called $q.m.$-solution (*solution in the quadratic mean* or *mean-square solution*) of the random differential equation (4.3) on the interval I, provided that for all $t \in I$ it holds that $f(X_t, t, \omega) \hat{=} F_t$ such that $F_t \in L_d^2$ and $\dot{X}_t \hat{=} F_t$.

If, additionally, $X_{t_0} \hat{=} X_0$ holds for a $(X_0, t_0) \in L_d^2 \times I$, then X_t is called $q.m.$-*solution with respect to the initial condition* (X_0, t_0). The $q.m.$-solutions of (4.3) are called *unique* if for each pair X_t, X_t^* of $q.m.$-solutions with respect to this initial condition it holds that $X_t \hat{=} X_t^*$.

The realisations $X_t(\omega)$ of a $q.m.$-solution do not need to be differentiable. Though under weak regularity conditions, $q.m.$-solutions are equivalent to $q.m.$-solutions with continuous realisations. This is, for instance, the case if F_t is $q.m.$-continuous on I due to item 8 from Sec. 4.3.

The following theorem gives a connection between $q.m.$-solutions and path-wise solutions:

Theorem 4.2 ($q.m.$-Solutions With Path-Wise Continuous Path-Wise Derivatives are Path-Wise Solutions). *Let $f(x, t, \omega)$ be path-wise continuous on $\mathbb{R}^d \times I$ and $\mathcal{A}$-measurable for every fixed $(x, t) \in \mathbb{R}^d \times I$. Assume that X_t is a $q.m.$-solution of (4.3) on I with a path-wise continuous derivative on I. Then, X_t is a path-wise solution of (4.3) on I, too.*

Proof. Following [45], p. 29, we have that property 12 from Sec. 4.3 implies the following for the path-wise derivative

$$\dot{X}_t \hat{=} f(X_t, t, \omega)\,, \qquad t \in I\,.$$

Thus X_t is a $\mathbb{P}$-solution and by Theorem 4.1 it is a path-wise solution of (4.3). □

Before you continue, make sure to answer the following questions:

Quiz: Section 4.3

Q1 Define the terms $q.m.$-continuous, $q.m.$-derivative, and $q.m.$-integral.

Q2 Why is the auto-correlation Γ_X of a second order stochastic process important for the analysis of $q.m.$ properties?

Q3 Prove property 1: I.e., show that X_t is $q.m.$-continuous if and only if $\Gamma_X(t, s)$ is continuous on $I \times I$.

Q4 Prove property 4: I.e., show that if X_t has a $q.m.$-derivative, then it is $q.m.$-continuous.

Q5 Prove property 12: I.e., show that if X_t is both $q.m.$-differentiable as well as path-wise differentiable, then its $q.m.$-derivative and its path-wise derivative are equivalent.

Compared with path-wise solutions, the $q.m.$-solutions have the disadvantage that they cannot be studied by simply transferring/ applying in a path-wise manner the results from the theory of deterministic ordinary differential equations. For instance, define $X_t := X_0 \exp(Y)$, where $Y \in L_1^2$ is a random variable the probability density of which is given by

$$p(s) = \begin{cases} \exp(-s) & \text{if} \quad s > 0 \\ 0 & \text{if} \quad s > 0 \end{cases} .$$

Then, X_t is not a $q.m.$-solution of $\dot{X}_t = YX_t$ for $t \geq 1$, because $\mathbb{E}(X_t^2) = \infty$ for $t > 1$.

On the other hand, the analytic realisations of the random differential equation's right hand side $f(x, t, \omega)$ may not be known or may be rather unfavorable, such that path-wise solutions do not exist. Though $q.m.$-solutions may exist in such a setting. Examples for such situations are white-noise driven stochastic differential equations for which Kiyoshi Itô (1915 – 2008) showed the existence of $q.m.$-solutions under certain Lipschitz- and growth conditions on $f(x, t, \omega)$, cf. [222]. In the following Sec. 4.5 we will give a short primer on Itô's ideas which essentially lead to a mean-square calculus.

Another example is given in [45], p. 30:

Example 4.4 (A Random Differential Equation With Known $q.m.$-Solution but Without Information About its Path-Wise Solutions, [45], p. 30). Let Z_t be a second order stochastic process that is $q.m.$-continuous on I. Then, a path-wise solution of the simple random differential equation $\dot{X}_t = Z_t$ is not known. Though due to property 6 from Sec. 4.3, the random differential equation $\dot{X}_t = Z_t$ has a uniquely determined $q.m.$-solution

$$X_t = X_0 + \int_{t_0}^{t} Z_\tau \mathrm{d}\tau$$

on I with respect to the initial condition $(X_0, t_0) \in L_d^2 \times I$.

The relationship between $q.m.$-solutions and path-wise solutions suggests that we can specialize results from the theory of differential equations on Banach-spaces to the space $\mathcal{L}_d^2$ and then apply them to path-wise solutions. Let $\tilde{X}_t \in \mathcal{L}_d^2$ denote the class of all elements of L_d^2 that are equivalent to X_t and let $\tilde{f}(\tilde{X}, t) \in \mathcal{L}_d^2$ be the class of ell elements of L_d^2 which are equivalent to $f(X_t, t, \omega)$. Then, X_t is a $q.m.$-solution of (4.3) if and only if $\tilde{X}_t$ is a solution of the differential equation

$$\frac{\mathrm{d}\tilde{X}_t}{\mathrm{d}t} = \tilde{f}(\tilde{X}_t, t),$$

in the Banach-space $\mathcal{L}_d^2$. Here, $\frac{\mathrm{d}\tilde{X}_t}{\mathrm{d}t}$ denotes the Frechet-derivative of $\tilde{X}_t$.

Well-known results on the existence, uniqueness and continuation of solutions of differential equations on Banach-spaces imply the following theorem:

Theorem 4.3 (Existence and Uniqueness of $q.m.$-Solutions). *Let $f(Y_t, t, \omega)$ be equivalent to a $q.m.$-continuous second order process on I for every $q.m.$-continuous process Y_t. Moreover, assume that for almost all $\omega \in \Omega$ the Lipschitz-condition*

$$\|f(x_1, t, \omega) - f(x_2, t, \omega)\| \;\leq\; L(t)\|x_1 - x_2\|\,, \tag{4.4}$$

holds, where $L(t)$ is a positive continuous function on I. Then, there is a unique $q.m.$-solution of (4.3) on I with respect to the initial condition $(X_0, t_0) \in L_d^2 \times I$.

Proof. Following [45], p. 30, let $J := (\alpha, \beta)$ be an arbitrary sub-interval of I such that $t_0 \in J$. Then (4.4) holds for $t \in J$ with a time-independent Lipschitz-constant $L := \max_{t\in J} L(t)$.

Due to our assumptions $f(X_0, t, \omega)$, $t \in I$, is equivalent to a second order $q.m.$-continuous process. Thus, $\mathbb{E}(\|f(X_0, t, \omega)\|^2)$ is continuous, which implies

$$\max_{i\in J} \mathbb{E}\left(\|f(X_0, t, \omega)\|^2\right) \;=\; M^2 \;<\; \infty\,.$$

Moreover, the Riemann-$q.m.$-integral $\int_{t_0}^{t} Y_\tau \mathrm{d}\tau$ is defined for any second order $q.m.$-continuous stochastic process Y_t.

Due to the equivalence between $q.m.$-solutions of random differential equations and solutions of differential equations in the Banach-space $\mathcal{L}_d^2$, the unique existence of an associated $q.m.$-solution of (4.3) on the interval $J_{t_1} := (\max\{\alpha, t_1 - c\}, \min\{\beta, t_1 + c\})$ with respect to the initial condition $(X_1, t_1) \in L_d^2 \times J$ follows. Here, $c := r(M + rL)^{-1}$ with an arbitrary fixed r.

Following the usual trail of thoughts for the continuation of solutions, the unique $q.m.$-solution on J_{t_0} with respect to the initial condition (X_0, t_0) can be extended uniquely on J. Thus, on each sub-interval $J \subset I$ there exists a unique $q.m.$-solution and consequently there exists a unique $q.m.$-solution on I. □

Example 4.5 ($q.m.$-Solution of a Linear Random Differential Equation, [45], p. 31). The conditions of Theorem 4.3 hold on $I = [0, \infty)$ for the linear random differential equation

$$\frac{\mathrm{d}X_t}{\mathrm{d}t} \;=\; A_t X_t \,+\, Z_t\,,$$

if Z_t is a second order $q.m.$-continuous stochastic process, the elements of the matrix A_t are path-wise continuous stochastic processes on I, and there is a continuous positive function $L(t)$ on I such that $\|A_t\| \overset{I}{\leq} L(t)$. The $q.m.$-continuity of a second order stochastic process Y_t implies the $q.m.$-continuity of the second order stochastic process $f(Y_t, t, \omega)$.

Without loss of generality, let us assume $t_0 < t < t + h < H$ and $\max_{t_0 \leq t \leq H} L^2(t) =: l^*$, then it holds that

$$\lim_{h\to 0} \mathbb{E}\left(\|A_{t+h}Y_{t+h} - A_t Y_t\|^2\right) \leq 2l^* \lim_{h\to 0} \mathbb{E}\left(\|Y_{t+h} - Y_t\|^2\right)$$
$$+ 2 \lim_{h\to 0} \mathbb{E}\left(\|(A_{t+h} - A_t)Y_t\|^2\right) = 2\mathbb{E}\left(\|(\lim_{h\to 0} A_{t+h} - A_t)Y_t\|^2\right) = 0,$$

where we applied the Lebesgue convergence theorem due to

$$\|(A_{t+h} - A_t)Y_t\|^2 \overset{I}{\leq} 2l^*\|Y_t\|^2, \quad \mathbb{E}(\|Y_t\|^2) < \infty.$$

Moreover,

$$\|A_t x_1 - A_t x_2\| \leq \|A_t\| \cdot \|x_1 - x_2\| \leq L(t)\|x_1 - x_2\|, \quad x_1, x_2 \in \mathbb{R}^d, \quad t \in I$$

holds for almost all $\omega \in \Omega$.

The following theorem establishes a relationship between $q.m.$-solutions and path-wise solutions:

Theorem 4.4 (Equivalence Between $q.m.$ and Path-Wise Solutions). *Let the following three conditions hold:*

1. *For almost all $\omega \in \Omega$ the function $f(x, t, \omega)$ is continuous on $\mathbb{R}^d \times I$, and for fixed $(x, t) \in \mathbb{R}^d \times I$ it is $\mathcal{A}$-measurable.*

2. *Let $L(t)$ be a continuous real positive function defined on I such that for almost all $\omega \in \Omega$ it holds that*

$$\|f(x_1, t, \omega) - f(x_2, t, \omega)\| \leq L(t)\|x_1 - x_2\|,$$

   where $x_1, x_2 \in \mathbb{R}^d$ and $t \in I$.

3. *Let $M(t)$ be a non-negative continuous function defined on I, and let $Z \in L_1^2$. Assume it holds for almost all $\omega \in \Omega$ that*

$$\|f(x, t, \omega)\| \leq M(t)\left(\|x + Z(\omega)\|\right),$$

   where $x \in \mathbb{R}^d$ and $t \in I$.

These ensure that there is a unique $q.m.$-solution of the random differential equation (4.3) on I with respect to the initial condition $(X_0, t_0) \in L_d^2 \times I$. On I this solution is equivalent to the unique path-wise solution for that initial condition.

Proof. Following [45], p. 32, we have that according to Theorem 3.2 there is a unique path-wise solution X_t on I such that $X_{t_0} \hat{=} X_0$. We will show that $X_t \in L_d^2$, $t \in I$, and that X_t is indeed a $q.m.$-solution, too.

Due to condition 1 of this theorem we have

$$X_t \stackrel{I}{=} X_0 + \int_{t_0}^{t} f(X_\tau, \tau, \omega) \mathrm{d}\tau \,. \tag{4.5}$$

Together with $u(t) := \int_{t_0}^{t} M(\tau)\mathrm{d}\tau$ and condition 3 we obtain

$$\|X_t\| \stackrel{I}{=} \|X_0\| + u(t)Z + \int_{t_0}^{t} M(\tau)\|X_\tau\|\mathrm{d}\tau \,.$$

Thus, by Gronwall's Lemma (Lemma 3.15) we get

$$\|X_t\| \stackrel{I}{=} (\|X_0\| + u(t)Z) \exp(u(t)) \tag{4.6}$$

which implies $\mathbb{E}(\|X_t\|^2) < \infty$ for all $t \in I$ because of

$$\mathbb{E}\left((\|X_0\| + u(t)Z)^2\right) \le 2\mathbb{E}\left(\|X_0\|^2\right) + 2u^2(t)\mathbb{E}(Z^2) < \infty \,.$$

Hence, $X_t \in L_d^2$.

Without loss of generality, let $t, t+h \in [t_0, H] \subset I$. Due to equations (4.5) and (4.6) and condition 3 it holds that

$$\|X_{t+h} - X_t\| \stackrel{I}{\le} \int_{t}^{t+h} M(\tau)\left(\|X_\tau\| + Z(\omega)\right)\mathrm{d}\tau \stackrel{I}{\le} m_H \cdot |h| \cdot (Q(\omega) + Z(\omega)) \,,$$

where $m_H := \max_{t_0 \le t \le H} M(t)$ and $Q(\omega) := (\|X_0\| + u(H)Z(\omega)) \exp(u(H))$.

If for any second order $q.m.$-continuous stochastic process Y_t, $t \in I$, the stochastic process $f(Y_t, t, \omega)$ is equivalent to a second order $q.m.$-continuous stochastic process G_t on I, then the conditions of Theorem 4.3 are satisfied and (4.3) has a unique $q.m.$-solution on I with respect to the initial condition (X_0, t_0).

Let $t, t+h \in [t_0, H] \subset I$. Lemma 3.10 and condition 3 imply there is a second order stochastic process G_t such that $f(Y_t, t, \omega) \stackrel{I}{=} G_t$.

Let $l^* := \max_{t_0 \le t \le H} L(t)$, then (analogously to example 4.5) it holds that

$$\begin{aligned} &\lim_{h\to 0} \mathbb{E}\left(\|G_{t+h} - G_t\|^2\right) \\ &\le 2l^* \lim_{h\to 0} \mathbb{E}\left(\|Y_{t+h} - Y_t\|^2\right) + 2\lim_{h\to 0} \mathbb{E}\left(\|f(Y_t, t+h, \omega - f(Y_t, t, \omega)\|^2\right) \\ &= 0 \,, \end{aligned} \tag{4.7}$$

where we applied the Lebesgue convergence theorem since

$$\|f(Y_t, t+h, \omega - f(Y_t, t, \omega)\|^2 \le 4m_H^2(\|Y_t\| + Z)^2 \,,$$

together with $\mathbb{E}(\|Y_t\|^2) < \infty$ and $\mathbb{E}(Z^2) < \infty$. Because of (4.7) the process G_t is $q.m.$-continuous.

Let G_t be a second order stochastic process such that $G_t \stackrel{I}{=} f(X_t, t, \omega)$. Then, the following inequality holds

$$V_h = \left\|h^{-1}(X_{t+h} - X_t) - G_t\right\| \stackrel{I}{\leq} 2m_H(Q + Z) = K.$$

Because X_t is a path-wise solution we have $\lim_{h\to 0} \|V_h(\omega)\|^2 = 0$ for almost all $\omega \in \Omega$. Since to $X_0 \in L^2_d$ and $Z \in L^2_1$, it holds that $\mathbb{E}(K^2) < \infty$. Thus, Lebesgue's convergence theorem can be applied, and it holds that

$$\lim_{h\to 0} \mathbb{E}\left(\|V_h\|^2\right) = \mathbb{E}\left(\lim_{h\to 0} \|V_h\|^2\right) = 0.$$

According to Definition4.3 the process X_t is thus a $q.m.$-solution of (4.3) and this shows the assertion. □

Example 4.6 ($q.m.$-Solutions of Linear Random Differential Equations, [45], p. 33). The conditions of Theorem 4.4 hold for a linear random differential equation

$$\frac{\mathrm{d}X_t}{\mathrm{d}t} = A_t(\omega)X_t + Z_t,$$

if the elements of A_t and the components of Z_t are path-wise continuous stochastic processes on I and there are continuous real functions $L(t)$, $N(t)$ on I as well as a $U \in L^2_1$ such that $\|A_t\| \stackrel{I}{\leq} L(t)$ and $\|Z_t\| \stackrel{I}{\leq} N(t)U$.

Finally, the following theorem describes the dependence of $q.m.$-solutions on initial conditions $(X_0, t_0) \in L^2_d \times I$ as well as parameters $\mu \in M$ occurring in the right hand side of the random differential equation, provided a certain Lipschitz-condition holds.

Theorem 4.5 (Dependence of $q.m.$-Solutions on Initial Conditions and Parameters). *Let $(X_0, t_0) \in L^2_d \times I$. For every $\bar{X} \in L^2_d$ and every $\mu \in M$ assume there exists a $q.m.$-solution $X_t(\bar{X}, \mu)$ of the random differential equation*

$$\frac{dX_t}{dt} = f(X_t, t, \omega, \mu)$$

on I with respect to the initial condition $(\bar{X}, t_0)$. For all $\mu \in M$ and all second order $q.m.$-continuous stochastic processes Y_t let $f(Y_t, t, \omega, \mu)$ be equivalent to a second order $q.m.$-continuous stochastic process.

Let the following Lipschitz-condition hold for a $\mu_0 \in M$ and for almost all $\omega \in \Omega$

$$\|f(x_1, t, \omega, \mu_0) - f(x_2, t, \omega, \mu_0)\| \leq L(t)\|x_1 - x_2\|,$$

where $x_1, x_2 \in \mathbb{R}^d$, $t \in I$ and let $L(t)$ be a non-negative real continuous function on I. Moreover, let the following additional inequality hold for all $X \in L^2_d$ and $t \in I$

$$\sup_{\mu \in M} \mathbb{E}\left(\|f(X,t,\omega,\mu_0) - f(X,t,\omega,\mu)\|\right) \leq \eta^2(t)\,,$$

where $\eta(t)$ is a non-negative function that is Lebesgue-integrable on every bounded sub-interval of I.

Then, for $t \in I$, $\mu \in M$ and $\bar{X} \in L^2_d$ the inequality

$$\begin{aligned}&\left(\mathbb{E}\left(\|X_t(X_0,\mu_0) - X_t(\bar{X},\mu)\|^2\right)\right)^{1/2}\\ &\leq \left(\mathbb{E}\left(\|X_0 - X\|^2\right)\right)^{1/2} \exp\left(\int_{t_0}^t L(\tau)d\tau\right) + \int_{t_0}^t \exp\left(\int_{t_0}^t L(s)ds\right)\eta(\tau)d\tau\end{aligned}$$

holds.

Proof. Following [45], p. 34, the assertion follows by applying Gronwall's Lemma (Lemma 3.15) to

$$u(t) := \left(\mathbb{E}\left(\|X_t(X_0,\mu_0) - X_t(\bar{X},\mu)\|^2\right)\right)^{1/2}.$$

$u(t)$ is continuous on I and because of property 7 from Sec. 4.3 it holds that

$$X_t(X,\mu) = X + \int_{t_0}^t f(X_t(X,\mu),\tau,\omega,\mu)\mathrm{d}\tau\,,$$

with $X \in L^2_d$ and $\mu \in M$. This leads to the inequalities

$$\begin{aligned}u(t) &\leq \left(\mathbb{E}\left(\|X_0 - \bar{X}\|^2\right)\right)^{1/2}\\ &\quad + \int_{t_0}^t \left(\mathbb{E}\left(\|f(X_\tau(X_0,\mu_0),\tau,\mu,\mu_0) - f(X_\tau(\bar{X},\mu),\tau,\omega,\mu)\|^2\right)\right)^{1/2}\mathrm{d}\tau\\ &\leq \left(\mathbb{E}\left(\|X_0 - \bar{X}\|^2\right)\right)^{1/2} + \int_{t_0}^t \eta(\tau)\mathrm{d}\tau + \int_{t_0}^t L(\tau)u(\tau)\mathrm{d}\tau\,.\end{aligned}$$

This shows the assertion. □

4.5 Excursion: A Primer on Itô's Stochastic Calculus

In the late 1930s/ 1940s Kiyoshi Itô was motivated to gain a deeper meaning of the underlying infinitesimal structure of Markov processes. His basic idea was to identify the tangents of the process and to reconstruct the process in a path-wise way from its tangents (see [99], pp. 2.).

For stochastic processes, the role of "straight lines" is played by processes whose increments are independent and identically distributed over time intervals of the same length. Such processes are named in honor of Paul Lévy, and in the one-dimensional continuous case the prototype of such a Lévy process is the Wiener process W_t.

Before you continue, make sure to answer the following questions:

Quiz: Section 4.4

Q1 Give the definition of a $q.m.$-solution.

Q2 Under which conditions is a $q.m.$-solution a path-wise solution? Prove this.

Q3 Give an example of a random differential equation which has a $q.m.$-solution but not a path-wise solution.

Q4 Under which conditions does a random differential equation have a unique $q.m.$-solution?

Q5 Apply this theorem to the scalar random differential equation $\dot{X}_t = a_t X_t - z_t$, where $a_t(\omega)$ and $z_t(\omega)$ are path-wise continuous scalar stochastic processes. Which additional assumptions are required to hold for a_t and z_t?

Q6 How do the $q.m.$-solution of random differential equations depend on initial conditions and parameters?

> "In the case of a diffusion it is therefore natural to say that a "tangent" of the Markov process in a state x should be an affine function of the Wiener process with coefficients depending on that state. Thus Itô was led to describe the infinitesimal behavior of the diffusion by a "stochastic differential equation" of the form
>
> $$\mathrm{d}X_t = H(X_t)\mathrm{d}W_t + a(X_t)\mathrm{d}t\,.$$
>
> [...] The second part of the program now consisted in solving the stochastic differential equation, i.e., constructing the trajectories of the Markov process in the form
>
> $$X_t(\omega) \;=\; x + \int_0^t H(X_s(\omega))\mathrm{d}W_s(\omega) + \int_0^t a(X_s(\omega))\mathrm{d}s\,.$$
>
> At this point a major difficulty arose" [99], p. 3.

In what remains, we follow [100] and [227] to define the definite stochastic integral

$$I(T) \;=\; \int_0^T f(t,\omega)\mathrm{d}W_t\,,$$

where W_t is a Wiener process and f a suitable stochastic function, and study its basic properties. One may define via integration by parts

$$I(T) = f(T)W_T - \int_0^T f'(t,\omega)W_t \mathrm{d}t \tag{4.8}$$

if f is absolutely continuous for each ω. This ansatz is called *Paley-Wiener-Zygmund integral*. However, if f is only continuous, or just integrable, this definition does not make sense.

Since W_t is nowhere differentiable with probability 1, the integral $\int_0^T f(t,\omega)\mathrm{d}W_t$ cannot be defined in the usual Lebesgue-Stieltjes sense. Moreover, the sample paths of W_t are not of finite variation on compact time intervals. Thus, as a consequence of the Banach-Steinhaus theorem the usual Riemann-Stieltjes integral cannot be applied either (cf. [212], pp. 43):

Proposition 4.7 (Riemann-Stieltjes). *Let* $\Pi := \{0 = t_1 < \cdots < t_n = 1\}$ *be a partition of* $[0,1]$ *with mesh size* $\mu(\Pi)$ *tending to zero. Suppose that the fixed function* g *is such that for all continuous functions* h *one has that*

$$S_\Pi(h) := \sum_{i=1}^{n-1} h(t_i)\left(g(t_{i+1}) - g(t_i)\right)$$

converges if $\mu(\Pi) \to 0$. *Then,* g *is of bounded variation.*

Proof. We view the S_Π as linear operators on the Banach space of continuous functions on $[0,1]$ endowed with the sup-norm $||\cdot||$. The operator norm $||\cdot||$ for a linear operator $S : X \to Y$ on Banach spaces $(X, |\cdot|_X)$ and $(Y, |\cdot|_Y)$ is defined as $||S|| := \sup_{x\in X\setminus\{0\}} \frac{|S(x)|_Y}{|x|_X}$. Note, that

$$|S_\Pi(h)| \le ||h|| \sum_{i=1}^{n-1} |g(t_{i+1}) - g(t_i)| = ||h|| \cdot V^1(g,\Pi)$$

holds, where $V^1(g,\Pi)$ denotes the variation of g over the partition Π. $V^1(g,\Pi) := \sum_{i=1}^{n-1} |g(t_{i+1}) - g(t_i)|$ is thus defined for each partition separately. The total variation $\sup_\Pi \sum_{i=1}^{n-1} |g(t_{i+1}) - g(t_i)|$ of g is the supremum over all such variations (with respect to all possible partitions Π), i.e., $\sup_\Pi V^1(g,\Pi)$. Hence, the operator norm $||S_\Pi||$ is less than $V^1(g,\Pi)$.

For any partition $\Pi = \{0 = t_1 < \cdots < t_n = 1\}$ we can find (for instance by linear interpolation) a continuous function h_Π (bounded by 1, i.e., $||h_\Pi|| = 1$) such that $h_\Pi(t_i) = \mathrm{sgn}(g(t_{i+1}) - g(t_i))$. For such an h_Π we have

$$S_\Pi(h_\Pi) = \sum_{i=1}^{n-1} |g(t_{i+1}) - g(t_i)| = V^1(g,\Pi).$$

It follows, that $||S_\Pi|| = V^1(g, \Pi)$.

By assumption, for any h we have that the sums $S_\Pi(h)$ converge if $\mu(\Pi) \to 0$, so that for any h the set with elements $|S_\Pi(h)|$ (for such Π) is bounded. By the Banach-Steinhaus theorem[1], also the $||S_\Pi||$ form a bounded set, i.e., $\sup_\Pi ||S_\Pi|| < \infty$. Hence, the variation $\sup_\Pi V^1(g, \Pi) = \sup_\Pi ||S_\Pi||$ is bounded, too. □

What Itô saw in [139] was that if one restricts the class of potential integrands $f(t, \omega)$ to those that are adapted to the underlying filtration of σ-algebras generated by the Wiener process, and if one restricts the choice of $\xi_i \in [t_i, t_{i+1}]$ to t_i, then one can use the independence of the increments of the Wiener process in a clever way to obtain the convergence of the sums to a limit which is a martingale, cf. [213].

4.5.1 Integration with Respect to White Noise

As a motivating example, let us make sense of integrals of the form

$$I(T) = \int_0^T W_t \mathrm{d}W_t \,. \tag{4.9}$$

we have already seen that the usual Lebesgue-Stieltjes (W_t is nowhere differentiable with probability 1) as well as the usual Riemann-Stieltjes (the sample paths of W_t are not of finite variation) interpretations fail. Moreover, as W_t is not absolutely continuous an integration by parts trick, like (4.8), does not work either.

Despite these roadblocks, if all functions involved in the integration (4.9) obeyed the usual rules of differential and integral calculus, we would expect that

- $I(T) = \frac{1}{2} W_T^2$ holds,
- independently of the sequence $\left(0 = t_0^{(m)} < t_1^{(m)} < \cdots < t_n^{(m)} = T\right)_{m \in \mathbb{N}}$ of partitions of the interval $[0, T]$, and
- independently of the choice of points in each position interval $[t_{i-1}^{(m)}, t_i^{(m)}]$ for the evaluation of the integral sums that define $I(T)$ (in a Riemann-Stieltjes sense).

[1] see [3], p. 202, [256], p. 141: Let X, Y be Banach spaces and $\mathcal{S}$ be a family of continuous linear operators from X to Y. Suppose that for all $h \in X$ the set $\{|S(h)| : S \in \mathcal{S}\}$ is bounded, i.e., $\sup\{|S(h)| : S \in \mathcal{S}\} < \infty$ for all $h \in X$. Then, the set $\{||S|| : S \in \mathcal{S}\}$ is bounded, too. (I.e., $\sup\{||S|| : S \in \mathcal{S}\} < \infty$.)

Let us therefore consider the equally-spaced partition $0 = t_0 < t_1 < \cdots < t_n = T$ such that $\Delta t := t_i - t_{i-1} = n^{-1}T$ (for all $i = 1, 2, \ldots n$) and choose a convex combination of the start point t_{i-1} and the end point t_i of each subinterval as evaluation value for the integrand function. In the manner of a Riemann-Stieltjes integral sum this leads to

$$
\begin{aligned}
S_n^\lambda &:= \sum_{i=1}^{n} \left((1-\lambda)W_{t_{i-1}} + \lambda W_{t_i}\right) \cdot \left(W_{t_i} - W_{t_{i-1}}\right) \\
&= (1-\lambda)\sum_{i=1}^{n} W_{t_{i-1}}\left(W_{t_i} - W_{t_{i-1}}\right) + \lambda \sum_{i=1}^{n} W_{t_i}\left(W_{t_i} - W_{t_{i-1}}\right) \quad (4.10)
\end{aligned}
$$

for all $\lambda \in [0,1]$. The method to split the integrand into a part weighted by $1-\lambda$ and another one weighted by λ and is motivated by the *stochastic λ-integrals*.

The first sum in (4.10) can be evaluated as follows:

$$
\begin{aligned}
\sum_{i=1}^{n} W_{t_{i-1}}\left(W_{t_i} - W_{t_{i-1}}\right) &= \tfrac{1}{2}\sum_{i=1}^{n}\left(W_{t_i}^2 - W_{t_{i-1}}^2 - \left(W_{t_i} - W_{t_{i-1}}\right)^2\right) \\
&= \tfrac{1}{2}\left(W_{t_n}^2 - W_{t_0}^2\right) - \tfrac{1}{2}\sum_{i=1}^{n}\left(W_{t_i} - W_{t_{i-1}}\right)^2 \\
&= \tfrac{1}{2}W_T^2 - \tfrac{1}{2}\Delta_n^2 ,
\end{aligned}
$$

as $W_0 = 0$. Hence, we have that only Δ_n^2 depends on the partition of $[0,T]$. Next, we show that Δ_n^2 converges in probability to T for $n \to \infty$.

Lemma 4.8. *For every $\varepsilon > 0$ it holds that*

$$
\lim_{n\to\infty} \mathbb{P}\left(\left|\Delta_n^2 - \mathbb{E}\left(\Delta_n^2\right)\right| \geq \varepsilon\right) = 0 . \quad (4.11)
$$

This type of convergence is called convergence in probability.

Proof. Equation (4.11) is a consequence of Chebysev's inequality (cf. Proposition 1.15) that reads in our setting as $\mathbb{P}\left(\left|\Delta_n^2 - \mathbb{E}\left(\Delta_n^2\right)\right| \geq \varepsilon\right) \leq \varepsilon^{-2}\mathbb{V}ar\left(\Delta_n^2\right)$. We have

$$
\begin{aligned}
\mathbb{V}ar\left(\Delta_n^2\right) &= \sum_{i=1}^{n} \mathbb{V}ar\left(\left(W_{t_i} - W_{t_{i-1}}\right)^2\right) \\
&= \sum_{i=1}^{n}\left(\mathbb{E}\left(\left(W_{t_i} - W_{t_{i-1}}\right)^4\right) - \left(\mathbb{E}\left(\left(W_{t_i} - W_{t_{i-1}}\right)^2\right)\right)^2\right) \\
&\leq \sum_{i=1}^{n}\mathbb{E}\left(\left(W_{t_i} - W_{t_{i-1}}\right)^4\right) \overset{(*)}{=} \frac{n}{\sqrt{2\pi\Delta t}}\int_{-\infty}^{\infty} x^4 \exp\left(-\frac{x^2}{2\Delta t}\right) \mathrm{d}x
\end{aligned}
$$

$$= \quad 3n\,(\Delta t)^2 \quad = \quad \frac{3T^2}{n} \quad \to \; 0\,, \qquad \text{as } n \to \infty\,,$$

as the Wiener increments $W_{t_i} - W_{t_{i-1}}$ are independent normally distributed random variables. The identity $(*)$ is due to the equal spacing of the partition points and the expression of the fourth moment of a normally distributed random variable by its distribution. Hence, the assertion (4.11) follows. □

As a consequence we have that

$$\lim_{n\to\infty} \Delta_n^2 \overset{\mathcal{P}}{=} \lim_{n\to\infty} \mathbb{E}\left(\Delta_n^2\right) \;=\; \lim_{n\to\infty} \sum_{i=1}^{n} \mathbb{E}\left(\left(W_{t_i} - W_{t_{i-1}}\right)^2\right) \;=\; T\,, \qquad (4.12)$$

holds, and hence

$$\lim_{n\to\infty} \sum_{i=1}^{n} W_{t_{i-1}}\left(W_{t_i} - W_{t_{i-1}}\right) \overset{\mathcal{P}}{=} \tfrac{1}{2} W_T \;-\; \tfrac{1}{2} T\,.$$

When we consider the squared Wiener increments and thus a kind of mean-square-limit, we can "tame" the unbounded variation of the Wiener Process and gain a proper definition of stochastic integrals. This can be applied to properly set-up the Itô stochastic integral, see, e.g., [222]. Another consequence of taking limits in the mean is that the limit no longer depends point-wise on the $\omega \in \Omega$, but holds for the whole Ω[2].

Note, we will set-up the stochastic Itô integral with respect to the limit in probability. Of course, as just indicated, a mean square limit will work too. Indeed, Itô's calculus is intrinsically a mean square calculus by virtue of Itô's formula.

Next, to the second term in (4.10):

$$\begin{aligned} \sum_{i=1}^{n} W_{t_i}\left(W_{t_i} - W_{t_{i-1}}\right) \;&=\; \tfrac{1}{2}\sum_{i=1}^{n}\left(W_{t_i}^2 - W_{t_{i-1}}^2 \;+\; \left(W_{t_i} - W_{t_{i-1}}\right)^2\right) \\ &=\; \tfrac{1}{2} W_T^2 \;+\; \tfrac{1}{2}\Delta_n^2 \;\overset{\mathcal{P}}{\longrightarrow}\; \tfrac{1}{2} W_T^2 \;+\; \tfrac{1}{2} T\,. \end{aligned} \qquad (4.13)$$

This finally leads to

$$\begin{aligned} \lim_{n\to\infty} S_n^{\lambda} \;&=\; \lim_{n\to\infty} \sum_{i=1}^{n}\left((1-\lambda) W_{t_{i-1}} + \lambda W_{t_i}\right)\cdot\left(W_{t_i} - W_{t_{i-1}}\right) \\ &\overset{\mathcal{P}}{=}\; \tfrac{1}{2} W_T \;+\; \left(\lambda - \tfrac{1}{2}\right) T\,. \end{aligned}$$

Obviously, different choices of the points in the partition intervals for the evaluation of the integrand lead to different results. It is apparent from these

[2] Recall: $\mathbb{E}(X(\omega)) = \int_\Omega X(\omega)\,\mathrm{d}\mathbb{P}(\omega)$.

calculations that the usual rules of differential and integral calculus are not suitable to handle stochastic integrals in a straightforward manner. To make sense of a stochastic integral in the way outlined above we have to define the choice of the evaluation rule upfront, i.e., the value of λ. Usually only two types of evaluation rules are widely used[3]: $\lambda = 0$ which leads to Itô's interpretation of the stochastic integral — and is the only possible choice such that the integral is a martingale — and $\lambda = \frac{1}{2}$ which leads to Stratonovich's interpretation of the stochastic integral — and is the only possible choice such that the rules of classical calculus remain valid.

Before you continue, make sure to answer the following questions:

Quiz: Section 4.5.1 – Integration with Respect to White Noise

Recall the essential steps of the above exposition and their key requirements:

Q1 Which of the following properties of the Riemann sums $\sum_{i=1}^{n} W_\xi(W_{t_i} - W_{t_{i-1}})$, with $\xi \in [t_{i-1}, t_i]$ are not valid in the limit $\max_i(t_i - t_{i-1})$?]

1. Independence of the sequence $\left(0 = t_0^{(m)} < t_1^{(m)} < \cdots < t_n^{(m)} = T\right)_{m \in \mathbb{N}}$ of partitions of $[0, T]$.
2. Independence of the choice of the evaluation point $\xi \in [t_{i-1}, t_i]$.

Q2 What additional conventions are set in stochastic integration theory to make-up for the roadblock/ roadblocks stated in Q1?

Q3 What are the values of $(t_i - t_{i-1})(W_{t_i} - W_{t_{i-1}})$ and $(W_{t_i} - W_{t_{i-1}})$ in probability as well as in expectation?

4.5.2 Introducing the 1D Itô & Stratonovich Stochastic Integral

Let W_t, $t \geq 0$, be a 1-dimensional Wiener process on a probability space $(\Omega, \mathcal{F}, \mathbb{P})$. Let $\mathcal{F}_t \subset \mathcal{F}$, $t \geq 0$, be an increasing family of σ-algebras in $\mathcal{F}$ (i.e., $\mathcal{F}_s \subset \mathcal{F}_t \subset \mathcal{F}$ if $s < t$) such that for all $t \geq 0$ it holds

- $\mathcal{A}(W_s : 0 \leq s \leq t) \subseteq \mathcal{F}_t$, and
- $\mathcal{A}(W_{t+s} - W_t : s \geq 0)$ is independent of $\mathcal{F}_t$.

[3] In [186], [187] and [260] a bunch of further rules varying from Itô's and Stratonovich's interpretation are considered.

One can take, for instance, the Wiener filtration $\mathcal{F}_t := \mathcal{A}(W_s : 0 \le s \le t)$ as such an increasing family and should informally think of $\mathcal{F}_t$ as "containing all information available at time t". $\mathcal{F}_t := \mathcal{A}(W_s : 0 \le s \le t)$ is the smallest family of σ-algebras with these properties. Here, $\mathcal{F}_t$ is generated by the Wiener process.

It is often necessary and desirable to augment $\mathcal{A}(W_s : 0 \le s \le t)$ with other events that are independent of $\mathcal{A}(W_s : 0 \le s < \infty)$, for instance, initial conditions. In the case of stochastic differential equations, we will usually take $\mathcal{F}_t := \mathcal{A}(W_s, X_0 : 0 \le s \le t)$, where X_0 is a random variable independent of $\mathcal{A}(W_s : 0 \le s < \infty)$.

The σ-algebra $\mathcal{A}(W_s : 0 \le s \le t)$ is called the *history* of the Wiener process up to (and including) time t. The σ-algebra $\mathcal{A}(W_{t+s} - W_t : s \ge 0)$ is the *future* of the Wiener process beyond time t.

A stochastic process $f(t,\omega)$ defined for $0 \le t \le T < \infty$ is called *adapted* with respect to $\mathcal{F}_t$ if for each $t \in [0,T]$, $f(t,\omega)$ is $\mathcal{F}_t$ measurable.

Stochastic processes $f(t,\omega)$ that are adapted to the Wiener filtration $\mathcal{F}_t$, are independent of the increments of the Wiener process $W_{t,\omega}$ "in the future". I.e., $f(t,\omega)$ is independent of $W_{t+s,\omega} - W_{t,\omega}$ for all $s > 0$. For instance, if $f(x)$ is an integrable deterministic function, then the functions $f(W_{t,\omega})$ and $\int_0^t f(W_{s,\omega})\mathrm{d}s$ are $\mathcal{F}_t$-adapted.

Definition 4.9 (The Class of Adapted Functions). We denote by $M_\omega^2[0,T]$ the class of $\mathcal{F}_t$-adapted stochastic processes $f(t,\omega)$ on the interval $[0,T]$ such that

$$\int_0^T \mathbb{E}\left(f^2(s,\omega)\right) \mathrm{d}s \;<\; \infty\,.$$

As motivated in Sec. 4.5.1, in order to give the limit of the Riemann sums for the stochastic integral meaning independent of an arbitrary choice of the evaluation points, we always have to clarify upfront which evaluation points we are considering. This is an additional rule for the stochastic integral. In principle, two such rules are commonly used:

- $\lambda = 0$: Evaluation at the start point of the partition intervals (Itô's interpretation).
- $\lambda = \frac{1}{2}$: Evaluation in the middle of the partition intervals (Stratonovich's interpretation).

We start with some first discussions on Itô's interpretation rule.

The Stochastic Itô Integral

Integration with respect to white noise is defined in this class $M_\omega^2[0,T]$ of stochastic processes adapted to the Wiener filtration. Itô's construction of the integral of a function $f(t,\omega) \in M^2_\omega[0,T]$ is similar to the λ-integral in Sec. 4.5.1 with $\lambda = 0$ and chooses the evaluation points at the start point of the partition intervals.

For any partition $0 = t_0 < t_1 < \cdots < t_n = T$, we form the *Itô sum*

$$S_n := \sum_{i=1}^{n} f(t_{i-1},\omega)\left(W_{t_i,\omega} - W_{t_{i-1},\omega}\right) . \tag{4.14}$$

Note, that the increment $W_{t_i,\omega} - W_{t_{i-1},\omega}$ is independent of $f(t_{i-1},\omega)$, because $f(t,\omega)$ is $\mathcal{F}_t$-adapted. One can show, see, e.g., [222], that for any sequence of partitions of the interval, such that $\max_i(t_i - t_{i-1}) \to 0$, the sequence $\{S_n(t,\omega)\}$ converges to the same limit, denoted

$$\int_0^T f(t,\omega)\mathrm{d}W_{t,\omega} \overset{\mathcal{P}}{:=} \lim_{\max_i(t_i - t_{i-1})\to 0} S_n \tag{4.15}$$

and called *Itô integral* of $f(t,\omega)$. Moreover, one can show that the convergence in (4.15) is uniform in t with probability one, i.e. on almost every trajectory of the Wiener process $W_{t,\omega}$, see, e.g., [222]. The Itô integral is a $\mathcal{F}_t$-adapted stochastic process in Ω. It takes different values on different realisations ω of the trajectories of the Wiener process. In particular, it is easy to show that for any integrable deterministic function f

$$\int_0^T f(t)\mathrm{d}W_t \sim \mathcal{N}\left(0, \int_0^T f(t)^2\mathrm{d}t\right)$$

holds.

The outstanding characteristic of the Itô integral is that it is a martingale: Among all λ-integrals in Sec. 4.5.1, Itô's interpretation is characterised by the fact that as a function of the upper limit, it is a martingale. This can be illustrated as follows (see [8], p. 61): By taking $X_t := \frac{1}{2}W_t^2 + (\lambda - \frac{1}{2})t$, then for $t \geq s$ we have almost surely that

$$\begin{aligned}
&\mathbb{E}(X_t \,|\, \mathcal{A}(X_u : u \leq s)) = \frac{1}{2}\mathbb{E}(W_t^2 \,|\, \mathcal{A}(\frac{1}{2}W_u^2 + (\lambda - \frac{1}{2})u : u \leq s)) + (\lambda - \frac{1}{2})t \\
&= \frac{1}{2}\mathbb{E}(\mathbb{E}(W_t^2 \,|\mathcal{A}(W_u : u \leq s)) \,|\, \mathcal{A}(\frac{1}{2}W_u^2 + (\lambda - \frac{1}{2})u : u \leq s)) + (\lambda - \frac{1}{2})t \\
&= \frac{1}{2}\mathbb{E}(\mathbb{E}(W_t^2 \,|\, \mathcal{A}(W_s : s \leq t)) \,|\, \mathcal{A}(\frac{1}{2}W_u^2 + (\lambda - \frac{1}{2})u : u \leq s)) + (\lambda - \frac{1}{2})t \\
&= \frac{1}{2}\mathbb{E}(t - s + W_s^2 \,|\, \mathcal{A}(\frac{1}{2}W_u^2 + (\lambda - \frac{1}{2})u : u \leq s)) + (\lambda - \frac{1}{2})t \\
&= \frac{1}{2}W_s^2 + \frac{1}{2}(t - s) + (\lambda - \frac{1}{2})t \\
&= X_s + \lambda(t - s) ,
\end{aligned}$$

where we have used the various properties of the conditional expectation and of a Wiener process. The process X_t is therefore a martingale[4], i.e.,

$$\mathbb{E}(X_t|\mathcal{A}(X_u : 0 \leq u \leq s)) \;=\; X_s \qquad \text{a.s.},$$

if and only if $\lambda = 0$, hence for Itô's choice of the evaluation points.

As we have already seen in Sec. 4.5.1

$$\int_0^T W_t \mathrm{d}t \;=\; \tfrac{1}{2}W_T^2 - \tfrac{1}{2}T\,.$$

Hence, Itô's integral leads to a result different from that implied by classical calculus. The correction term $-\frac{1}{2}T$ is essential for the martingale property of solutions of Itô integrals, though it also results in "new" rules of calculus. In particular, these new rules become apparent when discussing stochastic (Itô) differentials.

Stochastic Differentials & Itô's Formula

Given a scalar Wiener process W_t and a scalar at least twice differentiable function $g(x)$, what is $\mathrm{d}g(W_t)$? For this purpose, let us follow [224], p. 21, and consider the Taylor expansion of g

$$\begin{aligned} g(W_t + \mathrm{d}W_t) &= g(W_t) + g_x(W_t)\mathrm{d}W_t + \tfrac{1}{2}g_{xx}(W_t)(\mathrm{d}W_t)^2 + \ldots \\ \mathrm{d}g(W_t) &= g(W_t + \mathrm{d}W_t) - g(W_t) = g_x(W_t)\mathrm{d}W_t + \tfrac{1}{2}g_{xx}(W_t)(\mathrm{d}W_t)^2 + \ldots \end{aligned}$$

If W_t would be a deterministic function, the term $(\mathrm{d}W_t)^2$ could be neglected as a term of higher order. In the stochastic case, though,

$$\mathrm{d}W_t \;=\; W_{t+\mathrm{d}t} - W_t \qquad \Rightarrow \qquad (\mathrm{d}W_t)^2 \;=\; (W_{t+\mathrm{d}t} - W_t)^2$$

hold, and according to the properties of the Wiener process, we get with (4.11)

$$(\mathrm{d}W_t)^2 \overset{\mathcal{P}}{=} \mathbb{E}\left((\mathrm{d}W_t)^2\right) \;=\; \mathbb{E}\left((W_{t+\mathrm{d}t} - W_t)^2\right) \;=\; \mathrm{d}t\,.$$

Hence, the term $(\mathrm{d}W_t)^2$ is of first order in probability and cannot be neglected. In particular, we can formally set-up the following *multiplication table*

$$\mathrm{d}t \cdot \mathrm{d}t \;=\; 0\,, \qquad \mathrm{d}t \cdot \mathrm{d}W_t \;=\; 0\,, \qquad \mathrm{d}W_t \cdot \mathrm{d}W_t \;=\; \mathrm{d}t\,.$$

---

[4] To prove this martingale property for "arbitrary" integrands some effort has to be put into defining more precisely what conditions have to be imposed on the integrands. Illustratively, the integrands have to be such that for each time, there is no information needed that will be accessible only in a σ-algebra generated by some Wiener process in the future.

Thus, the formula

$$\mathrm{d}g(W_t) \;=\; g_x(W_t)\mathrm{d}W_t \;+\; \tfrac{1}{2}g_{xx}(W_t)\mathrm{d}t \qquad (4.16)$$

holds in the stochastic case. We will prove this rigorously via Itô's formula.

Below, we give the definition of a (Itô) stochastic differential, followed by citing a 1-dimensional version of Itô's formula in order to highlight the differences to classical differential and integral calculus.

Definition 4.10 (Stochastic Differential). Let X_t, $0 \le t \le T$, be a stochastic process such that for any $0 \le t_1 < t_2 \le T$

$$X_{t_2} \;-\; X_{t_1} \;=\; \int_{t_1}^{t_2} a(t)\mathrm{d}t \;+\; \int_{t_1}^{t_2} b(t)\mathrm{d}W_t\,,$$

where $\sqrt{|a|}, b \in M_\omega^2[0,T]$. Then, we say that X_t has a *stochastic differential* $\mathrm{d}X_t$ on $[0,T]$, given by

$$\mathrm{d}X_t \;=\; a(t)\mathrm{d}t \;+\; b(t)\mathrm{d}W_t\,.$$

Observe, that X_t is an $\mathcal{F}_t$-adapted stochastic function that belongs to $M_\omega^2[0,T]$.

Equation (4.16) and the chain rule of deterministic integral and differential calculus immediately imply Itô's celebrated formula (the analog to the chain rule in classical calculus). In its simplest form it reads as:

Lemma 4.11 (Itô's Formula (1D)). *Let $dX_t = a(t)dt + b(t)dW_t$ for $t \in [0,T]$ and $u : \mathbb{R} \times [0,T] \to \mathbb{R}$ be once continuously differentiable in $t \ge 0$ and twice continuously differentiable in $x \in \mathbb{R}$. Then, $u(X_t,t)$ has a stochastic differential given by*

$$\begin{aligned} du(X_t,t) \;&=\; \big(u_t(X_t,t) \,+\, a(t)u_x(X_t,t) + \tfrac{1}{2}b^2(t)u_{xx}(X_t,t)\big)\,dt \\ &\quad + b(t)u_x(X_t,t)dW_t\,, \end{aligned} \qquad (4.17)$$

Proof. See [100], pp. 80. □

This can be extended to functions $u : \mathbb{R}^n \to \mathbb{R}$, and leads us to the general formulation of *Itô's formula* involving 1-dimensional stochastic processes $X_t^{(1)}, \ldots, X_t^{(m)}$:

Theorem 4.6 (Itô's Formula (1D)). *Let $dX_t^{(i)} = a_i(t)dt + b_i(t)dW_t$, $i = 1,2,\ldots,m$, and let $u(x_1,\ldots,x_m,t)$ be once continuously differentiable in $t \ge$*

0 *and twice continuously differentiable in* $x = (x_1, \ldots, x_m) \in \mathbb{R}^m$. *Then,* $u(X_t^{(1)}, \ldots, X_t^{(m)})$ *has a stochastic differential given by*

$$du(X_t, t) = \left(u_t(X_t, t) + \sum_{i=1}^{m} a_i(t) u_{x_i}(X_t, t) + \frac{1}{2} \sum_{i,j=1}^{m} b_i(t) b_j(t) u_{x_i x_j}(X_t, t) \right) dt$$
$$+ \sum_{i=1}^{m} b_i(t) u_{x_i}(X_t, t) dW_t\,, \tag{4.18}$$

where $X_t = (X_t^{(1)}, \ldots, X_t^{(m)})^T$.

Proof. see [100], pp. 80 □

As a first example, we consider a very simplified model for financial markets, that can be seen as a crude approximation of the model given by Fischer Black and Myron Scholes in [28]: Let us assume, that the stock market is governed by regular events. Thus, it makes sense that the underlying stochastic differential equation may be a diffusion process like the geometric Brownian motion

$$\mathrm{d}u_t \;=\; a u_t \mathrm{d}t + b u_t \mathrm{d}W_t\,, \qquad u(0) = u_0$$

where a and b are real constants. Applying Itô's formula, with

$$\mathrm{d}X_t \;=\; (a - \tfrac{1}{2}b^2)\mathrm{d}t + b\mathrm{d}W_t\,, \qquad \text{and} \qquad u(x) \;=\; u_0 \exp(x)\,,$$

immediately proves that

$$u_t \;=\; u_0 \exp\left((a - \tfrac{1}{2}b^2)t + bW_t\right)$$

solves this linear stochastic differential equation.

In the financial context the parameter b is called volatility. The volatility is a simple but effective measure for what we in our daily life call "risk". For practical purposes, the model is valid only temporarily, as for instance, if the chief manager suddenly leaves or some new Stock regulations are implemented the parameters a and b may change, too. One way to actually see if the model is suitable, is to look at the quadratic variation of the logarithm of u_t. We have $\ln(u_t) = \ln(u_0) + (a - \frac{1}{2}b^2)t + bW_t$, hence, the quadratic variation of $\ln(u_t)$ at time t is

$$\langle \ln(u) \rangle_t \;:=\; \lim_{\max_i(t_i - t_{i-1}) \to 0} \sum_{i=1}^{n} (\ln(u_i) - \ln(u_{i-1}))^2 \;\overset{\mathcal{P}}{=}\; b^2 t\,.$$

It just means that we can approximate $b^2 t$ by $S := \sum_{k=1}^{n} (\ln(u_{k+1}) - \ln(u_k))^2$, where $0 = t_0 < t_1 < \cdots < t_n = t$. If S, as a function of time, follows a straight line, then our model is rather realistic and we can read off b^2 as the slope.

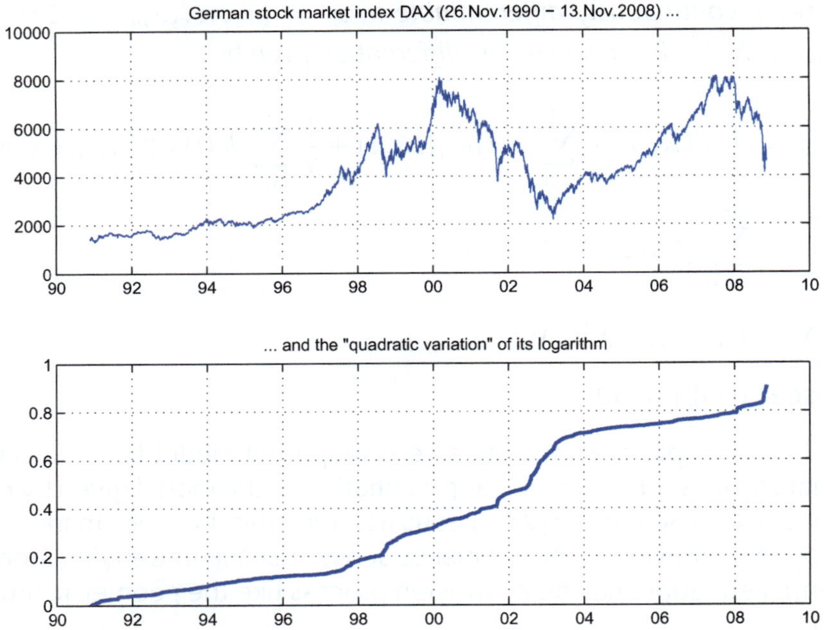

Figure 4.1. Values of the German stock market index from 26th of November 1990 until 13th of November 2008 (upper graph) and their corresponding quadratic variation (lower graph).

Figure 4.1 displays the values of the German stock market index DAX[5] over its corresponding quadratic variation (or, more precisely, the quadratic variation of the logarithmic model). We see that the model of the geometric Brownian motion is applicable for rather long time periods. For instance, one observes a slope $b^2 \approx 0.0038$ from January 2004 until December 2007. Though, in the case of "sudden" crashes of the DAX the slope of the quadratic variation changes dramatically indicating a tremendous increase of volatility and risk in the market.

We proceed, by studying another interpretation of stochastic integrals introduced by Ruslan Leont'evich Stratonovich (31.05.1930 – 13.01.1997), see [240]. Stratonovich's integral does not lead to martingales, though it allows us to keep the usual rules of differential and integral calculus.

5 source: http://de.finance.yahoo.com/

The Stochastic Stratonovich Integral

Another construction of a stochastic integral is similar to the λ-integral in Sec. 4.5.1 with $\lambda = \frac{1}{2}$. I.e., the choice of the evaluation points in according to a midpoint rule. For functions $f(X_t, t, \omega) \in M_\omega^2[0,T]$ and any partition $0 = t_0 < t_1 < \cdots < t_n = T$, the *stochastic Stratonovich integral* is given as

$$\int_0^T f(X_t, t) \circ \mathrm{d}W_t \overset{\mathcal{P}}{:=} \lim_{\max_i(t_i - t_{i-1}) \to 0} \sum_{i=1}^n f\left(\tfrac{1}{2}\left(X_{t_i} + X_{t_{i-1}}\right), t_{i-1}\right)\left(W_{t_i} - W_{t_{i-1}}\right), \tag{4.19}$$

where W_t is the 1-dimensional (standard) Wiener process, cf. [240]. The sum on the right hand side of equation (4.19) is called *Stratonovich sum.*

Note, that we write an "$\circ$" to indicate that the integral considered is of Stratonovich type, whereas integrals without "$\circ$" obey the Itô interpretation. Other authors use a "δ" instead of "$\circ$" to assign the Stratonovich increment "δW_t" contrary to the Itô increment "$\mathrm{d}W_t$", see, e.g., [170].

The Stratonovich integral is related to the Itô integral by the following correspondence formula, cf. [260].

Theorem 4.7 (The Wong-Zakai Correspondence). *Let X_t be a stochastic process with stochastic differential $dX_t = a(X_t, t)dt + b(X_t, t)dW_t$ and let $f(x,t) : \mathbb{R} \times [0,T] \to \mathbb{R}$ be continuously differentiable. Let $\sqrt{|a|}, b, f, f_x b \in M_\omega^2[0,T]$ and X_t, dX_t have a continuous realisation almost surely. Then, the Stratonovich and Itô interpretations can be converted into each other, as*

$$\int_0^T f(X_t, t) \circ dW_t = \int_0^T f(X_t, t) dW_t + \frac{1}{2}\int_0^T f_x(X_t, t) b(X_t, t) dt \qquad \text{a.s.} \tag{4.20}$$

in the sense that the left-hand side of (4.20) exists if and only if the right-hand side exists and they are equal.

Proof. Following [159], pp. 156, we show equation (4.20) by first considering each of the terms in the Stratonovich sums (4.19) separately. Applying the Mean Value Theorem we have

$$\begin{aligned} & f\left(\tfrac{1}{2}\left(X_{t_{j+1}} + X_{t_j}\right), t_j\right) - f\left(X_{t_j}, t_j\right) \\ &= \tfrac{1}{2} f_x\left(\tfrac{1}{2}\left((2-\theta_j)X_{t_{j+1}} + \theta_j X_{t_j}\right), t_j\right)\left(X_{t_{j+1}} - X_{t_j}\right) \end{aligned}$$

for some (random) $\theta_j \in [0,1]$ for $j = 1, \ldots, n$. Note, that $\theta_j = \theta_{j,\omega}$ depends on $\omega \in \Omega$. Since X_t has a stochastic differential

$$\begin{aligned} \Delta X_{t_j} &= X_{t_{j+1}} - X_{t_j} \\ &= a\left(X_{t_j}, t_j\right)\Delta t_j + b\left(X_{t_j}, t_j\right)\Delta W_{t_j} + \text{"terms of higher order"}, \end{aligned}$$

where $\Delta t_j = t_{j+1} - t_j$ and $\Delta W_{t_j} = W_{t_{j+1}} - W_{t_j}$ for $j = 1, \ldots, n$. Thus, each term in the Stratonovich sum equals

$$\begin{aligned}
& f\left(\tfrac{1}{2}\left(X_{t_{j+1}} + X_{t_j}\right), t_j\right) \Delta W_{t_j} \\
= \;& f(X_{t_j}, t_j)\Delta W_{t_j} \;+\; \tfrac{1}{2} f_x\left(\tfrac{1}{2}\left((2-\theta_j)X_{t_{j+1}} + \theta_j X_{t_j}\right), t_j\right) \Delta X_{t_j} \Delta W_{t_j} \\
= \;& f(X_{t_j}, t_j)\Delta W_{t_j} \;+\; \tfrac{1}{2} f_x\left(\tfrac{1}{2}\left((2-\theta_j)X_{t_{j+1}} + \theta_j X_{t_j}\right), t_j\right) b\left(X_{t_j}, t_j\right) \left(\Delta W_{t_j}\right)^2 \\
& + \tfrac{1}{2} f_x\left(\tfrac{1}{2}\left((2-\theta_j)X_{t_{j+1}} + \theta_j X_{t_j}\right), t_j\right) a\left(X_{t_j}, t_j\right) \Delta t_j \Delta W_{t_j} \\
& + \text{"terms of higher order"}\,.
\end{aligned}$$

Taking into account the continuity of the coefficients and the facts

$$\mathbb{E}\left(\left(\Delta W_{t_j}\right)^2\right) \;=\; \Delta t_j \qquad \text{and} \qquad \mathbb{E}\left(\Delta t_j \Delta W_{t_j}\right) \;=\; 0\,,$$

we obtain (4.20) in the mean square limit and thus in probability, as L^2-convergence implies convergence in probability. □

Hence, once the Itô integral is defined, and with it the right hand side of (4.20), the left hand side of (4.20) — the Stratonovich integral — is given, too.

In particular, the Itô and Stratonovich interpretation coincide if $f(x,t) = f(t)$ is independent of the state variable. Moreover, when the stochastic process X_t satisfies the Itô stochastic differential equation

$$\mathrm{d}X_t \;=\; a(X_t, t)\mathrm{d}t \;+\; b(X_t, t)\mathrm{d}W_t\,,$$

i.e., the respective Itô integral equation

$$X_t - X_0 \;=\; \int_0^T a(X_t, t)\mathrm{d}t \;+\; \int_0^T b(X_t, t)\mathrm{d}W_t\,,$$

we obtain from (4.20) with $f(X_t, t) \;\equiv\; b(X_t, t)$ that X_t is a solution of the Stratonovich stochastic differential equation

$$\mathrm{d}X_t \;=\; \underline{a}(X_t, t)\mathrm{d}t \;+\; b(X_t, t) \circ \mathrm{d}W_t\,,$$

where the modified drift term $\underline{a}(X_t, t)$ is defined by

$$\underline{a}(x,t) \;:=\; a(x,t) \;-\; \tfrac{1}{2} b(x,t) \frac{\partial}{\partial x} b(x,t)\,.$$

Note, that for a stochastic process X_t the diffusion terms $b(X_t, t)$ of its Itô and Stratonovich interpretation are equal, whereas the drift terms differ in these two interpretations in general (they are equal if $b(X_t, t) = b(t)$ is independent of X_t).

Example 4.12 (Itô vs. Stratonovich Solutions). Consider the linear stochastic differential equation

$$\mathrm{d}X_t \;=\; aX_t\mathrm{d}t \,+\, bX_t\mathrm{d}W_t\,, \tag{4.21}$$

where $a, b \in \mathbb{R} \setminus \{0\}$. Interpreted as an Itô equation the geometric Brownian motion

$$X_t^{(I)} \;=\; X_0 \exp\left(\left(a - \tfrac{1}{2}b^2\right)t + bW_t\right)$$

solves (4.21) for the initial value $X_0 \in \mathbb{R}$. Next consider (4.21) to be interpreted in the sense of Stratonovich, i.e.,

$$\mathrm{d}X_t \;=\; aX_t\mathrm{d}t \,+\, bX_t \circ \mathrm{d}W_t\,.$$

Applying the Wong-Zakai correspondence formula (4.20), this Stratonovich interpretation becomes the Itô differential equation

$$\mathrm{d}X_t \;=\; \left(a + \tfrac{1}{2}b^2\right)X_t\mathrm{d}t \,+\, bX_t\mathrm{d}W_t$$

which is solved (applying Itô's calculus) by

$$X_t^{(S)} \;=\; X_0 \exp\left(at + bW_t\right)\;;$$

exactly as we would have assumed by applying the rules of classical calculus to solve (4.21). Note, that both $X_t^{(I)}$ and $X_t^{(S)}$ do not change sign due to the exponential function.

Figure 4.2 displays and compares some simulations for the Itô solution $X_t^{(I)}$ and the Stratonovich solution $X_t^{(S)}$ of (4.21).

The major reason for using the Stratonovich integral is that it obeys — contrary to the Itô integral — the rules of classical calculus, which is extremely useful when dealing with stochastic differential equations on manifolds where one needs to perform transformations of the process and coordinate-changes regularly, see, e.g., [94], [170].

Let us illustrate this property of the Stratonovich integral by the following theorem, cf. [227], pp. 71. For simplicity of notation we restrict ourselves to functions of the (standard) Wiener process.

Theorem 4.8 (Stratonovich Calculus). *Let $F'(x) = f(x)$ and $G'(x) = g(x)$ be continuous functions. The Stratonovich integral satisfies the rules of deterministic differential and integral calculus:*

1. *The fundamental theorem of differential and integral calculus:*

$$\int_0^T f(W_s) \circ dW_s \;=\; F(W_s) \;-\; F(W_0)\,.$$

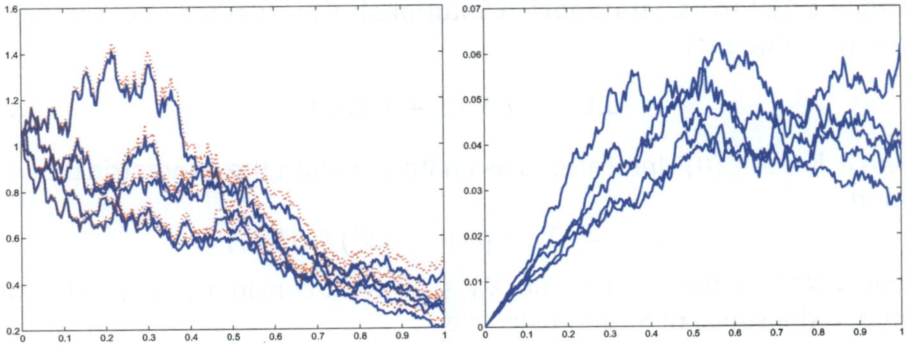

(a) (b)

Figure 4.2. Five simulation runs of (a) the Itô and Stratonovich solution of (4.21) and (b) their difference $X_t^{(S)} - X_t^{(I)}$, for $X_0 = 1$, $a = -1$, $b = 0.5$ and different random events $\omega \in \Omega$. In (a) the Itô solution $X_t^{(I)}$ is drawn by straight lines and the Stratonovich solution $X_t^{(S)}$ by dotted lines.

2. *The integration by parts formula:*

$$\int_0^T (F(W_t)g(W_t) + f(W_t)G(t)) \circ dW_s \;=\; F(W_t)G(W_T) - F(W_0)G(W_0)\,.$$

3. *The chain rule:*

$$\int_0^T f(G(W_t))g(W_t) \circ dW_s \;=\; F(G(W_T)) \;-\; F(G(W_0))\,.$$

Proof. We prove the first assertion (the remaining two follow analogously). For an arbitrary function $f \in \mathcal{C}^1(\mathbb{R})$ of a (standard) Wiener process W_t the Wong-Zakai correspondence formula (4.20) reads as

$$\int_0^T f(W_t) \circ \mathrm{d}W_t \;=\; \int_0^T f(W_t)\mathrm{d}W_t + \tfrac{1}{2}\int_0^T f'(W_t)\mathrm{d}t\,. \qquad (4.22)$$

Now, let F be the anti-derivative of f, i.e., $F'(x) = f(x)$ and $F''(x) = f'(x)$. Applying Itô's formula (4.17) to the thus transformed process $Y_t := F(W_t)$, we obtain

$$F(W_T) \;-\; F(W_0) \;=\; \int_0^T f(W_t)\mathrm{d}W_t + \tfrac{1}{2}\int_0^T f'(W_t)\mathrm{d}t\,,$$

which is the left hand side of (4.22). Hence, we get, that the Stratonovich integral reads as

$$\int_0^T f(W_t) \circ \mathrm{d}W_t \;=\; F(W_T) \;-\; F(W_0)\,,$$

as in classical calculus.

The remaining assertions will be proven in the problems. □

A further consequence of the Wong-Zakai correspondence formula is that the Stratonovich integral is not a martingale (which can be checked easily).

Before you continue, make sure to answer the following questions:

Quiz: Section 4.5.2 – Introducing Itô's & Stratonovich's Stochastic Integral (1D)

Q1 What is the definition of an adapted function with respect to a filtration $\mathcal{F}_t$.

Q2 How are the Itô and Stratonovich sums defined?

Q3 Give Itô's formula in one dimension and use it to compute the stochastic (Itô) differential of $\exp(at+bW_t)$ with respect to the time variable t $(a, b > 0)$.

Q4 Is there a relationship between the Itô and the Stratonovich stochastic integral? If so, how are they connected?

Q5 Under which conditions do Itô's and Stratonovich's interpretation of the stochastic integral lead to the same result?

Q6 Stratonovich's interpretation of the stochastic integral preserves the usual rules of calculus – give an illustrative example (e.g., the chain rule). What would Itô's interpretation give in your example?

4.6 Chapter's Summary

In this chapter two additional solution concepts for random differential equations were discussed: the $\mathbb{P}$-solutions and the $q.m.$-solutions. A $\mathbb{P}$-solution is a stochastic process that fulfills a given random differential equation with probability one, whereas a $q.m.$-solution is a second order stochastic process that again fulfills the given random differential equation with probability one such that the right hand side of this random differential equation is equivalent to a square integrable process. With respect to the set-up of $q.m.$-solutions essential characteristics of second order processes were discussed together with important results from mean-square calculus.

Besides introducing them and studying their properties we emphasized their connections to path-wise solutions in order to gain better insights into the latter notion of a solution. As we have seen, path-wise continuous differentiability can be viewed as the outstanding attribute of path-wise solutions which makes them accessible to the theory of (deterministic) ordinary differential equations. Under additional regularity assumptions, equivalence of these three concepts can be established (cf. Fig. 4.3).

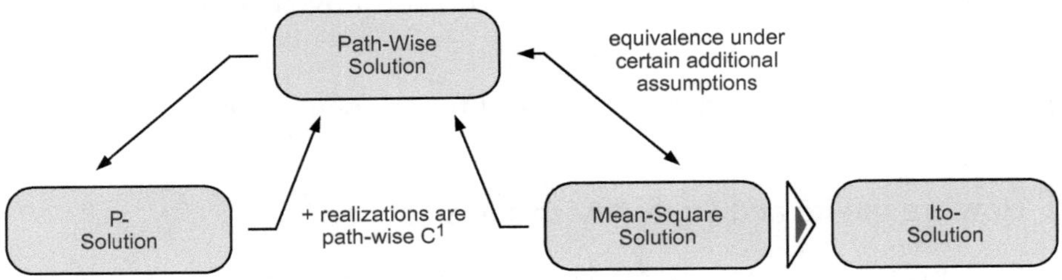

Figure 4.3. Interconnections between the three different solution concepts for random differential equations and their relation to solutions of Itô stochastic differential equations.

We had a brief excursion into Itô's stochastic integration theory highlighting the essential differences of integration with respect to white noise compared to integration with respect to a usual function. Utilizing the boundedness of the Wiener process's second variation we saw that Itô's calculus is actually a mean-square calculus and thus it is accessible via results from the theory of differential equations on Banach-spaces to the space $\mathcal{L}_d^2$.

Finally, the interpretation of the stochastic integral in the sense of Stratonovich allowed us to re-gain the classical rules of calculus for stochastic integration. Here, the Wong-Zakai correspondence gives an equivalence between Itô and Stratonovich stochastic integrals (under certain mild conditions). We refer to the literature for in depth discussions of stochastic integration with respect to white noise.

Problems

Classification: ☼ easy, ◷ easy with longer calculations, ☆ a little bit difficult, ⧉ challenging.

Exercise 4.13. [☼] **Existence of $\mathbb{P}$- or $q.m.$-Solutions – Part 1**
Let $a(\omega)$ be an exponentially distributed random variable. In example 3.6, we

have seen that the random initial value problem

$$\dot{X}_t = -a(\omega)X_t^2, \qquad X_0 = 1,$$

does not possess a path-wise solution. Discuss the existence (and uniqueness) of $\mathbb{P}$- or $q.m.$-solutions for this problem.

Exercise 4.14. [☼] **Existence of $\mathbb{P}$- or $q.m.$-Solutions – Part 2**
In example 3.6, we have seen that the differential equation with random initial conditions

$$\dot{x} = \sqrt{|x|}, \qquad \text{with the initial condition } X_0(\omega) = 0,$$

does not possess a path-wise solution. Discuss the existence (and uniqueness) of $\mathbb{P}$- or $q.m.$-solutions for this problem.

Exercise 4.15. [☼] **Properties of Second Order Processes – Part 1**
Prove properties 2, 3, 5 and 6. I.e.:

- Property 2: Let $X_{m,t}$ be a $q.m.$-continuous second order processes such that $q.m.\lim_{m\to\infty} X_{m,t} \hat{=} X_t$ holds uniformly for $t \in [a,b]$. Then, the limit process X_t is $q.m.$-continuous on $[a,b]$.

- Property 3: X_t has a $q.m.$-derivative Y_t if and only if the partial derivative $\frac{\partial^2}{\partial t \partial s}\Gamma_X(t,s)$ exists on $I \times I$. In this case $\Gamma_Y(t,s) = \frac{\partial^2}{\partial t \partial s}\Gamma_X(t,s)$. If the derivative exists for every element $(t,s) \in I \times I$ then it exists on $I \times I$, too.

- Property 5: Let $\Gamma_X(t,s)$ be continuous on $I \times I$, $g : I \to \mathbb{R}$ be a real function and assume that the Riemann-integral $V = \int_a^b \int_a^b g(t)g(s)\Gamma_X(t,s)\mathrm{d}s\mathrm{d}t$ exists. Then, the Riemann-$q.m.$-integral $Z := \int_a^b X_t g(t)\mathrm{d}t$ exists and $\mathbb{E}((Z^{(l)})^2) = V^{(l)}$ holds for every component $l = 1, 2, \ldots, d$.

- Property 6: If X_t is $q.m.$-continuous for all $t \geq t_0$, then $Z_t := \int_{t_0}^t X_\tau \mathrm{d}\tau$ is $q.m.$-differentiable and it holds that X_t is the $q.m.$-derivative of Z_t, i.e., $\dot{Z}_t \hat{=} X_t$.

Exercise 4.16. [☼] **Properties of Second Order Processes – Part 2**
Prove properties 7, 9, 10 and 11. I.e.:

- Property 7: Let the $q.m.$-derivative $\dot{X}_t$ of X_t be $q.m.$-continuous on I and $[a,b] \subset I$, then $\int_a^b \dot{X}\mathrm{d}t \hat{=} X_b - X_a$.

- Property 9: Let the random vectors $X_k \in L_d^2$ be normally distributed and assume their limit $q.m.\lim_{k\to\infty} X_k \hat{=} X$ exists. Then, X is normally distributed, too.

- Property 10: Let X_t be a Gaussian process such that $\Gamma_X(t,s)$ fulfills the assumptions in 3 and 5, respectively. Then its $q.m.$-derivative $\dot{X}_t$ and its Riemann-$q.m.$-integral $Z_t = \int_{t_0}^t X_\tau \mathrm{d}\tau$ are Gaussian processes, too such that $\Gamma_{\dot{X}}(t,s) = \frac{\partial^2}{\partial t \partial s}\Gamma_X(t,s)$ and $\Gamma_Z(t,s) = \int_{t_0}^t \int_{t_0}^s \Gamma_X(\tau,\sigma)\mathrm{d}\sigma \mathrm{d}\tau$, respectively.

- Property 11: Let $a(t)$ be a differentiable real function on I and assume that X_t has a $q.m.$-derivative $\dot{X}_t$ on I. Then, the $q.m.$-derivative of $a(t)\cdot X_t$ equals $\dot{a}(t)X_t + a(t)\dot{X}_t$. Recall, $\dot{a}(t)$ is the usual derivative with respect to t, whereas $\dot{X}_t$ has to be interpreted in the sense of a $q.m.$-derivative.

Exercise 4.17. [☼] Properties of Second Order Processes – Part 3
Prove properties 13, 14, 15 and 16. I.e.:

- Property 13: Let X_t be both $q.m.$-continuous as well as path-wise continuous. Then its $q.m.$-integral $\int_{t_0}^t X_\tau \mathrm{d}\tau$ and its path-wise integral $\int_{t_0}^t X_\tau \mathrm{d}\tau$ exist for $t_0, t \in I$, and are equivalent.

- Property 14: Let $\{X_t^k\}_{k\in\mathbb{N}}$ be a sequence of second order stochastic processes. Then there is a limit (stochastic) process X such that $q.m.\lim_{k\to\infty} X_t^k \hat{=} X_t$, $t \in I$, if and only if for each $t \in I$ and each $l \in \mathbb{N}$ the limit $\lim_{i,k} \mathbb{E}\left(X_t^{i(l)} X_t^{k(l)}\right)$ exists. Then, it holds that

$$\lim_{k\to\infty} \mathbb{E}\left(X_{t_1}^k \left(X_{t_2}^k\right)^T\right) = \mathbb{E}\left(X_{t_1} X_{t_2}^T\right) .$$

- Property 15: Assume $X_k \in L_d^2$, $X_k \hat{\geq} 0$ and $q.m.\lim_{k\to\infty} X_k = X$ for each $k = 1, 2, \ldots$. Then $X \hat{\geq} 0$ holds, too.

- Property 16: Let X_t be a stationary process the $q.m.$-derivative $\dot{X}_t$ of which exists. Then, $\mathbb{E}(X_t^T \dot{X}_t) = 0$.

Exercise 4.18. [☼] Some Applications of Itô's Formula

1. Use Itô's formula with the function $f(x) = x^3$ to show that $\int_0^t W_s^2 \mathrm{d}W_s = \frac{1}{3}W_t^3 - \int_0^t W_s \mathrm{d}s$.

2. Use Itô's formula to solve the following stochastic differential equation for the spot rate of interest:

$$\mathrm{d}r_t = a(b - r_t)\mathrm{d}t + \sigma \mathrm{d}W_t ,$$

   where $a, b > 0$ and $\sigma \in \mathbb{R}$.

Exercise 4.19. [☼] An Application of Itô's Calculus
Let $f : [a, b] \to \mathbb{R}$ be a deterministic function in $L^2[a, b]$, and $X_t := X_a + \int_a^t f(s)\mathrm{d}W_s$, with $a \leq t \leq b$. Show that the following holds:

$$\int_a^b f(t)X_t\mathrm{d}W_t = \tfrac{1}{2}\left(X_b^2 - X_a^2 + \int_a^b f(t)^2\mathrm{d}t\right) .$$

Exercise 4.20. [☼] Stratonovich's Integral is not a Martingale
Give an illustrative example that shows that the Stratonovich stochastic integral does not lead to a martingale in general.

Chapter 5

RDEs in Science & Engineering: Randomly Perturbed Flow Problems

In this chapter, we discuss flow problems in detail, in particular incompressible flow. These problems represent an important class of applications in computational science and engineering. The various possible random effects in the model, the geometry, the boundary conditions, or the parameters may be generalised to other flow scenarios, involving also coupled scenarios such as fluid-structure interaction or biofilm growth.

5.1 Key Concepts

Random effects certainly occur in continuum mechanical problems related to structural dynamics as described in Chap. 2 but are also present in fluid flows described by diffusive differential equations.

A first illustrative example for gas diffusion is given as follows:

Example 5.1 (Random Diffusion of a Radioactive Gas, cf. [202]). Following [202], radioactive gas transport is diffusion from contaminated ground into the atmosphere. We assume the ground and the atmosphere to be semi-infinite media with $x = 0$ as the boundary. Then the density $\rho(x,t)$ of the radioactive gas in the air is governed by the relations:

$$
\begin{aligned}
\frac{\partial \rho(x,t)}{\partial t} &= D\frac{\partial^2 \rho(x,t)}{\partial x^2} - k\rho(x,t)\,, \\
\left.\frac{\partial \rho(x,t)}{\partial x}\right|_{x=0} &= -\chi(t)\,, \qquad \rho(\infty,t) < \infty\,, \\
\rho(x,0) &= 0 \qquad x \geq 0\,,
\end{aligned}
$$

where $D = \text{const.}$ is the diffusion coefficient depending on the properties of the medium, $k = \text{const.}$, and χ is a continuous stochastic process.

As in the case of structural dynamics, PDE-based approaches are an important class of models to describe certain flow phenomena in science and engineering with spatial resolution. We restrict the detailed discussion of flow problems to incompressible and laminar flow with Newtonian fluids. To describe such flows mathematically, the incompressible *Navier-Stokes equations* (NSE) are typically used. Details on various aspects of the Navier-Stokes equations can be found in [86, 118, 119], e.g. Since this topic is a very broad field, we restrict the discussions to the aspects relevant for random effects.

When reading this chapter note the answers to the following questions

1. What do the incompressible Navier-Stokes equations look like?
2. What are basic approaches to solve such flow problems numerically?
3. Which type of boundary conditions are used?
4. Where do random effects enter into the scene?

as well as the following key concepts

1. Derivation of the Navier-Stokes equations
2. Random effects for incompressible flow scenarios

This chapter is structured as follows: We present the derivation and discretisation of the deterministic incompressible Navier-Stokes equations in Sec. 5.2 and 5.3, respectively. In Sec. 5.4, some possible random effects are discussed. We briefly extend the corresponding insights to other flow problems in Sec. 5.5. Finally, Section 5.6 wraps up the contents of this chapter.

5.2 Derivation of the Deterministic Navier-Stokes Equations

Assuming that the hypothesis of continuous distribution of material holds, we take the Eulerian point of view on a fluid (see Sec. 2.2.1), i.e. a liquid or gasiform material that may easily be deformed in contrast to solid bodies. In the following, we briefly derive the Navier-Stokes equations for incompressible flows, a system of partial differential equations. The goal is a mathematical description of unsteady fluid dynamics in a certain domain $\Omega \subset \mathbb{R}^d$, $d = 2, 3$, with boundary $\Gamma := \partial\Omega$ for a time interval $[t_0, T] \subset \mathbb{R}_0^+$. We use Reynolds' transport theorem as the tool for deriving the continuity equation and the momentum equations, briefly present the most important boundary conditions and finally summarise the incompressible Navier-Stokes equations in a compact form.

Reynolds' transport theorem describes the total time derivative of an integral of a time-dependent function $f \in \mathcal{C}^1(\Omega \times [t_0, T], \mathbb{R})$ over a time-dependent domain. For a continuously differentiable velocity of a fluid $\mathbf{u}(x, t) : \Omega \times [t_0; T] \to \mathbb{R}^d$, and for a constant volume Ω_c that matches the time-dependent volume Ω_t at time t, Reynolds' transport theorem states

$$\frac{d}{dt}\int_{\Omega_t} f(x,t)\,dx = \int_{\Omega_t\equiv\Omega_c}\frac{\partial f(x,t)}{\partial t}\,dx + \int_{\Omega_t\equiv\Omega_c}\left[\nabla f\cdot \mathbf{u} + f\,(\nabla\cdot\mathbf{u})\right]\,dx \tag{5.1}$$

$$= \int_{\Omega_t\equiv\Omega_c}\frac{\partial f(x,t)}{\partial t}\,dx + \int_{\Omega_t\equiv\Omega_c}\mathrm{div}\left(\ f(x,t)\,\mathbf{u}(x,t)\ \right)\,dx\,, \tag{5.2}$$

where div is the divergence operator. Several ways of proving this theorem can be found in [226], e.g. For the ease of notation, we do not distinguish between Ω_c and Ω_t in the formulas below.

The right-hand side of Eq. (5.2) consists of two parts: the convective term $\int_{\Omega_t}\mathrm{div}\left(\ f(x,t)\,\mathbf{u}(x,t)\ \right)dx$ corresponds to the variation of the integral with respect to the convective transport of the volume Ω_t, whereas $\int_{\Omega_t}\frac{\partial f(x,t)}{\partial t}\,dx$ represents the change of the integral over time with a fixed volume $\Omega_t\equiv\Omega_c$.

Equation (5.1) is still valid for tensor instead of scalar quantities f: We just have to replace the multiplication appearing in the divergence form (5.2) by a tensor product. The validity of this formula can directly been shown using the double contraction ":" known from tensor calculus (cf. [226]):

$$\begin{aligned}
\mathrm{div}\,(f\otimes\mathbf{u}) &= \nabla\,(f\otimes\mathbf{u}) : \underline{\underline{\mathrm{Id}}} = \left(\frac{\partial\,(f\otimes\mathbf{u})}{\partial x_k}\otimes e^k\right) : \underline{\underline{\mathrm{Id}}}\\
&= \left(\frac{\partial\,(f)}{\partial x_k}\otimes\mathbf{u}\otimes e^k\right) : \underline{\underline{\mathrm{Id}}} + \left(f\otimes\frac{\partial\,(\mathbf{u})}{\partial x_k}\otimes e^k\right) : \underline{\underline{\mathrm{Id}}}\\
&= \frac{\partial\,(f)}{\partial x_k}\otimes u_i e^i\otimes e^k : \delta_{lm}e^l\otimes e^m + f\otimes\frac{\partial\,(u_i e^i)}{\partial x_k}\otimes e^k : \delta_{lm}e^l\otimes e^m\\
&= \frac{\partial\,(f)}{\partial x_k}u_k \;+\; f\frac{\partial u_k}{\partial x_k} = \nabla f\cdot\mathbf{u} \;+\; f\,(\nabla\cdot\mathbf{u})\,,
\end{aligned}$$

where $\underline{\underline{\mathrm{Id}}}$ denotes the identity tensor of second order and where we use the Einstein summation convention for indices. Hence, Reynolds' transport theorem holds for tensor integrands f of arbitrary order:

$$\frac{d}{dt}\int_{\Omega_t} f(x,t)\,dx = \int_{\Omega_t}\frac{\partial f(x,t)}{\partial t}\,dx + \int_{\Omega_t}\mathrm{div}\,(f(x,t)\otimes\mathbf{u}(x,t))\,dx\,. \tag{5.3}$$

We now use Reynolds' transport theorem to derive the continuity equation. A basic assumption in fluid dynamics is the conservation of mass. In a fixed volume, the total mass must not change throughout a change in the state of the fluid. Mathematically, this can be represented via

$$\frac{d}{dt}\left(\int_{\Omega_t}\rho(x,t)\,dx\right) = 0, \tag{5.4}$$

where the density of the fluid is denoted by $\rho \in \mathbb{R}^+$. Applying Reynolds' transport theorem (5.2) to the left-hand side of Eq. (5.4) gives

$$0 = \int_{\Omega_t} \frac{\partial \rho(x,t)}{\partial t} \, dx + \int_{\Omega_t} \mathsf{div}\,(\rho(x,t)\mathbf{u}(x,t)) \; dx \,, \tag{5.5}$$

which holds for every arbitrary partial volume Ω_t. Applying the fundamental lemma of variational calculus, the integrand itself has to vanish:

$$0 = \frac{\partial \rho}{\partial t} + \mathsf{div}\,(\rho \mathbf{u}) = \frac{d}{dt}\rho + \rho(\nabla \cdot \mathbf{u}). \tag{5.6}$$

Eq. 5.6 is usually called the continuity equation. For incompressible flows, the density is time-independent. Thus, the total time derivative of the density $\frac{d}{dt}\rho$ vanishes for a specific location, and the continuity equation (5.6) reads

$$\nabla \cdot \mathbf{u} = 0. \tag{5.7}$$

The second important principle of conservation for fluid flow is the conservation of momentum. Applying Newton's second law of motion, the change of total momentum has to be equal to the sum of the applied external forces:

$$\frac{d}{dt}\left(\int_{\Omega_t} \rho(x,t)\mathbf{u}(x,t) \right) \, dx = \sum_j F_{a,j} \,, \tag{5.8}$$

where $F_{a,j}$ are either volume forces (denoted by $\mathbf{g}$), i.e. forces acting beyond fluid surfaces directly on the interior of the domain such as gravitation or electromagnetic forces, or surface forces, i.e. forces acting only on the surface of the fluid. The latter can be modelled by a stress tensor $\underline{\underline{\sigma}}$ of second order (denoted by double underlines). In fluid flows, no real stresses appear, though inter-molecular momentum exchange may be interpreted as stress.

To prepare the next step, we apply the divergence theorem for the external forces as a sum of surface and volume contributions,

$$\begin{aligned} \sum_j F_{a,j} &= \int_{\Omega_t} \rho(x,t)\, \mathbf{g}(x,t)\, dx + \int_{\partial\Omega_t} \underline{\underline{\sigma}}(x,t) \cdot \mathbf{n}\, ds \\ &= \int_{\Omega_t} \rho(x,t)\, \mathbf{g}(x,t)\, dx + \int_{\Omega_t} \mathsf{div}(\underline{\underline{\sigma}}(x,t))\, dx, \end{aligned} \tag{5.9}$$

where the outer normal vector of the boundary of the volume is represented by $\mathbf{n}$.

Now, Reynolds' transport theorem (5.3) in tensor form is applied to the temporal change of momentum in Eq. (5.8). The vector $\rho\mathbf{u}$ being a tensor of first order, we have

$$\frac{d}{dt}\left(\int_{\Omega_t}\rho(x,t)\mathbf{u}(x,t)\,dx\right)=\int_{\Omega_t}\left[\frac{\partial(\rho\mathbf{u})}{\partial t}(x,t)+\mathrm{div}\,(\rho\mathbf{u}\otimes\mathbf{u})\,(x,t)\right]\,dx. \tag{5.10}$$

Using Eqs. (5.8), (5.9), and (5.10) for the balance of momentum in the integral form, we get

$$\int_{\Omega_t}\left[\frac{\partial(\rho\mathbf{u})}{\partial t}(x,t)+\mathrm{div}\,(\rho\mathbf{u}\otimes\mathbf{u})\,(x,t)\right]\,dx=\int_{\Omega_t}\left[\rho(x,t)\,\mathbf{g}(x,t)+\mathrm{div}(\underline{\underline{\sigma}}(x,t))\right]\,dx. \tag{5.11}$$

Eq. (5.11) is valid for arbitrary volumes and, under the assumption of sufficient smoothness, gives

$$\frac{\partial(\rho\mathbf{u})}{\partial t}(x,t)+\mathrm{div}\,(\rho\mathbf{u}\otimes\mathbf{u})\,(x,t)=\rho\,\mathbf{g}(x,t)\;+\;\mathrm{div}(\underline{\underline{\sigma}}(x,t)). \tag{5.12}$$

Applying the divergence of tensors

$$\begin{aligned}\mathrm{div}\,(\rho\mathbf{u}\otimes\mathbf{u}) &= \nabla(\rho\mathbf{u}\otimes\mathbf{u}):\underline{\underline{\mathrm{Id}}} = \frac{\partial\left(\rho u_i e^i\otimes u_j e^j\right)}{\partial x_k}\otimes e^k:\delta_{lm}e^l\otimes e^m\\ &= \frac{\partial\rho}{\partial x_k}u_i e^i u_k\;+\;\rho\underbrace{\frac{\partial u_i}{\partial x_k}e^i u_k}_{(\mathbf{u}\cdot\nabla)\mathbf{u}}\;+\rho\underbrace{\frac{\partial u_k}{\partial x_k}}_{\nabla\cdot\mathbf{u}}u_i e^i\end{aligned}$$

and the continuity equation (5.7), we finally obtain the momentum equations for incompressible flow in divergence form

$$\rho\frac{\partial\mathbf{u}}{\partial t}(x,t)+\rho\,(\mathbf{u}\cdot\nabla)\,\mathbf{u}(x,t)=\rho\,\mathbf{g}(x,t)\;+\;\mathrm{div}(\underline{\underline{\sigma}}(x,t)). \tag{5.13}$$

All considerations above are independent of the specific fluid material. This information comes now into play via the definition of the stress tensor $\underline{\underline{\sigma}}$ which models the surface forces exerted on the fluid. In the following, we will always assume Newtonian fluids, i.e. fluids that fulfil two assumptions: linearity and isotropy. The former represents a linear dependency of the stress tensor on the velocity gradient $\nabla\mathbf{u}$ (cf. [168]), while the latter describes the fact that all stress forces in the fluid are acting independently of the specific direction in space. Typical examples for Newtonian fluids are water or air. Restriction to Newton fluids results in the following form of the stress tensor:

$$\underline{\underline{\sigma}}\;=\;-p\,\underline{\underline{\mathrm{Id}}}+\mu\underbrace{\left(\nabla u+(\nabla u)^T\right)}_{=2\varepsilon(u)}. \tag{5.14}$$

Hence, the stress tensor consists of a non-viscous pressure contribution and a viscous term $\varepsilon(\mathbf{u})$ which models the viscosity of the fluid subject to its local velocity in combination with the viscosity μ.

Applying this specific form of the stress tensor $\underline{\underline{\sigma}}$ as well as the rules for calculating the divergence of tensors in combination with the continuity equation (5.7), we obtain the momentum equations (5.13) explicitly as

$$
\begin{aligned}
\rho\frac{\partial \mathbf{u}}{\partial t} + \rho\,(\mathbf{u}\cdot\nabla)\,\mathbf{u} \quad &= \quad \rho\,\mathbf{g} + \mathrm{div}(\underline{\underline{\sigma}}) \quad = \quad \rho\,\mathbf{g} + \left(\frac{\partial \underline{\underline{\sigma}}}{\partial x_k}\otimes e^k\right) : \underline{\underline{\mathrm{Id}}} \\
&= \rho\,\mathbf{g} + \left(\frac{\partial\left(-p\,\delta_{ij}\,e^i\otimes e^j\right)}{\partial x_k}\otimes e^k\right) : \underline{\underline{\mathrm{Id}}} + \mu\left(\frac{\partial\left(\nabla u + (\nabla u)^T\right)}{\partial x_k}\otimes e^k\right) : \underline{\underline{\mathrm{Id}}} \\
&= \rho\,\mathbf{g} - \frac{\partial\left(p\,\delta_{ik}\,e^i\right)}{\partial x_k} + \mu\left(\frac{\partial}{\partial x_k}\left(\frac{\partial u_i}{\partial x_j}e^j\otimes e^i + \frac{\partial u_j}{\partial x_i}e^j\otimes e^i\right)\otimes e^k\right) : \left(\delta_{lm}e^l\otimes e^m\right) \\
&= \rho\,\mathbf{g} - \nabla p + \mu\left(\underbrace{\frac{\partial}{\partial x_k}\left(\frac{\partial u_k}{\partial x_j}\right)e^j}_{=\frac{\partial}{\partial x_j}\left(\frac{\partial u_k}{\partial x_k}\right)e^j = 0} + \frac{\partial}{\partial x_k}\left(\frac{\partial u_j}{\partial x_i}\right)e^j\right) \\
&= \rho\,\mathbf{g} - \nabla p + \mu\Delta\mathbf{u}\,.
\end{aligned}
\tag{5.15}
$$

With the continuity equation (5.7) and the momentum equations (5.15), we now have a description of the problem in the interior of the domain Ω of interest. In addition to that, we also need conditions on the boundary $\Gamma = \partial\Omega$ for all times of interest. We briefly summarise different types of boundary conditions (for details, see Gresho:[118],e.g.).

- **Velocity Dirichlet Conditions:** If velocity values are directly specified on a boundary, this conditions is denoted to be of Dirichlet-type. Classical examples are velocity inlet conditions and no-slip conditions. The former describe the full velocity data at inlet boundaries with a non-zero normal component of the velocity: $u_n = \mathbf{n}\cdot\mathbf{u} = \mathbf{n}\cdot\mathbf{w} \neq 0$. The latter model flow with fluid particles sitting directly next to solid walls inheriting directly the wall velocity $\mathbf{w}$ (often zero):

$$\mathbf{u}(x,t) = \mathbf{w}(x,t) \quad \text{on} \quad \Gamma\times[0;T]. \tag{5.16}$$

- **Force Conditions:** Force boundary conditions are usually used when the specific physical forces on the surface are known.

  The mathematical formulation,

$$\underline{\underline{\sigma}}\cdot\mathbf{n} = \mu\left(\nabla\mathbf{u} + (\nabla\mathbf{u})^T\right)\cdot\mathbf{n} - p\,\mathbf{n} = \mathbf{F}, \tag{5.17}$$

gives, for straight boundaries (or plane boundary faces in three dimensions),

$$\mu\left(\frac{\partial \mathbf{u}}{\partial n}+\nabla u_n\right)-p\,\mathbf{n}=\mathbf{F}. \tag{5.18}$$

One classical case are prescribed pressure values, for example in the scenario of a tube that connects two basins with different hydrostatic pressures that drive a flow through the tube. The pressure is fixed at the corresponding inlet and outlet boundaries that correspond to Neumann velocity boundaries.

- **Open Boundary Conditions:**

  Artificial boundaries are created whenever a larger or infinite domain is computationally limited to allow for a discretised solution. Specifying data on these parts is difficult since no information is available about the flow conditions outside the domain of interest. In [117], a nice survey on different approaches for these so-called open boundary conditions is given. However, there is presently no result concerning "best" or only "correct" conditions.

  The major problem (and simultaneously also the clue) is the coupling of velocity and pressure via the continuity equation $\nabla\cdot\mathbf{u}=0$. It is common to tackle open boundary conditions using force or pseudo-force conditions. In the case of 2D straight boundaries, the explicit formulation of (5.17) results in

$$2\mu\frac{\partial u_n}{\partial n}-p \;=\; \mathbf{n}\cdot\mathbf{F}=F_n \quad \text{(normal contribution)}, \tag{5.19}$$

$$\mu\left(\frac{\partial u_\tau}{\partial n}+\frac{\partial u_n}{\partial \tau}\right) \;=\; \tau\cdot\mathbf{F}=F_\tau \quad \text{(tangent contribution)}. \tag{5.20}$$

  The conditions are typically simplified by neglecting the term $(\nabla\mathbf{u})^T$ in Eq. (5.17), since the real force values are usually unknown, leading to

$$\mu\,\nabla\mathbf{u}\cdot\mathbf{n}-p\,\mathbf{n} \;=\; \mathbf{f}\,. \tag{5.21}$$

  Equations (5.19) and (5.20) then have the form

$$\mu\frac{\partial u_n}{\partial n}-p \;=\; f_n \quad \text{(normal contribution)} \tag{5.22}$$

$$\mu\frac{\partial u_\tau}{\partial n} \;=\; f_\tau \quad \text{(tangent contribution)}, \tag{5.23}$$

  where the pseudo force **f** represents only a part of the full physical force **F**. By setting the tangent part (5.20) of the pseudo force **f** to zero, one obtains the condition $\frac{\partial u_\tau}{\partial n}=0$ which is frequently used at outlets (cf. [119]).

Hence, using open boundary conditions in this "do-nothing" form corresponds to Neumann boundary conditions for the velocity.

- **Periodic Boundary Conditions:** The periodic boundary conditions are used to eliminate boundaries in the domain and realise "infinite domains". Thus, the basic idea of these type of boundary conditions is that there is no real boundary. Boundary points are considered as usual inner points (with the corresponding equations) where just a part of the surrounding is located at the other end of the domain. Hence, leaving the domain in this "torus world" results in a reentry on the other side, and no specific data is prescribed on such boundaries.

Of course, a combination of different boundary conditions applied to different parts of the boundary Γ is possible. Different conditions for different velocity components may even be used for the same location x (see [118] p. 383 for a nice visualisation). An example for such an overlapped boundary condition are slip-wall conditions modelling non-viscous walls: Tangentially to the wall, the flow does not encounter viscous resistance, but normally to the wall, no flow is allowed. This corresponds to a combination of velocity Dirichlet conditions (normal part) and velocity Neumann conditions (tangent part).

In many cases, however, the type of boundary is identical for all components of the velocity. If this holds for the complete domain, the boundary consists of disjoint Dirichlet and Neumann parts:

$$\Gamma = \Gamma_D \cup \Gamma_N \,.$$

Putting it altogether, for all stated boundary conditions, a specification of the type of velocity conditions is necessary, which fixes inherently also the pressure boundary conditions. The pressure is not fixed independently by itself but is treated as an auxiliary variable that assures a solenoidal flow field. This is an important observation that translates also to the numerical discretisation.

Finally, since the Navier-Stokes equations are also time-dependent, we need initial conditions: Again, it is sufficient to fix velocity conditions: the pressure reacts accordingly. The initial velocity values can't, however, be chosen in an arbitrary manner but have to respect the conditions

$$\nabla \cdot \mathbf{u}_0 = 0 \qquad \text{in } \Omega, \tag{5.24}$$

$$\mathbf{n} \cdot \mathbf{u}_0 = \mathbf{n} \cdot \mathbf{w}_0 \qquad \text{on } \Gamma_D \,. \tag{5.25}$$

The complete system of the Navier-Stokes equations for incompressible flow with corresponding boundary conditions has the following compact form:

- continuity equation

$$\nabla \cdot \mathbf{u}(x,t) = 0 \qquad \forall\,(x,t) \in \overline{\Omega} \times [0;T]\,, \tag{5.26}$$

- momentum equations

$$\rho\frac{\partial \mathbf{u}}{\partial t} + \rho\,(\mathbf{u}\cdot\nabla)\,\mathbf{u} + \nabla p - \mu\,\Delta\mathbf{u} = \rho\mathbf{g} \quad \forall\,(x,t) \in \Omega \times [0;T]\,, \tag{5.27}$$

- boundary conditions

$$\mathbf{u}(x,t) = \mathbf{w}(x,t) \quad \forall\,(x,t) \in \Gamma_D \times (0;T]\,, \tag{5.28}$$

$$[\mu\,\nabla\mathbf{u}\cdot\mathbf{n} - p\,\mathbf{n}]\,(x,t) = \mathbf{f}(x,t) \quad \forall\,(x,t) \in \Gamma_N \times (0;T]\,, \tag{5.29}$$

- initial conditions

$$\nabla \cdot \mathbf{u}(x, t_0 = 0) = 0 \qquad \forall\, x \in \Omega\,, \tag{5.30}$$

$$\mathbf{n} \cdot \mathbf{u}(x, t_0 = 0) = \mathbf{n}\cdot\mathbf{w}_0 \qquad \forall\, x \in \Gamma_D\,. \tag{5.31}$$

Before you continue, make sure to answer the following questions:

Quiz: Section 5.2

Q1 Are the incompressible Navier-Stokes equations linear or non-linear?

Q2 What are the four boundary conditions for NSE scenarios?

Q3 Which random effects may appear for incompressible flow problems? List at least three!

5.3 Numerical Solution of the Navier-Stokes Equations

The goal of computational approaches to the Navier-Stokes Equations is to compute approximate solutions $\mathbf{u}_h, p_h$ of $\mathbf{u}(x,t), p(x,t)$ on discrete points in space and time. Using this data, other quantities of interest—such as the lift or drag force coefficients of an object in flow—can be computed via postprocessing. As is usual for PDEs, we need to discretise the underlying space-time domain and the operators in the Navier-Stokes equations. For the former, a classical mesh for the domain Ω and a 1D grid window in time is typically used (see Sec. 2.3.1). For the operator discretisation in Eq. (5.26)–(5.27), all common approaches such as finite differences (FD, see [119], e.g.), finite volumes (FV, cf. [172], e.g.), and finite elements (FE, see [118], e.g.) can of course be used, resulting in a set of semi-discrete equations (still continuous in time):

$$\begin{aligned} A\dot{\mathbf{u}}_h + D\mathbf{u}_h + C(\mathbf{u}_h)\mathbf{u}_h - M^T p_h &= f, \\ M\mathbf{u}_h &= 0, \end{aligned}$$

A variety of approaches exist of how to combine the spatial discretisation of the incompressible NSE with integration in time: explicit methods, fully implicit discretisations, stabilised formulations, etc. The crucial point in the discretisation is how to respect the continuity equation [1].

We will briefly discuss one classical approach: the Chorin projection scheme (developed in [58] and independently in [249]). The basic idea of this scheme is to decouple the solution of the pressure and velocity fields in two separate calculation steps, similar to a predictor-corrector approach. First, an intermediate velocity—typically called F—is computed which does not yet fulfil the algebraic mass conservation constraint. Second, the pressure is calculated using a Poisson-type equation obtained from the continuity equation and is used to project the intermediate velocity onto a space of discretely divergence-free velocity fields and then to update the velocities accordingly.

Starting with the semi-discrete Navier-Stokes equations, we derive the continuity equation (5.32) once with respect to time:

$$M\dot{\mathbf{u}}_h = 0. \tag{5.32}$$

We now formally invert the momentum equations with respect to $\dot{\mathbf{u}}_h$ and insert them into (5.32). This results in the discrete pressure Poisson equation (PPE),

$$\underbrace{\left(MA^{-1}M^T\right)}_{=:\,Q} p_h = MA^{-1}[-f + \underbrace{D\mathbf{u}_h + C(\mathbf{u}_h)\mathbf{u}_h}_{=F}] - \dot{g}, \tag{5.33}$$

where Q represents a discrete analogue of the continuous Laplacian, and $\dot{g} := M\dot{\mathbf{u}}_h$ vanishes everywhere except on boundaries due to prescribed velocity data.

To avoid the solution of an extra linear system of equations indicated by A^{-1}, one typically uses mass lumping. The mass matrix contributions are accumulated only in diagonal entries of A (often called row summing), and we

[1] Note that recently continuously divergence-free discretisations via finite elements have gained more attention (see [118, 198, 96]). While the development of such elements is not straightforward, one advantage of these schemes is the simultaneous conservation of momentum and energy which classical discretisations do not provide (cf. [118]).

just need to invert scalar values separately. Hence, just one linear system of equations for the pressure has to be solved in each time step, with lower dimension than the whole system which involves momentum and continuity equations. This assures a (discretely) solenoidal flow field, while the velocities are updated explicitly. This approach is often called semi-implicit.

Applying an explicit one-step method (see Sec. 7.3) with the corresponding update function Φ_f fixes the time levels of the discretised velocity degrees of freedom $\mathbf{u}_h$ in (5.33). The necessary computations for the Chorin projection scheme for the explicit Euler time integration (cf. Sec. 7.3.1) is shown in Algorithm 1. Note that for the explicit Runge-Kutta method instead of forward Euler, one needs to perform a cycle of the computations in Algorithm 1 for each step vector.

Algorithm 1 Explicit time stepping with the Chorin projection scheme.

```
while t < T_end do
    calculate new time step size
    calculate intermediate velocities F
    assemble PPE right-hand side
    solve PPE
    calculate ∇p_h
    update u_h
    t = t + Δt_n
end while
```

To get a brief impression of how typical benchmark scenarios for incompressible flow and corresponding discrete solutions look like, we present results for the well-known laminar flow around a cylinder in a channel described in detail in [251]. Figure 5.1 shows the two-dimensional geometric setup of the cylinder benchmark scenario. The 2D "cylinder" is located near the inlet of a channel where a fixed, parabolic velocity profile is prescribed (i.e. Dirichlet boundary conditions), while open boundary conditions are applied at the outlet. The coefficients for drag and lift forces onto the obstacle serve as the main "hard" reference data with respect to the accuracy of the obtained solution.

The following results are taken from [197]; the approximate solutions have been computed with a finite element space discretisation using classical Q1Q0 elements combined with the explicit Euler time integration using the Chorin projection scheme. In Fig. 5.1, an example of a steady-state flow around the cylinder with Reynolds number Re $= 20$ is given: The pressure contour lines are visualised in black over the coloured velocity values. The steady double vortex behind the cylinder is visible from the low to negative velocity values there. Fig. 5.2 shows one example of an adaptive Cartesian

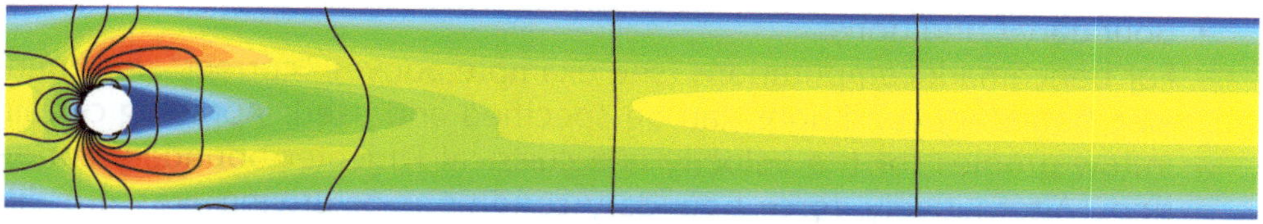

Figure 5.1. Simulation results for the DFG benchmark "laminar flow around a cylinder" [251] at $\mathrm{Re} = 20$. The contour lines of the pressure are visualised over the horizontal velocity distribution.

grid using refinement boxes to steer the level of resolution (i.e. the size of the mesh) in different regions of the flow. This grid results in a total number of 88,857 cell and vertex degrees of freedom. The drag and lift coefficients of $c_d = 5.680, c_l = 0.0150$ agree relatively well with the reference data of $c_d = 5.580, c_l = 0.0107$, showing a relative error of about 1.8 and 18% for this still rather coarse resolution of the grid.

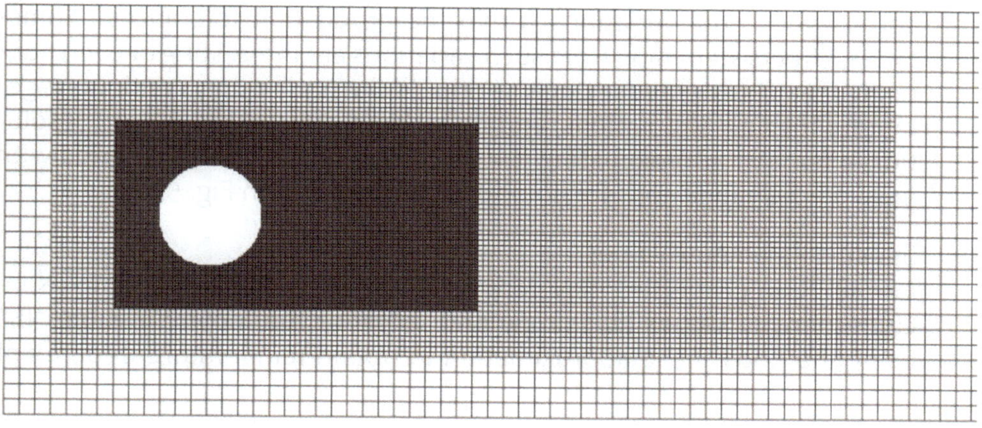

Figure 5.2. Visualisation of an adaptive Cartesian grid for the benchmark flow around a cylinder 2D–1 (cf. [251]).

5.4 Random Perturbation of Incompressible Flow

Different kind of uncertainties may be observed or desired to resolve in the case of incompressible fluid flow. Random effects may arise due to statistically inferred input parameters on different levels:

- Boundary Conditions:
  For benchmark computations (see example above), "correct" deterministic boundary conditions can be specified and used: Applying certain inflow profiles for the velocity is a classical Dirichlet boundary condition. Alternatively, force boundary conditions or driving pressure differences combined with velocity Neumann conditions may be specified (see Sec. 5.2).

  For real flow scenarios, however, specifying the "correct" conditions is actually possible only in rare cases. Measured data from experiments or similar are inherently error-prone and, thus, already carry uncertainty.

- Geometric Uncertainty:
  Several possibilities of geometric uncertainty exist. First, a specific geometry appearing in nature—such as riverbeds etc.—will always only be approximated by a geometric model (generated via CAD representations or measurements). Second, even if the exact geometry is known as in the case of flow around technical components modeled in CAD (a wing profile or car body, e.g.), the construction of actual parts will surely introduce certain geometric variations within the production tolerances. Sometimes, effects stemming from such variations are interesting or even non-negligible. Besides these uncertainties in the underlying, continuous geometry for flow simulations, additional ones are created by grid resolution effects in approximating the geometry (using Cartesian grids with pure brick-shaped cells for smooth boundaries, as in Fig. 5.2 e.g.).

- Specific Parameter Values:
  Certain fluid properties may vary inside the flow domain. Examples include density values (in the generalisation of the incompressible NSE using the Boussinesq approximation to obtain temperature-dependent

Before you continue, make sure to answer the following questions:

Quiz: Section 5.3

Q1 Which different spatial discretisation schemes may be used for the Navier-Stokes equations?

Q2 What is the basic idea of the Chorin projection?

Q3 What is the bulk of the computational costs for the numerical solution of the incompressible Navier-Stokes equations?

buoyancy effects) or viscosity data, where a minor mixture of fluids in realistic setups (such as nuclear reactor safety computations) affects the overall flow properties.

The above-mentioned effects get combined with other effects in coupled problems such as fluid-structure interaction scenarios or chemical reactions in flow problems, where the incompressible Navier-Stokes equations are one of several physical problems to be modeled and solved.

Before you continue, make sure to answer the following questions:

Quiz: Section 5.4

Q1 What are the three categories of random effects mentioned above?

Q2 List at least one concrete example for each category with respect to the Navier-Stokes equations!

5.5 Extension to Other Flow Problems

Many of the above mentioned random effects are not restricted to the incompressible Navier-Stokes equations in the laminar regime but can directly be generalised to other flow scenarios; we briefly list four possibilities that directly come to the mind. Magneto-hydrodynamical (MHD) simulations may encounter additional stochastic effects via the magnetic field data. Compressible flows around airfoils depend on the detailed geometry data of the wing profile which hardly ever comes perfectly out of production. Furthermore, far-field effects stemming from (imperfect) boundary conditions might affect the whole flow field. In the case of the shallow-water equations, statistical effects may arise in the bathymetry (i.e. the geometrical deepness of the sea ground). Finally, turbulent flow simulations naturally incorporate statistical effects intrinsically, but additional variations in material properties (wall roughness etc.) or boundary conditions may be considered that could have large macroscopic effects. And of course modeling effects play an important role when it comes to large eddy simulations (LES) of turbulent flow scenarios.

This last example widens the focus on a completely different category of uncertainty effects: Uncertainties due to model inadequacy.

- Inadequate Model Simplification:
  Currently, no major doubt exists in the community that the Navier-Stokes equations are the correct approach to model the corresponding flows. However, the above-mentioned example of the Boussinesq already raises the question of how well this simplification approximates real temperature-dependent (and energy-conserving) flow scenarios. When it comes to upscaling approaches (for instance Darcy flow for porous-media flows used in reservoir simulations, e.g.), these modeling effects become more important.

- Unknown Relevant Factors for more Complex Flows:
  Sometimes, it may not even be fully clear, what exactly happens in flows on certain length or time scales. Certain effects are not (yet) measurable with current state-of-the-art sensors.

Before you continue, make sure to answer the following questions:

Quiz: Section 5.5

Q1 List at least two other flow problems besides the incompressible NSE where random effects may play a role!

Q2 Which additional category of random effects comes newly into play compared to Sec. 5.4?

5.6 Chapter's Summary and Outlook

In this chapter, we discussed flow problems in more detail, in particular we analysed incompressible flow. These problems represent an important class of applications in computational science and engineering. The various possible random effects in the model, geometry, boundary conditions, or parameters may be generalised to other flow scenarios, involving coupled scenarios such as fluid-structure interaction.

The equations of fluid flow can be coupled with those of biological growth to enable the simulation and control of so-called biofilms, see Fig. 5.3 (a). Biofilms are an aggregate of microorganisms in which cells adhere to each other on a surface, e.g. the interior wall of a tube. Obviously, such biofilms reduce the speed of fluids in the tubes in which they grow and they may cause serious health problems when they serve as a medium for infectious bacteria, like Legionella, cf. [195]. Biofilms in marine engineering systems, such as pipelines of the offshore oil and gas industry can lead to substantial corrosion problems (see [228]). Corrosion is mainly due to abiotic factors; however, at least 20% of corrosion is caused by microorganisms that are attached to the metal subsurface (i.e. microbially-influenced corrosion), see figure 5.3 (b). Such corrosion problems add a third component to the constituting equations, namely the description of electro-chemical processes that are responsible for corrosion. Clearly, this gives rise to a plethora of new, mathematically challenging and industrially relevant applications for random and stochastic (partial/ ordinary) differential equations.

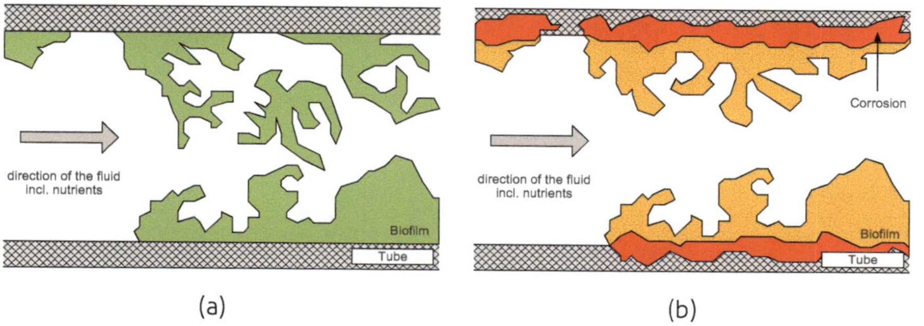

Figure 5.3. Sketches of biofilms in tubes filled with some fluid: (a) fluid interaction with biological processes and (b) fluid interaction with biological and electro-chemical processes leading to corrosion of the tube's walls.

Part II

The Path-Wise Deterministic Setting

The mathematician's patterns, like the painter's or the poet's must be beautiful; the ideas, like the colours or the words must fit together in a harmonious way. Beauty is the first test: there is no permanent place in this world for ugly mathematics.

GODFREY HAROLD HARDY *(1877 - 1947)*

Chapter 6

Recap: Theory of Ordinary Differential Equations (ODEs)

This chapter serves as a holistic introduction to the theory of ordinary differential equations (without singularities). After providing some preliminaries, integral curves in vector fields are discussed, i.e., ordinary differential equations $\dot{x} = F(t, x)$. We start with continuous right hand sides F and their ε-approximate solutions as well as the Peano-Cauchy existence theorem and its implications. Our discussion continues with Lipschitz-continuous functions and the existence and uniqueness theorem of Picard-Lindelöf. In particular, we analyse maximal integral curves, define the three types of maximal integral curves that can occur in autonomous systems and show the transformation of a d-th order equation into a first order system. We deal with the existence of solutions in the extended sense where the right hand side function may be discontinuous on a set of Lebesgue-measure zero. Caratheodory's existence theorem and its implications are studied together with maximum and minimum solutions. Then, we study the broad class of linear ordinary differential equations by discussing the unique existence of their solutions and their explicit construction. Applications of the theory focus on first integrals and oscillations in the deterministic pendulum and the Volterra-Lotka system. Finally, we provide a first glance into the existence, uniqueness and extension of solutions of ordinary differential equations on infinite-dimensional Banach spaces.

6.1 Key Concepts

As we have seen in Chap. 3, random (ordinary) differential equations can be interpreted path-wise as (deterministic) ordinary differential equations. Thus, it pays-off to recall the basic properties of these deterministic equations, and, as we will see, it is interesting to compare the results and proof-techniques we utilized in Chap. 3 with those that are relevant for (deterministic) ordinary differential equations.

Before diving into the details, let us summarize the key results of this chapter: Let $J \subseteq \mathbb{R}$, $U \subseteq \mathbb{R}^d$ ($d \in \mathbb{N}$) and $\Lambda \subseteq \mathbb{R}^m$ ($m \in \mathbb{N}_0$) be open subsets and F a smooth function on the Cartesian product of these to $\mathbb{R}^d$, i.e., $F \in \mathcal{C}^k(J \times U \times \Lambda, \mathbb{R}^n)$ for some $1 \leq k \leq \infty$. An *ordinary differential equation*

(ODE) is an equation of the form

$$\dot{x} = F(t, x, \lambda), \tag{6.1}$$

where the dot indicates the differentiation with respect to the time variable $t \in J$, $x \in U$ is a vector of state variables and $\lambda \in \Lambda$ is a vector of parameters. If $d > 1$, the equation (6.1) is called a *system of d first order (ordinary) differential equations* and the differentiation is considered component-wise. In particular, during the study of bifurcation/ change of stability scenarios (cf. Sec. 8.3) the dependence on parameters will become essential.

For fixed $\lambda \in \Lambda$ a *solution* (or *trajectory*) of (6.1) is a function $\varphi : J_0 \subset J \to U$, J_0 open, given by $t \mapsto \varphi(t)$ such that the following holds for all times $t \in J_0$:

$$\frac{\mathrm{d}\varphi}{\mathrm{d}t} = F(t, \varphi(t), \lambda).$$

The fundamental issues in ODE theory concern existence, uniqueness, continuity (with respect to the parameters and the initial conditions) and extensibility of solutions. Fortunately, all these issues are resolved, provided the right-hand side function F has enough regularity: Every initial value problem

$$\dot{x} = F(t, x, \lambda) \quad \text{with initial conditions} \quad x(t_0) = x_0 \in U \tag{6.2}$$

has a unique solution[1] which is smooth with respect to the initial conditions and parameters[2]. Moreover, the solution of an initial value problem can be extended with respect to time until it either reaches the boundary ∂U of the domain of definition $U \in \mathbb{R}^d$ of the differential equation or blows up to infinity[3].

Let us be more precise: If there is some finite T such that $\lim_{t\to T} \|\varphi(t, x_0, \lambda)\| \to \infty$, we say that the solution *blows up in finite time*. Such solutions will leave every compact subset of $U \subseteq \mathbb{R}^d$ (see [162], p. 143).

Moreover, every d-th order differential equation, $d > 1$, is equivalent to a system of d first order differential equations, see [12], p. 105 or [162], p. 148.

Finally, as we have seen in Chap. 4, (deterministic) ordinary differential equations on Banach spaces play a crucial role in understanding the mean-square solutions of random differential equations. Because the Euclidean $\mathbb{R}^d$ is the "simplest" Banach space, we will basically gain the same results as

---

1 See, for instance, [56], p. 3, theorem 1.2, or [12], p. 93, corollary 3. In [162], pp. 140, the uniqueness and existence statements are provided for locally Lipschitz-continuous right-hand sides.

2 See, for instance, [56], p. 3, theorem 1.3, or [12], p. 93, corollary 4.

3 See, for instance, [56], p. 3, theorem 1.4, and [162], pp. 143. Again, [162] provides his statements for locally Lipschitz-continuous right-hand sides.

cited above, though the methods for establishing them will take the infinite-dimensional nature of arbitrary Banach spaces into account.

When reading this chapter note the answers to the following key concepts

1. Vector fields, local Lipschitz-continuity, and integral curves.
2. Estimating bounds for the solution of ordinary differential equations by Gronwall's lemma.
3. The theorems of
   (a) Cauchy-Peano (local existence w.r.t. continuous right-hand sides),
   (b) Picard-Lindelöf (local existence and uniqueness w.r.t. Lipschitz-continuous right-hand sides),
   (c) Caratheodry (local existence of solutions in the extended sense),

   and their applications.
4. ε-approximate solutions.
5. Life-time and extension of maximal integral curves.
6. Classification of maximal integral curves in autonomous/ time-independent systems.
7. Maximum and minimum solutions of ordinary differential equations in the extended sense.
8. Fundamental system and matrix for homogeneous linear ordinary differential equations.
9. Method of the variation of constant.
10. First integrals and oscillations (deterministic pendulum equation and Volterra-Lotka equation).
11. Existence, uniqueness and extension of solutions of ordinary differential equations on infinite-dimensional Banach spaces.

as well as the following questions

1. How are vector fields and integral curves connected?
2. What is the method of successive approximation?
3. How and under which conditions can solutions be extended beyond a given open interval of existence?

4. What does the lemma of Arzela-Ascoli state?
5. How can we use MATLAB to symbolically solve ordinary differential equations and display the solutions graphically?
6. How can a d-th order equation be transformed into a first order system?
7. What are solutions in the extended sense?
8. How are the solution of a inhomogeneous linear ordinary differential equations constructed?

The main sources for this chapter are [162], Chap. 4, for the numerics-friendly introduction of ordinary differential equations on $\mathbb{R}^d$ as integral curves of vector fields, as well as [82], Chap. 1, for the brief discussion of ordinary differential equations on arbitrary Banach spaces.

This chapter is structured as follows: First, Sec. 6.2 equips us with the essential information about vector fields and their representation as well as the notion of local Lipschitz continuity and Gronwall's lemma. Next integral curves in vector fields are discussed, i.e., ordinary differential equations $\dot{x} = F(t, x)$. We start in Sec. 6.3 by analyzing ordinary differential equations with continuous right hand side F and studying ε-approximate solutions as well as the Peano-Cauchy existence theorem and its implications. Then, in Sec. 6.4, we continue our discussion with Lipschitz-continuous functions F and the existence and uniqueness theorem of Picard-Lindelöf. It is here that we give the MATLAB commands for symbolically solving ordinary differential equations and analyse maximal integral curves. In particular, we give the three types of maximal integral curves that can occur in autonomous systems and show the transformation of a d-th order equation into a first order system. Sec. 6.5 deals with existence of solutions in the extended sense where the right hand side function may be continuous except for a set of Lebesgue-measure zero. Caratheodory's existence theorem and its implications are studied together with maximum and minimum solutions. Section 6.6 analyses the broad class of linear ordinary differential equations by discussing the unique existence of their solutions and explicitly constructing them. Applications are given in Sec. 6.7, where we discuss in particular the deterministic pendulum and the Volterra-Lotka system. Sec. 6.8 rounds up our discussion by providing a first glance into the existence, uniqueness and extension of solutions of ordinary differential equations on infinite-dimensional Banach spaces. Finally, Sec. 6.9 wraps up the contents of this chapter.

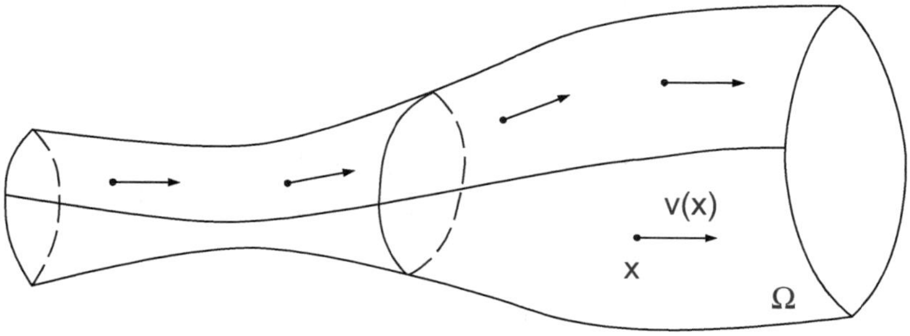

Figure 6.1. Sketch of the geometric/ physical interpretation of a vector field on a domain Ω.

6.2 Preliminaries

Following [162], pp. 131, vector fields occur naturally in many physics and engineering contexts. Here, we recall their basic definitions and properties together with some illustrative examples. We lay the foundations for the discussion of integral curves in vector fields, i.e. the study of ordinary differential equations.

6.2.1 Vector Fields and Their Representation

Definition 6.1 (Vector Field & $\mathcal{C}^k$ Vector Field). A *vector field* v on a set $\Omega \subset \mathbb{R}^d$ is a mapping $v : \Omega \to \mathbb{R}^d$ which associates a vector $v(x) \in \mathbb{R}^d$ to each element $x \in \Omega$. If v is a $\mathcal{C}^k$-mapping, we call v a $\mathcal{C}^k$ *vector field*.

Geometrically, a vector field v is interpreted by attaching the vector $v(x)$ to each point $x \in \Omega$. This means, that all pairs $(x, v(x)) \in \mathbb{R}^d \times \mathbb{R}^d$ are formed. Physically, a vector field is often interpreted as the velocity field of a stationary, i.e. time-independent, (fluid) flow, where $v(x)$ is the velocity vector at the position $x \in \Omega$, cf. Fig. 6.1.

Example 6.2 (Constant Fields, cf. [162], p. 131). *Constant (vector) fields* are defined by $v(x) := v \in \mathbb{R}^d$ for all $x \in \Omega$. In this chapter we will interpret a vector $v \in \mathbb{R}^d$ as a constant vector field, and denote this by $v : v(x) = v$.

Example 6.3 (Central Fields, cf. [162], p. 131). Let $I \subset \mathbb{R}$ be an interval. On the shell of a ball centered at the origin of the Euclidean $\mathbb{R}^d$, *central (vector) fields* are defined by

$$v(x) \; := \; a\left(\|x\|\right) \cdot x\,, \qquad \text{where } a : I \to \mathbb{R} \text{ is a function}\,,$$

with $x \in \mathbb{R}^d$. For instance, on $\mathbb{R}^3 \setminus \{0\}$ the gravitational field is given as

$$v(x) := -\frac{x}{\|x\|^3}.$$

Example 6.4 (Rotation Fields, cf. [162], p. 131). Let $I \subset \mathbb{R}$ be an interval. On a circle centered at the origin of the Euclidean plane $\mathbb{R}^2$, *rotation (vector) fields* are defined by

$$v(x) := a(\|x\|) \cdot \begin{pmatrix} -x_2 \\ x_1 \end{pmatrix}, \qquad \text{where } a : I \to \mathbb{R} \text{ is a function},$$

with $x = (x_1, x_2)^T \in \mathbb{R}^2$.

Example 6.5 (Gradient Fields, cf. [162], p. 132). Let $f : \Omega \to \mathbb{R}$ be a differentiable function on an open subset of the Euclidean $\mathbb{R}^d$. The *gradient (vector) field* $\nabla f : \Omega \to \mathbb{R}^d$ of f is defined as

$$v : x \mapsto v(x) := \nabla f(x).$$

Examples of these just discussed four vector fields are displayed in Fig. 6.2 with the aid of MATLAB's quiver function. E.g., the plot of the gradient field is generated via

```
[X,Y] = meshgrid(-2:.5:2);
%
Z = X.*exp(-X.^2 - Y.^2);
[DX,DY] = gradient(Z,.2,.2);
%
quiver(X,Y,DX,DY,'-k')
```

The vector fields $\eta_1, \eta_2, \ldots, \eta_d : \Omega \to \mathbb{R}^d$ are called a *basis of the vector fields on* Ω, if the vectors $\eta_1(x), \eta_2(x), \ldots, \eta_d(x)$ form a basis of $\mathbb{R}^d$ for every point $x \in \mathbb{R}^d$. In this setting, any further vector field $v : \Omega \to \mathbb{R}^d$ can be expressed as a linear combination of the vector fields $\eta_1, \eta_2, \ldots, \eta_d$ with the aid of uniquely defined functions $a_1, a_2, \ldots, a_d : \Omega \to \mathbb{R}$:

$$v(x) = \sum_{i=1}^{d} a_i(x)\eta_i(x), \qquad x \in \Omega.$$

Next, we extend the notion of a derivative along a vector to the notion of a derivative with respect to a vector field:

Definition 6.6 (Derivative with Respect to a Vector Field). Let $f : \Omega \to \mathbb{R}^m$ be a differentiable mapping on an open subset $\Omega \subset \mathbb{R}^d$, and $v : \Omega \to \mathbb{R}^d$ a vector field. Then, the derivative of f at $x \in \Omega$ in direction of the vector $v(x)$

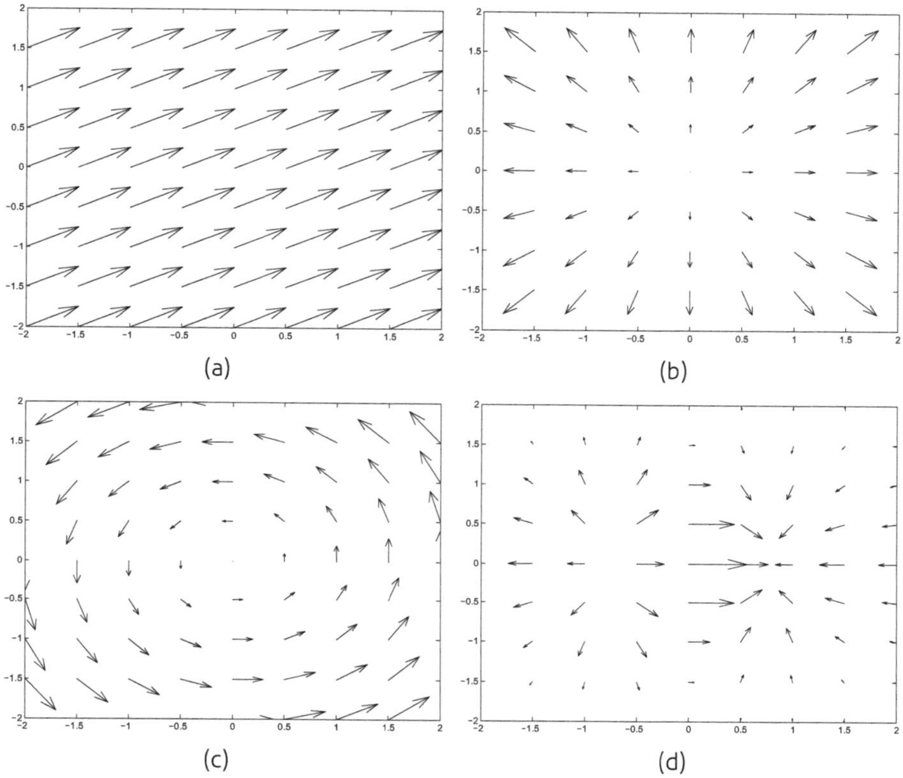

Figure 6.2. Velocity plots for (a) the constant vector field $v(x,y) = (2,1)^T$, (b) the central field $v(x,y) = (x,y)^T$, (c) the rotation field $v(x,y) = (-y,x)$ and (d) the gradient field $v(x,y) = \nabla f(x,y)$, where $f(x,y) = x \cdot \exp\left(-x^2 - y^2\right)$.

is called the *derivative of f at x with respect to the vector field v* and denoted by $\partial_v f(x)$; i.e.,

$$\partial_v f(x) \;:=\; \partial_{v(x)} f(x) \;=\; \lim_{t \to 0} \frac{f\left(x + tv(x)\right) - f(x)}{t}\,.$$

For a scalar function $f : \Omega \to \mathbb{R}$ the notation $\partial_h f(x) = \mathrm{d}f(x)h = f'(x)h$ for $h \in \mathbb{R}^d$ immediately leads to

$$\partial_v f(x) \;=\; \mathrm{d}f(x)v(x) \;=\; f'(x)v(x) \;=\; \langle \nabla f(x), v(x) \rangle\,.$$

In the context of Lyapunov functions, this special derivative is called *directional derivative*. It plays the essential role in establishing (asymptotic) stability of an equilibrium point as we will see in Chap. 8. If $a_1, a_2, \ldots, a_d$: $\Omega \to \mathbb{R}$ are the component functions of the vector field v, i.e. $v(x) =$

$(a_1(x), a_2(x), \ldots, a_d(x))^T$, it follows that

$$\partial_v f(x) \;=\; \sum_{i=1}^{d} a_i(x) \frac{\partial f(x)}{\partial x_i} \,.$$

Finally, let us discuss some MATLAB commands that can be used to compute (partial) derivatives. Symbolic manipulations in MATLAB, like the symbolic calculation of partial derivatives, are actually performed by the program MAPLE. The following lines of code illustrate how MATLAB can be used for symbolic differentiation, cf. [177]. E.g., let us consider the differentiation of $f(k,x) = k \cdot \sin(x^3)$ first with respect to x

```
>> syms x k
>> f = k*sin(x^3);
>> diff(f,z)
ans = 3*k*(x^2)*cos(x^3)
```

and then with respect to k

```
>> df = diff(f,k)
df = sin(x^3)
```

Finally, we substitute a numerical value for x:

```
>> subs(df,x,0.5)
ans = 0.1246
```

6.2.2 Technical Requirements

Let $\mathbb{K} = \mathbb{R}$ or $\mathbb{K} = \mathbb{C}$. In the following we consider the space $\mathbb{K}^d$ to be equipped with the l_∞-norm

$$\|x\| \;:=\; \|x\|_\infty \;:=\; \max_{i=1,\ldots,d} |x_i| \,, \qquad \text{for all } x = (x_1, x_2, \ldots, x_d)^T \in \mathbb{K}^d \,.$$

The corresponding subordinate operator norm on the space $\mathbb{K}^{d\times d}$ of $d \times d$-matrices is the row sum norm

$$\|A\| \;:=\; \|A\|_\infty \;:=\; \max_{i=1,\ldots,d} \sum_{k=1}^{d} |a_{ik}| \,, \qquad \text{for all } A = (a_{ij})_{i,j=1,\ldots,d} \in \mathbb{K}^{d\times d} \,.$$

Next, let $U \subset \mathbb{R} \times \mathbb{K}^d$ be an open set and $t \in \mathbb{R}$ as well as $x \in \mathbb{K}^d$. The mapping

$$F : \left\{ \begin{array}{rcl} U & \to & \mathbb{K}^m \\ (t,x) & \mapsto & F(t,x) \end{array} \right.$$

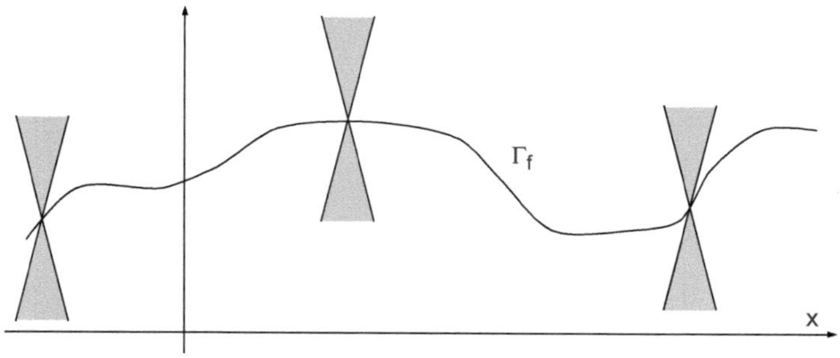

Figure 6.3. Illustration of the double cone property of the graph Γ_f of a Lipschitz-continuous function f. In the case of a differentiable function f, the slope of the double cone is limited by the maximal slope of f.

is called *Lipschitz-continuous with respect to* x if it is continuous and if there is a constant $L \geq 0$ such that

$$\|F(t,x) - F(t,x')\| \leq L\|x - x'\|. \tag{6.3}$$

holds for all elements $(t,x),(t,x') \in U$. Geometrically, for a Lipschitz continuous function f, there is a double cone whose vertex can be translated along the graph of f, so that the graph always remains entirely outside the cone, see Fig. 6.3.

F is called *locally Lipschitz-continuous with respect to* x if for every element $(t_0, x_0) \in U$ there is a subset $U_0 \subset U$ such that the restriction $F_{|U_0}$ of F on U_0 is Lipschitz-continuous with respect to x. The concept of local Lipschitz continuity and its separation from (global) Lipschitz continuity are illustrated in Figure 6.4.

The following lemma describes a large class of locally Lipschitz-continuous mappings:

Lemma 6.7 (Locally Lipschitz-Continuous Mappings). *Let $U \subset \mathbb{R} \times \mathbb{K}^d$ be an open set and $F : U \to \mathbb{K}^m$ be partially differentiable with respect to $x_1, x_2, \ldots, x_d$ at each element $(t,x) \in U$, such that all partial derivatives $\partial_{x_1}F(t,x)$, $\partial_{x_2}F(t,x)$, ..., $\partial_{x_d}F(t,x)$ are continuous on U. Then, F is locally Lipschitz-continuous with respect to x.*

More rigorous: F is Lipschitz-continuous with respect to x on every compact subset $Q = I \times K$ of U, where $I \subset \mathbb{R}$ is an interval and $K \subset \mathbb{K}^d$ is a convex set.

Proof. Following, [162], p. 138, let $F = (F_1, F_2, \ldots, F_m)^T$ and denote by M the maximum of the operator norms $\|\partial_{x_i}F_k\|_Q$ on Q for $i = 1,2,\ldots,d$ and $k = 1,2,\ldots,m$. Then, $L := nM$ is an upper bound for the norm of $F'(x)$, $x \in Q$, with $n := \max\{d, m\}$, and thus the required Lipschitz-constant. □

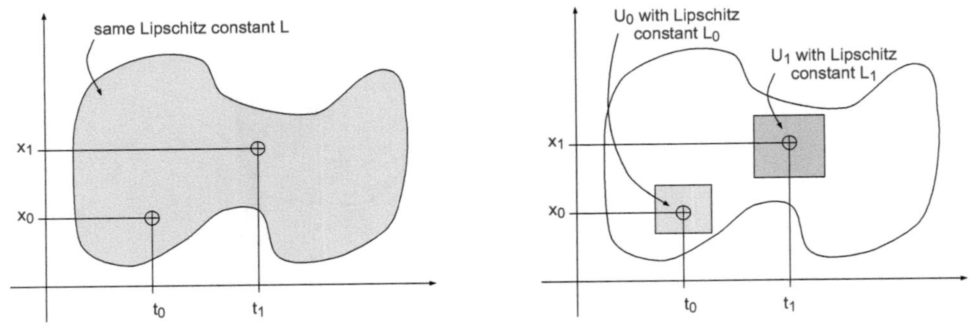

Figure 6.4. Illustration of (global) Lipschitz-continuity (left), where one and the same Lipschitz-constant L is valid within the whole domain U, and of local Lipschitz-continuity (right), where specific Lipschitz-constants L_0, L_1, ... are assigned to different to different sub-sets $U_0, U_1, \ldots$.

When studying ordinary differential equations $\dot{x} = F(t, x)$ one commonly considers the corresponding integral equation. I.e., we require a notion for the integral of a continuous mapping $f : [a, b] \to \mathbb{K}^d$. Let $f_1, f_2, \ldots, f_d$ be the component functions of f, then the integral of f is defined as

$$\int_a^b f(t)\mathrm{d}t := \left(\int_a^b f_1(t)\mathrm{d}t\,, \int_a^b f_2(t)\mathrm{d}t\,, \ldots, \int_a^b f_d(t)\mathrm{d}t \right)^T .$$

and it holds that

$$\left\| \int_a^b f(t)\mathrm{d}t \right\| \le \left| \int_a^b \|f(t)\|\, \mathrm{d}t \right| .$$

The proof of this inequality is straightforward, cf. [162], p. 139: Let f_i be that component of f for which

$$\left\| \int_a^b f(t)\mathrm{d}t \right\| = \max_{i=1,\ldots,d} \left| \int_a^b f_i(t)\mathrm{d}t \right| = \left| \int_a^b f_i(t)\mathrm{d}t \right| .$$

For this component the following holds,

$$\left| \int_a^b f_i(t)\mathrm{d}t \right| \le \left| \int_a^b |f_i(t)|\, \mathrm{d}t \right| \le \left| \int_a^b \|f(t)\|\, \mathrm{d}t \right|$$

which shows the assertion.

Finally, Gronwall's lemma will be used frequently in the discussions that follow:

Lemma 6.8 (Gronwall's Lemma). *Let $g : I \to \mathbb{R}$ be a continuous function on the interval I such that $g \geq 0$. Moreover, for one t_0 and all t in I this functions satisfies the inequality*

$$g(t) \leq A\left|\int_{t_0}^{t} g(s)ds\right| + B,$$

where $A, B \geq 0$ are constant. Then,

$$g(t) \leq B \exp\left(A|t-t_0|\right)$$

holds for all $t \in I$.

Proof. Following [162], p. 140, we show the assertion for $t > t_0$ as the case $t < t_0$ is analogously. The assertion needs to be proven just in a neighborhood of the point t where $g(t) > 0$. There, the assumed inequality reads as

$$\frac{\dot{G}(t)}{G(t)} \leq A, \quad \text{and thus } \frac{\mathrm{d}}{\mathrm{d}t}\ln\left(G(t)\right) \leq A,$$

where we set $G(t) := A\left|\int_{t_0}^{t} g(s)\mathrm{d}s\right| + B$. Thus, $G(t) \leq G(t_0)\exp\left(A|t-t_0|\right)$ holds. The assertion follows due to $g \leq G$. □

Before you continue, make sure to answer the following questions:

Quiz: Section 6.2

Q1 Give the definition of a vector field together with a geometric/ physical interpretation.

Q2 Give some illustrative examples of vector fields.

Q3 Give the definition of the derivative $\partial_v f(x)$ of a function f at a point x w.r.t. a vector field v.

Q4 Let $f(x, y) = x \cdot y$ and $v(x, y) = (-y, x)^T$. What is $\partial_v f(x, y)$?

Q5 Give the definition of a locally Lipschitz-continuous function.

Q6 Give an example of a locally Lipschitz-continuous function that is not globally Lipschitz.

Q7 State Gronwall's lemma and sketch its proof.

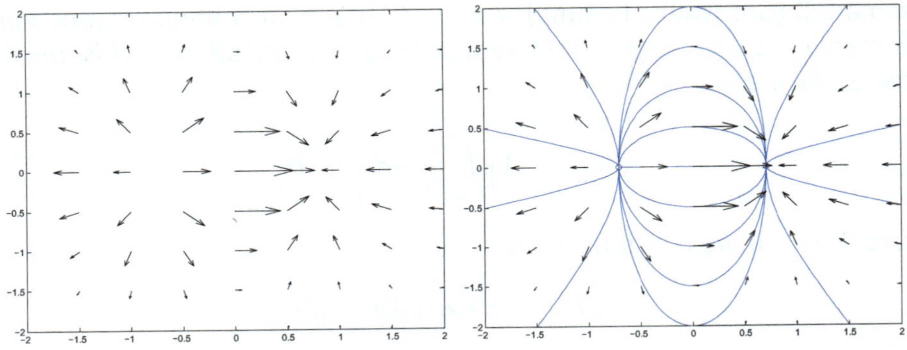

(a) (b)

Figure 6.5. Comparison of the velocity plot (a) and integral curves (b) for the gradient field $v(x, y) = \nabla f(x, y)$, where $f(x, y) = x \cdot \exp\left(-x^2 - y^2\right)$. In (b) we see that the velocity field is at any point tangential to the integral curve.

6.3 Integral Curves in Vector Fields: Ordinary Differential Equations – Part I

Following [162], pp. 137, an *integral curve* in a vector field $v : \Omega \to \mathbb{R}^d$ is a differentiable curve $\varphi : I \to \Omega$ defined on an interval $I \subset \mathbb{R}$ such that

$$\dot{\varphi}(t) = v\left(\varphi(t)\right)$$

holds for all $t \in I$. Interpreting v as the velocity field of a fluid, then the integral curve can be viewed as the trajectory of a particle transported by the fluid.

Example 6.9 (Integral Curves in a Rotation Field, cf. [162], p. 137). The conditions for an integral curve $\varphi = (\varphi_1, \varphi_2)$ in the rotation field $v(x, y) = (-y, x)$ are

$$\dot{\varphi}_1(t) = -\varphi_2(t)\,, \quad \text{and} \quad \dot{\varphi}_2(t) = \varphi_1(t)$$

or, in complex variables, $\dot{\varphi} = i\varphi$. Its solutions are the curves $\varphi(t) = c\exp(it)$, where $c \in \mathbb{C}$ is a constant.

Since $\dot{\varphi} = v(\varphi)$ we know that the velocity field v is at any point tangential to the integral curve, see Fig. 6.5. This observation immediately implies a numerical scheme for the approximation of integral curves based on the velocity field and known as *Euler's method* (see Chap. 7):

$$\varphi_{n+1}^{(h)} = \varphi_n^{(h)} + h \cdot v(\varphi_n^{(h)})\,.$$

In order to compute an update $\varphi_{n+1}^{(h)}$ we start at an initial point $\varphi_n^{(h)}$ and advance in the direction of the velocity field $v(\varphi_n^{(h)})$ at $\varphi_n^{(h)}$ with a step size h. For small values of h this ansatz is equivalent to the difference quotient approximation of the derivative $\dot{\varphi}$ and we hope that as $h \to 0$ the sequence of approximations $\{\varphi_n^{(h)}\}$ tends towards the exact solution φ of the (ordinary) differential equation $\dot{\varphi} = v(\varphi)$.

Of course, this method can only work if there is indeed a solution of $\dot{\varphi} = v(\varphi)$. An indication of the problems we face is given by the following example:

$$\dot{\varphi} = -\varphi^2 .$$

It is clear that a solution of this equation that passes through the point $(t, x) = (1, 1)$ is given by $\varphi(t) = t^{-1}$. However, this solution does not exist at $t = 0$. This shows that any general existence theorem will necessarily have to be of local nature and existence on a larger domain can only be established under additional conditions.

Let D be a domain in the (t, x) plane and suppose that F is a real-valued function such that $F \in \mathcal{C}(D)$. The central problem is to find a differentiable function φ defined on a real t interval I such that

(i) $(t, \varphi(t)) \in D$ for all $t \in I$, and

(ii) $\dot{\varphi}(t) = F(t, \varphi(t))$ for all $t \in I$.

This problem implies an *ordinary differential equation of the first order*, and denoted by

$$\dot{x} = F(t, x) . \tag{6.4}$$

If such an interval I and function φ exist, then φ is called a solution of the differential equation $\dot{x} = F(t, x)$ on I. Clearly, $\varphi \in \mathcal{C}^1(I)$ on account of (ii).

The first local existence result we state is the theorem of Cauchy-Peano. As we have seen in Chap. 3, it can be easily transfered to the setting of random differential equations. Following [64], pp. 3, the proof of this result proceeds in two stages: First, by construction we will show the existence of an "approximate" solution (in a sense to be made precise below), then we will prove that there is a sequence of these approximate solutions which tends towards the actual solution.

6.3.1 Approximate Solutions & Prerequisites for the Proof of the Cauchy-Peano Existence Theorem

Following [64], pp. 3, let F be a real-valued continuous function on a domain D in the (t, x) plane. An *ε-approximate solution* of $\dot{x} = F(t, x)$ on an interval I is a function $\varphi \in \mathcal{C}(I)$ such that

(i) $(t, \varphi(t)) \in D$ for all $t \in I$,

(ii) $\varphi \in \mathcal{C}^1(I)$, except possibly for a finite set of time points $S \subset I$, where $\dot{\varphi}$ may have simple discontinuities[4], and

(iii) $|\dot{\varphi}(t) - F(t, \varphi(t))| \leq \varepsilon$ for all $t \in I \setminus S$.

Any function $\varphi \in \mathcal{C}$ that satisfies the property (ii) on I is said to have a piecewise continuous derivative on I, and this is denoted by $\varphi \in \mathcal{C}^1_p(I)$.

Let $F \in \mathcal{C}$ be bounded on the rectangle

$$R \; := \; \{(t,x) \in D \, : \, |t - \tau| \leq a \, , \, |x - \xi| \leq b \, , \, a, b > 0\}$$

centered at the point (τ, ξ), and let

$$M \; := \; \max_{(t,x) \in \mathbb{R}} \{F(t,x)\} \, , \quad \text{as well as} \quad \alpha \; = \; \min\left(a \, , \, \frac{b}{M}\right) \, .$$

Theorem 6.1 (Local Existence of an ε-Approximate Solution). *Let $F \in \mathcal{C}(R)$. Given any $\varepsilon > 0$, there exists an ε-approximate solution φ of $\dot{x} = F(t,x)$ on $|t - \tau| \leq \alpha$ such that $\varphi(\tau) = \xi$.*

Proof. Following [64], pp. 3, let $\varepsilon > 0$ be given. Without loss of generality, an ε-approximate solution will be constructed for the interval $[\tau, \tau + \alpha]$ (an analogue construction can be carried out for $[\tau - \alpha, \tau]$). This approximate solution will consist of a polygonal path starting at (τ, ξ), i.e., a finite number of straight-line segments joined end to start.

Since $F \in \mathcal{C}(R)$, it is uniformly continuous, and hence, for the given ε, there exists a δ_ε such that

$$\left|F(t,x) - F(\tilde{t}, \tilde{x})\right| \; \leq \; \varepsilon \tag{6.5}$$

if $(t,x), (\tilde{t}, \tilde{x}) \in R$, and $|t - \tilde{t}| \leq \delta_\varepsilon$ as well as $|x - \tilde{x}| \leq \delta_\varepsilon$. Next, we partition the interval $[\tau, \tau + \alpha]$ into n segments

$$\tau \; = \; t_0 \; < \; t_1 \; < \; \ldots \; < \; t_n \; < \; \tau + \alpha$$

such that

$$\max |t_k - t_{k-1}| \; \min\left\{\delta_\varepsilon \, , \, \frac{\delta_\varepsilon}{M}\right\} \, . \tag{6.6}$$

From the initial point (τ, ξ) construct a straight-line segment with slope $F(\tau, \xi)$ proceeding to the right of τ until it intersects the line $t = t_1$ at some point (t_1, x_1). This segment must lie inside the triangular region T bounded by the lines issuing from (τ, ξ) with slopes M and $-M$, and the line $t = \tau + \alpha$, cf. Fig. 6.6. This follows immediately from the definition of α and the fact

[4] A function f is said to have a simple discontinuity at a point c if the right and left limits of f at c exist but are not equal.

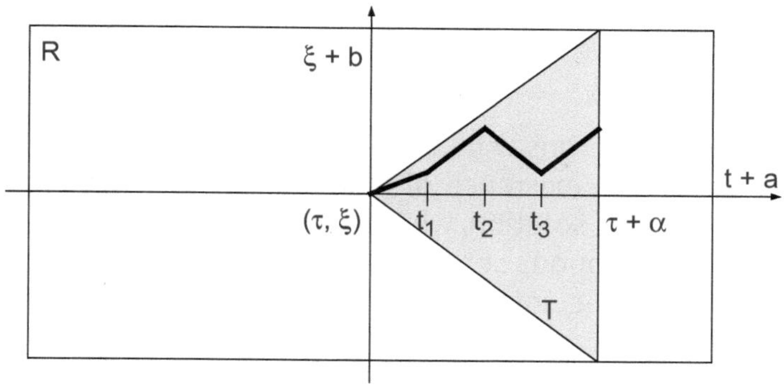

Figure 6.6. Construction of the straight-line segments as used in the proof of Theorem 6.1, cf. [64], p. 4.

that $|f(t,x)| \leq M$. In particular, the constructed segment actually meets the line $t = t_1$ in T. At the point (t_1, x_1) construct to the right of t_1 a straight-line segment with slope $F(t_1, x_1)$ up to the intersection with $t = t_2$, say at (t_2, x_2). If we continue in this fashion, in a finite number of steps the resultant path φ will meet the line $t = \tau + \alpha$. Further, the path will lie completely within T.

This φ is the required ε-approximate solution. Analytically, it may be expressed as

$$\begin{cases} \varphi(\tau) &= \xi \\ \varphi(t) &= \varphi(t_{k-1}) + F(t_{k-1}, \varphi(t_{k-1}))(t - t_{k-1}), \qquad t_{k-1} < t \leq t_k \end{cases} \tag{6.7}$$

for $k = 1, \ldots, n$.

From the construction of φ it is clear that $\varphi \in \mathcal{C}_p^1([\tau, \tau+\alpha])$, and that

$$|\varphi(t) - \varphi(\tilde{t})| \leq M|t - \tilde{t}|, \qquad \text{for all } t, \tilde{t} \in [\tau, \tau+\alpha]. \tag{6.8}$$

If t is such that $t_{k-1} < t < t_k$, then (6.8) together with (6.6) imply that $|\varphi(t) - \varphi(t_{k-1})| \leq \delta_\varepsilon$. But from (6.7) and (6.5),

$$|\dot{\varphi}(t) - F(t, \varphi(t))| = |F(t_{k-1}, \varphi(t_{k-1})) - F(t, \varphi(t))| \leq \varepsilon.$$

This shows that φ is an ε-approximate solution, as asserted. □

The construction of Theorem 6.1 is sometimes used as a practical means for finding the an approximate solution. In fact, what has been found is really a set of points $(t_k, \varphi(t_k))$ and these are joined by line segments. The points, by (6.7), satisfy the difference equation

$$x_k - x_{k-1} = (t_k - t_{k-1}) F(t_{k-1}, x_{k-1}).$$

This is a formulation of the Euler scheme that might be used on digital computers. It reduces to

$$x_k = x_{k-1} + hF(t_{k-1}, x_{k-1})$$

for equally spaced partitions with $h = t_k - t_{k-1}$ for $k = 1, \ldots, n$.

In order to prove the existence of a sequence of approximate solutions that tend to the analytic solution, where the only hypothesis is $f \in \mathcal{C}(R)$, the notion of an equi-continuous set of functions is required. A set of functions $G = \{F\}$ defined on a real interval I is said to equi-continuous on I if, given any $\varepsilon > 0$, there exists a $\delta_\varepsilon > 0$, independent of $F \in G$ and also $t, \tilde{t} \in I$ such that

$$\left|f(t) - f(\tilde{t})\right| < \varepsilon \,, \text{whenever } |t - \tilde{t}| < \delta_\varepsilon \,.$$

The fundamental property of such sets of functions required for the proof of Cauchy-Peano's existence theorem is given in the following lemma:

Lemma 6.10 (Lemma of Arzela-Ascoli). *On a bounded interval I, let $G = \{F\}$ be an infinite, uniformly bounded, equi-continuous set of functions. Then G contains a sub-sequence $\{F_n\}_{n\in\mathbb{N}} \subset G$ which is uniformly convergent on I.*

Proof. Following [64], pp. 5, let $\{r_k\}_{k\in\mathbb{N}}$ be the rational numbers in I enumerated in some order. The set of numbers $\{F(r_1)\}$, $F \in G$, is bounded, and hence there is a sequence of distinct functions $\{F_{n1}\}$, $F_{n1} \in G$, such that the sequence $\{F_{n1}(r_1)\}$ is convergent. Similarly, the set of numbers $\{F_{n1}(r_2)\}$ has a convergent sub-sequence $\{F_{n2}(r_2)\}$. In this way, we can define an infinite set of functions $F_{nk} \in G$, $n, k = 1, 2, \ldots$, which have the property that $\{F_{nk}\}$ converges at $r_1, r_2, \ldots, r_k$. Define F_n to be the function F_{nn}. Then the diagonal sequence $\{F_n\}$ is the required sequence which is uniformly convergent on I.

Clearly, $\{F_n\}$ converges at each of the rationals on I. Thus, given any $\varepsilon > 0$ and any rational number $r_k \in I$, there exists an integer $N_\varepsilon(r_k)$ such that

$$|F_n(r_k) - F_m(r_k)\| < \varepsilon \,, \qquad \text{for all } n, m > N_\varepsilon(r_k) \,.$$

For the given ε there is a δ_ε, independent of $t, \tilde{t}$ and $F \in G$ such that

$$\left|F(t) - F(\tilde{t})\right| < \varepsilon \,, \qquad \left|t - \tilde{t}\right| < \delta_\varepsilon \,.$$

Next, partition the interval I into a finite number of sub-intervals $I_1, I_2, \ldots, I_p$ such that the length of the largest sub-interval is less than δ_ε. For each I_k choose a rational number $\tilde{r}_k \in I_k$. If $t \in I$, then t is in some I_k, and hence

$$\begin{aligned} |F_n(t) - F_m(t)| &\leq |F_n(t) - F_n(\tilde{r}_k)| + |F_n(\tilde{r}_k) - F_m(\tilde{r}_k)| + |F_m(\tilde{r}_k) - F_m(t)| \\ &< 3\varepsilon \,, \end{aligned}$$

provided that $n, m > \max\{N_\varepsilon(\tilde{r}_1), \ldots, N_\varepsilon(\tilde{r}_p)\}$. This proves the uniform convergence of the sequence of the sequence $\{F_n\}$ on I. □

6.3.2 The Cauchy-Peano Existence Theorem

Theorem 6.2 (Cauchy-Peano's Existence of Solutions). *Let $F \in \mathcal{C}(R)$, then there exists a solution $\varphi \in \mathcal{C}^1$ of $\dot{x} = F(t,x)$ on $|t-\tau| \leq \alpha$ for $\varphi(\tau) = \xi$.*

Proof. Following [64], pp. 6, let $\{\varepsilon_n\}_{n\in\mathbb{N}}$ be a monotonously decreasing sequence of positive real numbers tending to zero as $n \to \infty$. By Theorem 6.1, for each ε_n there is an ε_n-approximate solution, φ_n, of $\dot{x} = F(t,x)$ on $|t-\tau| \leq \alpha$ such that $\varphi(\tau) = \xi$. Let us choose one such solution φ_n for each ε_n. From (6.8) it follows

$$\left|\varphi_n(t) - \varphi_n(\tilde{t})\right| \leq M\left|t - \tilde{t}\right| . \tag{6.9}$$

Applying (6.9) to $\tilde{t} = \tau$, we readily see, since $|t-\tau| \leq b/M$, that the sequence $\{\varphi_n\}$ is uniformly bounded by $|\xi| + b$. Moreover, (6.9) implies that $\{\varphi_n\}$ is an equi-continuous set. By the lemma of Arzela-Ascoli (Lemma 6.10) there is a sub-sequence $\{\varphi_{n_k}\}_{k\in\mathbb{N}}$ of $\{\varphi_n\}$ converging uniformly on $[\tau-\alpha, \tau+\alpha]$ towards a limit function φ, which must be continuous since each φ_n is continuous. (Indeed, it follows from (6.9) that $|\varphi(t) - \varphi(\tilde{t})| \leq M|t-\tilde{t}|$.)

This limit function φ is a solution of $\dot{x} = F(t,x)$ which meets the required specifications. To see this, we write the relation defining φ_n as an ε_n-approximate solution in an integral form, as follows:

$$\varphi_n(t) = \xi + \int_\tau^t \left(F(s, \varphi_n(s)) + \Delta_n(s)\right) \mathrm{d}s , \tag{6.10}$$

where $\Delta_n(t) = \dot{\varphi}_n(t) - F(t, \varphi_n(t))$ at those points where $\dot{\varphi}_n$ exists, and $\Delta_n(t) = 0$ otherwise. Because φ_n is an ε_n-approximate solution, $|\Delta_n(t)| \leq \varepsilon_n$ holds. Since F is uniformly continuous on the rectangle R, and $\varphi_{n_k} \to \varphi$ uniformly on $[\tau-\alpha, \tau+\alpha]$, as $k \to \infty$, it follows that $F(t, \varphi_{n_k}(t)) \to F(t, \varphi(t))$ uniformly on $[\tau-\alpha, \tau+\alpha]$, as $k \to \infty$. Replacing n by n_k in (6.10), we obtain

$$\varphi_n(t) = \xi + \int_\tau^t F(s, \varphi(s)) \mathrm{d}s \tag{6.11}$$

with $k \to \infty$. But from (6.11), and upon differentiation, $\dot{\varphi}(t) = F(t, \varphi(t))$, for $F(t, \varphi(t))$ is a continuous function. It is clear from this that φ is a solution of $\dot{x} = F(t,x)$ on $|t-\tau| \leq \alpha$ of class $\mathcal{C}$. □

In general, the choice of a sub-sequence of $\{\varphi_n\}$ in the above proof is necessary, for there are a polygonal paths $\{\varphi_n\}$ which diverge everywhere on a whole interval about $t = \tau$ as $\varepsilon_n \to 0$, see problem 6.40.

If we assume that a solution of $\dot{x} = F(t,x)$ through (τ, ξ), if it exists, is unique, then every sequence of polygonal paths $\{\varphi_n\}$ for which $\varepsilon_n \to 0$ must converge on $|t-\tau| \leq \alpha$, and hence uniformly, to a solution, for $\{\varphi_n\}$ is an equi-continuous set on $|t-\tau| \leq \alpha$. Let us prove this assertion by reducio ad absurdum: Suppose this would be wrong. Then there would be a sequence of polygonal paths $\{\varphi_n\}$ divergent at some point on $|t-\tau| \leq \alpha$. This implies the existence of at lest two sub-sequences of $\{\varphi_n\}$ tending to different limit functions. Both will be solutions, and this leads to a contradiction. Therefore, if uniqueness is assured, the choice of a sub-sequence in the Cauchy-Peano existence theorem (Theorem 6.2) is unnecessary.

It can happen that the choice of a sub-sequence is unnecessary even though uniqueness is not satisfied. The following example illustrates this.

Example 6.11 (Existence of an Infinite Number of Solutions, cf. [64], p. 7). The ordinary differential equation

$$\dot{x} = x^{1/3} \tag{6.12}$$

has an infinite number of solutions starting at $(0,0)$ which exist on $[0,1]$. For any c such that $0 \leq c \leq 1$, the function φ_c defined by

$$\begin{aligned} \varphi_c(t) &= 0, \qquad 0 \leq t \leq c, \\ \varphi_c(t) &= \left(\tfrac{2}{3}(t-c)\right)^{3/2}, \qquad c < t \leq 1, \end{aligned}$$

is a solution of (6.12) on $[0,1]$.

If the construction of Theorem 6.1 is applied to equation (6.12), one finds that the only polygonal path starting at the point $(0,0)$ is φ_1. This shows that this method can not, in general, give all solutions of $\dot{x} = F(t,x)$.

Theorem 6.3 (Existence of a Local Solution). *Let $F \in C$ on a domain D in the (t,x) plane, and suppose (τ,ξ) is any point in D. Then there exists a solution φ of $\dot{x} = F(t,x)$ on some t interval containing τ in its interior.*

Proof. Following [64], p. 7, since D is open, there exists an $r > 0$ such that all points, whose distance from (τ,ξ) is less than r, are contained in D. Let R be any closed rectangle containing (τ,ξ), and contained in this open circle of radius r. Then the Cauchy-Peano existence theorem (Theorem 6.2) applied to $\dot{x} = F(t,x)$ on R gives the required result. □

Application of the Euler method does not always lead to the expected results, see Fig. 6.7.

This and questions of uniqueness and global existence motivate the study of ordinary differential equations from theoretical and numerical points of view; see Chap. 7 for the discussion of numerical schemes.

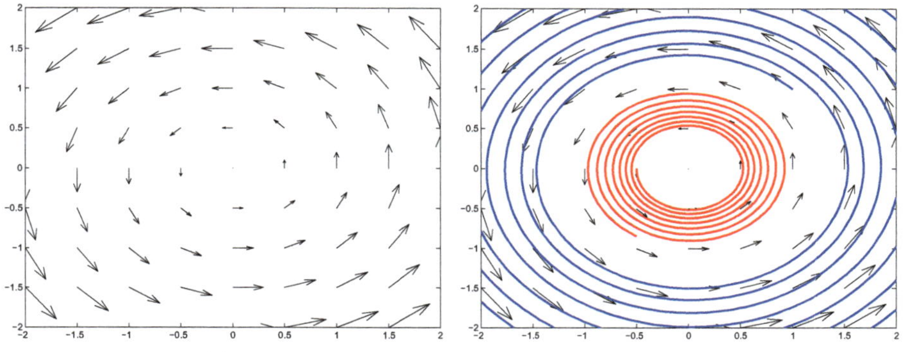

(a) (b)

Figure 6.7. Comparison of the velocity plot (a) and the Euler approximation with $h = 0.03$ (b) for the rotation field $v(x, y) = (-y, x)$. We see a spiraling out of the approximate solutions in contrast to the periodic orbits we derived as the real solutions in example 6.9.

Before you continue, make sure to answer the following questions:

Quiz: Section 6.3

Q1 Give the definition of an integral curve.

Q2 What is an ε-approximate solution of an ordinary differential equation?

Q3 What can you say about the local existence of ε-approximate solutions?

Q4 Sketch the proof of the assertion you used in Q3.

Q5 State the lemma of Arzela-Ascoli and sketch its proof.

Q6 State the Cauchy-Peano existence theorem.

Q7 Sketch the proof of Cauchy-Peano's existence theorem.

Q8 Give an example of an ordinary differential equation that has infinitely many solutions.

Q9 What can you say about the existence of local solutions?

Q10 Sketch the proof of the assertion you used in Q9.

6.4 Integral Curves in Vector Fields: Ordinary Differential Equations – Part II

In this chapter the existence of integral curves is studied in the broader setting of *time-dependent vector fields*, termed as *continuous dynamical systems*. Dynamical systems will be studied from a general point of view in Chap. 8. For now, a (continuous) dynamical system is a continuous mapping $S : U \to \mathbb{K}^d$ on a open set $U \subset \mathbb{R} \times \mathbb{K}^d$, where $\mathbb{K} = \mathbb{R}$ or $\mathbb{K} = \mathbb{C}$. Elements in U are denoted by the pair (t, x) with "time" $t \in \mathbb{R}$ and "position" $x \in \mathbb{K}^d$.

A *solution* or *integral curve* of the vector field S is a differentiable curve $\varphi : I \to \mathbb{K}^d$ on an interval $I \subset \mathbb{R}$ such that $(t, \varphi(t)) \in U$ and

$$\dot{\varphi}(t) \;=\; S\left(t, \varphi(t)\right) , \qquad \text{for all } t \in I .$$

It is common to write this last identity in the form of a time-dependent *first order ordinary differential equation* with right-hand side $F(t, x)$ (instead of $S(t, x)$):

$$\dot{x} \;=\; F(t, x) ,$$

or, equivalently as the *first order system of ordinary differential equation*

$$\begin{aligned} \dot{x}_1 &= F_1(t, x_1, x_2, \ldots, x_d) , \\ &\vdots \\ \dot{x}_d &= F_d(t, x_1, x_2, \ldots, x_d) . \end{aligned}$$

Often, an *initial condition* $(t_0, x_0) \in U$ is given and a specific solution φ is demanded such that $\varphi(t_0) = x_0$. This setting is called the *initial value problem* (IVP) and formally denoted by

$$\dot{x} \;=\; F(t, x) , \quad x(t_0) \;=\; x_0 .$$

For a given ordinary differential equation $\dot{x} = F(t, x)$, its corresponding integral equation is gained by applying the fundamental theorem of calculus:

Lemma 6.12 (Equivalence of Ordinary Differential Equations and Integral Equations). *Let $F : U \to \mathbb{K}^d$ be continuous on the open set $U \subset \mathbb{R} \times \mathbb{K}^d$. A continuous function $\varphi : I \to \mathbb{K}^d$ on the interval $I \subset \mathbb{R}$ with $(t, \varphi(t)) \in U$ solves the initial value problem*

$$\dot{\varphi}(t) \;=\; F\left(t, \varphi(t)\right) , \quad \varphi(t_0) \;=\; x_0 ,$$

for all $t \in I$ if and only if

$$\varphi(t) \;=\; \varphi(t_0) + \int_{t_0}^{t} F\left(s, \varphi(s)\right) ds$$

holds for all $t \in I$.

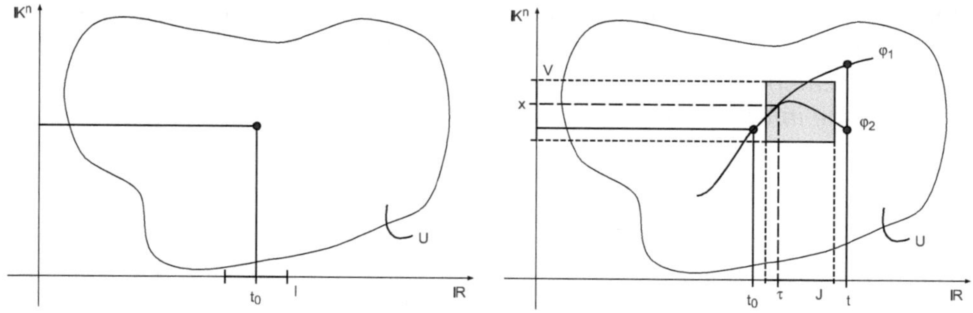

Figure 6.8. Sketch of the situation discussed in the proof of Theorem 6.4.

This essential reformulation enables us to work with continuous functions instead of differentiable ones and has advantages when considering limit processes such as, for example, in the construction of solutions of ordinary differential equations.

6.4.1 Local Existence & Uniqueness of Solutions

Theorem 6.4 (Uniqueness of Integral Curves). *Let the dynamical system $F : U \to \mathbb{K}^d, U \subset \mathbb{R} \times \mathbb{K}^d$ be locally Lipschitz continuous in x. If two integral curves $\varphi_1, \varphi_2 : I \to \mathbb{K}^d$ of F coincide in at a time $t_0 \in I$, then $\varphi_1(t) \equiv \varphi_2(t)$ for all $t \in I$.*

Proof. Following [162], pp. 140, let $I' \subset I$ be the set of times $t \in I$ for which $\varphi_1(t) = \varphi_2(t)$ holds. Due to continuity I' is closed in I, and we will show that I' is open in I as well.

Let $t_0 \in I'$ and $J \times V \subset U$ be a neighborhood of $(t_0, \varphi_1(t_0))$ in which F is locally Lipschitz-continuous with Lipschitz constant L, cf. Fig. 6.8. Let $\psi := \varphi_2 - \varphi_1$. Because of $\psi(t_0) = 0$ we get with Lemma 6.12

$$\|\psi(t)\| \le \left| \int_{t_0}^{t} \|F(s, \varphi_2(s)) - F(s, \varphi_1(s))\| \, \mathrm{d}s \right| \le L \cdot \left| \int_{t_0}^{t} \|\psi(s)\| \, \mathrm{d}s \right|$$

for all $t \in J \cap I$. By Gronwall's lemma (Lemma 6.8) we have that $\psi \equiv 0$ on $J \cap I$, i.e., $J \cap I \subset I'$. Finally, since I' is not empty and I is connected we have that $I' \equiv I$. □

Remark 6.13 (The Lipschitz-Condition is Essential). The uniqueness theorem for integral curves does not hold without the Lipschitz condition on F in general. For instance, consider the scalar initial value problem $\dot{x} = \sqrt{|x|}$, $x_0 = 0$. This has infinitely many solutions on $\mathbb{R}$. Some of which are $x(t) \equiv 0$ and $x(t) = \frac{1}{4}\mathrm{sign}(t) \cdot t^2$.

Theorem 6.5 (Local Existence; Picard-Lindelöf). *Let the dynamical system $F : U \to \mathbb{K}^d$ be locally Lipschitz continuous in x on an open set $U \subset \mathbb{R} \times \mathbb{K}^d$. Then, for every element $(t_0, x_0) \in U$ there is an interval $I_\delta(t_0) = (t_0 - \delta, t_0 + \delta)$, $\delta > 0$, on which the initial value problem*

$$\dot{x} \ = \ F(t, x)\,, \qquad x(t_0) \ = \ x_0\,, \tag{6.13}$$

has a unique solution.

More precisely: Let $Q := \overline{I_a(t_0)} \times \overline{K_b(x_0)}$ be an arbitrary compact cuboid in U on which F is Lipschitz-continuous in x with Lipschitz-constant L. Let $\delta \leq a$ be a positive number such that $\delta \|F\|_Q \leq b$ and $\delta L < 1$. Then, the initial value problem (6.13) has a unique solution φ on $I_\delta(t_0)$. This solution lies in $\overline{K_b(x_0)}$ since we have

$$\|\varphi(t) - x_0\| \ \leq \ b\,, \qquad \text{for all } t \in I_\delta(t_0)\,, \tag{6.14}$$

and it is the limit of the sequence $(\varphi_k)_{k \in \mathbb{N}}$ of Picard-Lindelöf iteration functions *which are defined as*

$$\varphi_0(t) \ := \ x_0\,, \qquad \varphi_{k+1}(t) \ := \ x_0 + \int_{t_0}^{t} F\left(s, \varphi(s)\right) ds\,. \tag{6.15}$$

The sequence $(\varphi_k)_{k \in \mathbb{N}}$ converges uniformly on $I_\delta(t_0)$.

Proof. Following [162], pp. 141, it is sufficient to construct a continuous function $\varphi : I_\delta(t_0) \to \mathbb{K}^d$ such that φ satisfies the inequality (6.14) for all $t \in I_\delta(t_0)$ as well as the integral equality

$$\varphi(t) \ = \ x_0 + \int_{t_0}^{t} F\left(s, \varphi(s)\right) \mathrm{d}s\,. \tag{6.16}$$

We interpret (6.16) as a fixed point identity: Therefore, let $\mathcal{M}$ be the space of all continuous functions $\psi : I_\delta(t_0) \to \mathbb{K}^d$ such that

$$\|\psi(t) - x_0\| \ \leq \ b\,, \qquad \text{for all } t \in I_\delta(t_0)\,,$$

and P be that mapping, that maps a function $\psi \in \mathcal{M}$ to $P\psi : I_\delta(t_0) \to \mathbb{K}^d$ by

$$(P\psi)(t) \ := \ x_0 + \int_{t_0}^{t} F\left(s, \varphi(s)\right) \mathrm{d}s\,.$$

$P\psi$ is continuous and satisfies the inequality

$$\|(P\psi)(t) - x_0\| = \left\| \int_{t_0}^{t} F\left(s, \psi(s)\right) \mathrm{d}s \right\| \leq \left| \int_{t_0}^{t} \|F\left(s, \psi(s)\right)\| \,\mathrm{d}s \right| \leq \delta \|F\|_Q \leq b\,.$$

Thus, for $\psi \in \mathcal{M}$ we also have $P\psi \in \mathcal{M}$.

With the aid of P, the integral equality (6.16) reads as the fixed point identity $P\varphi = \varphi$. In order to apply the Banach fixed point theorem, we introduce a metric on $\mathcal{M}$ as follows: For $\psi_1, \psi_2 \in \mathcal{M}$ we define

$$d(\psi_1, \psi_2) := \sup \{\|\psi_1(t) - \psi_2(t)\| : t \in I_\delta(t_0)\} .$$

In this metric the sequence $(\psi_k)_{k\in\mathbb{N}}$ converges in $\mathcal{M}$ if and only if it converges uniformly on $I_\delta(t_0)$. As $\overline{K_b(x_0)}$ is bounded it follows that $(\mathcal{M}, d)$ is a complete metric space (i.e. a Banach space). Moreover, $P : \mathcal{M} \to \mathcal{M}$ is a contraction because

$$\begin{aligned} d(P\psi_1, P\psi_2) &= \sup_I \left\| \int_{t_0}^t (F(s, \psi_1(s)) - F(s, \psi_2(s)))\, \mathrm{d}s \right\| \\ &\le \sup_I \left| \int_{t_0}^t \|F(s, \psi_1(s)) - F(s, \psi_2(s))\|\, \mathrm{d}s \right| \\ &\le \sup_I \left| \int_{t_0}^t L\, \|\psi_1(s) - \psi_2(s)\|\, \mathrm{d}s \right| \\ &\le \delta \cdot L \cdot d(\psi_1, \psi_2), \quad \text{with } \delta L < 1 . \end{aligned}$$

Thus, Banach's fixed point theorem can be applied and there exists a unique function $\varphi \in \mathcal{M}$ such that $P\varphi = \varphi$. Here, φ lies in $\overline{K_b(x_0)}$ and solves the initial value problem (6.13). □

Example 6.14 (Construction of a Solution, cf. [162], pp. 142). Let us apply the Picard-Lindelöf iteration for the rotation field $v(x, y) = (-y, x)$. We apply it to the initial value problem

$$\begin{aligned} \dot{x} &= -y, \quad \text{with } x(0) = 1 , \\ \dot{x} &= x, \quad \text{with } y(0) = 0 . \end{aligned}$$

We obtain

$$\begin{aligned} \varphi_1(t) &= \begin{pmatrix} 1 \\ 0 \end{pmatrix} + \int_0^t \begin{pmatrix} 0 \\ 1 \end{pmatrix} \mathrm{d}s = \begin{pmatrix} 1 \\ t \end{pmatrix} \\ \varphi_2(t) &= \begin{pmatrix} 1 \\ 0 \end{pmatrix} + \int_0^t \begin{pmatrix} -s \\ 1 \end{pmatrix} \mathrm{d}s = \begin{pmatrix} 1 - \frac{1}{2}t^2 \\ t \end{pmatrix} \\ \varphi_3(t) &= \begin{pmatrix} 1 \\ 0 \end{pmatrix} + \int_0^t \begin{pmatrix} -s \\ 1 - \frac{1}{2}s^2 \end{pmatrix} \mathrm{d}s = \begin{pmatrix} 1 - \frac{1}{2}t^2 \\ t - \frac{1}{3!}t^3 \end{pmatrix} \\ &\vdots \end{aligned}$$

$$\varphi_{2k+1}(t) = \begin{pmatrix} 1 - \frac{1}{2!}t^2 + - \cdots + (-1)^k \frac{1}{(2k)!} t^{2k} \\ t - \frac{1}{3!}t^3 + - \cdots + (-1)^k \frac{1}{(2k+1)!} t^{2k+1} \end{pmatrix}.$$

The components of φ_{2k+1} are the Taylor polynomials of the sine function and the cosine function, respectively, of order $2k+1$. In the limit $k \to \infty$ the solution of the initial value problem results as

$$\varphi(t) = \begin{pmatrix} \cos(t) \\ \sin(t) \end{pmatrix}.$$

6.4.2 Interlude: Solving ODEs Symbolically with MATLAB

For the symbolic solution of differential equations with MATLAB, we can use the dsolve command, cf. [134]. In order to calculate the solution of the scalar (ordinary) differential equation $\dot{x} = t \cdot x$ type:

```
>> x = dsolve('Dx = t*x', 't')
x = C1*exp(1/2*t^2)
```

MATLAB uses capital D to indicate the derivative and requires that the entire equation appears in single quotes. By default, MATLAB assumes t to be the

Before you continue, make sure to answer the following questions:

Quiz: Section 6.4 – Part I

Q1 How are integral curves and solutions of ordinary differential equations related?

Q2 In what respect are ordinary differential equations and integral equations equivalent?

Q3 Under which conditions are integral curves unique?

Q4 Sketch the proof of your assertion from Q3.

Q5 State the theorem of Picard-Lindelöf.

Q6 Sketch the proof of the Picard-Lindelöf theorem. What is the essential step in this proof?

Q7 Apply the Picard-Lindelöf iteration to solve the initial value problem $\dot{x} = tx$, $x(0) = 1$.

independent variable, so we could have used the command x = dsolve('Dx = t*x') without ambiguity.

If we want to use the same equation a number of times, we may define it as a variable:

```
>> eqn1 = 'Dy = x*y'
>> y = dsolve(eqn1, 'x')
y = C1*exp(1/2*x^2)
```

To solve the corresponding initial value problem $y'(x) = x \cdot y$ with respect to $y(1) = 1$, we use

```
>> y = dsolve(eqn1, 'y(1)=1', 'x')
y = 1/exp(1/2)*exp(1/2*x^2)
```

or

```
>> inits1 = 'y(1)=1';
>> y = dsolve(eqn1, inits1, 'x')
y = 1/exp(1/2)*exp(1/2*x^2)
```

Now that we have solved the differential equation we may want to plot the solution to get a rough idea of its behavior. Here, we immediately run into two problems: (i) the expression y we get from MATLAB is not suited for array operations and (ii) y is a symbol or symbolic object. Following [134], the first of these obstacles is straightforward to fix by applying the vectrorize command. For the second, we employ the eval command, that evaluates or executes text strings that constitute valid MATLAB commands. Hence, we can use

```
>> x = linspace(0,1,20);
>> z = eval( vectorize(y) );
>> plot(x,z)
```

to gain the result shown in Fig. 6.9 (a).

Consider now the solution of the second order equation

$$y''(x) + 8y'(x) + 2y(x) = \cos(x), \qquad y(0) = 0, \quad y'(0) = 1.$$

Then, the following, self-explanatory MATLAB code solves the equation and draws Fig. 6.9 (b):

```
>> eqn2 = 'D2y + 8*Dy + 2*y = cos(x)';
>> inits2 = 'y(0) = 0, Dy(0) = 1';
>> y = dsolve(eqn2, inits2, 'x')
y = 1/65*cos(x) + 8/65*sin(x)
   + (-1/130+53/1820*14^(1/2))*exp((-4+14^(1/2))*x)
   - 1/1820*(53+14^(1/2))*14^(1/2)*exp(-(4+14^(1/2))*x)
>> x = linspace(0,1,20);
>> z = eval( vectorize(y) );
>> plot(x,z)
```

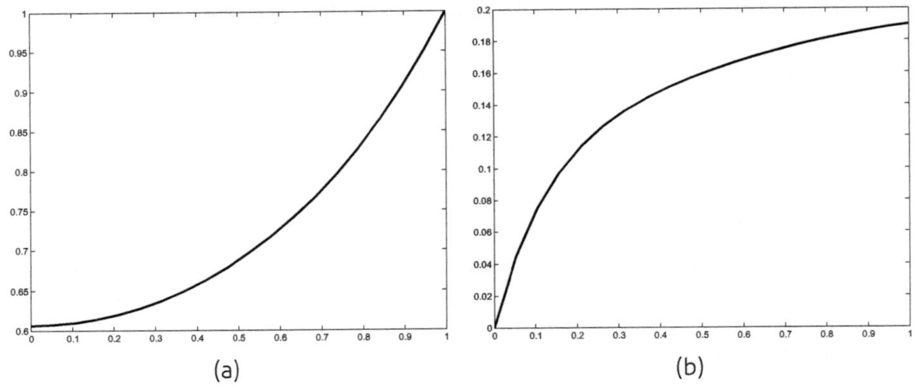

Figure 6.9. (a) Graph of the solution y of $y'(x) = x \cdot y$ with respect to $y(1) = 1$ and (b) graph of the solution y of $y''(x) + 8y'(x) + 2y(x) = \cos(x)$ with respect to $y(0) = 0$, $y'(0) = 1$.

Finally, we to solve and plot solutions to the linear system

$$\dot{x} = x + 2y - z\,, \tag{6.17}$$
$$\dot{y} = x + z\,, \tag{6.18}$$
$$\dot{z} = 4x - 4y + 5z\,, \tag{6.19}$$

of three coupled equation, cf. [134]. We first find the general solution analogously to the scalar case. Here, we just have to brace each equation in its own pair of single quotation marks:

```
>> [x, y, z] = dsolve('Dx = x + 2*y - z', 'Dy = x + z',
          'Dz = 4*x - 4*y + 5*z')
x = -C1*exp(3*t) + 2*C1*exp(2*t) + 2*C2*exp(2*t) - 2*C2*exp(t)
  - 1/2*C3*exp(3*t) + 1/2*C3*exp(t)
y = C1*exp(3*t) - C1*exp(2*t) + 2*C2*exp(t) - C2*exp(2*t)
  + 1/2*C3*exp(3*t) - 1/2*C3*exp(t)
z = -4*C1*exp(2*t) + 4*C1*exp(3*t) - 4*C2*exp(2*t)
  + 4*C2*exp(t) - C3*exp(t) + 2*C3*exp(3*t)
```

Of course, if we use MATLAB to double check analytic solutions of this system, we have to keep in mind that MATLAB's choice of the constants C_1, C_2 and C_3 probably would not correspond to our own choice. E.g., we might have $C := -2C_1 + \frac{1}{2}C_3$ such that the coefficients of $\exp(t)$ in the expression for x are combined. Fortunately, there is no such ambiguity when the initial values are assigned.

To solve a corresponding initial value problem, e.g. with $x(0) = 1$, $y(0) = 2$, and $z(0) = 3$ we can use

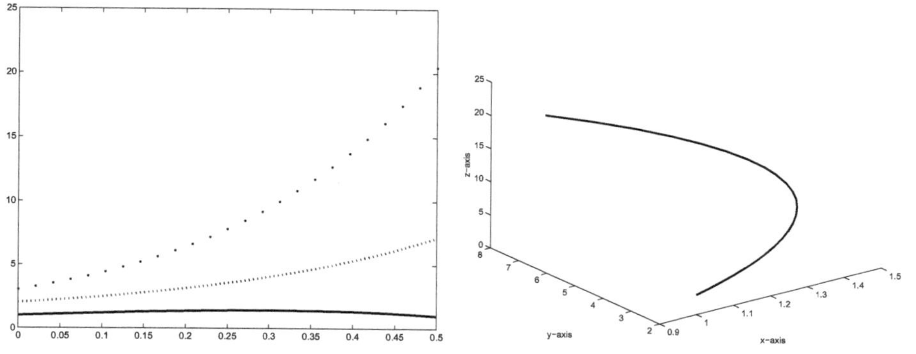

Figure 6.10. (a) Graph of the solutions $x(t)$, $y(t)$ and $z(t)$ of the system (6.17)-(6.19) over time with respect to the initial conditions $x(0) = 1$, $y(0) = 2$, and $z(0) = 3$, and (b) solution trajectory in the x-y-z-space starting at $(x(0), y(0), z(0))^T = (1, 2, 3)^T$.

```
>> inits3 = 'x(0) = 1, y(0) = 2, z(0) = 3';
>> [x, y, z] = dsolve('Dx = x + 2*y - z', 'Dy = x + z',
           'Dz = 4*x - 4*y + 5*z', inits3)
x = -5/2*exp(3*t) + 6*exp(2*t) - 5/2*exp(t)
y = 5/2*exp(3*t) - 3*exp(2*t) + 5/2*exp(t)
z = -12*exp(2*t) + 10*exp(3*t) + 5*exp(t)
```

Finally, we plot this solution to obtain the results displayed in Fig. 6.10:

```
>> t = linspace(0, .5, 25);
>> xx = eval( vectorize(x) );
>> yy = eval( vectorize(y) );
>> zz = eval( vectorize(z) );
>> plot(t, xx, '-k', t, yy, ':k', t, zz, '.k')
```

and

```
plot3(xx, yy, zz,'-k')
```

Compared to the velocity plots shown earlier, we immediately see that Figures 6.9 and 6.10 provide less (global) information. Here, we just see one integral curve, but cannot estimate the behavior of nearby integral curves (at least without solving the differential equations with different initial conditions again). This global point of view on vector/ solution fields will be discussed in depth in Chap. 8 on deterministic dynamical systems.

The other observation we should make here is that we cannot estimate the behavior of the integral curves beyond the time-frames in which we plotted them. This leads to questions about extensibility and maximality of integral curves.

6.4.3 Maximal Integral Curves

Definition 6.15 (Maximal Integral Curve). A integral curve $\varphi : I \to \mathbb{K}^d$ of the dynamical system $F : U \to \mathbb{K}^d$ through the point $(t_0, \varphi(t_0))$ is called *maximal* if for every other integral curve $\psi : J \to \mathbb{K}^d$ through this point it holds that: $J \subset I$ and $\psi = \varphi_{|J}$.

Lemma 6.16 (Uniqueness of a Maximal Integral Curve). *If the dynamical system $F : U \to \mathbb{K}^d$ is locally Lipschitz-continuous with respect to x uniformly in t, then every initial value problem*

$$\dot{x} = F(x,t), \qquad x(t_0) = x_0$$

has one and only one maximal solution.

Proof. Following [162], p. 143, let I be the union of all intervals I_α on which the initial value problem has a solution φ_α, where α is an element of a suitable index set. For $t \in I$ choose a I_α such that $t \in I_\alpha$ and set $\varphi(t) := \varphi_\alpha(t)$.

Let I_β be another interval such that $t \in I_\beta$ and $[t_0, t] \subset I_\alpha \cap I_\beta$. Due to the uniqueness Theorem 6.4 we obtain $\varphi_\alpha(t) = \varphi_\beta(t)$. Thus, $\varphi(t)$ is defined independently from I_α. Of course, $\varphi(t) : I \to \mathbb{K}^d$ is a maximal curve. □

The following theorem gives an essential assertion about the intervals of definition for maximal integral curves:

Theorem 6.6 (Finite Life-Time of Solutions). *A maximal integral curve which has finite life-time leaves every compact set.*

More precisely: Let $\varphi : (\alpha, \beta) \to \mathbb{K}^d$ be a maximal integral curve of the dynamical system $F : U \to \mathbb{K}^d$ which is locally Lipschitz-continuous in x uniformly in t. If $\beta < \infty$, then for any compact set $K \subset U$ and every $\varepsilon > 0$ there is a $\tau \in (\beta - \varepsilon, \beta)$ such that $(\tau, \varphi(\tau)) \notin K$. An analogous assertion holds for $\alpha > -\infty$.

Proof. Following [162], p. 143, assume that for all $t \in (\beta - \varepsilon, \beta)$ it holds that $(t, \varphi(t)) \in K$. First, we claim that φ can be continued continuously on $(\alpha, \beta]$. Therefore it suffices to show that φ is uniformly continuous on $(\beta - \varepsilon, \beta)$. This follows from the fact that when $M := \|F\|_K$ for all $t_1, t_2 \in (\beta - \varepsilon, \beta)$, the following inequality holds:

$$\|\varphi(t_2) - \varphi(t_1)\| = \left\| \int_{t_1}^{t_2} \dot{\varphi}(s) \mathrm{d}s \right\| \leq \left| \int_{t_1}^{t_2} \|F(s, \varphi(s))\| \, \mathrm{d}s \right| \leq M \cdot |t_2 - t_1| \, .$$

Let the continuous continuation of φ on $(\alpha, \beta]$ be denoted by $\tilde{\varphi}$. Next, we show that $\tilde{\varphi}$ is an integral curve of F, too. Because K is closed, we have

$(\beta, \tilde{\varphi}(\beta)) \in K$ and thus in U. Moreover, we have for arbitrary $t, t_0 \in (\alpha, \beta)$ that

$$\tilde{\varphi}(t) = \tilde{\varphi}(t_0) + \int_{t_0}^{t} F(s, \tilde{\varphi}(s))\, \mathrm{d}s\,. \tag{6.20}$$

Due to the continuity of $\tilde{\varphi}$ on $(\alpha, \beta]$, equality (6.20) holds for $t = \beta$ also. Hence, the function $\tilde{\varphi} : (\alpha, \beta] \to \mathbb{K}^d$ solves the differential equation $\dot{x} = F(x, t)$. This contradicts that $\varphi : (\alpha, \beta) \to \mathbb{K}^d$ is a maximal solution. □

For a dynamical system on the product structure $I \times \Omega$ the following important result holds:

Corollary 6.17 (Leaving Every Compact Set). *Let $\varphi : (\alpha, \beta) \to \mathbb{K}^d$ be a maximal integral curve of the dynamical system $F : I \times \Omega \to \mathbb{K}^d$ which is locally Lipschitz-continuous in x uniformly in t. If β is not the right boundary point of the interval I, then for every compact subset $J \subset \Omega$ and for every $\gamma \in (\alpha, \beta)$ there is a $t \in (\gamma, \beta)$ such that $\varphi(t) \notin K$. An analogue result holds for α.*

If φ lies completely in a compact subset of Ω, then φ is defined on the whole of I.

Proof. Following [162], p. 144, let $[\gamma, \beta] \times K$ be a compact subset of $I \times \Omega$. Then, Theorem 6.6 can be applied to this subset. □

There are however vector fields defined everywhere on $\mathbb{R} \times \mathbb{K}^d$ with perfect differentiability properties, which do not have a solution defined for the whole of $\mathbb{R}$.

Example 6.18 (A Solution of an ODE with Blow-Up in Finite Time, cf. [162], p. 144). Consider $\dot{x} = 1 + x^2$ on $\mathbb{R} \times \mathbb{R}$. The solutions $\varphi_c(t) = \tan(t - c)$ on the intervals $I_{\pi/2}(c)$ are already solutions on maximal intervals: a solution that would be defined on an interval of length $> \pi$ coincides with φ_c on a certain interval $I_{\pi/2}(c)$ due to the uniqueness Theorem 6.4. This is impossible because of $|\varphi_c(t)| \to \infty$ for $t \to c \pm \pi/2$.

Let $\dot{x} = 1 + x^2$ be the velocity of a point moving in the line $\mathbb{R}$. Then, its velocity increases in a proportion that is faster than the distance $|x|$, and consequently this point escapes to infinity in finite time.

On the other hand, if $\dot{x}$ increases at most proportionally to $|x|$ then the escape to infinity requires infinite time.

An analogous result of escape to infinity in infinite time is obtained in the general case of linearly bounded mappings:

Definition 6.19 (Linearly Bounded Mapping). A mapping $F : I \times \mathbb{K}^d \to \mathbb{K}^d$ is called *linearly bounded* if there are continuous functions $a, b : I \to \mathbb{R}$ such that

$$\|F(t,x)\| \leq a(t)\|x\| + b(t)\,.$$

holds for all $(t,x) \in I \times \mathbb{K}^d$

Theorem 6.7 (Life-Time of Linearly Bounded Integral Curves). *Every maximal integral curve φ of a linearly bounded and locally Lipschitz-continuous dynamical system $F : I \times \mathbb{K}^d \to \mathbb{K}^d$ in x uniformly in t is defined everywhere on I.*

Proof. Following [162], p. 144, let $(\alpha, \beta) \subset I$ be the interval of definition of φ. Assume that β is not the right boundary point of I, then φ would be unbounded on $[t_0, \beta)$ for an arbitrary $t_0 \in (\alpha, \beta)$. Next, from

$$\varphi(t) = \varphi(t_0) + \int_{t_0}^{t} F\left(s, \varphi(s)\right) \mathrm{d}s$$

it follows that

$$\|\varphi(t)\| \leq \|a\|_{[t_0,\beta]} \cdot \int_{t_0}^{t} \|\varphi(s)\|\, \mathrm{d}s + \|\varphi(t_0)\| + \|b\|_{[t_0,\beta]} \cdot |\beta - t_0|\,.$$

Due to this estimate, φ must be bounded because of Gronwall's lemma on $[t_0, \beta]$. This is a contradiction, and thus β is already the right boundary point. □

6.4.4 Maximal Integral Curves in Time-Independent Vector Fields

Let $v : \Omega \to \mathbb{K}^d$ be a vector field on an open subset $\Omega \subset \mathbb{K}^d$. The corresponding ordinary differential equation $\dot{x} = F(x,t) = v(x)$ is termed an *autonomous ordinary differential equation*, and Ω is called its *phase space*. If v is (locally) Lipschitz-continuous with respect to x, then F has the same property with respect to x. Far-reaching importance has the result that every maximal integral curve of a locally Lipschitz-continuous vector field is either constant, periodic or has no retuning point. To show this assertion, we rely on two simple yet essential remarks on time shifts for integral curves.

Proposition 6.20 (Remarks on Time Shifts). *Let $\varphi : I \to \Omega$ be a maximal integral curve of a locally Lipschitz-continuous vector field $v : \Omega \to \mathbb{K}^d$. Then the following holds:*

1. *For every $c \in \mathbb{R}$ the function $\varphi_c : I + c \to \Omega$, $\varphi_c(t) := \varphi(t-c)$ is also a maximal integral curve of v.*
2. *Let $\psi : J \to \Omega$ be a maximal integral curve such that $\psi(s) = \varphi(r)$ for a times $s \in J$ and $r \in I$, respectively. Then $J = I + s - r$ and $\psi = \varphi_{s-r}$ holds.*

Before you continue, make sure to answer the following questions:

Quiz: Section 6.4 – Part II

Q1 What is a maximal integral curve?

Q2 Under which conditions does a dynamical system has a unique maximal solution?

Q3 Sketch the proof of the assertion you used in Q2.

Q4 State the following result precisely: "a maximal integral curve which has finite life-time leaves every compact set".

Q5 Sketch the proof of the assertion you used in Q4.

Q6 Give an example for an ordinary differential equation with blow-up solutions.

Q7 What is a linearly bounded mapping?

Q8 What can you say about the life-time of linearly bounded integral curves?

Q9 Sketch the proof of the assertion you used in Q8.

Proof. Following [162], p. 145, the first assertion follows from

$$\dot{\varphi}_c(t) \;=\; \dot{\varphi}(t-c) \;=\; v\left(\varphi(t-c)\right) \;=\; v\left(\varphi_c(t)\right) .$$

Finally, due to the maximality of both φ_{s-r} and ψ the identity $\varphi_{s-r}(s) = \varphi(r) = \psi(s)$ implies $J \subset I + s - r \subset J$ as well as $\varphi_{s-r} = \psi$. This shows the second assertion. □

Though, φ and φ_c may be different curves due to their time parametrisation, their trajectories in Ω are identical. If φ solves the initial value problem $\dot{x} = v(x)$ with initial condition $x(t_0) = x_0$, then φ_{t_0} solves the initial value problem $\dot{x} = v(x)$ with initial condition $x(0) = x_0$. This correspondence is often used to normalize the initial time of an integral curve to zero.

As a consequence of 2 we have that the trajectories of the maximal integral curves form a disjoint partition of the phase space Ω, termed the *phase portrait*.

Theorem 6.8 (The Three Types of Integral Curves). *Let v be a locally Lipschitz-continuous vector field on Ω. Up to time shifts, there is a unique maximal integral curve through every point of Ω, and for these integral curves exactly one of the following cases holds:*

1. *For at least one $t_0 \in I$ it holds that $\dot{\varphi}(t_0) = 0$. Then $I = \mathbb{R}$, and φ is constant where $\varphi(t)$ is a root of v.*

2. *For all $t \in I$ it holds that $\dot{\varphi} \neq 0$, and φ has a returning point, i.e., $\varphi(r) = \varphi(s)$ for suitable $r, s \in I$, $r \neq$. Then, $I = \mathbb{R}$ and φ is periodic such that $\varphi(t+p) = \varphi(t)$ with $p := s - r$ for all $t \in \mathbb{R}$.*

3. *For all $t \in I$ it holds that $\dot{\varphi} \neq 0$, and φ has no returning point.*

Proof. Following [162], p. 146, let φ and ψ be maximal integral curves through $x_0 \in \Omega$, i.e., $\varphi(r) = \psi(s)$ for suitable r and s, respectively. Then, according to the result 2 of the remark on time shifts we have $\psi = \varphi_{s-r}$. Next to the classification:

1. If $\dot{\varphi}(t_0) = 0$ then $x_0 = \varphi(t_0)$ is a root of v, because $v(\varphi(t_0)) = \dot{\varphi}(t_0)$. Thus, the constant function $\psi : \mathbb{R} \to \Omega$, $\psi(t) = x_0$ solves the initial value problem $\dot{x} = v(x)$, $x(t_0) = x_0$, either. Due to the maximality of φ assertion 1 follows.

2. Next, let $\varphi(s) = \varphi(r)$ where $p := s - r \neq 0$. According to result 2 of the remark on time shifts we have $I = I + p$ and $\varphi = \varphi_p$.

This shows the assertions. □

According to case 1 the roots of v are exactly the critical points of integral curves of the vector fields and at the same moment the trajectories of the constant integral curves. Correspondingly, they are also called *critical points* or *equilibrium points* of the vector field.

Theorem 6.9 (Addendum on Periodic Functions). *A periodic continuous function $\varphi : \mathbb{R} \to \mathbb{K}^d$ is either constant or there is a number $p > 0$ such that $\varphi(s) = \varphi(p)$ holds if and only if $t - s = k \cdot p$ for a $k \in \mathbb{Z}$.*

Proof. Following [162], p. 146, let A be the set of all periods of φ, i.e., all numbers a such that $\varphi(t+a) = \varphi(t)$ for all $t \in \mathbb{R}$. The continuity of φ implies that A is a closed sub-group of $\mathbb{R}$. Thus the assertion follows from the following statement:

Statement: Every closed sub-group $A \subset \mathbb{R}$ is either identical to 0, or $\mathbb{R}$ or $\mathbb{Z} \cdot p$ for a $p \neq 0$.

We begin to prove this statement by assuming $A \neq 0$ and $A \neq \mathbb{R}$. First, we show that $A^+ := A \cap \mathbb{R}_+$ has a smallest element. If this is not the case, then for every $\varepsilon > 0$ there would be some $a \in A^+$ where $a < \varepsilon$, then for every $x \in \mathbb{R}$ there is a $k \in \mathbb{Z}$ such that $ka \leq x < (k+1)a$, i.e., $|x - ka| < \varepsilon$. Since A is closed we have $x \in A$, and thus $A = \mathbb{R}$, which is a contradiction to our initial assumption. Next, let p be the smallest element of A^+. For any $a \in A$ there is a $k \in \mathbb{Z}$ such that $kp \le a < (k+1)p$, i.e., $0 \le a - kp < p$. If $a - kp \ne 0$, then $a - kp$ is in A^+, and is smaller than p, leading to a contradiction. □

Constant and periodic integral curves of a vector field are among most interesting cases of orbits; the integral curves without returning points can be considered as the most common. All three types can occur in the same vector field as the following example illustrates:

Example 6.21 (Phase Portrait, cf. [162], pp. 146). Consider the planar autonomous system

$$\begin{pmatrix} \dot{x} \\ \dot{y} \end{pmatrix} = \begin{pmatrix} y \\ -x \end{pmatrix} + (1 - x^2 - y^2) \begin{pmatrix} x \\ y \end{pmatrix} = v(x, y)\,, \qquad (6.21)$$

on $\mathbb{R}^2$. Up to time shifts there is a unique solution through every point of $\mathbb{R}^2$. The origin $(0,0)$ is the only root of v and thus the curve $t \mapsto (0,0)$ is the only constant maximal integral curve. Every other non-constant maximal integral curve hence lies in $\mathbb{R}^2 \setminus (0,0)$.

We can construct such non-constant maximal integral curves by changing to polar coordinates

$$\begin{aligned} x(t) &= r(t) \cdot \cos(\varphi(t))\,, \\ y(t) &= r(t) \cdot \sin(\varphi(t))\,, \end{aligned}$$

where $r > 0$ and φ are continuously differentiable functions. With this ansatz we see that the solution curves of (6.21) in $\mathbb{R}^2 \setminus (0,0)$ are equivalent to the solutions of the following system

$$\dot{r} = r(1 - r^2)\,, \qquad \dot{\varphi} = 1\,.$$

This is an example of an ordinary differential equation with separated variables.

Thus, we can discuss the one-dimensional solutions for r and φ separately and then put them together to obtain the whole two-dimensional picture. Typical solution of $\dot{r} = r(1 - r^2)$ are:

1. the constant solution $r = 1$,

2. the strictly monotonously increasing solution $r : \mathbb{R} \to (0,1)$ with

$$r(t) = \sqrt{\frac{1}{1 + \exp(-2t)}}$$

where $\lim_{t \to -\infty} r(t) = 0$ and $\lim_{t \to \infty} r(t) = 1$, and

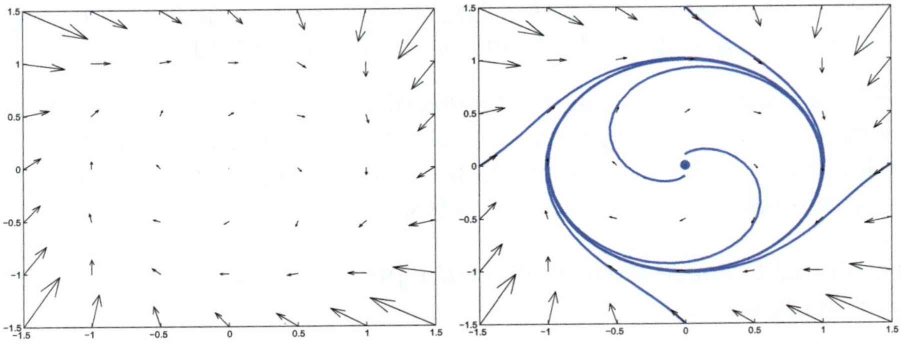

(a) (b)

Figure 6.11. Velocity plot (a) and sketch of system's (6.21) phase portrait (b) showing the existence of an equilibrium, and a periodic solution as well as regular solutions with no returning points.

3. the strictly monotonously decreasing solution $r : \mathbb{R}^+ \to (1, \infty)$ with

$$r(t) = \sqrt{\frac{1}{1 - \exp(-2t)}}$$

where $\lim_{t \downarrow 0} r(t) \to \infty$ and $\lim_{t \to \infty} r(t) = 1$.

By time shifts we obtain all maximal solutions of $\dot{r} = r(1 - r^2)$, $r > 0$, from these three types.

Finally, we combine these solutions with the solution $\varphi(t) = t$ of $\dot{\varphi} = 1$ to obtain the solutions of (6.21): In the case 1 this leads to periodic solutions and in the cases 2 and 3 to spirals that tend towards the periodic solution for $t \to \infty$, see Fig. 6.11. All further solutions of (6.21) are obtained by time shifts.

6.4.5 Systems of 1st Order

Besides ordinary differential equations of the form $\dot{x} = F(t, x)$ that involve only first-order derivatives, equations must be considered that contain higher-order derivatives. An example for such a higher-order ordinary differential equation is the oscillation equation $\ddot{x} = -x$. It can be transformed to a system of first-order by setting $x_1 := x$ and $x_2 := \dot{x}_1$. This leads to the system

$$\ddot{x} = -x\,, \quad \Rightarrow \quad \begin{cases} \dot{x}_1 = x_2 \\ \dot{x}_2 = -x_1 \end{cases}$$

which is related to the rotation field $v(x_1, x_2) = (x_2, -x_1)$.

Let $f : I \times \Omega \to \mathbb{K}$ be a scalar valued function on an interval I and a open set $\Omega \subset \mathbb{K}^d$. The general every d-th order ordinary differential equation

$$x^{(d)} = f\left(t, x, \dot{x}, \dots, x^{(d-1)}\right), \tag{6.22}$$

can be transformed into a d-dimensional system of first-order equations

$$\left.\begin{aligned} \dot{x}_1 &= x_2 \\ \dot{x}_2 &= x_3 \\ &\vdots \\ \dot{x}_{d-1} &= x_d \\ \dot{x}_d &= f(t, x_1, x_2, \dots, x_d) \end{aligned}\right\} \text{ or in vector notation: } \dot{x} = F(t,x) \begin{pmatrix} x_2 \\ \vdots \\ x_d \\ f(t,x) \end{pmatrix}. \tag{6.23}$$

The function $\varphi : I \to \mathbb{K}^d$, $\varphi = (\varphi_1, \varphi_2, \dots, \varphi_d)$ is an integral curve of the system (6.23) if and only if it has the form $\varphi = (\varphi_1, \dot{\varphi}_1, \dots, \varphi_1^{(d-1)})$, where $\varphi_1 : I \to \mathbb{K}$ is a solution of (6.22).

One may recognize that F is locally Lipschitz-continuous in x uniformly in t if and only f has this property. Thus, the existence and uniqueness theorems for systems of first-order equations are also existence and uniqueness theorems for higher-order ordinary differential equations. For higher-order equations one unusually takes $x(t_0)$ together with the values $\dot{x}(t_0), \dots, x^{d-1}(t_0)$ of the first $d-1$ derivatives as initial conditions at the initial time t_0.

Before you continue, make sure to answer the following questions:

Quiz: Section 6.4 – Part III

Q1 Give a classification of the integral curves that can occur in locally Lipschitz-continuous autonomous dynamical systems.

Q2 Sketch and discuss the phase portrait of

$$\dot{r} = r(1 + r^2), \quad \text{and} \quad \dot{\varphi} = 1.$$

Q3 Rewrite the second-order damped Duffing oscillator equation $\ddot{x} + \dot{x} + x + ax^3 = 0$, $a > 0$, as a system of first order equations.

Q4 Rewrite the forth-order equation $x^{(4)} + ax^3 + b\dot{x} + c\ddot{x} = 0$, $a, b, c > 0$, as a system of first order equations.

6.5 Solutions of ODEs in the Extended Sense

Following [64], Chap. 2, we now discuss the extension of the notion of a solution of an ordinary differential equation $\dot{x} = f(t,x)$ in the sense that the solution function in the extended sense matches the right-hand side $f(t,x)$ except on a set of Lebesgue-measure zero.

We have already seen that if the right hand side f is a continuous function on some (t,x)-domain D, then the (ordinary) differential equation $\dot{x} = f(t,x)$ together with the initial condition $x(t_0) = x_0$ is equivalent to the integral equation

$$x(t) = x_0 + \int_{t_0}^{t} f(s, x(s))\mathrm{d}s\,. \tag{6.24}$$

In other words, if φ is a solution of $\dot{x} = f(t,x)$ on some interval I for which $\varphi(t_0) = x_0$, then $x(t) = \varphi(t)$ will satisfy (6.24) on I and conversely.

Clearly, the integral in (6.24) makes sense for many functions f that are not continuous. Recall that the continuity of f guarantees that a solution of $\dot{x} = f(t,x)$ is continuously differentiable. Thus, if a continuously differentiable function is not demanded (like in the case of jump processes), the continuity restriction on f can be relaxed.

Suppose f is a real-valued (not necessarily continuous) function defined on some sub-set S of the (t,x)-space. Then, we can extend the notion of solutions to the differential equation $\dot{x} = f(t,x)$ by addressing the following problem: Find an absolutely continuous function φ defined on a real time interval I such that

1. $(t, \varphi(t)) \in S$ for all $t \in I$,
2. $\dot{\varphi}(t) = f(t, \varphi(t))$ for all $t \in I$, except on a set of Lebesgue-measure zero.

If such an interval I and a function φ exist, then φ is said to be the *solution of* $\dot{x} = f(t,x)$ *in the extended sense on* I. Notice that the absolute continuity of a solution guarantees the existence of $\dot{\varphi}$ almost everywhere (i.e., except on a set of Lebesgue-measure zero), so that requirement 2 makes sense.

If $f \in \mathcal{C}(S)$ and φ is a solution of $\dot{x} = f(t,x)$ in the extended sense on I, then we can immediately conclude $\dot{\varphi} \in \mathcal{C}(I)$ from requirement 2, and therefore the more general notion of the differential equation and its solution in the extended sense reduces to the common definition of an ordinary differential equation when $f \in \mathcal{C}(S)$. Usually, it will be clear from the context which meaning is attached to $\dot{x} = f(t,x)$ and its solution φ, and hence it will rarely be necessary to add the phrase "in the extended sense".

6.5.1 The Theorem of Caratheodory

As regards the existence of solution of $\dot{x} = f(t,x)$ in the extended sense on I, Constantin Caratheodory (1873 - 1950) has proved the following quite general theorem under the assumption that f be bounded by a Lebesgue-integrable function of t. Following [64], pp. 43, the proof will be carried out in one space dimension only; it will be clear what modifications are required in the case of a system of ordinary differential equations, see problem 6.50. R will denote the rectangle

$$R := \left\{(t,x) \in \mathbb{R}^2 : |t-\tau| \le a\,,\ |x-\xi| \le b\right\}\,,$$

where (τ,ξ) is a fixed given point in the (t,x)-plane and a, b are positive real numbers.

Theorem 6.10 (Theorem of Caratheodory). *Let f be defined on R, and suppose it is measurable in t for each fixed x as well as continuous in x for each fixed t. If there exists a Lebesgue-integrable function m on the interval $|t-\tau| \le a$ such that*

$$|f(t,x)| \le m(t)\,, \qquad \text{for all } (t,x) \in R\,, \tag{6.25}$$

then there exists a solution φ of $\dot{x} = f(t,x)$ in the extended sense on some interval $|t-\tau| \le \beta$ ($\beta > 0$), satisfying $\varphi(\tau) = \xi$.

Proof. Following [64], pp. 43, let us consider the case $t \ge \tau$ as the situation is similar when $t \le \tau$. If M is defined by

$$\begin{cases} M(t) = 0 & \text{for all } t < \tau \\ M(t) = \displaystyle\int_\tau^t m(s)\mathrm{d}s & \text{for all } \tau \le t \le \tau + a \end{cases}, \tag{6.26}$$

then it is clear that M is continuous, non-decreasing (as $m \ge 0$ by (6.25)), and $M(\tau) = 0$. Hence, $(t, \xi \pm M(t)) \in R$ for some interval $\tau \le t \le \tau + \beta \le \tau + a$, where β is some positive constant.

Choose any $\beta > 0$ for which this is true, and define the Picard approximations φ_j, $j = 1, 2, \ldots$, by

$$\begin{cases} \varphi_j(t) = \xi & \text{for all } \tau \le t \le \tau + \beta/j \\ \varphi_j(t) = \xi + \displaystyle\int_\tau^{t-\beta/j} f(s, \varphi_j(s))\mathrm{d}s & \text{for all } \tau + \beta/j < t \le \tau + \beta \end{cases}. \tag{6.27}$$

Clearly, φ_1 is defined on $\tau \le t \le \tau + \beta$, for it is the constant ξ. For any fixed $j \ge 1$, the first formula in (6.27) defines φ_j on $\tau \le t \le \tau + \beta/j$, and since $(t, \xi) \in R$ for $\tau \le t \le \beta/j$, the second formula in (6.27) defines φ_j as a continuous function on the interval $\tau + \beta/j < t \leq \tau + 2\beta/j$. Further, on this latter interval

$$|\varphi_j(t) - \xi| \leq M\left(t - \tfrac{\beta}{j}\right), \tag{6.28}$$

by virtue of (6.25) and (6.26).

Assume that φ_j is defined on $\tau \leq t \leq \tau + k\beta/j$ for $1 < k < j$. Then the second formula of (6.27) defines φ_j for $\tau + k\beta/j < t \leq \tau + (k+1)\beta/j$ since knowledge of the measurable integrand is only required on $\tau \leq t \leq \tau + k\beta/j$. Also, on $\tau + k\beta/j < t \leq \tau + (k+1)\beta/j$, the function φ_j satisfies (6.28), because of (6.25) and (6.26).

Therefore, by induction, (6.27) defines all φ_j as continuous functions on $\tau \leq t \leq \tau + \beta$ which satisfy

$$\begin{cases} \varphi_j(t) = \xi & \text{for all } \tau \leq t \leq \tau + \beta/j \\ |\varphi_j(t) - \xi| \leq M\left(t - \frac{\beta}{j}\right) & \text{for all } \tau + \beta/j < t \leq \tau + \beta \end{cases} . \tag{6.29}$$

If t_1 and t_2 are any two points in the interval $[\tau, \tau + \beta]$, then on account of (6.25), (6.26) and (6.27),

$$|\varphi_j(t_1) - \varphi_j(t_2)| \leq \left| M\left(t_1 - \tfrac{\beta}{j}\right) - M\left(t_2 - \tfrac{\beta}{j}\right) \right| . \tag{6.30}$$

Since M is continuous on $[\tau, \tau + \beta]$, it is uniformly continuous there. This implies by (6.30), that the set $\{\varphi_j\}$ is uniformly bounded on $[\tau, \tau + \beta]$. Consequently it follows by the lemma of Arzela-Ascoli (Lemma 6.10) that there exists a sub-sequence $\{\varphi_{j_k}\}$ which converges uniformly on $[\tau, \tau + \beta]$ to a continuous limit function φ as $k \to \infty$.

From (6.25),

$$|f(t, \varphi_{j_k}(t))| \leq m(t), \qquad \text{for all } \tau \leq t \leq \tau + \beta,$$

and since f is continuous in x for fixed t,

$$f(t, \varphi_{j_k}(t)) \to f(t, \varphi(t)), \qquad \text{for } k \to \infty,$$

for every fixed $t \in [\tau, \tau + \beta]$. Therefore, Lebesgue's dominated convergence theorem may be applied to give

$$\lim_{k \to \infty} \int_\tau^t f(s, \varphi_{j_k}(s)) \mathrm{d}s = \int_\tau^t f(s, \varphi(s)) \mathrm{d}s \tag{6.31}$$

for any $t \in [\tau, \tau + \beta]$. But,

$$\varphi_{j_k}(t) = \xi + \int_\tau^t f(s, \varphi_{j_k}(s)) \mathrm{d}s - \int_{t-\beta/j}^t f(s, \varphi_{j_k}(s)) \mathrm{d}s,$$

where it is clear that the latter integral vanishes as $k \to \infty$. Hence, letting $k \to \infty$, and applying (6.31), it follows that

$$\varphi(t) = \xi + \int_\tau^t f(s, \varphi(s)) \mathrm{d}s \,,$$

from which the assertion of the theorem follows at once. □

It is interesting to remark that the original approximations (6.27) must converge to a solution in the case where a unique solution is known. This situation does not apply for ordinary successive approximations as the following example shows.

Example 6.22 (Continuity of f is Not Sufficient for the Convergence of the Successive Approximations, cf. [64], p. 53). Let the right-hand side f of an ordinary differential equation $\dot{x} = f(t, x)$ be defined by

$$f(t,x) := \begin{cases} 0 & \text{for } t = 0 \text{ and } -\infty < x < +\infty \\ 2t & \text{for } 0 < t \le 1 \text{ and } -\infty < x < 0 \\ 2t - 4x/t & \text{for } 0 < t \le 1 \text{ and } 0 \le x \le t^2 \\ -2t & \text{for } 0 < t \le 1 \text{ and } t^2 < x < +\infty \,. \end{cases}$$

On the region $0 \le t \le 1$, $-\infty < x < +\infty$, this function f is continuous and bounded by the constant 2. For the initial point $(\tau, \xi) = (0, 0)$, the successive approximations

$$\begin{aligned} \varphi_0(t) &= \xi \\ \varphi_{m+1}(t) &= \xi + \int_\tau^t f(s, \varphi_m(s)) \mathrm{d}s \,, \qquad \text{for all } m = 0, 1, 2, \ldots \end{aligned}$$

become, for $0 \le t \le 1$

$$\varphi_0(t) = 0 \,, \quad \varphi_{2m-1}(t) = t^2 \,, \quad \text{and} \quad \varphi_{2m}(t) = -t^2 \,, \quad \text{for all } m = 1, 2, \ldots \,.$$

The sequence $\{\varphi_m(t)\}$ has two cluster values for each $t \neq 0$, and hence the successive approximations do not converge. Note, neither of the two convergent sub-sequences $\{\varphi_{2m-1}\}$ and $\{\varphi_{2m}\}$ converge to a solution, for

$$\dot{\varphi}_{2m-1}(t) = 2t \neq f(t, t^2) \,, \quad \text{and} \quad \dot{\varphi}_{2m}(t) = -2t \neq f(t, -t^2) \,.$$

6.5.2 Maximum & Minimum Solutions

For the scalar case it can be shown that all solutions of $\dot{x} = f(t,x)$ (in the extended sense) issuing from an initial point (τ, ξ) can be bracketed between two special solutions, the *maximum* and the *minimum solutions*. Let f be defined on the rectangle R as in Caratheodory's theorem (Theorem 6.10).

If φ_M is a solution of $\dot{x} = f(t,x)$ passing through (τ, ξ), existing on some interval I containing τ, with the property that every other solution φ of $\dot{x} = f(t,x)$ passing through (τ, ξ) and existing on I is such that

$$\varphi(t) \;\le\; \varphi_M(t)\,, \qquad \text{for all } t \in I\,,$$

then φ_M is called a *maximum solution* of $\dot{x} = f(t,x)$ on I passing through (τ, ξ). Similarly, if φ_m is a solution of $\dot{x} = f(t,x)$ on an interval I for which $\varphi_m(t) = \xi$, and such that

$$\varphi(t) \;\ge\; \varphi_m(t)\,, \qquad \text{for all } t \in I\,,$$

holds for every other solution φ of $\dot{x} = f(t,x)$ on I for which $\varphi(t) = \xi$, then φ_m is called *minimum solution* of $\dot{x} = f(t,x)$ on I passing through (τ, ξ). Clearly, the functions φ_M and φ_m, if they exist, must be unique.

The existence of φ_M and φ_m will now be demonstrated under the Caratheodory assumptions.

Theorem 6.11 (Existence of a Maximum & Minimum Solution). *Let the hypothesis of Caratheodory's theorem (Theorem 6.10) be satisfied. Then there exists a maximum solution φ_M and a minimum solution φ_m of $\dot{x} = f(t,x)$ on $|t-\tau| \le \beta$ passing through (τ, ξ).*

Proof. Following [64], pp. 45, we will prove the existence of φ_M on $[\tau, \tau+\beta]$. Now any solution φ of $\dot{x} = f(t,x)$ passing through the point (τ, ξ) must satisfy

$$\varphi(t) \;=\; \xi + \int_\tau^t f(s, \varphi(s))\mathrm{d}s \tag{6.32}$$

as far as it exists, and from (6.32) it follows that

$$|\varphi(t_1) - \varphi(t_2)| \;\le\; |M(t_1) - M(t_2)| \tag{6.33}$$

for any two points t_1, t_2 where φ exists, where M is defined by (6.26).

Since M is continuous, (6.33) implies, by the Cauchy criterion for convergence, that the solution φ can be continued, if necessary, to the entire interval $[\tau, \tau+\beta]$, making use of the Caratheodory existence theorem. Thus, all solutions of $\dot{x} = f(t,x)$ passing through (τ, ξ) exist on $[\tau, \tau+\beta]$, and must satisfy (6.33) there. From the uniform continuity of M on $[\tau, \tau+\beta]$, it follows from

(6.33) that the set of all solutions $\{\varphi\}$ of $\dot{x} = f(t,x)$ on $[\tau, \tau+\beta]$ is an equicontinuous set, that is, given any $\varepsilon > 0$, there exists a $\delta_\varepsilon > 0$, independent of t and the solution φ, such that

$$\left|\varphi(\hat{t}) - \varphi(\tilde{t})\right| < \varepsilon, \qquad \text{whenever } |\hat{t} - \tilde{t}| < \delta_\varepsilon, \tag{6.34}$$

and $\hat{t}, \tilde{t}$ are in $[\tau, \tau+\beta]$.

Let Φ be the function defined by

$$\Phi(t) = \sup_{t \in [\tau, \tau+\beta]} \{\varphi(t)\}$$

taken over all solutions φ of $\dot{x} = f(t,x)$ on $[\tau, \tau+\beta]$ passing through (τ, ξ). Clearly, Φ exists on $[\tau, \tau+\beta]$ and is continuous (and hence uniformly continuous) there. Thus, for any given $\varepsilon > 0$ there is a $\delta_\varepsilon > 0$ such that not only (6.34) is true for this δ_ε, but also for $\hat{t}, \tilde{t}$ in $[\tau, \tau+\beta]$,

$$\left|\Phi(\hat{t}) - \Phi(\tilde{t})\right| < \varepsilon, \qquad \text{whenever } |\hat{t} - \tilde{t}| < \delta_\varepsilon. \tag{6.35}$$

It will be shown that Ψ is a solution of $\dot{x} = f(t,x)$ satisfying $\Psi(\tau) = \xi$, and if φ_M is defined to be Φ, it is clear that this φ_M will satisfy the requirements of the theorem on $[\tau, \tau+\beta]$. For a given $\varepsilon > 0$, choose δ_ε such that (6.34) and (6.35) hold. Subdivide the interval $[\tau, \tau+\beta]$ into n intervals by the points $\tau = t_0 < t_1 < t_2 < \cdots < t_n = \tau + \beta$ in such a way that

$$\max(t_{i+1} - t_i) < \delta_\varepsilon .$$

For every t_i, $i = 0, 1, \ldots, n-1$, choose a solution φ_i of $\dot{x} = f(t,x)$ passing through (τ, ξ) so that

$$0 \leq \Phi(t_i) - \varphi_i(t_i) < \varepsilon$$

and for $i \geq 1$

$$\varphi_i(t_i) - \varphi_{i-1}(t_i) \geq 0 .$$

This is possible from the definition of Φ.

Now, for the given ε, define the function φ_ε as follows: Let

$$\varphi_\varepsilon(t) = \varphi_{n-1}(t), \qquad \text{for } t_{n-1} \leq t \leq t_n = \tau + \beta .$$

If $\varphi_{n-1}(t_{n-1}) > \varphi_{n-2}(t_{n-1})$, define φ_ε to the left of t_{n-1} as φ_{n-1} up to the point τ_{n-2}, if it exists, in (t_{n-2}, t_{n-1}) nearest t_{n-1} such that

$$\varphi_\varepsilon(\tau_{n-2}) = \varphi_{n-1}(\tau_{n-2}) = \varphi_{n-2}(\tau_{n-2}) .$$

If τ_{n-2} exists, define $\varphi_\varepsilon(t) := \varphi_{n-2}(t)$ for $t_{n-2} \leq t < \tau_{n-2}$. If τ_{n-2} does not exist, define φ_ε on $[t_{n-2}, t_{n-1})$ as φ_{n-1}. If $\varphi_{n-1}(t_{n-1}) = \varphi_{n-2}(t_{n-1})$, define

$\varphi_\varepsilon(t) := \varphi_{n-2}(t)$ on $[t_{n-2}, t_{n-1})$. Continuing in this way, one can define a solution φ_ε of $\dot{x} = f(t,x)$ on $[\tau, \tau+\beta]$ passing through the point (τ, ξ), obtained by patching together solutions of $\dot{x} = f(t,x)$, and having the property

$$0 \leq \Phi(t_i) - \varphi_\varepsilon(t_i) < \varepsilon, \qquad \text{for all } i = 0, 1, \ldots, n. \tag{6.36}$$

Since the variation of Φ and φ_ε in each interval $[t_i, t_{i+1}]$ is less than ε, by (6.34) and (6.35), there results from (6.36)

$$0 \leq \Phi(t) - \varphi_\varepsilon(t) < 3\varepsilon, \qquad \text{for all } \tau \leq t \leq \tau + \beta. \tag{6.37}$$

Letting $\varepsilon = 1/m$, $m = 1, 2, \ldots$, one obtains a sequence $\dot{\varphi}_{1/m}$ of solutions which by (6.37) converges uniformly to Φ on $[\tau, \tau+\beta]$. From this fact and an application of Lebesgue's dominated convergence theorem to (6.32) with φ replaced by $\varphi_{1/m}$ it follows that

$$\Phi(t) = \xi + \int_\tau^t f(s, \Phi(s))\mathrm{d}s, \qquad \text{for all } \tau \leq t \leq \tau + \beta,$$

that is, Φ is a solution of $\dot{x} = f(t,x)$ satisfying $\Phi(\tau) = \xi$, and from its definition it is the maximum solution φ_M on $[\tau, \tau+\beta]$. □

Theorem 6.12 (Continuation of Solutions). *In a domain D of the (t,x) plane let the function f be defined, measurable in t for fixed x and continuous in x for fixed t. Let there exist an integrable function m such that $|f(t,x)| \leq m(t)$ for $(t,x) \in D$. Then, given a solution φ of $\dot{x} = f(t,x)$ for $t \in (a,b)$, it is the case that $\varphi(b-0)$ exists and if $(b, \varphi(b-0)) \in D$ then φ can be continued over $(a, b+\delta]$ for some $\delta > 0$. A similar result holds at a.*

Thus the solution φ can be continued up to the boundary of D. Moreover, the same continuation is valid for a maximum solution φ_M or a minimum solution φ_m.

Proof. The proof is very similar to that of the continuation Theorem 6.6, see problem 6.51, too. □

Corollary 6.23 (Backwards-Continuation of Solutions via the Maximum and Minimum Solution). *Let the hypotheses of Caratheodory's theorem (Theorem 6.10) be satisfied and let φ_M and φ_m, the maximum and minimum solution through (τ, ξ), exist over $[\tau, \tau+\beta]$, where $\beta \leq a$. Then for any c satisfying $\varphi_m(\tau+\beta) < c < \varphi_M(\tau+\beta)$ there is at least one solution φ through (τ, ξ) for $\tau \leq t \leq \tau+\beta$ and with $\varphi(\tau+\beta) = c$.*

Proof. Following [64], p. 47, we start with the solution through $(\tau+\beta, c)$ and continue it to the left. It need not leave the region $\varphi_m(t) \leq x \leq \varphi_M(t)$, $\tau \leq t \leq \tau+\beta$, since it can always be continued back along one of these extreme solutions if it meets one of them. Thus, it can be continued back to (τ, ξ). □

Theorem 6.13 (Continuation of the Maximum Solution). *Let the hypotheses of the continuation theorem (Theorem 6.12) be valid, and suppose the maximum solution $\varphi_{M,\xi}$ of $\dot{x} = f(t,x)$ through (τ, ξ), exist over an interval $[\tau, \tau+\alpha]$. Then there exists a $\delta > 0$ such that $\dot{x} = f(t,x)$ has a maximum solution $\varphi_{M,\eta}$ for each η, $\xi \leq \eta < \xi + \delta$ on $[\tau, \tau+\alpha]$ with $\varphi_{M,\eta}(\tau) = \eta$.*

Moreover, $\varphi_{M,\eta} \to \varphi_{M,\xi}$ as $\eta \to \xi + 0$, uniformly over $[\tau, \tau+\alpha]$.

Proof. Following [64], pp. 47, we have by Theorem 6.11 that $\varphi_{M,\eta}$ certainly exists over some interval with the left end point τ, if $\eta - \xi$ is small enough. From the definition of the maximum solution, it follows readily that, for $\tilde{\eta} > \eta > \xi$,

$$\varphi_{M,\tilde{\eta}}(t) \;\geq\; \varphi_{M,\eta}(t) \;\geq\; \varphi_{M,\xi}(t)\,.$$

Thus $\varphi_{M,\eta}$ is monotone non-decreasing in η and is bounded from below. Therefore, for each t on some interval $[\tau, \tau+\beta]$, there exists

$$\Phi(t) \;=\; \varphi_{M,\xi+0}(t) \;\geq\; \varphi_{M,\xi}(t)\,. \tag{6.38}$$

Since $\varphi_{M,\eta}$ satisfies (6.33),

$$|\Phi(t_1) - \Phi(t_2)| \;\leq\; |M(t_1) - M(t_2)|\,,$$

so that Φ is continuous. From

$$\varphi_{M,\eta}(t) \;=\; \eta + \int_\tau^t f(s, \varphi_{M,\eta}(s))\mathrm{d}s$$

it follows on letting $\eta \to \xi + 0$ that

$$\Phi(t) \;=\; \xi + \int_\tau^t f(s, \Phi(s))\mathrm{d}s\,.$$

But this implies Φ is a solution of $\dot{x} = f(t,x)$ through (τ, ξ). Thus, by (6.38), $\Phi(t) = \varphi_{M,\xi}(t)$ over $[\tau, \tau+\beta]$. The uniformity of the convergence of $\varphi_{M,\eta}$ to $\varphi_{M,\xi}$ follows from the equi-continuity of $\varphi_{M,\eta}$ in t, as proved by (6.33).

The above argument is clearly valid over the range of existence of Φ on $[\tau, \tau+\alpha]$. Suppose that for some $t_0 \leq \tau+\alpha$ and for every small $h > 0$, Φ exists over $[\tau, t_0 - h]$ but not over $[\tau, t_0 + h]$. Then for any given $\varepsilon > 0$ there exists a $\delta_\varepsilon > 0$ such that

$$|\varphi_{M,\eta}(t_0 - \varepsilon) - \varphi_{M,\xi}(t_0 - \varepsilon)| \;\leq\; \varepsilon\,, \tag{6.39}$$

if $0 \leq \eta - \xi < \delta_\varepsilon$.

Let the region H be the set of points (t, x) which satisfy the inequalities

$$|t - t_0| \;\leq\; \gamma\,, \quad \text{and} \quad |x - \varphi_{M,\xi}(t_0 - \gamma)| \;\leq\; \gamma + M(t) - M(t_0 - \gamma)\,.$$

By choosing γ small enough, $H \subset D$. Any solution φ of $\dot{x} = f(t, x)$ which starts on the left vertical side $t = t_0 - \gamma$ of H (i.e., $|\varphi(t_0 - \gamma) - \varphi_{M,\xi}(t_0 - \gamma)| \leq \gamma$) will, by (6.33), remains in H as t increases. Thus any such solution can be continued to $t_0 + \gamma$.

By choosing ε in (6.39) so that $\varepsilon = \gamma$, it follows that for $0 < \eta - \xi < \delta_\varepsilon$ the solutions $\varphi_{M,\eta}$ can be continued to $t_0 + \gamma$. This implies the existence of Φ over $[\tau, t_0 + \gamma]$, which contradicts the assumption on t_0. Thus, $t_0 > \tau + \alpha$ and therefore Φ exists over $[\tau, \tau + \alpha]$. □

Before you continue, make sure to answer the following questions:

Quiz: Section 6.5

Q1 What is meant by an ordinary differential equation in the extended sense and its solution?

Q2 State the theorem of Caratheodory.

Q3 Sketch the proof of the theorem of Caratheodory and explain why one has to be cautious about the convergence of successive approximations.

Q4 Give the definition of a maximum and a minimum solution.

Q5 What can you say about the existence of a maximum and a minimum solution?

Q6 Sketch the proof of the assertion you used in Q5.

Q7 What can you say about the continuation of solutions beyond a time interval (a, b)?

Q8 State the conditions that allow you to re-construct a solution if the maximum and minimum solutions are known such that all three of these solutions have the same initial point.

Q9 Sketch the proof of the assertion you used in Q8.

6.6 Linear Ordinary Differential Equations

Linear ordinary differential equations play an essential role in science and engineering. They occur on the one hand when superposition comes into play and serve, on the other hand, for the wide majority of non-linear problems as suitable approximations. As we will see in our study of linearized stability (in Sec. 8.4) for specific non-linear systems the linearisation, i.e., linear ordinary differential equation, around its equilibria already contains all required (local) information about this system.

Moreover, in Chaps. 12 and 13, we will utilise the methods for determining the deterministic solutions of linear ordinary differential equations for setting-up solution formulas for linear random differential equations.

Linear ordinary differential equations are of the form

$$\begin{aligned}\dot{x}_1 &= a_{11}(t)x_1 + \cdots + a_{1d}(t)x_n + b_1(t)\\ &\vdots\\ \dot{x}_d &= a_{d1}(t)x_1 + \cdots + a_{dd}(t)x_n + b_d(t)\end{aligned}$$

or in matrix-vector notation

$$\dot{x} = A(t)x + b(t)\,,$$

where $A : I \to \mathbb{K}^{d\times d}$ and $b : I \to \mathbb{K}^d$ are given mappings on a interval $I \subset \mathbb{R}$. The linear ordinary differential equation is called *homogeneous* if $b \equiv 0$ and *non-homogeneous* with *inhomogeneity* b otherwise.

6.6.1 Existence & Uniqueness of Solutions

Obviously, linear systems are Lipschitz-continuous and linearly bounded as long as A and b do not have jumps. Thus, following [162], pp. 149, existence and uniqueness of solutions is straightforward:

Theorem 6.14 (Existence and Uniqueness). *Let $A : I \to \mathbb{K}^{d\times d}$ and $b : I \to \mathbb{K}^d$ be continuous, then every initial value problem*

$$\dot{x} = A(t)x + b(t)\,, \quad x(t_0) = x_0$$

has a unique solution which is defined on the whole of $I \subset \mathbb{R}$.

Proof. Following [162], p. 149, we have that $F(t,x) := A(t)x + b(t)$ is linearly bounded. Moreover, on every compact interval $J \subset I$ this function F on $J \times \mathbb{K}^{d\times d}$ is Lipschitz-continuous with respect to x uniformly in t with Lipschitz-constant $L := \max_{t\in J} \|A(t)\|$. □

Corollary 6.24 (Existence of a Unique Solution). *Let $a_0, a_1, \ldots, a_{d-1}, b : I \to \mathbb{K}$ be continuous functions and $x_0, x_1, \ldots, x_{d-1} \in \mathbb{K}$ be given numbers. Then, the initial value problem of order $d-1$*

$$x^{(n)} = \sum_{k=0}^{d-1} a_k(t)x^{(k)} + b(t)\,, \quad x^{(k)} = x_k\,, \quad k = 0,1,\ldots,d-1\,,$$

has a unique solution on the whole of $I \subset \mathbb{R}$.

Corollary 6.25 (The Solution Space of Linear Homogeneous ODEs). *1. The set $\mathcal{L}$ of all solution of the linear homogeneous ordinary differential equation $\dot{x} = A(t)x$ defined on I is a d-dimensional $\mathcal{K}$-vector space.*

2. *Any d solutions $\varphi_1, \varphi_2, \dots, \varphi_d : I \to \mathbb{K}^n$ form a basis of $\mathcal{L}$ if and only if the vectors $\varphi_1(t), \varphi_2(t), \dots, \varphi_d(t)$ form a basis of $\mathbb{K}^d$ for at least one $t \in I$ (and thus for all $t \in I$).*

Proof. Following [162], p. 149, we first see that every linear combination $c_1\varphi_1 + c_2\varphi_2 + \cdots + c_k\varphi_k$ of solutions $\varphi_1, \varphi_2, \dots, \varphi_k \in \mathcal{L}$ is again a solution of the homogeneous equation $\dot{x} = A(t)x$, thus $\mathcal{L}$ is a vector space. In order to determine its dimension we study the initial value homomorphism $\alpha_{t_0} : \mathcal{L} \to \mathbb{K}^d$, $\alpha_{t_0}\varphi := \varphi(t_0)$ for a $t_0 \in I$. Due to the existence theorem α_{t_0} is surjective and due to the uniqueness theorem it is injective, too. Thus, $\mathcal{L}$ is d-dimensional. This establishes the first assertion.

For the second assertion it is sufficient to note that the initial value homomorphism α_{t_0} is an isomorphism for every $t \in I$. Thus it transforms a basis into another one. □

A basis $\varphi_1, \varphi_2, \dots, \varphi_d$ of the solution space $\mathcal{L}$ for the homogeneous equation $\dot{x} = A(t)x$ is called a *fundamental system*. Collecting $\varphi_1, \varphi_2, \dots, \varphi_d$ together leads to the matrix valued mapping

$$\Phi := (\varphi_1, \varphi_2, \dots, \varphi_d) : I \to \mathbb{K}^{d\times d}, \qquad t \mapsto (\varphi_1(t), \varphi_2(t), \dots, \varphi_d(t)) .$$

This mapping Φ is called *fundamental matrix* of $\dot{x} = Ax$ and, by construction, it obeys the equation

$$\dot{\Phi}(t) = A(t)\Phi(t) .$$

According to part two of Corollary 6.25 the mapping Φ is invertible. Let $\varphi \in \mathcal{L}$ be any other solution, then it can be written as a linear combination $\varphi = c_1\varphi_1 + c_2\varphi_2 + \cdots + c_d\varphi_d$ of the elements of the fundamental system and scalars $c_i \in \mathbb{K}$, $i = 1, 2, \dots, d$. Thus, with $c := (c_1, c_2, \dots, c_d)^T$ we have

$$\varphi(t) = \Phi(t)c, \qquad c \in \mathbb{K}^d .$$

If $\Psi = (\psi_1, \psi_2, \dots, \psi_d)$ is another fundamental matrix, then it holds that $\Psi = \Phi C$, where C is an invertible matrix.

To show Liouville's theorem on the deformation of volumes in phase space under the solutions of homogeneous linear differential equations, we require a lemma on the differentiation of determinants:

Lemma 6.26 (Differentiation of a Determinant). *Let $\Phi : I \to \mathbb{K}^{d\times d}$ be a continuously differentiable matrix valued mapping with columns $\varphi_1, \varphi_2, \ldots, \varphi_d$ on an interval I. Then it holds that*

$$\frac{d}{dt}\det(\Phi(t)) = \sum_{i=1}^{d} \det\left(\varphi_1, \ldots, \varphi_{i-1}, \dot{\varphi}_i, \varphi_{i+1}, \ldots, \varphi_d\right)\big|_t \, .$$

Proof. Following [162], p. 151, we consider the function

$$f : I^n \to \mathbb{K}, \qquad f(t_1, t_2, \ldots, t_d) := \det\left(\varphi_1(t_1), \varphi_2(t_2), \ldots, \varphi_d(t_d)\right) .$$

According to the definition of partial differentiation along with the linearity of the determinant function in each column, we obtain

$$\frac{\partial}{\partial t_i} f(t_1, t_2, \ldots, t_d) = \det\left(\varphi_1(t_1), \ldots, \varphi_{i-1}(t_{i-1}), \dot{\varphi}_i(t_i), \varphi_{i+1}(t_{i+1}), \ldots, \varphi_d(t_d)\right) .$$

Thus, f is continuously partial differentiable. Because of $\det(\Phi(t)) = f(t, t, \ldots, t)$ the assertion follows with the aid of the chain rule. □

Theorem 6.15 (Theorem of Liouville). *Let $\Phi I \to \mathbb{K}^{d\times d}$ be a fundamental matrix of $\dot{x} = A(t)x$, then $\det(\Phi)$ obeys the ordinary differential equation*

$$\frac{d}{dt}\det(\Phi(t)) = tr(A(t)) \cdot \det(\Phi(t)) \tag{6.40}$$

on I.

Proof. Following [162], p. 150, we show first that (6.40) holds at any point $t \in I$ where $\Phi(t) = \mathbb{I}$ is the identity matrix. Let $\varphi_1, \varphi_2, \ldots, \varphi_d$ be the columns of Φ. By construction $\varphi_i(t) = e_i$ is the ith canonical basis vector and $\dot{\varphi}_i(t) = A(t)e_i$ holds. With Lemma 6.26 we have

$$\frac{\mathrm{d}}{\mathrm{d}t}\det\left(\Phi(t)\right) = \sum_{i=1}^{d} \det\left(e_1, \ldots, e_{i-1}, A(t)e_i, e_{i+1}, \ldots, e_d\right) = \mathrm{tr}\left(A(t)\right) .$$

This already shows the assertion for t and Φ such that $\Phi(t) = \mathbb{I}$. The general case will now be reduced to this one. For arbitrary fixed $t \in I$ we consider the fundamental matrix $\Psi := \Phi \cdot C$, where $C := \Phi^{-1}$. Thus, for t and Ψ we can apply the above result and obtain

$$\frac{\mathrm{d}}{\mathrm{d}t}\det(\Psi(t)) = \mathrm{tr}(A(t)) \cdot \det(\Psi(t)) .$$

By definition of Ψ the assertion follows for Φ, too. □

Let us return to the inhomogeneous ordinary differential equation

$$\dot{x} \;=\; A(t)x \;+\; b(t)\,.$$

As in linear algebra, one can show that every solution of this equation can be gained by adding a special solution, called the *particular solution*, to a solution of the homogeneous equation $\dot{x} = A(t)x$. Intuitively, you can think of the solution space of an inhomogeneous equation as an affine space shifted from a linear sub-vector space (the solution space of the homogeneous equation) by the inhomogeneity $b(t)$. If one knows the fundamental system of the homogeneous equation, then a particular solution can be constructed by the method of variation of the constant:

Theorem 6.16 (Variation of the Constant). *Let Φ be the fundamental matrix of the homogeneous equation $\dot{x} = A(t)x$, then*

$$x_p(t) \;:=\; \Phi(t)\cdot c(t) \quad \textit{with} \quad c \;:=\; \int \Phi^{-1}(s)b(s)ds$$

is a solution of the inhomogeneous equation $\dot{x} = A(t)x + b(t)$.

Proof. Following [162], p. 151, we have

$$\dot{x}_p \;=\; \dot{\Phi}c + \Phi\dot{c} \;=\; A\Phi c + \Phi\Phi^{-1}b \;=\; Ax_p + b\,,$$

which shows the assertion. □

6.6.2 Construction of Solutions for Linear ODEs with Constant Matrices

Following [162], pp. 152, the Picard-Lindelöf iteration for the initial value problem

$$\dot{x} \;=\; Ax \qquad x(0) \;=\; x_0$$

with a constant matrix A leads to

$$\varphi_1(t) \;=\; x_0 + \int_0^t Ax_0 \mathrm{d}s \;=\; (\mathbb{I} + At)\,x_0\,,$$

$$\varphi_2(t) \;=\; x_0 + \int_0^t A\,(\mathbb{I} + As)\,x_0 \mathrm{d}s \;=\; \left(\mathbb{I} + At + \tfrac{1}{2}A^2t^2\right)x_0$$

$$\vdots$$

$$\varphi_k(t) \;=\; \left(\mathbb{I} + At + \tfrac{1}{2}A^2t^2 + \cdots + \tfrac{1}{k!}A^kt^k\right)x_0\,.$$

For $k \to \infty$, we have

$$\varphi(t) = \exp(At)x_0 \,, \tag{6.41}$$

where $\exp(At)$ is the matrix valued exponential function that obeys the ordinary differential equation $\frac{d}{dt}\exp(At) = A\exp(At)$. This allows us to verify immediately that $\varphi(t) = \exp(At)x_0$ solves the given initial value problem.

The use of MATLAB's matrix exponential is straightforward, cf. [115], chapter 3. E.g., for the constant matrix

$$A = \begin{pmatrix} 1 & 1 \\ -2 & 4 \end{pmatrix}$$

we can calculate $\exp(A)$ as follows:

```
>> A = sym([1 1; -2 4]);
>> expm(A)
ans =
[   2*exp(2)-exp(3),      exp(3)-exp(2)]
[ -2*exp(3)+2*exp(2),  -exp(2)+2*exp(3)]
```

In particular, we can immediately obtain the solution of $\dot{x} = Ax$ with initial condition $x_0 = (-4, 2)$:

```
>> A = sym([1 1; -2 4]);
>> syms t
>> expm(A*t)*[-4; 2]
ans =
[   6*exp(3*t) - 10*exp(2*t)]
[ -10*exp(2*t) + 12*exp(3*t)]
```

The representation (6.41) of the solution as $\varphi(t) = \exp(At)x_0$ is well suited for theoretical considerations, like discussions on the geometric structure of the solution.

Example 6.27 (Solutions on a Sphere, cf. [162], pp. 152). If A is real and anti-symmetric, then every solution φ of $\dot{x} = Ax$ lives on a sphere around the origin, i.e.,

$$\|\varphi(t)\|_2 = \|\varphi(0)\|_2 \qquad \text{for all } t \in \mathbb{R} \,.$$

This can be seen as follows: if A is real and anti-symmetric then $\exp(At)$ is orthogonal: $\exp(At)\,(\exp(At))^T = \exp((A + A^T)t) = \mathbb{I}$.

Substituting the elements $v_1, v_2, \ldots, v_n$ of an arbitrary basis of the $\mathbb{K}^n$ into (6.41) in place of x_0, we obtain a basis of the solution space $\mathcal{L}$ of the homogeneous equation $\dot{x} = Ax$. For instance the canonical basis vectors $e_1, e_2, \ldots, e_d$ lead to the columns of $\exp(At)$ as a basis of $\mathcal{L}$.

Usually, the computation of exp(At) is tiresome. Therefore, in order to obtain a basis of $\mathcal{L}$ we choose another path based on the Jordan normal form of A.

The easiest and most important situation occurs when A has d linearly independent eigenvectors, as in the case of A being a real symmetric matrix, or A having d distinct eigenvalues.

Lemma 6.28 (A Basis of $\mathcal{L}$ Consisting of Eigenvectors of A). *Let v be an eigenvector of A and λ be its corresponding eigenvalue, then*

$$\varphi_v : \mathbb{R} \to \mathbb{K}^d, \qquad \varphi_v(t) := \exp(\lambda t) v,$$

solves the initial value problem $\dot{x} = Ax$, $x(0) = v$. Let $v_1, v_2, \ldots, v_d$ be linearly independent eigenvectors of A and $\lambda_1, \lambda_2, \ldots, \lambda_d$ be their corresponding eigenvalues, then $\varphi_{v_1}, \varphi_{v_2}, \ldots, \varphi_{v_d}$ form a fundamental system.

Proof. Following [162], p. 153, we have that φ_v is a solution of the homogeneous equation, because

$$\dot{\varphi}_v = \lambda \exp(\lambda t) v = \exp(\lambda t) A v = A\varphi_v .$$

Moreover, the solutions $\varphi_{v_1}, \varphi_{v_2}, \ldots, \varphi_{v_d}$ form a basis of $\mathcal{L}$ because their values $\varphi_{v_1}(0), \varphi_{v_2}(0), \ldots, \varphi_{v_d}(0)$ form a basis of $\mathbb{K}^d$. □

If A does not have a full set of d linearly independent eigenvectors, as in the case of multiple eigenvalues, then a fundamental system can be constructed with the aid of generalized eigenvectors.

Definition 6.29 (Generalized Eigenvector). A vector $v \in \mathcal{C}^d$, $v \neq 0$ is called *generalized eigenvector of the matrix A corresponding to the eigenvalue λ*, if there is a natural number s such that

$$(A - \lambda\mathbb{I})^s v = 0 .$$

The smallest number s is called *degree* of v.

The generalized eigenvectors of degree one are the eigenvectors them self. If v is a generalized eigenvector of degree s, then the vectors

$$v_s := v, \quad v_{s-1} := (A - \lambda\mathbb{I}) v, \quad \ldots \quad v_1 := (A - \lambda\mathbb{I})^{s-1} v$$

are generalized eigenvectors of degree s, $s-1$, ..., 1, respectively. v_1 is an eigenvector and v_i, $i = 2, 3, \ldots, s$ is a solution of the equation $(A - \lambda\mathbb{I}) v_i = v_{i-1}$.

Example 6.30 (Generalized Eigenvectors of a 3×3 Matrix, cf. [162], p. 153). For the following matrix A the number 1 is an eigenvalue of algebraic multiplicity 3 but having only one eigenvector, namely e_1:

$$A = \begin{pmatrix} 1 & 2 & 3 \\ 0 & 1 & 2 \\ 0 & 0 & 1 \end{pmatrix} .$$

Moreover,

$$\begin{aligned} (A - \lambda\mathbb{I})\, e_2 &= 2e_1 , & (A - \lambda\mathbb{I})^2\, e_2 &= 0 \\ (A - \lambda\mathbb{I})\, e_3 &= 3e_1 + 2e_2 , & (A - \lambda\mathbb{I})^2\, e_3 &= 4e_1 , & (A - \lambda\mathbb{I})^3\, e_3 &= 0 . \end{aligned}$$

i.e., e_s is a generalized eigenvector of degree s, $s = 1, 2, 3$.

The theorem on the Jordan normal form of matrices from Linear Algebra implies

Theorem 6.17 (A Basis Consisting of Generalized Eigenvectors). *For any matrix $A \in \mathcal{C}^{d\times d}$ there is a basis of $\mathcal{C}^d$ consisting of generalized eigenvectors that contains for any k-fold eigenvalue λ exactly k generalized eigenvectors $v_1, v_2, \dots, v_k$, where the degree of v_s is at most s.*

A basis $h_1, h_2, \dots, h_d$ of generalized eigenvectors gives rise to the fundamental system $\exp(At)h_1$, $\exp(At)h_2$, $\dots$, $\exp(At)h_d$ of the homogeneous equation $\dot{x} = Ax$. Next, we analyse the construction of such a solution

$$\varphi_v : \mathbb{R} \to \mathbb{C}^d , \qquad \varphi_v(t) := \exp(At)v ,$$

where v is a generalized eigenvector of degree s corresponding to the eigenvalue λ:

$$\exp(At)\, v = \exp(\lambda\mathbb{I}t) \exp((A - \lambda\mathbb{I})\, t)\, v = \exp(\lambda t) \cdot \sum_{k=0}^{\infty} \tfrac{1}{k!} (A - \lambda\mathbb{I})^k\, t^k v .$$

Since $(A - \lambda\mathbb{I})^k\, v = 0$ for $k \geq s$, the above infinite sum reduces to a finite one, and we obtain

$$\varphi_v(t) = \exp(\lambda t)\, p_v(t) , \quad \text{where} \quad p_v(t) = \sum_{k=0}^{s-1} \tfrac{1}{k!} (A - \lambda\mathbb{I})^k\, t^k v . \tag{6.42}$$

Here, p_v is a polynomial of degree less or equal to $s - 1$ the coefficients $\frac{1}{k!} (A - \lambda\mathbb{I})^k\, v$ of which are vectors in $\mathcal{C}^d$. For $s = 1$ we have $p_v(t) = v$.

Let us summarize the construction methodology in the following theorem:

Theorem 6.18 (Blueprint for the Construction of Fundamental System for $\dot{x} = Ax$). *The following procedure leads to a fundamental system of the linear homogeneous equation $\dot{x} = Ax$:*

1. *Determine all distinct eigenvalues $\lambda_1, \lambda_2, \dots, \lambda_r$ of A and their algebraic multiplicities $k_1, k_2, \dots, k_r$, where $k_1 + k_2 + \dots + k_r = d$ holds.*

2. *For each eigenvalue λ_ρ, $\rho = 1, 2, \dots, r$, with algebraic multiplicity k_ρ construct k_ρ solutions of $\dot{x} = Ax$ by*

   - *first determining the corresponding generalized eigenvectors $v_1, v_2, \dots, v_k$, where v_s has degree less or equal to s, and*
   - *then calculating, according to (6.42), these solutions*

$$\varphi_{v_s}(t) = \exp(\lambda_\rho t)\, p_{v_s}(t)\,, \qquad s = 1, 2, \dots, k_\rho\,.$$

The obtained d solutions form a fundamental system of $\dot{x} = Ax$.

Example 6.31 (Fundamental System for a 3×3 Matrix, cf. [162], p. 154). As seen in example 6.30: A has the 3-fold eigenvalue 1 with corresponding eigenvector e_1. e_2 is a generalized eigenvector of degree 2 and e_3 is a generalized eigenvector of degree 3. According to our construction blueprint we thus get the solutions

$$\varphi_1(t) = \exp(t)\, e_1 = \exp(t) \begin{pmatrix} 1 \\ 0 \\ 0 \end{pmatrix}$$

$$\varphi_2(t) = \exp\left(\mathbb{I} + (A - \mathbb{I})\, t\right) e_2 = \exp(t) \begin{pmatrix} 2t \\ 1 \\ 0 \end{pmatrix},$$

$$\varphi_3(t) = \exp\left(\mathbb{I} + (A - \mathbb{I})\, t + \tfrac{1}{2}(A - \mathbb{I})^2 t^2\right) e_3 = \exp(t) \begin{pmatrix} 3t + 2t^2 \\ 2t \\ 1 \end{pmatrix}.$$

Here, $\varphi_1(t), \varphi_2(t), \varphi_3(t)$ form a fundamental system.

6.7 First Integrals & Oscillations

Following [162], pp. 157, first information about the trajectories of integral curves can be gained from first integrals. A *first integral* is a of a vector field

Before you continue, make sure to answer the following questions:

Quiz: Section 6.6

Q1 What can you say about the existence and uniqueness of the solutions of $\dot{x} = Ax + b$, where $x, b \in \mathbb{R}^d$ and $A \in \mathbb{R}^{d\times d}$?

Q2 Give the definition of a fundamental system and of a fundamental matrix.

Q3 State the theorem of Liouville and sketch its proof.

Q4 Describe the general construction method for a solution of $\dot{x} = Ax + b$, where $x, b \in \mathbb{R}^d$ and $A \in \mathbb{R}^{d\times d}$.

Q5 Solve

$$\begin{pmatrix} \dot{x}_1 \\ \dot{x}_2 \end{pmatrix} = \begin{pmatrix} -8 & 3 \\ -18 & 7 \end{pmatrix} \begin{pmatrix} x_1 \\ x_2 \end{pmatrix} + \exp(-x) \begin{pmatrix} 5 \\ 12 \end{pmatrix}.$$

$v : \Omega \to \mathbb{R}^d$, $\Omega \subset \mathbb{R}^d$ is a $\mathcal{C}^1$-function $E : \Omega \to \mathbb{R}$ the derivative of which vanishes along v:

$$\partial_v E(x) = \sum_{i=1}^{d} v_i(x) \partial_i E(x) = 0. \tag{6.43}$$

For instance the function $E : \mathbb{R}^2 \to \mathbb{R}$, $E(x, y) = x^2 + y^2$ is a first integral of the rotation field $v : \mathbb{R}^2 \to \mathbb{R}^2$, $v(x, y) = (-y, x)$.

The importance of a first integral E is that it takes a constant value on each integral curve φ of the vector field due to the following identity

$$\frac{\mathrm{d}}{\mathrm{d}t} (E \circ \varphi)(t) = E'(\varphi(t))\, \dot{\varphi}(t) = E'(\varphi(t))\, v(\varphi(t)) = \partial_v E(\varphi(t)) = 0.$$

Theorem 6.19 (Integral Curves and Level Sets of First Integrals). *Every integral curve of v lies on a level set of E.*

In the above example of a rotation field we have that every integral curve $\neq 0$ lies on a circle $E(x, y) = x^2 + y^2 = r^2$

There is no general method to determine the first integral of a given vector field. Though, a set of methodological approaches enable the computation of first integrals in specific concrete cases, see Sec. 6.7.3. In physical applications conservation laws, like the conservation of energy, help to establish first integrals, see Sec. 6.7.1 and 6.7.2.

Before we continue to discuss specific examples we give two further remarks important for the study of level sets of first integrals, cf. [162], p. 160:

1. A non-empty level set $E^{-1}(c)$ for a regular value $c \in \mathbb{R}$ is a $(d-1)$-dimensional sub-manifold on $\Omega \subset \mathbb{R}^d$. For an element $x \in \Omega$ at which the value $E(x)$ is regular the condition (6.43) can be interpreted in the sense that $v(x)$ is a tangential vector to the level set through x (i.e., $v(x)$ lies in the kernel of $\mathrm{d}E(x)$).

2. Let E be a $\mathcal{C}^2$-function and x_0 a non-degenerate critical point, i.e., a point such that $E'(x_0) = 0$ and non-degenerate Hessian $E''(x_0)$. The shape of the level sets in the vicinity of x_0 is determined by the eigenvalues of $E''(x_0)$ up to diffeomorphisms according to the Lemma of Morse (see problem 6.53).

   We cite a special case of this lemma which has particular importance here: Let all eigenvalues of $E''(x_0)$ be positive, then there is a diffeomorphism $h : K \to \Omega_0$ of a ball $K \subset \mathbb{R}^d$ around 0 to an environment $\Omega_0 \subset \Omega$ of x_0 such that

   $$E \circ h(\xi) \;=\; E(x_0) + \xi_1^2 + \cdots + \xi_d^2 \,.$$

   Thus: Let $E'(x_0) = 0$ and $E''(x_0) > 0$, then there is an environment $\Omega_0 \subset \Omega$ of the point x_0 such that every integral curve φ of v through a point $x \in \Omega_0$ lies in a manifold $\sum \subset \Omega_0$ which is diffeomorphic to a $(d-1)$-sphere and has infinite life time.

Remark 6.32 (Vector Fields on Manifolds). First integrals naturally give rise to the study of vector fields on manifolds. A *vector field* or *tangential field on a manifold* $M \subset \mathbb{R}^d$ is a mapping v that assigns a tangential vector $v(x)$ to each element of $x \in M$. An integral curve to such a field is a curve $\varphi : I \to M$ such that $\dot{\varphi}(t) = v(\varphi(t))$ for all $t \in I$. If, for instance, M is a level set of a first integral of the vector field $V : \Omega \to \mathbb{R}^d$ on the open set $\Omega \subset \mathbb{R}^d$, then $v := V_{|M}$ is a vector field on M.

Finally, we give two examples of first integrals in systems from physics and one from biology/ ecology.

6.7.1 Application 1: The General Oscillation Equation

Following [162], p. 157, let $U : \Omega \to \mathbb{R}$ be a $\mathcal{C}^1$-function (a potential) on an open set $\Omega \subset \mathbb{R}^d$ which has an isolated minimum at x_0. The equation

$$\ddot{x} \;=\; -\nabla U^T(x)\,,$$

i.e., the first-order system

$$\frac{\mathrm{d}}{\mathrm{d}t}\begin{pmatrix} x \\ y \end{pmatrix} = \begin{pmatrix} y \\ -\nabla U^T(x) \end{pmatrix} =: v(x,y)\,, \tag{6.44}$$

has the energy function $E : \Omega \times \mathbb{R}^d \to \mathbb{R}$, given by

$$E(x,y) := \tfrac{1}{2}\sum_{i=1}^{d} y_i^2 + U(x)$$

as a first integral, because

$$\partial_v E(x,y) = y^T \nabla U^T(x) - \nabla U(x) y = 0\,.$$

Any solution $\varphi = (x,y) = (x,\dot{x})$ of the system (6.44), where x is a solution of $\ddot{x} = -\nabla U^T(x)$, lies in a level set of E such that $E(x,\dot{x})$ is constant with the constant being determined by the initial condition $(x(0),\dot{x}(0))$ (conservation of energy).

6.7.2 Application 2: The Deterministic Pendulum

Since the congenial insight of Johannes Kepler (1571 - 1630), Santorio Santorio (Sanctorius, 1561 - 1636) and Galileo Galilei (1564 - 1642) to time the pulse of a human via a pendulum, see [196], pp. 20, this easy to physically realise but hard to analyse device has gained enormous interest, both from the purely mathematical as well as from the applied point of view. Following [29][5], let us briefly discuss the properties of the deterministic (mathematical) *pendulum*, and its wide ranging applications.

Figure 6.12 (a) shows a simple sketch of the pendulum. Here, ϑ represents the angle that the pendulum makes with the vertical, l and m are its length and mass respectively, and g is the acceleration of gravity ($9.81\frac{m}{s^2}$ at the surface of the earth). The equation of motion of the pendulum is[6]

$$\ddot{\vartheta} + \omega^2 \sin(\vartheta) = 0 \quad \Leftrightarrow \quad \left\{ \begin{array}{rcl} \dot{x} & = & -\omega^2 \sin(\vartheta) \\ \dot{\vartheta} & = & x \end{array} \right., \tag{6.45}$$

where $\omega^2 := \frac{g}{l}$ is the angular forcing frequency. Approximating the nonlinear term $\sin(\vartheta)$ by ϑ in this equation for small values of ϑ leads to the (simple) *harmonic oscillator*

$$\ddot{\vartheta} + \omega^2 \vartheta = 0 \quad \Leftrightarrow \quad \left\{ \begin{array}{rcl} \dot{x} & = & -\omega^2 \vartheta \\ \dot{\vartheta} & = & x \end{array} \right.. \tag{6.46}$$

---

5 See, [47] for an extremely nice discussion of "the rigid pendulum—an antique but evergreen physical model", too.

6 See, e.g., [83], pp. 184, or [116], pp. 243, for a derivation of the pendulum's equation of motion by means of physical considerations, i.e., Newtons law of motion.

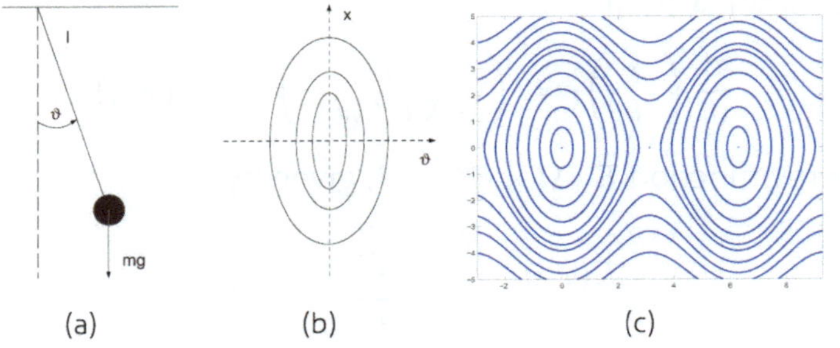

Figure 6.12. Sketch of the (mathematical) pendulum (a), the phase portrait of the (simple) harmonic oscillator (b) and numerical simulation of the pendulum's phase portrait for $\omega = 2$ (c). In (b) and (c), the abscissa-axis displays the angle variable ϑ and the ordinate-axis the angular velocity $\dot{\vartheta} = x$.

By direct substitution, we immediately verify that

$$\vartheta(t) \;=\; A\cos(\omega t) \,+\, B\sin(\omega t) \tag{6.47}$$

is a solution for the harmonic oscillator (6.46), whereby the constants A and B are determined by the initial conditions: $A = \vartheta(0)$ and $B = \omega^{-1}\dot{\vartheta}(0)$. Moreover, the rate of change (velocity) of the harmonic oscillator is given by

$$\dot{\vartheta}(t) \;=\; B\omega\cos(\omega t) \,-\, A\omega\sin(\omega t)$$

and a brief calculation shows that the trajectories of the harmonic oscillators obey the equations of an ellipse in phase space, parameterized by the time variable t, as sketched in Fig. 6.12 (b), i.e.,

$$\vartheta^2(t) \,+\, \omega^{-2}\dot{\vartheta}^2(t) \;=\; A^2 \,+\, B^2\,.$$

See also [5], pp. 1, for a discussion of the dynamical properties of the harmonic oscillator[7]. In particular, we immediately have that the origin is stable, but not asymptotically stable (see problem 8.25).

It is surprising that, unlike the equation for the harmonic oscillator, there is no closed form solution of the pendulum equation (6.45) analogous to the identity (6.47). A solution gained via a separation of variables ansatz, can be expressed in terms of elliptic integrals

$$T \;=\; 4\omega\int_0^{\pi/2} \frac{\mathrm{d}\vartheta}{\sqrt{1 \,-\, k^2\sin^2(\vartheta)}}\,,$$

[7] See, e.g., [5] for a discussion of various refinements/ generalisations of the (simple) deterministic oscillator model, and see [111] for the discussion of the stochastically disturbed counterparts.

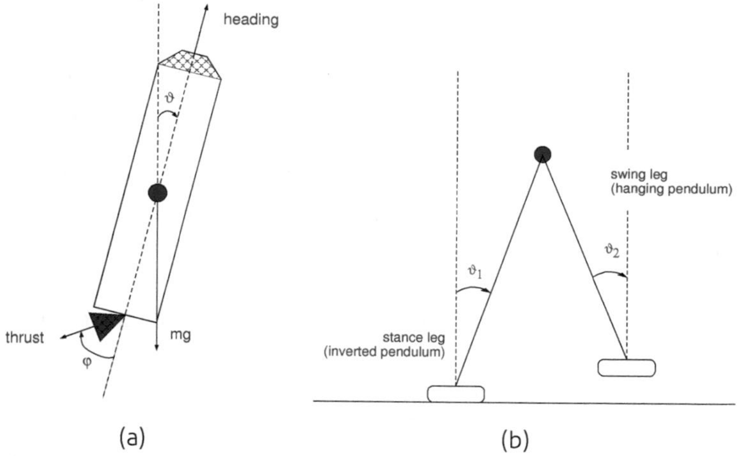

Figure 6.13. Applications of the pendulum system, cf. [29]: (a) a rocket acts like an inverted pendulum at the end of a thrust vectored motor; (b) the swing and stance legs of a bipedal walker can be modelled as a coupled pendulum.

where T denotes the oscillation/ periodic time, see, e.g., [116], pp. 245, [131], pp. 50, or [112], pp. 1.

The numerically generated phase portrait in Fig. 6.12 (c) illustrates that the trajectories can be described by the total energy E of the pendulum

$$E = \tfrac{1}{2}\dot{\vartheta}^2 - \omega^2 \cos(\vartheta) .$$

Recall that E is constant along solutions[8], i.e., a first integral of the motion, as

$$\dot{E} = \dot{\vartheta}\ddot{\vartheta} + \omega^2\dot{\vartheta}\sin(\vartheta) = \dot{\vartheta}\left(\ddot{\vartheta} + \omega^2 \sin(\vartheta)\right) = 0 ,$$

and therefore each trajectory in Fig. 6.12 (c) displays a level curve of the energy, see, e.g., [12, p. 147] for a detailed analysis of the phase portrait.

The mathematical pendulum system is interesting in its own right as a problem in dynamics and control, see, e.g., [29]. However, its importance is more than academic as many practical engineering systems can be approximately modelled as pendulum systems:

A rocket acts like an inverted pendulum at the end of a thrust vectored motor where active control is required to maintain proper attitude during ascent. Figure 6.13 (a) shows a diagram of a rocket whose pitch angle ϑ can be controlled during ascent by varying the angle φ of the thrust vector. The pitch dynamics of the rocket can be approximated by a controlled pendulum.

8 This is intuitively clear as we know from elementary physics that, without friction, the energy of the pendulum is constant.

In Biomechanics, the pendulum is often used to model bipedal walking, see Fig. 6.13 (b). In bipedal robots the *stance leg* that is in contact with the ground is often modelled as an inverted pendulum while the *swing leg* behaves as a freely swinging pendulum, suspended from the hip. In fact, studies of human postural dynamics and locomotion have found many similarities with pendulum dynamics. Measurements of muscle activity during walking indicate that the leg muscles are active primarily during the beginning of the swing phase after which they shut off and allow the leg to swing through like a pendulum. Nature has taught us to exploit the natural tendency of the leg to swing like a pendulum during walking, which may partially account for the energy efficiency of walking[9].

Quiet standing requires control of balance, and *postural sway* results from stretch reflexes in the muscles, which are a type of local feedback stabilisation of the inverted pendulum dynamics involved in standing, cf., e.g., [54], or [230].

There are many other examples of problems in engineering systems or the natural sciences where pendulum dynamics provides useful insight, including stabilisation of overhead (gantry) cranes, roll stabilisation of ships and trucks, and slosh control of liquids or the analysis of earthquakes, cf., e.g., [132]. Thus, a study of the pendulum and pendulum-like systems is an excellent starting point to understand issues in nonlinear dynamics.

In real live application there is often some forcing present altering the pendulum's equation of motion (6.45). Such a *forced pendulum* is described by

$$\ddot{\vartheta} + \omega^2 \sin(\vartheta) = \rho \sin(t) \quad \Leftrightarrow \quad \begin{cases} \dot{x} = -\omega^2 \sin(\vartheta) + \rho \sin(t) \,, \\ \dot{\vartheta} = x \,, \end{cases}$$

where the term $\rho \sin(t)$ ($\rho \neq 0$) is a 2π-periodic external forcing term constantly providing energy to the pendulum, cf. [2], pp. 54, or [56], pp. 384, for a general formulation/ discussion in terms of forced and damped oscillators[10]. Moreover, [39] and [209], for instance, discuss further modifications of the (simple) mathematical pendulum, like an *elastic pendulum* with an elastic instead of a inelastic rod (pp. 51, 132)[11], a *double pendulum*, where the mass

9 This is in particular true for penguins who – because of their short feet – have no other choice than to "swing" in order to move forward on solid ground.

10 A damping term $c\dot{\vartheta}$, corresponding to friction at the pivot, with friction constant $c \neq 0$, can be integrated into the pendulum dynamics in the following way:

$$\ddot{\vartheta} + c\dot{\vartheta} + \omega^2 \sin(\vartheta) = 0 \,.$$

See, e.g., [2], pp. 54, [56], pp. 384, or [49]. Moreover, e.g., [246], discusses the oscillation-rotation attractors in damped and forced pendulum systems.

11 See, e.g., [74] for a discussion of the occurrence of deterministic chaos in the elastic pendulum system.

is replaced by a second pendulum (pp. 136), or a pendulum with a freely moving point of suspension (pp. 219). Recently the *reaction wheel pendulum* attained much attention (here, the mass at the end of the pendulum rod is replaced by a independently rotating flywheel); for a comprehensive treatment of this pendulum type, see [29].

When the suspension point of the pendulum is forced to undergo rapid vertical periodic oscillations, it can be shown that the inverted pendulum position is stable — see, for instance, [173], [48], [185] or [50].

6.7.3 Application 3: The Volterra-Lotka System

Following [162], pp. 158, the system

$$\dot{x} = a(y) \cdot x\,, \qquad \dot{y} = -b(x) \cdot y\,, \tag{6.48}$$

can be interpreted in terms of the time evolution of two interacting biological or economic populations $x(t)$ and $y(t)$. At time t the growth rates $\dot{x}/x$ and $\dot{y}/y$ are determined by the current potential $\int_{y_0}^{y} a(\eta)/\eta \mathrm{d}\eta$ and $\int_{x_0}^{x} b(\xi)/\xi \mathrm{d}\xi$ respectively of the parameter functions.

Such a model was first applied by the Italian mathematician Vito Volterra (1860 – 1940) and the US-American biologist Alfred Lotka (1880 – 1949) to understand the fluctuations of sharks $x(t)$ and sardines $y(t)$ in the Mediterranean sea after World War I by using linear functions for $a(y)$ and $-b(x)$.

Here, we assume that $a, b : \mathbb{R} \to \mathbb{R}$ are strictly monotonically decreasing $\mathcal{C}^1$-functions each having a unique positive root

$$b(\xi) = 0\,, \quad a(\eta) = 0\,, \qquad \xi, \eta \in \mathbb{R}^+\,.$$

We will show that all positive solutions of (6.48) are periodic. First,though, some preparatory conclusions to set the scene:

1. The only non-negative equilibria are the origin $(0, 0)$ and the mixed species point $(\xi, \eta) \in \mathbb{R}^2_+$.

2. System (6.48) has the solutions $(c\exp(a(0)t), 0)$ and $(0, c\exp(-b(0)t))$, where $c \in \mathbb{R}$. These lie on the x- and y-axis respectively. Due to the existence and uniqueness theorem this implies that every solution φ starting with $\varphi(0) \in \mathbb{R}^2_+$ in the positive quadrant stays in this quadrant for all times. We say that the positive quadrant is invariant, cf. Sec. 8.2.1.

3. There exists a first integral on $\mathbb{R}^2_+$ which has the form

   $$E(x, y) = F(x) + G(y)\,.$$

   The corresponding condition for a first integral to vanish along solutions then reads as

   $$a(y)xF'(x) - b(x)yG'(y) = 0\,,$$

and is satisfied, for instance, if $F'(x) = -b(x)/x$ and $G'(y) = a(y)/y$. Thus, we define

- F as the anti-derivative of $-b(x)/x$ on $\mathbb{R}^+$ such that $F(\xi) = 0$, and
- G as the anti-derivative of $a(y)/y$ on $\mathbb{R}^+$ such that $G(\eta) = 0$.

We immediately recognize that F is strictly monotonously decreasing on $(0, \xi]$ and strictly monotonously increasing on $[\xi, \infty)$, and that G exhibits the same properties with η in the place of ξ.

Moreover, $F(u), G(u) \to \infty$ for $u \to 0$ as well as for $u \to \infty$. This implies that E has an isolated minimum in (ξ, η) with $E(\xi, \eta) = 0$. Every level set $E^{-1}(\alpha)$, $\alpha > 0$, is a compact sub-set of the positive quadrant $\mathbb{R}^2_+$ and intersects with each of the lines $x = \xi$ and $y = \eta$ at exactly two points, see Fig. 6.14 for the description of these intersection points A_0, A_1, A_2, and A_3.

4. Let $\varphi = (x, y)$ be the maximal integral curve with $\varphi(0) = A_0 \in E^{-1}(\alpha)$, $\alpha > 0$. φ lies in $E^{-1}(\alpha)$, i.e., in a compact set, and thus is defined for all $t \in \mathbb{R}$. Next, we show that there is a $t_1 \in (0, \infty)$ such that

   - x is strictly monotonously decreasing in $[0, t_1]$, whereas y is strictly monotonously increasing in $[0, t_1]$, and
   - $\varphi(t_i) = A_1$.

   To see this, let $t_1 := \sup\{t : x(t) > \xi \text{in the whole of } [0, t]\}$. According to the second differential equation $\dot{y} > 0$ thus holds in $[0, t_1)$, and y is hence strictly monotonously increasing. This implies $y(t) > y(0) = \eta$ on $(0, t_1)$. Consequently, due to the first differential equation $\dot{x} < 0$ holds, and x is strictly monotonously decreasing.

   Next, we show that $t_1 < \infty$. Therefore, choose an $\varepsilon \in (0, t_1)$. For $t \in (\varepsilon, t_1)$ it holds that $y(t) \geq y(\varepsilon) > \eta$ and thus $a\,(y(t)) \leq a\,(y(\varepsilon)) =: \alpha < 0$. According to the first differential equation $x(t) \leq \exp\left(\alpha(t - t_0)\right) x(\varepsilon)$ holds for these values of t. Because of this and the fact $x(t) > \xi > 0$ it has to hold that $t_1 < \infty$.

   The definition of t_1 it immediately implies now that $x(t_1) = \xi$ and $\varphi(t_1) = A_1$.

5. As above in conclusion 4 one can show the existence of parameter points $t_1 < t_2 < t_3 < t_4$ such that $\varphi(t_k) = A_k$, $k = 2, 3, 4$. In particular, one has with $T = t_4$ that
   $$\varphi(T) = A_4 = A_0 = \varphi(t_0)$$
   holds.

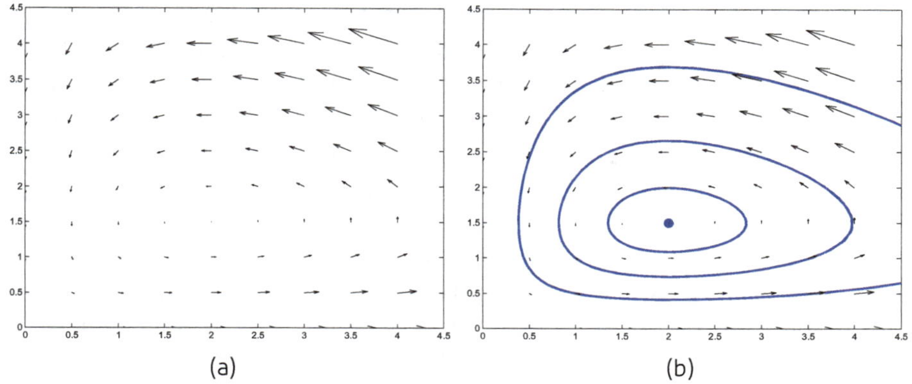

Figure 6.14. Velocity plot (a) and sketch of the phase portrait (b) of the Volterra-Lotka system $\dot{x} = (3 - 2y)x$ and $\dot{y} = (x - 2)y$ showing the existence of an equilibrium and periodic solutions.

Altogether, this shows the following theorem:

Theorem 6.20 (Solution of the Volterra-Lotka System). *Every maximal integral curve φ with $\varphi(0) \in \mathbb{R}^2_+$ lies for all times in this quadrant and is periodic.*

Before you continue, make sure to answer the following questions:

Quiz: Section 6.7

Q1 Give the definition of a first integral.

Q2 How are first integrals and integral curves connected?

Q3 Give the ordinary differential equation that describes the deterministic pendulum. Sketch and discuss its phase portrait.

Q4 Give the ordinary differential equation that describes the Volterra-Lotka system. Sketch and discuss its phase portrait.

6.8 Ordinary Differential Equations on Banach Spaces

Following [82], Chap. 1, we end our discussion on deterministic ordinary differential equations with Lipschitz-type conditions for the existence and uniqueness of deterministic ordinary differential equations on arbitrary Banach spaces. These Lipschitz-type conditions do not lead to the most refined results for such equations, though they are similar to what was discussed in the $\mathbb{K}^d$-setting of the previous sections. The advantage is that they are already sufficient to understand a large class of problems related to mean-square solutions of random differential equations as introduced in Chap. 4.

From now on we let X be a Banach space over $\mathbb{K}$, i.e., a complete normed vector space over $\mathbb{K}$. Completeness in this regard means that for every Cauchy sequence $\{x_n\}_{n=1}^{\infty} \subset X$ there exists an element $x \in X$ such that $\lim_{n\to\infty} x_n = x$ with respect to the chosen norm. Moreover, let $D \subset X$ such that $x_0 \in D$, $J = [0, a] \subset \mathbb{R}$ an interval and $f : I \times D \to X$ a function. In this section, we are interested in continuously differentiable functions

$$x : [0, \delta] \to D, \quad \text{for some } \delta \in (0, a]$$

such that

$$\dot{x} = f(t, x) \quad \text{on } [0, \delta] \text{ and } x(0) = x_0 . \tag{6.49}$$

Such a function x is called a (local) solution of the differential equation (6.49).

6.8.1 Existence & Uniqueness of Solutions

The following facts may be proved as in the case of $X = \mathbb{K}^d$, for instance by means of successive approximation with the aid of Picard-Lindelöf sequences:

If f is continuous and satisfies the Lipschitz-condition

$$|f(t, x) - f(t, y)| \le L\,|x - y| ,$$

for $t \in I$ and $x, y \in D$, then (6.49) has a unique solution on I, provided $D = X$.

If D is the ball $\overline{\mathbb{B}_r(x_0)} = \{x : |x - x_0| \le r$ and f is Lipschitz-continuous (as above), then (6.49) has a unique solution on $[0, \delta]$, where $\delta := \min\{a, r/M\}$ and $M := \sup\{|f(t, x)| : t \in I, x \in D\}$. Note, due to the lemma of Riesz, the ball $\overline{\mathbb{B}_r(x_0)}$ is no longer compact in infinite dimensions, and this is contrary to what we are used to in the finite dimensional spaces $\mathbb{K}^d$, cf. [85].

Assume that D is open and f is locally Lipschitz-continuous, i.e., to each $(t, x) \in I \times D$ there exists an $\eta = \eta(t, x) > 0$ and $L = L(t, x) > 0$ and a neighborhood U_x of x such that $|f(t, u) - f(t, v)| \le L\,|u - v|$ for $s \in I \cap [t, t+\eta]$ and $u, v \in U_x$. Then, (6.49) has a unique solution, defined either on the whole interval J or only on a subinterval $[0, \delta)$ which is maximal with respect to the extension of solutions (see Sec. 6.8.2).

By means of the simple results just mentioned it is easy to construct approximate solutions for (6.49) in the case where f is continuous. Following [82], pp. 4, we show at first that such an f may be approximated by locally Lipschitz-continuous functions.

Lemma 6.33 (Approximation by Lipschitz-Continuous Functions). *Let X, Y be Banach spaces, $\Omega \subset X$ open and $f : \Omega \to Y$ continuous. Then, to each $\varepsilon > 0$ there exists a locally Lipschitz-continuous function $f_\varepsilon : \Omega \to Y$ such that $\sup_\Omega |f(x) - f_\varepsilon(x)| \leq \varepsilon$.*

Proof. Following [82], p. 5, let $U_\varepsilon(x) := \{y \in \Omega : |f(y) - f(x)| < \frac{1}{2}\varepsilon\}$, $x \in \Omega$, then we have that $U_\varepsilon(x)$ is open and $\Omega = \cup_{x\in\Omega} U_\varepsilon(x)$.

Next, let $\{V_\lambda : \lambda \in \Lambda\}$ be a locally finite refinement of $\{U_\varepsilon(x) : x \in \Omega\}$, i.e., an open cover of Ω such that each $x \in \Omega$ has a neighborhood $V(x)$ with $V(x) \cap V_\lambda \neq 0$ only for finitely many $\lambda \in \Lambda \subset \Omega$, and such that to each $\lambda \in \Lambda$ there exists a $x \in \Omega$ with $V_\lambda \subset U_\varepsilon(x)$.

Define $\alpha_\lambda : X \to \mathbb{R}$ by

$$\alpha_\lambda(x) := \begin{cases} 0 & \text{for } x \neq V_\lambda \\ \rho(x, \partial V_\lambda) & \text{for } x \in V_\lambda \end{cases},$$

where $\rho(x, A) = \inf\{|x - y| : y \in A\}$. Let

$$\phi_\lambda(x) := \left(\sum_{\mu\in\Lambda} \alpha_\mu(x)\right)^{-1} \alpha_\lambda(x) \qquad \text{for } x \in \Omega .$$

Since α_λ is Lipschitz-continuous on X and $\{V_\lambda : \lambda \in \Lambda\}$ is locally finite, ϕ_λ is locally Lipschitz-continuous on Ω. For every $\lambda \in \Lambda$ we choose some $a_\lambda \in V_\lambda$ and we define

$$f_\varepsilon(x) := \sum_{\lambda\in\Lambda} \phi_\lambda(x) f(a_\lambda) \qquad \text{for } x \in \Omega .$$

We have $f_\varepsilon(x)$ locally Lipschitz-continuous in Ω and

$$|f_\varepsilon(x) - f(x)| = \left|\sum_{\lambda\in\Lambda} \phi_\lambda(x) \left(f(a_\lambda) f(x)\right)\right| \leq \sum_{\lambda\in\Lambda} \phi_\lambda(x) \left|f(a_\lambda) f(x)\right| .$$

Now, suppose $\phi_\lambda(x) \neq 0$. Then, $x \in V_\lambda \subset U_\varepsilon(x_0)$ for some $x_0 \in \Omega$ and $a_\lambda \in V_\lambda$, hence $|f(a_\lambda) - f(x)| \leq \varepsilon$ and therefore

$$|f_\varepsilon(x) - f(x)| \leq \varepsilon \sum_{\lambda\in\Lambda} \phi_\lambda(x) = \varepsilon ,$$

which shows the assertion. □

The following result on continuous extensions of continuous mappings may be proved along similar lines (cf. problem 6.59):

Lemma 6.34. *Let X, Y be Banach spaces, $\Omega \subset X$ closed and $f : \Omega \to Y$ continuous. Then there is a continuous extension $\tilde{f} : X \to Y$ such that $\tilde{f}(X) \subset \text{conv}(f(\Omega))$, where $\text{conv}(f(\Omega))$ denotes the convex hull of $f(\Omega)$.*

Now, we can find approximate solutions of (6.49):

Theorem 6.21 (Approximate Solutions). *Let $I = [0, a]$ be an interval, $D = \overline{\mathbb{B}_r(x_0)} \subset X$ be the ball around $x_0 \in X$ with radius $r > 0$, and $f : I \times D \to X$ be continuous and bounded by $|f(t,x)| \le M$ on $I \times D$. Moreover, let $\varepsilon > 0$ and $a_\varepsilon := \min\{a, r/(M+\varepsilon)\}$. Then there exists a continuously differentiable function $x_\varepsilon : [0, a_\varepsilon] \to D$ such that*

$$\dot{x}_\varepsilon = f(t, x_\varepsilon) + y_\varepsilon(t)\,, \quad x_\varepsilon = x_0\,, \quad \textit{and} \quad |y_\varepsilon(t)| \le \varepsilon\,, \quad \textit{on } [0, a_\varepsilon]\,.$$

which approximates the exact solution of $\dot{x} = f(t,x)$ up to an error of ε.

Proof. Following [82], p. 6, we have by Lemma 6.34 that f has a continuous extension $\tilde{f} : \mathbb{R} \times X \to X$ such that $|\tilde{f}(t,x)| \le M$ everywhere. By Lemma 6.33, there exists a locally Lipschitz-continuous approximation $\tilde{f}_\varepsilon : \mathbb{R} \times X \to X$ such that

$$\left|\tilde{f}_\varepsilon(t,x) - \tilde{f}(t,x)\right| \le \varepsilon\,,$$

in particular,

$$\left|\tilde{f}_\varepsilon(t,x) - f(t,x)\right| \le \varepsilon \qquad \text{on } J \times D\,,$$

and

$$\left|\tilde{f}_\varepsilon(t,x)\right| \le M + \varepsilon\,.$$

Let x_ε be the solution of $\dot{x} = \tilde{f}_\varepsilon(t,x)$, $x(0) = x_0$. This solution exists on $[0, a_\varepsilon)$ and satisfies $\dot{x}_\varepsilon = f(t, x_\varepsilon) + y_\varepsilon(t)$ with

$$|y_\varepsilon(t)| = \left|\tilde{f}_\varepsilon(t, x_\varepsilon) - f_\varepsilon(t, x_\varepsilon)\right| \le \varepsilon\,.$$

This shows the assertion. □

6.8.2 Extension of Solutions

In the case $\dim(X) < \infty$ and f is continuous on $J \times X$, it is well known that every solution x of (6.49) either exists on J or only on some sub-interval $[0, \delta_x)$ and then blows-up, $\lim_{t \to \delta_x} |x(t)| \to \infty$.

However, in the case of $\dim(X) = \infty$ one can find a continuous function $f : [0, \infty) \times X \to X$ such that (6.49) has a (unique) solution x on $[0, 1)$ only (and not for any further time), where x remains bounded. In particular $\lim_{t \to 1} x(t)$ does not exist. This is possible since a non-linear continuous mapping needs not to map bounded sets into bounded sets:

Example 6.35 (Counter-Example for the Extensibility of Solutions, cf. [82], pp. 7). Let $\dim(X) = \infty$. Then, there exists an infinite dimensional closed subspace $X_0 \subset X$ equipped with a Schauer base, i.e., there are sequences $\{e_i\}_{i\in\mathbb{N}} \subset X_0$ and $\{e_i^*\}_{i\in\mathbb{N}} \subset X_0^*$ (X_0^* is the dual space of X_0) such that $x \in X_0$ has the unique representation

$$x = \sum_{i\geq 1} \langle x \,,\, e_i^* \rangle e_i$$

(see, e.g., [176], p. 10). For every $i \in \mathbb{N}$ we assume $|e_i| = 1$ and consider the initial value $x_0 = e_1 \in X_0$. It is enough to construct an $f : [0,\infty) \times X_0 \to X_0$ since f has a continuous extension $\tilde{f} : [0,\infty) \times X \to X_0$ by Lemma 6.34, and therefore every solution of (6.49) with $\tilde{f}$ is already in X_0.

Let $0 < t_1 < t_2 < \cdots < 1$ be a partition of $(0,1)$ such that $\lim_{i\to\infty} t_i = 1$, and χ_i be the characteristic function of $[t_i, t_{i+1}]$. Set $\phi(t) := \max\{0, 2t-1\}$, $\alpha_1(t) \equiv 1$ and $\alpha_i \in C^1(\mathbb{R})$ for $i \geq 2$ such that

$$\alpha_i(t) \begin{cases} = 1 & \text{on } [t_i, t_{i+1}]\,, \\ \in (0,1) & \text{on } \left(\frac{1}{2}(t_{i-1}+t_i), t_i\right) \cup \left(t_{i+1}, \frac{1}{2}(t_{i+1}+t_{i+2})\right), \\ = 0 & \text{otherwise}\,. \end{cases}$$

By means of these functions we define

$$f(t,x) := \sum_{i\geq 2} \left(\phi\left(\langle x, e_{i-1}^*\rangle\right)\chi_{i-1}(t) + \phi\left(\langle x, e_{i+1}^*\rangle\right)\chi_{i+1}(t)\right)\dot{\alpha}_i(t)e_i\,.$$

Since $\langle x, e_i^*\rangle \to 0$ as $i \to \infty$, one can see that f is locally Lipschitz-continuous on $[0,\infty) \times X_0$. For

$$x(t) = \sum_{i\geq 1} \alpha_i(t)e_i$$

we have $x_0 = e_1$ and

$$\dot{x}(t) = \sum_{i\geq 2} \dot{\alpha}_i(t)e_i = f(t, x(t)) \qquad \text{on } [0,1)\,.$$

Hence, x is the unique solution of (6.49). Since $x(t_i) = e_1 + e_{i-1} + e_i$ and

$$x\left(\tfrac{1}{2}(t_i + t_{i+1})\right) = e_1 + e_i \qquad \text{for } i \geq 3\,,$$

the limit $\lim_{t\to 1} x(t)$ does not exist. Finally, $|x(t)| \leq 3$ on $[0,1)$.

6.8.3 Linear Equations

We finish by discussing a simple situation from [82], pp. 8, in which the results of Sec. 6.8.1 apply: the linear problem

$$\dot{x} \;=\; A(t)x + b(t)\,, \qquad x(0) \;=\; x_0\,, \tag{6.50}$$

where $A : I \to L(X)$ ($L(X) = B(X, X)$ is the space of bounded linear operators from X to itself) and $b : J \to X$ are continuous. Let $R(t, s)$ be the solution of the initial value problem on $L(X)$

$$\dot{U} \;=\; A(t)U \quad \text{with} \quad U(s) \;=\; id_X\,.$$

Then, $R(t, s)x_0$ is the solution on J of

$$\dot{x} \;=\; A(t)x \quad \text{with} \quad x(s) \;=\; x_0\,.$$

Using this fact, we can verify that

- $R(t, s) = R(t, \tau)R(\tau, s)$ for any $t, \tau, s \in I$,
- $R(t, s)$ is a homeomorphism of X onto X,
- $R(t, s)^{-1} = R(s, t)$, and
- $(t, s) \mapsto R(t, s)$ is continuous.

Therefore

$$x(t) \;=\; R(t, 0)x_0 + \int_0^t R(t, s)b(s)\text{d}s$$

is the unique solution of (6.50). In particular, if $A(t) \equiv A \in L(X)$, then

$$x(t) \;=\; \exp(At)x_0 + \int_0^t \exp\left(A(t-s)\right)b(s)\text{d}s\,. \tag{6.51}$$

This fact can be exploited in connection with countable systems of ordinary differential equations where an infinite dimensional matrix $A = (a_{ij})$ defines a bounded linear operator on the sequence space under consideration. For example (see [82], p. 9), let us discuss

$$X \;:=\; \ell^1 \;=\; \left\{ x \in \mathbb{R}^{\mathbb{N}} \,:\, \sum_{j\geq 1} |x_i| < \infty \right\}$$

and suppose that A satisfies

$$\sup_j \sum_{i\geq 1} |a_{ij}| \;<\; \infty\,.$$

Obviously, $A \in L(\ell^1)$ and therefore the solution of (6.50) is (6.51).

Now, again consider $X = \ell^1$, though we assume that A satisfies

$$\sup_i \sum_{j \geq 1} |a_{ij}| < \infty .$$

Then, A is defined on $x \in \ell^1$ but Ax may not belong to ℓ^1. Therefore, we can not have a solution of (6.50) for every $x_0 \in \ell^1$. Nevertheless, we may proceed in the following two directions: On the one hand we may restrict A to its proper domain $D(A) = \{x \in \ell^1 : Ax \in \ell^1\}$ and ask whether (6.50) has a solution at least if x_0 and $b(t)$ belong to $D(A)$. Results of this type can be found, for instance, in [82], Chap. 8. On the other hand we may ask whether (6.50) always has a solution at least in some Banach space larger than the one under consideration ("generalized solutions"). In the present example it is easy to answer this question: Since

$$\ell^1 \subset \ell^\infty = \left\{ x \in \mathbb{R}^{\mathbb{N}} : \sup_i |x_i| \leq \infty \right\}$$

and $A \in L(\ell^\infty)$, equation (6.50) has a unique solution in ℓ^∞ and (6.51) is valid.

In general, the condition that A be bounded from X to some larger Banach space Y such that X is continuously embedded in Y is not sufficient for existence. The following example illustrates this.

Example 6.36 (Non-Existence of Solutions After Embedding, cf. [82], p. 10). Let $A = (a_{ij})$, where $a_{1j} = 1$ for $j \geq 2$, $a_{i1} = 1$ for $i \geq 2$ and $a_{ij} = 0$ otherwise. Obviously, A is bounded from ℓ^1 into ℓ^∞. If x is a solution of the initial value problem $\dot{x} = Ax$, $x(0) = x_0$, then

$$x_i(t) = x_{0i} + \int_0^t x_1(s)\mathrm{d}s \qquad \text{for } i \geq 2$$

and

$$\dot{x}_1(t) = \sum_{i \geq 2} x_i(t) .$$

Hence, $x_1(t) \equiv 0$ and $\sum_{i \geq 2} x_{01}(t) = 0$. In particular, there is no solution if $x_0 \in \ell^1$ and $x_{01} \neq 0$.

A positive result in this direction is the following theorem, where we assume in particular, that A is bounded from the small space to a whole family of larger spaces.

Theorem 6.22 (Existence of a Unique Solution After Embedding). *Let (X_s), $0 \leq \alpha \leq s \leq \beta$ be a scale of Banach spaces such that $X_s \subset X_{s'}$ for $s < s'$ and $|x|_{s'} \leq |x|_s$ for $x \in X_s$. Suppose further*

1. *$A : I \to L(X_s, X_{s'})$ is continuous for every pair (s, s') with $\alpha \leq s' < s \leq \beta$,*

2. *$|A(t)|_{L(X_s,X_{s'})} \leq M/(s - s')$ for some constant M independent of s, s' and t, and*

3. *$x_0 \in X_\beta$ and $b : I \to X_\beta$ continuous.*

Then, for every $s \in [\alpha, \beta)$, equation (6.50) has a solution $x : [0, \delta(\beta - s)) \to X_s$, where $\delta = \min\{a, 1/(M \times e)\}$. The solution is uniquely determined for $s \in (\alpha, \beta)$, and

$$|x(t) - x_0|_{X_s} \;=\; \left(|x_0|_{X_\beta} + \frac{\beta - s}{M} \max_{[0,t]} |b(\tau)|_\beta\right) \cdot \frac{M \cdot e \cdot t}{\beta - s - M \cdot e \cdot t} \,. \qquad (6.52)$$

Proof. Following [82], p. 11, we start with the existence part by considering the successive approximations

$$\begin{aligned} x_0(t) &\equiv x_0 \\ x_k(t) &= x_0 + \int_0^t (A(\tau)x_{k-1}(\tau) + b(\tau))\, \mathrm{d}\tau \qquad \text{for } k \geq 1 \,. \end{aligned}$$

By induction, $x_k(t) \in X_s$ for every $s \in [\alpha, \beta)$ and $k \geq 0$. Let

$$M_t \;:=\; |x_0|_\beta + \frac{\beta - s}{M} \cdot \max_{[0,t]} |b(\tau)|_\beta \,.$$

We claim

$$|x_k(t) - x_{k-1}(t)|_{X_s} \;\leq\; M_t \cdot \left(\frac{M \cdot e \cdot t}{\beta - s}\right)^k \qquad \text{for } k \geq 1 \,.$$

We have

$$|x_1(t) - x_0|_{X_s} \;\leq\; t \cdot \left(\frac{M}{\beta - s}|x_0|_{X_\beta} + \max_{[0,t]} |b(\tau)|_{X_\beta}\right) \;\leq\; M_t \cdot \frac{M \cdot e \cdot t}{\beta - s} \,.$$

If the inequality holds for k then

$$\begin{aligned} |x_{k+1}(t) - x_k(t)| &\leq \int_0^t |A(\tau)\,(x_k(\tau) - x_{k-1}(\tau))\,|_{X_s} \mathrm{d}\tau \\ &\leq \frac{M}{\varepsilon} \int_0^t |x_k(\tau) - x_{k-1}(\tau)|_{X_{s+\varepsilon}} \mathrm{d}\tau \\ &\leq M_t \cdot \left(\frac{M \cdot e \cdot t}{\beta - s - \varepsilon}\right)^k \cdot \frac{M}{\varepsilon} \cdot \frac{t^{k+1}}{k+1} \,. \end{aligned}$$

With $\varepsilon = \frac{\beta-s}{k+1}$ we obtain

$$|x_{k+1}(t)-x_k(t)| \leq M_t \cdot \left(\frac{M \cdot e \cdot t}{\beta - s}\right)^{k+1} \cdot e^{-1} \cdot \left(1+\frac{1}{k}\right)^k \leq M_t \cdot \left(\frac{M \cdot e \cdot t}{\beta - s}\right)^{k+1} .$$

Hence,

$$x_k \rightarrow x(t) := x_0 + \sum_{k \geq 1} (x_k(t) - x_{k-1}(t))$$

uniformly on every closed subinterval of $[0, \delta(\beta-s))$, and (6.52) holds. Therefore,

$$x(t) = x_0 + \int_0^t (A(\tau)x(\tau) + b(\tau))\,\mathrm{d}\tau \qquad \text{on } [0, \delta(\beta - s)) .$$

Now, $A(t)x(t)$ is continuous, since $x(t)$ is continuous on $[0, \delta(\beta - s - \varepsilon))$ with values in $X_{s+\varepsilon}$ and $t \rightarrow A(t) \in L(X_{s+\varepsilon}, X_s)$ is continuous for every $\varepsilon \in (0, \beta - s)$. Hence, x is a solution of (6.50).

Finally, we address the uniqueness part: Let $s \in (\alpha, \beta)$, and $x : [0, \eta] \rightarrow X_s$ satisfy $\dot{x} = A(t)x$, $x(0) = 0$. Then, $N := \{t : x(t) = 0\}$ is closed. But N is also open in $[0, \eta]$. To prove this, let $t_0 \in N$ and $s' < s$. As in the proof of the existence we obtain

$$|x(t)|_{X_{s'}} \leq M_2 \cdot \left(\frac{M \cdot e \cdot |t - t_0|}{s - s'}\right)^k \qquad \text{for } k \geq 1 ,$$

by induction, where $M_2 := \max\{|x(t)|_{X_s} : t \in [0, \eta]\}$. Hence, $|x(t)|_{X_{s'}} = 0$ for $t \in [0, \eta]$ and $|t - t_0| \leq (M \cdot e)^{-1}(s - s')$. Since N is open and closed, we have $N = [0, \eta]$. □

The next example illustrates this theorem.

Example 6.37 (Moments of the Diffusion Equation, cf. [82], p. 12). Consider the diffusion equation

$$u_t = u_{xx} + axu_x + bx^2u$$

with the initial condition $u(0, x) = \phi(x)$. If u is interpreted as probability density, e.g. of a particle undergoing Brownian motion, i.e. $u(t, x) \geq 0$, and

$$\int_{-\infty}^{\infty} u(t, x)\mathrm{d}x = 1 ,$$

then one may be interested in the moments

$$u_n(t) = \int_{-\infty}^{\infty} u(t, x)x^n\mathrm{d}x .$$

We now multiply the (partial) differential equation by x^n and integrate over $\mathbb{R}$. Assuming that partial integration is justified, we obtain the countable system

$$\begin{aligned} \dot{u}_n &= n(n-1)u_{n-2} - a(n+1)u_n + bu_{n+2}\,, \\ u_n(0) &= \int_{-\infty}^{\infty} \phi(x)x^n \mathrm{d}x\,, \end{aligned}$$

where $u_{-1}(t) \equiv 0$ and $u_0(t) \equiv 1$. Let A be the corresponding infinite matrix and consider the scale of embedded Banach spaces

$$X_s = \left\{ u \in \mathbb{R}^{\mathbb{N}} : |u|_{X_s} = \sum_{j\geq 1} |u_j| \exp(js)(j!)^{-1/2} < \infty \right\} \quad \text{for } 0 < \alpha \leq s \leq \beta\,.$$

Obviously, $|u|_{X_{s'}} \leq |u|_{X_s}$ for $s' < s$ and $u \in X_s$. A simple calculation yields

$$|A|_{L(X_s, X_{s'})} \leq M \cdot (s-s')^{-1}\,, \qquad \text{with } M := \exp(2\beta)\cdot(1+|a|+|b|)\,.$$

Hence, the moments of ϕ are in X_β, and the moments of u are in X_s for $0 \leq t \leq (M\cdot e)^{-1}(\beta - s)$.

We may replace the linear right hand side $A(t)x + b(t)$ of Theorem 6.22 by a non-linear function f such that $f : I \times X_s \to X_{s'}$ is continuous and

$$|f(t,x) - f(t,y)|_{X_{s'}} = \frac{M}{s-s'} \cdot |x-y|_{X_s} \qquad \text{for } x, y \in X_s \text{ and } t \in I\,.$$

Further generalisations and applications are given, e.g., in [53].

6.9 Chapter's Summary

With Sec. 6.2, we started our study of deterministic ordinary differential equations (without singularities) by first collecting essential background information about vector fields and their representation as well as local Lipschitz continuity and Gronwall's lemma.

After these preliminaries we discussed continuous integral curves in vector fields, i.e. the solutions of ordinary differential equations $\dot{x} = F(t,x)$ with F being at most continuous. In Sec. 6.3, we analysed continuous right hand side equations and studied ε-approximate solutions as well as the Peano-Cauchy existence theorem and its implications. Then, in Sec. 6.4, we continued our discussion for Lipschitz-continuous functions F and the existence and uniqueness theorem of Picard-Lindelöf. It is here that we gave the MATLAB commands for symbolically solving ordinary differential equations and analysed maximal integral curves. In particular, we gave the three types of

Before you continue, make sure to answer the following questions:

Quiz: Section 6.8

Let X, Y be Banach spaces and $\Omega \subset X$ be open.

Q1 Describe the concept of an ordinary differential equation on Banach spaces. What does one mean when speaking about a solution in this context?

Q2 Can every continuous function $f : \Omega \to Y$ be approximated by a locally Lipschitz-continuous function?

Q3 Sketch the proof of the assertion you used in Q2.

Q4 Under which conditions does an ε-approximate solution of the exact solution of an ordinary differential equation on Banach spaces exist?

Q5 Sketch the proof of the assertion you used in Q4.

Q6 What can you say about the extension of solutions? Illustrate your assertion.

Q7 Under which conditions does unique solution of a linear ordinary differential equation on Banach spaces exist?

Q8 Sketch the proof of the assertion you used in Q7.

maximal integral curves that can occur in autonomous systems and showed the transformation of a d-th order equation into a first order system.

In Sec. 6.5, we studied the existence of solutions in the extended sense where the right hand side function may be continuous except for a set of Lebesgue-measure zero. Caratheodory's existence theorem and its implications were analysed together with maximum and minimum solutions.

Section 6.6 returns to Lipschitz-continuous right hand sides and discusses the broad class of linear ordinary differential equations: We studied the unique existence of their solutions and explicitly constructed them.

Applications of the theory were given with respect to first integrals and oscillations in Sec. 6.7. Here, we discussed in particular the deterministic pendulum and the Volterra-Lotka system.

Of course, there is a variety of great books on deterministic ordinary differential equations and their applications, for instance, [38], [15], [128], or [167] provide a wealth of background examples from the fields of biology, chemistry, engineering and the sciences. In particular, [231] covers lots of historic material related to the evolution of the mathematical study of ordinary differential equations.

Finally, based on [82], Chap. 1, Sec. 6.8 rounded up our discussion by providing a first glance into the existence, uniqueness and extension of solutions of ordinary differential equations on infinite dimensional Banach spaces.

Problems

Classification: ☼ easy, ⊕ easy with longer calculations, ☆ a little bit difficult, ◫ challenging.

Exercise 6.38. [☼] Identifying Integral Curves
Match the solution curves in Fig. 6.15 to the ODEs below.

1. $\dot{x} = x^2 - 1$.
2. $\dot{x} = t^{-2}x - 1$.
3. $\dot{x} = \sin(t)\sin(x)$.
4. $\dot{x} = \sin(tx)$.
5. $\dot{x} = 2t + x$.
6. $\dot{x} = \sin(3t)$.

Exercise 6.39. [☼] The Velocity Field of the 3-Dimensional Rigid Rotation
Sketch the velocity field $v : \mathbb{R}^3 \to \mathbb{R}^3$, $v(x) := \omega \times x$ of a rigid rotation in the Euclidean $\mathbb{R}^3$ with the vectorial rotation speed $\omega \in \mathbb{R}^3$ and compute its divergence.

Exercise 6.40. [☆] Convergence of Polygonal Approximations
Let

$$\dot{x} = |x|^{-3/4} + t\sin(\pi/t)\,, \qquad x(0) = 0\,.$$

Show that if polygonal approximate solutions are set-up as in Theorem 6.1 they need not converge as $\varepsilon \to 0$.

Hint: Consider $t \geq 0$ and let $t_k = k\delta$, $k = 0, 1, 2, \ldots$, where $\delta = (n + 1/2)^{-1}$ for some large n. If n is even, show that the polygonal solution $\varphi_n(t)$ satisfies $\varphi_n(\delta) = 0$, $\varphi_n(2\delta) = \delta^2$, $\varphi_n(3\delta) > \frac{1}{2}\delta^{3/2}$. Once $\varphi_n(t) \geq t^{3/2}/6$ it stays there as long as $t < 1/2000$. Indeed, for $t \geq 4\delta$ and as long as $\varphi_n(t) \geq t^{3/2}/6$, $\dot{\varphi}_n(t) > \varphi_n^{1/2}(t - \delta) - t > \frac{1}{2}(t - \delta)^{3/2} - t > t^{3/5}/10$. Since $t^{3/5}/10 > (\mathrm{d}/\mathrm{d}t)(t^{3/2}/6)$, the result follows. If n is odd, $\varphi_n(t) < -t^{3/2}/6$ for $3\delta < t < 1/2000$.

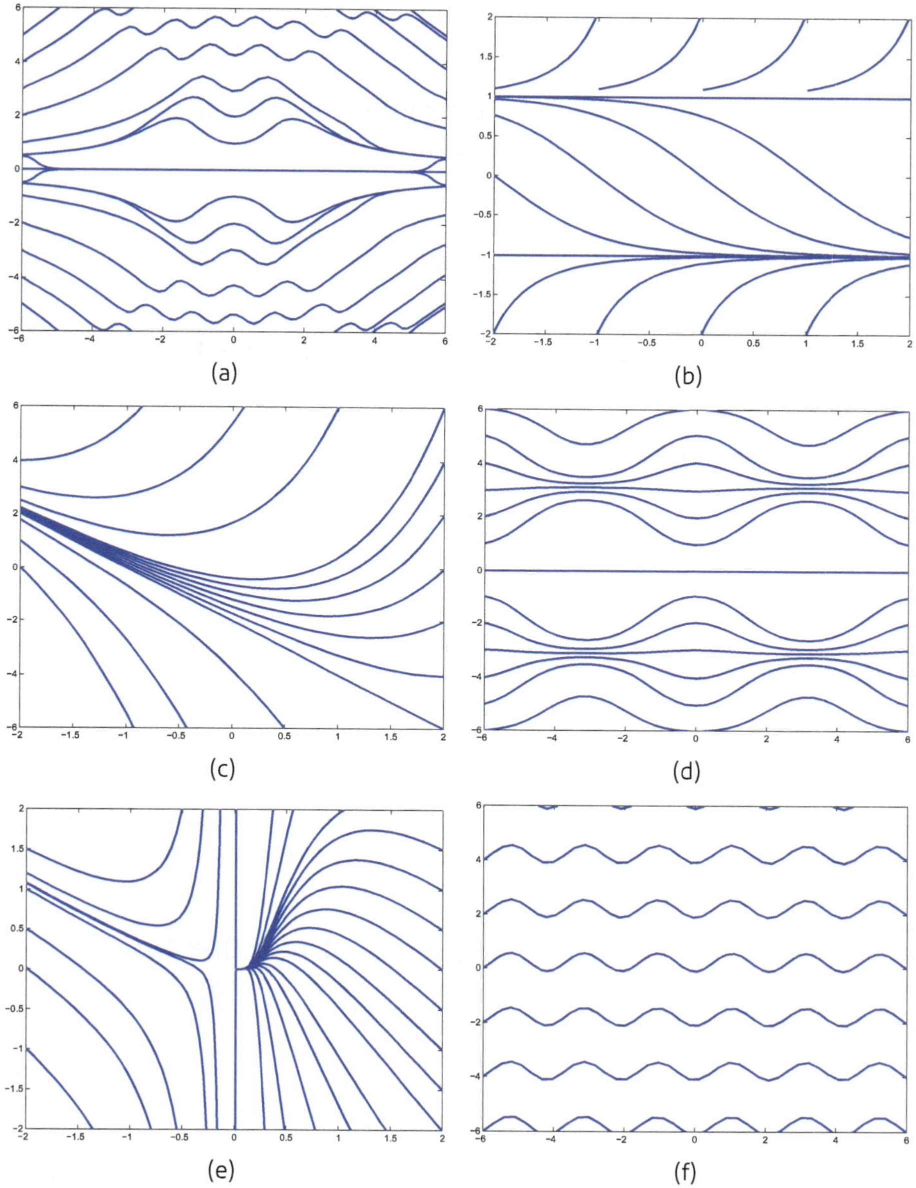

Figure 6.15. Integral curve plots for problem 6.38.

Exercise 6.41. [☼] **Application of the Picard-Lindelöf Method – Pt. 1**
Find the exact solution of the initial value problem

$$\dot{x} = x^2, \quad x(0) = 1.$$

Starting with $x_0(t) = 1$, apply the Picard-Lindelöf iteration to calculate $x_1(t)$, $x_2(t)$ and $x_3(t)$, and compare these results with the exact solution.

Exercise 6.42. [☼] **Application of the Picard-Lindelöf Method – Pt. 2**
Find the exact solution of the initial value problem

$$\dot{x} = 2t\,(1+x)\,, \quad x(0) = 0\,.$$

Starting with $x_0(t) = 0$, calculate $x_1(t)$, $x_2(t)$, $x_3(t)$ and $x_4(t)$, and compare these results with the exact solution.

Exercise 6.43. [☼] **Application of the Picard-Lindelöf Method – Pt. 3**
It is instructive to see how the Picard-Lindelöf iteration works with a choice of the initial approximation other than the constant function $x_0(t) = x_0$. Apply the method to the initial value problem

$$\dot{x} = t + x\,, \quad x(0) = 1\,.$$

with

1. $x_0(t) = \exp(t)$,
2. $x_0(t) = 1 + t$, and
3. $x_0(t) = \cos(t)$.

Exercise 6.44. [☼] **Existence of Solutions**
Show that every initial value problem with the differential equation $\dot{x} = t \cdot |\sin(tx)|$ has exactly one solution that is defined on the whole of $\mathbb{R}$. The solutions with $x(0) \neq 0$ do not have zeros.

Exercise 6.45. [☼] **Solution of a Homogeneous Linear ODE-System**
A transport mechanism allows the exchange between two reservoirs of salts in homogeneous solution. This system is described by

$$\begin{aligned} \dot{x}_1 &= -k_1 x_1 + k_2 x_2\,, \\ \dot{x}_2 &= k_1 x_1 - k_2 x_2\,, \end{aligned}$$

where x_i denotes the amount of salt in reservoir i, $i = 1, 2$, and the k_i's are constant exchange rates. Compute the solution with respect to the initial conditions $x_1(0) = s_1$ and $x_2(0) = s_2$, and discuss its behavior for $t \to \infty$.

Exercise 6.46. [☼] **Solution of an Inhomogeneous Linear ODE-System – Pt. 1**
Give the general real solution of the following two-dimensional inhomogeneous system of linear differential equations

$$\frac{\mathrm{d}}{\mathrm{d}t}\begin{pmatrix} x(t) \\ y(t) \end{pmatrix} = \begin{pmatrix} 1 & -1 \\ 0 & 2 \end{pmatrix}\begin{pmatrix} x(t) \\ y(t) \end{pmatrix} + \begin{pmatrix} \sin(t) \\ \cos(t) \end{pmatrix}\,,$$

where $x, y : \mathbb{R} \to \mathbb{R}$ and $t \in \mathbb{R}$.

Exercise 6.47. [☼] **Solution of an Inhomogeneous Linear ODE-System – Pt. 2**
Let $v \in \mathcal{C}^d$ be an eigenvector of the matrix $A \in \mathbb{R}^{d\times d}$ corresponding to the eigenvalue $\lambda \in \mathcal{C}$. Show that for the inhomogeneous differential equation

$$\dot{x} = Ax + \exp(\omega t)\, v$$

a particular solution is obtained by the ansatz

$$x = \beta \exp(\omega t)\, v\,, \quad \text{if } \omega \neq \lambda \quad \text{and} \quad x = \beta t \exp(\omega t)\, v\,, \quad \text{if } \omega = \lambda\,.$$

In each case determine the constant β. Finally, compute the general solution of the differential equation

$$\dot{x} = \begin{pmatrix} 2 & 0 & 1 \\ 0 & 2 & 0 \\ 0 & 1 & 3 \end{pmatrix} x + \exp(2t) \begin{pmatrix} 2 \\ 0 \\ 1 \end{pmatrix}.$$

Exercise 6.48. [☼] **Integral Curves in the Plane**
Determine all constant and all periodic integral curves of the vector field $v : \mathbb{R}^2 \to \mathbb{R}^2$ with $v(0,0) = (0,0)^T$ and

$$v(x,y) = \begin{pmatrix} y \\ -x \end{pmatrix} + (x^2+y^2) \sin\left(\frac{1}{\sqrt{x^2+y^2}}\right) \begin{pmatrix} x \\ y \end{pmatrix},$$

if $(x,y) \neq (0,0)$. Moreover, sketch the remaining integral curves qualitatively.

Exercise 6.49. [☼] **Equivalence of Integral Curves**
Let $v : \Omega \to \mathbb{R}^d$ be a continuous vector field on an open set $\Omega \subset \mathbb{R}^d$, and let $\alpha : \Omega \to \mathbb{R} \setminus \{0\}$. Show that the traces/ trajectories of the integral curves of the two vector fields v and αv coincide.

More precisely: If ψ is a integral curve of the vector field αv with $\psi(\tau_0) = x_0$ and if $\tau = \tau(t)$ is a time transformation defined by

$$\int_{\tau_0}^{\tau} \alpha(\phi(s))\, \mathrm{d}s = t - t_0\,,$$

then $\varphi := \psi \circ \tau$ is an integral curve of the vector field v with $\varphi(\tau_0) = x_0$.

Exercise 6.50. [☼] **Transporting the Theorem of Caratheodory into Multiple Dimensions**
Formulate and proof the theorem of Caratheodory (Theorem 6.10) for systems of ordinary differential equations (in the extended sense).

Exercise 6.51. [☼] **Proving the Continuity Theorem for Solutions in the Extended Sense**
Following the proof of the continuation Theorem 6.6 show that Theorem 6.12 holds for solutions in the extended sense: I.e., for solutions in the extended sense sow the following:

> In a domain D of the (t, x) plane let the function f be defined, measurable in t for fixed x and continuous in x for fixed t. Let there exist an integrable function m such that $|f(t,x)| \leq m(t)$ for $(t,x) \in D$. Then, given a solution φ of $\dot{x} = f(t,x)$ for $t \in (a,b)$, it is the case that $\varphi(b-0)$ exists and if $(b, \varphi(b-0)) \in D$ then φ can be continued over $(a, b+\delta]$ for some $\delta > 0$. A similar result holds at a.
>
> Thus the solution φ can be continued up to the boundary of D. Moreover, the same continuation is valid for a maximum solution φ_M or a minimum solution φ_m.

Exercise 6.52. [☼] **On the Form of the Fundamental Matrix**
Determine the form of the fundamental matrix (often called *Wronski-matrix*) of the first order system associated to a d-th order linear ordinary differential equation.

Exercise 6.53. [☆] **Lemma of Morse**
In the vicinity of a stationary point a a $\mathcal{C}^2$-function $f : \mathbb{R}^d \to \mathbb{R}$ with non-degenerate Hessian $f''(a)$ can be approximated by the quadratic form $(x-a)^T f''(a)(x-a)$. The lemma of Morse tells us that f itself is locally and in suitable coordinates this quadratic form. This lemma is essential for the mathematical field of differential topology.

Lemma 6.54 (Lemma of Morse). *Let $f : U \to \mathbb{R}$ be a $\mathcal{C}^2$-function in a vicinity U of $0 \in \mathbb{R}^d$ such that $f(0) = 0$, $f'(0) = 0$ and non-degenerate Hessian $f''(0)$. Then there is a diffeomorphism $\varphi : U_0 \to V$ of an environment U_0 of 0 to an environment V of 0 such that*

$$f \circ \varphi^{-1}(y) = y_1^2 + \cdots + y_k^2 - \left(y_{k+1}^2 + \cdots + y_d^2\right) .$$

Proof the lemma of Morse.

Hint: Start by constructing a symmetric matrix $A(x) \in \mathbb{R}^{d\times d}$ the coefficients of which are $\mathcal{C}^\infty$-functions such that

$$f(x) = \tfrac{1}{2} x^T A(x) x , \qquad A(0) = f''(0) .$$

For x near 0 the matrix $A(x)$ is invertible. Now set $B(x) := A(0)A(x)^{-1}$. Then $B(0) = E$. With the help of a power series constrict a differentiable mapping

$Q : U_0 \to \mathbb{R}_s^{n\times n}$ such that $Q^2 = B$. Because $BA = AB^T$ we have $QA = AQ^T$ and with $S := Q^{-1}$ it follows $A = SA(0)S^T$. Moreover, define $\psi(x) := S(x)^T$, and thus we obtain

$$f(x) \;=\; \tfrac{1}{2}\psi(x)^T A(x)\psi(x)\,.$$

Due to $\psi'(0) = E$ the function ψ has an inversion φ at 0. With this, we get

$$(f \circ \varphi)\,(y) \;=\; \tfrac{1}{2}y^T A(0)y\,.$$

Finally, this last quadratic form can be transformed to principal axis form by a linear transformation.

Exercise 6.55. [☼] **On Second Order Systems**
For the solutions of a homogeneous second order linear differential equation with continuous coefficients on the interval I show the following:

1. Every solution different from the null-solutions has simple zeros only, and the set of its zeros does not have a cumulation point in I.
2. If (φ, ψ) is a fundamental system, then there is a zero of ψ between any two zeros of φ. This result is sometimes called *separation theorem*.

   Hint: $\varphi\dot{\psi} - \psi\dot{\varphi}$ has no zeros.

Exercise 6.56. [☼] **On the Volterra-Lotka System**
Show that the Volterra-Lotka system (6.48) has also first integrals in the three quadrants $\mathbb{R}^+ \times \mathbb{R}^-$, $\mathbb{R}^- \times \mathbb{R}^+$, and $\mathbb{R}^- \times \mathbb{R}^-$. Has the Volterra-Lotka system periodic solutions in one of these three quadrants?
Hint: Discuss whether the level sets of the first integrals are compact.

Exercise 6.57. [☼] **Oscillations of a Spherical Buoy**
A spherical buoy of radius r floats half submerged in water. If it is suppressed slightly, a restoring force equal to the weight to the displaced water presses it upward; and if it is then released, it will bob up and down. Find the period of oscillation if the friction of the water is neglected.

Exercise 6.58. [☼] **Oscillations Under Gravity**
Suppose that a straight tunnel is drilled through the earth through any two points on the surface. If tracks are laid, then – neglecting friction – a train placed in the tunnel at one end will roll under the earth under its own weight, stop at the other end, and return. Show that the time required for a complete round trip is the same for all such tunnels, and estimate its value.

Exercise 6.59. [✩] **Continuous Extensions of Continuous Mappings**
Prove Lemma 6.34: Let X, Y be Banach spaces, $\Omega \subset X$ closed and $f : \Omega \to Y$ continuous. Then there is a continuous extension $\tilde{f} : X \to Y$ such that $\tilde{f}(X) \subset \text{conv}(f(\Omega))$, where $\text{conv}(f(\Omega))$ denotes the convex hull of $f(\Omega)$.

Exercise 6.60. [☼] Filling the Gaps in Example 6.37
Example 6.37 is rather fast-paced. Fill the gaps of its outline.

Chapter 7

Recap: Simulation of Ordinary Differential Equations

The relevant aspects of the numerical simulation of ordinary differential equations are discussed. Classical explicit one-step methods such as the explicit Euler or Runge-Kutta schemes are presented before motivating implicit approaches for stiff ODEs. A variety of example implementations show the behaviour of the different schemes applied to different initial value problems. The brief discussion of the Newmark family of schemes and of symplectic methods widens the scope of this chapter to approaches that are typically not treated in course curricula but that provide useful features worth being on the radar in the context of RODE simulations.

7.1 Key Concepts

Many methods exist for obtaining analytical solutions of ordinary differential equations. As we have seen in Chap. 6, these approaches are primarily limited to special—and typically simple—types of ODEs. When applicable, they produce a solution in the form of a nice analytic formula. In practical problems, we frequently encounter ODEs that are not amenable to solution by special methods, and a numerical approximation for solutions must be found. Moreover, even if an analytical solution can be obtained, a numerical solution of the ordinary differential equation may be preferable in certain cases, especially if the analytical solution is very complicated and, thus, computationally expensive to evaluate.

Numerical methods for ordinary differential equations may be derived in various ways. One approach is to use the link between ordinary differential equations and integration (see [55, pp. 299], e.g.). Integrating both sides of the initial value problem (see Eq. (7.2) below) from s to $s+h$, we obtain

$$\int_s^{s+h} \dot{y}(t)\mathrm{d}t = \int_s^{s+h} f(t, y(t))\mathrm{d}t\,,$$

and get the integral equation of the corresponding ODE on $[s, s+h]$ as

$$y(s+h) = y(s) + \int_s^{s+h} f(t, y(t))\mathrm{d}t\,. \qquad (7.1)$$

Approximating the integral with some numerical integration scheme results in formulae for the approximate solution of ODEs. Thus, a strong relation between numerical quadrature and numerical integration for solving ordinary differential equations exists. In the case when the approximation of $\int_s^{s+h} f(t, y(t))dt$ does not involve the value $y(s+h)$ (the value one is trying to approximate for the next time step), the corresponding methods are called *explicit methods*. Vice-versa, if $y(s+h)$ is in the formula, we call these *implicit methods*. Such schemes typically imply the solution of a non-linear system of equations in each time step which makes these methods computationally more expensive. Certain circumstances or problems may require the use of implicit schemes.

While reading this chapter, note the answers to the following questions

1. How can ODE systems be treated numerically?
2. What categories of problems exist and which specific methods do they require?
3. Why do so many different integration schemes exist? How do they differ?

as well as the following key concepts:

1. Consistency of ODE integration schemes,
2. Convergence of ODE integration schemes,
3. Explicit vs. implicit methods.

The remainder of this chapter has the following structure: We present a result concerning the condition of initial value problems in Sec. 7.2. In Sec. 7.3, a survey on classical one-step methods is given. We briefly discuss stiff ODEs as a motivation for implicit schemes and present the implicit Euler method and the trapezoidal rule in Sec. 7.4. The family of Newmark methods in Sec. 7.5 constructed for second-order systems in structural dynamics is an alternative to formulating the problem as a system of ODEs. In Sec. 7.6, a brief excursion leads us to symplectic methods, a category of non-standard schemes which possess certain useful conservation properties. Finally, Section 7.7 summarises the contents of this chapter.

7.2 General Aspects of Numerical Solution of ODEs

For the sake of compatibility with classical literature in the field of numerics for ODEs, we now alter the notation in the definition of the initial value problem (IVP) (6.2)):

$$\dot{y} \;=\; f(t, y(t)), \quad t \in [t_0, t_e] \qquad (7.2)$$

$$y(t_0) \quad = \quad y_0 \,, \tag{7.3}$$

for $f : [t_0, t_e] \times \mathbb{R}^m \mapsto \mathbb{R}^m$ being sufficiently smooth. Note that we omit the vector of parameters λ of Eq. (6.2) in this definition since it does not have any impact on the numerical schemes derived below. The consideration of first-order systems is sufficient since every system of higher order can be reformulated to formally be of first order (see Sec. 6.4.5). Hence, numerical schemes that are able to deal with systems of equations may also be applied to ODEs containing higher-order derivatives.

Following [216], the following theorem shows how much two solutions of the ODE system can differ depending on the difference in their initial values. This corresponds to analysing the condition of the initial value problem.

Theorem 7.1. *For two solutions y, z of the ODE* (7.2) *(with different initial values) and for all t, t_0, it holds:*

$$||y(t) - z(t)|| \le ||y(t_0) - z(t_0)|| \cdot e^{L \cdot |t - t_0|} \tag{7.4}$$

with $0 < L \in \mathbb{R}$.

Proof. Let $t_0 < t$. $D(t) := ||y(t) - z(t)||$ is a continuous scalar function. Hence, $E(t) := e^{-Lt} \int_{t_0}^{t} D(t) dt$ is continuously differentiable. Therefore, it holds

$$D(t) = \left(e^{Lt} E(t)\right)' = L e^{Lt} E(t) + e^{Lt} E'(t) \,. \tag{7.5}$$

□

From (7.1) with $s = t_0$ for $y(t)$ and $z(t)$, from the triangle inequality and the Lipschitz continuity (6.3), we have

$$D(t) \le ||y_0 - z_0|| + \int_{t_0}^{t} ||f(\tau, y(\tau)) - f(\tau, z(\tau))|| d\tau \le ||y_0 - z_0|| + \int_{t_0}^{t} L D(\tau) d\tau \,. \tag{7.6}$$

Plugging (7.5) into (7.6) (left) and using the definition of E (right), we obtain

$$\begin{aligned} L e^{Lt} E(t) + L e^{Lt} E'(t) &\le ||y_0 - z_0|| + L e^{Lt} E(t) \,, \\ E'(t) &\le ||y_0 - z_0|| L e^{-Lt} , \\ E(t) = \int_{t_0}^{t} E'(\tau) d\tau &\le ||y_0 - z_0|| \left(e^{-L t_0} - e^{-Lt}\right) \,. \end{aligned}$$

Note that $E(t_0) = 0$ by definition. Using these estimates of the right-hand side of (7.5) results in

$$D(t) \le ||y_0 - z_0|| \cdot \left(e^{L(t - t_0)} - 1 + 1\right) \,. \tag{7.7}$$

In case of $x < x_0$, one uses $-x$ as independent variable and $-f$ as right-hand side of the corresponding ODE. Eq. (7.4) represents an upper bound for the error (or condition).

For numerical solutions, we encounter inexact initial values due to round-off errors and/or discretisation errors. Hence, the reliability of solutions measured via (7.4) diminishes with $t_e >> t_0$ due to the exponential term. More accurate error representations exist but are more complex to derive and evaluate.

In the following sections, different discretisation methods for ODEs will be discussed. All of them rely on a discretisation of the time interval under consideration, denoted by $t_0 < t_1 < \ldots < t_n = t_e$; $h_n = t_{n+1} - t_n$ is called the time step size which we assume to be equidistant in the following (i.e. $h_i = h \; \forall i$). For the numerical integration of ODEs, we need to compute approximate solutions $y^n \approx y(t_n)$. All methods—in the following and in general—can be formulated using an update function Φ_f:

$$y^{n+1} \quad = \quad y^n + h \cdot \Phi_f(h, y_h), \quad n = 0, 1, \ldots \,. \tag{7.8}$$

Note that Φ_f in (7.8) needs to somehow evaluate the right-hand side f of the ODE (7.2) and depends on the time step size h and a vector y_h representing k previously computed approximations (i.e. $y_h = (y^n, y^{n-1}, \ldots, y^{n-k+1})$).

The field is now prepared to discuss some methods in detail. We first present explicit one-step methods before tackling implicit schemes.

Before you continue, make sure to answer the following questions:

Quiz: Section 7.2

Q1 What is the basic idea to derive numerical schemes for ODEs using the relation (7.1)?

Q2 Why are implicit schemes computationally more expensive?

Q3 What is the problem of the upper bound for the condition in Eq. (7.4) for large values of t?

Q3 What does the update function Φ of a numerical scheme generally depend on?

7.3 Explicit One-Step Methods for ODEs

The class of one-step methods uses data of only one previous time step t_n to compute an approximate solution of the next time step t_{n+1}. Hence, in the explicit case, the update function Φ_f of (7.8) depends on y^n only, and we may write the update in a more specific form:

$$y^{n+1} = y^n + \Phi_f\left(h_n, t_n, y^n, f(t_n, y^n)\right). \tag{7.9}$$

We now present three well-known representatives of explicit one-step methods for ODEs: the explicit Euler method, Heun's method, and the classical Runge-Kutta scheme. We show MATLAB functions used to compare the three methods in Sec. 7.3.4.

7.3.1 Explicit Euler Method

The explicit Euler method, also known as the forward Euler (FE) method, is a straightforward approach to compute numerical approximations of ODEs.

One way of motivating this first-order scheme is to approximate the time derivative of the initial value problem (7.2) using forward finite differences

$$\dot{y} \doteq \frac{y^{n+1} - y^n}{h_n} \tag{7.10}$$

for time steps n and $n+1$. Combining this with a linear approximation of the unknown function y at time step t_{n+1} using the derivative evaluated at the previous time step t_n results in the explicit Euler scheme:

$$y^{n+1} = y^n + h_n \cdot f(t_n, y^n). \tag{7.11}$$

The update function of the explicit Euler method as a one-step method in the sense of Eq. (7.9) is clearly visible.

In terms of relation to numerical quadrature schemes, the explicit Euler method can be interpreted as the left rectangle quadrature rule applied to (7.1) in Sec. 7.2, i.e.

$$\int_s^{s+h} f(t, y(t))\mathrm{d}t \approx f(s, y(s)) \cdot h,$$

Explicit Euler is the computationally cheapest ODE solver in terms of memory and evaluations of the system function f. This holds per time step and, thus, is not a statement on efficiency in terms of accuracy per cost, for that we would have to consider the approximation order and possible restrictions on the time step size h.

MATLAB Example 7.1 integrateEuler.m: Implementation of the explicit Euler method with equidistant time stepping.

```
function err = integrateEuler(t0, tend, y0, fhandle, N, yRef)

h = 1.0 * (tend-t0) / N;
t = t0:h:tend;
noTimeSteps = length(t);

y = y0;
for n=2:noTimeSteps
    y = y + h * fhandle(t(n-1),y);
end
err = y - yRef;
```

One possible implementation of the explicit Euler scheme is given in MATLAB Example 7.1. We use equidistant time steps of size h to successively approximate the solution at N time steps. The return value holds the difference to an analytical reference solution (for problems where this exists, see problem (7.24) in Sec. 7.3.4 below).

7.3.2 Heun's Method

For explicit Euler, we used a function evaluation of f at y^n. The information required to compute the next time step only relies on the last time step with a certain "delay" in new tangent information. One idea to improve this is to use averaged tangent information by incorporating information closer to the new time step t_{n+1}. Since explicit schemes are simple and efficient, we do not want to use $f(t_{n+1}, y^{n+1})$ (as this would result in an expensive non-linear system of equations to be solved in each time step). Since the explicit Euler method approximates exactly this value, we can use the corresponding formula in the function f. The resulting scheme is Heun's method:

$$y^{n+1} = y^n + \frac{h_n}{2} \cdot [f(t_n, y^n) + f(t_{n+1}, y^n + h_n \cdot f(t_n, y^n))] . \quad (7.12)$$

Note that we invest an additional evaluation of f in each time step. For simple problems, this does not represent a major overhead, but when it comes to more complex scenarios, this may introduce substantial additional costs (in the case of ODE integration inside PDEs, a function evaluation involves the computation of all discrete spatial operators on the whole grid). The advantage of Heun's method over the explicit Euler scheme is one order of con-

MATLAB Example 7.2 integrateHeun.m: Implementation of Heun's method with equidistant time stepping.

```
function err = integrateHeun(t0, tend, y0, fhandle, N, yRef)

h = 2.0 * (tend-t0) / N;
t = t0:h:tend;
noTimeSteps = length(t);

y = y0;
for n=2:noTimeSteps
    step1 = h * fhandle(t(n-1),y);
    step2 = h * fhandle(t(n),  y+step1);
    y = y + (step1 + step2)/2.0;
end
err = y - yRef;
```

sistency[1]. MATLAB Example 7.2 shows an implementation of Heun's method.

7.3.3 Explicit Runge-Kutta Schemes

As with the explicit Euler and Heun's method, Runge-Kutta schemes belong to the class of one-step methods. The basic idea is to update the current value y^n with a linear approximation similar to Heun's method but using an estimated slope consisting of an average of s instead of 2 individual slopes:

$$y^{n+1} = y^n + h_n \cdot \sum_{j=1}^{s} b_j Y'_{nj}. \tag{7.13}$$

The step vectors Y_{ni} and corresponding slopes Y'_{ni} are recursively defined as

$$Y_{ni} = y^n + h_n \sum_{j=1}^{s} a_{ij} Y'_{nj}, \quad Y'_{ni} = f(t_n + c_i h_n, Y_{ni}) \qquad \forall i = 1, \ldots \tag{7.14}$$

The classical Runge-Kutta scheme (RK4) has the following explicit form

$$\begin{aligned}
Y_1 &= y^n, & Y'_1 &= f(t_n, Y_1),\\
Y_2 &= y^n + \tfrac{h_n}{2} Y'_1, & Y'_2 &= f(t_n + h_n/2, Y_2),\\
Y_3 &= y^n + \tfrac{h_n}{2} Y'_2, & Y'_3 &= f(t_n + h_n/2, Y_3),\\
Y_4 &= y^n + h_n Y'_3, & Y'_4 &= f(t_{n+1}, Y_4),
\end{aligned}$$

[1] i.e. error reduction capability of the method) with decreasing time steps, see Sec. 7.3.4.

MATLAB Example 7.3 integrateRK4.m: Implementation of the 4th-order Runge-Kutta scheme with equidistant time stepping.

```
function err = integrateHeun(t0, tend, y0, fhandle, N, yRef)

h = 4.0 * (tend-t0) / N;
t = t0:h:tend;
noTimeSteps = length(t);

y = y0;
for n=2:noTimeSteps
    step1 = h * fhandle(t(n-1),          y);
    step2 = h * fhandle(t(n-1)+h/2.0,y+step1/2.0);
    step3 = h * fhandle(t(n-1)+h/2.0,y+step2/2.0);
    step4 = h * fhandle(t(n),          y+step3);
    y = y + (step1 + 2*step2 + 2*step3 + step4)/6.0;
end
err = y - yRef;
```

$$y^{n+1} = y^n + h_n \left(\frac{1}{6} Y_1' + \frac{1}{3} Y_2' + \frac{1}{3} Y_3' + \frac{1}{6} Y_4' \right) .$$

The corresponding implementation is given in MATLAB Example 7.3. Note that step$i$ corresponds to Y_i'.

Consistency conditions exist that allow us to specify the parameters a_{ij}, b_i, and c_i for a desired order of consistency p of a Runge-Kutta method (see Sec. 7.3.4 for definition and details on consistency). These conditions can be derived via Taylor expansion or Butcher trees (cf. [46]). A mnemonic structure called *Butcher tableau* may be used to arrange the corresponding coefficients:

$$\begin{array}{c|c} c & \mathcal{A} \\ \hline & b \end{array} := \begin{array}{c|cccc} c_1 & a_{11} & a_{12} & \dots & a_{1s} \\ c_2 & a_{21} & a_{22} & \dots & a_{2s} \\ \vdots & \vdots & \vdots & \ddots & \vdots \\ c_s & a_{s1} & a_{s2} & \dots & a_{ss} \\ \hline & b_1 & b_2 & \dots & b_s \end{array} .$$

An explicit Runge-Kutta method is characterised by a lower triangular matrix $\mathcal{A}$ in the Butcher tableau (often skipping the zero entries). Implicit schemes are possible but will not be discussed in this context. The price for a higher order p are higher storage requirements (the slopes Y_{ni}' have to be available to perform the final update) as well as higher computational costs: s increases with the approximation order p and, thus, more evaluations of the right-hand side function f are required (see discussion in Sec. 7.3.2).

7.3.4 Consistency & Convergence

Now that we have seen some particular classical methods, we cover the aspects of consistency and convergence. Note that the following definitions hold not only for the case of explicit one-step methods but also for general time discretisations.

Definition 7.1 (Local Discretisation Error). The *local discretisation error* of a numerical integration scheme with update function Φ_f for an arbitrary point of reference ξ, η is given by

$$\tau(h, \xi, \eta) \quad = \quad \frac{1}{h}\left[z(\xi + h) - z(\xi)\right] - \Phi_f(h, z) \tag{7.15}$$

where $z(.)$ is the solution of the local initial value problem $\dot{z} = f(., z), z(\xi) = \eta$.

This definition is rather abstract and generally has no direct benefit for the user of a method. However, we will need it to demonstrate important properties of the methods from the numerical point of view. For explicit one-step (or even s-step) methods, the local discretisation error τ has a certain interpretation. Since all arguments in the vector y_h in $\Phi_f(h, y_h)$ are freely selectable (as it is an explicit scheme), $\Phi_f(h, z)$ is exactly the increment $(y^{n+1} - z(\xi))/h$ that the method will compute for exact input data $y^n = z(\xi)$. Hence, (7.15) simplifies to

$$\tau(h, \xi, \eta) = (z(\xi + h) - y^{n+1})/h \; .$$

The local discretisation error is the error purely stemming from the step of ξ to $\xi + h$ without any previous error contributions. In particular, for our one-step methods, we get

$$h_h \cdot \tau(h_n, t_n, y^n) = z(t_n + h_n) - y^{n+1} \; . \tag{7.16}$$

The local discretisation error motivates the definition of consistency of a method:

Definition 7.2 (Consistency). A method with update function Φ_f is called *consistent* if the local discretisation error τ goes to zero (uniformly in ξ and η) if $h \to 0$:

$$||\tau(h, \xi, \eta)|| \leq \sigma(h) \quad \text{and} \quad h \to 0 \quad \Rightarrow \quad \sigma(h) \to 0 \tag{7.17}$$

A method has the *order of consistency* p if $\sigma(h) = \mathcal{O}(h^p)$.

Consistency is a minimal property of reasonable numerical methods: If we knew and used the analytical solution of the IVP in the method, our scheme should not create approximate solutions that are far off. Of course, we would like to construct methods that are not only consistent but also tend to compute in every point a decent approximation of the solution which converges to the real one. To do so, we need the notion of the global discretisation error:

Definition 7.3 (Global Discretisation Error). The *global discretisation error* is the difference between the discrete approximation and the exact solution at a specific time $t \neq t_0$:

$$e(h,t) := y^n(h) - y(t) \tag{7.18}$$

where $nh = t - t_0$.

With $e(h,t)$, we now have a direct measure of the accuracy of the method (i.e. the error from an intuitive point of view) depending on a given mesh resolution h. A "correct" method should provide a vanishing global discretisation error as the mesh gets finer:

Definition 7.4 (Convergence). A method with update function Φ_f is called *convergent* on $[t_0, t_e]$ if

$$h \to 0 \quad \Rightarrow \quad e(h,t) \to 0 \tag{7.19}$$

uniformly for all $t \in [t_0, t_e]$. The *order of convergence* is the largest value p for which

$$||e(h,t)|| \leq s(t)h^p \quad \text{holds with a limited function} \quad s : [t_0, t_e] \to \mathbb{R}\,. \tag{7.20}$$

For explicit one-step methods, the following relation between consistency and convergence (and the respective orders) exists (see [216]):

Theorem 7.2. *Let $y(t)$ be the solution of the initial value problem* (7.2)–(7.3) *and y^i the approximate solution computed with an explicit one-step method via* (7.9),

$$\begin{aligned} y^0 &= y_0, \\ y^{n+1} &= y^n + h_n \cdot \Phi_f(h_n, y^n), \\ h_n &= t_{n+1} - t_n, \qquad n = 0, 1, \ldots \end{aligned}$$

For

$$||f(t,y) - f(t,z)|| \leq L||y - z|| \quad \text{and} \quad ||\tau(h,\xi,\eta)|| \leq \sigma(h)$$

it follows that

$$||y(t_k) - y^k|| \leq \sigma(h_{\text{max}}) \cdot \left(e^{L\cdot|t_k - t_0|} - 1\right)/L \quad \text{for all } k\,.$$

Proof. W.l.o.g. let $t_0 < t_k$ and all step sizes h_n be positive. We introduce the solutions $z_n(t)$ of the differential equation with the initial values $z_n(t_n) = y^n, n = 0, \dots, k$. In particular, $z_0(t) = y(t)$ holds. Furthermore, due to (7.16), we have

$$h_n \tau(h_n, t_n, y^n) = z_n(t_{n+1}) - y^{n+1} = z_n(t_{n+1}) - z_{n+1}(t_{n+1})\,. \tag{7.21}$$

From Eq. (7.21) we get

$$||z_n(t_{n+1}) - z_{n+1}(t_{n+1})|| \le h_n \sigma(h_n) \le h_n \sigma(h_{\max})\,. \tag{7.22}$$

We now use Theorem 7.4, indicating how far two neighbouring solutions z_n and z_{n+1} can drift apart within the distance $t_{k+1} - t_{n+1}$, to obtain

$$||z_n(t_{k+1}) - z_{n+1}(t_{n+1})|| \le ||z_n(t_{n+1}) - z_{n+1}(t_{n+1})|| \cdot e^{L\cdot|t_{k+1}-t_{n+1}|}\,. \tag{7.23}$$

Hence, we have the following estimates for all k:

$$\begin{aligned}
||y(t_k) - y^k|| &= ||z_0(t_k) - z_k(t_{t_k})|| \le \sum_{n=0}^{k-1} ||z_n(t_k) - z_{n+1}(t_k)|| \\
&\overset{(7.23)}{\le} \sum_{n=0}^{k-1} ||z_n(t_{n+1}) - z_{n+1}(t_{n+1})|| e^{L\cdot|t_k - t_{n+1}|} \\
&\overset{(7.22)}{\le} \sigma(h_{\max}) \sum_{n=0}^{k-1} h_n \cdot e^{L\cdot|t_k - t_{n+1}|} \\
&\le \sigma(h_{\max}) \sum_{n=0}^{k-1} \int_{t_n}^{t_{n+1}} e^{L\cdot|t_k - t|} dt = \sigma(h_{\max}) \int_{t_0}^{t_k} e^{L\cdot|t_k-t|} dt\,.
\end{aligned}$$

□

The essence of Theorem 7.2 is the proportionality of the global discretisation error (on the left) to the local discretisation error (on the right). The constant $\left(e^{L\cdot|t_k - t_0|} - 1\right)/L$ does not depend on the time step size h_n of the discretisation. Hence, the analysis of an explicit one-step method is transferred to the task of computing $\tau(h, \xi, \eta) = \tau \cdot h^p + \dots$ via Taylor expansion. The resulting order of the three methods mentioned above is as follows:

explicit Euler:	$p = 1$,	$\tau(h, \xi, \eta) = \mathcal{O}(h)$,	$e(h,t) = \mathcal{O}(h)$,
Heun's method:	$p = 2$,	$\tau(h, \xi, \eta) = \mathcal{O}(h^2)$,	$e(h,t) = \mathcal{O}(h^2)$,
Runge-Kutta4:	$p = 4$,	$\tau(h, \xi, \eta) = \mathcal{O}(h^4)$,	$e(h,t) = \mathcal{O}(h^4)$,

The following example (taken from [216]) demonstrates the properties of these three explicit methods. We consider the initial value problem

$$\dot{y} = \sqrt{t^2 + y^2}, \quad y(0) = 0\,, \tag{7.24}$$

N	explicit Euler $h = 2/N$	explicit Heun $h = 4/N$	Runge-Kutta 4th order $h = 8/N$
4	−0.9812	−0.2433 1354	0.1755 6871 99280
8	−0.5676	−0.0792 5875	0.0229 1221 36475
16	−0.3098	−0.0219 2620	0.0021 6896 12998
32	−0.1626	−0.0056 4107	0.0001 7903 63352
64	−0.0834	−0.0014 1657	0.0000 1369 86432
128	−0.0422	−0.0003 5365	0.0000 0100 30986
256	−0.0213	−0.0000 8825	0.0000 0007 15003
512	−0.0107	−0.0000 2203	0.0000 0000 50064
1024	−0.0053	−0.0000 0550	0.0000 0000 03463
2048	−0.0027	−0.0000 0138	0.0000 0000 00239
4096	−0.0013	−0.0000 0034	0.0000 0000 00018
reduction rate:	1/2	1/4	1/16

Table 7.1. Global discretisation error $e(h, t)$ of the Euler, Heun, and Runge-Kutta method for the initial value problem (7.24) at $t = 2$: Comparison of different resolutions N shows the corresponding orders of convergence $p = 1, 2, 4$.

with Lipschitz constant $L = 1$ integrated on the time interval [0,2]. The analytical solution of (7.24) at $t = 2$ is 2.5886 0752 6700 (see Problem 7.6).

Table 7.1 shows the global discretisation error $e(h, t)$ at $t = 2$. Each row holds the results of the three different methods for comparable computational effort (i.e. N function evaluations of the right-hand side f). In each column, the corresponding order of convergence is visible. MATLAB Example 7.4 holds the script ivp_explicit to produce the results of Tab. 7.1 by using the right-hand side function f_explicit (of MATLAB Example 7.5) and the three explicit methods of MATLAB Examples 7.1, 7.2, and 7.3.

To summarise, explicit one-step methods are typically robust and reliable, relatively easy to implement and analyse, and particularly well suited for variable time step sizes (cf. [216]). However, there are certain types of ODE systems for which explicit methods are not well suited. In such cases, implicit methods are more useful and will be analysed in the following section.

7.4 Implicit Methods

We start this section on implicit methods with a brief motivating example ODE that is difficult to approximate with explicit methods. For the solution of

MATLAB Example 7.4 ivp_explicit.m: Script for the initial value problem (7.24).

```
t0 = 0;
tend = 2;
y0 = 0;

fhandle = @f_explicit;
yRef = 2.588607526700;
for i=2:12
    N = 2^i;
    errEuler = integrateEuler(t0, tend, y0, fhandle, N, yRef);
    errHeun  = integrateHeun( t0, tend, y0, fhandle, N, yRef);
    errRK4   = integrateRK4(  t0, tend, y0, fhandle, N, yRef);
    fprintf('%4d & %1.4f && %1.8f && ...
        %1.13f\n',N,errEuler,errHeun,errRK4);
end
```

MATLAB Example 7.5 f_explicit.m: Right-hand side function for the initial value problem (7.24).

```
function [dy]=f(t,y)

dy = sqrt(t^2+y^2);
```

the model IVP

$$\dot{y} = -0.5 \cdot y\,, \tag{7.25}$$
$$y(0) = 1\,, \tag{7.26}$$

Fig. 7.1 shows a number of approximate solutions of the explicit Euler method for different time step sizes ranging from $h = 6$ to $h = 1$. For comparison, we also plot the analytic solution. We observe that the approximations with coarser step sizes "explode" in the sense that they do not follow the analytic solution in a stable manner. Some additional effects take place besides the accuracy issues for coarse time steps: stability features of numerical schemes for ODEs impose a restriction on the time step size. Conditions to obtain stable solutions with explicit methods exist which are often called Courant-Friedrichs-Lewy (CFL) conditions (see [67]). Obviously, these stability issues did not appear in the problems presented before (no oscillations or explosions in Tab. 7.1, e.g., despite relatively coarse time step sizes). Stability issues are, therefore, not (only) a problem of a specific numerical method

Before you continue, make sure to answer the following questions:

Quiz: Section 7.3

Q1 What is the basic idea to derive the explicit Euler method to numerically solve ODEs?

Q2 By which factor is the error reduced for each halfing of h_n if you apply a method of the following order?

- first order ($\mathcal{O}(h_n)$),
- second order ($\mathcal{O}\left(h_n^2\right)$),
- third order ($\mathcal{O}\left(h_n^3\right)$),
- fourth order ($\mathcal{O}\left(h_n^4\right)$)

Q3 What is the order of consistency of the explicit Euler method, Heun's method, and the Runge-Kutta-4 scheme?

Q4 What criteria exist to determine convergence of a one-step method?

Q5 What is the difference between consistency and convergence?

Q6 What is the advantage of higher-order methods such as the classical Runge-Kutta scheme compared to the explicit Euler method, e.g.? What are the additional costs?

Q7 What is a Butcher tableau?

but mainly related to the particular problem: the underlying equation. This motivates the notion of *stiff ODEs* that we present in the following section.

7.4.1 Stiff ODEs

Unluckily, there is no hard mathematical definition of stiffness. However, one may define it in a soft manner:

Definition 7.5 (Stiff Equations). An ordinary differential equation is called *stiff* (in a certain interval of integration) if all explicit one-step methods applied to it with a stable time step size provide very small local errors (in that interval).

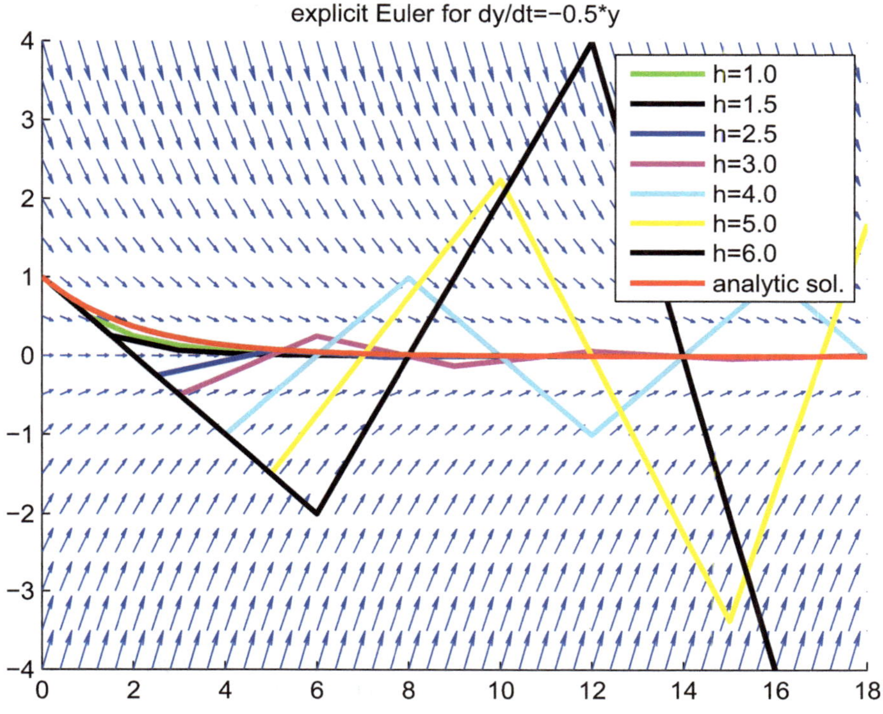

Figure 7.1. The explicit Euler method applied to the example IVP $\dot{y} = -0.5 \cdot y$ for different time step sizes h. For coarser step sizes, the instability effects are clearly visible.

This definition states that the time step size of explicit one-step methods applied to stiff equations is limited by stability, not by accuracy. Furthermore, it implies that stiff equations lead to instabilities if explicit methods do not obey the stability (or CFL) conditions.

Examples of stiff equations include damped mechanical systems, parabolic partial differential equations (such as the Navier-Stokes Equations, cf. Sec. 5.2), or chemical reaction kinetics. Solutions to such systems typically show a certain behaviour over time: A transient phase—typically in the beginning—shows large changes in the solution which stagnates in the constant phase. A simple scalar example of this type is the Dahlquist test equation

$$\dot{y}(t) = \lambda \cdot y(t) , \quad \lambda \in \mathbb{C} . \tag{7.27}$$

The analytical solution of (7.27) is given by

$$y(t) = y_0 \cdot e^{\lambda \cdot t} \tag{7.28}$$

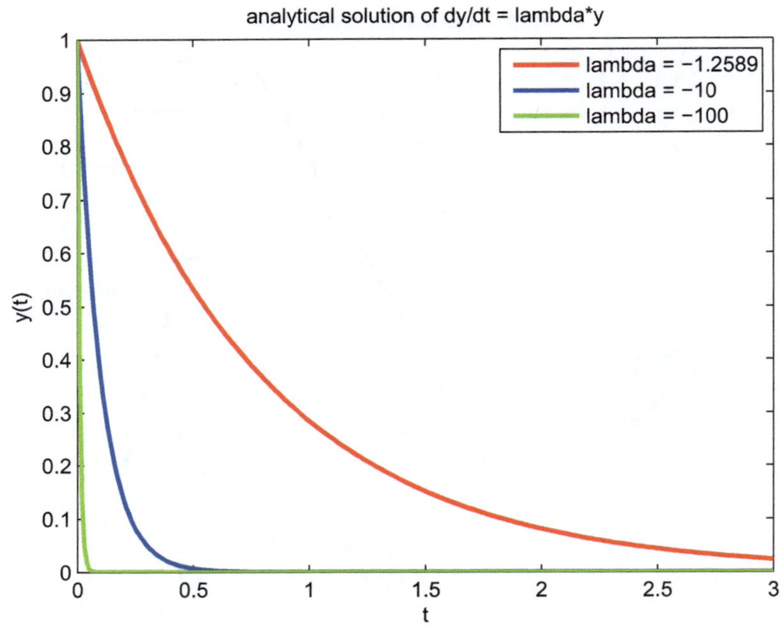

Figure 7.2. Analytical solutions for the stiff IVP $\dot{y} = \lambda \cdot y, y_0 = 1$ for $\lambda = -1.2589, -1, -100$. The transient phase ends at about $3.0, 0.6, 0.1$, respectively.

and is bounded for all $y_0 \in \mathbb{C}, t \geq t_0$ iff Re$\{\lambda\} \leq 0$. The solution for three different real-valued λ is visualised in Fig. 7.2; the transient phases (up to values of t of about 3, 0.6, and 0.1) are clearly observable. Note that also the initial value problem (7.25) above is of the Dahlquist test equation type with $\lambda = -0.5$. The notion of stability of a method is crucial but also complex. Many different variants exist and are related to stability functions and stability regions. We do not discuss these aspects in detail but refer to classical literature such as [125, 84].

Now that we know a category of stiff test equations, it is time to present the remedy to the associated stability problems: implicit methods.

7.4.2 Implicit Euler Method

The simplest implicit scheme is the implicit Euler method. In fact, the only change to the explicit Euler method is the point of evaluation of the IVP right-hand side f: Instead of t_n, f is evaluated at time t_{n+1} and at y^{n+1},

$$y^{n+1} = y^n + h_n \cdot f(t_{n+1}, y^{n+1}) . \tag{7.29}$$

The appearance of y^{n+1} in the right-hand side of (7.29) has a huge impact on the necessary solution procedures. In the (common) case of a non-linear function f, a non-linear system of equations has to be solved in (7.29) in each time step. Only for rare exceptions, y^{n+1} can be found analytically[2]; typically, numerical approaches such as the Newton method have to be applied. We do not go into the details of solving non-linear systems of equations, since this would represent at least another separate chapter, but rely instead on existing implementations in MATLAB.

MATLAB Example 7.6 holds an implementation of the implicit Euler method with equidistant time stepping[3]. MATLAB Example 7.7 defines the non-linear function F to be solved in each time step. This function used inside is fsolve, MATLAB's standard method for non-linear system solves.

In Sec. 7.4.4, we apply the implicit Euler to the initial value problem (7.24) of Sec. 7.3.4. The order of consistency (and convergence) for the implicit Euler scheme is $p = 1$, as in the explicit case. We are, therefore, interested in generating higher-order methods.

7.4.3 Trapezoidal Rule

A variety of higher-order implicit schemes exist such as Adams-Moulton methods, implicit Runge-Kutta schemes, but we will only briefly present the trapezoidal rule (which is also called Crank-Nicholson and actually represents the simplest scheme of the Adams-Moulton family). The trapezoidal rule gets its name from the corresponding quadrature rule applied to (7.1) in Sec. 7.2:

$$\int_s^{s+h} f(t, y(t))\mathrm{d}t \approx \tfrac{1}{2}\left(f(s, y(s)) + f(s+h, y(s+h))\right) \cdot h\,,$$

This results in the following integration scheme

$$y^{n+1} = y^n + \frac{h_n}{2} \cdot \left(f(t_n, y^n) + f(t_{n+1}, y^{n+1})\right) . \tag{7.30}$$

Note that—as for explicit schemes—the higher order ($p = 2$ in this case) comes at the price of more evaluations of f. Due to the appearance of y^{n+1} in the right-hand side of (7.30), the trapezoidal rule is fully implicit. The corresponding implementation given in MATLAB Example 7.8 is nearly identical to the implicit Euler except for the data passed to the evaluation of the non-linear function F (see MATLAB Example 7.9): We now need additional parameters t_n and y^n. Note that we have lowered the default tolerance for

---

2 Already for the scalar example of $f(t, y(t)) = -log(y(t))$, we can not solve the corresponding non-linear system (7.29) by hand.

3 Note that the call of optimset just suppresses additional default MATLAB output but does not change the functionality of the solver settings.

MATLAB Example 7.6 integrateEulerImpl.m: Implementation of the implicit Euler method with equidistant time stepping.

```
function err = integrateEulerImpl(t0, tend, y0, fhandle, N, yRef)

h = 1.0 * (tend-t0) / N;
t = t0:h:tend;
noTimeSteps = length(t);

y = zeros(1,noTimeSteps);
y(1) = y0;
for n=1:noTimeSteps-1
    yold  = y(n);
    yInit = y(n);
    %trigger solution of non-linear system
    opts = optimset('Diagnostics','off', 'Display','off');
    ynew = fsolve(@(x) ...
        nonlinearSysIE(t(n+1),x,yold,h,fhandle),yInit,opts);
    y(n+1) = ynew;
end
err = y(end) - yRef;
```

MATLAB Example 7.7 nonlinearSysIE.m: Implementation of the non-linear function F of the implicit Euler method.

```
function [F]=nonlinearSysIE(t,y,yold,h,fhandle)

F = yold - y + h*fhandle(t,y);
```

the stopping criteria to 10^{-11} to prepare the application examples in the next section.

7.4.4 Consistency & Convergence

We apply the implicit Euler and the trapezoidal rule to the problem (7.24). The resulting global discretisation error $e(h,t)$ at $t = 2$ is given in Tab. 7.2 and this shows the order of convergence of both methods: The implicit Euler scheme is of first order, whereas the trapezoidal rule is a second-order scheme. Note that the absolute values of e are of the same order as in case of the explicit schemes.

In addition to the respective schemes and the general problem implementation given in MATLAB Examples 7.6, 7.7, 7.8, 7.9, and 7.5, we used the code of MATLAB Example 7.10 to run the comparison for the given problem. When

MATLAB Example 7.8 integrateTR.m: Implementation of the trapezoidal rule with equidistant time stepping.

```
function err = integrateTR(t0, tend, y0, fhandle, N, yRef)

h = 2.0 * (tend-t0) / N;
t = t0:h:tend;
noTimeSteps = length(t);

y = zeros(1,noTimeSteps);
y(1) = y0;
for n=1:noTimeSteps-1
    yold  = y(n);
    yInit = y(n);
    %trigger solution of non-linear system
    opts = optimset('Diagnostics','off', ...
        'Display','off','TolFun',1e-11);
    ynew = fsolve(@(x) ...
        nonlinearSysTR(t(n+1),t(n),x,yold,h,fhandle),yInit,opts);
    y(n+1) = ynew;
end
err = y(end) - yRef;
```

MATLAB Example 7.9 nonlinearSysTR.m: Implementation of the non-linear function F of the implicit trapezoidal rule.

```
function [f]=nonlinearSysTR(t,told,y,yold,h,fhandle)

f = yold - y + h/2.0* (fhandle(told,yold) + fhandle(t,y));
```

applying the implicit scheme to the Dahlquist test example (7.25), we can get stable approximate solutions for large time step sizes as Fig. 7.3 shows. Of course, the global discretisation error for the coarse time step sizes of $h = 6.0$ and $h = 5.0$ is visible in the first time steps, but the approximate solutions do not blow up or oscillate—rather they converge in time to the steady-state solution zero. Hence, implicit methods are a remedy (and sometimes the only choice) for stiff equations if one cannot afford small time step sizes which explicit schemes require. These implicit schemes come at the price of a non-linear system of equations to be solved in each time step; this considerably increases the computational cost per time step (as in the above scalar examples, compare the runtimes). Certain cases also exist where the numerical solution of the non-linear system even fails for different reasons. Which category of methods—explicit or implicit—to choose heavily depends on the

N	implicit Euler $h = 2/N$	trapezoidal rule $h = 4/N$
4	2.6125	0.3387 7052
8	0.8856	0.0723 0838
16	0.3851	0.0174 5608
32	0.1812	0.0043 2708
64	0.0880	0.0010 7949
128	0.0434	0.0002 6973
256	0.0215	0.0000 6742
512	0.0107	0.0000 1686
1024	0.0054	0.0000 0421
2048	0.0027	0.0000 0105
4096	0.0013	0.0000 0026
reduction:	1/2	1/4

Table 7.2. Global discretisation error $e(h, t)$ of the implicit Euler and trapezoidal rule for the initial value problem (7.24) at $t = 2$: Comparison of different resolutions N shows the corresponding order of convergence $p = 1, 2$.

MATLAB Example 7.10 ivp_implicit.m: Script for the initial value problem (7.24) solved with implicit Euler and trapezoidal rule.

```
t0 = 0;
tend = 2;
y0 = 0;

fhandle = @f_explicit;
yRef = 2.588607526700;
for i=2:12
    N = 2^i;
    errIE = integrateEulerImpl(t0, tend, y0, fhandle, N, yRef);
    errTR = integrateTR( t0, tend, y0, fhandle, N, yRef);
    fprintf('%4d & %1.4f && %1.8f\n',N,errIE,errTR);
end
```

underlying problem and the available ressources and software. There are cases where it is hard to predict, but the properties of the different schemes can typically be summarised as follows:

- explicit schemes
    - \+ cheap time steps

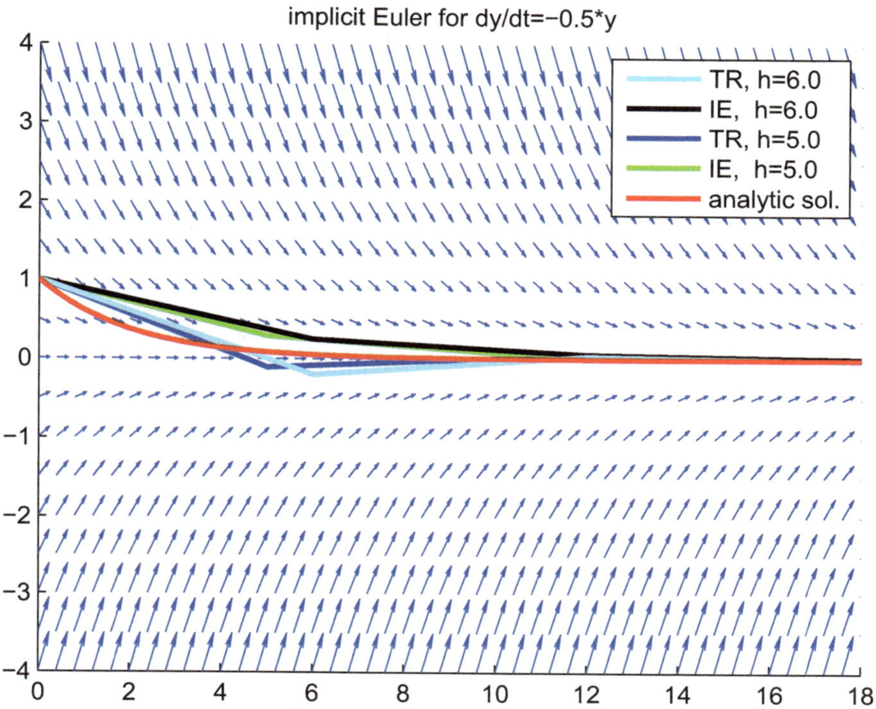

Figure 7.3. Solutions for problem (7.25) obtained with implicit Euler and trapezoidal rule for $h = 5.0$ and $h = 6.0$. Note the different behaviour of the implicit methods compared to the explicit ones in Fig. 7.1.

+ robust
− many time steps necessary

- implicit schemes
  − expensive time steps
  ? robust
  + less time steps necessary .

7.5 Excursion: The Newmark Scheme

An additional class of methods is the Newmark scheme (see [201, 24],e.g.). This approach differs from all methods covered so far in the underlying ODE equations which are tackled. Classical integration schemes are applied to

Before you continue, make sure to answer the following questions:

Quiz: Section 7.4

Q1 What is the characterisation of stiff ODEs?

Q2 What is the major price to pay to use implicit methods for ODE integration?

Q3 What is the advantage of implicit schemes?

Q4 Which order of consistency do the implicit Euler and the trapezoidal rule possess, respectively?

first-order systems of the form of (7.2), while the Newmark scheme is based on the second-order formulation of a given system:

$$M \cdot \ddot{y}(t) + C \cdot \dot{y}(t) + K \cdot y(t) = F(t, y(t)) , \tag{7.31}$$

$$y(t_0) = y_0 , \tag{7.32}$$

$$\dot{y}(t_0) = dy_0 . \tag{7.33}$$

The Newmark method was developed for the solution of structural dynamics problem which are exactly of the type (7.31). Since our main application example of moving buildings is described by structural mechanics, the Newmark approach is relevant, and we briefly describe the most important aspects following [261].

The basic idea of the Newmark method is to represent the velocity and displacement at the new time step t_{n+1} via

$$\dot{y}^{n+1} = \dot{y}^n + h \cdot \left[(1-\gamma)\ddot{y}^n + \gamma\ddot{y}^{n+1}\right] , \tag{7.34}$$

$$y^{n+1} = y^n + h\dot{y}^n + h^2 \cdot \left[\left(\frac{1}{2} - \beta\right)\ddot{y}^n + \beta\ddot{y}^{n+1}\right] . \tag{7.35}$$

Inserting (7.34)–(7.35) into the equations of motion (7.31) for time step t_{n+1} allows us to solve for the new accelerations $\ddot{y}^{n+1}$. Reinserting the resulting $\ddot{y}^{n+1}$ into (7.34)–(7.35) yields the new velocity and displacement. Since we are primarily interested in the displacement y^{n+1} as our quantity of interest, we reformulate the formulas (7.34) and (7.35) as:

$$\dot{y}^{n+1} = \frac{\gamma}{\beta h}(y^{n+1} - y^n) - \frac{\gamma - \beta}{\beta}\dot{y}^n - \frac{\gamma - 2\beta}{2\beta}h\ddot{y}^n , \tag{7.36}$$

$$\ddot{y}^{n+1} = \frac{1}{\beta h^2}(y^{n+1} - y^n) - \frac{1}{\beta h}\dot{y}^n - \frac{1 - 2\beta}{2\beta}\ddot{y}^n . \tag{7.37}$$

Inserting (7.36) and (7.37) into (7.31) directly results in a formula allowing us to compute y^{n+1}.

The implementation of the Newmark scheme for a system without damping[4] is given in MATLAB Example 7.11 using coefficients k_i to shorten the notation. The update of $\mathrm{dy\_new}:= \dot{y}^{n+1}$ and $\mathrm{ddy\_new}:= \ddot{y}^{n+1}$ is identical to (7.36) and (7.37). The update of the displacements y is obtained by solving a linear system of equations involving the system matrices. This variant of the Newmark scheme represents an explicit variant where the known state of motion at the beginning of a time step (i.e. at t_n) is extrapolated to the end of the step (i.e. to t_{n+1}).

To cross-check whether our implementation of the Newmark scheme works correctly, we apply it to a very simple test problem:

$$\ddot{y}(t) + \cdot y(t) = 0\,, \tag{7.38}$$

$$y(0) = 0\,, \tag{7.39}$$

$$\dot{y}(0) = 1\,, \tag{7.40}$$

where we set $M = 1$, $K = 1$, and $F = 0$ in (7.31) and use $t_e = 1$. The analytical solution of this problem is $y(t) = sin(t)$. The global discretisation error at $t_e = 1$ is listed for different resolutions in Tab. 7.3. It clearly shows the order $p = 2$ of the Newmark scheme.

resolution		global discretisation error
N	$h = 2/N$	$e(h, t = 1)$
4	0.500000	0.0110 2116
8	0.250000	0.0027 9917
16	0.125000	0.0007 0258
32	0.062500	0.0001 7582
64	0.031250	0.0000 4397
128	0.015625	0.0000 1099
256	0.007813	0.0000 0275
512	0.003906	0.0000 0069
1024	0.001953	0.0000 0017
reduction:		1/4

Table 7.3. Global discretisation error $e(h, t)$ of the Newmark method for the initial value problem (7.38)–(7.40) at $t = 1$: Comparison of different resolutions N (or h) shows the corresponding order of convergence $p = 2$.

In fact, the Newmark scheme defines a whole family of methods: The coefficients β and γ have to be chosen and this choice is not without effects. It

[4] i.e. with $C = 0$

MATLAB Example 7.11 integrateNewmark.m: Implementation of the Newmark method.

```
% solve for system: M*y''+ K*y = F
function y = integrateNewmark(M,K,F, y0, dy0, t0, tend, N)
h = 2.0 * (tend-t0) / N;
t = t0:h:tend;
noTimeSteps = length(t);
% coefficients
beta  = 1/4;
gamma = 1/2;
k1 = 1/h^2/beta;
k2 = 1/h/beta;
k3 = 1/2/beta - 1;
k4 = gamma/beta - 1;
k5 = gamma*h/2/beta - h;
k6 = gamma/h/beta;
% initialise
y   = zeros(length(y0),noTimeSteps); y(1,:) = y0;
dy  = dy0;
ddy = M\(F(1)-K*y0);
% construct system matrix
A = k1*M + K;
for i = 2:noTimeSteps
    % solve for y
    b = F(i) + M*(k1*y(:,i-1) + k2*dy + k3*ddy);
    y_new = A\b;
    % solve for ddy
    ddy_new = y_new*k1 - y(:,i-1)*k1 - dy*k2 - ddy*k3;
    % solve for dy
    dy_new  = y_new*k6 - y(:,i-1)*k6 - dy*k4 - ddy*k5;
    % update
    y(:,i) = y_new;
    dy     = dy_new;
    ddy    = ddy_new;
end
return
```

steers the order of the method as well as its category (explicit or implicit) and stability properties. Note that for a choice of parameters resulting in implicit Newmark schemes, the order of the method drops from 2 to 1. Alternative methods exist that lead to the family of generalized-α methods. These methods are based on the Newmark scheme but provide even more parameters to choose. We do not go into the details of the different variants but close this section by repeating that for structural dynamics problems, the Newmark scheme may be a valid alternative.

Before you continue, make sure to answer the following questions:

Quiz: Section 7.5

Q1 What is the basic difference in the underlying ODE system for the application of the Newmark scheme compared to methods discussed in the previous sections?

Q2 Verify that $y(t) = sin(t)$ really is the analytical solution of (7.38)–(7.40).

Q3 What is the order of convergence of the Newmark method?

Q4 Is there just one Newmark method? Why?

7.6 Excursion: Symplectic Methods

In this section, we briefly present symplectic methods, a special type of methods from the category of geometric integration schemes (see [123]). To motivate symplectic schemes, we consider a simple example of population dynamics, a Volterra-Lotka model taken from [123]. In Sec. 6.7.3, this model has been presented in a general form. We now use a particular variant for a predator-prey scenario where $u(t)$ denotes the population of the predators and $v(t)$ of the prey:

$$\dot{u} = u(v-2) \,, \tag{7.41}$$

$$\dot{v} = v(1-u) \,. \tag{7.42}$$

For the numerical solutions of this system, we implemented the right-hand-side function f of the IVP in MATLAB Example 7.12. The code for the explicit and implicit Euler method, given in MATLAB Example 7.13 and 7.14, is different than the code in Sec. 7.3.1 and 7.4.2 only with respect to the dimension of the system: we are now dealing with a 2D instead of a scalar system.

Figures 7.4 (a) and 7.4 (b) show the results of the explicit and implicit Euler for our Volterra-Lotka example using 125 time steps of size $h = 0.12$ and initial values $y_0 = (2,2)^T$ and $y_0 = (4,8)^T$, respectively. Knowing that the continuous solutions of Volterra-Lotka systems stay on orbits (sketched in black in the figures), the numerical solutions (blue) obviously do not behave nicely: The explicit Euler solution blows up whereas the implicit Euler approximation gets contracted. In fact, this behaviour is not a problem of coarse time steps, but an inherent feature of the schemes. So, are there methods that produce approximate solutions that do respect the underlying periodicity?

MATLAB Example 7.12 f_VolterraLotka.m: Implementation of the right-hand side f of the Volterra-Lotka system (7.41)–(7.42).

```
function [dy]=f_VolterraLotka(t,y)
dy = y;
dy(1) = y(1) * (y(2) - 2);
dy(2) = y(2) * (1 - y(1));
```

MATLAB Example 7.13 eulerExplicitVL.m: Implementation of the explicit Euler method with equidistant time stepping for the Volterra-Lotka system (7.41)–(7.42).

```
function y = eulerExplicitVL(y0, fhandle, h, noTimeSteps)

y = zeros(2,noTimeSteps+1);
y(:,1) = y0;
for n=1:noTimeSteps
    y(:,n+1) = y(:,n) + h * fhandle(h*n,y(:,n));
end
```

MATLAB Example 7.14 eulerImplicitVL.m: Implementation of the implicit Euler method with equidistant time stepping for the Volterra-Lotka system (7.41)–(7.42).

```
function y = eulerImplicitVL(y0, fhandle, h, noTimeSteps)

y = zeros(2,noTimeSteps+1);
y(:,1) = y0;
for n=1:noTimeSteps
    yold  = y(:,n);
    yInit = y(:,n);
    %trigger solution of non-linear system
    ynew = fsolve(@(x) nonlinearSysIE(h*n,x,yold,h,fhandle),yInit);
    y(:,n+1) = ynew;
end
```

The positive answer to this question is visible from Fig. 7.4 (c) which shows the approximate solution of the *symplectic Euler method* using the same time steps as above with initial values $y_0 = (2,2)^T$, $y_0 = (4,2)^T$, and $y_0 = (6,2)^T$.

MATLAB Example 7.15 shows an implementation of the symplectic Euler method which is identical to the implicit Euler variant except for the definition of the non-linear function F to be solved in each time step. The code

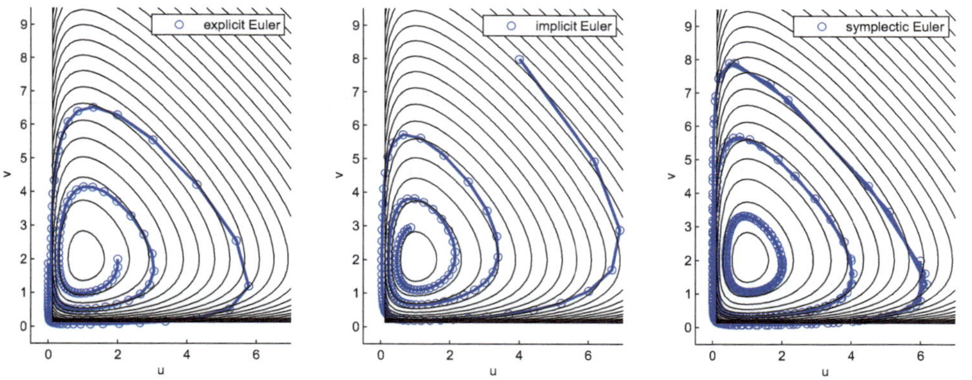

Figure 7.4. Approximate solutions of the Volterra-Lotka system (7.41)–(7.42) with 125 time steps of size $h = 0.12$: explicit Euler (a) with initial value $y_0 = (2,2)^T$, implicit Euler (b) using $y_0 = (4,8)^T$, and symplectic Euler with $y_0 = (2,2)^T$, $y_0 = (4,2)^T$, and $y_0 = (6,2)^T$, respectively.

for this function is given in MATLAB Example 7.16; it shows that the effective non-linearity only exists for the first of the two components u, v. Hence, the symplectic Euler method is an implicit scheme that treats part of the right-hand side of the IVP in an explicit manner.

We complete the discussion of the implementations for this system with MATLAB Example 7.17 holding all relevant code to generate the pictures of Fig. 7.4. In particular, the function $\mathrm{plotSolutionVL}$ computes the field $E = \ln u - u + 2\ln v - v$ which is a first integral[5]. First integrals have been defined in Sec. 6.7.

In fact, a *symplectic scheme* is a scheme that preserves the values of a first integral of a given problem over time. Another example of a first integral is the Hamiltonian of the mathematical pendulum discussed in Sec. 6.7.2:

$$E \; = \; H(p,q) \; = \; p^2/2 - \cos q \, .$$

So in this case, a symplectic scheme preserving E for the pendulum actually preserves the total energy of the system. This conservation of energy is an important feature for Hamiltonian Systems such as molecular dynamics or celestial mechanics. When discretising the underlying system and integrating in time, one should still respect this physical constraint; this is exactly the motivation for symplectic integration schemes.

[5] E is obtained by dividing Eqs. 7.41 and 7.42 by each other.

MATLAB Example 7.15 eulerSymplecticVL.m: Implementation of the symplectic Euler method with equidistant time stepping for the Volterra-Lotka system (7.41)–(7.42).

```
function y = eulerSymplecticVL(y0, fhandle, h, noTimeSteps)

y = zeros(2,noTimeSteps+1);
y(:,1) = y0;
for n=1:noTimeSteps
    yold  = y(:,n);
    yInit = y(:,n);
    %trigger solution of non-linear system
    ynew = fsolve(@(x) nonlinearSysSE(h*n,x,yold,h,fhandle),yInit);
    y(:,n+1) = ynew;
end
```

MATLAB Example 7.16 nonlinearSysSE.m: Implementation of the non-linear function F of the symplectic Euler method.

```
function [f]=nonlinearSysSE(t,y,yold,h,fhandle)

f = yold - y + h*fhandle(t,[y(1);yold(2)]);
```

We finish this section by giving a brief outlook on symplectic properties of Runge-Kutta schemes of [123]:

Theorem 7.3. *If the coefficients of a Runge-Kutta method satisfy*

$$b_i a_{ij} + b_j a_{ji} = b_i bj \quad \forall i, j = 1, \ldots, s \tag{7.43}$$

then it is symplectic.

So one can check whether a given Runge-Kutta scheme has the symplectic property. Using the representation of explicit Euler, implicit Euler, trapezoidal rule, and RK4 as Runge-Kutta schemes of Sec. 7.7, it is straightforward to show that all these methods are not symplectic (independent of their explicit or implicit nature!). It actually turns out (see [123]) that the sufficient conditions (7.43) for symplectic methods are also necessary conditions.

We close this excursion with a final note on the numerical integration of the Navier-Stokes equations (see Chap. 5.3): In order to provide a numerical conservation of energy, a symplectic integration method may be interesting, though typically this approach is not used.

MATLAB Example 7.17 plotSolutionVL.m: Visualisation of numerical solutions of the Volterra-Lotka system (7.41)–(7.42) compared with analytical orbits.

```
function plotSolutionVL(legendString, varargin)
fh=figure;
hold on;
for i=1:length(varargin)
    y = varargin{i};
    plot(y(1,:),y(2,:),'bo');
    plot(y(1,:),y(2,:),'b','LineWidth',2);
end
axis equal; axis([-.5 7 -.5 9.5]);
xlabel('u'); ylabel('v');
legend(legendString,'Location','NorthEast')
%plot level curves!
h=.125;
urange = 0:h:7;
vrange = 0:h:9.5;
[U,V] = meshgrid(urange,vrange);
E = log(U) - U + 2*log(V) - V;
v = -10.5:0.35:-0;
contour(U,V,E,v,'k');
hold off;
%print to file
set(fh,'PaperPositionMode','Auto')
saveas(fh,legendString,'epsc')
```

Before you continue, make sure to answer the following questions:

Quiz: Section 7.6

Q1 Why are symplectic methods interesting? What is their advantage?

Q2 Recall the definition of a first integral.

Q3 Which application scenarios may directly profit from symplectic methods?

Q4 What do the conditions for symplectic Runge-Kutta schemes look like? Are they sufficient or necessary conditions?

7.7 Chapter's Summary

In the sections above, we encountered a variety of different schemes for the numerical integration of ODE systems. In particular, we presented the corresponding orders of consistency and convergence of the different methods. Pros and cons of the methods with respect to stiff and non-stiff IVPs have been discussed: Explicit one-step methods are typically robust and reliable, easy to implement and analyse, and computationally cheap. However, they fail for stiff ODE problems where implicit schemes are the methods of choice allow for large time step sizes without stability problems. The price we pay for these advantages is a loss of robustness and higher computational efforts necessary to compute a single time step.

Note that this categorisation of methods—being a Runge-Kutta, Adams-Moulton, or Newmark scheme—is not unique. For instance, implicit Runge-Kutta schemes (not discussed above) can become implicit Euler and trapezoidal rule for a certain choices of parameters[6]. In fact, a Butcher tableau can be set up for various methods that we presented in this chapter. We demonstrate this by representing the explicit Euler, implicit Euler, trapezoidal rule (or Crank-Nicholson), and the classical Runge-Kutta scheme of fourth order in corresponding Butcher tableaus:

$$
\begin{array}{c|c} 0 & 0 \\ \hline & 1 \end{array}\,, \qquad
\begin{array}{c|c} 1 & 1 \\ \hline & 1 \end{array}\,, \qquad
\begin{array}{c|cc} 0 & & \\ 1 & \frac{1}{2} & \frac{1}{2} \\ \hline & \frac{1}{2} & \frac{1}{2} \end{array}\,, \qquad
\begin{array}{c|cccc} 0 & & & & \\ \frac{1}{2} & \frac{1}{2} & & & \\ \frac{1}{2} & 0 & \frac{1}{2} & & \\ 1 & 0 & 0 & 1 & \\ \hline & \frac{1}{6} & \frac{1}{3} & \frac{1}{3} & \frac{1}{6} \end{array} \tag{7.44}
$$

We emphasise that for the numerical integration in the context of PDEs, the order of consistency of an ODE method has to fit the order of the spatial discretisation scheme in use. Since the overall error is the sum of both spatial and temporal discretisation errors, the choice of the ODE scheme is no longer independent.

This chapter is far from an exhaustive source for numerics of ODEs. We did not cover multi-step methods which reuse information previously computed in former time steps rather than additional data such as Runge-Kutta schemes. Extrapolation methods (see [239]) provide very high accuracy as necessary in celestial mechanics simulations. Furthermore, similar to the explicit Runge-Kutta methods, implicit variants can be derived and visualised via Butcher tableaus. We discussed the notion of stability only very briefly ignoring a variety of different definitions (some of them being particularly

6 Note that the same can be done for the Adams-Moulton and Newmark families.

relevant for multi-step methods). Finally, we did not tackle the problem of adaptive time stepping which can provide means to reduce the computational costs by a constant factor for certain problems. Hence, we recommend to the interested reader the following literature [124, 125, 239, 84, 231] for details on these topics. We are now well prepared for following tasks of ODE integration in the context of RODE simulations. In particular, we are going to analyse derived random variants of certain (explicit) schemes in Chap. 14.

Problems

Classification: ☼ easy, ⌚ easy with longer calculations, ☆ a little bit difficult, ⧉ challenging.

Exercise 7.6. [☼] Analytical Solution of an IVP
Compute the analytical solution $y(t)$ of the initial value problem (7.24),

$$\dot{y} \;=\; \sqrt{t^2 + y^2}, \quad y(0) \;=\; 0\,,$$

to show that $y(t=2)$ =2.5886 0752 6700.

Exercise 7.7. [☆] Explicit & Implicit Euler Analytically Applied to a Stiff ODE
Consider the following system of two coupled ODEs with initial conditions

$$\begin{array}{rcll} \dot{x}(t) & = & -20x - 19y\,, & \text{with } x(0) = 2 \\ \dot{y}(t) & = & -19x - 20y\,, & \text{with } y(0) = 0 \end{array} . \tag{7.45}$$

The analytical solution of (7.45) is

$$\begin{array}{rcl} x(t) & = & \exp(-39t) + \exp(-t) \\ y(t) & = & \exp(-39t) - \exp(-t) \end{array} .$$

The component $\exp(-39t)$ quickly becomes negligible as t increases, starting at 0. The solution is then approximately given by

$$x(t) \;=\; -y(t) \;=\; \exp(-t)\,,$$

and this function is smooth and decreasing to zero. One would assume that in almost any numerical solution, a large step size could be used. However, see what is the result examining the **explicit Euler** scheme:

1. Apply the explicit Euler to the IVP (7.45) with a general step size h.
2. Derive a closed form for the resulting difference equations so that x^n and y^n are represented as polynomials in h with degree n.

3. What restriction for h has to be fulfilled to guarantee that the numerical solution converges to zero (and thus imitates the actual solution in long-term behaviour)?

Now, realise a similar approach for the **implicit Euler** scheme:

1. Apply the implicit Euler to the IVP (7.45) with a general step size h.
2. Derive a closed form for the resulting difference equations (possible for this simple non-linear case).
3. Solve the resulting 2×2 linear system of equations analytically.
4. Reformulate the system solution such that a closed form of the approximate solution at time step n using the initial value appears. Is there any restriction on the time step size h for the implicit Euler method as a necessary condition to converge to zero with increasing t?

Exercise 7.8. [⌚] **Population Dynamics: Explicit Methods**
We examine the following ordinary differential equation describing the dynamics of the population of a certain species:

$$\dot{p} = \left(1 - \frac{p}{10}\right) \cdot p \tag{7.46}$$

with initial condition

$$p(0) = 1. \tag{7.47}$$

The analytical solution is given by

$$p(t) = \frac{10}{1 + 9e^{-t}}.$$

We use this rather simple equation with a known exact solution to examine the properties of different numerical methods in the following.

1. Use MATLAB to plot the function $p(t)$ in a graph.
2. Consider a general initial value problem

   $$\dot{y} = f(y), \quad y(0) = y_0.$$

   Implement and apply the following explicit numerical methods with variable step size h_n and final time t_e:

   1) explicit Euler method,
   2) Heun's method,
   3) 4th-order Runge-Kutta method

as a MATLAB functions depending on the right hand side $f(y)$, the initial value y_0, the time step size h_n and the final time t_e. The output of the function shall be a vector containing all computed approximate values for y.

3. For each of the three methods implemented, compute approximate solutions for equation (7.46) with initial conditions (7.47), end time $t_e = 5$, and with time step sizes $h_n = 1, \frac{1}{2}, \frac{1}{4}, \frac{1}{8}$. For each case, compute the approximation error

$$E = \sqrt{\frac{h_n}{5} \sum_k (p_k - p_{k,exact})^2},$$

where p_k denotes the approximation and $p_{exact,k}$ the exact solution at $t = h_n \cdot k$.

Plot your solutions in one graph per method (together with the given solution from above) and write down the errors in the tables below.

4. For each of the three methods, determine the factor by which the error is reduced if the step size h_n is halved. Write down the results in the tabular below.

5. In general, we do not know the exact solution of an equation we have to solve numerically (otherwise, we would not have to use a numerical method, in fact ;)). To anyhow guess the accuracy of a method, we can use the difference between our best approximation (the one with the smallest time step h_n) and the other approximations:

$$\tilde{E} = \sqrt{\frac{h_n}{5} \sum_k (p_k - p_{k,best})^2},$$

where p_k denotes the approximation with time step h_n, $p_{best,k}$ the best approximation at $t = h_n \cdot k$.

Compute $\tilde{E}$ for all time steps and methods used, write down the results in the tables below, and compare them to the exact error. What do you observe?

explicit Euler method				
h_n				
error				
error red.				
error app.				

Heun's method				
h_n				
error				
error red.				
error app.				

classical Runge-Kutta scheme (RK4)				
h_n				
error				
error red.				
error app.				

Exercise 7.9. [☼] **Population Dynamics: Implicit Methods**
We examine a similar population dynamics model as in Problem 7.8 but with different parameters

$$\dot{p} = 7\left(1 - \frac{p}{10}\right) \cdot p \tag{7.48}$$

and with a different initial condition

$$p(0) = 20. \tag{7.49}$$

The analytical solution of (7.48)–(7.49) is given by

$$p(t) = \frac{200}{20 - 10e^{-7t}}.$$

1. Plot the function $p(t)$ in a graph.
2. Reuse the explicit Euler method and the Heun's method implemented in Problem 7.8 to compute approximate solutions for equation (7.48) with initial conditions (7.49), end time $t_e = 5$, and $h_n = 1, \frac{1}{2}, \frac{1}{4}, \ldots, \frac{1}{32}$.

   Plot your solutions in one graph per method (together with the given analytical solution). Plot the function in the range $t \in [0, 5]$ and $p \in [0, 20]$.
3. Implement the following implicit numerical methods with variable step size h_n and end time t_e

   - 1) implicit Euler method,
   - 2) trapezoidal rule,

   for the solution of the initial value problem

   $$\dot{y} = f(y), \quad y(0) = y_0$$

   as a function of the right-hand side f, the first derivative of the right-hand side with respect to y, initial value y_0, the stepsize h_n and the end time t_e. The output of the function is a vector containing all computed approximate values for y. Use an accuracy limit of 10^{-4} for the Newton iteration in each time step. Stick to the signatures of the functions from Problem 7.8 as far as possible.
4. For both methods implemented, compute—as far as possible— approximate solutions for equation (7.48) with initial conditions (7.49) and with time steps $h_n = \frac{1}{2}, \frac{1}{4}, \frac{1}{8}, \frac{1}{16}, \frac{1}{32}$. Plot your solutions in one graph per method (together with the given analytical solution).

5. Compare the results of the implicit methods to those computed with the explicit methods:

   Compute the approximation error

   $$E = \sqrt{\frac{h_n}{5} \sum_k (p_k - p_{k,exact})^2}$$

   for each $h_n = \frac{1}{2}, \frac{1}{4}, \frac{1}{8}, \frac{1}{16}, \frac{1}{32}$, where p_k denotes the approximation of $p(h_n \cdot k)$, $p_{exact,k}$ the exact values of p at $t = h_n \cdot k$.

   Collect the results in the table below.

6. For each of the following methods, determine the factor by which the error is reduced if the step size h_n is halved.

   1) explicit Euler,
   2) Heun's method,
   3) implicit Euler,
   4) trapezoidal rule,

   Write down the results in the table below. Is this consistent with what you expected? Why?

7. In addition to accuracy, we examine an additional aspect of 'quality' of a method: the *stability*. Descriptively spoken, stability denotes the applicability of a method for varying parameters, whereas at least results similar to the exact/correct solution have to be achieved (In particular, unphysical oscillations should not occur). With this heuristic definition, decide for which of the used values for h_n each of the four examined methods is stable (in the case of our problem).

   Mark stable cases by a cross in the last tabular. Try to find a simple criterion to determine whether a solution is stable or not and write the result to the MATLAB console.

explicit Euler					
h_n	$\frac{1}{2}$	$\frac{1}{4}$	$\frac{1}{8}$	$\frac{1}{16}$	$\frac{1}{32}$
error					
error red.					

Heun's method					
h_n	$\frac{1}{2}$	$\frac{1}{4}$	$\frac{1}{8}$	$\frac{1}{16}$	$\frac{1}{32}$
error					
error red.					

implicit Euler					
h_n	$\frac{1}{2}$	$\frac{1}{4}$	$\frac{1}{8}$	$\frac{1}{16}$	$\frac{1}{32}$
error					
error red.					

trapezoidal rule					
h_n	$\frac{1}{2}$	$\frac{1}{4}$	$\frac{1}{8}$	$\frac{1}{16}$	$\frac{1}{32}$
error					
error red.					

Stable cases				
	explicit Euler	Heun	implicit Euler	TR
$h_n = \frac{1}{2}$				
$h_n = \frac{1}{4}$				
$h_n = \frac{1}{8}$				
$h_n = \frac{1}{16}$				
$h_n = \frac{1}{32}$				

8. Which type of methods (explicit/implicit) would you choose for

    - the initial value problem from Problem 7.8,
    - the current initial value problem (7.48)–(7.49)?

    Give a reason why you do or do not choose a certain type of methods in the corresponding cases!

Exercise 7.10. [☼] **Simulating Solutions and Phase Space Diagrams**
For $t \in \mathbb{R}$ consider the forced linear oscillator differential equation

$$\ddot{y}(t) + 2cd\dot{y}(t) + d^2 y(t) = f(t), \qquad x \in \mathcal{C}^2(\mathbb{R}, \mathbb{R}), \tag{7.50}$$

with $f(t) = e^{-0.1t}\sin(10t)$ and constants $c = 0.64$ and $d = 15.56$[7].

1. Formulate (7.50) as a first-order system in time.

2. Solve this system numerically with the explicit Euler method for the initial condition $y(0) = 3$ and step sizes of 10^{-2} as well as 10^{-4}. Plot the solution against the time variable $t \in [0, 10]$, and in the y-$\dot{y}$-phase plane.

3. Solve the first-order system numerically with the implicit Euler method for the initial condition $y(0) = 3$ and step sizes of 10^{-2} as well as 10^{-4}. Plot the solution against the time variable $t \in [0, 10]$, and in the y-$\dot{y}$-phase plane.

4. As usual, the energy of this oscillator is given by

$$E(\dot{y}, y, t) = \tfrac{1}{2}(\dot{y}(t))^2 + \tfrac{1}{2}d^2 (y(t))^2 .$$

   Compare the energies obtained for the numerical solutions with the explicit and implicit Euler method by plotting the (discrete) energy over time.

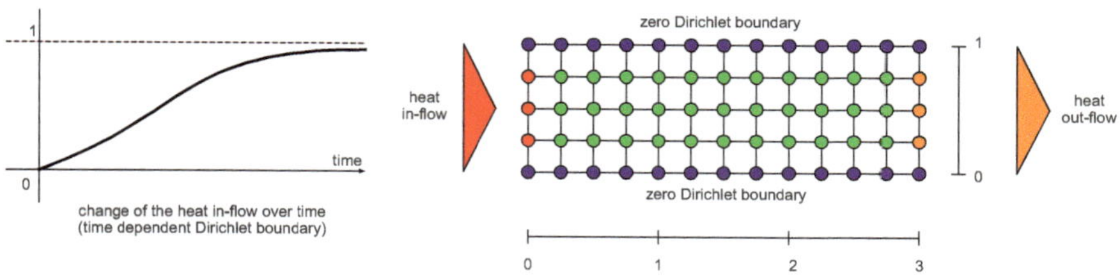

Figure 7.5. Sketch of the situation described in Problem 7.11.

Exercise 7.11. [☆] **Computer Exercise: Solving the 2D Heat Equation**
In this exercise, we combine the spatial FD discretisation presented in Chap. 2 with the time discretisation. To not confuse the notation, the time step size is now denoted by Δt. Figure 7.5 sketches a heat conduction problem on $(x, y) \in [0, 3] \times [0, 1]$ with initial condition $u(0, x, y) = 0$, fixed zero Dirichlet boundary conditions $u(t, x, y) = 0$ for $(x, y) \in [0, 3] \times \{0\}$ and

7 In the case of f being Gaussian white noise, this equation actually models an earthquake excitation for firm soil conditions according to Kanai and Tajimi's work (see Sec. 3.2.2).

$(x, y) \in [0, 3] \times \{1\}$, von Neumann boundary conditions for $(x, y) \in \{3\} \times [0, 1]$ that account for an heat out-flow as well as a time dependent Dirichlet boundary condition $u(t, x, y) = f(t) = \frac{2}{\pi}\,\text{arctan}(t)$ for $(x, y) \in \{0\} \times [0, 1]$ which describes a continuous heating up to a saturation point.

Use a coarse regular Cartesian grid with $h_x = h_y = 0.25$.

1. Give the precise form of the von Neumann boundary conditions in this case.
2. Set-up the propagation matrix of this heat conduction problem of each time step including the boundary conditions.
3. Argue why the choice of $\Delta t = 0.1$ is not a good idea. What would be a smarter choice for Δt?
4. Write a computer program that solves the above described heat conduction problem for $\Delta t = 0.03$ and $\Delta t = 0.001$. Plot the solution for $t \in \{0, 0.5, 1, 5\}$ in the x-y-plane.

Chapter 8

Deterministic Dynamical Systems and Stability of Solutions

We start our brief review on deterministic dynamical systems by introducing the fundamental concepts of (continuous) dynamical systems, long-time behavior, invariance and attraction. This prepares us to analyze stability in the sense of Lyapunov. First, we use Lyapunov-functions for proving (asymptotic) stability in non-linear systems. Next, we analyze the correspondence between the stability properties of non-linear systems and their linearisations. Here, we give the theorem of Hartman and Grobman, a classification of equilibria in planar systems with respect to their stability properties as well as techniques for the determination of the position of Lyapunov exponents of a linear system, like the Routh-Hurwitz criterion of the Lozinskii-measure method.

8.1 Key Concepts

A dynamical system $(X, \mathbb{T}, S_t)$ can be considered as an evolution in the phase space X with respect to time $t \in \mathbb{T}$. If $\mathbb{T} = \mathbb{R}_0^+$ or $\mathbb{T} = \mathbb{R}$ the dynamical system is called continuous, whereas for $\mathbb{T} = \mathbb{N}_0$ or $\mathbb{T} = \mathbb{Z}$ it is called discrete. In this chapter, we focus on continuous dynamical systems that result from ordinary differential equations. Finally, S_t maps the state of the system at time $t_0 = 0$ to the state at time t. This map is called semi-flow or flow map.

The main sources of the following sections are the very nicely written monographs [257], [12], [162] and [56], the lecture notes [189] and in particular [106].

When reading this chapter note the answers to the following questions

1. What is a dynamical system?
2. What does stability mean, and what is the basin of attraction of an asymptotically stable equilibrium?
3. What is a Lyapunov-function?
4. What does the theorem of Hartman-Grobman state and under which conditions can it be applied?

as well as the following key concepts

1. (Continuous) dynamical systems and stability (in the sense of Lyapunov).

2. The principle of linearized stability.
3. The Routh-Hurwitz criterion and the Lozinskii-measure method.

This chapter is structured as follows: Sec. 8.2 introduces the fundamental notions and concepts of (continuous) dynamical systems, long-time behavior, invariance and attraction. This paves the ground to analyze stability in the context of dynamical systems. Hence, Sec. 8.3 gives the definition of (Lyapunov) stability and studies the use of Lyapunov-functions for proving (asymptotic) stability in non-linear systems. The correspondence between the stability properties of non-linear systems and their linearisation is analyzed in Sec. 8.4. We give the theorem of Hartman and Grobman, a classification of equilibria in planar system with respect to their stability properties as well as the techniques for the determination of the position of Lyapunov exponents of a linear system, like the Routh-Hurwitz criterion of the Lozinskii-measure method. Finally, Sec. 8.5 wraps up the contents of this chapter.

8.2 Continuous Dynamical Systems from Ordinary Differential Equations

Let us now use ordinary differential equations to generate continuous dynamical systems: First, we define a dynamical system $(X, \mathbb{R}_0^+, S_t)$ on the complete metric space/ Banach space (X, dist), where dist denotes the usual metric/ distance.

Definition 8.1 (Continuous Dynamical System/ Equilibrium Point/ Periodic Orbit). The triple $(X, \mathbb{T}, S_t)$ is called a *(continuous) dynamical system*, if $\mathbb{T} = \mathbb{R}_0^+$ or $\mathbb{T} = \mathbb{R}$, then X is a Banach space, and $S_t : X \to X$ is a family of mappings for $t \in \mathbb{R}_0^+$ such that the

1. the *flow map* $S_t : X \to X$ is defined and $S_t(x)$ is continuous with respect to both the *time variable* $t \in \mathbb{T}$ and the *space variable* $x \in X$.
2. the pair $(\{S_t\}_{t\in\mathbb{T}}, \circ)$ forms a *semi-group*, i.e., $S_0 = id_X$ and $S_{t+s} = S_t \circ S_s$ for all $t, s \in \mathbb{T}$, where "$\circ$" denotes the usual composition of mappings.

The underlying space X is called the *phase space* of the dynamical system. A point $x^* \in X$ is called *equilibrium (point)* if $S_t x^* = x^*$ for all times $t \in \mathbb{T}$. Moreover, if $S_T x = x$ for a $T > 0$, then the set $\{S_t x \in \mathbb{X} : t \in [0; T]\}$ is called the *periodic orbit* of the point $x \in X$ and the minimal $T > 0$ with this property is the so-called *period* of the periodic orbit.

Remark 8.2 (Groups & Flow-Property). If S_t is bijective for all $t \in \mathbb{T}$, i.e., if $\mathbb{T} = \mathbb{R}$, then we can define $S_{-t} :=(S_t)^{-1}$ for all $t \geq 0$ and then the pair

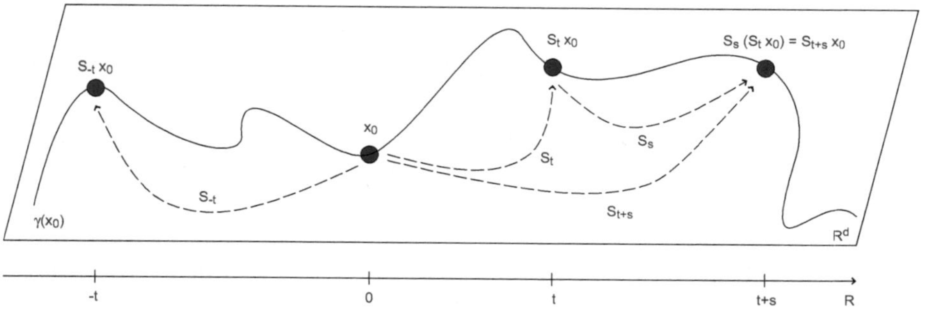

Figure 8.1. Flow property of S_t on the phase space $\mathbb{R}^d$.

$(\{S_t\}_{t\in\mathbb{T}}, \circ)$ forms a *group*, i.e., $S_0 = id_X$ and $S_{t+s} = S_t \circ S_s$ for all $t, s \in \mathbb{T}$. This so-called *flow property* of S_t on the phase space X is illustrated in Fig. 8.1.

Such (continuous) dynamical systems that have the flow-property may be defined by ordinary differential equations: Let $f \in \mathcal{C}^k(\mathbb{R}^d, \mathbb{R}^d)$, $1 \le k \le \infty$[1]. If the initial value problem of the autonomous equation

$$\dot{x} = f(x) \qquad \text{with initial conditions} \qquad x(0) = x_0 \tag{8.1}$$

has a unique solution $x(t)$ for all $x_0 \in \mathbb{R}^d$, which exists for all times $t \in \mathbb{R}$, then the unique solution $x(t)$ of (8.1) induces a dynamical system, This is done by defining the flow map S_t as

$$S_t x_0 := x(t) .$$

Because of the existence and uniqueness of solutions, $(\{S_t\}_{t\in\mathbb{R}}, \circ)$ is a group. We will call dynamical systems induced by linear ordinary differential equations *linear (dynamical) systems* and those induced by non-linear ordinary differential equations *non-linear (dynamical) systems.*

Remark 8.3 (ODEs That do not Induce a Dynamical System). The ordinary differential equation $\dot{x} = x^2$ does not define a dynamical system in our sense: The solution of the initial value problem is

$$x(t) = \frac{1}{x_0^{-1} - t}$$

which is defined only up to $t_{max} := x_0^{-1}$. We can however, define it for $X = \mathbb{R}_0^-$.

[1] Contrary to the discussion of ordinary differential equations, we will now suppress the notation of the parameters. But note, that the parameters may play a crucial role in stability analysis. Furthermore, we set $t_0 = 0$.

Moreover, non-autonomous ordinary differential equations $\dot{x} = f(t,x)$ do not define a semi-group. Though, by extending the non-autonomous system $\dot{x} = f(t,x)$ by the trivial identity $\dot{t} = 1$ such a system can always be transformed to an autonomous one $\dot{y} = F(y)$, where $y = (t,x)^T$ and $F(y) := F(t,x) = (1, f(t,x))^T$. Thus, we can view autonomous system as general. One disadvantage of this approach is that we lose information about attractors as $t \to \infty$.

Example 8.4 (Dynamical Systems from Partial Differential Equations, cf. [106], pp. 3). Let us consider the following heat-conduction equation for $u = u(t,x)$, $x \in (0,\pi)$

$$\begin{cases} u_t &= \alpha u_{xx} \\ u(0,x) &= u_0(x) \\ u(t,0) &= u(t,\pi) \;=\; 0\,, \quad \text{for all } t \geq 0 \end{cases},$$

and let us set $X = L^2(0,\pi)$. Since this problem has a unique solution for all $u_0 \in L^2(0,\pi)$ the map $S_t u_0 := u(t,\cdot)$ defines a dynamical system on X. For instance, if we set $u_0 := \sum_{k=1}^{\infty} a_k \sin(kx)$, then $u(x,t) = \sum_{k=1}^{\infty} a_k \exp(-k^2\alpha t)\sin(kx)$ is the solution.

8.2.1 Long-time Behavior, Invariance and Attraction

The time evolution of a starting point $x \in \mathbb{R}^d$ under the flow map S_t, $t \in \mathbb{R}$ is characterised by its orbit or trajectory $\gamma(x)$:

$$\gamma(x) := \{S_t x : t \in \mathbb{R}\}\,.$$

Concerning just the positive time evolution, the *positive orbit* of x is defined as $\mathcal{O}^+ = \cup_{t\geq 0} S_t x \subset \mathbb{R}^d$.

We can ensure the existence of orbits for all $t \in \mathbb{R}$, even if the system $\dot{x} = f(x)$ has a blow-up in finite time, by rescaling the original differential equation: Consider the modified system

$$\dot{x} = \frac{f(x)}{1 + \|f(x)\|^2} =: g(x)\,,$$

where $g \in \mathcal{C}^k(\mathbb{R}^d, \mathbb{R}^d)$ if $f \in \mathcal{C}^k(\mathbb{R}^d, \mathbb{R}^d)$. Since g is obtained from f by multiplication with a positive scalar the modified system and the original one have the same positive orbits, only the velocity is different. Thus, dynamically speaking the two systems exhibit the same properties. Since $\|g(x)\| \leq \frac{1}{2}$, solutions of the modified system exist for all $t \in \mathbb{R}$.

The long-time behavior of dynamical systems is described by the concepts of "α-limit sets" and "ω-limit sets", of "invariance" and "attraction":

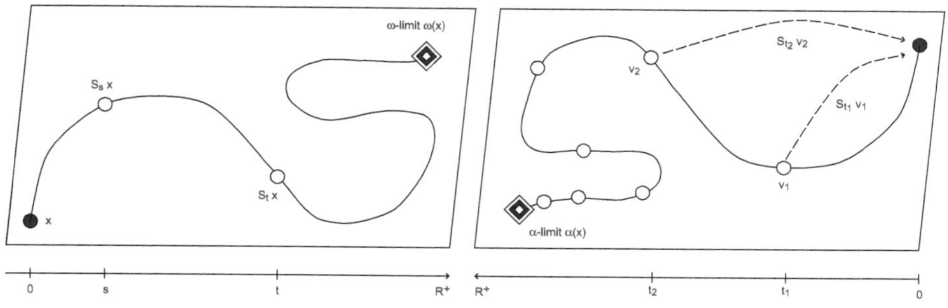

Figure 8.2. Illustrations of the concepts of an ω- (left) and α-limit point (right).

Definition 8.5 (α- and ω-Limit Set/ Point). Let $(\mathbb{R}^d, \mathbb{R}, S_t)$ be a dynamical system. For a subset $B \subset \mathbb{R}^d$, we define the *ω-limit set* $\omega(B)$ and the *α-limit set* $\alpha(B)$ as

$$w \in \omega(B) \iff \exists t_n \overset{n\to\infty}{\longrightarrow} \infty, x_n \in B : S_{t_n} x_n \overset{n\to\infty}{\longrightarrow} w$$
$$v \in \alpha(B) \iff \exists t_n \overset{n\to\infty}{\longrightarrow} -\infty, x_n \in B : S_{t_n} x_n \overset{n\to\infty}{\longrightarrow} v\,.$$

Analogously, the *ω-limit set* $\omega(x) := \omega(\{x\})$ and the *α-limit set* $\alpha(x) := \alpha(\{x\})$ of a point $x \in \mathbb{R}^d$ are defined as

$$w \in \omega(x) \iff \exists t_n \overset{n\to\infty}{\longrightarrow} \infty : S_{t_n} x \overset{n\to\infty}{\longrightarrow} w$$
$$v \in \alpha(x) \iff \exists t_n \overset{n\to\infty}{\longrightarrow} -\infty : S_{t_n} x \overset{n\to\infty}{\longrightarrow} v\,.$$

Figure 8.2 illustrates this definition of an α- and ω-limit point. Later on, for random dynamical systems we will define the concept of a pullback limit that can be considered as a combination of both α- and ω-limits.

Definition 8.6 (Invariance). Let $(\mathbb{R}^d, \mathbb{R}, S_t)$ be a dynamical system and $B \subset \mathbb{R}^d$.

1. B is called *positively invariant* if $S_t B \subset B$ for all $t \geq 0$.
2. B is called *negatively invariant* if $S_t B \supset B$ for all $t \geq 0$.
3. B is called *invariant*, if $S_t B = B$ for all $t \geq 0$.

Note that for positively invariant sets B, their closure $\overline{B}$ and interior $int(B)$ are positively invariant, too. We will show this in problem 8.19.

To give the concept of "attraction", we first have to define the half-distance $dist(A, B)$ of two sets[2] $A, B \subset \mathbb{R}^d$:

$$\text{dist}(A, B) := \sup_{x \in A} \text{dist}(x, B) := \sup_{x \in A} \inf_{y \in B} \|x - y\| .$$

Definition 8.7 (Attraction of Points). Let $(\mathbb{R}^d, \mathbb{R}, S_t)$ be a dynamical system, $A, B \subset \mathbb{R}^d$ with A compact. We say A *attracts points of* B *(uniformly)*, if $dist(S_t p, A) \to 0$ for $t \to \infty$ holds for all points $p \in B$.

8.3 Lyapunov Stability

The most fundamental concept in dynamical systems is that of stability. Actually, the question of stability (of our solar system) and its discussion by Henri Poincare gave birth to this discipline, see [23]. Today, the most common notion of stability for deterministic systems dates back to A. Lyapunov's work. This "Lyapunov stability" is illustrated in Fig. 8.3:

Definition 8.8 (Stability and Asymptotic Stability in the Sense of Lyapunov). Let $(\mathbb{R}^d, \mathbb{R}, S_t)$ be a dynamical system and I be an invariant set.

Before you continue, make sure to answer the following questions:

Quiz: Section 8.2

Q1 Give the precise definition of a dynamical system as discussed above.

Q2 For $x \in \mathbb{R}^d$, let the candidates for flow maps be defined as

$$S_t x := xe^{-t}, \quad S_t x := xt, \quad S_t x := x(t+1), \quad \text{and} \quad S_t x := x .$$

Do these examples define dynamical systems? Sketch their behavior.

Q3 Consider the differential equation $\dot{x} = x^2$, $x \in \mathbb{R}$. Give its explicit solution. Does this equation define a dynamical system?

Q4 Determine all equilibria of the dynamical system given by $\dot{x} = \sin(x)$, $x \in \mathbb{R}$. Sketch the behavior of this system in phase space.

---

[2] Because $dist(A, B) = 0$ only implies $A \subset B$ we can only speak of a "half-distance" instead of a "distance".

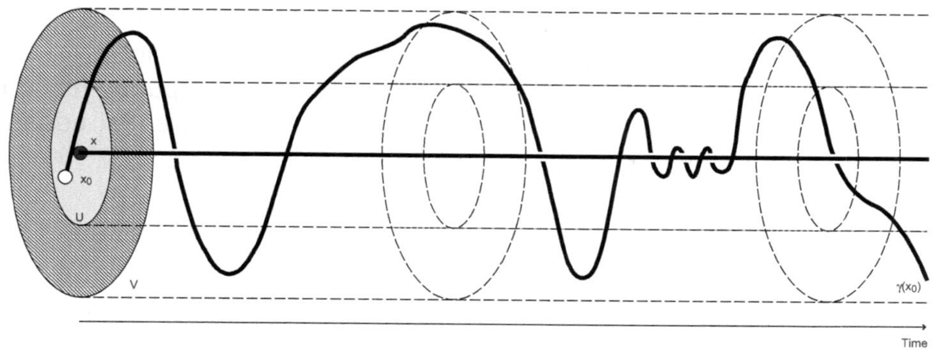

Figure 8.3. Illustration of the concept of stability in the sense of Lyapunov.

1. I is called *stable*, if for each neighborhood $\mathcal{V}$ of I there is a neighborhood $\mathcal{U}$ of I such that $S_t\mathcal{U} \subset \mathcal{V}$ for all times $t \geq 0$.

2. I is called *asymptotically stable*, if I is stable and there is a neighborhood $\mathcal{W}$ of I such that I attracts points of $\mathcal{W}$.

Note, for a stable invariant set I it is required only that trajectories stay "near" I, i.e., they are allowed to meander around I without exceeding a certain distance from I. Asymptotic stability on the other hand says that trajectories approach I.

By means of so-called *Lyapunov-functions*[3] (*Lyapunov's second method*) the Lyapunov stability of a set or an equilibrium can be checked easily.

8.3.1 The Method of Lyapunov-Functions

In the theory of deterministic dynamical systems Lyapunov-functions[4] —although hard to find and/ or construct—play a rather important role to prove (asymptotic) stability. Definitions and (some) applications for the classical method are, e.g., very well explained in [27], pp. 139, [244], pp. 23, [56],

3 ...see, e.g., [106], section 2.2, [6], pp. 24 for Lyapunov- and Chetaev-functions, [27], pp. 134 for $\mathcal{C}^1$-Lyapunov-functions and pp. 166 for non-continuous Lyapunov-functions, or [107] and [108] for the numerical treatment of Lyapunov-functions with radial basis functions.

4 Aleksandr Mikhailovich Lyapunov's (1857 - 1918) main work considered the theory of rotating liquids. E.g., he showed that a sufficient condition for stability is that the second and higher variations of the potential energy are positive—cited according to http://www-groups.dcs.st-and.ac.uk/~history/.

pp. 23, [206][5] which utilizes sums of squares decompositions or [108][6] where radial basis functions are applied to construct Lyapunov-functions.

Classically a function $V \in \mathcal{C}(\mathbb{R}^d, \mathbb{R})$ is called *a strict Lyapunov-function*[7] for an equilibrium point $x_0 \in \mathbb{R}^d$ of the deterministic differential equation $\dot{x} = f(x)$, if the following two conditions hold

- $V'(x) := \langle \nabla V(x), f(x) \rangle < 0$ for all $x \in K \setminus \{x_0\}$ and
- $V(x) > V(x_0)$ for all $x \in K \setminus \{x_0\}$,

where $K \subset \mathbb{R}^d$ is some neighborhood of x_0. The function V', called the *orbital derivative*, is the derivative of V along a solution of $\dot{x} = f(x)$. Thus, V is decreasing along solutions of the underlying differential equation.

The existence of such a function is for instance guaranteed by the following theorem attributed to N. N. Krasovskii, where $\mathbb{R}^d_K := \{x \in \mathbb{R}^d : \|x\| < K\}$ with $K > 0$:

Theorem 8.1 (Krasovskii's Existence Theorem of a Lyapunov-Function). *Let $x(t; x_0, t_0)$ be a solution of the ordinary differential equation $\dot{x} = f(x,t)$ with respect to the initial conditions $t_0 \in I$ and $x_0 = x_{t_0}$, where $f : \mathbb{R}^d_K \times I \to \mathbb{R}^d$ is continuous and has bounded partial derivatives $f_{x_i} = \frac{\partial}{\partial x_i} f$, $i = 1, 2, \ldots, d$. Moreover, assume that the growth-bound*

$$\|x(t; x_0, t_0)\| \leq b\|x_0\| \exp(-a(t - t_0)), \qquad \text{for } t \geq t_0 \text{ and } \|x_0\| < H$$

holds for a $H \leq b^{-1}K$ with $a, b > 0$.

Then, there exists a continuously differentiable function $V(x,t) : \mathbb{R}^d_H \times I \to \mathbb{R}$ such that

1. $c_1\|x\|^2 \leq V(x,t) \leq c_2\|x\|$,
2. $V_t(x,t) + V'(x,t) \leq -c_3\|x\|$, *and*
3. $\|(V_{x_1}(x,t), \ldots, V_{x_d}(x,t))\| \leq c_4\|x\|$,

for $x \in \mathbb{R}^d_H$ and $t \in I$, and positive constants c_1, c_2, c_3 and c_4 (with $V_t = \frac{\partial}{\partial ti} V$). Here, $K = H = \infty$ may hold, too.

For an asymptotically stable equilibrium, one can imagine the corresponding Lyapunov-function as a kind of "cup" over the phase space, as sketched in Fig. 8.4.

5 In [206], Lyapunov-functions are constructed by applying the sums of square decomposition. For details on the methodology of sums of squares, see the work of P. A. Parrilo, in particular his PhD thesis [208].

6 In [108], for the first time a method to construct Lyapunov-functions by radial basis functions is developed and rigorously proven.

7 cf. [108], p. 18.

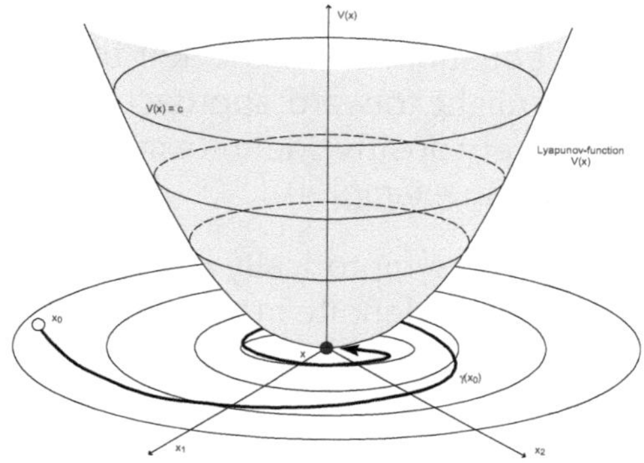

Figure 8.4. Lyapunov-function $V : \mathbb{R}^2 \to \mathbb{R}$ over a two-dimensional phase space with characteristic orbits in the phase space and projection of the Lyapunov-function's level curves into the phase space, see [209], p. 101.

Theorem 8.2 (Lyaponov-Function Stability Theorem). *Let V be a Lyapunov-function on a neighborhood of N of an equilibrium x_0, then the equilibrium point is stable.*

Proof. Following [106], pp. 20, there is an ε_0 such that $N_{\varepsilon_0} := \{x : \text{dist}(x, x_0) < \varepsilon_0\}$ fulfills $N_{\varepsilon_0} \subset N$. Fix an $\varepsilon \leq \varepsilon_0$ and let $\mu := \min\{V(x) : \text{dist}(x, x_0) = \varepsilon\}$. Since V is a continuous function with $V(x) = 0$ if and only if $x = x_0$, there is a $\delta > 0$ (and $\delta \leq \varepsilon$) such that $V(x) < \mu$ for all $x \in N_\delta := \{x : \text{dist}(x, x_0) < \delta\}$.

Next, we show via reducio ad absurdum that $x \in N_\delta$ implies $S_t x \in N_\varepsilon$. Therefore, let us assume that this is false. Then there is a time $t_0 \geq 0$ such that $\text{dist}(S_{t_0} x, x_0) = \varepsilon$ and $\text{dist}(S_t x, x_0) < \varepsilon$ for all $t \in [0, t_0)$. Since $V(S_t x)$ is non-decreasing for $t \in [0, t_0)$, $V(x) < \mu$ implies $V(S_{t_0} x) < \mu$. This is a contradiction to $\mu := \min\{V(x) : \text{dist}(x, x_0) = \varepsilon\}$ since $\text{dist}(S_{t_0} x, x_0) = \varepsilon$. □

Suppose we would like to investigate the stability of the linear system $\dot{x} = Ax$ using a Lyapunov-function. If we set $V(x) := x^T P x$ with a positive definite matrix P, we get

$$\dot{V}(x) = \dot{x}^T P x + x^T P \dot{x} = x^T \left(A^T P + PA\right) x \, .$$

We define the matrix Q by the *Lyapunov equation*

$$A^T P + PA =: -Q \, ,$$

and if Q is positive definite then the origin in the linear system $\dot{x} = Ax$ is globally asymptotically stable. There are basically two options for obtaining the pair (P, Q):

- *Option 1*: Choose an arbitrary positive definite matrix P and compute Q via the Lyapunov equation. Then check if this Q is positive definite. Unfortunately, this straight forward approach is not a smart, as if A is just stable, then not every positive definite matrix P will imply a positive definite matrix Q (give an example!).

- *Option 2*: If A is stable or asymptotically stable, any positive definite matrix Q will lead to a positive definite matrix P via the Lyapunov equation. Thus, set $Q = \mathbb{I}$, and then solve the Lyapunov equation for P.

In particular, we have the following theorem:

Theorem 8.3 (Asymptotic Stability via the Lyapunov Equation). *All eigenvalues of a matrix $A \in \mathbb{R}^{d\times d}$ lie in the left half plane if and only if for any given symmetric positive definite matrix Q there is a unique positive definite matrix P satisfying the Lyapunov equation.*

Proof. See [183], pp. 98. □

Note that P and Q are related by the identity

$$P = \int_0^\infty \exp(A^T t) Q \exp(At) \mathrm{d}t \,.$$

However, this equation is seldom used to obtain P, and P is instead directly computed from the Lyapunov equation.

MATLAB's version of the Lyapunov equation reads as $A^T P + PA =: -Q$ and is available via the command lyap(A, Q). Thus, if we want to determine the matrix P for the linear system

$$\begin{pmatrix} \dot{x} \\ \dot{y} \end{pmatrix} = \begin{pmatrix} 0 & 1 \\ -6 & -5 \end{pmatrix} \begin{pmatrix} x \\ y \end{pmatrix}$$

with $Q = \mathbb{I}$, we can apply the following:

```
>> A = [0, 1; -6, -5];
>> Q = eye(2);
>> P = lyap(A', Q)
P = 1.1167    0.0833
    0.0833    0.1167
>> eig(P)
ans =    0.1098
         1.1236
```

A calculation by hand leads to

$$P = \begin{pmatrix} 67/60 & 1/12 \\ 1/12 & 7/60 \end{pmatrix} .$$

8.3.2 La Salle's Principle & its Implications

A Lyapunov-function can either be a strict Lyapunov-function, i.e., the orbital derivative is strictly less than zero except on x_0, or its orbital derivative can be zero everywhere. In this case, the Lyapunov-function is called a first integral (cf. Sec. 6.7) and solutions lie on the level sets of the function. Recall, a *first integral* is a function $V \in \mathcal{C}^1(\mathbb{R}^d, \mathbb{R})$ such that its orbital derivative vanishes everywhere, i.e., $V'(x) = 0$. Thus, solutions have constant values V and lie on level sets of V.

Theorem 8.4 (La Salle's Principle). *Let C be a compact, positively invariant and non-empty set and $V : C \to \mathbb{R}$ a $\mathcal{C}^1$-function such that $V'(x) \leq 0$ for all $x \in C$. Then $\omega(x) \subset \{y \in U : V'(y) = 0\}$ for all $x \in C$.*

Proof. Following [106], p. 21, let $x \in C$ and $z \in \omega(x)$. We assume that $V'(z) < 0$. Then there is a $\tau > 0$ such that $V(S_\tau) < V(z)$. Since $z \in \omega(x)$ there is a sequence $\{t_n\}$ with $t_n \to \infty$ for $n \to \infty$ such that $S_{t_n}x \to z$. We can choose the sequence such that $t_{n+1} - t_n > \tau$ for all $n \in \mathbb{N}$. We have $V(S_{t_n}x) \leq V(S_{\tau+t_n}x)$ because $V'(y) \leq 0$ for all $y \in C$. By continuity we have $S_{\tau+t_n}x \to S_\tau z$ and thus $V(S_{\tau+t_n}x) \to V(S_\tau z)$. Hence, $V(z) \leq V(S_\tau z)$ which is a contradiction. □

La Salle's principle can be used to show asymptotic stability for non-linear (autonomous or periodic) systems:

Theorem 8.5 (Asymptotic Stability Theorem). *Let V be a strict Lyapunov-function on a neighborhood N of an equilibrium x_0. More generally, let V be*

Before you continue, make sure to answer the following questions:

Quiz: Section 8.3 – Part I

Q1 Give the definition of a Lyapunov-function and illustrate the concept of a Lyapunov-function geometrically.

Q2 State Krasovskii's theorem for the existence of a Lyapunov-function.

Q3 Let V be a Lyapunov-function on a neighborhood of an equilibrium x_0. What can you say about the stability of x_0?

Q4 Sketch the proof of the assertion you used in Q3.

Q5 How can you construct a Lyapunov-function for the linear system $\dot{x} = Ax$ and what do you have to keep in mind when doing this?

a Lyapunov-function and assume that no positive orbit is contained in the set $M := \{x \in N \setminus x_0 : V'(x) = 0\}$*. Then* x_0 *is asymptotically stable.*

If moreover $K \subset N$ *is a compact positively invariant set, then* K *is contained in the basin of attraction, i.e.,* $K \subset \mathcal{A}$*.*

Proof. Following [106], p. 22, we first have that if V is a strict Lyapunov-function, then $M = \{\}$ by definition.

x_0 is stable by Theorem 8.2. There is an $\varepsilon > 0$ such that $N_\varepsilon := \{x : \text{dist}(x, x_0) < \varepsilon_0\}$ fulfills $\overline{N_\varepsilon}$ is compact and $\overline{N_\varepsilon} \subset N$. Let $\mu := \min_{x : \text{dist}(x,x_0)=\varepsilon} V(x)$. By the continuity of V and the condition that $V(x) = 0$ if and only if $x = x_0$, we have $\mu > 0$. Next, let us set $\tilde{K} := \{x \in \overline{N_\varepsilon} : V(x) \leq \mu\} \subset N$. This set is compact, positively invariant and a neighborhood of x_0. By La Salle's Principle (Theorem 8.4) we have $\omega(x) \subset \{x \in N : V'(x) = 0\} = I \dot{\cup} M$ for all $x \in K$. Assume, in order to obtain a contradiction, there is a $z \in \omega(x) \cap M$, then $\gamma^+(z) \in \omega(x) \cap M$, too, because x_0 is invariant. (We know $\gamma^+ \subset \omega(x) \subset I \dot{\cup} M$. If $S_t z = x_0$ for a $t \geq 0$, then also $z = x_0$ by the invariance — note that we are in $\mathbb{R}^d$ — and this is a contradiction to $z \in M$.) Thus, this is a contradiction to the assumption and hence $\omega(x) = x_0$.

We still have to show that $\omega(x) = x_0$ implies $S_t x \to x_0$. We know that x_0 is stable. Let us assume the opposite, i.e., there is a $\varepsilon > 0$ and there are times $t_n \to \infty$ such that $\text{dist}(S_{t_n} x, x_0) \geq 2\varepsilon$. But, since x_0 is stable, for this ε there is a $\delta > 0$ such that $\text{dist}(y, x_0) < \delta$ implies $\text{dist}(S_t y, x_0) \leq \varepsilon$ for all $t \geq 0$. Because of $\omega(x) = x_0$ there hence is a T_0 such that $S_{T_0} x \in N_\delta := \{x : \text{dist}(x, x_0) < \delta\}$ and by the stability $S_{T_0+t} x \in N_\varepsilon$ for all $t \geq 0$ which is a contradiction to the assumption.

The same argumentation holds for the compact and positively invariant set K. □

Finally, we define a uniformly unbounded function, which tends uniformly to infinity as x tends to infinity. More precisely we define:

Definition 8.9 (Uniformly Unbounded Function). A continuous function $V : N \to \mathbb{R}$ is called *uniformly bounded* on N if given any $\alpha > 0$ there is a compact set $K \subset N$, $K \neq N$, such that $V(x) \geq \alpha$ for all $x \in N \setminus K$.

For example

$$V(x, y) = y^2 + \frac{x^2}{1 + x^2}$$

is not uniformly unbounded on $\mathbb{R}^2$. Choose $\alpha = 2$ and assume there was a compact and hence bounded set $K \subset \mathbb{B}_R(0)$, $R > 0$, such that $V(x) \geq 2$ for all $x \neq K$. But this does not hold for the point $(R+1, 0)$, since $V(R+1, 0) \leq 1$ which is a contradiction to the assumption.

Theorem 8.6 (On the Basin of Attraction). *Let $V : N \to \mathbb{R}$ be an uniformly unbounded strict Lyapunov-function, where N is a neighborhood of the equilibrium x_0. Then $N \subset \mathcal{A}(x_0)$.*

Proof. Following [106], p. 24, we have that since V is uniformly unbounded on N for any $\alpha > 0$ the set $K_\alpha := \{x \in N : V(x) \leq \alpha\}$ is compact. It is positively invariant because V is a (strict) Lyapunov-function. By Theorem 8.5 we have $K_\alpha \in A(x_0)$. $N = \cup_{n \in \mathbb{N}} K_n$ and thus $N \subset A(x_0)$. □

8.4 Structural Stability & Linearisation

Lyapunov-stability is a local concept of a set or an equilibrium point. Thus, especially for equilibrium points, the linearized dynamics around them are helpful for studying the stability of complicated non-linear systems. This works if the orbits of the non-linear system and that of its linearisation (e.g. via Taylor-expansion) are topological conjugate to each other which is the case if the equilibrium is hyperbolic as we will see in the following.

8.4.1 The Principle of Linearized Stability & the Theorem of Hartman-Grobman

In order to study the stability of an equilibrium $x^* \in \mathbb{R}^d$, $d \in \mathbb{N}$, of a dynamical system, it is often useful to examine the *(local) linearized system* around x^*

Before you continue, make sure to answer the following questions:

Quiz: Section 8.3 – Part II

Q1 State La Salle's principle and sketch its proof.

Q2 Which conditions do you need to show asymptotic stability with the aid of Lyapunov-functions?

Q3 Sketch the proof of the assertion you used in Q2.

Q4 Give the definition of a uniformly unbounded function.

Q5 What can you say about the basin of attraction of a uniformly unbounded Lyapunov-function?

Q6 Sketch the proof of the assertion you used in Q5.

(this is called *Lyapunov's first method*). A non-linear system is linearized as follows: Let us consider the dynamical system which comes from the ordinary differential equation $\dot{x} = f(x)$, $f \in \mathcal{C}^2(\mathbb{R}^d, \mathbb{R}^d)$. First, on a neighborhood $\mathcal{U}$ of x^* a transformation $T : \mathcal{U} \subset \mathbb{R}^d \to \mathbb{R}^d$ is defined as $T(x) := \tilde{x} := x - x^*$ (hence the equilibrium is mapped into the origin). Second, the function f is changed correspondingly to a function $\tilde{f}$ of $\tilde{x}$. Third, a first order Taylor expansion is performed on the thus transformed right-hand side $\tilde{f}$ with the new equilibrium $\tilde{x}^* = 0$. Fourth, the higher order terms are neglected. This procedure leads to the linear system

$$\dot{x} = f(x) \qquad \rightsquigarrow \qquad \dot{\tilde{x}} = A\tilde{x} \,,$$

where A is the derivative $\mathrm{D}\tilde{f}(0)$ of $\tilde{f}$ with respect to the transformed space variables $\tilde{x}$. Let us now concentrate on linear systems (and suppress the tilde-symbol).

Since the solutions of the linear initial value problem

$$\dot{x} = Ax \qquad \text{with initial conditions} \qquad x(0) = x_0$$

are given by $x(t) = e^{At}x_0$, the long-time behavior of the solutions (or, more precisely, its components) is determined by the eigenvalues of $A \in \mathbb{R}^{d\times d}$ (these eigenvalues are often called *Lyapunov exponents* of the equilibrium). We can conclude the stability of the equilibrium $x^* = 0$ (by applying these long-time behavior considerations on solutions in a suitable neighborhood of $x^* = 0$):

Theorem 8.7 (Stability of the Origin in Linear Systems). *Consider the dynamical system induced by the linear ordinary differential equation $\dot{x} = Ax$, $A \in \mathbb{R}^{d\times d}$.*

1. *If Re$(\lambda) > 0$ for an eigenvalue λ of A or if Re$(\lambda) = 0$ for an eigenvalue λ of A with higher algebraic than geometric multiplicity, then the origin is unstable.*

2. *If all eigenvalues λ of A have non-positive real part (i.e., Re$(\lambda) \leq 0$) and all eigenvalues with Re$(\lambda) = 0$ are semi-simple[8], then the origin is stable.*

3. *If Re$(\lambda) < 0$ for all eigenvalues λ of A, then the origin is asymptotically stable.*

Proof. See [106], Sec. 2.1, or [253, p. 74]. □

In particular, a suitable neighborhood of the origin can be divided into three invariant subspaces, which are the span of all generalized eigenvectors

8 ..., i.e., their algebraic and geometric multiplicity are equal.

to the eigenvalues λ of A such that their real parts share the same sign: the *unstable subspace* is spanned by those with $\text{Re}(\lambda) > 0$, the *stable subspace* by those with $\text{Re}(\lambda) < 0$ and the *center subspace* by those with $\text{Re}(\lambda) = 0$.

Figure 8.5 sketches the various types of dynamics of planar systems with respect to the stability properties of the equilibria therein, see [188], p. 140.

Finally, we can extend these concepts to non-linear systems, at least locally, and only for those equilibria for which the matrix of the linearized system has no eigenvalues with vanishing real part (i.e., no center subspace is present)[9].

Definition 8.10 (Hyperbolic Equilibrium). Let x^* be an equilibrium of the ordinary differential equation $\dot{x} = f(x)$, $f \in \mathcal{C}^1(\mathbb{R}^d, \mathbb{R}^d)$. Then x^* is called *hyperbolic*, if $\text{Re}(\lambda) \neq 0$ for all eigenvalues λ of $\text{D}f(x^*)$. If not, x^* is called *non-hyperbolic*.

The next theorem tells us that locally around hyperbolic equilibria, the dynamics of the linearized and non-linear system are equivalent.

Theorem 8.8 (Hartman-Grobman's Theorem of Linearized Stability). *Consider the dynamical system given by the autonomous differential equation* $\dot{x} = f(x)$ *on* $\mathbb{R}^d$ *with* $f \in \mathcal{C}^2(\mathbb{R}^d, \mathbb{R}^d)$ *and* $x^* \in \mathbb{R}^d$ *being an equilibrium point of this system. Let* $\Phi(t; x)$ *denote the flow of this system and* $\Psi(t; x)$ *the flow of the linearized system* $\dot{x} = Df(x^*)(x - x^*)$. *If* x^* *is hyperbolic, then there is an open subset* $\mathcal{U} \subset \mathbb{R}^d$ *such that* $x^* \in \mathcal{U}$, *and a homeomorphism*[10] $h : \mathcal{U} \to h(\mathcal{U}) \subset \mathbb{R}^n$ *such that* $\Psi(t; h(x)) = h(\Phi(t; x))$ *whenever* $x \in \mathcal{U}$.

Proof. See [106], Sec. 2.3, or [56, pp. 305–315]. □

This local stability conserving relationship between linear and non-linear systems allows us to extend the concepts of the stable, unstable and center subspace to non-linear dynamical systems. The corresponding *(local) stable, unstable* and *center manifold* of the non-linear system are invariant and tangent to their counterpart subspaces. The stable and unstable manifolds are uniquely determined and have the same dimension as their counterparts[11].

Though Hartman-Grobman's theorem of linearized stability does not recommend anything for the non-hyperbolic case, i.e., the case of eigenvalues with vanishing real part. One can use center manifold methods[12] or Lyapunov-functions (Lyapunov's second method) to gain more information.

---

9 A more general approach is given by so-called normal forms where not only the linear approximation of the nonlinear system is considered but also higher order approximations, see, e.g., [194].

10 Recall: A *homeomorphism* is a bijective continuous mapping with continuous inverse.

11 ..., see, e.g. [106], Sec. 2.3, or [244], p. 159 Th. 2.6.5 and p. 160 Th. 2.6.6.

12 ...see [106], Sec. 4.1, or [6, pp. 56], e.g.,

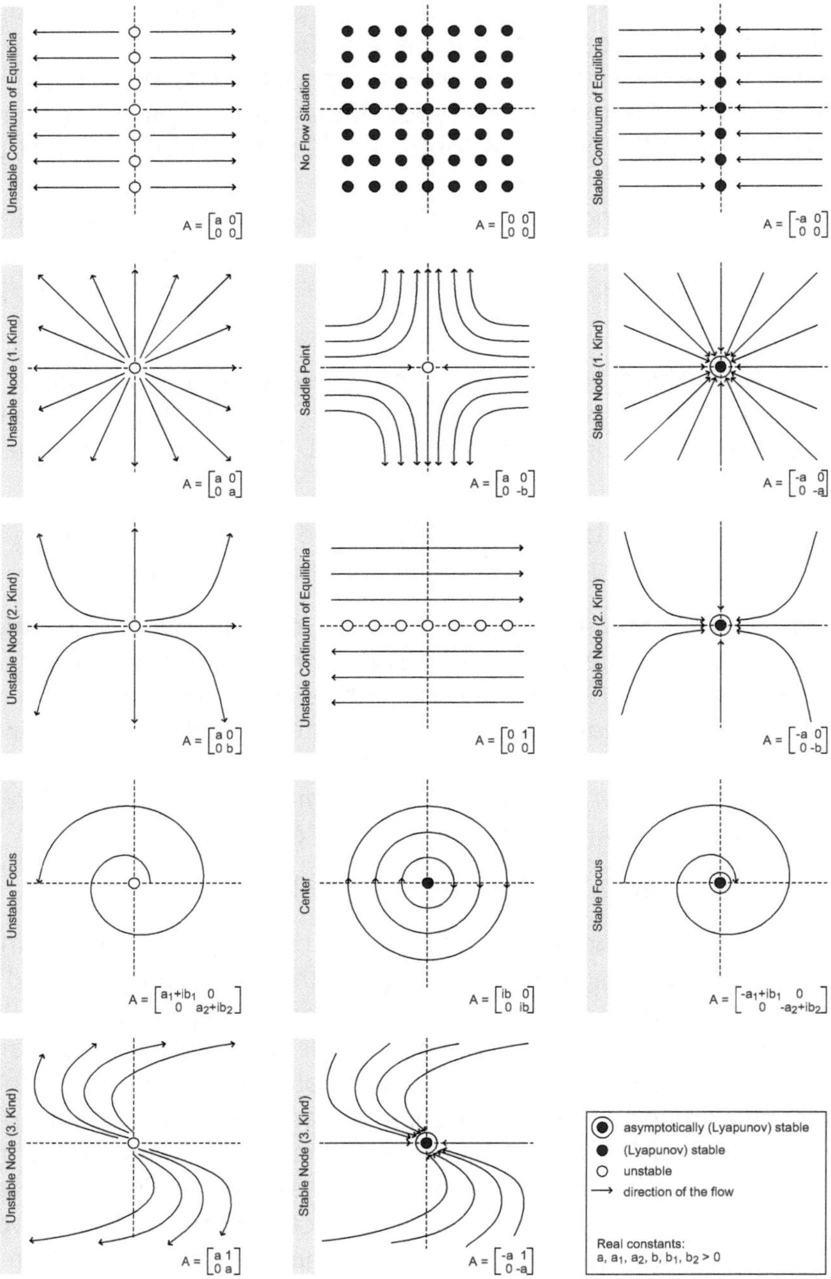

Figure 8.5. Graphical classification of equilibria in planar systems with respect to their stability properties, see [188], p. 140.

In the remainder of this section we discuss two methods that allow us to decide if the spectrum $\sigma(A)$ of a $d \times d$-matrix A lies in the left half-plane, i.e., if the Lyapunov-exponents of A lead to (linear) stability. These methods are known as the Routh-Hurwitz stability criterion and the application of the Lozinskii-measure.

8.4.2 The Routh-Hurwitz Stability Criterion

We follow [102], pp. 172, for the outline of the Routh-Hurwitz stability criterion (see [102], pp. 190) and its improvement, the Lienard-Chipart stability criterion (see [102], pp. 220).

For the derivation of the Routh-Hurwitz criterion, we consider the polynomial

$$f(z) = a_0 z^n + b_0 z^{n-1} + a_1 z^{n-2} + b_1 z^{n-3} + \ldots, \qquad (a_0 \neq 0),$$

Before you continue, make sure to answer the following questions:

Quiz: Section 8.4 – Part I

Q1 What can you say about the stability of the origin at linear systems with respect to the sign of the real parts of the eigenvalues of the system matrix?

Q2 Give some representative phase space plots for the different types of stability properties/ combinations that can occur in planar systems.

Q3 What is a hyperbolic equilibrium?

Q4 State the theorem of Hartman and Grobman.

Q5 Determine the stability of the null-solution of the damped non-linear pendulum

$$\dot{x} = y, \qquad \text{and} \qquad \dot{y} = -\sin(x) - 0.1y.$$

with real coefficients $a_0, b_0, \ldots$. The following square matrix of order n is called *Hurwitz matrix*

$$H := \begin{pmatrix} b_0 & b_1 & b_2 & \ldots & b_{n-1} \\ a_0 & a_1 & a_2 & \ldots & a_{n-1} \\ 0 & b_0 & b_1 & \ldots & b_{n-2} \\ 0 & a_0 & a_1 & \ldots & a_{n-2} \\ 0 & 0 & b_0 & \ldots & b_{n-3} \\ & \vdots & & \ddots & \vdots \end{pmatrix}, \qquad \text{where} \quad \begin{cases} a_k = 0 & \text{if} \quad k > \left[\frac{n}{2}\right] \\ b_k = 0 & \text{if} \quad k > \left[\frac{n-1}{2}\right] \end{cases}.$$

We will transform the Hurwitz matrix successively to upper triangular form by Gaussian elimination. We confine our attention to the regular case, where $b_0 \neq 0$, $c_0 := a_1 - a_0 b_1 b_0^{-1} \neq 0$, $d_0 := b_1 - b_0 c_1 c_0^{-1} \neq 0$, etc. First, after multiplying the odd-numbered rows by $a_0 b_0^{-1}$ we subtract these from the corresponding rows of even number. Thus, the rows of even number have entries $c_0, c_1, \ldots$, where $c_k = 0$ if $k > \left[\frac{n}{2}\right] - 1$. Next, after multiplying the rows of even number by $b_0 c_0^{-1}$ we subtract these from the corresponding rows of odd number. This leads to

$$H \rightsquigarrow \begin{pmatrix} b_0 & b_1 & b_2 & \ldots & b_{n-1} \\ 0 & c_0 & c_1 & \ldots & c_{n-2} \\ 0 & b_0 & b_1 & \ldots & b_{n-2} \\ 0 & 0 & c_0 & \ldots & c_{n-3} \\ 0 & 0 & b_0 & \ldots & b_{n-3} \\ & \vdots & & \ddots & \vdots \end{pmatrix} \rightsquigarrow \begin{pmatrix} b_0 & b_1 & b_2 & \ldots & b_{n-1} \\ 0 & c_0 & c_1 & \ldots & c_{n-2} \\ 0 & 0 & d_0 & \ldots & d_{n-3} \\ 0 & 0 & c_0 & \ldots & c_{n-3} \\ 0 & 0 & d_0 & \ldots & d_{n-4} \\ & \vdots & & \ddots & \vdots \end{pmatrix}.$$

Applying this procedure further, we ultimately arrive at an upper triangular nth order matrix

$$R := \begin{pmatrix} b_0 & b_1 & b_2 & \ldots \\ 0 & c_0 & c_1 & \ldots \\ 0 & 0 & d_0 & \ldots \\ \vdots & & & \ddots \end{pmatrix},$$

which we shall call the *Routh matrix*.

Definition 8.11 (Equivalence of Matrices by Minors). Two $n \times n$-matrices A and B are said to be *equivalent* if and only if for any $1 \leq k \leq n$ the corresponding kth order minors on the first k rows of these matrices are equal, i.e. (in the notation of permutations)

$$A \begin{bmatrix} 1 & 2 & \ldots & k \\ i_1 & i_2 & \ldots & i_k \end{bmatrix} = B \begin{bmatrix} 1 & 2 & \ldots & k \\ i_1 & i_2 & \ldots & i_k \end{bmatrix},$$

for $1 \leq i_k \leq n$ and $k = 1, 2, \ldots, n$.

Since, the kth order minors on the first k rows ($k = 1, 2, \dots, n$) do not change their values if one subtracts from any rows another row multiplied by an arbitrary constant, the Hurwitz and Routh matrices are equivalent:

$$H\begin{bmatrix} 1 & 2 & \dots & k \\ i_1 & i_2 & \dots & i_k \end{bmatrix} = R\begin{bmatrix} 1 & 2 & \dots & k \\ i_1 & i_2 & \dots & i_k \end{bmatrix},$$

for $1 \le i_k \le n$ and $k = 1, 2, \dots, n$.

The equivalence of the Hurwitz and Routh matrices permits us to express all of the elements of R in terms of the minors of the Hurwitz matrix H, and therefore in terms of coefficients of the given polynomial $f(z)$. We obtain

$$\begin{array}{llllll} H\begin{bmatrix} 1 \\ 1 \end{bmatrix} = b_0 , & H\begin{bmatrix} 1 & 2 \\ 1 & 2 \end{bmatrix} = b_0 c_0 , & H\begin{bmatrix} 1 & 2 & 3 \\ 1 & 2 & 3 \end{bmatrix} = b_0 c_0 d_0 , \\ H\begin{bmatrix} 1 \\ 2 \end{bmatrix} = b_1 , & H\begin{bmatrix} 1 & 2 \\ 1 & 3 \end{bmatrix} = b_0 c_1 , & H\begin{bmatrix} 1 & 2 & 3 \\ 1 & 2 & 4 \end{bmatrix} = b_0 c_0 d_1 , \\ H\begin{bmatrix} 1 \\ 3 \end{bmatrix} = b_2 , & H\begin{bmatrix} 1 & 2 \\ 1 & 4 \end{bmatrix} = b_0 c_2 , & H\begin{bmatrix} 1 & 2 & 3 \\ 1 & 2 & 5 \end{bmatrix} = b_0 c_0 d_2 , \\ \vdots & \vdots & \vdots \quad . \end{array}$$

The successive principal minors of the Hurwitz matrix H are usually called the *Hurwitz determinants*, and denoted by

$$\Delta_1 = H\begin{bmatrix} 1 \\ 1 \end{bmatrix} = b_0 ,$$

$$\Delta_2 = H\begin{bmatrix} 1 & 2 \\ 1 & 2 \end{bmatrix} = \det\begin{pmatrix} b_0 & b_1 \\ a_0 & a_1 \end{pmatrix} = a_1 b_0 - a_0 b_1 = b_0 c_0 ,$$

and so on to

$$\Delta_n = H\begin{bmatrix} 1 & 2 & \dots & n \\ 1 & 2 & \dots & n \end{bmatrix} = \det\begin{pmatrix} b_0 & b_1 & b_2 & \dots & b_{n-1} \\ a_0 & a_1 & a_2 & \dots & a_{n-1} \\ 0 & b_0 & b_1 & \dots & b_{n-2} \\ 0 & a_0 & a_1 & \dots & a_{n-2} \\ & \vdots & & \ddots & \end{pmatrix} .$$

In the regular case all of the quantities $b_0, c_0, d_0, \dots$ are defined and different from zero. This is characterised by the inequalities

$$\Delta_1 = b_0 \neq 0, \quad \Delta_2 = b_0 c_0 \neq 0, \quad \Delta_3 = b_0 c_0 d_0 \neq 0, \quad \dots \quad \Delta_n \neq 0 .$$

In particular,

$$b_0 = \Delta_1 , \quad c_0 = \frac{\Delta_2}{\Delta_1} , \quad d_0 = \frac{\Delta_3}{\Delta_2} , \quad \dots .$$

Theorem 8.9 (Routh-Hurwitz Theorem). *In the regular case the number k of roots of the real polynomial $f(z) = a_0 z^n + \ldots$ which lie in the right half-plane is given by the formula*

$$k = V\left(a_0, \Delta_1, \frac{\Delta_2}{\Delta_1}, \frac{\Delta_3}{\Delta_2}, \ldots, \frac{\Delta_n}{\Delta_{n-1}}\right)$$

or equivalently

$$k = V(a_0, \Delta_1, \Delta_3, \ldots) + V(1, \Delta_2, \Delta_4, \ldots),$$

where $V(x_1, \ldots, x_n)$ denotes the number of sign changes in the sequence $x_1, \ldots, x_n$.

Proof. The proof is based in Routh's criterion, cf. [102], pp. 172. □

The special case in which all roots of $f(z)$ lie in the left have-plane $\mathrm{Re}(z) < 0$ follows from the Routh-Hurwitz Theorem for $k = 0$.

Corollary 8.12 (Routh-Hurwitz Stability Criterion). *In the regular case, in order for all roots of the real polynomial $f(z) = a_0 z^n + \ldots$ ($a_0 \neq 0$) to have negative real parts, it is necessary and sufficient that the inequalities*

$$a_0\Delta_1 > 0,\ \Delta_2 > 0,\ a_0\Delta_3 > 0,\ \Delta_4 > 0,\ \ldots \begin{cases} a_0\Delta_n > 0 & \text{for } n \text{ odd} \\ \Delta_n > 0 & \text{for } n \text{ even} \end{cases} \tag{8.2}$$

hold. If $a_0 > 0$ these conditions reduce to $\Delta_i > 0$ for $i = 1, 2, \ldots, n$.

A real polynomial $f(z)$ the coefficients of which satisfy the conditions in (8.2) is a real polynomial whose roots all have negative real parts, and is called *Hurwitz polynomial.*

If we introduce the usual notation for the coefficients of the polynomial

$$f(z) = \alpha_n z^n + \alpha_{n-1} z^{n-1} + \alpha_{n-2} z^{n-2} + \ldots + \alpha_0, \qquad \text{with } \alpha_n > 0,$$

then the Routh-Hurwitz conditions my be re-written in the regular case as the following determinantal inequalities

$$\Delta_1 = \alpha_{n-1} > 0, \quad \Delta_2 = \begin{pmatrix} \alpha_{n-1} & \alpha_{n-3} \\ \alpha_n & \alpha_{n-2} \end{pmatrix} > 0,$$

$$\Delta_3 = \begin{pmatrix} \alpha_{n-1} & \alpha_{n-3} & \alpha_{n-5} \\ \alpha_n & a_{n-2} & \alpha_{n-4} \\ 0 & \alpha_{n-1} & \alpha_{n-3} \end{pmatrix} > 0,$$

and so on to

$$\Delta_n = \begin{pmatrix} \alpha_{n-1} & \alpha_{n-3} & \alpha_{n-5} & \cdots \\ \alpha_n & a_{n-2} & \alpha_{n-4} & \cdots \\ 0 & \alpha_{n-1} & \alpha_{n-3} & \cdots \\ & & & \ddots \end{pmatrix} > 0 .$$

If the determinant conditions $\Delta_i > 0$ $(i = 1, 2, \ldots, n)$ are satisfied, then the polynomial $f(z)$ may be written as a product of α_n by factors of the form $z + \gamma_2$ and $z^2 + \gamma_1 z + \gamma_0$ with $\beta_0, \beta_1, \beta_2 > 0$. Thus, all coefficients of $f(z)$ must be positive, i.e. $\alpha_i > 0$ $(i = 1, 2, \ldots, n)$.

In contrast to the determinant conditions $\Delta_i > 0$ $(i = 1, 2, \ldots, n)$, the coefficient conditions $\alpha_i > 0$ $(i = 1, 2, \ldots, n)$ are necessary but not sufficient for all the roots of $f(z)$ to lie in the left half-plane. However, once the coefficient conditions $a_i > 0$ $(i = 1, 2, \ldots, n)$ are satisfied the determinant conditions $\Delta_i > 0$ $(i = 1, 2, \ldots, n)$ are no longer independent. For instance, when $n = 4$ the Routh-Hurwitz conditions reduce to the single inequality $\Delta_3 > 0$, when $n = 5$ to the pair of inequalities $\Delta_2 > 0$ and $\Delta_4 > 0$, and when $n = 6$ to the pair of inequalities $\Delta_3 > 0$ together with $\Delta_5 > 0$.

Following [102], pp. 221, this circumstance was studied by the French mathematicians Lienard and Chipart and led them, in 1914, to the discovery of stability criteria different from the Routh-Hurwitz criterion.

Theorem 8.10 (The Lienard-Chipart Stability Criteria). *Necessary and sufficient conditions for the real polynomial $f(z) = \alpha_n z^n + \alpha_{n-1} z^{n-1} + \alpha_{n-2} z^{n-2} + \cdots + \alpha_0$ with $\alpha_n > 0$ to have only roots with negative real parts may be expressed in any of the four following forms*

1. $\alpha_0, \alpha_2, \alpha_4, \cdots > 0$ *and* $\Delta_1, \Delta_3, \cdots > 0$,
2. $\alpha_0, \alpha_2, \alpha_4, \cdots > 0$ *and* $\Delta_2, \Delta_4, \cdots > 0$,
3. $\alpha_0, \alpha_1, \alpha_3, \cdots > 0$ *and* $\Delta_1, \Delta_3, \cdots > 0$, *or*
4. $\alpha_0, \alpha_1, \alpha_3, \cdots > 0$ *and* $\Delta_2, \Delta_4, \cdots > 0$.

Proof. An elegant proof is based on the so-called Cauchy index and Hankel sequences, see [102], pp. 221. □

Clearly, these four Lienard-Chipart stability criteria have an advantage over the Routh-Hurwitz conditions in that they involve about half as many determinant inequalities.

From Theorem 8.10 it follows that for a real polynomial in which all coefficients (or even only some of them, namely $\alpha_0, \alpha_2, \alpha_4, \ldots$ or $\alpha_0, \alpha_1, \alpha_3, \ldots$) are positive, the Routh-Hurwitz determinant inequalities $\Delta_i > 0$ $(i = 1, 2, \ldots, n)$

are not independent. In particular: the positivity of the Hurwitz determinants of odd order implies that of the Hurwitz determinants of even order, and conversely.

8.4.3 The Lozinskii-Measure and Stability

In this exposition on the Lozinskii-Measure we follow [174]: Let $\mathbb{K}^{d\times d}$ be the linear space of $d \times d$-matrices with real entries ($\mathbb{K} = \mathbb{R}$) or complex entries $\mathbb{K} = \mathbb{C}$. Every $A \in \mathbb{K}^{d\times d}$ will be identified with the linear operator on $\mathbb{K}^d$ it represents and let $\wedge$ denote the exterior product on $\mathbb{K}^d$. For an integer $1 \le k \le d$ the *kth additive compound matrix* $A^{[k]}$ of A is a linear operator on the kth exterior space $\wedge^k\mathbb{K}^d$ with respect to the canonical basis of this kth exterior product space $\wedge^k\mathbb{K}^d$. The kth additive compound matrix is given as

$$A^{[k]}(u_1 \wedge \cdots \wedge u_k) = \sum_{i=1}^{k} u_1 \wedge \cdots \wedge Au_i \wedge \dots u_k\,,$$

where $u_1 \wedge \cdots \wedge u_k$ is a decomposable element. By linear extension, $A^{[k]}$ is defined over the whole $\wedge^k\mathbb{K}^d$.

The entries of $A^{[k]}$ are linear relations of those of $A = (a_{ij})$. For any integer $i = 1, \dots, \binom{d}{k}$, let $(i) = (i_1, \dots, i_k)$ be the ith member in the lexicographic ordering of integer k-tuples such that $1 \le i_1 < \cdots < i_k \le d$. Then the entry in the ith row and the jth column of $Z = (z_{ij}) = A^{[k]}$ is

$$z_{ij} = \begin{cases} a_{i_1,i_1} + \cdots + a_{i_k,i_k}\,, & \text{if } (i) = (j)\,, \\ (-1)^{r+s} a_{j_r,i_s}\,, & \text{if exactly one entry } i_s \text{ of } (i) \text{ does not occur} \\ & \text{in } (j) \text{ and } j_r \text{ does not occur in } (i)\,, \\ 0\,, & \text{if } (i) \text{ differs from } (j) \text{ in two or more entries}\,. \end{cases}$$

As special cases, we have $A^{[1]} = A$ and $A^{[d]} = \mathrm{tr}(A)$. The second (additive) compound matrix $A_d^{[2]}$ for an $d \times d$-matrix $A = (a_{ij})$ is given in Table 8.1 for $d = 2, 3, 4$.

For the purpose of connecting our results on compound matrices to the stability of the original matrix a spectral property of $A^{[k]}$ is required: Let $\sigma(A) = \{\lambda_i : i = 1, \dots, d\}$, then the spectrum of $A^{[k]}$ reads as $\sigma(A^{[k]}) = \{\lambda_{i_1} + \cdots + \lambda_{i_k} : 1 \le i_1 < \cdots < i_k \le d\}$.

Let $\|\cdot\|$ denote a vector norm on $\mathbb{K}^d$ as well as the corresponding operator norm it induces on $\mathbb{K}^{d\times d}$.

Definition 8.13 (The Lozinskii-Measure). The *Lozinskii-Measure* (or *logarithmic norm*) μ on $\mathbb{K}^{d\times d}$ with respect to $\|\cdot\|$

$$\mu(A) = \lim_{h\to 0+} \frac{\|\mathbb{I} + hA\| - 1}{h}\,, \qquad (8.3)$$

$$A_2^{[2]} = a_{11} + a_{22} = \mathrm{tr}(A)$$

$$A_3^{[2]} = \begin{pmatrix} a_{11}+a_{22} & a_{23} & -a_{13} \\ a_{32} & a_{11}+a_{33} & a_{12} \\ -a_{31} & a_{21} & a_{22}+a_{33} \end{pmatrix}$$

$$A_4^{[2]} = \begin{pmatrix} a_{11}+a_{22} & a_{23} & a_{24} & -a_{13} & -a_{14} & 0 \\ a_{32} & a_{11}+a_{33} & a_{34} & a_{12} & 0 & -a_{14} \\ a_{42} & a_{43} & a_{11}+a_{44} & 0 & a_{12} & a_{13} \\ -a_{31} & a_{21} & 0 & a_{22}+a_{33} & a_{34} & -a_{24} \\ -a_{41} & 0 & a_{21} & a_{43} & a_{22}+a_{44} & a_{23} \\ 0 & -a_{41} & a_{31} & -a_{42} & a_{32} & a_{33}+a_{44} \end{pmatrix}$$

Table 8.1. The second (additive) compound matrix $A_d^{[2]}$ for an $d \times d$-matrix $A = (a_{ij})$ for $d = 2, 3, 4$, cf. [174], p. 263.

for $A \in \mathbb{K}^{d\times d}$.

A Lozinskii-measure $\mu(A)$ dominates the stability modulus $s(A)$ as the following lemma states:

Lemma 8.14 (A Lozinskii-Measure Dominates the Stability Modulus). *Let $A \in \mathbb{K}^{d\times d}$, $|\cdot|$ be a norm on $\mathbb{K}^{d\times d}$, and μ a Lozinskii-measure. Then $s(A) \le \mu(A) \le |A|$.*

Proof. See [66] □

The Lozinskii-measure of $A = (a_{ij})$ with respect to the three most common vector norms $\|x\|_\infty = \sup_i\{|x_i| : i = 1, \dots d\}$, $\|x\|_1 = \sum_{i=1}^d |x_i|$ and $\|x\|_2 = \left(\sum_{i=1}^d (x_i)^2\right)^{1/2}$ on $\mathbb{K}^d$ are, cf. [66],

$$\mu_\infty(A) = \sup_{i=1,\dots,d} \left\{ \mathrm{Re}(a_{ii}) + \sum_{k=1,\, k\neq i}^{d} |a_{ik}| \right\},$$

$$\mu_1(A) = \sup_{k=1,\dots,d} \left\{ \mathrm{Re}(a_{kk}) + \sum_{i=1,\, i\neq k}^{d} |a_{ik}| \right\},$$

and

$$\mu_2(A) = s\left(\tfrac{1}{2}\left(A + A^*\right)\right),$$

respectively, where A^* denotes the Hermitian of A. If A is real symmetric, then $\mu_2(A) = s(A)$. In particular, for a real matrix A, conditions $\mu_\infty(A) < 0$ or $\mu_1(A) < 0$ can be interpreted as $a_{ii} < 0$ for $i = 1, \ldots, d$.

Let $P \in \mathbb{K}^{d\times d}$ be invertible. We define the norm $\|\cdot\|_P$ via $\|x\|_P := \|Px\|$ and denote the corresponding Lozinskii-measure by μ_P. Then we obtain immediately from the definition of the Lozinskii-measure:

Lemma 8.15. *Let $P \in \mathbb{K}^{d\times d}$ be invertible. Then $\mu_P(A) = \mu(PMP^{-1})$ holds for any Lozinskii-measure μ.*

This allows us to interpret the modulus of stability as an infimum of over all Lozinskii-measures.

Proposition 8.16. *For any $A \in \mathbb{K}^{d\times d}$ it holds that*

$$s(A) \;=\; \inf\left\{\mu(A) \;:\; \mu \text{ is a Lozinskii-measure on } \mathbb{K}^{d\times d}\right\}. \qquad (8.4)$$

Proof. Following [174], p. 252, we first prove the case when $\mathbb{K} = \mathbb{C}$. The relation (8.4) obviously holds for diagonalisable matrices by Lemma 8.15 and the Lozinskii-measures μ_∞ and μ_1. Furthermore, the infimum in (8.4) can be achieved if a A is diagonalisable. The general case can be shown based on this observation, the fact that any A can be approximated by diagonalisable matrices in $\mathbb{C}^{d\times d}$, and the continuity of $\mu(\cdot)$, which is implied by the property $|\mu(A) - \mu(B)| \le \|A - B\|$.

Next, let $A \in \mathbb{R}^{d\times d}$, then $s(A) = \inf\{\mu_{\mathbb{C}}(A)\}$, where $\mu_{\mathbb{C}}$ are Lozinskii-measures with respect to the vector norms $\|\cdot\|_{\mathbb{C}}$ on $\mathbb{C}^d$. When restricted to $\mathbb{R}^d$, each $\|\cdot\|_{\mathbb{C}}$ induces a vector norm $\|\cdot\|_{\mathbb{R}}$ on $\mathbb{R}^d$. Let $\mu_{\mathbb{R}}$ be the corresponding Lozinskii-measures. Then $\mu_{\mathbb{C}}(A) \ge \mu_{\mathbb{R}}(A)$ holds for $A \in \mathbb{R}^{d\times d}$, since $\|A\|_{\mathbb{C}} \ge \|A\|_{\mathbb{R}}$ holds for the induced norms. This shows that (8.4) is valid for the case when $\mathbb{K} = \mathbb{R}$, too. □

For the proof of Proposition 8.16 we see that if $\mathbb{K} = \mathbb{C}$, then

$$s(A) \;=\; \inf\left\{\mu_\infty\left\{PMP^{-1}\right) \;:\; P \in \mathrm{GL}(d, \mathbb{C})\right\},$$

where $\mathrm{GL}(d, \mathbb{C})$ is the group of invertible complex matrices. The same relations holds if μ_∞ is replaced by μ_1. Similar relations do not hold for $\mathbb{K} = \mathbb{R}$.

Finally, we have

Corollary 8.17 (Equivalence of a Negative Lozinskii-Measure and the Negative Stability Modulus). *Let $A \in \mathbb{K}^{d\times d}$. Then $s(A) < 0 \Leftrightarrow \mu(A) < 0$ for some Lozinskii-measure μ on $\mathbb{K}^{d\times d}$.*

Let $A \in \mathbb{R}^{d\times d}$ in the following.

Theorem 8.11. *For $s(A) < 0$ it is sufficient and necessary that $s(A^{[2]}) < 0$ and $(-1)^d \det(A) > 0$.*

Proof. Following [174], p. 253, we have that by the spectral property of $A^{[2]}$, the condition $s(A^{[2]}) < 0$ implies that at most one of the eigenvalues of A can have a non-negative real part. We may thus assume that all of the eigenvalues are real. It is then simple to see that the existence of one and only one non-negative eigenvalue is precluded by the condition $(-1)^d \det(A) > 0$. □

Theorem 8.11 and Corollary 8.17 lead to the following result.

Theorem 8.12 (Stability & the Lozinskii-Measure). *Let $(-1)^d \det(A) > 0$. Then A is stable if and only if $\mu(A^{[2]}) < 0$ for some Lozinskii-measure μ on $\mathbb{R}^{N\times N}$ with $N = \binom{d}{2}$.*

Example 8.18 (Stability of a Parameter-Dependent Matrix, see [174], p. 253). We show that the 3×3-matrix

$$A(t) = \begin{pmatrix} -1 & -t^2 & -1 \\ t & -t-1 & t \\ t^2 & 1 & -t^2-1 \end{pmatrix}$$

is stable for all $t > 0$. From Table 8.1 we read off the second (additive) compound matrix

$$A^{[2]}(t) = \begin{pmatrix} -2-t & -t & 1 \\ 1 & -2-t^2 & -t^2 \\ -t^2 & t & -2-t-t^2 \end{pmatrix}.$$

$A^{[2]}(t)$ is diagonally dominant in its rows. Hence, let μ be the Lozinskii-measure with respect to the norm $\|x\|_\infty = \sup\{|x_1|, |x_2|, |x_3|\}$. Then $\mu(A^{[2]}(t)) = -1 < 0$. Moreover, $\det(A(t)) = -2t^5 - 3t^3 - 2t^2 - t - 1 < 0$ for $t > 0$. The stability of $A(t)$ follows from Theorem 8.12.

8.5 Chapter's Summary

We started our review on deterministic dynamical systems by introducing the fundamental notions and concepts of (continuous) dynamical systems, long-time behavior, invariance and attraction. This paved the way towards analysing stability in the sense of Lyapunov. First, the use of Lyapunov-functions for proving (asymptotic) stability in non-linear systems was studied. Next, we analyzed the correspondence between the stability properties of non-linear systems and their linearisations. Here, we gave the theorem of

Hartman and Grobman, a classification of equilibria in planar systems with respect to their stability properties, as well as the techniques for the determination of the position of Lyapunov exponents of a linear system, like the Routh-Hurwitz criterion of the Lozinskii-measure method.

Problems

Classification: ☼ easy, ◷ easy with longer calculations, ☆ a little bit difficult, ▣ challenging.

Exercise 8.19. [☼] On Invariance

Show the assertion stated in Definition 8.6, i.e., that for positively invariant sets B, their closure $\overline{B}$ and interior $int(B)$ are positively invariant, too.

Exercise 8.20. [☼] Stability via Lyapunov-Functions

1. Show that the zero solution of the system

$$\dot{x} = -x - xy^2\,, \qquad \dot{y} = -y - x^2y\,,$$

is globally asymptotically stable, by guessing a suitable Lyapunov-function.

2. Investigate the stability of the zero solution of

$$\dot{x} = -xy - x\,, \qquad \dot{y} = y^3 - xy^3 + xy - y\,,$$

by using the function $V(x, y) = -x - \ln(1 - x) - y - \ln(1 - y)$ locally around $(x, y) = (0, 0)$.

Before you continue, make sure to answer the following questions:

Quiz: Section 8.4 – Part II

Q1 State the Routh-Hurwitz stability criterion.

Q2 Give the definition of the Lozinskii-measure.

Q3 What can you say about stability of the null-solution using the Lozinskii-measure.

Q4 Sketch the proof of the assertion you used in Q3.

Exercise 8.21. [☼]-[⊕] **What Really Happened at the Paris Peace Talks**
The original plan developed by Henry Kissinger and Le Duc Tho to settle the Vietnamese war is described below. It was agreed that 1 million South Vietnamese ants and 1 million North Vietnamese ants would be placed in the backyard of the Presidential palace in Paris and be allowed to fight it out for a long period of time. If the South Vietnamese ants destroyed nearly all the North Vietnamese ants, then South Vietnam would retain control of all of its land. If the North Vietnamese ants were victorious, then North Vietnam would take over all of South Vietnam. If they appeared to be fighting to a standoff then South Vietnam would be partitioned according to the proportion of ants remaining. Now, the South Vietnamese ants, denoted by S, and the North Vietnamese Ants, denoted by N, compete against each other according to the following differential equations:

$$\frac{dS}{dt} = \frac{1}{10}S - \frac{1}{20}S \times N \tag{8.5}$$
$$\frac{dN}{dt} = \frac{1}{100}N - \frac{1}{100}N^2 - \frac{1}{100}S \times N \tag{8.6}$$

Note that these equations correspond to reality since South Vietnamese ants multiply much more rapidly than the North Vietnamese ants, but the North Vietnamese ants are much better fighters. The battle began at 10:00 sharp on the morning of May 19, 1972, and was supervised by a representative of Poland and a representative of Canada. At 2:43 p.m. on the afternoon May 21, the representative of Poland , being unhappy with the progress of the battle, slipped a bag of North Vietnamese ants into the backyard, but he was spotted by the eagle eyes of the representative of Canada. The South Vietnamese immediately claimed a foul and called off the agreement, thus setting the stage for the protracted talks that followed in Paris. The representative of Poland was hauled before a judge in Paris for sentencing. The judge, after making some remarks about the stupidity of the South Vietnamese gave the Polish representative a very light sentence. Justify mathematically the judges's decision.

Hint:

1. Show that the lines $N = 2$ and $N + S = 1$ divide the first quadrant into three regions (see Fig. 8.6) in which dS/dt and dN/dt have fixed signs.

2. Show that every solution $S(t)$, $N(t)$ of (8.5)-(8.6) which start in region I or region III must eventually enter region II.

3. Show that every solution $S(t)$, $N(t)$ of (8.5)-(8.6) which start in region II must remain there for all future time.

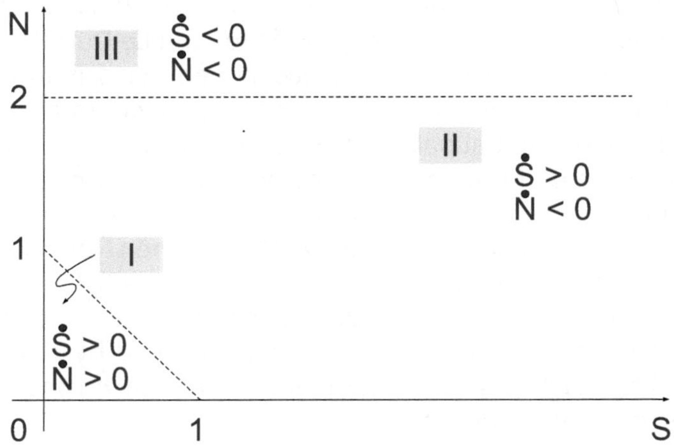

Figure 8.6. Sketch of the phase space corresponding to problem 8.21.

4. Conclude from 3 that $S(t) \to \infty$ for all solutions $S(t)$, $N(t)$ of (8.5)-(8.6) with $S(t_0)$ and $N(t_0)$ positive. Conclude, too, that $N(t)$ has a finite limit (≤ 2) as $t \to \infty$.

5. To prove that $N(t) \to 0$, observe that there exists t_0 such that $dN/dt \leq -N$ for $t \geq t_0$. Conclude from these inequality that $N(t) \to 0$ as $t \to \infty$.

Exercise 8.22. [☼] **Application of the Routh-Hurwitz Method**
Apply the Routh-Hurwitz method to determine the location of all roots of the following polynomials:

1. $p(x) = 3x + 5$,

2. $p(x) = -2x^2 - 5x - 100$,

3. $p(x) = 523x^2 - 57x + 189$,

4. $p(x) = (x^2 + x - 1)(x^2 + x + 1)$, and

5. $p(x) = x^3 + 5x^2 + 10x - 3$.

Exercise 8.23. [☼] **Implementation of the Routh-Hurwitz Algorithm**
Implement the Routh-Hurwitz method for determining the location of the roots of a polynomial of degree 4 within the complex plane in MATLAB. Use the MATLAB root finding methods to plot the roots in the complex plane.

1. Test your code against the analytic results derived in problem 8.22.

2. Apply your code to determine the stability of the zero solution of the linear oscillator differential equation

$$\ddot{x}(t) + 2cd\dot{x}(t) + d^2x(t) = 0\,, \qquad x \in \mathcal{C}^2(\mathbb{R}, \mathbb{R})\,,$$

with constants $c = 0.64$ and $d = 15.56$.

3. Finally, consider the parameters c and d from part b) to be arbitrary, i.e., $(c, d) \in \mathbb{R}^2$. Define a function depending on c and d that is positive if the origin is asymptotically stable in exercise b), zero if the origin is stable and negative if the origin is unstable. Plot the graph of this function against the c-d-plane.

Exercise 8.24. [☼] Application of the Lozinskii-Measure
In an article by Rupp and Scheurle a reduced model for fish-jellyfish interactions is proposed, where fish being assumed to represent the dominant predatory species feeding on jellyfish. This model is given by the following set of coupled non-linear ordinary differential equations

$$\dot{x} = \left(c + \frac{y}{1+y}\right)x - x^2\,, \quad \text{and} \quad \dot{y} = \beta\left(1 - d\frac{x}{1+y}\right)y\,,$$

where $c \in \mathbb{R}$ and $\beta, d > 0$ are parameters, and x denotes the fish population whereas y stands for the jellyfish population, respectively.

1. Show that the origin is an equilibrium and give the linearisation of the equations at this point.

2. Apply the method of Lozinskii-Measures to determine (linearised) stability of the origin.

Exercise 8.25. [☼] Linearized and Non-Linear Stability
Consider the pendulum-system

$$\dot{x} = y\,, \qquad \dot{y} = -\sin(x)\,.$$

1. Determine all equilibria.

2. Determine their stability with respect to the linearized system.

3. What can we say about the stability with respect to the original non-linear system? *Hint:* Find a suitable Lyapunov-function.

Exercise 8.26. [☼] Stability of Randomly Chosen Systems, cf. [242], pp. 143
Suppose we pick a linear system at random: what's the probability that the origin will be, say, an unstable spiral?

1. To be more specific, consider the system

$$\dot{x} = Ax\,, \qquad \text{where} \quad A = \begin{pmatrix} a & b \\ c & d \end{pmatrix}.$$

   Suppose we pick the entries a, b, c, d independently and at random from a uniform distribution on the interval $[-1, 1]$. Find the probabilities of all the different kinds of equilibria.

2. To check your answers (or if you hit an analytic road block), try the Monte Carlo Method. Generate millions of random matrices on your computer and have the machine count the relative frequency of saddles, unstable spirals, etc.

3. Are the answers the same if you use a normal distribution instead of a uniform distribution?

Part III

Efficient Data Structures & the Propagation of Random Excitations

Chapter 9
Fourier-Transform

Le plus court chemin entre deux vérités dans le domaine réel passe par le domaine complexe.
JACQUES SALOMON HADAMARD

The basic aspects of the continuous and the discrete Fourier transform are discussed in this chapter, with the focus on the latter including various MATLAB examples. The famous Fast Fourier Transform is derived. We briefly present the trigonometric variants of the discrete Fourier transform related to symmetry properties of the underlying input data; these trigonometric transforms allow us to realise fast Poisson solvers on Cartesian grids as needed in the workshop problem (cf. Chap. 18).

9.1 Key Concepts

This chapter contains a brief survey of continuous and discrete Fourier transforms. The continuous Fourier transform is summarised in a compact manner for two reasons. First, to provide the basic relations to the discrete counterpart. Second, to prepare several aspects necessary for the analysis of noise spectra in Chap. 10. In contrast, the discrete Fourier transform and corresponding variants are discussed in more detail. The focus is on applications for PDE problems (see Chap. 2) such as the ones described in Sec. 9.6 and used in the workshop problem (see Chap. 18). Note that the discrete (and in particular the Fast) Fourier transform are essential tools for the realisation of other approaches to PDE discretisations such as spectral methods (see [51], e.g.).

A nice Fourier analysis applet is provided by Fu-Kwun Hwang[1]: By defining the amplitudes of 16 sinusoidal functions $\sin(kf)$ and $\cos(kf)$, $k = 0, 1, 2, \ldots, 15$, for a given frequency f, you can compose your own Fourier series and play the sound that the corresponding acoustic wave realises. Paul Falstad presents another really great applet on his website[2].

From the point of view of mathematics, the Fourier transform is a function representing the amplitude and phase at each frequency. Hence, it is gener-

---

1 http://www.phy.ntnu.edu.tw/ntnujava/index.php?topic=17 (NTNUJAVA Virtual Physics Laboratory, Java Simulations in Physics)

2 http://www.falstad.com/fourier/

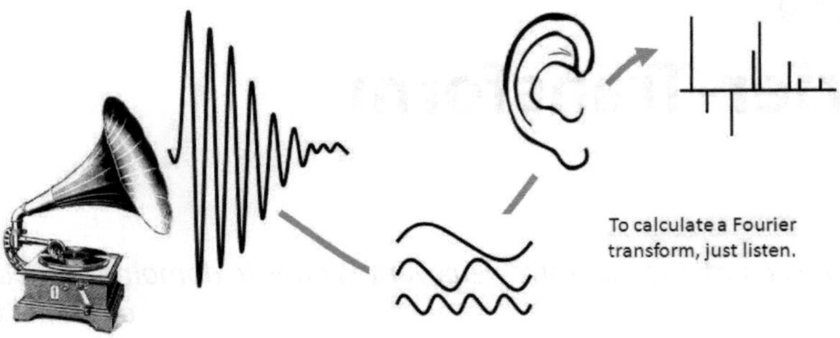

Figure 9.1. To calculate a Fourier transform, just listen, cf. [221].

ally necessary to specify two real-valued numbers (or one complex number) at each frequency. The complex representation is quite popular as it invites the use of complex analysis (see [105], e.g.) which often provides easier results. The theory of complex analysis and the Fourier transform have become indispensable in the numerical calculations underlying the analysis of mechanical vibrations, the design of electrical circuits, and the study of wave propagation both in the macro as well as in the quantum cosmos ([36]).

Referring to the acoustic examples mentioned above, one sees that time-based waves can be decomposed into sums of sinusoidal functions, each with unique amplitude, frequency, and phase. This process converts a rather complicated time-dependent function (the sound wave) into a more manageable sum of sine functions. The amplitude (and phase) of each sine term creates the power spectrum, which is the response of the original wave in the frequency domain. This conversion is called the Fourier transform. For the example of a square wave, this time-to-frequency-domain-conversion concept is illustrated in Fig. 9.2.

Figures 9.2 (a)–9.2 (f) show partial sums with only frequencies 1, 3, 5, 7, 31, until 131, respectively. Hence, the concrete time-dependent signals are for $t \in [0, 2\pi]$

$$
\begin{aligned}
y_1(t) &= \sin(t) \\
y_3(t) &= \sin(t) + \frac{1}{3}\sin(3t) \\
y_5(t) &= \sin(t) + \frac{1}{3}\sin(3t) + \frac{1}{5}\sin(5t) \\
y_7(t) &= \sin(t) + \frac{1}{3}\sin(3t) + \frac{1}{5}\sin(5t) + \frac{1}{7}\sin(7t) \\
y_{31}(t) &= \sin(t) + \frac{1}{3}\sin(3t) + \frac{1}{5}\sin(5t) + \frac{1}{7}\sin(7t) + \ldots + \frac{1}{31}\sin(31t)
\end{aligned}
$$

$$y_{131}(t) = \sin(t) + \frac{1}{3}\sin(3t) + \frac{1}{5}\sin(5t) + \frac{1}{7}\sin(7t) + \ldots + \frac{1}{131}\sin(131t)$$

The frequency and amplitude information from Fourier transform are visualised in Fig. 9.2 (g)–(i) for the input data y_1, y_7, and y_{131}, respectively. We obtain exactly the correct frequency information (where the peaks are) together with the correct amplitudes (the fractions in front of each sine term) of the corresponding input signal.

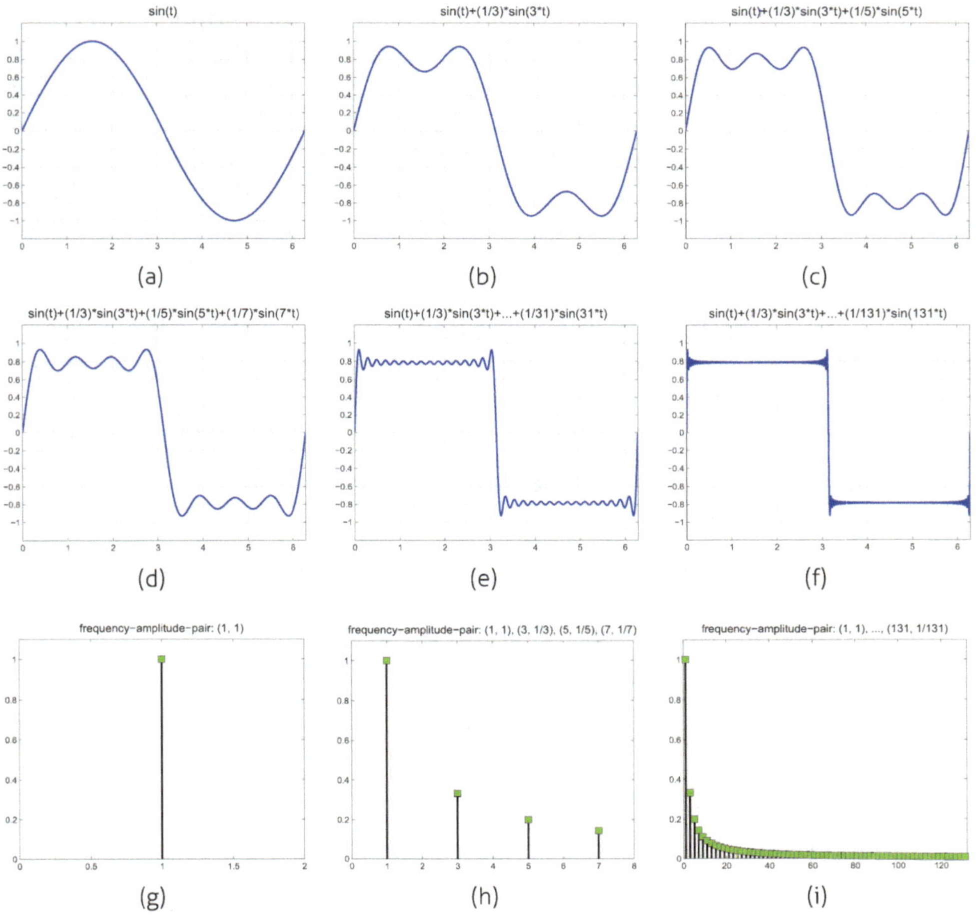

Figure 9.2. (a)–(f): the sum of considered sine functions uses terms up to frequency 1, 3, 5, 7, 31, and 131; the rectangle function is better and better approximated. (g)–(i): the frequency-amplitude plots for the signal of (a), (d), and (f), respectively.

Of course, we can also assign the notes of a particular instrument (a guitar, e.g.) to a precisely defined power spectrum that enables us to distinguish (for example) a guitar from a violin. A prominent example where this technique has successfully been applied is in the analysis of the opening chord in the Beatles' "Hard Day's Night" (1964)

"It's been a hard day's night
And I've been working like a dog".

Everyone who tried to replay this sound by guitar would give up in desperation. Why is that? About 40 years (!) after the publication of the song, Jason Brown of Dalhousie's Department of Mathematics applied the Fourier transform to this example. His group discovered that the frequencies did not match the assumed instrumentation: "George played a 12-string Rickenbacker, Lennon had his six string, Paul had his bass [...] none of them quite fit what I found," he explains. "Then the solution hit me: it wasn't just those instruments. There was a piano in there as well, and that accounted for the problematic frequencies"[3]. This is a simple but illustrative example of how the Fourier transform can simplify generally complex problems by changing the point of view to the frequency domain.

Note that in the following, we will use both notations $\overline{z}$ and z^* for representing the conjugate complex $x - iy$ of a complex number $z = x + iy$, depending on which variant is easier to write down.

When reading this chapter note the answers to the following questions

1. What is one common representation for the continuous Fourier transform?

2. What is one common representation for the discrete Fourier transform?

3. How does the Fast Fourier transform work?

as well as the following key concepts

1. MATLAB scripts for different variants of the discrete Fourier transform,

2. Variants of the discrete Fourier transform,

3. Creating a fast solver for the Poisson equation via the discrete sine transform.

This chapter is structured as follows: In Sec. 9.2, the concept and notation of the continuous Fourier transform is presented. Section 9.3 contains the motivation of the discrete Fourier transform via a trigonometric interpolation problem as well as the derivation of the forward and inverse transformation formulas. We present the backbone of Fourier transform algorithms, the Fast

[3] See http://www.mathstat.dal.ca/~brown/AHDNjib.html for the discussion of the sound wave and its spectrum as well as some interviews with J. Brown.

Fourier transform, in Sec. 9.4. In Sec. 9.5, different variants of the discrete Fourier transform are derived using symmetry properties of the underlying data before applying the acquired Fourier knowledge to the problem of solving the Poisson equation in Sec. 9.6. Finally, Section 9.7 wraps up the contents of this chapter.

9.2 The Continuous Fourier Transform

Relying on [162] for notation and results, we briefly summarise the continuous Fourier transform for two reasons. First, basic continuous definitions and results are relevant for the aspects discussed in Sec. 10. Second, the continuous representation serves as a motivation for the discrete Fourier transform tackled in Sec. 9.3.

Denoting the vector space of complex-valued Lebesgue-integrable functions on the $\mathbb{R}^d$, $d \in \mathbb{N}$, by $\mathcal{L}^1(\mathbb{R}^d)$ and the scalar product of two vectors $a, b \in \mathbb{R}^d$ by $< a, b >$, we make the following

Definition 9.1 (Continuous Fourier Transform). Let $f \in \mathcal{L}^1(\mathbb{R}^d)$. Then, the *Fourier transform* of f is the function $\widehat{f} : \mathbb{R}^d \to \mathbb{C}$ with

$$\widehat{f}(x) := \frac{1}{(2\pi)^{d/2}} \int_{\mathbb{R}^d} f(t) e^{-i<x,t>} dt, \qquad x \in \mathbb{R}^d$$

The function $\widehat{f}$ is often called the *continuous spectrum* of f. In the case of scalar functions and transforms ($d = 1$), the continuous Fourier transform directly states

$$\widehat{f}(x) = \frac{1}{\sqrt{2\pi}} \int_{\mathbb{R}} f(t) e^{-ixt} dt, \qquad x \in \mathbb{R} . \tag{9.1}$$

Note that the scaling factor, as below in the discrete case, may be used in different variants. We stick to the symmetric form, using $\sqrt{2\pi}^{-1}$ in the scalar case. This motivates the following example.

Example 9.2. Fourier transform of $f(t) = e^{-at^2}, a > 0$
Inserting the definition (9.1) results in

$$\widehat{f}(x) = \frac{1}{\sqrt{2\pi}} \int_{-\infty}^{\infty} e^{-at^2} e^{-ixt} dt = \frac{1}{\sqrt{2\pi}} e^{-x^2/4a} \int_{-\infty}^{\infty} e^{-a(t+ix/2a)^2} dt .$$

With the Cauchy integral theorem and some minor calculations (see [162]), we obtain

$$\int_{-\infty}^{\infty} e^{-a(t+ix/2a)^2} dt = \int_{-\infty}^{\infty} e^{-at^2} dt = \sqrt{\frac{\pi}{a}} ,$$

and, hence,

$$\widehat{f}(x) \quad = \quad \frac{1}{\sqrt{2a}} e^{-x^2/4a} \, .$$

In particular, for $a = 1/2$, the Fourier transform of the Gaussian is identical to the original function: $\widehat{f} = f$. This result also holds in the higher-dimensional case $d > 1$ for $f(t) = e^{-\|t\|^2/2}$ (see [162]).

If we know the continuous spectrum $\widehat{f}$ of a given function f, is it possible to transform that spectrum back to the original function (i.e. applying an inverse Fourier transform)? The positive answer to this question gives the following theorem.

Theorem 9.1. *Let $f \in \mathcal{L}^1(\mathbb{R}^d)$ be a function whose Fourier transform $\widehat{f}$ is also in $\mathcal{L}^1(\mathbb{R}^d)$. Then,*

$$f(t) \quad = \quad \frac{1}{(2\pi)^{d/2}} \int_{\mathbb{R}^d} \widehat{f}(x) e^{i<x,t>} dx$$

holds almost everywhere.

For the proof, see [162]. The scaling factor of this inverse transform is the same $(1/\sqrt{2\pi})$ as in the forward mode due to our symmetric choice. Note that Theorem 9.1 indicates that

$$\widehat{\widehat{f}}\,(t) = f(-t) \, ,$$

i.e. that we can compute the forward and the inverse Fourier transform in a very similar manner applying some pre- and/or postprocessing. We will also encounter this feature in the context of discrete Fourier transforms below. Translating, once more, the inverse Fourier transform of Theorem 9.1 to the scalar case results in

$$f(t) \quad = \quad \frac{1}{\sqrt{2\pi}} \int_{-\infty}^{\infty} \widehat{f}(x) e^{ixt} dx, \qquad t \in \mathbb{R} \, .$$

9.3 The Discrete Fourier Transform

9.3.1 Definition of the DFT and IDFT

In the following, we always assume complex input data for the Fourier transform if not indicated differently.

Definition 9.3 (Discrete Fourier Transform). For a vector of N complex numbers $(f_0, \ldots, f_{N-1})$, the *discrete Fourier transform (DFT)* is given by the vector $(F_0, \ldots, F_{N-1})$, where

$$F_k = \frac{1}{N} \sum_{n=0}^{N-1} f_n e^{-i2\pi nk/N} .$$

The DFT may be interpreted in different ways. Typically, it is considered as an approximation of the coefficients of the continuous Fourier series. A second possibility is to interpret the DFT as a trigonometric interpolation. Using the assumptions

- choose N ansatz functions $g_k(x) := e^{ikx}$ with $kx \in [0, 2\pi]$, $k = 0, \ldots, N-1$,
- assume N supporting points: $x_n := 2\pi n/N$, $n = 0, \ldots, N-1$,
- let f_n, $n = 0, \ldots, N-1$ be N interpolation values,

we may formulate the trigonometric interpolation problem as:

Find N weights F_k such that at all supporting points x_n

$$f_n = \sum_{k=0}^{N-1} F_k g_k(x_n) \qquad \Leftrightarrow \qquad f_n = \sum_{k=0}^{N-1} F_k e^{i2\pi nk/N} \tag{9.2}$$

holds.

The interpretation of the DFT as an interpolation problem consists of finding a representation of the signal f_n—or o a function $f(x)$—of the form $f(x) = \sum_{k=0}^{N-1} F_k g_k(x)$. Due to Euler's formula, the ansatz functions g_k are combined sine and cosine functions:

$$e^{ikx} = \cos(kx) + i \sin(kx) .$$

The F_k are, thus, called *Fourier coefficients*:

- k represents the wave number
- the value of F_k represents the amplitude of the corresponding frequency

The Fourier transform provides a *frequency spectrum*. Its application is useful when a problem is easier to solve in the *frequency domain* than in the *spatial domain* (see Fig. 9.3 for a visualisation). For a first hands-on example of what the Fourier coefficients represent, try Problem 9.5.

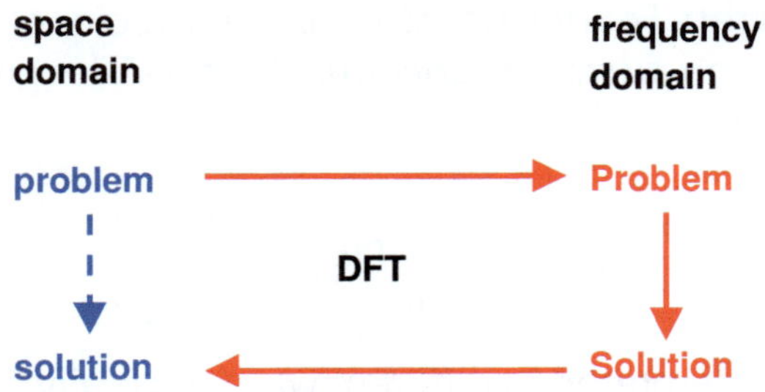

Figure 9.3. Visualisation of the alternative solution path via the frequency domain: For certain problems, this approach is advantageous and even efficient despite the overhead due to transformations in between the different domains.

The interpolation problem 9.2 actually represents a linear system of equations:

$$f_n = \sum_{k=0}^{N-1} F_k \omega_N^{nk}, \quad \text{for all } n = 0, \dots, N-1\,,$$

where $\omega_N := e^{i2\pi/N}$, i.e. $\omega_N^{nk} = e^{i2\pi nk/N}$. Denoting the vectors of the f_n and F_k as $\mathbf{f} := (f_0, \dots, f_{N-1})$ and $\mathbf{F} := (F_0, \dots, F_{N-1})$, the linear system of equations can be formulated in matrix-vector notation as

$$\mathbf{W}\mathbf{F} = \mathbf{f}\,, \tag{9.3}$$

where the entries of the matrix $\mathbf{W}$ are given by $W_{nk} := \omega_N^{nk}$. Note that this formulation would imply the solution of a dense linear system of equations of dimension N in order to compute the Fourier coefficients F_k. The complexity of algorithms such as Gaussian elimination or LU decomposition is of order $\mathcal{O}(N^3)$, hence too expensive for efficient calculations on considerably large input vectors $\mathbf{f}$. Luckily, it is possible to simplify this task. The matrix $\mathbf{W}$ has the form

$$\mathbf{W} = \begin{pmatrix} \omega_N^0 & \omega_N^0 & \omega_N^0 & \dots & \omega_N^0 \\ \omega_N^0 & \omega_N^1 & \omega_N^2 & \dots & \omega_N^{(N-1)} \\ \omega_N^0 & \omega_N^2 & \omega_N^4 & \dots & \omega_N^{2(N-1)} \\ \vdots & \vdots & \vdots & & \vdots \\ \omega_N^0 & \omega_N^{(N-1)} & \omega_N^{2(N-1)} & \dots & \omega_N^{(N-1)(N-1)} \end{pmatrix}$$

and obviously is symmetric: $\mathbf{W} = \mathbf{W}^T$. Furthermore,

$$\mathbf{W}\left(\mathbf{W}^T\right)^* = \mathbf{W}\mathbf{W}^H = N\mathbf{I} \tag{9.4}$$

holds (with H denoting the Hermitian), since

$$[\mathbf{W}\mathbf{W}^H]_{kl} = \sum_{j=0}^{N-1} \omega_N^{kj} \left(\omega_N^{lj}\right)^* = \sum_{j=0}^{N-1} \omega_N^{(k-l)j} = \begin{cases} N & \text{if} \quad k = l \\ 0 & \text{if} \quad k \neq l. \end{cases}$$

From Eq. 9.4, it follows that

$$\mathbf{W}^{-1} = \frac{1}{N}\mathbf{W}^H , \tag{9.5}$$

i.e. we have a direct representation of the inverse of the matrix $\mathbf{W}$ as

$$\mathbf{W}^{-1} = \frac{1}{N} \begin{pmatrix} \omega_N^0 & \omega_N^0 & \omega_N^0 & \dots & \omega_N^0 \\ \omega_N^0 & \omega_N^{-1} & \omega_N^{-2} & \dots & \omega_N^{-(N-1)} \\ \omega_N^0 & \omega_N^{-2} & \omega_N^{-4} & \dots & \omega_N^{-2(N-1)} \\ \vdots & \vdots & \vdots & & \vdots \\ \omega_N^0 & \omega_N^{-(N-1)} & \omega_N^{-2(N-1)} & \dots & \omega_N^{-(N-1)(N-1)} \end{pmatrix}$$

Hence, the vector $\mathbf{F}$ of the Fourier coefficients can directly be computed as a matrix-vector product:

$$\mathbf{F} = \frac{1}{N}\mathbf{W}^H\mathbf{f} \qquad \text{or} \qquad F_k = \frac{1}{N}\sum_{n=0}^{N-1} f_n\omega_N^{-nk} .$$

The corresponding computational effort is now $\mathcal{O}(N^2)$ and is significantly lower than that of Eq. 9.3. We are going to see an even more efficient variant in Sec. 9.4.

Until now, we derived a way to compute the Fourier coefficients F_k out of a given discrete input vector f_n (DFT). If problems of the type depicted in Fig. 9.3 have to be tackled, the corresponding backward mode or inverse Fourier transform (IDFT) is necessary.

Definition 9.4 (Inverse Discrete Fourier Transform). The *inverse discrete Fourier Transform* (IDFT) of the vector $(F_0, \dots, F_{N-1})$ is given by the vector $(f_0, \dots, f_{N-1})$, where

$$f_n = \sum_{k=0}^{N-1} F_k e^{i2\pi nk/N} .$$

As the names already indicate, the DFT and IDFT are inverse operations (see Problem 9.6):

$$F_k = \frac{1}{N}\sum_{n=0}^{N-1} f_n e^{-i2\pi nk/N} , \qquad\qquad f_n = \sum_{k=0}^{N-1} F_k e^{i2\pi nk/N} .$$

operation	forward	inverse
continuous Fourier transform	$\widehat{f}(x) = \frac{1}{\sqrt{2\pi}} \int_{\mathbb{R}} f(t) e^{-ixt} dt$	$f(t) = \frac{1}{\sqrt{2\pi}} \int_{-\infty}^{\infty} \widehat{f}(x) e^{ixt} dx$
discrete Fourier transform	$F_k = \frac{1}{N} \sum_{n=0}^{N-1} f_n e^{-i2\pi nk/N}$	$f_n = \sum_{k=0}^{N-1} F_k e^{i2\pi nk/N}$

Table 9.1. Summary of the one-dimensional continuous and discrete Fourier Transform in forward and inverse mode.

$$\mathsf{F} = \mathrm{DFT}(\mathrm{IDFT}(\mathsf{F})) \qquad \text{or} \qquad \mathsf{f} = \mathrm{IDFT}(\mathrm{DFT}(\mathsf{f}))\,.$$

Using again the vectors $\mathsf{f} := (f_0, \ldots, f_{N-1})$ and $\mathsf{F} := (F_0, \ldots, F_{N-1})$, one can formulate the DFT and IDFT in matrix-vector notation as

$$\mathsf{F} = \frac{1}{N}\mathsf{W}^H \mathsf{f}\,, \qquad \mathsf{f} = \mathsf{W}\mathsf{F}\,.$$

In particular, from Eq. (9.35) of Problem 9.6 follows that an algorithm for computing either the DFT or the IDFT is sufficient to treat both transforms with the help of some minor postprocessing. Note that the f_n and the F_k are discretely periodic since $\omega_N^{nk} = \omega_N^{n(k+N)} = \omega_N^{(n+N)k}$:

$$f_{n+N} = f_n, \qquad F_{k+N} = F_k \qquad \forall k, n \in \mathbb{Z}\,.$$

The formulas for the discrete and continuous Fourier transform in forward and inverse mode are summarised in Tab. 9.1

We emphasize that for the discrete Fourier transform—similar to the continuous Fourier transform—a lot of different variants for formulating the (I)DFT exist as a variety of possible choices concerning:

- the scaling factor $\frac{1}{N}$ in the IDFT instead of the DFT; alternatively a factor $\frac{1}{\sqrt{N}}$ in DFT and IDFT;
- switched signs in the exponent of the exponential function in the DFT and IDFT;
- the notation of j for the imaginary unit (electrical engineering).

MATLAB Example 9.1 dft_naive.m: Direct version of DFT.

```
function F = dft_naive(f)

s = size(f);
N = s(1);
F = zeros(s);
for k = 0:N-1
    for n = 0:N-1
        F(k+1,:) = F(k+1,:) + f(n+1,:) * exp(-1i*2*pi*k*n/N);
    end
end
F = F/N;
```

MATLAB Example 9.2 dft_twiddle.m: DFT using twiddle factors.

```
function F = dft_twiddle(f)

s = size(f);
N = s(1);
F = zeros(s);
%omega(n+1) = omega^n
omega = exp(1i*2*pi * (0:N-1)' / N);
for k = 0:N-1
    F(k+1,:) = sum(bsxfun(@times, f , omega.^(-k)));
end
F = F/N;
```

9.3.2 MATLAB Examples for DFT and IDFT

Depending on the programming language of a given library, an additional index shift may appear. For instance, the MATLAB formulation of DFT and IDFT in our notation read

$$F_{k+1} = \frac{1}{N}\sum_{n=0}^{N-1} f_{n+1}e^{-i2\pi nk/N} \qquad k = 0,\ldots,N-1 \tag{9.6}$$

$$f_{n+1} = \sum_{k=0}^{N-1} F_{k+1}e^{i2\pi nk/N} \qquad n = 0,\ldots,N-1 \tag{9.7}$$

since array indexing starts with one instead of zero in MATLAB.

A direct implementation of the DFT using explicit loops in MATLAB is given in MATLAB Example 9.1. This direct implementation is not very efficient due to two facts: First, the direct usage of loops in MATLAB is known to be slower than vectorised notation. Second, the full computation of the powers of ω_N is computationally expensive since all calculations are performed inside each loop iteration. MATLAB Example 9.2 shows a more efficient variant tackling the above two issues. In particular, the so-called *twiddle factors* ω_N^n are precomputed and reused in the loop. Of course, these optimisations in the implementation do not change the (bad) complexity of the algorithm. In Sec. 9.4, an improvement of the overall algorithm is presented.

Note that MATLAB uses a different scaling than Eq. (9.6)–(9.7) for its built-in methods fft and ifft:

$$F_{k+1} = \sum_{n=0}^{N-1} f_{n+1} e^{-i2\pi nk/N} \qquad k = 0, \ldots, N-1 \tag{9.8}$$

$$f_{n+1} = \frac{1}{N} \sum_{k=0}^{N-1} F_{k+1} e^{i2\pi nk/N} \qquad n = 0, \ldots, N-1 \tag{9.9}$$

Hence, one needs to scale the results of fft by the factor $1/N$ to compare it with dft_naive or dft_twiddle.

Note further that both MATLAB Example 9.1 and MATLAB Example 9.2 support not only vector but also matrix input data[4]. This is relevant for the higher-dimensional examples in the following subsection.

9.3.3 DFT in Higher Dimensions

For certain applications such as image compression or fast Poisson solvers (see Sec. 9.6), higher-dimensional discrete Fourier transforms may have to be solved. Luckily, these higher-dimensional cases can be transformed into sequences of one-dimensional DFTs. Thus, no new algorithm has to be developed or implemented. For example, the two-dimensional variant of the DFT for a squared problem ($N = N_1 = N_2$) has the form

$$\widetilde{F}_{kl} = \underbrace{\frac{1}{N^2}}_{=:A} \sum_{n=0}^{N-1} \sum_{m=0}^{N-1} f_{nm} e^{-i2\pi nk/N} e^{-i2\pi mk/N}$$

which can directly be rewritten as

$$\frac{1}{N} \sum_{n=0}^{N-1} \underbrace{\left(\frac{1}{N} \sum_{m=0}^{N-1} f_{nm} e^{-i2\pi mk/N} \right)}_{:= \widehat{F}_{nl}} e^{-i2\pi nk/N} ,$$

4 In particular, bsxfun(@times, f , omega.^(−k)) is identical to the formulation f .* ... omega.^(−k) applied to all columns of the matrix f.

where we identify the nesting of the 1D DFTs: For each n, $\widehat{F}_{nl}$ are computed via N 1D transforms. Of course, such reformulations also work for other dimensions ($d > 2$) and for non-squared problems with $A = N_1 N_2, N_1 \neq N_2$.

In MATLAB, we may reuse existing 1D implementations of DFT and simply call them in a nested manner to get higher-dimensional DFT results. For the 2D case, we may first 1D-transform all rows and then all columns to get the 2D-transform:

```
F = dft_twiddle(dft_twiddle(f).').';
```

Note that the above example uses the column-wise DFT computation of a matrix input f supported in dft_twiddle; the transpositions are necessary to obtain the correct results and ordering. We emphasize that the column-wise computation does not represent the most efficient but rather the most convenient way for calculating a 2D DFT.

Before you continue, make sure to answer the following questions:

Quiz: Section 9.3

Q1 Are the Fourier coefficients real- or complex-valued numbers, in general?

Q2 Concerning the computational effort: When will it be advantageous to solve a problem in the frequency domain instead of in the spatial domain where it is typically formulated? Do you know an (easy) example in applied mathematics?

Q3 What is the difference in complexity between the computation of the N Fourier coefficients via the linear system of equations and the matrix-vector product, respectively?

Q4 Show that the following properties hold for $\omega_N = e^{i2\pi/N}$:

$$\begin{aligned} \sum_{k=0}^{n-1} \overline{\omega_N}^{kl} &= \begin{cases} n, & \text{for } l = 0 \\ 0, & \text{for } l = 1, \ldots, n-1 \end{cases} \\ \sum_{k=0}^{n-1} \omega_N^{kl} \, \overline{\omega_N}^{kj} &= \begin{cases} n, & \text{for } l = j \\ 0, & \text{for } l \neq j \end{cases} . \end{aligned}$$

Which trick is useful here?

9.4 The Fast Fourier Transform

The computational complexity of the discrete Fourier transform in its original form, as derived in Sec. 9.3, is quadratic. The corresponding effort is too high for most applications. However, more efficient formulations are possible that need only $\mathcal{O}(N \log(N))$ operations; such DFT formulations are typically called *fast Fourier transforms* (FFT). We are going to present the most common FFT variant in more detail (proposed by James Cooley and John W. Tukey [65]).

Since the DFT and the IDFT are equivalent in the sense that

$$\mathsf{DFT}(f) = \frac{1}{N}\overline{\mathsf{IDFT}(\overline{f})}$$

holds (see Problem 9.6), it is sufficient to have an efficient algorithm and implementation for only one of the two transforms. Therefore, we are going to restrict the considerations for a fast algorithm to the IDFT case in the following.

9.4.1 FFT Idea

The basic idea of the fast Fourier transform is to apply a divide-and-conquer approach. Assuming that the length of the data and coefficient vector $\mathbf{x} := \mathbf{f}$ and $\mathbf{X} := \mathbf{F}$ is a power of two (i.e. $N = 2^p$), it is possible to treat the even and odd indices separately in the IDFT formulation: For $n = 0, 1, \ldots, \frac{N}{2} - 1$ it holds

$$x_n = \sum_{k=0}^{N-1} X_k \omega_N^{nk} = \sum_{k=0}^{\frac{N}{2}-1} \left(X_{2k}\omega_N^{2nk} + X_{2k+1}\omega_N^{(2k+1)n} \right) .$$

Setting $Y_k := X_{2k}$ and $Z_k := X_{2k+1}$ and using $\omega_N^{2nk} = \omega_{N/2}^{nk}$, we obtain a sum of two IDFT on $\frac{N}{2}$ coefficients:

$$x_n = \sum_{k=0}^{N-1} X_k \omega_N^{nk} = \underbrace{\sum_{k=0}^{\frac{N}{2}-1} Y_k \omega_{N/2}^{nk}}_{:= y_n} + \omega_N^n \underbrace{\sum_{k=0}^{\frac{N}{2}-1} Z_k \omega_{N/2}^{nk}}_{:= z_n} . \qquad (9.10)$$

Applying the same even vs. odd separation for the indices $\frac{N}{2}, \ldots, N-1$ results in

$$x_{n+\frac{N}{2}} \quad = \quad y_{n+\frac{N}{2}} + \omega_N^{(n+\frac{N}{2})} z_{n+\frac{N}{2}}$$

Since $\omega_N^{(n+\frac{N}{2})} = -\omega_N^n$, and since y_n and z_n have a period of $\frac{N}{2}$, we obtain the so-called *butterfly scheme*:

$$x_n \quad = \quad y_n + \omega_N^n z_n$$

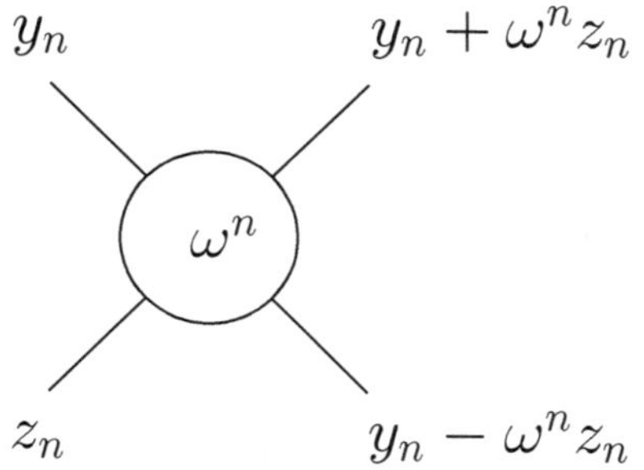

Figure 9.4. Visualisation of the butterfly scheme (9.11).

$$x_{n+\frac{N}{2}} = y_n - \omega_N^n z_n \,. \tag{9.11}$$

In Fig. 9.4, our visualisation of the butterfly scheme illustrates the inspiration for this name.

If we interpret the separation into even and odd contributions as a function call to compute the IDFT on a smaller set of parameters, we may formulate it in the following manner

$$\begin{aligned}
(x_0, x_1, \ldots, \ldots, x_{N-1}) &= \mathsf{IDFT}(X_0, X_1, \ldots, \ldots, X_{N-1}) \\
&\Downarrow \\
(y_0, y_1, \ldots, y_{\frac{N}{2}-1}) &= \mathsf{IDFT}(X_0, X_2, \ldots, X_{N-2}) \\
(z_0, z_1, \ldots, z_{\frac{N}{2}-1}) &= \mathsf{IDFT}(X_1, X_3, \ldots, X_{N-1}) \,,
\end{aligned}$$

The separation into even and odd contributions that we investigated for the initial overall vector **x** may, of course, be used in an recursive manner on the resulting smaller vectors **y** and **z** as well. Since $N = 2^p$, p such recursion levels are possible up to the point where only a single scalar value remains to be computed. The resulting scheme is visualised in Fig. 9.5 for a vector length of $N = 4$ (i.e. $p = 2$ levels of recursion). Note that the algorithm consists of two phases:

1. resorting of input data;
2. combination following the "butterfly scheme".

The computational effort $C(N)$ ($N = 2^p$) is given by the following recursion equation

$$C(N) = \begin{cases} \mathcal{O}(1) & \text{for} \quad N = 1 \\ \mathcal{O}(N) + 2C\left(\frac{N}{2}\right) & \text{for} \quad N > 1 \end{cases} \tag{9.12}$$

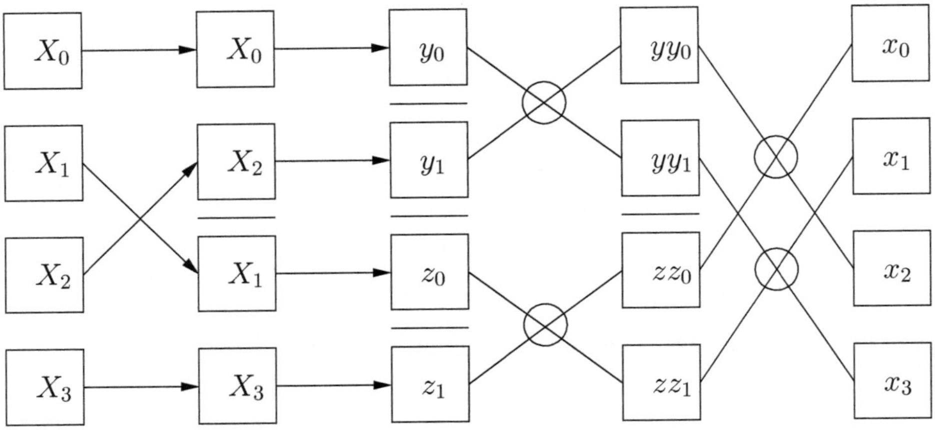

Figure 9.5. Visualisation of the FFT algorithm: sorting phase (left) and combination phase with the butterfly operator (right) for a vector length of $N = 4$. The horizontal bars indicate the borders of the odd/even sorted data.

Applying Eq. 9.12 once more recursively results in

$$\begin{aligned} C(N) &= \mathcal{O}(N) + 2C\left(\tfrac{N}{2}\right) \\ &= \mathcal{O}(N) + 2\mathcal{O}(\frac{N}{2}) + 4C\left(\tfrac{N}{4}\right) \end{aligned}$$

Since $N = 2^p$, we need to perform, in total, $p = \frac{\log N}{\log 2}$ such recursion steps. Hence, we obtain an overall effort $C(N) = \mathcal{O}(N \log N)$ as desired.

9.4.2 MATLAB Examples: Recursive and Iterative FFT

MATLAB example 9.3 implements the FFT in a recursive manner. Again, the algorithm is formulated in MATLAB syntax for array indexing starting at one instead of zero. This implementation works nicely and delivers (at least for smaller tests) the correct result compared with MATLAB's built-in FFT. However, there are several drawbacks of this recursive variant. Recursive function calls may be relatively expensive (independent of the programming language). Furthermore, additional storage is necessary (intermediate results an input vectors have to be stored somewhere: Compare the vectors x, x0, and x1 in MATLAB Example 9.3). These intermediate arrays are not necessary to store the input data (which is already available in the very first input vector) but appear mainly due to the sorting phase. Hence, a direct approach for the sorting would enable a simple, iterative algorithm without additional memory requirements.

MATLAB Example 9.3 ifft_rec.m: Recursive FFT. Note that the notation odd-/even here refers to formula (9.10), not to MATLABs indexing starting with one instead of zero.

```
function x = ifft_rec(X)

if length(X) == 1
    x = X;
else
    N = length(X);
    omega = exp(1i * 2 * pi * (0:N/2-1) / N)';

    x0 = ifft_rec(X(1:2:end)); % IFFT on even terms
    x1 = ifft_rec(X(2:2:end)); % IFFT on odd  terms

    x = zeros(size(X));
    x(1:N/2)       = x0 + omega .* x1; % butterfly
    x(N/2+1:end) = x0 - omega .* x1;
end
```

If we cross-check the indices of the original array components and compare them with the resulting reordered indices, the sorting into even and odd contributions results in an effect called bit reversal. This effect is obvious for a binary representation of the corresponding indices for the small example presented in Tab. 9.2 with $N = 8$: For each component X_k, the decimal and binary representation of the corresponding index are indicated without and with brackets, respectively. An index in the sorted field (right) has the reversed (i.e. mirrored) binary representation compared to the original index (left). Note that, on a given level of recursion, the even indices are always sorted into the upper half of the sub-array, whereas the odd indices are put into the lower half. Hence, the distinction between even/odd is based on the least significant bit while the distinction between upper/lower is visible from the most significant bit of the binary representation of an index.

In MATLAB Example 9.4, one variant of a bit reversal-induced sorting is presented Note that the bit reversal needs $\mathcal{O}(p) = \mathcal{O}(\log N)$ operations. Therefore, the overall sorting results also in a complexity of $\mathcal{O}(N \log N)$ (directly visible from the loops in MATLAB Example 9.4). Actually, the sorting is non-negligible: It may consume up to 10–30 % of the CPU time necessary for the overall FFT! Luckily, MATLAB provides a built-in method to efficiently compute the bit reversal: bitrevorder(). Certain digital signaling processors implement bit reversal in hardware for even higher performance (see [104], e.g.). The bit reversal together with the sorting may be used in order to formulate

$X_{0(000)}$		$X_{0(000)}$		$X_{0(000)}$
$X_{1(001)}$		$X_{2(010)}$	$\rightarrow$	$X_{4(100)}$
$X_{2(010)}$		$X_{4(100)}$		$X_{2(010)}$
$X_{3(011)}$	$\rightarrow$	$X_{6(110)}$		$X_{6(110)}$
$X_{4(100)}$		$X_{1(001)}$		$X_{1(001)}$
$X_{5(101)}$		$X_{3(011)}$	$\rightarrow$	$X_{5(101)}$
$X_{6(110)}$		$X_{5(101)}$		$X_{3(011)}$
$X_{7(111)}$		$X_{7(111)}$		$X_{7(111)}$

Table 9.2. Bit reversal for a coefficient vector of length $N = 8$ throughout the FFT sorting: For each component X_k, the decimal and binary representation of the indices are indicated without and with brackets, respectively.

MATLAB Example 9.4 bit_reversal.m: Sorting of a vector using bit reversal.

```
function y = bit_reversal(x)

y = x;
N = length(x);
p = log2(N);

for n=0:N-1
    j=0;
    m=n;
    for i=0:p-1 % Compute p-valued bit reversal of n in j
        j = 2*j + mod(m,2);
        m = floor(m/2);
    end
    if (j>n)    % exchange (idx shift by 1 due to matlab idx)
        y(j+1) = x(n+1); y(n+1) = x(j+1); %
    end
end
```

an in-place variant of the recursive FFT. Alternatively, we can use this sorting for an iterative variant, as MATLAB Example 9.5 shows.

MATLAB Example 9.5 ifft_iter.m: Iterative FFT.

```
function x = ifft_iter(X)

N = length(X);
x = X(1+bitrevorder(0:N-1));

for L=2.^(1:log2(N))                                    % L=2,4,8,...,N
    omega = exp(1i * 2 * pi * (0:L/2-1) / L)'; % omega_L^j
    for K=0:L:N-1
        % skip additional j-loop via vector comput.: j=0..L/2-1
        left  = 1+K       : K+L/2;
        right = 1+K+L/2 : K+L;

        z = omega .* x(right);
        x(right) = x(left) - z;
        x(left)  = x(left) + z;
    end
end
```

9.4.3 Outlook: FFT Variants and Libraries

A variety of different FFT variants exists. Obviously, two- or three-dimensional DFT can be realised via corresponding FFTs. Furthermore, the basic Cooley-Tukey algorithm needs to be modified if the length of the input vector is not a power of two. A division into three parts—also called radix-3 variant—is more or less straightforward (cf. [135],e.g.) allowing for $N = 3^p$. Similar approaches are of course possible for other—in particular prime—numbers. Besides, a decomposition for $N = N_1 N_2$ via

$$x_n = \sum_{k=0}^{N-1} X_k \omega_N^{nk} = x_{n_1+n_2N_1} = \sum_{k_2=0}^{N_2-1} \left[\left(\sum_{k_1=0}^{N_1-1} X_{k_1N_2+k2} \omega_{N_1}^{n_1k_1} \right) \omega_N^{n_1k_2} \right] \omega_{N_2}^{n_2k_2}$$

shows the possibility to break down this general problem quite arbitrarily.

MATLAB's built-in FFT methods (such as fft, fft2 for 2D, and fftn for nD) rely on the FFTW[5] library. The following non-complete list of features makes FFTW a very popular FFT package: FFTW is

- free software;
- fast; in particular, cache efficiency is achieved (compared to other publicly available software and competitive to vendor versions of FFT);

5 "Fastest Fourier Transform in the West" http://www.fftw.org/

- portable to different hardware and operating systems;
- supporting a variety of code variants (multi-dimensional FFT, arbitrary vector lengths, etc.).

For the FFT of a given vector length N (or size in more than one dimension), FFTW creates a so-called *plan* which mainly holds the decomposition of N as a product of (prime) numbers as described above. Often, several such plans exist: For instance, $N = 4 = 2^2$ may be tackled by one radix-4 variant or two radix-2 calls. These variants may behave differently depending on the underlying hardware (CPU, memory layout) or software (compilers, libraries, operating system). Therefore, the FFTW library has the ability to experimentally determine the fastest computational method (i.e. determine the best plan) for a particular input data on a given system. These performance results of different plans can be stored in a database for future runs of FFTs of the same size and dimensionality.

Of course, a variety of other FFT packages are available which differ with respect to different properties. We skip the detailed discussion and are now going to focus on different variants of Fourier transforms that are important in the context of solving PDEs numerically.

Before you continue, make sure to answer the following questions:

Quiz: Section 9.4

Q1 Why is the FFT better than the DFT for actual computations?

Q2 What is the difference of the FFT implementations presented in Sec. 9.4.2 compared to the built-in method fft in MATLAB? This is important to know if you want to compare or cross-check results of your/an implementation.

Q3 What variants are available to compute the FFT for vectors of length different from a power of two?

9.5 Variants of Fourier Transforms via Symmetry Properties

In this section, we present several discrete transforms related to the DFT via symmetry conditions for the periodic continuation of the underlying input data f. The following types of symmetry are relevant and lead to corresponding transforms:

SYMMETRY TYPE	INPUT	TRANSFORM
real symmetry	$f_n \in \mathbb{R}$	Real-valued DFT (RDFT)
odd symmetry	$f_n = -f_{-n}$	Discrete Sine Transform (DST)
even symmetry	$f_n = f_{-n}$	Discrete Cosine Transform (DCT)

For the sake of simplicity, we switch the index representation of the DFT of length N to a centred variant around zero resulting in indices $n = -\frac{N}{2} + 1, \ldots, 0, \ldots, \frac{N}{2}$. The basic idea for all three types of symmetry is: If we take an input of the specific kind of symmetry, what is the form of the DFT applied to this specific case?

We will cover the real symmetry now and the odd and even symmetry in the following two subsections. Applying the real symmetry $f_n \in \mathbb{R}\ \forall n$ (i.e. $f_n^* := \overline{f_n} = f_n$), we may directly deduce from Def. 9.3

$$F_k = \frac{1}{N} \sum_{n=-\frac{N}{2}+1}^{\frac{N}{2}} f_n e^{-i2\pi nk/N} = \frac{1}{N} \sum_{n=-\frac{N}{2}+1}^{\frac{N}{2}} f_n \left(\cos\left(\frac{2\pi nk}{N}\right) - i \sin\left(\frac{2\pi nk}{N}\right)\right) .$$

The following properties are, hence, straightforward:

- $\text{Re}\left\{F_k\right\} = \frac{1}{N} \sum_{n=-\frac{N}{2}+1}^{\frac{N}{2}} f_n \cos\left(\frac{2\pi nk}{N}\right)$,

  $\text{Im}\left\{F_k\right\} = -\frac{1}{N} \sum_{n=-\frac{N}{2}+1}^{\frac{N}{2}} f_n \sin\left(\frac{2\pi nk}{N}\right)$

- Only N independent, real coefficients are necessary, since

$$F_k^* = \frac{1}{N} \sum f_n^* \left\{\omega_N^{-nk}\right\}^* = \frac{1}{N} \sum f_n \omega_N^{-n(-k)} = F_{-k} .$$

So we have the following mapping from N real values f_n to N real-valued coefficients and vice versa (even though the full coefficients F_k are, of course, still complex):

$$\left\{f_{-\frac{N}{2}+1}, \ldots, \quad f_0, \quad \ldots, \ f_{\frac{N}{2}}\right\}$$

$$\text{DFT} \ \Downarrow \quad \Uparrow \ \text{IDFT}$$

$$\left\{F_0, \text{Re}\{F_1\}, \text{Im}\{F_1\}, \quad \ldots, \quad \text{Re}\{F_{\frac{N}{2}-1}\}, \text{Im}\{F_{\frac{N}{2}-1}\}, F_{\frac{N}{2}}\right\}$$

Since we know that one direction (forward or inverse mode) of the DFT is sufficient to compute both, we shall restrict the further considerations in this section to the IDFT. Besides, we of course insert the above symmetry in the coefficients: $F_{-k} = F_k^*$. For a DFT of overall length $2N$ (again, for the sake of simplicity and to avoid the fractions), this results in

$$\begin{aligned}
f_n &= \sum_{k=-N+1}^{N} F_k e^{i2\pi nk/2N} \\
&= F_0 + \sum_{k=1}^{N-1} \left(F_k e^{i2\pi nk/2N} + F_{-k} e^{-i2\pi nk/2N}\right) + F_N e^{i2\pi nN/2N} \\
&= F_0 + \sum_{k=1}^{N-1} \left(F_k e^{i2\pi nk/2N} + \left\{F_k e^{i2\pi nk/2N}\right\}^*\right) + F_N e^{i\pi n} \\
&= F_0 + 2\sum_{k=1}^{N-1} \text{Re}\left\{F_k e^{i2\pi nk/2N}\right\} + F_N e^{i\pi n} \\
&= F_0 + 2\sum_{k=1}^{N-1} \left(\text{Re}\{F_k\} \cos\left(\frac{\pi nk}{N}\right) - \text{Im}\{F_k\} \sin\left(\frac{\pi nk}{N}\right)\right) + F_N \cos(\pi n) \;.
\end{aligned}$$

Now, in order to obtain a formulation with purely real-valued coefficients, we set $a_k := 2\,\text{Re}\{F_k\}$ and $b_k := -2\,\text{Im}\{F_k\}$ (but $a_0 := \text{Re}\{F_0\}$ and $a_N := \text{Re}\{F_N\}$) to get:

$$f_n = a_0 + \sum_{k=1}^{N-1} \left(a_k \cos\left(\frac{\pi nk}{N}\right) + b_k \sin\left(\frac{\pi nk}{N}\right)\right) + a_N \cos(\pi n) \qquad (9.13)$$

Equation (9.13) defines the *real inverse discrete Fourier transform* using the (real-valued) representation for F_k:

$$a_k = \frac{1}{N} \sum_{n=-N+1}^{N} f_n \cos\left(\frac{\pi nk}{N}\right), \qquad b_k = \frac{1}{N} \sum_{n=-N+1}^{N} f_n \sin\left(\frac{\pi nk}{N}\right) \;.$$

Having a definition of the (I)RDFT at hand, the question becomes how to compute it efficiently (i.e. similar to the FFT for the (I)DFT). The direct computation of a real-valued DFT using the complex FFT (with real-valued input data) is inefficient since we would spoil effort on N redundant components (due to the symmetry). Several better choices are possible:

- compute two real DFTs from one complex FFT;

- compute a real DFT of length $2N$ from one complex FFT of length N;
- define and implement a "compact" real FFT: use the symmetry of the data directly in the algorithm.

We do not go into the details of these variants but turn our interest to the odd and even symmetry. For several fields of application, including data compression (such as JPEG) or solving PDEs numerically, modified transforms for real-valued input data are advantageous. The DFT always assumes a periodic continuation of the input vector resulting in (possibly large) jumps at borders, i.e. the end of the input vector that is continued with its beginning. However, if we make different assumptions about the continuation of the discrete input data at the boundaries, we obtain other transforms which are often called *discrete trigonometric transforms* (DTT; see [184], e.g.). We distinguish two different aspects of the continuation at boundaries of a given data vector f:

- *symmetry*: odd ($f_n = -f_{-n}$) or even ($f_n = f_{-n}$) symmetry of the data.
- *mirror* vs. *copy*: The axis or point of symmetry can be put directly on the first/last data entry (*mirror*) or at the center of the interval between this entry and the one before/after (*copy*). See Fig. 9.6 for some examples.

Since two boundaries exist (the beginning and end of the input vector), we have four possibilities concerning symmetry (denoted by E and O) and another four (independent) possibilities concerning mirror/copy (abbreviated by M and C):

$$
\begin{aligned}
\text{E} &: \text{ even}\\
\text{O} &: \text{ odd}\\
\text{M} &: \text{ mirror}\\
\text{C} &: \text{ copy}\,.
\end{aligned}
$$

This results in a total of 16 variants for the symmetric-periodic sequences which are visualised in Fig. 9.6 and 9.7 together with their corresponding transforms (and their abbreviations for symmetry and mirror/copy such as MEME etc.). Note that eight of the DTT variants belong to the discrete sine transform (DST) and another eight to the discrete cosine transform (DCT) where the beginning boundary (at index zero) always shows a fixed odd and even symmetry, respectively. DCT I and DST I show mirroring at both boundaries (cf. Fig. 9.6 (a) and 9.6 (b)), whereas DCT II and DST II provide copying (see Fig. 9.6 (c) and 9.6 (d)). Note that mirroring in combination with an odd symmetry results in zero (averaged) values. The most important variants are the four first ones of each category: DCT I–IV and DST I–IV which are summarised in Fig. 9.6. The survey of the DTT variants in Fig. 9.6 and 9.7 shows an additional feature: The 16 different members of the DTT "family" for a given length N have different numbers of actual vector lengths: $5 \times N - 1, 10 \times N$ and $1 \times N + 1$ (compare the number of the red data points).

In the following two subsections, we are going to have a detailed look at one representative for DST and DCT, respectively. Again, the basic idea is: If we take an input of a specific kind of symmetry, how does the DFT applied to this specific case look like? Using the resulting formulas, we may then apply them to arbitrary input data as a new category of transforms with different results and properties than the DFT.

9.5.1 The Discrete Sine Transform

We derive the DST version I by directly plugging the symmetry and mirror/-copy conditions (odd and mirror at both ends each, cf. Fig. 9.6 (b)) into the DFT of length $2N$, for the sake of simplicity in the notation. Hence, we assume

$$\begin{aligned} f_{-N+1}, \dots, f_N &\in \mathbb{R}, \\ f_{-n} &= -f_n, \\ f_0 = f_N = f_{-N} &= 0 \end{aligned}$$

The DFT then has the following form ($k = -N+1, \dots, N$):

$$\begin{aligned} F_k &= \frac{1}{2N} \sum_{n=-N+1}^{N} f_n \omega_{2N}^{-nk} \\ &= \frac{1}{2N} \left(\underbrace{f_0}_{=0} + \sum_{n=1}^{N-1} \left(f_n \omega_{2N}^{-nk} + f_{-n} \omega_{2N}^{nk} \right) + \underbrace{f_N}_{=0} \omega_{2N}^{-Nk} \right) \\ &= \frac{1}{2N} \sum_{n=1}^{N-1} f_n \left(\omega_{2N}^{-nk} - \omega_{2N}^{nk} \right) = \frac{-i}{N} \sum_{n=1}^{N-1} f_n \sin\left(\frac{\pi n k}{N} \right). \qquad (9.14) \end{aligned}$$

Note that due to Eq. (9.14) we have the same (odd) symmetry in the coefficients F_k:

$$F_{-k} = \frac{-i}{N} \sum_{n=1}^{N-1} f_n \sin\left(\frac{\pi n(-k)}{N} \right) = \frac{-i}{N} \sum_{n=1}^{N-1} f_n \left(-\sin \frac{\pi n k}{N} \right) = -F_k \, .$$

Hence, only half of the indices are needed to be computed to obtain all relevant data. In particular, $k = 1, \dots, N-1$ is sufficient, since

$$F_0 = \frac{-i}{N} \sum_{n=1}^{N-1} f_n \sin(0) = 0, \qquad F_{\pm N} = \frac{-i}{N} \sum_{n=1}^{N-1} f_n \sin\left(\frac{\pi \pm N(-k)}{N} \right) = 0 \, .$$

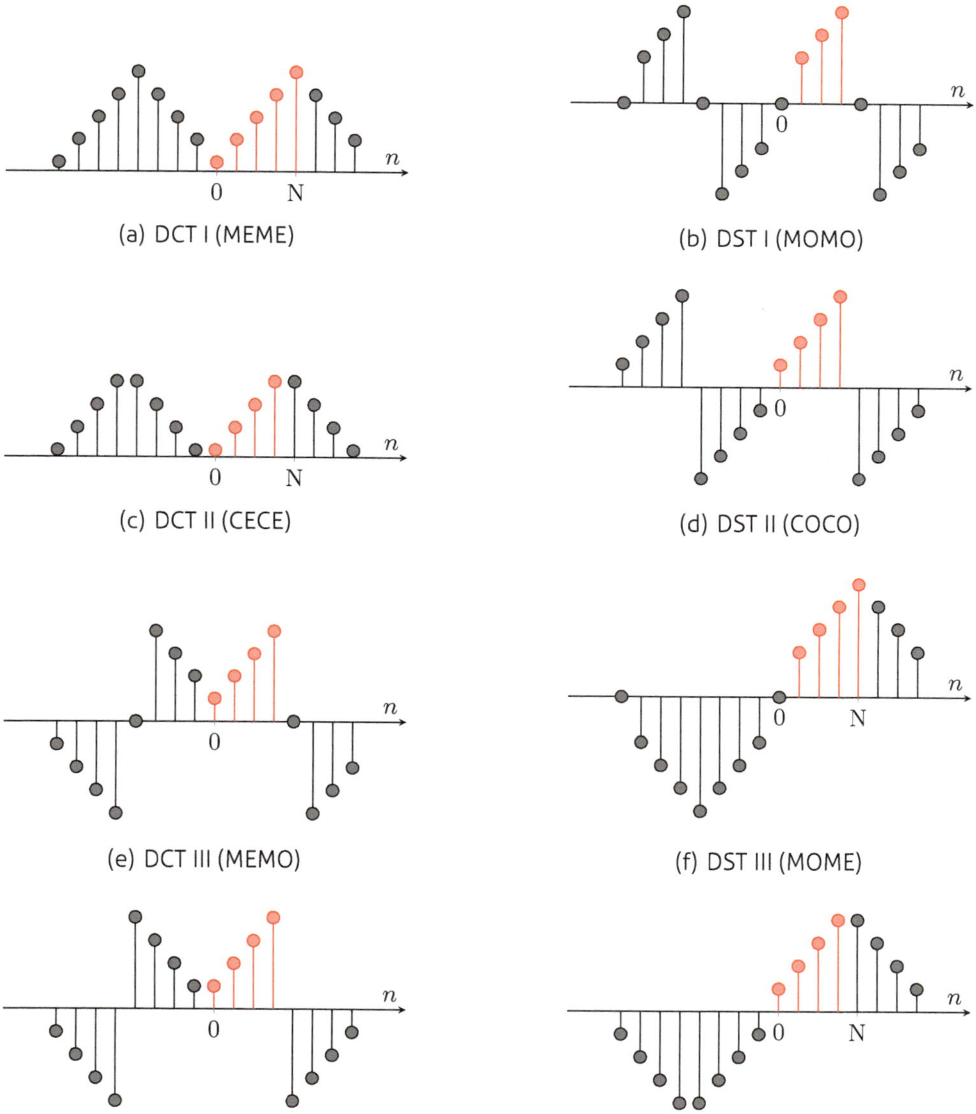

(a) DCT I (MEME)

(b) DST I (MOMO)

(c) DCT II (CECE)

(d) DST II (COCO)

(e) DCT III (MEMO)

(f) DST III (MOME)

(g) DCT IV (CECO)

(h) DST IV (COCE)

Figure 9.6. Survey of the first four variants of each DCT (left) and DST (right). The fix odd and even symmetry, resp., is clearly visible in the centre around zero, whereas the symmetry in the continuation as well as the mirror/copy properties at the boundaries vary.

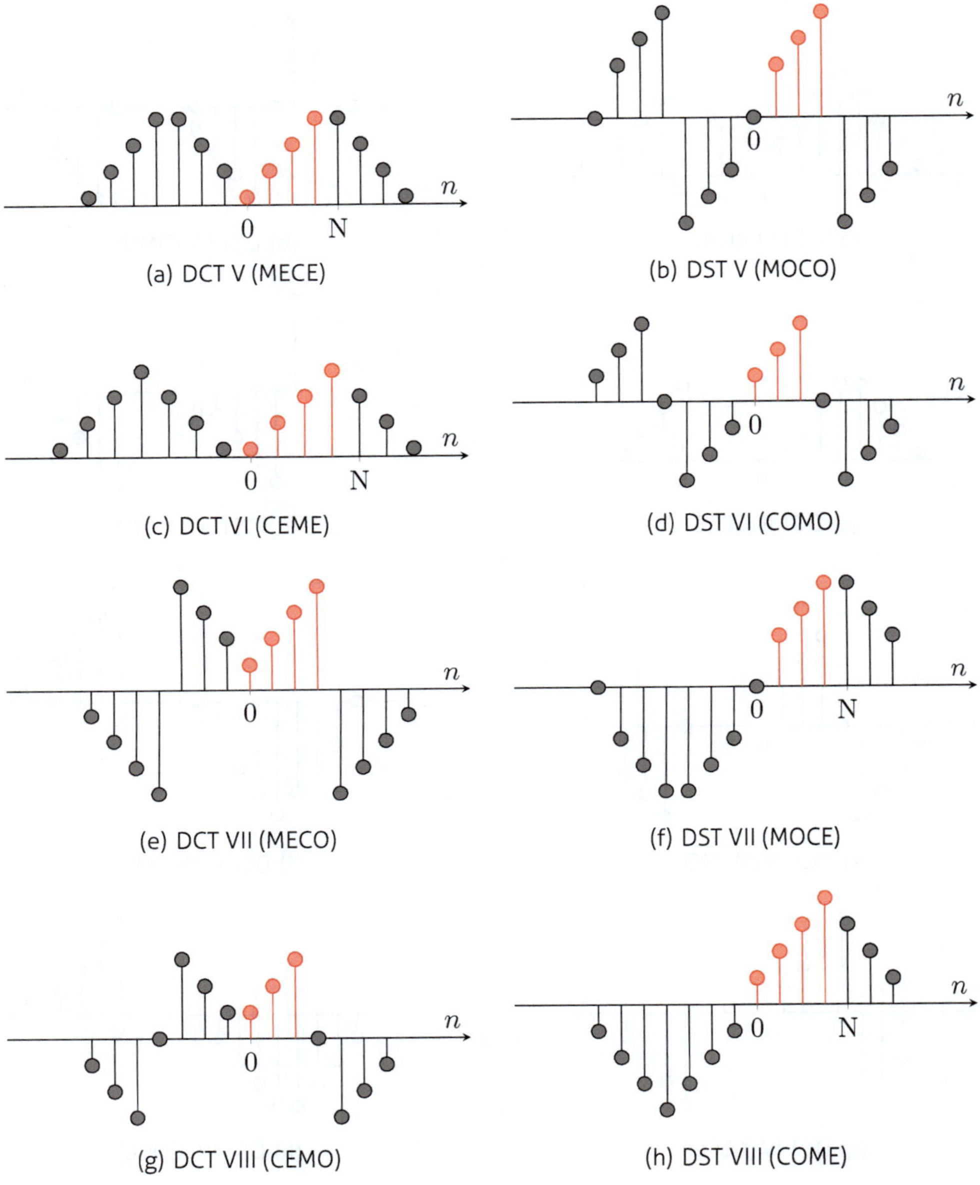

Figure 9.7. Survey of the second four variants of each DCT (left) and DST (right). Again, the fix odd and even symmetry, resp., is clearly visible in the centre around zero, whereas the symmetry in the continuation as well as the mirror/copy properties at the boundaries vary.

Analogous calculations for the IDFT result in

$$f_n = 2i \sum_{k=1}^{N-1} F_k \sin\left(\frac{\pi n k}{N}\right), \quad n = 1, \dots N-1\,. \tag{9.15}$$

From Eq. (9.14) and (9.15), we obtain the forward and inverse mode of the real-valued DST using $\widehat{F}_k := iF_k$:

$$\widehat{F}_k = \frac{1}{N} \sum_{n=1}^{N-1} f_n \sin\left(\frac{\pi n k}{N}\right), \quad k = 1, \dots, N-1\,, \tag{9.16}$$

$$f_n = 2 \sum_{k=1}^{N-1} \widehat{F}_k \sin\left(\frac{\pi n k}{N}\right) \quad n = 1, \dots N-1\,. \tag{9.17}$$

Note that there is no additional symmetry assumed in the data or in the coefficients. Hence, the DST represents a new category of transforms compared to DFT. The equivalence of DST and DFT only holds for antisymmetric input data. The definition and derivation of the DST leads to one possible variant of implementation for its computation using pre- and postprocessing:

1. use the input vector f to generate a $2N$ vector x with odd symmetry

$$x_{-n} = -x_n \qquad \text{for } n = 1, \dots N-1$$
$$x_0 = x_N = 0$$

2. compute the coefficients X_k via the fast, real-valued FFT on the vector x;
3. postprocessing: $\widehat{X}_k = -\text{Im}\{X_k\}$ for $k = 1, \dots, N-1$.
4. if necessary: scaling

Besides the pre-/postprocessing approach, direct compact algorithms exist and are used (in the FFTW library [101], e.g.). Note that variants of the DST (I) formulation exist (different with respect to index shifts or starting indices, etc.) that are all equivalent to Eq. (9.16)-(9.17). In Sec. 9.6.2, the DST derived above is applied to a specific PDE problem.

9.5.2 The Discrete Cosine Transform

Similar to the DST I from the previous section, we are now going to derive the discrete cosine transform of a particular variant: The DCT II is sometimes called "the DCT" (see [1]), since it is probably the most prominent trigonometric transform due to the fact that it is the workhorse of image compression

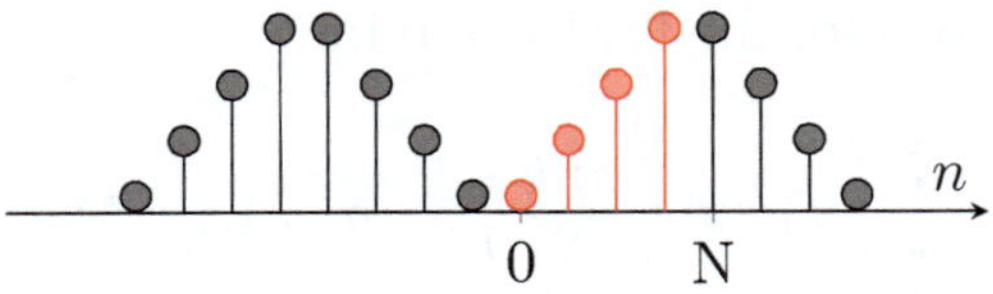

Figure 9.8. Visualisation of the DCT II with the copy even symmetry at both ends of the data vector (CECE).

(jpeg, mpeg, etc.). For the purpose of interpreting image data, the symmetric-periodic continuation corresponding to DCT II (see Fig. 9.8 for a visualisation) is necessary: discrete image data stored in pixels are treated by typically 8×8 blocks of a two-dimensional DCT II. Since the underlying image function is, by default, smooth at the block borders, the even symmetry (no jumps!) in combination with the copying (CECE) is the variant of choice for the continuation at the borders.

Assuming the CECE symmetry in the underlying data f of length $2N$ (again for simplicity) translates to

$$\begin{aligned} f_n &= f_{2N-n-1}, \qquad n = 0, \ldots, N\,, \\ f_N &= f_{N-1}\,, \\ f_0 &= f_{-1} = f_{2N} = F_{2N-1}\,. \end{aligned}$$

We now formulate the DFT (of length 2N for simplicity) for the above symmetry:

$$\begin{aligned} F_k &= \frac{1}{2N}\sum_{n=0}^{2N-1} f_n \omega_{2N}^{-nk} = \frac{1}{2N}\sum_{n=0}^{N-1} f_n \omega_{2N}^{-nk} + \frac{1}{2N}\sum_{n=N}^{2N-1} f_n \omega_{2N}^{-nk} \\ &\overset{n \rightsquigarrow 2N-n-1}{=} \frac{1}{2N}\sum_{n=0}^{N-1} f_n \omega_{2N}^{-nk} + \frac{1}{2N}\sum_{n=N-1}^{0} f_{2N-n-1}\omega_{2N}^{-(2N-n-1)k} \\ &= \frac{1}{2N}\sum_{n=0}^{N-1} f_n \omega_{2N}^{-nk} + \frac{1}{2N}\sum_{n=0}^{N-1} f_n \omega_{2N}^{k}\omega_{2N}^{nk} \underbrace{\omega_{2N}^{-2Nk}}_{=e^0=1} \\ &= \frac{1}{2N}\sum_{n=0}^{N-1} f_n \left(\omega_{2N}^{-nk} + \omega_{2N}^{k}\omega_{2N}^{nk}\right) \\ &= \frac{1}{2N}\omega_{2N}^{k/2}\sum_{n=0}^{N-1} f_n \left(\underbrace{\omega_{2N}^{-k/2}\omega_{2N}^{-nk}}_{=:z} + \underbrace{\omega_{2N}^{k/2}\omega_{2N}^{nk}}_{=:\bar{z}}\right) \qquad (9.18) \end{aligned}$$

Applying the relation $z + \bar{z} = 2\text{Re}\{z\}$ for complex numbers, Eq. (9.18) reads

$$F_k = \underbrace{\omega_{2N}^{k/2} \frac{1}{N} \sum_{n=0}^{N-1} f_n \cos\left(\frac{\pi}{N}(n+\frac{1}{2})k\right)}_{=:\widetilde{F}_k}, \qquad k = 0, \dots, 2N-1 \quad (9.19)$$

Defining $\widetilde{F}_k$ via $F_k = \omega_{2N}^{k/2} \widetilde{F}_k$ results in the classical definition of the DCT II (often called quarter-wave DCT due to the shift of a quarter wave length $1/2$). Note that the following properties hold:

$$\begin{aligned} \widetilde{F}_{-k} &= \widetilde{F}_k, \qquad k = 0, \dots, N-1 \quad \text{(due to the symmetry of cos)}, \\ \widetilde{F}_N &= \frac{1}{N} \sum_{n=0}^{N-1} f_n \cos\left(\frac{\pi N}{N}(n+\frac{1}{2})k\right) = 0 \,. \end{aligned}$$

Hence, it is sufficient to compute $\widetilde{F}_k$ for the N indices $k = 0, 1, \dots, N-1$. Analogous calculations for the inverse DFT result in

$$f_n = \widetilde{F}_0 + 2 \cdot \sum_{k=1}^{N-1} \widetilde{F}_k \cos\left(\frac{\pi}{N} n(k+\frac{1}{2})\right), \quad n = 1, \dots N-1 \,. \qquad (9.20)$$

To summarise, we obtained the forward and inverse DCT II from Eqs. (9.19) and (9.20) as

$$\begin{aligned} \widetilde{F}_K &= \frac{1}{N} \sum_{n=0}^{N-1} f_n \cos\left(\frac{\pi}{N}(n+\frac{1}{2})k\right) & k = 0, \dots, N-1 \,, \quad (9.21) \\ f_n &= \widetilde{F}_0 + 2 \cdot \sum_{k=1}^{N-1} \widetilde{F}_k \cos\left(\frac{\pi}{N} n(k+\frac{1}{2})\right), & n = 1, \dots, N-1 \,. (9.22) \end{aligned}$$

As in the case of the DST, variants of the DCT II formulation exist, but they are all equivalent to Eq. (9.21)-(9.22). Note that in contrast to DFT, both forward and inverse mode of the DCT II (and actually of all DCT variants) are real-valued transforms with real-valued coefficients. As for the DST, there is no additional symmetry assumed in the data or in the coefficients. Hence, DCT may be applied to arbitrary data such as pixel values in image compression. DCT can efficiently be implemented in software as well as hardware (the latter being important for fast image processing). Direct transforms similar to the FFT exist, known as fast cosine transforms (FCT; cf. [101], e.g.).

9.6 Solving the Poisson Equation

In this section, the continuous Fourier transform and the discrete cosine transform are applied to the model problem of solving the Poisson equation. Both the continuous and the discrete approach are designed as exercises for the reader.

9.6.1 Fourier's Method for Partial Differential Equations

In the following, we present a set of problems leading to Fourier's method deriving analytical solutions for general partial differential equations with particular emphasis on the heat equation.

Let $(V, \langle\cdot,\cdot\rangle)$ be a (complex) pre-Hilbert space and $A : V \to V$ a linear map. We assume that $\dim V = n < \infty$ and $\{v_1, \ldots, v_n\}$ is an orthonormal basis of eigenvectors of A corresponding to the eigenvalues $\{\lambda_1, \ldots, \lambda_n\}$.

Step 1:

Consider the following initial value problem for a linear ordinary differential equation on V:

$$\frac{d}{dt} w = Aw\,, \qquad w(0) = g \in V\,. \tag{9.23}$$

Prove that if $a_i := \langle g, v_i\rangle$, then $g = \sum_{i=1}^{n} a_i v_i$ holds, and

$$w(t) = \sum_{i=1}^{n} a_i e^{\lambda_i \cdot t} v_i \tag{9.24}$$

is the unique solution of (9.23).

Formulate the analogous statement for the initial value problem

$$\frac{d^2}{dt^2} w = Aw\,, \qquad w(0) = g, \quad \frac{d}{dt} w(0) = h \in V. \tag{9.25}$$

Before you continue, make sure to answer the following questions:

Quiz: Section 9.5

Q1 Are the coefficients of DCT II complex or real-valued? Why?

Q2 Derive the IDST I formula (9.17) using the IDFT for the properties of the coefficients $\widehat{F}_k$.

Q3 Derive the IDCT II formula (9.22) using the IDFT for the properties of the coefficients $\widetilde{F}_k$.

Why is the extra term "$\frac{d}{dt}w(0) = h$" needed, here?

Step 2 (Fourier's Method for the Wave Equation):

Let us now search for solutions of the wave equation with respect to periodic boundary conditions:

$$\begin{cases} u_{tt} - u_{xx} = 0 & \text{in } [0, 2\pi] \times (0, \infty), \\ u = g, u_t = h & \text{on } [0, 2\pi] \times \{0\}, \\ D_x^\alpha u(t, 0) = D_x^\alpha u(t, 2\pi) & \text{for } t \in [0, \infty), |\alpha| \leq 2. \end{cases} \tag{9.26}$$

Define the pre-Hilbert space $V := \{v \in \mathcal{C}^2([0, 2\pi], \mathbb{C}) \mid D^\alpha v(0) = D^\alpha v(2\pi) \text{for } |\alpha| \leq 2\}$ with scalar product $\langle f_1, f_2 \rangle := \frac{1}{2\pi} \int \overline{f_1(x)} f_2(x) \mathrm{d}x$.

The goal is to solve (9.26) by interpreting it as an ordinary differential equation (9.25) in V with the operator

$$A := \frac{d^2}{dx^2} : V \subset L^2[0, 2\pi] \to L^2[0, 2\pi].$$

To this end, we assume $g, h \in V$.

1. Show that $\{e^{ikx}\}_{k \in \mathbb{Z}}$ is an orthonormal system of eigenvectors of A.
2. Show that this system is complete in the sense that each $v \in V$ can be written as $v(x) = \sum_{k \in \mathbb{Z}} c_k \, e^{ikx}$ (the series converges in the norm induced by $\langle \cdot, \cdot \rangle$).
3. Determine the corresponding eigenvalues.

Now, consider the wave equation (9.26) with initial conditions such that for $0 < x < 2\pi$ it holds that $g(x) = x^2(x - 2\pi)^2$ and $h \equiv 0$. Verify that g can be extended to an element of V. Apply the ansatz from step 1)—formally, as we do not know initially if the series we are dealing with converges—with respect to the basis $\{e^{ikx}\}_{k \in \mathbb{Z}}$ and the initial conditions g and h. Prove that this leads to a solution and plot this solution.

Step 3 (Fourier's Method for the Heat Equation – Periodic Boundary Conditions):

Calculate the solution for the following heat equation on the interval $\Omega = (0, 2\pi) \subset \mathbb{R}_+$ and plot it:

$$\begin{cases} u_t - u_{xx} = 0 & \text{for } t > 0 \text{ and } x \in \overline{\Omega}, \\ u(0; x) = x^2(x - 2\pi)^2 & \text{for } x \in \overline{\Omega}, \\ \partial_x^k u(t; 0) = \partial_x^k u(t; 2\pi) & \text{for } t \geq 0 \text{ and } k = 0, 1, 2. \end{cases} \tag{9.27}$$

Step 4 (Fourier's Method for the Heat Equation – Neumann Boundary Conditions):

Consider the following one-dimensional von Neumann boundary problem for the heat equation:

$$\begin{cases} u_t - u_{xx} = 0 & \text{in } (0,\infty) \times [0,\pi], \\ u = g & \text{on } \{0\} \times [0,\pi], \\ u_x(0,t) = u_x(\pi,t) = 0 & \text{for } t \geq 0, \end{cases} \tag{9.28}$$

The aim of this exercise is to construct explicit solutions of (9.28) for suitable initial data g. Therefore, we conduct the following steps:

1. Choose a suitable pre-Hilbert space V such that (9.28) becomes a linear ODE there and formulate this ODE.

2. Choose a suitable orthonormal basis of V of eigenvectors of the linear operator A and obtain the corresponding eigenvalues.

   Hint: Eigenvectors $v_k \in V$ of A are directly obtained via $Av_k = \lambda v_k$ (boundary conditions!).

3. Make an ansatz for the solution of (9.28). Under which conditions does this ansatz lead to a solution? Justify your assertions.

4. Prove that if $u(t;x) \in C^{1,2}_{t,x}(\overline{\mathbb{R}_+} \times \overline{\Omega})$ is a solution of (9.28), then

   $$u(t;x) \;\to\; C \quad \text{uniformly with } t \to \infty\,.$$

   Identify the constant $C \in \mathbb{R}$.

5. Solve (9.28) for $g(x) = x$ and plot the solution.

   Hint: The solution will not be smooth everywhere on the boundary.

Step 5 (Fourier's Method for the Heat Equation – 2D):
Let $0 < a$, $0 < t$ and $0 < x < l_1$, $0 < y < l_2$. Determine the solution of the two-dimensional heat conduction problem

$$u_t = a^2 \Delta u$$

subject to

$$u(t,0,y) = u(t,x,0) = 0 \qquad u(t,l_1,y) = u(t,x,l_2) = 0$$

and $u(0,x,y) = f(x,y)$ for $0 < x < l_1$, $0 < y < l_2$.

Hint: The ansatz $u(t,x,y) = T(t) \cdot X(x) \cdot Y(y)$ combined with Fourier's methods may be helpful.

We refer to [13] for a more detailed discussion of the Fourier method applied to PDEs and plenty of examples as well as exercises.

9.6.2 Fast Poisson Solver

We now switch our perspective to a discrete point of view. Using the discrete sine transform (DST) for the Poisson problem

$$\Delta u(x) = f(x), \qquad x \in \Omega := [0,1]^2\,, \tag{9.29}$$
$$u(x) = 0, \qquad x \in \Gamma := \partial\Omega\,, \tag{9.30}$$

a fast alternative solver for the resulting linear system of equations can be constructed.

a) **1D Problem**

In the one-dimensional case, the discretisation of Eq. (9.29) with centred finite differences results in the following linear system of equations

$$u_{n+1} - 2u_n + u_{n-1} = f_n, \quad n = 1, \ldots, N-1\,, \tag{9.31}$$

where $u_1, \ldots, u_{N-1}$ represent the unknowns and f_n is defined as $f_n := h^2 \cdot f(x_n)$ using the mesh size h. Due to the Dirichlet zero boundary conditions $u_0 = u_N = 0$, no equations for index 0 and N exist and no contribution in the discretised right hand side f_1 and f_{N-1} appear.

Now, we use the (inverse) discrete sine transform (DST) of u and f

$$u_n = 2\sum_{k=1}^{N-1} U_k \sin\frac{\pi n k}{N}, \qquad f_n = 2\sum_{k=1}^{N-1} F_k \sin\frac{\pi n k}{N}$$

and insert these terms into the system of equations (9.31).

i) Show that the DST coefficients U_k depend on the F_k in the following way

$$U_k = \frac{F_k}{2\cos\frac{\pi k}{N} - 2} \qquad \text{for} \quad k = 1, \ldots, N-1\,. \tag{9.32}$$

Hence, the U_k can be retrieved directly from the F_k, without solving a linear system of equations!

ii) Formulate an algorithm which solves the system of equations (9.31) efficiently by using the dependency (9.32) and the fast sine transform(s).

b) **2D Problem**

In the two-dimensional case, the FD discretisation of Eq. (9.29) results in the following linear system of equations

$$u_{n,m+1}+u_{n+1,m}-4u_{n,m}+u_{n-1,m}+u_{n,m-1}=f_{n,m}, \quad n,m=1,\ldots,N-1, \tag{9.33}$$

again with homogeneous Dirichlet boundary conditions and the scaling $f_{n,m} := h^2 \cdot f(x_n, y_m)$ using the mesh size $h_x = h_y = h$.

Similar to 1D, we use the 2D (inverse) discrete sine transform (DST) of u and f

$$u_{n,m} = 2\sum_{k=1}^{N-1}\sum_{l=1}^{N-1} U_{k,l} \sin\frac{\pi nk}{N} \sin\frac{\pi ml}{N},$$
$$f_{n,m} = 2\sum_{k=1}^{N-1}\sum_{l=1}^{N-1} F_{k,l} \sin\frac{\pi nk}{N} \sin\frac{\pi ml}{N}$$

and insert these terms into the linear system of equations (9.33).

i) Show that the DST coefficients $U_{k,l}$ depend on the $F_{k,l}$ in the following way

$$U_{k,l} = \frac{F_{k,l}}{2\cos\frac{\pi k}{N} + 2\cos\frac{\pi l}{N} - 4} \qquad \text{for} \quad k,l=1,\ldots,N-1\,. \tag{9.34}$$

Hence, the $U_{k,l}$ can be retrieved directly from the $F_{k,l}$, without solving a linear system of equations!

ii) Formulate a 2D algorithm which solves the system of equations (9.33) efficiently by using the dependency (9.34) and the fast sine transform(s).

9.7 Chapter's Summary

This chapter summarises important aspects of Fourier transforms. The repetition of basic aspects of the continuous Fourier transform paves the way for analysing noise spectra in Chap. 10.

The focus of this chapter has been put on the discrete Fourier transform. We derived the workhorse of all computational Fourier transforms, the FFT in the Cooley-Tukey form. The 16 different discrete trigonometric transforms (DST and DCT) have been derived using the simple means of symmetry and symmetry points in the underlying data vectors. These transforms allow for, besides other applications, the direct solution of discrete Poisson(-like) scenarios and are used in the workshop project (cf. Chap. 18).

Of course, many interesting aspects of continuous and discrete Fourier transforms have been skipped. For further reading, we recommend [163] (including plenty of historic materials) and the corresponding exercise collection [164], [105] for applications concerning filtering and extensions to wavelets, and [101] for various aspects concerning the efficient implementation.

Problems

Classification: ☼ easy, ◷ easy with longer calculations, ☆ a little bit difficult, ◫ challenging.

Exercise 9.5. [☼] **Signals in Space and Frequency Domain**
Which of the three complex signals s_1, s_2, s_3 over time in Fig.9.9 (upper images) belongs to which set of discrete Fourier coefficients f_1, f_2, f_3 (lower images)?

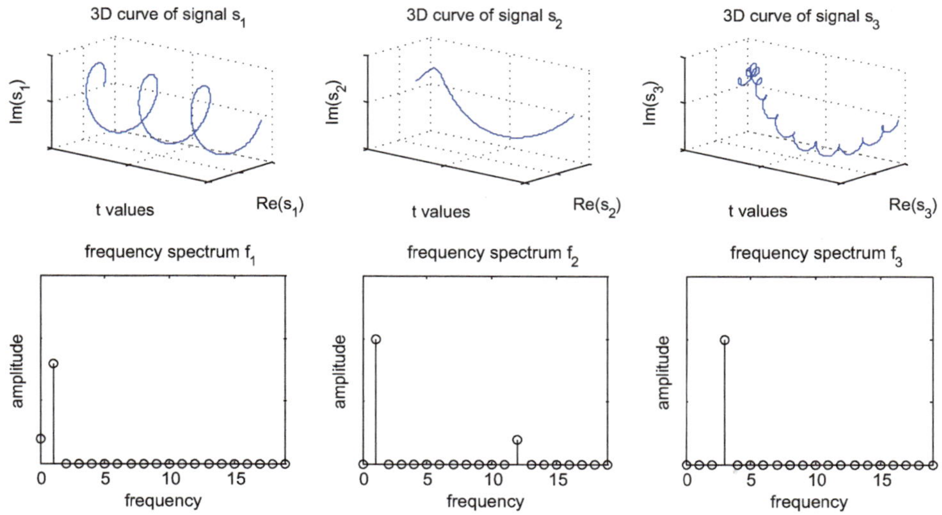

Figure 9.9. Visualisation of the signals s_1, s_2, and s_3 and the frequencies f_1, f_2, and f_3 relevant for Problem 9.5.

Exercise 9.6. [☼] Equivalence of DFT and IDFT
Show the following equivalence

$$\mathsf{DFT}(f) = \frac{1}{N}\,\overline{\mathsf{IDFT}(\bar{f})}\,. \tag{9.35}$$

Exercise 9.7. [☼] Linearity of DFT and IDFT
Show that the DFT and IDFT are linear, i.e. that the following holds for $\alpha, \beta \in \mathbb{C}$

$$\begin{aligned} \mathsf{DFT}(\alpha f + \beta g) &= \alpha\,\mathsf{DFT}(f) + \beta\,\mathsf{DFT}(g) \\ \mathsf{IDFT}(\alpha f + \beta g) &= \alpha\,\mathsf{IDFT}(f) + \beta\,\mathsf{IDFT}(g)\,. \end{aligned}$$

Chapter 10

Noise Spectra and the Propagation of Oscillations

We start this chapter with the basic definitions related to the spectral representation of stationary and periodic stochastic processes. Based on these, we study the notions of energy, power and spectral density. Moreover, we give several examples for colored noise processes. Next, we discuss the frequency domain method for response analysis and the concept of linear filters. In particular, we apply this method to our problem of multi-storey excitation due to seismic impacts.

10.1 Key Concepts

We introduced white noise as a vanishing mean process with δ-distributed auto-correlation function. This process is named by analogy to white light which has equal energy per cycle (hertz), and thus a flat frequency spectrum (as white noise). Consequently, stochastic processes with a special non-flat energy spectrum are called colored noise processes. They can, for instance, be generated from white noise by application of linear transformations, often named linear filters. Two examples of such a transformation are our earthquake excitation models, the Kanai-Tajimi filter or Clough-Penzien filter, discussed in Sec. 3.2.2. There, Fig. 3.6 (b) shows the spectral representation of the Kanai-Tajimi filter's output with white noise input. Other examples for colored noise processes are the Ornstein-Uhlenbeck process that we applied to transform stochastic differential equations into random differential equations or power-law noises which we discuss in this chapter.

Despite understanding the "colors" of stochastic processes by passing to the frequency domain via the Fourier transform, the propagation of oscillations can be studied directly in the frequency domain. This gives us another tool for the description of ground motion impact on multi-storey buildings.

When reading this chapter note the answers to the following questions

1. What can we say about the spectral representation of stationary and periodic processes?
2. What are power law noises?
3. How does the frequency domain method for response analysis work?

4. What are linear (time invariant) filters, and how can they be used to generate colored noise?

as well as the following key concepts

1. The energy and power spectral densities of stationary processes, and
2. The (complex) frequency response function.

This chapter is structured as follows: In Sec. 10.2, we give the basic definitions and implications related to the spectral representation of stationary and periodic stochastic processes. Based on these, we study the notions of energy, power and spectral density in Sec. 10.3. Moreover, we give several examples for colored noise processes. Next, in Sec. 10.4, we discuss the frequency domain method for response analysis and the concept of linear filters. In particular, we apply this method to our problem of multi-storey excitations due to seismic impacts. Finally, Sec. 10.5 wraps up the contents of this chapter.

10.2 Spectral Properties of Stationary & Periodic Processes

In this section, we provide some non-trivial applications of continuous Fourier analysis, namely the spectral representation theorem in its scalar version (based on [91]) as well as essential spectral properties of multi-dimensional stationary and periodic processes (based on [45], chapter 3.1).

10.2.1 Stochastic Integration & the Spectral Representation Theorem

Let us start with some definitions: Let $L^2(\Omega, \mathcal{F}, \mathbb{P})$ be the space of all real-valued square integrable random variables defined on the probability space $(\Omega, \mathcal{F}, \mathbb{P})$. Then, we define the pairing $\langle X, Y\rangle := \mathbb{E}(XY) = \mathbb{C}ov(X,Y)$. Then, $\langle\cdot,\cdot\rangle$ is just a semi-product as $\langle X, X\rangle = 0$ if $\mathbb{P}(X=0) = 1$ that can be extended to a scalar product by considering equivalence classes $X \sim Y$ if $\mathbb{P}(X=Y) = 1$. In particular, the space of all such equivalence classes is a Hilbert space.

An *orthogonal-increment process* on $\Pi := [-\pi, \pi]$ is a complex-valued stochastic process $\{Z(\lambda), \lambda \in [-\pi,\pi]\}$ such that

- $\langle Z(\lambda), 1\rangle = \mathbb{E}(Z(\lambda)) = 0$, for all $\lambda \in [-\pi,\pi]$,
- $\|Z(\lambda)\|^2 = \mathbb{E}(Z(\lambda)\overline{Z(\lambda)}) < \infty$, for all $\lambda \in [-\pi,\pi]$, and
- $\langle Z(\lambda_2) - Z(\lambda_1), Z(\mu_2) - Z(\mu_1)\rangle = 0$, for $(\lambda_1,\lambda_2] \cap (\mu_1,\mu_2] = \{\}$.

The process $\{Z(\lambda), \lambda \in [-\pi, \pi]\}$ is called *right-continuous* if

$$\|Z(\lambda+\delta) - Z(\lambda)\|^2 = \mathbb{E}\left(|Z(\lambda+\delta) - Z(\lambda)|^2\right) \to 0, \qquad \text{as } \delta \to 0.$$

Next, let $Z := \{Z(\lambda), \lambda \in [-\pi, \pi]\}$ be a right-continuous orthogonal-increment process on $[-\pi, \pi]$. Then Z has a unique distribution function F such that

- $F(\lambda) = 0$ for $\lambda \leq -\pi$ and $F(\lambda) = F(\pi)$ for $\lambda \geq \pi$, and
- $F(\mu) - F(\lambda) = \|Z(\mu) - Z(\lambda)\|^2 = \mathbb{V}ar\,(Z(\mu) - Z(\lambda))$.

Let $Z := \{Z(\lambda), \lambda \in [-\pi, \pi]\}$ be an orthogonal-increment process with distribution function F. For functions $f \in L^2(F)$, which are functions such that

$$\int_\Pi f(\lambda) \mathrm{d}F(\lambda) < \infty,$$

the stochastic integral

$$I(f) = \int_\Pi f(\lambda) \mathrm{d}Z(\lambda)$$

is defined in the usual three-step way (also called stochastic induction, cf. [259]):

1. First let us consider primitive functions/ step functions

$$f(\lambda) := \sum_{i=1}^{n} \alpha_i \mathbb{I}_{(\lambda_i, \lambda_{i+1}]}(\lambda),$$

   with $\alpha_i \in \mathbb{C}$, $-\pi < \lambda_1 < \lambda_2 < \cdots < \lambda_n \leq \pi$ and set

$$I(f) := \sum_{i=1}^{n} \alpha_i \left(Z(\lambda_{i+1}) - Z(\lambda_i)\right).$$

2. Note, that the isometry

$$\langle I(f), I(g) \rangle_{L^2(\Omega, \mathcal{F}, \mathbb{P})} = \langle f, g \rangle_{L^2(F)}$$

   holds.

3. The set of primitive functions is dense in $L^2(F)$. Thus for any $f \in L^2(F)$ there exists a sequence of primitive functions f_n such that $\|f_n - f\| \to 0$. Since

$$\|I(f_n) - I(f_m)\|_{L^2(\Omega, \mathcal{F}, \mathbb{P})} = \|f_n - f_m\|_{L^2(F)} \to 0,$$

   the sequence $I(f_n)$ converges in norm and we are ready to finally define

$$I(f) := \lim_{n \to \infty} I(f_n).$$

Finally, let x_t be a (weakly) stationary process. Then, there is a right-continuous orthogonal-increment process $Z_x := \{Z_x(\lambda), \lambda \in \Pi = [-\pi, \pi]\}$ with distribution function F_x such that

$$x_t = \int_\Pi \exp(i\lambda t) \mathrm{d} Z_x(\lambda), \qquad \text{with probability one}.$$

In particular, if $Y_x = \{Y_x(\lambda), \lambda \in \Pi\}$ and $Z_x = \{Z_x(\lambda), \lambda \in \Pi\}$ are two such processes, then

$$\mathbb{P}(Y_x(\lambda) = Z_x(\lambda), \lambda \in \Pi) = 1.$$

Moreover, the auto-covariance function of x_t is given by

$$\gamma(\tau) = \int_\Pi \exp(i\lambda\tau) \mathrm{d} F_x(\lambda).$$

In this context, F is called *spectral distribution function* of x_t. Under the additional assumption that F_x is absolutely continuous with density f_x, then

$$F_x(\lambda) = \int_{-\pi}^{\lambda} f_x(\mu) \mathrm{d}\mu,$$

and f_x is called the *spectral density function* of x_t. It satisfies

$$\gamma(\tau) = \int_\Pi f_x(\lambda) \exp(i\lambda\tau) \mathrm{d}\lambda.$$

Next, we extend this body of knowledge to the multi-dimensional setting.

10.2.2 Stationary & Periodic Processes

As defined in Chap. 1, a stochastic process X_t, $t \in I$, is called stationary if it is a second order process, if its mean $m_x(t)$ is constant on I and its covariance function depends only on the difference $(t_2 - t_1)$ between two points in time: $C_X(t_1, t_2) = C_X(t_2 - t_1)$. The covariance function $C_X(t)$ is also called the *correlation function*.

Thus, if the matrix function $K(t_1, t_2)$ from Lemma 1.27 depends only on $t_2 - t_1$ and if $C(\tau) = C(t_1, t_1 + \tau)$ is continuous on I and if finally $m(t) = m$ is constant, than the Gaussian process that exists due to Lemma 1.27 as well as every other process with mean m and covariance function $C(t_1, t_2)$ is a stationary mean-square continuous stochastic process according to property 1 from Sec. 4.3. On the other hand, every covariance function $C(t_1, t_2)$ of a stationary mean-square continuous process satisfies the conditions of Lemma 1.27, and the correlation function $C(\tau)$ is continuous.

Lemma 10.1 (Spectral representation & Spectral Distribution). *The elements of the correlation function of a stationary mean-square continuous process X_t have the spectral representation*

$$C_{k,l}^X(\tau) = \int_{-\infty}^{\infty} \exp(i\lambda\tau)\, dF_{k,l}^X(\lambda)\,, \quad \textit{for}\; k,l = 1,2,\ldots,d\,, \tag{10.1}$$

where the complex matrix function

$$F_X = \left(F_{k,l}^X\right)_{k,l=1,2,\ldots,d}$$

is called the the spectral distribution of X_t. It has the following properties:

1. *The matrix $F_X(\lambda_2) - F_X(\lambda_1)$ is non-negative definite for arbitrary $\lambda_1 < \lambda_2$.*
2. *The diagonal elements $F_{k,k}^X$ are real functions and it holds that*

$$\mathrm{tr}\,(F(+\infty) - F(-\infty)) < \infty$$

holds.

If there are complex functions $\Phi_{k,l}^X$ on $\mathbb{R}$ such that

$$F_{k,l}^X(\lambda) = \int_{-\infty}^{\lambda} \Phi_{k,l}^X(\rho)\mathrm{d}\rho\,, \quad \lambda \in \mathbb{R}\,,$$

then $\Phi^X = \left(\Phi_{k,l}^X\right)_{k,l=1,2,\ldots,d}$ is called the *spectral density of X_t*. A sufficient condition for the existence of a spectral density is

$$\int_{-\infty}^{\infty} \|C_X(\tau)\|\,\mathrm{d} < \infty\,.$$

Lemma 10.2 (Mean-Square Differentiability of Mean-Square Continuous Processes). *A stationary mean-square continuous process X_t is mean-square differentiable if and only if*

$$\int_{-\infty}^{\infty} \lambda^2 dF_{k,k}^X(\lambda) < \infty\,, \quad k = 1,2,\ldots,d\,, \tag{10.2}$$

holds. $\dot{\overline{X}}_t$ is also a stationary process.

Lemma 10.3 (Spectral Distribution of ``Taylor"-Sums). *Let $\overline{X}_t^{(k)}$ denote the k-th mean-square continuous mean-square derivative of the one-dimensional second order stationary stochastic process X_t and let Z_t be defined as*

$$Z_t := \sum_{k=0}^{m} b_k \overline{X}_t^{(k)}\,, \quad t \in I\,, \quad m \geq 0\,,$$

where b_k, $k = 1, 2, \dots, m$, are real coefficients. Then, the spectral distribution of Z_t is

$$F_Z(\lambda) = \int_{-\infty}^{\lambda} \left| \sum_{k=0}^{m} b_k (i\rho)^k \right|^{-2} dF_X(\rho) . \tag{10.3}$$

Moreover it holds that

$$m_Z(t) = \sum_{k=0}^{m} b_k \frac{d^k}{dt^k} m_X(t) , \quad t \in I .$$

Let us give some examples of one-dimensional correlation functions and their corresponding spectral densities:

1. $C_\tau = C \exp(-\rho|\tau|)$, $C, \rho > 0$, and

$$\varphi(\lambda) = \frac{D}{\rho^2 + \lambda^2} , \quad D = \frac{C\rho}{\pi} .$$

2. $C_\tau = C \exp(-\rho|\tau|) \cos(\beta\tau)$, $C, \rho > 0$, and

$$\varphi(\lambda) = \frac{D(\lambda^2 + b^2)}{\lambda^4 + 2a\lambda^2 + b^4} , \quad D = \frac{C\rho}{\pi} ,$$

with $b = \sqrt{\rho^2 + \beta^2}$, $a = \rho^2 - \beta^2$.

3. Let $D, b > 0$ and $\sqrt{b} + a > 0$

$$\varphi(\lambda) = \frac{D}{\lambda^4 + 2a\lambda^2 + b}$$

or with $\sqrt{b} = \omega^2$ and $\sqrt{b} + a = 2\rho^2$

$$\varphi(\lambda) = \frac{D}{(\lambda^2 - \omega^2) + 4\rho^2\lambda^2} ,$$

respectively.

(a) $\omega^2 - \rho^2 = -\beta^2 < 0$

$$C(\tau) = \frac{D\pi}{4\rho\beta_1\omega^2} \left((\rho + \beta_1) \exp\left(-(\rho - \beta_1)|\tau|\right) - (\rho - \beta_1) \exp\left((\rho + \beta_1)|\tau|\right) \right) ,$$

(b) $\omega^2 = \rho^2$

$$C(\tau) = \frac{D\pi}{2\rho^3} \exp\left(-\rho|\tau|\right) \left(1 + \rho|\tau|\right) ,$$

(c) $\omega^2 - \rho^2 = \beta^2 > 0$

$$C(\tau) = \frac{D\pi}{2\rho\omega^2} \exp\left(-\rho|\tau|\right) \left(\cos\left(\beta\tau\right) + \frac{\rho}{\beta} \sin\left(\beta|\tau|\right)\right).$$

A second order stochastic process X_t, $t \in I$, is called *θ-periodic*, if

$$m_X(t_1 + \theta) = m_X(t_1), \quad C_X(t_1 + \theta, t_2 + \theta) = C_X(t_1, t_2), \quad t_1, t_2 \in I,$$

holds. A second order stochastic process is stationary if and only if it is θ-periodic for an arbitrary $\theta > 0$. If the stochastic process X_t is the mean-square limit m.s.-$\lim_{k\to\infty} X_t^{(k)} \to X_t$ of a sequence $\{X_t^{(k)}\}_{k=1,2,\ldots}$ of θ-periodic processes, then X_t is θ-periodic, too.

In the proceeding let I_0 be either $\mathbb{R}$ itself or an interval $[t_0, \infty)$. A stochastic process X_t, $t \in I_0$, (which may not be of second order) is called strictly θ-periodic if all finite dimensional distribution functions of the form

$$F\left(x_1, x_2, \ldots, x_k\,;\, t_1 + \tau, t_2 + \tau, \ldots, t_k + \tau\right), \quad \text{for } t_1, t_2, \ldots, t_k \in I$$

are θ-periodic, i.e., periodic functions in τ with an identical period θ. If the processes $X_t^{(k)}$, $k = 1, 2, \ldots$, are strictly θ-periodic and if X_t is a stochastic process such that $\lim_{k\to\infty} X_t^{(k)} \hat{=} X_t$, $t \in I_0$. Then, X_t is strictly θ-periodic.

Lemma 10.4 (Strictly θ-Periodic Processes from θ-Periodic Processes). *Let X_t, $t \in I$, be a strictly θ-periodic stochastic process, and let $y : I_0 \times \mathbb{R}^d \to \mathbb{R}^m$ be θ-periodic for every fixed $x \in \mathbb{R}^m$ and $\mathbb{R}^d$-measurable for fixed $t \in I_0$. Then, $Y_t = y(t, X_t)$ is a strictly θ-periodic stochastic process.*

For instance, if $X_t = (X_t^{(1)}, X_t^{(2)}, \ldots, X_t^{(k)})$ with m-dimensional vector processes $X_t^{(i)}$, then the linear combination $\sum_{i=1}^{k} a_i X_t^{(i)}$ is strictly θ-periodic.

X_t, $t \in I_0$, is a strictly θ-periodic m-dimensional vector process if and only if the function

$$\psi(t) = \mathbb{E}\left(\varphi\left(X_{t+t_1}, X_{t+t_2}, \ldots, X_{t+t_r}\right)\right)$$

is θ-periodic for every $r \in \mathbb{N}$, and for every bounded continuous function φ defined on $\mathbb{R}^{m\cdot r}$ as well as for arbitrary $t_\rho \in I_0$, $\rho = 1, 2, \ldots, r$.

A stochastic process X_t, $t \in I$, is called *strictly stationary* if all finite-dimensional distribution functions of the form $F(x_1, \ldots, x_k; t_1 + \tau, \ldots, t_k + \tau)$ are independent from τ. A stochastic process is strictly stationary if and only if it is strictly θ-periodic for every $\theta > 0$.

Before you continue, make sure to answer the following questions:

Quiz: Section 10.2

Q1 Explain how a meaningful stochastic integration can be defined.

Q2 Give the definition of the auto-covariance/ auto-correlation function of a stationary process.

Q3 Give the definition of the spectral distribution function and the spectral density function.

Q4 What can you say about the spectral representation and spectral density of a mean-square continuous process?

Q5 What is a stationary process and what is a strictly stationary process?

Q6 Under which conditions is a mean-square continuous stationary process differentiable? Is the result stationary again? Why?

Q7 What is a periodic process and what is a strictly periodic process?

10.3 Energy & Power Spectral Density, and Examples for Colored Noise

Following [203], let us first recall the fundamental notions relevant for the frequency domain interpretation of stochastic processes[1]:

The energy spectral density describes how the *energy*

$$\int_{-\infty}^{\infty} |x(t)|^2 \mathrm{d}t\,.$$

of a signal or a time series $x(t)$ is distributed with frequency. The energy spectral density is most suitable for pulse-like signals (or transients) that have a finite total energy. In this case, Parseval's theorem gives us an alternative expression for the energy of $x(t)$ in terms of its Fourier transform $\hat{x}(\omega)$:

$$\int_{-\infty}^{\infty} |x(t)|^2 \mathrm{d}t = \frac{1}{2\pi}\int_{-\infty}^{\infty} |\hat{x}(\omega)|^2 \mathrm{d}\omega\,,$$

where ω denotes the angular frequency. Since the integral on the right-hand side is the energy of the signal, the integrand $|\hat{x}(\omega)|^2$ can be interpreted as a

[1] Cf. http://en.wikipedia.org/wiki/Spectral_density

density function describing the energy per unit frequency contained in the signal at frequency ω. In light of this, the *energy spectral density* of a signal $x(t)$ is defined as

$$S_{xx}(\omega) = |\hat{x}(\omega)|^2 = \left| \int_{-\infty}^{\infty} x(t)e^{-i\omega t} \mathrm{d}t \right|^2 .$$

Consider the following example for the energy spectral density to connect this concept to physical intuition[2]:

Example 10.5 (Energy Spectral Density of an Electrical Pulse). Suppose $V(t)$ represents the potential (in volts) of an electrical pulse propagating along a transmission line of impedance Z, and suppose the line is terminated with a matched resistor (so that all of the pulse energy is delivered to the resistor and none is reflected back). By Ohm's law, the power delivered to the resistor at time t is equal to $V(t)^2/Z$, so the total energy is found by integrating $V(t)^2/Z$ with respect to time over the duration of the pulse. To find the value of the energy spectral density $S_{xx}(\omega)$ at frequency ω, one could insert between the transmission line and the resistor a bandpass filter which passes only a narrow range of frequencies ($\Delta\omega$, say) near the frequency of interest and then measure the total energy $E(\omega)$ dissipated across the resistor. The value of the energy spectral density at ω is then estimated to be $E(\omega)/\Delta\omega$. In this example, since the power $V(t)^2/Z$ has units of $V^2\Omega^{-1}$, the energy $E(\omega)$ has units of $V^2 s\Omega^{-1} = J$, and hence the estimate $E(\omega)/\Delta\omega$ of the energy spectral density has units of JHz^{-1}, as required. In many situations, it is common to forgo the step of dividing by Z so that the energy spectral density instead has units of $V^2 sHz^{-1}$.

The above definition of energy spectral density is most suitable for pulse-like signals for which the Fourier transforms exist. For continuous signals that describe, for example, stationary physical processes, it makes more sense to define a power spectral density (PSD), which describes how the power of a signal or time series is distributed over the different frequencies, as in the simple example given previously. Here, *power* can be the actual physical power, or more often the squared value of the signal. The total power P of a signal $x(t)$ is the following time average:

$$P := \lim_{T\to\infty} \frac{1}{2T} \int_{-T}^{T} x(t)^2 \mathrm{d}t .$$

In analysing the frequency content of the signal $x(t)$, one might like to compute the ordinary Fourier transform $\hat{x}(\omega)$; however, for many signals of interest this Fourier transform does not exist. Because of this, it is advantageous to

[2] Cited from http://en.wikipedia.org/wiki/Spectral_density, visited at the 9th of March 2013.

work with a truncated Fourier transform $\hat{x}_T(\omega)$, where the signal is integrated only over a finite interval $[0, T]$:

$$\hat{x}_T(\omega) = \frac{1}{\sqrt{T}} \int_0^T x(t) e^{-i\omega t} \mathrm{d}t .$$

Then the *power spectral density* can be defined as

$$S_{xx}(\omega) := \lim_{T\to\infty} \mathbb{E}\left(|\hat{x}_T(\omega)|^2\right) ,$$

where we explicitly have

$$\begin{aligned} \mathbb{E}\left(|\hat{x}_T(\omega)|^2\right) &= \mathbb{E}\left(\tfrac{1}{T}\int_0^T x^*(t)e^{i\omega t}\mathrm{d}t \int_0^T x(t')e^{-i\omega t'}\mathrm{d}t'\right) \\ &= \tfrac{1}{T}\int_0^T\int_0^T \mathbb{E}\left(x^*(t)x(t')\right) e^{i\omega(t-t')}\mathrm{d}t\,\mathrm{d}t' . \end{aligned}$$

Using such formal reasoning, one may already guess that for a stationary random process, the power spectral density $f(\omega)$ and the autocorrelation function

$$\gamma(\tau) := \mathbb{E}\left(X(t)X(t+\tau)\right)$$

of the signal should form a Fourier transform pair. Provided that $\gamma(\tau)$ is absolutely integrable, then

$$S_{xx}(\omega) = \int_{-\infty}^{\infty} \gamma(\tau)\exp(-i\omega\tau)\mathrm{d}\tau = \hat{\gamma}(\omega) .$$

A deep theorem by Norbert Wiener and Aleksandr Khinchin (the Wiener-Khinchin theorem) makes sense of this formula for any wide-sense stationary process under weaker hypotheses: γ does not need to be absolutely integrable, it only needs to exist. But the integral can no longer be interpreted as usual. The formula also makes sense if interpreted as involving distributions (in the sense of Laurent Schwartz, not in the sense of a statistical cumulative distribution function) instead of functions. If γ is continuous, Bochner's theorem can be used to prove that its Fourier transform exists as a positive measure, whose distribution function is F (but not necessarily as a function and not necessarily possessing a probability density). Many authors use this equality to actually define the power spectral density, cf. [218].

The power of the signal in a given frequency band $[\omega_1, \omega_2]$ can be calculated by integrating over positive and negative frequencies,

$$\int_{\omega_1}^{\omega_2} \left(S_{xx}(\omega) + S_{xx}(-\omega)\right) \mathrm{d}\omega = F(\omega_2) - F(-\omega_2) ,$$

where F is the integrated spectrum whose derivative is S_{xx}.

One of the results of Fourier analysis is Parseval's theorem which states that the area under the energy spectral density (ESD) curve is equal to the area under the square of the magnitude of the signal, the total energy:

$$\int_{-\infty}^{\infty} |f(t)|^2 \, \mathrm{d}t \;=\; \int_{-\infty}^{\infty} ESD(\omega)\mathrm{d}\omega \,.$$

Proposition 10.6 (Properties of the Power Spectral Density). *Some properties of the power spectral density include:*

- *The spectrum of a real valued process is an even function of frequency:* $S_{xx}(-\omega) = S_{xx}(\omega)$.
- *If the process is continuous and purely indeterministic, the auto-covariance function can be reconstructed by using the Inverse Fourier transform*
- *The power spectral density describes the distribution of the variance over frequency. In particular,*

$$\mathbb{V}ar(X_n) \;=\; \gamma_0 \;=\; 2\int_0^{1/2} S_{xx}(\omega)d\omega \,.$$

- *The power spectral density is a linear function of the auto-covariance function in the sense that if* γ *is decomposed into two functions* $\gamma(\tau) = \alpha_1\gamma_1(\tau) + \alpha_2\gamma_2(\tau)$*, then* $f = \alpha_1 S_{xx,1} + \alpha_2 S_{xx,2}$.

Proof. See problem 10.10 for the proof. □

Moreover, the integrated spectrum or power spectral distribution $F(\omega)$ is defined as

$$F(\omega) \;=\; \int_{-\infty}^{\omega} S_{xx}(\omega')\mathrm{d}\omega' \,.$$

Given two signals $x(t)$ and $y(t)$, each of which possess power spectral densities $S_{xx}(\omega)$ and $S_{yy}(\omega)$, it is possible to define a cross-spectral density given by

$$S_{xy}(\omega) \;=\; \lim_{T\to\infty} \mathbb{E}\left(\left(F_x^T(\omega)\right)^* F_y^T(\omega)\right) .$$

The cross-spectral density (or "cross power spectrum") is thus the Fourier transform of the cross-correlation function.

$$S_{xy}(\omega) = \int_{-\infty}^{\infty} R_{xy}(t)e^{-j\omega t}\mathrm{d}t = \int_{-\infty}^{\infty}\left(\int_{-\infty}^{\infty} x(\tau)\cdot y(\tau+t)d\tau\right)\mathrm{exp}(-j\omega t)\mathrm{d}t \,,$$

where $R_{xy}(t)$ is the cross-correlation of $x(t)$ and $y(t)$.

By an extension of the Wiener-Khinchin theorem, the Fourier transform of the cross-spectral density $S_{xy}(\omega)$ is the cross-covariance function. In light of this, the power spectral density is seen to be a special case of the cross-spectral density for $x(t) = y(t)$.

10.3.1 A More Realistic Model for Brown's Observation: The Ornstein-Uhlenbeck Process

Recall Browns observations on pollen particles suspended in a liquid and the properties a mathematical model for such particles should posses. Let us focus on the 1-dimensional situation and assume that a Brownian particle obeys Newton's second law of motion, i.e.,

$$\ddot{Y} = m^{-1}F(\dot{Y}, Y, t) = m^{-1}\left(-\tilde{a}\dot{Y} + \tilde{b}\xi_t\right),$$

where $Y_t : [0,T] \to \mathbb{R}$ is the position process and $X_t := \dot{Y}_t : [0,T] \to \mathbb{R}$ is the velocity process of the particle with mass $m > 0$. The force $F(x,t) : \mathbb{R} \times [0,T] \to \mathbb{R}$ acting on the particle is assumed to be the sum of a friction force $f_1(x,t) = -\tilde{a}x$ with friction coefficient $\tilde{a} > 0$ (damping) and a random diffusion force $f_2(x,t) = \tilde{b}\xi_t$ with diffusion coefficient $\tilde{b} \in \mathbb{R} \setminus \{0\}$. The stochastic driving process ξ_t is assumed to be uncorrelated with mean zero, in particular, we assume $\xi_t = w_t$ to be the white noise process.

With Gaussian initial conditions Y_0 and Y_1 the corresponding system of first order Itô stochastic differential equations becomes

$$\mathrm{d}\begin{pmatrix} Y_t \\ X_t \end{pmatrix} = \begin{pmatrix} X_t\mathrm{d}t \\ -aX_t\mathrm{d}t + b\mathrm{d}W_t \end{pmatrix}, \qquad \mathrm{d}\begin{pmatrix} Y(0) \\ X(0) \end{pmatrix} = \mathrm{d}\begin{pmatrix} Y_0 \\ Y_1 \end{pmatrix},$$

with $a := m^{-1}\tilde{a}$ and $b := m^{-1}\tilde{b}$. Hence, once X_t is known Y_t follows by integration:

$$Y_t = Y_0 + \int_0^t X_t \mathrm{d}t.$$

The velocity process X_t is called *Ornstein-Uhlenbeck process*[3], and its defining stochastic differential equation

$$\mathrm{d}X_t = -aX_t\mathrm{d}t + b\mathrm{d}W_t \quad \text{with} \quad X(0) = X_0, \tag{10.4}$$

where, for convenience, $X_0 := Y_1$ is set, has the solution

$$X_t = e^{-at}X_0 + b\int_0^t e^{-a(t-s)}\mathrm{d}W_t, \qquad t \geq 0,$$

3 See, e.g., [200] for a comparison of the Ornstein-Uhlenbeck process with the process studied by Einstein in [92].

as is straightforward to verify. If we assume $X_0 = Y_1$ to be normally distributed, we gain from the explicit solution that X_t is also Gaussian for all times $t \geq 0$.

In particular, $\mathbb{E}(X_t) = e^{-at}\mathbb{E}(X_0)$ and

$$\begin{aligned}
\mathbb{E}(X_t^2) = \mathbb{E}\left(e^{-2at}X_0^2 + 2be^{-at}X_0\int_0^t e^{-a(t-s)}\mathrm{d}W_s + b^2\left(\int_0^t e^{-a(t-s)}\mathrm{d}W_s\right)^2\right) \\
= e^{-2at}\mathbb{E}\left(X_0^2\right) + 2be^{-2at}\mathbb{E}\left(X_0^2\right)\mathbb{E}\left(\int_0^t e^{-a(t-s)}\mathrm{d}W_s\right) + b^2\int_0^t e^{-2a(t-s)}\mathrm{d}s \\
= e^{-2bt}\mathbb{E}\left(X_0^2\right) + \frac{b^2}{2a}\left(1 - e^{-2bt}\right).
\end{aligned}$$

Thus, the variance $Var(X_t) = \mathbb{E}(X_t^2) - \mathbb{E}^2(X_t)$ is given by

$$Var(X_t) = e^{-2bt}Var\left(X_0\right) + \frac{b^2}{2a}\left(1 - e^{-2bt}\right),$$

assuming of course, $Var(X_0) < \infty$. For any such initial condition X_0, we therefore have

$$\mathbb{E}(X_t) \overset{t\to\infty}{\longrightarrow} 0 \qquad \text{and} \qquad Var(X_t) \overset{t\to\infty}{\longrightarrow} \frac{b^2}{2a}.$$

From the explicit form of the solution, we see that the distribution of X_t approaches $\mathcal{N}(0, \frac{b^2}{2a})$ as $t \to \infty$. This means, regardless of the initial distribution, the solution of the stochastic differential equation for large times "settles down" into a Gaussian distribution whose variance $\frac{b^2}{2a}$ represents a balance between the random disturbing force $b\mathrm{d}W_t$ and the frictional damping force $-aX_t$ in each time increment.

Finally, back to the position process Y_t. We have

$$\begin{aligned}
\mathbb{E}(Y_t) = \mathbb{E}(Y_0) + \int_0^t \mathbb{E}(X_s)\mathrm{d}s = \mathbb{E}(Y_0) + \int_0^t e^{-as}\mathbb{E}(Y_1)\mathrm{d}s \\
= \mathbb{E}(Y_0) + \frac{1 - e^{-at}}{a}\mathbb{E}(Y_1),
\end{aligned}$$

and a somewhat lengthy calculation shows

$$Var(Y_t) = \ldots = Var(Y_0) + \frac{b^2}{a^2}t + \frac{b^2}{2a^3}\left(-3 + 4e^{-at} - e^{-2at}\right).$$

We refer to [110] for the simulation of Ornstein-Uhlenbeck noise.

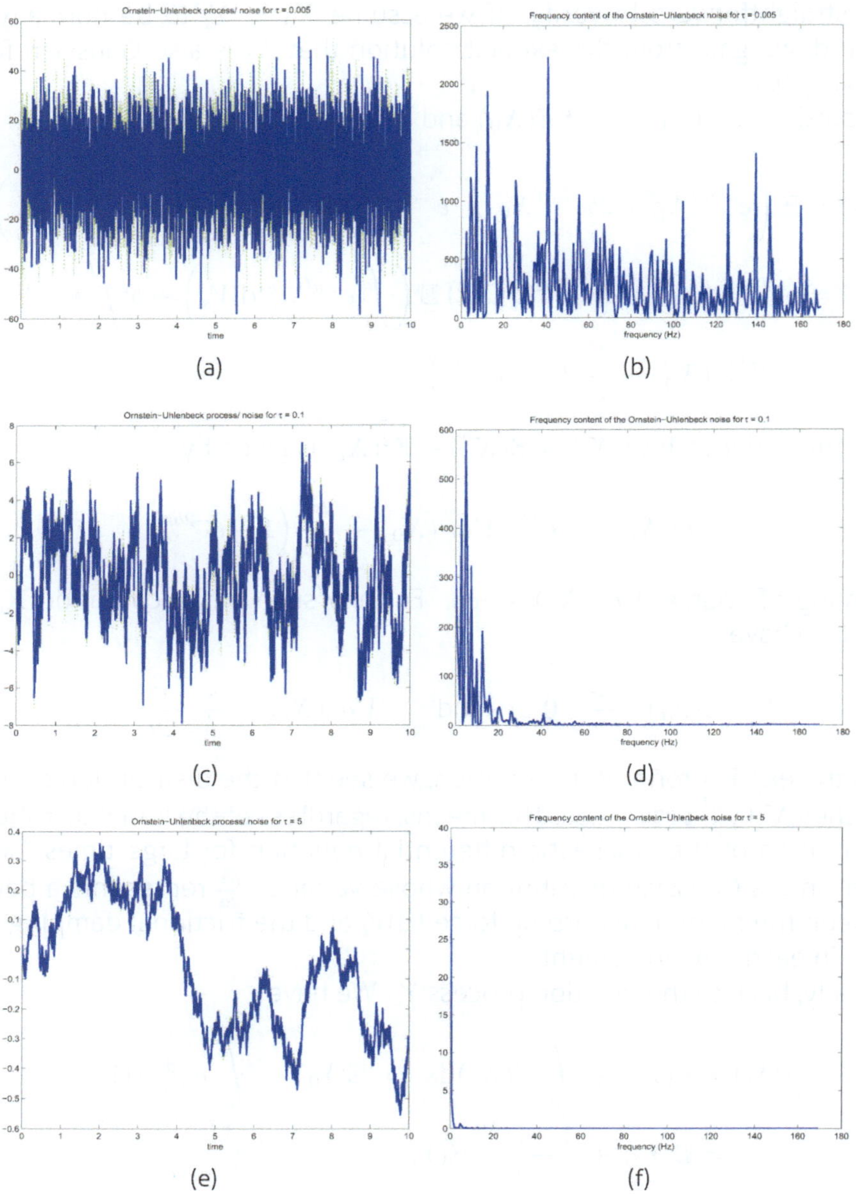

Figure 10.1. Three simulation runs for the Ornstein-Uhlenbeck process with different values of the correlation time τ (left hand side) and their corresponding power spectrum (right hand side). In (a), (b) we have $\tau = 0.005$, in (c), (d) $\tau = 0.1$ and in (e), (f) $\tau = 5$.

10.3.2 Power-Law Noise & its Simulation

Following [182], [236] and [127] it is reasonable to assume that many instances of noise occurring in nature have frequencies f that scale according to an inverse power law $f^{-\beta}$. In this setting, white noise is a special case with $\beta = 0$ where all frequencies contribute equally to the power spectrum, while colored noises are dominated by frequencies in a certain range. Specifically, red/ pink noise ($\beta = 1$) is dominated by low frequency (or long-period) cycles and the underlying stochastic process has autocorrelated increments. In the time domain, this produces an increased probability of having long runs of above (or below) average conditions; more intense slopes have been termed brown ($\beta = 2$). Recently, [252] investigated 152 (temperature) data sets of maritime, coastal and inland terrestrial environments to consolidate this hypothesis: it was shown that noise in terrestrial environments tends to be white ($\beta < 0.5$)[4], whereas noise in maritime environments (coastal and sea surface) tend to be red/ pink ($\beta \approx 1$) or brown ($\beta \approx 2$). See, [73], p. 970, for a summary of geophysical data showing the range of spectral exponents β found in nature. This suggests that dynamics of stochastic phenomena taking place in the human body, for instance, might be better modeled by red/ pink or even brown noise instead of white noise.

The importance of colored noise for biological systems in general is emphasised [237], [219] and [145], who showed that population persistence and extinction are influenced by noise color; though the effect of noise may be masked by periodic or chaotic dynamics, cf. [90]. Moreover, [169] predicted an increased risk of extinction in populations experiencing red/ pink noise relative to those experiencing white noise, based upon the simple observation that in red/ pink noise long runs of "poor" conditions for survival are more likely than they are in white noise. We refer to [252] for further literature sources supporting this hypothesis. Finally, [73] shows that very long population persistence times are more likely for black noise ($\beta = 3$) than for red/ pink one.

We refer to [190], [178] for the simulation of power-law noise.

10.4 The Frequency Domain Method for Response Analysis

The goal of this section is to shed some light on the propagation of external excitations by transforming the problem from the time domain to the frequency domain. In this domain, the propagation problem becomes a multipli-

4 Analyzing lakes and rivers, [80] found white or at most pink $\beta \approx 0.5$ spectra, which suggests that these systems are strongly coupled to the terrestrial environment.

cation of the excitation spectrum with a suitable function, called the *complex frequency response function*, see Fig. 10.2.

Following [57], pp. 851, as an introduction, let us consider the single-degree of freedom system damped by viscous effects

$$m\ddot{u} + c\dot{u} + ku = p(t)\,, \tag{10.5}$$

subject to an external force $p(t)$. From the physical point of view $\omega_n := \sqrt{m^{-1}k}$ denotes the natural frequency of the vibration, and $\zeta = c(2m\omega_n)^{-1}$ the damping ratio of the system. The displacement (or deformation) with respect to $p(t) = \sin(\omega t)$ is

$$u^s(t) = \frac{1}{k}\frac{(1-(\omega_n^{-1}\omega)^2)\sin(\omega t) - (2\zeta\omega_n^{-1}\omega)\cos(\omega t)}{(1-(\omega_n^{-1}\omega)^2)^2 + (2\zeta\omega_n^{-1}\omega)^2}\,,$$

and with respect to $p(t) = \cos(\omega t)$ it is

$$u^c(t) = \frac{1}{k}\frac{(1-(\omega_n^{-1}\omega)^2)\cos(\omega t) + (2\zeta\omega_n^{-1}\omega)\sin(\omega t)}{(1-(\omega_n^{-1}\omega)^2)^2 + (2\zeta\omega_n^{-1}\omega)^2}\,.$$

Before you continue, make sure to answer the following questions:

Quiz: Section 10.3

Q1 Give the definition of the energy and the energy spectral density of a signal.

Q2 Give the definition of the power and the power spectral density of a signal as well as the connection between the power spectral density and the auto-correlation of a signal.

Q3 How are the energy spectral density and the power spectral density connected?

Q4 Give some properties of the power spectral density connected and show how they can be derived from the definition of the power spectral density.

Q5 Why can the Ornstein-Uhlenbeck process be considered as better model for Brown's pollen experiment than the Wiener process?

Q6 Why is the Ornstein-Uhlenbeck process an example for colored noise?

Q7 Give some examples of power law noises.

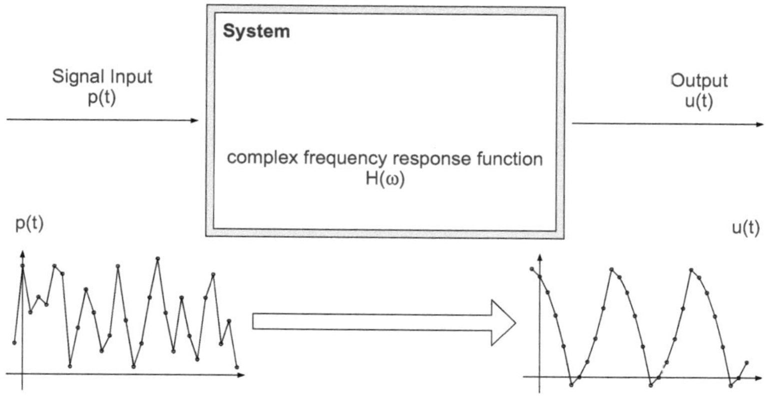

Figure 10.2. Schematic signal input-output diagram.

Thus, the steady state response of the system (10.5) will be harmonic motion at the forcing frequency, ω, which can be expressed in complex-number notation as

$$u(t) \;=\; u^c(t) + iu^s(t) \;=\; H(\omega)e^{i\omega t},$$

where $H(\omega)$ is the so-called *complex frequency response function* that takes $e^{i\omega t}$ as an input and results in $u(t)$, the solution of (10.5) with $p(t) = e^{i\omega t}$.

To determine the unknown complex frequency response function $H(\omega)$ in $u(t) = H(\omega)e^{i\omega t}$, we differentiate this expression for $u(t)$, i.e.,

$$\dot{u} \;=\; i\omega H(\omega)e^{i\omega t}, \qquad \ddot{u} \;=\; -\omega^2 H(\omega)e^{i\omega t},$$

and plug it into the system equation (10.5):

$$(-\omega^2 m + i\omega c + k)H(\omega)e^{i\omega t} \;=\; e^{i\omega t}.$$

Canceling the $e^{i\omega t}$ term from both sides of the equation gives $H(\omega)$ as a rational function in ω:

$$\begin{aligned} H(\omega) \;&=\; \frac{1}{-\omega^2 m + i\omega c + k} \;=\; \frac{1}{k}\frac{1}{(1-(\omega_n^{-1}\omega)^2) + i(2\zeta\omega_n^{-1}\omega)} \\ &=\; \frac{1}{k}\frac{(1-(\omega_n^{-1}\omega)^2) - i(2\zeta\omega_n^{-1}\omega)}{(1-(\omega_n^{-1}\omega)^2)^2 + (2\zeta\omega_n^{-1}.\omega)^2} \end{aligned}$$

This method is easily generalized to suitable scalar differential operators $\mathcal{L}$ and periodic input signals $p(t)$ the Fourier series of which exists over the interval $[0,T]$ such that we have

$$\mathcal{L}u \;=\; p(t) \;=\; \sum_{k=-\infty}^{\infty} \hat{p}_k e^{i(k\omega_0)t}, \quad \hat{p}_k \;:=\; \tfrac{1}{T}\int_0^T p(t)e^{-i(k\omega_0)t}\mathrm{d}t, \; k \in \mathcal{Z},$$

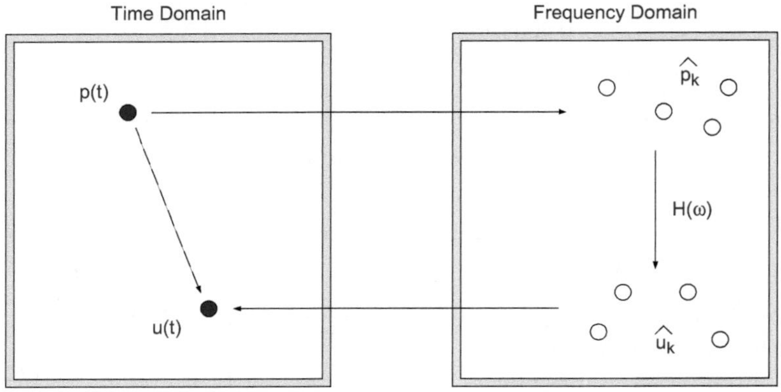

Figure 10.3. Solution of a response problem with respect to periodic excitation $p(t)$ by transformation to the frequency domain, cf. [57], p. 857.

with the base frequency ω_0. Thus, for periodic input signals p the solution u reads as

$$u(t) = \sum_{k=-\infty}^{\infty} H(k\omega_0)\hat{p}_k e^{i(k\omega_0)t},$$

where the complex frequency response function $H(\omega)$ can be derived from $\mathcal{L}$ as in the just discussed case of a single-degree of freedom system with viscous damping (10.5).

This procedure is known in the engineering literature as the *frequency domain method for the analysis of structural response to periodic excitation.* It is shown schematically in Fig. 10.3: The excitation $p(t)$ is transferred from the time domain to the frequency domain, where it is described by its Fourier coefficients $\hat{p}_k$, $k \in \mathbb{Z}$. The response to the kth harmonic is then simply given by $u_k = H(k\omega_0)\hat{p}_k$ in the frequency domain. Adding the responses to all harmonic excitations gives the response $u(t)$ in the time domain.

Example 10.7 (An Undamped Structure Subject to Rectangular Pulses, cf. [57], p. 857). Consider an undamped structure subject to a rectangular pulse, i.e.

$$m\ddot{u} + ku = p(t) = \begin{cases} p_0 & 0 \le t \le T/2 \\ -p_0 & T/2 \le t \le T \end{cases},$$

with amplitude $p_0 > 0$. We determine its solution via the frequency domain method. The Fourier coefficients of p read as

$$\hat{p}_k = \tfrac{1}{T}\int_0^T p(t)e^{-i(k\omega_0)t}\mathrm{d}t = \frac{-p_0}{ik\omega_0 T}\left(e^{-i(k\omega_0)t}\Big|_0^{T/2} - e^{-i(k\omega_0)t}\Big|_{T/2}^{T} \right).$$

To evaluate the exponential term we note that $\omega_0 T = 2\pi$. Therefore,

$$e^{ik\omega_0 T/2} = e^{ik\pi} = \begin{cases} +1 & k \text{ even} \\ -1 & k \text{ odd} \end{cases},$$

and $e^{-ik\omega T} = e^{-i2k\pi} = 1$. Thus,

$$\hat{p}_k = \frac{-p_0 i}{k\pi}, \quad \text{for } k \text{ odd}, \qquad \text{and} \qquad \hat{p}_k = 0 \text{ for } k \text{ even},$$

as we would expect from an odd function like $p(t)$.

Next, the frequency response function of our undamped system is

$$H(k\omega_0) = \frac{1}{k}\frac{1}{1-\beta_k^2}, \qquad \text{with} \qquad \beta_k := \tfrac{1}{\omega_n} k\omega_0, \quad \text{and} \quad \omega_n := \sqrt{m^{-1}k},$$

and hence

$$u(t) = \frac{-2p_0 i}{\pi} \sum_{k=-\infty}^{\infty} \tfrac{1}{k} \frac{1}{1-\beta_k^2} e^{-i(k\omega_0)t}.$$

Finally, we extend this method to arbitrary excitations $p(t)$ represented by their Fourier-Transforms $\hat{p}(\omega)$

$$p(t) = \frac{1}{2\pi}\int_{-\infty}^{\infty} \hat{p}(\omega) e^{i\omega t} \mathrm{d}\omega, \quad \text{and} \quad \hat{p}(\omega) = \int_{-\infty}^{\infty} p(t) e^{-i\omega t} \mathrm{d}t.$$

The response of a linear system to an arbitrary excitation $p(t)$ can be determined by combining the responses to individual harmonic excitation terms. The response of the system to the excitation $\hat{p}(\omega)e^{i\omega t}$ is given by $H(\omega)\hat{p}(\omega)e^{i\omega t}$. Superposing the responses to all harmonic terms results in the total response

$$u(t) = \tfrac{1}{2\pi}\int_{-\infty}^{\infty} H(\omega)\hat{p}(\omega)e^{i\omega t} \mathrm{d}\omega.$$

This is known as the *frequency-domain method for analysis of structural response to arbitrary excitations.*

10.4.1 Propagation of Excitations

Of course, we can consider the output of a system as the input for a new system and so on. Due to our frequency domain ansatz, the computation of the result becomes relatively easy as the new spectral representation is the product to the complex frequency response functions of the old and the new system together with the Fourier transform of the initial input excitation.

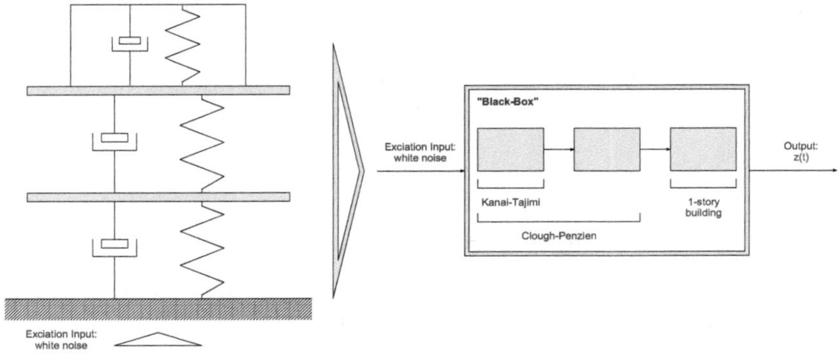

Figure 10.4. Sketch of a damped one-storey building with mass m subject to the Clough-Penzien excitation and corresponding input-output system for the frequency-domain method for analysis of structural response to arbitrary excitations.

For illustration purposes, we apply this procedure to determine the displacement $z(t)$ of a viscously damped one-storey building with mass m subject to the Clough-Penzien excitation, cf. (3.3). Thus, en passant we derive both, the power spectrum of the Kanai-Tajimi S_{TK} as well as that of the Clough-Penzien filter S_{CP} as used in Sec. 3.2.2.

The system we now consider is given by the following three-dimensional set of differential equations, cf. Fig. 10.4:

$$\ddot{x} + 2ab\dot{x} + b^2 x = -w_t \qquad (10.6)$$

$$\ddot{y} + 2cd\dot{y} + d^2 x = \ddot{x} + w_t \qquad (10.7)$$

$$m\ddot{z} + 2ef\dot{z} + f^2 x = \ddot{y} \qquad (10.8)$$

where w_t is a Gaussian white noise process with constant frequency function $S_{w_t}(\omega) = \hat{w}_t = D_0$ and $\ddot{x} + w_t$ is the Kanai-Tajimi excitation and $\ddot{y}(t)$ the Clough-Penzien excitation as introduced in Sec. 3.2.2.

With the complex frequency response transfer function

$$H_0(\omega; a, b) = \frac{1}{b^2 - \omega^2 + 2iab\omega}$$

we have that the solution of (10.6) reads as

$$x(t) = \frac{1}{2\pi} \int_{-\infty}^{\infty} \underbrace{-D_0 \frac{1}{b^2 - \omega^2 + 2iab\omega}}_{=: S_x(\omega)} e^{i\omega t} \mathrm{d}\omega \,,$$

and thus, that the Fourier transform $H_1(\omega)$ of the excitation $\ddot{x} + w_t$ is

$$H_x(\omega) = -\omega^2 S_x(\omega) + D_0 = D_0 \frac{\omega^2 + b^2 - \omega^2 + 2iab\omega}{b^2 - \omega^2 + 2iab\omega}$$

$$= \quad D_0 \frac{b^4 + (4a^2-1)b^2\omega^2 - 2iab\omega^3}{(b^2-\omega^2)^2 + 4a^2b^2\omega^2}$$

(recall: two-fold differentiation with respect to time in the time domain becomes multiplication by $-\omega^2$ in the frequency domain).

Finally, the displacement z of the roof of this one-storey building is given by:

$$z(t) \quad = \quad \frac{1}{2\pi} \int_{-\infty}^{\infty} H_y(\omega) \frac{m^{-1}f^2 - \omega^2 - 2iem^{-1}f\omega}{(m^{-1}e^2-\omega^2)^2 + 4e^2m^{-2}f^2\omega^2} e^{i\omega t} d\omega \,.$$

10.4.2 Linear Filter

The just described method is a special application of a concept called a *linear filter*: Following [91], a linear filter is an operator L that transforms a stochastic process or signal x_t, the input, into another stochastic process or signal y_t, the output:

$$y_t \quad = \quad Lx_t \,, \qquad t \in \mathbb{T} \,.$$

with $\mathbb{T} = \mathbb{Z}$ (time-discrete case) or $\mathbb{T} = \mathbb{R}$ (time-continuous case) depending on the context.

Definition 10.8 (Linear Time Invariant Filter). Let $\mathcal{E}$ be the space of all stationary stochastic processes. A *linear time invariant filter* L is a function $L : \mathcal{E} \to \mathcal{E}$ with the following three properties:

1. *scale preservation*: $L(\alpha x_t) = \alpha L(x_t)$,
2. *superposition*: $L(x_t + y_t) = L(x_t) + L(y_t)$, and
3. *time invariance*: $L(B^k x_t) = B^k L(x_t)$ for all $k \in \mathbb{Z}$, where B is the back-shift operator which acts on x_t as $B^k x_t = x_{t-k}$.

Important time-discrete examples for linear time invariant filters are

- *smoothing filter*:

$$y_t \quad = \quad Lx_t \;:=\; \tfrac{1}{4}x_{t+1} + \tfrac{1}{2}x_t + \tfrac{1}{4}x_{t-1} \,, \qquad t \in \mathbb{Z} \,,$$

  which reduces the difference between adjacent observations.

- *differencing filter*:

$$y_t \quad = \quad Lx_t \;:=\; x_t - x_{t-1} \,.$$

When linear time invariant filters are applied to a complex exponential, the output is also a complex exponential with exactly the same frequency multiplied by a factor that depends on the frequency and the linear filter.

Proposition 10.9 (Transformation Property of Time Invariant Filters & the Transfer Function). *Let L be a linear time invariant filter, and $e_\omega(t) := \exp(i\omega t)$ be the complex exponential function. Then, e_ω is an eigenfunction of L with eigenvalue $A(\omega) := L(e_\omega(0))$.*

The function $A(\omega)$ is called the transfer function *of the filter L.*

Proof. In particular due to time invariance, we have that

$$L(B^k e_\omega) \;=\; L(\exp(-i\omega k)e_\omega) \;=\; \exp(-i\omega k)L(e_\omega) \;=\; B^k L(e_\omega)\,.$$

For $k = t$ it follows that

$$L(e_\omega(t)) \;=\; \exp(i\omega t)L(e_\omega(0)) \;=\; A(\omega)e_\omega(t)\,,$$

which shows the assertion. □

If x_t has the spectral representation

$$x_t \;=\; \int_{\mathbb{R}} \exp(i\omega t)\mathrm{d}Z_x(\lambda)$$

with spectral distribution F_x and $A \in L^2(F_x)$, then

$$y_t \;=\; L(x(t)) \;=\; \int_{\Pi} A(\lambda)\exp(i\lambda t)\mathrm{d}Z_x(\lambda)\,,$$

and

$$F_y(\lambda) \;=\; \int_{-\pi}^{\lambda} |A(\mu)|^2\,\mathrm{d}F_x(\mu)\,.$$

Let L be a time invariant linear filter with transfer function $A(\lambda)$ such that $A \in L^2(F_x)$ with Fourier series $A(\lambda) = \sum_{k\in\mathbb{Z}} a(k)\exp(-i\lambda k)$, then

$$y_t \;=\; L(x_t) \;=\; \sum_{k\in\mathbb{Z}} a(k)x_{t-k}\,.$$

The function a is called *impulse response function* of the filter L as $y_t = a(t)$ holds for $x_t = \delta_0$.

Colored noise can be generated by passing the white noise through a shaping filter, and the response of the colored noise can be varied by adjusting the parameters of the shaping filter, usually a low-pass filter (cf., e.g., [193] for further backgrounds on low pass filters). A low-pass filter is a (discrete) time invariant linear filter with transfer function

$$A(\lambda) \;=\; \begin{cases} 1 & \text{if } \lambda \le \lambda_0 \\ 0 & \text{if } \lambda > \lambda_0 \end{cases}$$

for some $\lambda_0 \in (0, \pi)$. This type of filter cuts off the higher elements of the spectrum and lets just the low elements of the spectrum pass.

For the simulation of colored noise we follow a MATLAB Example introduced by Mathuranathan Viswanathan[5] in analogy to the visualisation of white noise given in Sec. 3.2.1:

MATLAB Example 10.1. visColor.m: Generating the visualisation of colored noise in Fig. 10.5.

```
a = 0.9;              % low pass filter parameter
L = 50 ;              % number of samples for auto-correlation
Fs = 1000;            % sampling rate
Fc = 10;              % carrier frequency for the dummy signal
t = 0:1/Fs:2;         % time base
variance = 1;         % variance of white noise
% generate a dummy signal - later on this can be replaced
% by the signal that we are interested in
signal=5*sin(2*pi*Fc*t);
% Generate Gaussian white noise with mean 0 & variance 1
x = sqrt(variance)*randn(1, length(signal));
% Colored Noise Generation with the aid of a first order
% low pass filter y(n)=a*y(n-1)+(1-a)*x(n), with
% filter trasfer function Y(Z) = X(Z)*(1-a)/(1-aZ^-1)
subplot(3,1,1);
   [y zf] = filter(1-a,[1 -a],x);
   coloredNoise = y;
   plot(t,coloredNoise);
   title('\bf{Colored Noise}');
   xlabel('Time (s)'); ylabel('Amplitude');
% Calculate auto-correlation of the colored noise, L is the
% number of samples used in auto-correlation calculation
subplot(3,1,2);
   [coloredNoiseCov,lags] = xcov(coloredNoise,L);
   stem(lags,coloredNoiseCov/max(coloredNoiseCov));
   title('\bf{Normalized AutoCovariance of Colored Noise}');
   xlabel('Lag [samples]');
% Frequency domain representation of noise
subplot(3,1,3);
   NFFT = 2^nextpow2(length(coloredNoise));
   coloredNoiseSpectrum = fft ...
      (coloredNoise,NFFT)/length(coloredNoise);
   f = Fs/2*linspace(0,1,NFFT/2+1);
   stem(f,2*abs(coloredNoiseSpectrum(1:NFFT/2+1)))
   title('\bf{Frequency Domain representation of Colored Noise}');
   xlabel('Frequency (Hz)'); ylabel('|Y(f)|');
```

These three plots are displayed in Fig. 10.5. We can clearly see the non-flat spectrum in the frequency-domain.

---

5 cf. http://www.gaussianwaves.com/2012/05/colored-noise-generation-in-matlab/

10.5 Chapter's Summary

We started this chapter with the basic definitions and implications related to the spectral representation of stationary and periodic stochastic processes. Based on this exposition, we studied the notions of energy, power and spectral density. Moreover, we gave several examples for colored noise processes like the Ornstein-Uhlenbeck process or power-law noise processes. Next, we discussed the frequency domain method for response analysis and the concept of linear filters. In particular, we applied this method to our problem of multi-storey excitation due to seismic impacts and the propagation of these impacts through wireframe structures.

Problems

Classification: ☼ easy, ◷ easy with longer calculations, ☆ a little bit difficult,

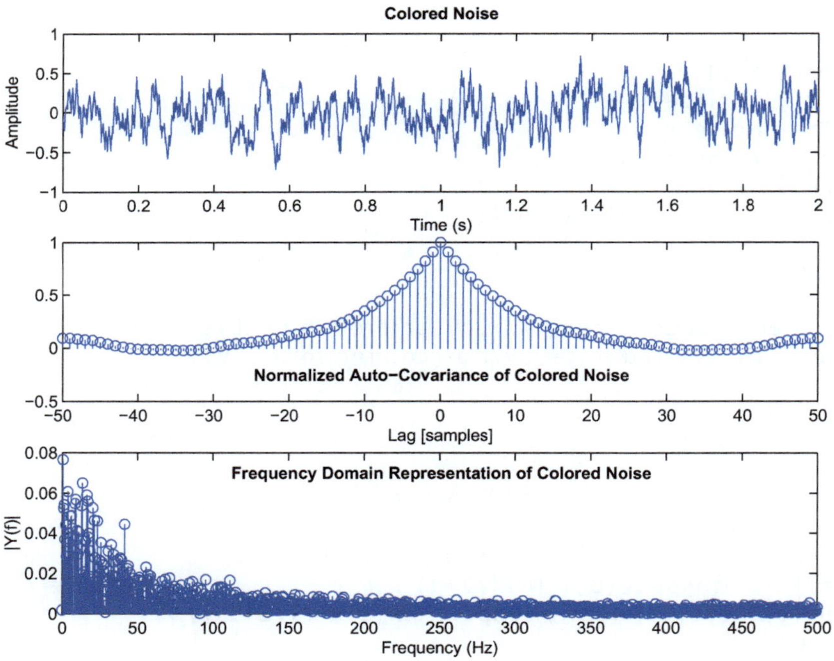

Figure 10.5. Properties of colored noise process: (top) sample path in the time-domain, (middle) normalized auto-correlation and (bottom) frequency representation.

Before you continue, make sure to answer the following questions:

Quiz: Section 10.4

Q1 Give the definition of the (complex) frequency response function.

Q2 Explain the frequency domain method for response analysis.

Q3 Give the definition of a linear time invariant filter L. What are the eigenfunctions and eigenvalues of such a filter?

Q4 Explain why linear (time invariant) filters can be used to generate colored noise.

challenging.

Exercise 10.10. [☼] Proving the Properties of the Power Spectral Density
Prove the properties of the power spectral density given in proposition 10.6. I.e., show the following assertions:

1. The spectrum of a real valued process is an even function of frequency: $S_{xx}(-\omega) = S_{xx}(\omega)$.
2. If the process is continuous and purely indeterministic, the auto-covariance function can be reconstructed by using the Inverse Fourier transform
3. The power spectral density describes the distribution of the variance over frequency. In particular, $\mathbb{V}ar(X_n) = \gamma_0 = 2\int_0^{1/2} S_{xx}(\omega)\mathrm{d}\omega$.
4. The power spectral density is a linear function of the auto-covariance function in the sense that if γ is decomposed into two functions $\gamma(\tau) = \alpha_1\gamma_1(\tau) + \alpha_2\gamma_2(\tau)$, then $f = \alpha_1 S_{xx,1} + \alpha_2 S_{xx,2}$.

Exercise 10.11. [☼] Energy & Power Spectrum of Seismic Excitations
Explicitly calculate the energy and power spectral densities of the output of the Kanai-Tajimi and the Clough-Penzien filter, as discussed in Sec. 3.2.2.

Exercise 10.12. [☼] Application of the Frequency Domain Method – Part 1
Determine the frequency response function of the system

$$\dot{x} + 5x = 4\dot{p} + 12p\,,$$

with the input $p(t) := 20\sin(4t)$.

Exercise 10.13. [☼] **Application of the Frequency Domain Method – Part 2**
Suppose a sinusoidal force $p(t) = P_0 \sin(\omega t)$ is applied to a mechanical system with

$$m\ddot{x} + b\dot{x} + kx = p(t)$$

as its equation of motion. Assuming that the displacement x is measured from the equilibrium position, find the frequency response function of this system.

Chapter 11

Space Filling Curves for Scientific Computing

The subject of space-filling curves has fascinated mathematicians for over a century [...] Working in this area is like skating on the edge of reason.
HANS SAGAN

In this chapter, the basic concepts, definitions, and properties of space-filling curves are introduced through the examples of the Hilbert and Peano curve. We briefly present three different categories of applications that motivate the use of these special curves in the context of simulations. Two variants for the construction of (discrete iterations of) the curves are explained in detail such that the reader is in the position to use space-filling curves for a concrete task such as ordering Cartesian mesh cells as used in the workshop problem (see Chaps. 17 and 18).

11.1 Key Concepts

This chapter may be used as a compact introduction to space-filling curves. We heavily rely on the approach taken in [19] while restricting properties and proofs to the absolute minimum necessary for a basic understanding. We take steps towards a hands-on approach with MATLAB.

The basic idea of space-filling curves is to construct mappings from intervals to higher-dimensional domains such that these mapping are still continuous and surjective. This astonishing feature of the cardinality of higher-dimensional domains did not fit into the mathematical framework of that time and contributed to the nickname "monster curves". We will discover that there is nothing monstrous about these curves, but that they can be helpful. In particular, these curves provide a clever approach to ordering unknowns in spatially discrete meshes discussed in Chap. 2.

When reading this chapter, note the answers to the following questions

1. What are space-filling curves?
2. What are space-filling curves useful for (in the context of Scientific Computing)?
3. What useful properties do (certain) space-filling curves possess?

as well as the following key concepts

1. The Hilbert and Peano curve,
2. Grammar-based construction of space-filling curves,
3. Arithmetisation of space-filling curves.

This chapter is structured as follows: We present the basic idea and definitions of space-filling curves in Sec. 11.2 before motivating the usage in Sec. 11.3 by different types of applications. In Sec. 11.4, two approaches for constructing space-filling curves in a discrete sense are explained: the construction via grammars and arithmetisations. Finally, Section 11.5 summarises the contents of this chapter.

11.2 The Concept of Space-filling Curves

In the following, $I \subset \mathbb{R}$ represents a compact interval, typically the unit interval $I := [0, 1]$.

Definition 11.1 (Direct Image). Let $f : I \to \mathbb{R}^n$. Then, $f_*(I) := \{f(t) \in \mathbb{R}^n \mid t \in I\}$ is called the *direct image* of the mapping f.

Definition 11.2 (Curve). For a continuous mapping $f\colon I \to \mathbb{R}^n$, the direct image $f_*(I)$ is called a *curve*. $x = f(t), t \in I$, is called *parametric representation* of the curve.

Note that in the definitions above, $\mathbb{R}^n$ may be replaced by any Euclidian vector space (providing a norm and scalar product).

Definition 11.3 (Space-filling Curve). Given a continuous mapping $f\colon I \to \mathbb{R}^n$, $n \geq 2$, then the corresponding curve $f_*(I)$ is called a *space-filling curve*, if $J_n\left(f_*(I)\right) > 0$.

The Jordan content J_n (area or volume for $n = 2$ or 3) denotes a Jordan-measurable subset of $\mathbb{R}^n$ (see [162]).

Often, the direct image of a space-filling curve is the unit hypercube: $Q := f_*(I) = [0, 1]^n$. If we assume f to be surjective, then $f_*(I)$ is a space-filling curve if the area (volume) of Q is positive.

What is an example of a space-filling curve? We will have a closer look at a very prominent example: the so-called Hilbert curve in two dimensions. The idea of how to construct such a curve is simple and uses recursion starting on the unit square $Q = [0, 1]^2$:

1. Split the quadratic domain into 4 congruent subsquares.
2. Find a space-filling curve for each subdomain; each of those curves is a scaled and rotated or reflected variant of the original curve.

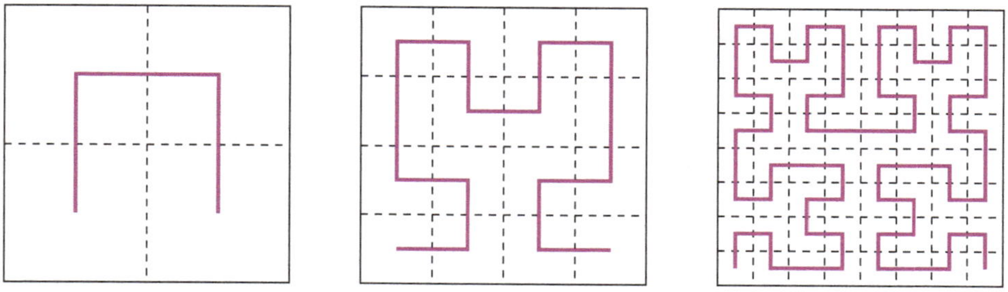

Figure 11.1. The first three discrete iterations of the 2D Hilbert curve.

3. Join the four subcurves in a suitable way; i.e. choose the reflection/rotation such that continuity is preserved.

4. Repeat the above steps on each of the 4 subsquares.

Figure 11.1 shows the resulting first three discrete iterations of the Hilbert curve: The pink polygonal line connects the centres of the subsquares of each level of recursion. Discrete iterations visualise the reflection and/or rotation applied to the curve when going from one level to the next.

In order to provide a thorough definition of a space-filling curve in general and of the Hilbert curve in particular, the above-mentioned rotations and/or reflections need to be determined. This will be done in Sec. 11.4. The construction of a real curve and the corresponding continuous mapping is realised via splitting of the interval $I = [0, 1]$ into 4 subintervals, mapping each subinterval to one subsquare (in a specific manner), and carrying out these steps ad infinitum. The nested subintervals converge to a point in I and the corresponding nested subsquares converge to a point in Q.

Definition 11.4 (Hilbert curve). The (2D) *Hilbert curve* $h : I \to Q$ is defined by the following steps:

- Each parameter $t \in I = [0, 1]$ is contained in a sequence of intervals

$$I \supset [a_1, b_1] \supset \ldots \supset [a_n, b_n] \supset \ldots ,$$

  where each interval results from a division-by-four of the previous interval.

- Each such sequence of intervals can be uniquely mapped to a corresponding sequence of 2D intervals (subsquares)

- The 2D sequence of intervals converges to a unique point q in $q \in Q := [0, 1] \times [0, 1]$ – q is defined as $h(t)$.

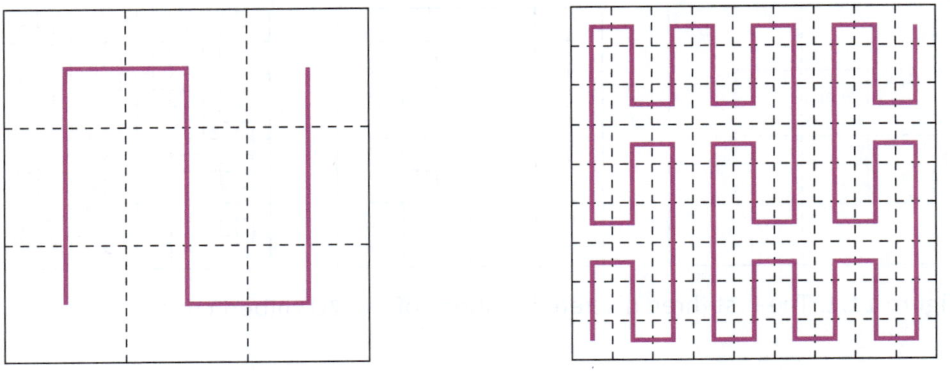

Figure 11.2. The first two discrete iterations of the 2D Peano curve with a vertical switch-back pattern.

Theorem 11.1. *The Hilbert curve $h : I \to Q$ is a space-filling curve.*

Proof. Three basic features of the Hilbert curve need to be proven: the mapping property of h, the surjectivity of the mapping, and the continuity of the curve. The proof is given in [19]. □

Another prominent example of a space-filling curve is the Peano curve. It is historically the first space-filling curve that has been discovered. For the definition of the curve in our sense, we use the same approach as for the Hilbert curve, i.e. defining discrete iterations and then going to the limit of infinitely many recursions. In Fig. 11.2, the first two discrete iterations of (one variant of) the Peano curve are visualised. The division by three in each dimension is an obvious difference from the Hilbert curve. The so-called *switch-back patterns* appear in Fig. 11.2 in the vertical direction resulting in an s-like shape; other variants of the curve exist with pure horizontal or mixed horizontal and vertical switch-back patterns. In contrast to the unique 2D Hilbert curve, there are 272 unique variants of switch-back type (see [223]) of the Peano curve. The proof of surjectivity and continuity of the Peano curve is straightforward and analogous to the one of the Hilbert curve.

The above examples represent two-dimensional space-filling curves. Of course, corresponding space-filling curves can be constructed for higher dimensions $n > 2$, too (cf. [19]).

To close this general section on space-filling curves, we briefly list an—incomplete—history of important events in this field:

1877: Georg Cantor constructs a bijective mapping from the unit interval $[0,1]$ onto the unit square $[0,1]^2$.
1879: Eugen Netto proves that a bijective mapping $f\colon I \to Q \subset \mathbb{R}^n$ can not be continuous, i.e. cannot be a curve, simultaneously (if Q has a smooth boundary).
1886: Camille Jordan introduces a rigorous definition of curves.
1890: Giuseppe Peano constructs the first space-filling curves.
1890: Hilbert provides a geometric construction of Peano's curve and introduces a new example: the Hilbert curve.
1904: Lebesgue curve
1912: Sierpinski curve

Despite the abstract definition and appearance of the space-filling curves, they—and in particular their discrete iterations—are advantageous for a number of different applications in Scientific Computing.

Before you continue, make sure to answer the following questions:

Quiz: Section 11.2

Q1 Which two properties does a curve need to be space-filling?

Q2 What is the difference between discrete iterations and the continuous representation of a space-filling curve?

Q3 What point in the unit square $[0,1]^2$ easily illustrates that the Hilbert curve (as every space-filling curve) is not a bijective mapping?

Q4 How many variants of the Hilbert curve and the Peano curve, respectively, exist in two dimensions?

11.3 Applications of Space-filling Curves

Space-filling curves are one way of serialising (i.e. ordering) multi-dimensional data. One example of such data are the grid cells in the discrete iterations in Fig. 11.1 and 11.2: The curve—represented by the polygonal line—steps into the cells one by one, ordering them in a unique way. If we interpret the grid cells as a Cartesian grid used for the discretisation of PDEs or as an array of pixels in image processing, two prominent fields of application are directly clear. Space-filling curves are relatively easy to construct and provide an inherent serial order of higher-dimensional data together with not optimal but good locality properties.

In the context of PDEs, three different categories of applications exist, each on a different level of granularity of the underlying problem (cf. [26]):

1. *High Level*: Parallelisation
   Use space-filling curves to divide a given discrete domain (number of cells) onto a set of processors. In case of a static distribution (done once in the beginning of the computation), this is done only once. For dynamic load balancing, i.e. re-distributing data (and, thus, work) when changes appear at runtime, the space-filling curve ordering gives information about which directions are possible to shift the work: the beginning or end (or both) of the local part of the curve on the given processor. This is a classical (in the sense of historical) field of application of (standard) space-filling curves that typically provide a nice (even if non-optimal) volume to surface ratio which is crucial for computational vs. communication costs.

2. *Intermediate Level*: Numbering of unknowns
   When simulating phenomena modelled via PDEs, sparse (non-)linear systems of equations typically have to be solved. The behaviour of algorithms (such as fill in for solvers or preconditioners) strongly depends on the order of the unknowns. This observation led to classical algorithms such as Cuthill-McKee (see [76]) that reorder the degrees of freedom to improve the solution of the underlying linear system of equations. The space-filling curves are an alternative approach to reorder data that is not geometrically ordered or is ordered in a non-advantageous manner. Hence, the curves provide a (relatively cheap) approach, again with perhaps not optimal but decent features, at least with respect to cost-benefit ratio (cf. [26], e.g.).

3. *Low Level*: Cache optimisation
   Ordering mesh items of (PDE) simulations in a clever way (i.e. providing temporal and spatial locality of the data access) allows for an efficient usage of the memory hierarchy in modern computer architectures. Again, space-filling curves provide good locality features and have successfully been used in various applications (see [121, 20, 254, 197], e.g.).

In all three categories, the curve provides a "string of pearls" of ordered items. The "good but non-optimal" locality of space-filling curves is due to their Hölder continuity (cf. [19]) and has to be interpreted in the sense that there may be algorithms that provide better results but typically are (much) more complicated and/or costly. Thus, space-filling curves represent a good compromise and nice heuristics.

For the workshop project (see part V), we only use the second category of application—the numbering of the unknowns—since neither parallelisation nor cache effects can be tackled within the context of this contribution. We motivate the usage of SFC enumerations by checking the properties of the system matrix A of the elastic problem (2.21). For levels of refinement p=3,...,6, a corresponding regular Cartesian grid with $2^p \times 2^p$ vertices is created on a square domain, and A is constructed accordingly. For demonstration purposes, we computed the complete LU decomposition of A without pivoting. In Tab. 11.1, the bandwidth of A as well as the relative fill-in of $L+U$ are presented for the five different enumerations described in Fig. 18.3: a default row-wise or column-wise lexicographic enumeration, a diagonal ordering of the unknowns, a meander-type enumeration iterating back and forth in a row-wise manner, and finally the Hilbert curve ordering. We observe that the Hilbert enumeration shows by far the largest bandwidth, but due to the fact that only a few entries realise these high values, the fill-in of $L+U$ is not extensive but actually competitive with the other enumeration types;

type \ level	3	4	5	6
lexico-row	73	273	1057	4161
lexico-col	73	273	1057	4161
diagonal	79	287	1087	4223
meander	79	287	1087	4223
Hilbert	116	468	1876	7508
dimension of A	128	512	2048	8192

type \ level	3	4	5	6
lexico-row	49.37	50.44	50.47	50.31
lexico-col	49.78	50.91	50.79	50.49
diagonal	50.39	51.13	50.85	50.50
meander	49.36	50.44	50.48	50.31
Hilbert	47.56	49.54	50.05	50.12

Table 11.1. Bandwidth of the original matrix A (left) and number of nonzeros of the LU decomposition of A (right; in percent) for the five different types of enumeration in a Cartesian grid of given level of refinement p=3–6. The size of A is equal to $2 \times 2^p \times 2^p$.

In Fig. 11.3, the sparsity pattern of A and $L+U$ are shown for the diagonal and the Hilbert ordering for a refinement p=4: The bandwidth of A and the fill-in effect are clearly visible, as well as the comparable orders of magnitude of fill-in for these two very different enumerations.

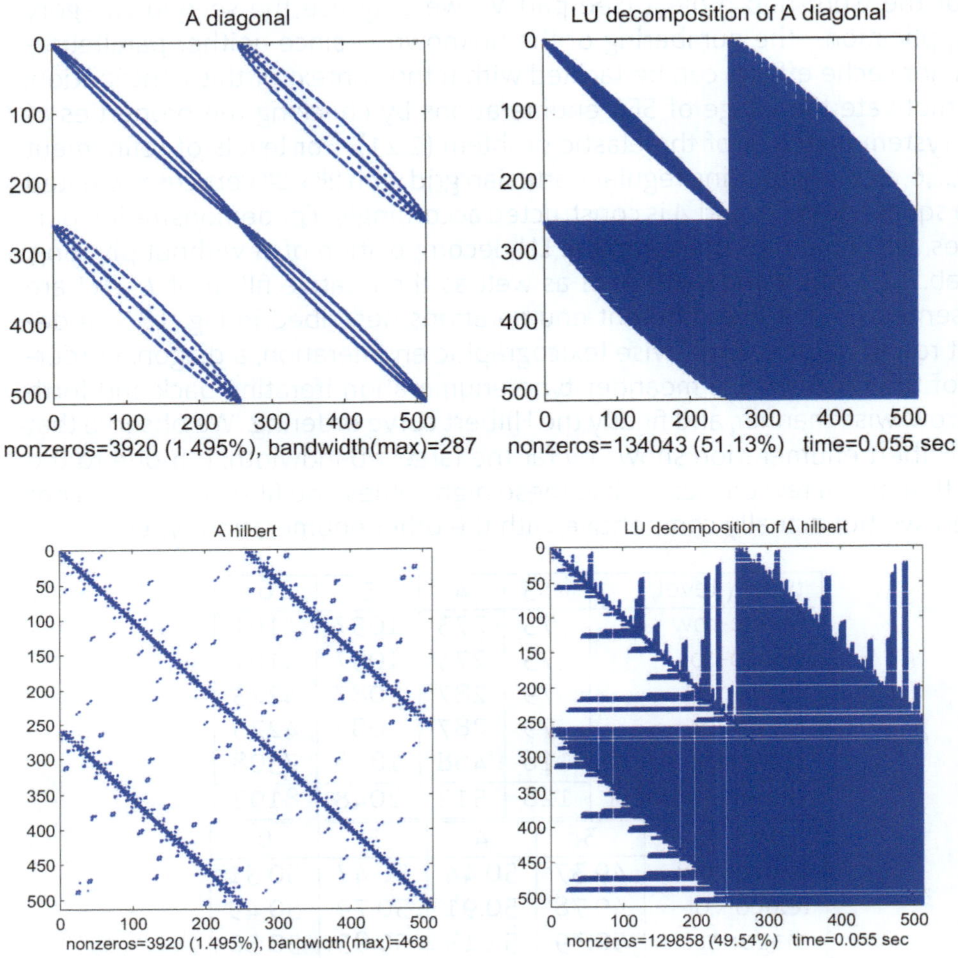

Figure 11.3. Sparsity pattern of the original system matrix A (left) and of its LU decomposition (right; fill-in) for the diagonal and Hilbert SFC enumeration of the degrees of freedom for the level of refinement p=4.

Besides the mere storage results, necessary runtimes for the decomposition are also relevant. In Fig. 11.4, the corresponding fill-in percentage as well as the runtime for the decomposition are visualised. We observe competitive results for the Hilbert curve ordering which represents, for this level of refinement, the best choice. Table 11.1 shows that this is not always the case for this concrete application.

It is hard to beat the row-wise lexicographic ordering on strictly regular grids, however SFC enumerations typically outperform other enumerations

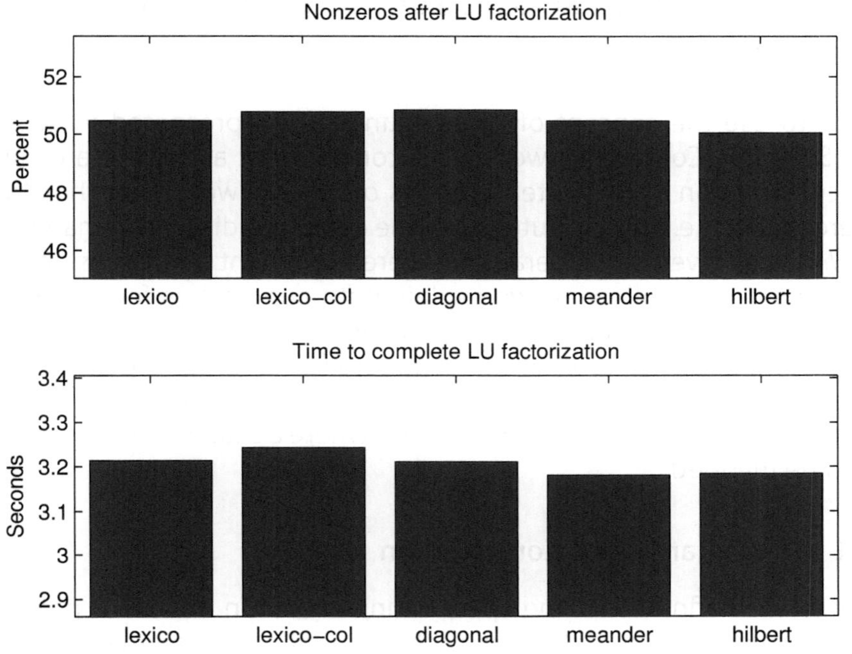

Figure 11.4. Properties of the LU decomposition of A using the five different enumerations: percentage of non-zeros(top) and runtime (bottom) for the level of refinement p=5.

in cases of adaptive grids due to their locality properties (in the context of recursive refinement and respective enumeration problems).

Before you continue, make sure to answer the following questions:

Quiz: Section 11.3

Q1 Why are space-filling curves frequently used for ordering purposes?

Q2 What other kind of application for SFCs comes to your mind?

11.4 Computational Construction of Space-filling Curves

In order to use the concept of space-filling curves presented in Sec. 11.2 within Scientific Computing, we need a concept that allows for a computational construction of (discrete iterations of) SFCs. Two approaches for the "forward mode", i.e. the computation of the corresponding coordinates of the image $h(t)$ of a given curve parameter t, are prominent: grammars and arithmetisations. In the following, we explain both approaches for the concrete example of the Hilbert curve and of the Peano curve. A grammar-based construction of a SFC will result in the complete (discrete iteration of a) curve. Hence, this approach is always useful if the full curve is needed, whereas the arithmetisation is advantageous in the case where a selection of single points need to be mapped.

11.4.1 Grammar-based Construction

The recursive definition of the space-filling curves in Sec. 11.2 provides a method for construction inherently, which is called a grammar. Individual discrete iterations of a curve are generated by replacing parent patterns (so-called non-terminal symbols) by a sequence of child patterns (a combination of non-terminal and terminal symbols) which are rotated and/or reflected variants of the original pattern. We just need to know or construct which pattern will be replaced in which manner; this corresponds to the production rules of a grammar. There is one additional condition: We have to replace all non-terminal symbols at once (i.e. at the same time) during a refinement step. The resulting grammars using this rule are often called *L-systems* (Lindenmayer systems).

The Hilbert Curve

In order to get a better feeling of how grammar-based constructions of SFCs look, we provide one variant of construction of the Hilbert curve[1]. In Fig. 11.5, the four basic patterns A, B, C, and H are visualised, together with their replacements on the next finer level. The green arrows, denoted by the symbols $\leftarrow, \uparrow, \rightarrow, \downarrow$ in text form, represent the so-called *terminal* symbols (i.e. they are not going to be replaced) and "glue" together the subcurves (or iterations).

1 Note that several variants of a grammar formulation exist for the unique Hilbert curve; one alternative are the so-called turtle grammars. The basic idea of all grammar variants is similar to the approach presented here.

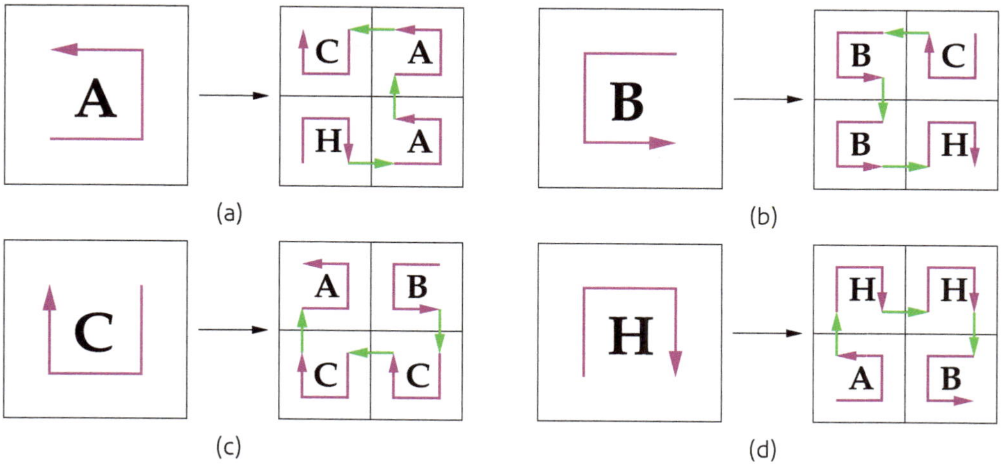

Figure 11.5. Towards a grammar for the Hilbert curve: The four basic patterns (or non-terminal symbols) A, B, C, and H of the curve together with the corresponding replacement patterns. The green arrows are the terminal symbols $\{\leftarrow, \uparrow, \rightarrow, \downarrow\}$.

The *production rules* can be formulated as

$$\begin{aligned} H &\longleftarrow A \uparrow H \rightarrow H \downarrow B \\ A &\longleftarrow H \rightarrow A \uparrow A \leftarrow C \\ B &\longleftarrow C \leftarrow B \downarrow B \rightarrow H \\ C &\longleftarrow B \downarrow C \leftarrow C \uparrow A \end{aligned}$$

where the symbol $\longleftarrow$ indicates the replacement of the pattern on the left-hand side by the combination of patterns on the right-hand side. Note that the pattern H represents the start pattern of the recursive grammar construction.

One variant of implementing an algorithm for the complete grammar construction is given in the MATLAB examples 11.1, 11.2, and 11.3 which contain the recursive production rules, the plotting of the terminal symbols (arrows), and the overall setup and creation of the curve, respectively. In Fig. 11.6, the resulting Hilbert curve of level (or depth) three is visualised.

Since we did not specify production rules that result in terminal symbols only, the grammar productions will not stop inherently but have to be stopped at a given level (depth)provided by the user (or environment) that is applying the space-filling curve. In the case of a direct correspondence of PDE mesh cells and cells of the discrete iterates of a space-filling curve, the stopping criterion is, of course, straightforward.

MATLAB Example 11.1 The four production rules H, A, B, and C of the Hilbert grammar.

```
% production rules for the Hilbert grammar
function H(depth)
if(depth > 0)
    A(depth - 1); up();
    H(depth - 1); right();
    H(depth - 1); down();
    B(depth - 1);
end
return

function A(depth)
if(depth > 0)
    H(depth - 1); right();
    A(depth - 1); up();
    A(depth - 1); left();
    C(depth - 1);
end
return

function B(depth)
if(depth > 0)
    C(depth - 1); left();
    B(depth - 1); down();
    B(depth - 1); right();
    H(depth - 1);
end
return

function C(depth)
if(depth > 0)
    B(depth - 1); down();
    C(depth - 1); left();
    C(depth - 1); up();
    A(depth - 1);
end
return
```

The Peano Curve

For the Peano curve, a similar standard approach for a constructive description via a grammar can be used. Of course, we now need different production rules, since we subdivide each square into nine child squares. Again, four basic patterns and corresponding production rules are necessary. In Fig. 11.7, the four patterns are shown together with their (graphical) replacement.

MATLAB Example 11.2 Creation of coordinates for the terminal symbols.

```
% adding coordinates for vert. line up
function up()
    global hy; global x; global y;
    x(end+1)=x(end);
    y(end+1)=y(end) + hy;
    return
```

```
% adding coordinates for vert. line down
function down()
    global hy; global x; global y;
    x(end+1)=x(end);
    y(end+1)=y(end) - hy;
    return
```

```
% adding coordinates for horiz. line left
function left()
    global hx; global x; global y;
    x(end+1)=x(end) - hx;
    y(end+1)=y(end);
    return
```

```
% adding coordinates for horiz. line left
function right()
    global hx; global x; global y;
    x(end+1)=x(end) + hx;
    y(end+1)=y(end);
    return
```

We do not formulate the production rules or the corresponding MATLAB algorithm explicitly since this represents one of the exercise problems at the end of this chapter. Instead, we are now going to describe an alternative approach to construct image values of the space-filling curve mapping: the so-called arithmetisation.

11.4.2 Arithmetisation

The basic idea of the arithmetisation for space-filling curves is to be able to map a single parameter t directly to its image $h(t)$ (or a discrete approximation

MATLAB Example 11.3 standardHilbert.m: setting of parameters, recursive grammar creation and plotting of the Hilbert curve.

```
% create and plot the Hilbert curve
function standardHilbert(depth)
numberOfCellsPerDim = 2.0^(depth);
sizeX = 1.0; sizeY = 1.0;
global hx; global hy;
hx = sizeX/numberOfCellsPerDim;
hy = sizeY/numberOfCellsPerDim;
global x; global y;
x = [0.5] * hx; y = [0.5] * hy;

% start of recursion
H(depth);

% plot of the curve
figure;
plot(x,y,'r');
grid on; axis([0 sizeX 0 sizeY]);
set(gca,'XTick',[0:hx:sizeX],'YTick',[0:hy:sizeY]);
xlabel('x'); ylabel('y');
title(['Hilbert curve for depth ' num2str(depth)]);

return
```

of it) without constructing the whole curve in all other parts of the image domain.

Note that the arithmetisation represents, in addition, a tool to create a (pseudo-) inverse mapping $\tilde{h}^{-1}$ for each space-filling curve h. A full inverse mapping h^{-1} is not possible because of the non-injectivity of space-filling curves proven by Netto. The inverse mapping $\tilde{h}^{-1}$ can be used to compute the curve parameter t for different coordinates in the two-dimensional domain. We will not go into the details of the (pseudo-)inverse mapping but now present the arithmetisation of the Hilbert and Peano curve, respectively.

The Hilbert Curve

An important ingredient for the arithmetisation of the Hilbert curve is the representation of the given curve parameter t in a different number system.

The construction of the Hilbert curve relies on the subdivision by four of both the image domain $Q = [0,1]^2$ and the parameter interval $I = [0,1]$. In particular, each curve parameter $t \in I$ can be represented as an interval sequence by choosing one of the four subdivided intervals in each recur-

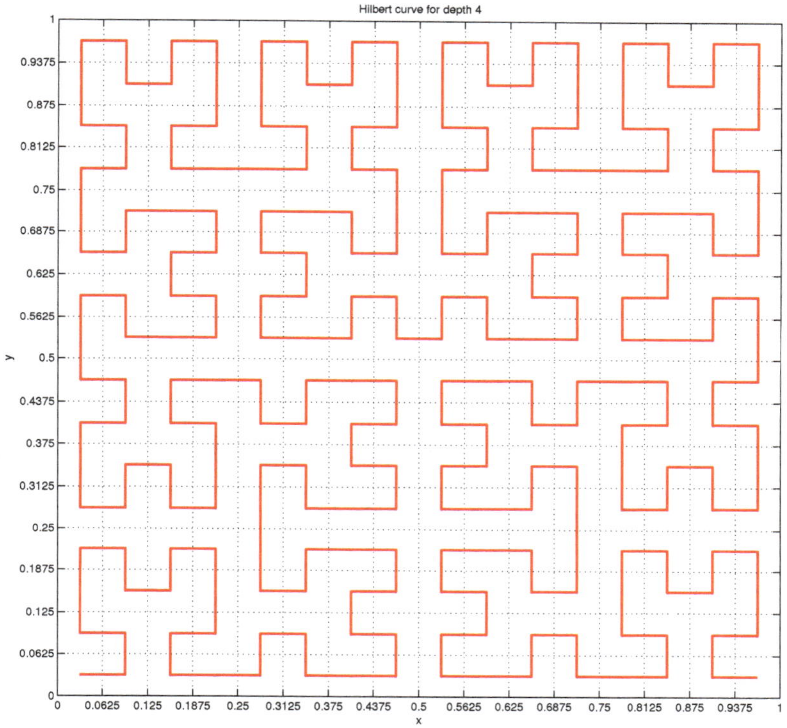

Figure 11.6. Hilbert curve of level 4 created via standardHilbert(4).

sion step. If we rewrite the boundaries of these subintervals in a quaternary representation instead of a decimal one, each border translates to one digit $\{0,\ldots,3\}$ of the quaternary system:

$$\left[\tfrac{1}{4},\tfrac{2}{4}\right] = [0_4.1, 0_4.2]\,, \quad \left[\tfrac{3}{4},1\right] = [0_4.3, 1_4.0]$$

Furthermore, every subsquare of the target domain Q contains a scaled, translated, and rotated or reflected Hilbert curve (self similarity of the curve). Defining four basic mappings $H_i, i \in \{0,\ldots,3\}$, that perform exactly these transformations, results in

$$\begin{aligned} H_0 &:= \begin{pmatrix} 0 & \frac{1}{2} \\ \frac{1}{2} & 0 \end{pmatrix}\begin{pmatrix} x \\ y \end{pmatrix}, & H_1 &:= \begin{pmatrix} \frac{1}{2} & 0 \\ 0 & \frac{1}{2} \end{pmatrix}\begin{pmatrix} x \\ y \end{pmatrix} + \begin{pmatrix} 0 \\ \frac{1}{2} \end{pmatrix}, \\ H_2 &:= \begin{pmatrix} \frac{1}{2} & 0 \\ 0 & \frac{1}{2} \end{pmatrix}\begin{pmatrix} x \\ y \end{pmatrix} + \begin{pmatrix} \frac{1}{2} \\ \frac{1}{2} \end{pmatrix}, & H_3 &:= \begin{pmatrix} 0 & -\frac{1}{2} \\ -\frac{1}{2} & 0 \end{pmatrix}\begin{pmatrix} x \\ y \end{pmatrix} + \begin{pmatrix} 1 \\ \frac{1}{2} \end{pmatrix}. \end{aligned} \tag{11.1}$$

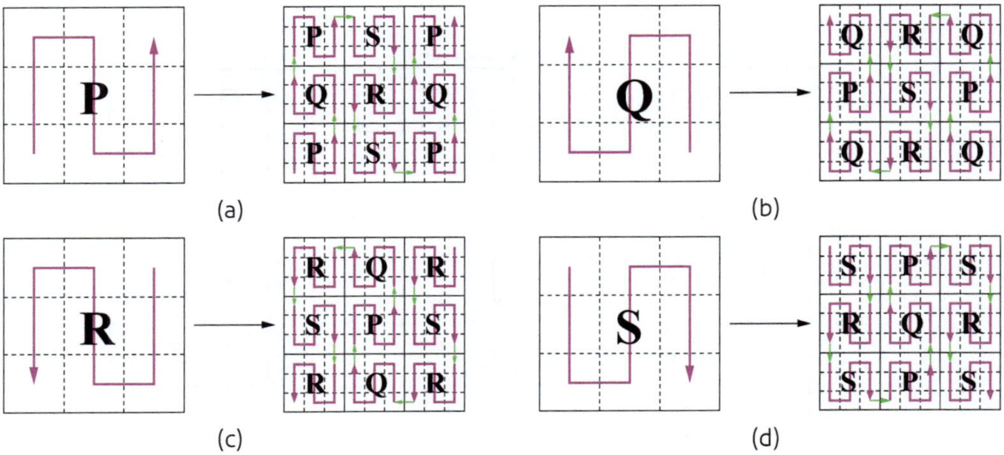

Figure 11.7. Towards a grammar for the Peano curve: The four basic patterns (or non-terminal symbols) P, Q, R, and S of the curve together with the corresponding replacement patterns (image courtesy of M. Bader [19]). The green arrows represent the terminal symbols $\{\leftarrow, \uparrow, \rightarrow, \downarrow\}$.

It is easy to see why the H_i have exactly the given form by comparing what happens in the first recursion step in Fig. 11.1 (or even better in Fig. 11.5 (d)):

- The scaling by two that appears in each H_i corresponds to the division of Q by two in each dimension.
- The translation changes the origin of the corresponding subsquare accordingly.
- The rotation or reflection can be computed directly applying simple 2D geometric transformations (see Problem 11.7).

Every subdivision of a parameter interval corresponds to applying one of the mappings H_i to the existing curve in a recursive manner. Which H_i has to be chosen is determined by the number i of the subinterval that contains the curve parameter t (i.e. by the quaternary digit of t of the corresponding level). Therefore, we may use the following recursive approach:

$$h(\underbrace{0_4.q_1q_2q_3q_4\ldots}_{=:\,t}) = H_{q_1} \circ h(\underbrace{0_4.q_2q_3q_4\ldots}_{=:\,\tilde{t}}). \tag{11.2}$$

The curve parameter $\tilde{t} = 0_4.q_2q_3q_4\ldots$ in the right-hand side of Eq. 11.2 is the relative parameter in the subinterval $[0_4.q_1, 0_4.(q_1+1)]$. The coordinates $h(\tilde{t})$ represent the relative position of the curve point in the corresponding subsquare Q_1. The mapping H_{q_1} transforms $h(\tilde{t})$ to its correct position in the unit square Q. Expanding recursion (11.2) leads to:

$$h(0_4.q_1q_2q_3q_4\ldots) = H_{q_1} \circ H_{q_2} \circ H_{q_3} \circ H_{q_4} \circ \cdots \tag{11.3}$$

Therefore, the basic idea for the construction of the arithmetic representation reads:

1. Find a quaternary representation of the curve parameter t.
2. Use the quaternary coefficients to determine the required sequence of operations.

One variant of an algorithm for the arithmetisation of the Hilbert curve is presented in the MATLAB example 11.4. The quaternary digits q of each recursion level are computed by $\lfloor 4t \rfloor$ since multiplying t by four results in shifting the quaternary point by one entry to the right:

$$4 \cdot t \;=\; 4 \cdot 0_4.q_1q_2q_3q_4\ldots \;=\; (q_1.q_2q_3q_4\ldots)_4$$

Hence, by cutting off the integer part of $4t$, we obtain the resulting remaining parameter $\tilde{t}$ of Eq. 11.2 relative to the corresponding subinterval; the variable r plays the role of $\tilde{t}$ in our example. Note that the stopping criterion for the recursion is a given maximum depth of the recursion tree. If the maximum depth is not necessary due to trailing zeros in the quaternary representation of the curve parameter t, unnecessary recursions are avoided. Furthermore, a value of $t = 1$ is not allowed in this example since $\lfloor 4t \rfloor = 4$ which will result in the error message. This exceptional case does not represent a problem because we know by construction that the exit point of the continuous Hilbert curve at $t = 1$ is $h(t) = (1, 0)$.

It is easy to show (see [19]) that the values of $h(t)$ defined by the arithmetisation are independent of the concrete form of the quaternary representation that may exist for certain values of t:

- For a finite number of non-zero quaternary digits in t, we may add further zeros which do not change the coordinates of $h(t)$, since $\binom{0}{0}$ is a fixed point of H_0, i. e. $H_0\binom{0}{0} = \binom{0}{0}$.
- Similar to representing values such as 1.0 by $0.9\bar{9}$ in the decimal system, one could change the representation of

  $$t = 0_4.q_1\ldots q_n \qquad \text{to} \qquad \tilde{t} = 0_4.q_1\ldots q_{n-1}(q_n - 1)333\ldots$$

  in the quaternary system (assuming $q_n \neq 0$ for the sake of simplicity). Then

  $$h(0_4.q_1\ldots q_n) = h(0_4.q_1\ldots q_{n-1}(q_n - 1)333\ldots),$$

  since

  $$H_{q_n} \circ \lim_{n\to\infty} H_0^n = H_{q_n - 1} \circ \lim_{n\to\infty} H_3^n\,.$$

MATLAB Example 11.4 hilbertArith.m: arithmetisation of the Hilbert curve.

```
function [x,y]=hilbertArith(t, depth)

if depth == 0
    x = 0; y = 0;
else
    q = floor(4*t);
    r = 4*t - q;           % shorter t in quaternary repr.
    if r == 0 && q == 0 % avoid unnecessary recursions
        x = 0; y = 0;
    else
        [x1,y1] = hilbertArith(r,depth-1); % recursive call
        switch q
            case 0
                x =  y1/2.0;      y =  x1/2.0;
            case 1
                x =  x1/2.0;      y =  y1/2.0+0.5;
            case 2
                x =  x1/2.0+0.5; y =  y1/2.0+0.5;
            case 3
                x = -y1/2.0+1.0; y = -x1/2.0+0.5;
            otherwise error('should never appear!');
        end
    end
end

return
```

The Peano Curve

The arithmetic definition of the Peano curve follows the same idea as for the Hilbert curve. This time, we subdivide the parameter interval into nine subintervals. Hence, we rely on the nonal representation (with basis nine) of the parameter t,

$$t = 0_9.n_1n_2n_3n_4\ldots\ .$$

The corresponding image point of the Peano curve can then be represented by the following sequence of mappings

$$p(0_9.n_1n_2n_3n_4\ldots) \quad = \quad P_{n_1} \circ P_{n_2} \circ P_{n_3} \circ P_{n_4} \circ \cdots \begin{pmatrix} 0 \\ 0 \end{pmatrix}, \qquad (11.4)$$

where each P_{n_i} is one of the nine basic mappings

$$P_2\begin{pmatrix}x\\y\end{pmatrix}=\begin{pmatrix}\frac{1}{3}x+0\\\frac{1}{3}y+\frac{2}{3}\end{pmatrix},\quad P_3\begin{pmatrix}x\\y\end{pmatrix}=\begin{pmatrix}\frac{1}{3}x+\frac{1}{3}\\-\frac{1}{3}y+1\end{pmatrix},\quad P_8\begin{pmatrix}x\\y\end{pmatrix}=\begin{pmatrix}\frac{1}{3}x+\frac{2}{3}\\\frac{1}{3}y+\frac{2}{3}\end{pmatrix},$$

$$P_1\begin{pmatrix}x\\y\end{pmatrix}=\begin{pmatrix}-\frac{1}{3}x+\frac{1}{3}\\\frac{1}{3}y+\frac{1}{3}\end{pmatrix},\quad P_4\begin{pmatrix}x\\y\end{pmatrix}=\begin{pmatrix}-\frac{1}{3}x+\frac{2}{3}\\-\frac{1}{3}y+\frac{2}{3}\end{pmatrix},\quad P_7\begin{pmatrix}x\\y\end{pmatrix}=\begin{pmatrix}-\frac{1}{3}x+1\\\frac{1}{3}y+\frac{1}{3}\end{pmatrix},$$

$$P_0\begin{pmatrix}x\\y\end{pmatrix}=\begin{pmatrix}\frac{1}{3}x+0\\\frac{1}{3}y+0\end{pmatrix},\quad P_5\begin{pmatrix}x\\y\end{pmatrix}=\begin{pmatrix}\frac{1}{3}x+\frac{1}{3}\\-\frac{1}{3}y+\frac{1}{3}\end{pmatrix},\quad P_6\begin{pmatrix}x\\y\end{pmatrix}=\begin{pmatrix}\frac{1}{3}x+\frac{2}{3}\\\frac{1}{3}y\end{pmatrix}.$$

Before you continue, make sure to answer the following questions:

Quiz: Section 11.4

Q1 What are the basic ingredients of a grammar for space-filling curves?

Q2 Grammar of the Hilbert curve: Using the start pattern H, the first level of replacement when applying the grammar is represented by the string $H \longleftarrow A \uparrow H \rightarrow H \downarrow B$. How does the second level of replacement look like (as a string)?

Q3 How could the production rules in MATLAB example 11.1 be changed to rules where the last level of recursion can be avoided (think of else statements)?

Q4 What is the value of the Hilbert curve coordinates at $t = 0.5$?

Q4 Use the arithmetisation of the Hilbert curve to manually compute the coordinates $h(t = \frac{1}{4})$.

11.5 Chapter's Summary

In this chapter, the concept of space-filling curves has been presented for the examples of the Hilbert and Peano curve together with a list of possible variants of applications which is not close to complete. We have seen how to construct space-filling curves (or their discrete iterations) via grammars and arithmetisation and can apply this knowledge to the concrete problems of ordering Cartesian mesh cells specified in the workshop project (see Chaps. 17 and 18).

Note that for the sake of simplicity only, we restricted all aspects of SFCs to two-dimensions. Of course, three- and even higher-dimensional versions of the Hilbert and Peano curve exist and can be computed using similar approaches (cf. [19]). Furthermore, a variety of other space-filling curves are available which fit into the concept and construction but differ slightly with respect to certain features.

Problems

Classification: ☼ easy, ⌚ easy with longer calculations, ☆ a little bit difficult,
challenging.

Exercise 11.5. [☼] **Usage of different construction variants for SFC**
We covered two variants for constructing space-filling curve images: the grammar-based construction and the arithmetisation. Which of the two may be advantageous in which context?

Exercise 11.6. [⌚] **MATLAB Exercise: Grammar for the Hilbert-Moore curve**
In Fig. 11.8, the first three discrete iterations of a modified Hilbert curve a visualised: the so-called Hilbert-Moore curve. In contrast to the usual Hilbert curve, this curve has its (continuous) start and end point at $x = (0.5, 0)$, i.e. it is a closed curve. Derive a grammar for the Hilbert-Moore curve (hint: take care of a different start pattern compared to the usual Hilbert curve).

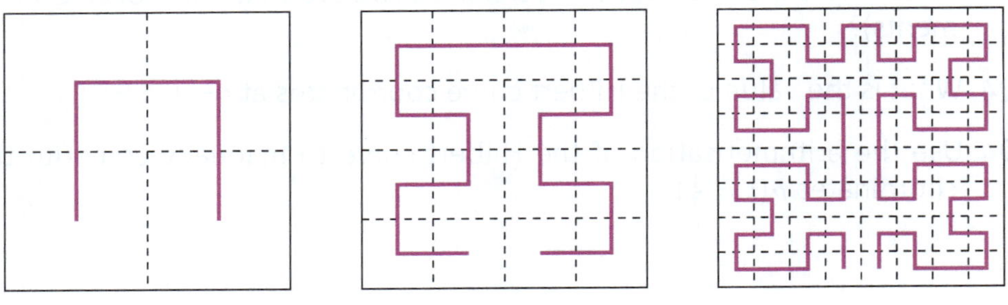

Figure 11.8. Three discrete iterations of the Hilbert-Moore curve.

Exercise 11.7. [⌚] **Arithmetisation of the Hilbert curve**
Derive the basic mappings H_1, H_2, H_3, H_4 of Eq. (11.1) via combining full 2D rotation and/or mirroring transformations of the form

$$\begin{pmatrix} \cos(\alpha) & -\sin(\alpha) \\ \sin(\alpha) & \cos(\alpha) \end{pmatrix}, \begin{pmatrix} \pm 1 & 0 \\ 0 & \pm 1 \end{pmatrix},$$

where α represents the angle of rotation. Hint: Take care of the order in case of a combination of rotation and mirroring.

Exercise 11.8. [☆] **MATLAB Exercise: Grammar for the Peano curve**
Based on the visualisation of the first discrete iterations of the Peano curve in Fig. 11.7(a) to (d), create and implement the full grammar for the Peano curve. You may reuse the functions $\mathrm{up}()$, $\mathrm{down}()$, $\mathrm{left}()$, and $\mathrm{right}()$ of MATLAB example 11.2.

Exercise 11.9. [📖] **Hölder Continuity of the Peano curve**
A d-dimensional space-filling curve can be seen as a mapping $f(x) : \mathbb{R} \to \mathbb{R}^d$. Thus, the Hölder continuity for this mapping is defined as

$$||f(x) - f(y)|| \leq c \cdot |x - y|^{\frac{1}{d}}.$$

Show that the two-dimensional Peano curve is Hölder continuous, with respect to the 2-norm. How large is c?

Part IV

Path-Wise Solutions of Random Differential Equations and Their Simulation

Determinism, like the Queen of England rules
– but does not govern.
MICHAEL VICTOR BERRY *(1941 -)*

Chapter 12

Linear RODEs with Stochastic Inhomogeneity

As in the discussion of deterministic ordinary differential equations, linear random differential equations can be studied with relative ease and exhibit a wealth of interesting phenomena. We start the discussion of the solutions of linear random differential equations by first assuming the randomness to be located in the inhomogeneity only. Important considerations include the stochastic characteristics of the solution process as well as periodic and stationary solution types. In particular, we give first stability conditions with respect to which solutions converge towards periodic or stationary ones. Stochastic effects in the coefficients will be studied in Chap. 13.

12.1 Key Concepts

In this section, we follow [45], pp. 38, to study the random (ordinary) differential equation

$$\dot{X}_t = A(t)X_t + Z_t\,, \tag{12.1}$$

where $A(t) : I \to \mathbb{R}^{d\times d}$ is a continuous and Z_t is a d-dimensional stochastic process. This type of random differential equation has particular significance in technical applications. Dynamical systems are often well approximated by linear non-random equations that account for random input data. Thus, we discuss this type of equation thoroughly and in particular its special case of the Kanai-Tajimi model (3.7):

$$\begin{pmatrix} \dot{z}_1 \\ \dot{z}_2 \end{pmatrix} = \begin{pmatrix} 0 & -1 \\ \omega_g^2 & -2\zeta_g\omega_g \end{pmatrix} \begin{pmatrix} z_1 \\ z_2 \end{pmatrix} + \begin{pmatrix} O_t \\ (1-2\zeta_g\omega_g)O_t \end{pmatrix}.$$

We shall also discuss second order random differential equations with constant coefficients and stationary random forcing Z_t as in the Kanai-Tajimi excitation of wireframe structures.

When reading this chapter note the answers to the following key concepts

1. Common covariance of a particular solution process and the driving process,
2. Asymptotic un-correlation of the common covariance,
3. Path-wise/ mean-square/ strict θ-periodicity,

4. Existence of path-wise/ mean-square/ strict θ-periodic and stationary solutions,

5. Convergence towards periodic and stationary solutions,

as well as the following questions

1. How does the general solution formula of (12.1) for path-wise and mean-square solutions look like?

2. Which differential equations do the mean and the common covariance obey?

3. How are periodic and stationary solutions connected?

4. Under which conditions do solutions converge towards a θ-periodic (or stationary) path-wise/ mean-square solution?

5. What results do we get for higher-order random differential equations?

This chapter is structured as follows: We start by giving the general formula for path-wise and mean-square solutions of (12.1) in Sec. 12.2 and by discussing the stochastic properties (expectation, auto-covariance, common covariance, etc.) of the solution process. Then, in Sec. 12.3, we focus our attention on path-wise/ mean-square periodic and path-wise/ mean-square stationary solutions. We study the conditions that lead to their existence as well as the convergence properties of other solutions towards these periodic and stationary ones. Next, in Sec. 12.4, the results we obtained so far are applied to higher-order linear random differential equations with constant coefficients and stationary inhomogeneity. Finally, Sec. 12.5 wraps up the contents of this chapter.

12.2 The General Solution Formula

As we have seen in Chap. 6, the solutions on I of the deterministic linear homogeneous (ordinary) differential equation

$$\dot{x}(t) \;=\; A(t)x(t) \tag{12.2}$$

form a d-dimensional vector space. Combining the d linearly independent solutions as columns in a matrix leads to the fundamental matrix $\Phi(t)$ corresponding to (12.2), with $\Phi(t_0) = \mathbb{I}$. The fundamental matrix satisfies the matrix differential equation

$$\dot{\Phi}(t) \;=\; A(t)\Phi(t)\,. \tag{12.3}$$

Now, recall a property we frequently used in Chap. 3: For any path-wise continuous d-dimensional stochastic process Z_t on I and for any continuous vector-valued function $f(t, x)$ on $I \times \mathbb{R}^d$ there is a stochastic process Y_t such that

$$Y_t \stackrel{I}{=} \int_a^t f(\tau, Z_\tau) \mathrm{d}\tau ,$$

where $a \in \mathbb{R}$ or $a = -\infty$. In the latter case, we assume the integral $\int_{-\infty}^t f(\tau, Z_\tau) \mathrm{d}\tau$ is finite for almost all $\omega \in \Omega$ and all $t \in I$.

Theorem 12.1 (Closed Solution Formulas for $\dot{X}_t = A(t)X_t + Z_t$). *Given the random (ordinary) differential equation (12.1), the following assertions hold:*

1. *If the stochastic process Z_t is path-wise continuous on I, then*

$$X_t = \Phi(t)X_0 + \Phi(t) \int_{t_0}^t \Phi^{-1}(\tau) Z_\tau d\tau \tag{12.4}$$

   is the unique path-wise solution of (12.1) with respect to the initial condition $(X_0, t_0) \in S_d \times I$.

2. *If Z_t is a mean-square continuous stochastic process of degree two, then*

$$X_t = \Phi(t)X_0 + \Phi(t) ⨍_{t_0}^t \Phi^{-1}(\tau) Z_\tau d\tau \tag{12.5}$$

   is the unique mean-square solution of (12.1) with respect to the initial condition $(X_0, t_0) \in S_d \times I$.

3. *If Z_t is a mean-square and path-wise stochastic process of degree two, then the path-wise solutions and mean-square solutions of (12.1) with the initial condition $(X_0, t_0) \in L_d^2 \times I$ are equivalent.*

Proof. Following [45], pp. 40, we have that part 1 is obtained from (12.4) via path-wise differentiation, and uniqueness follows from Theorem 3.2. Part 2 follows via mean-square differentiation from (12.5) and by taking into account the properties 6 and 11 from Sec. 4.3. Uniqueness of the solution follows from Theorem 4.3. Finally, part 3 follows from parts 1 and 2 with property 12 from paragraph 4.3. □

If the matrix $A(t) \equiv A$ is constant, then $\Phi(t)$ has the form $\Phi(t) = \exp(A(t - t_0))$ and the path-wise solution (12.4) reads as

$$X_t = \exp(A(t - t_0)) X_0 + \int_{t_0}^t \exp(A(t - \tau)) Z_\tau \mathrm{d}\tau . \tag{12.6}$$

Analogously, we obtain the mean-square solution (12.5) in this case as

$$X_t = \exp\left(A(t-t_0)\right) X_0 + \int_{t_0}^{t} \exp\left(A(t-\tau)\right) Z_\tau \mathrm{d}\tau . \qquad (12.7)$$

12.2.1 Stochastic Properties of Path-Wise Solution

If the fundamental matrix $\Phi(t)$ is known, then the general form of the path-wise solutions of (12.1) is given by equation (12.4). In general, the functional form of the realisations of Z_t is not known and thus the integration necessary for an evaluation of (12.4) can not be carried out. On the other hand, in view of probability theory and stochastic analysis, equation (12.1) can be solved once the probability distribution of X_t is determined.

For instance, in order to get the one-dimensional probability distribution of X_t from (12.4), we have to obtain the probabilities

$$\mathbb{P}\left(X_t \in B\right) = \mathbb{P}\left(\omega \in \Omega : X_t = \Phi(t)X_0 + \Phi(t)\int_{t_0}^{t} \Phi^{-1}(\tau)Z_\tau \mathrm{d}\tau \in B\right),$$

$B \in \mathcal{B}^d$, based on the distribution corresponding to the process (X_0, Z_τ), $\tau \in [t_0, t]$. The mathematical difficulties of this task are obvious, and thus one

Before you continue, make sure to answer the following questions:

Quiz: Section 12.2 – Part I

Q1 Given the inhomogeneous random differential equation $\dot{X}_t = A(t)X_t + Z_t$ with a path-wise continuous driving process Z_t. How does the general solution for X_t look like?

Q2 Derive this formula from previous results.

Q3 Given the inhomogeneous random differential equation $\dot{X}_t = A(t)X_t + Z_t$ with a mean-square continuous driving process Z_t. How does the general solution for X_t look like?

Q4 Derive this formula from previous results.

Q5 How do the two formulas of Q1 and Q3 simplify if $A(t) \equiv A$ is constant?

Q6 Under which conditions are the unique path-wise and the unique mean-square solution of $\dot{X}_t = A(t)X_t + Z_t$ equivalent?

Q7 Derive your assertion from Q6 from previous results.

may, in general, try to determine certain characteristic properties of the solution instead, like the mean and the covariance, based on the characteristic properties of the driving process Z_t.

We start with the study of the characteristic properties of the particular solutions

$$Y_t = \Phi(t) \int_{t_0}^{t} \Phi^{-1}(\tau) Z_\tau \mathrm{d}\tau \,, \tag{12.8}$$

and

$$Y_t = \Phi(t) ⨍_{t_0}^{t} \Phi^{-1}(\tau) Z_\tau \mathrm{d}\tau \,, \tag{12.9}$$

respectively, as these often lead to the characteristic properties of $X_t = \Phi(t)X_0 + Y_t$ in an easy manner.

First, we formulate a technical lemma that can be proven analogously to the Theorems 2.2.5 and 2.2.7 in [87]:

Lemma 12.1. *Let H and G be intervals in $\mathbb{R}$ and $\mathbb{R}^m$, respectively, as well as $K := H \times G$. Let g be a function defined on $K \times \mathbb{R}^d$ and $\mathcal{B}_K \times \mathcal{B}^d$-measurable. Assume that X_t is a path-wise continuous process on K and that $h(t) = \mathbb{E}(g(t, X_t))$ exists for all $t \in K$. Then, h is a $\mathcal{B}_K$- measurable function.*

Moreover, if

$$\int_D \mathbb{E}\left(\left\|g\left((t_1,t_2), X_{(t_1,t_2)}\right)\right\|\right) dt_1 < \infty$$

additionally holds for $D \in \mathcal{B}_H$, $t_2 \in G$, then for almost all $\omega \in \Omega$ it holds that

$$\int_D g\left((t_1,t_2), X_{(t_1,t_2)}\right) dt_1 < \infty \,,$$

as well as

$$\mathbb{E}\left(\int_D g\left((t_1,t_2), X_{(t_1,t_2)}\right) dt_1\right) = \int_D \mathbb{E}\left(g\left((t_1,t_2), X_{(t_1,t_2)}\right)\right) dt_1 \,.$$

Next, we calculate the expectation of the particular solution Y_t. In the case of a path-wise solution (12.8) we assume that the vector-valued process Z_t is path-wise continuous, that the expectation of each of its components exists and that

$$\int_{t_0}^{t} \mathbb{E}\left(\|Z_\tau\|\right) \mathrm{d}\tau < \infty \,, \qquad t \in I \,, \tag{12.10}$$

holds. In the case of a mean-square solution (12.9) we assume that Z_t is a second order mean-square continuous vector-valued process. Due to Lemma 12.1 and the interchangeability of taking the expectation and mean-square integration, we get in both cases

$$\mathbb{E}\left(Y_t\right) = \Phi(t) \int_{t_0}^{t} \Phi^{-1}(\tau) \mathbb{E}(Z_\tau) \mathrm{d}\tau \,. \tag{12.11}$$

The expectation of the path-wise solution of (12.1) with respect to the initial condition $(X_0, t_0) \in L_d^1 \times I$ and the expectation of the mean-square solution of (12.1) with respect to the initial condition $(X_0, t_0) \in L_d^2 \times I$, respectively, is a solution (in the extended sense in the case of the path-wise solution) of the ordinary (deterministic) differential equation

$$\frac{\mathrm{d}m_X(t)}{\mathrm{d}t} = A(t)m_X(t) + m_Z(t)$$

with respect to the initial condition $m_X(t_0) = \mathbb{E}(X_0)$.

For the computation of the covariance function of Y_t we assume that, in the case of a path-wise solution, Z_t is a second order path-wise continuous stochastic process and

$$\int_{t_0}^{t_1}\int_{t_0}^{t_2} \mathbb{E}\left(\|Z_{\tau_1}Z_{\tau_2}\|\right) \mathrm{d}\tau_1 \mathrm{d}\tau_2 < \infty, \quad t_1, t_2 \in I \tag{12.12}$$

holds, and in the case of a mean-square solution, we assume that Z_t is mean-square continuous on I, respectively. Taking Lemma 12.1 into account in the first case and

$$\mathbb{E}\left(\left(⨍_{t_0}^{t_1} Z_{\tau_1}\mathrm{d}\tau_1\right)\left(⨍_{t_0}^{t_2} Z_{\tau_2}\mathrm{d}\tau_2\right)\right) = \int_{t_0}^{t_1}\int_{t_0}^{t_2} \mathbb{E}\left(Z_{\tau_1}Z_{\tau_2}^T\right) \mathrm{d}\tau_1 \mathrm{d}\tau_2$$

in the second case, respectively, then we obtain in both cases

$$C_Y(t_1, t_2) = \Phi(t_1)\left(\int_{t_0}^{t_1}\int_{t_0}^{t_2} \Phi^{-1}(\tau_1)C_Z(\tau_1, \tau_2)\Phi^{-T}(\tau_2)\mathrm{d}\tau_1\mathrm{d}\tau_2\right)\Phi^T(t_2)\,. \tag{12.13}$$

Illustratively, the common covariance function

$$C_{YZ}(t_1, t_2) = \mathbb{E}\left(Y_{t_1}Z_{t_2}^T\right) - m_Y(t_1)m_Z^T(t_2)$$

of Y_t and Z_t characterises the relation between the input process and the output process. For its computation we assume in the case of path-wise solutions that Z_t is a second order path-wise continuous stochastic process and

$$\int_{t_0}^{t_1} \mathbb{E}\left(\left\|Z_\tau Z_{t_2}^T\right\|\right) \mathrm{d}\tau < \infty, \quad t_1, t_2 \in I\,, \tag{12.14}$$

holds. In the case of a mean-square solution, we assume that Z_t is mean-square continuous on I. It then follows that

$$C_{YZ}(t_1, t_2) = \Phi(t_1)\int_{t_0}^{t_1} \Phi^{-1}(\tau)C_Z(\tau, t_2)\mathrm{d}\tau\,. \tag{12.15}$$

Thus, for fixed t_2 the common covariance function $C_{YZ}(t_1, t_2)$ is a solution (in the extended sense in the case of the path-wise solution) of the deterministic matrix differential equation

$$\frac{\mathrm{d}C_{YZ}(t_1, t_2)}{\mathrm{d}t_1} = A(t)C_{YZ}(t_1, t_2) + C_Z(t_1, t_2)$$

with respect to the initial condition $C_{YZ}(t_0, t_2) = 0$.

With equation (12.15) we obtain conditions for the asymptotic decay of the common covariance function of Y_t and Z_t for increasing time distances. For instance, if Y_t and Z_t are stationary processes with positive variance then such decay conditions are given by asymptotic un-correlation.

Theorem 12.2 (Asymptotic Un-Correlation). *We formulate the following conditions:*

1. *Z_t is a second order path-wise continuous stochastic process satisfying (12.14).*
2. *Z_t is a mean-square continuous stochastic process.*
3. $\int_{t_0}^{t} \|\Phi(t)\Phi^{-1}(\tau)\| d\tau \leq H < \infty, t \geq t_0$.
4. $\lim_{(t-\tau)\to\infty} \|\Phi(t)\Phi^{-1}(\tau)\| = 0$.
5. $\lim_{(t-\tau)\to\infty} \|C_Z(\tau, t)\| = 0$.

Asymptotic un-correlation

$$\lim_{(t_1-t_2)\to\infty} C_{YZ}(t_1, t_2) = 0$$

holds in the case of a path-wise solution Y_t (12.8) provided conditions 1 together with 3 - 5 are satisfied, and in the case of a mean-square solution Y_t (12.9) provided conditions 2 together with 3 - 5 are satisfied, respectively.

Proof. Following [45], pp. 43, let us assume $t_2 \geq t_1$ and $\varepsilon > 0$ to be given. Together with $t_2 < t^* < t_1$ and condition 3 we get from equation (12.15) the following estimate

$$\begin{aligned} \|C_{YZ}(t_1, t_2)\| \leq\ & (t^* - t_0) \sup_{\tau\in[t_0,t^*]} \left\|\Phi(t_1)\Phi^{-1}(\tau)\right\| \cdot \|C_Z(\tau, t_2)\| \\ & + H \sup_{\tau\in[t^*,t_1]} \|C_Z(\tau, t_2)\| \ . \end{aligned}$$

Due to condition 5 we can choose $t^* > t_2$ such that the second term in (12.16) is less than $\varepsilon/2$ for all $t_1 > t^*$. Then we choose $t_\varepsilon > t^* > 0$ such that the first term is less than $\varepsilon/2$ for all $t_1 > t_\varepsilon$, too, because of condition 4. For every $\varepsilon > 0$ there thus is a $t_\varepsilon > t_2$ such that $\|C_{YZ}(t_1, t_2)\| < \varepsilon$, $t_1 > t_\varepsilon$. This shows the assertion. □

Before you continue, make sure to answer the following questions:

Quiz: Section 12.2 – Part II

Q1 Let X_t be a path-wise (mean-square) solution of $\dot{X}_t = A(t)X_t + Z_t$, where Z_t is a path-wise (mean-square) continuous stochastic process. Which differential equation does $\mathbb{E}(X_t)$ obey?

Q2 How would things change in Q1 if X_t would be a path-wise (mean-square) solution of $\dot{X}_t = A(t, \omega)X_t + Z_t$, where the entries of $A(t, \omega)$ are stochastic processes, too?

Q3 How is the common covariance C_{XZ} of a solution X_t of $\dot{X}_t = A(t)X_t + Z_t$ and its driving process Z_t defined?

Q4 Which differential equation does the common covariance C_{XZ} from Q3 satisfy?

Q5 Under which conditions does C_{XZ} decay to zero for $t \to \infty$?

Q6 Sketch the proof of the assertion you used in Q5.

12.2.2 The Special Case of a Gaussian Inhomogeneity

If the inhomogeneous forcing process Z_t of the random differential equation (12.1) is a Gaussian process, then the path-wise solution (12.8) and the mean-square solution (12.9) are Gaussian processes, too. As the probability distribution of a Gaussian process is determined by its mean and covariance we get the corresponding distribution of $Y_t \stackrel{I}{=} \int_{t_0}^{t} f(\tau, Z_\tau) d\tau$ in this case by applying (12.11) and (12.13).

The $2d$-dimensional process $W_t = (Y_t, Z_t)$ is a Gaussian process, too, the probability distribution of which is given by

$$m_W(t) = (m_Y(t), m_Z(t)) , \tag{12.16}$$

and

$$C_W(t_1, t_2) = \begin{pmatrix} C_Y(t_1, t_2) & C_{YZ}(t_1, t_2) \\ C_{ZY}(t_1, t_2) & C_Z(t_1, t_2) \end{pmatrix} . \tag{12.17}$$

Theorem 12.3 (Stochastic Characteristics of the Process $W_t = (Y_t, Z_t)$). *Let Z_t be a path-wise continuous Gaussian process with the properties (12.10), (12.12) and (12.14) or a mean-square continuous Gaussian process, respectively. Let Y_t be the path-wise solution (12.8) or the mean-square solution (12.9) of (12.1), respectively. Then $W_t = (Y_t, Z_t)$ is a Gaussian process the mean and covariance of which are given by (12.11), (12.13), (12.15), (12.16), and (12.17).*

Proof. Following [45], pp. 43, we have that the path-wise solution (12.8) or the mean-square solution (12.9), respectively, can be considered by definition to be a path-wise or mean-square integral, respectively, over a path-wise continuous or mean-square continuous integrand, respectively, as the limit in probability 1 or as mean-square limit, respectively, of Riemannian sums of the form

$$S_r(t) \;=\; \Phi(t) \sum_{i=1}^{N_r} \Phi^{-1}(\tau_{r,i}) Z_{\tau_{r,i}} \left(\tau_{r,i+1} - \tau_{r,i}\right) .$$

With Z_t the composed process $(S_r(t), Z_t)$ is a Gaussian process and thus its limit $W_t = (Y_t, Z_t)$ is a Gaussian process, too. □

Before you continue, make sure to answer the following questions:

Quiz: Section 12.2 – Part III

Let X_t be the solution of

$$\dot{X}_t \;=\; \begin{pmatrix} 1 & 0 \\ 0 & 1 \end{pmatrix} X_t + \begin{pmatrix} 0 \\ 1 \end{pmatrix} W_t ,$$

where W_t is the 1-dimensional standard Wiener process.

Q1 Determine the mean $\mathbb{E}(X_t)$ and the auto-covariance $C_X(t_1, t_2)$ of the solution process X_t.

Q2 Let $L_t := (X_t, W_t)$. Determine the mean $m_L(t)$ and the covariance $C_L(t_1, t_2)$ of the Gaussian process L_t.

12.3 Periodic and Stationary Solutions

In this section, we study random differential equations of the form

$$\dot{X}_t \;=\; A(t) X_t + Z_t , \qquad (12.18)$$

where $A(t) : I \to \mathbb{R}^{d\times d}$ is a matrix of periodic continuous functions and Z_t a d-dimensional periodic stochastic process. Under certain conditions we will construct a periodic path-wise solution or mean-square solution, respectively, of (12.18) towards which all other solutions converge asymptotically. As an immediate consequence of this discussion we gain a stationary solution towards which all other solutions converge asymptotically in the special case of a constant coefficient matrix $A(t) \equiv A$ and a stationary forcing process Z_t.

12.3.1 Existence of Periodic and Stationary Solutions

First note that the fundamental matrix $\Phi(t)$ of the homogeneous system

$$\dot{x}(t) = A(t)x(t)\,, \tag{12.19}$$

where $A(t)$ is continuous and θ-periodic, is of the form

$$\Phi(t) = P(t)\cdot \exp(Rt)\,,$$

where $P(t)$ is a non-singular 2θ-periodic matrix function and R is a constant matrix[1]. The eigenvalues of R are called *characteristic exponents of* $A(t)$. Moreover, for every $\alpha \in \mathbb{R}$ the matrix function $\Phi(t)\Phi^{-1}(t-\alpha)$ is θ-periodic. For a constant matrix A the fundamental matrix reads as $\Phi(t) = \exp(At)$ and the eigenvalues of A are the characteristic exponents.

Theorem 12.4 (Path-Wise θ-Periodic Solutions - Part I). *Let the following three conditions be satisfied:*

1. *$A(t)$ is a continuous θ-periodic matrix function defined on I.*
2. *Z_t is a path-wise continuous process on I.*
3. *$\int_{\infty}^{t_0} \mathbb{E}\left(\left\|\exp(-P\tau)P^{-1}(\tau)Z_\tau\right\|\right) d\tau < \infty$.*

Then

$$Y_t^0 = P(t)\int_{-\infty}^{t} \exp\left(R(t-\tau)\right)P^{-1}(\tau)Z_\tau d\tau \tag{12.20}$$

is a path-wise solution of (12.18). Moreover, if

4. *Z_t is a strictly θ-periodic process,*

then $W_t^0 = (Y_t^0, Z_t)$ and thus Y_t^0 are θ-periodic processes.

Proof. Following [45], pp. 44, we first note that due to condition 3, and according to Lemma 12.1

$$\gamma = \int_{-\infty}^{t_0} \exp\left(-R\tau\right)P^{-1}(\tau)Z_\tau \mathrm{d}\tau\,, \qquad \text{for all } \omega\in\Omega\,, \tag{12.21}$$

[1] Cf. Theorem 3.5.1 from [64].

and thus $Y_t^0 = \Phi(t)\gamma + Y_t$, for all $t \in I$, are defined and finite. Because of Y_t the function Y_t^0 is a path-wise solution of (12.18), too, see equations (12.4) and (12.8). Due to the continuity of the integrand function we have that Y_t^0 is for almost all $\omega \in \Omega$ a limit of Riemannian sums:

$$Y_t^0 = \int_0^\infty \Phi(t)\Phi^{-1}(t-\alpha)Z_{t-\alpha}\mathrm{d}\alpha \stackrel{I}{=} \lim_{r\to\infty} S_t^r ,$$

with the sums S_t^r given by

$$S_t^r := \sum_{i=1}^{N_r} \Phi(t)\Phi^{-1}(t-\alpha_{r,i})Z_{t-\alpha_{r,i}}\left(\alpha_{r,i+1}-\alpha_{r,i}\right) .$$

Next, for every r, the process $W_t^r = (S_t^r, Z_t)$ is of the form $W_t^r = y(t, V_t^r)$, where

$$V_t^r = (Z_{t-\alpha_{r,1}}, Z_{t-\alpha_{r,2}}, \ldots, Z_{t-\alpha_{r,N_r}}, Z_t)$$

is a strictly θ-periodic process and $y(t,v)$ is a θ-periodic function in t for fixed v. Thus, for every r, the process W_t^r itself and hence $W_t \stackrel{I}{=} \lim_{r\to\infty} W_t^r$ are both strictly θ-periodic. This shows the second assertion. □

Corollary 12.2 (Strict θ-Periodicity). *Let V_t be a stochastic process such that (Z_t, V_t) is strictly θ-periodic. Then, analogously to the proof of Theorem 12.4 one can show that (Y_t^0, Z_t, V_t) is strictly θ-periodic.*

Corollary 12.3 (Sufficient Requirements for Condition 3). *Let the conditions 1, 2 and 4 hold together with*

5. *All characteristic exponents λ_k of the matrix $A(t)$ have negative real parts, i.e., $\max_k Re(\lambda_k) < -\rho < 0$.*

Then,

$$\int_0^\theta \mathbb{E}\left(\|Z_\tau\|\right) < \infty \qquad (12.22)$$

is sufficient for the fulfillment of condition 3. This follows from the estimates

$$\|\exp(tR)\| \leq c \cdot \exp(-\rho t) ,$$

and

$$\begin{aligned} &\textstyle\int_{-\infty}^{t_0} \mathbb{E}\left(\left\|\exp(-R\tau)\, P^{-1}(\tau)\right\|\right) d\tau \\ &\leq \max_{0\leq\tau\leq 2\theta} \left\|P^{-1}(\tau)\right\| \textstyle\sum_{k=0}^{\infty} c \cdot \exp\left((t_0 - 2k\theta)\rho\right) \cdot \int_{t_0-2(k+1)\theta}^{t_0-2k\theta} \mathbb{E}\left(\|Z_\tau\|\right) d\tau \\ &= K \cdot \exp(t_0\rho) \cdot \textstyle\sum_{k=0}^{\infty} \left(\exp(-2\rho\theta)\right)^k < \infty . \end{aligned}$$

If $\mathbb{E}\left(\|Z_t\|\right)$ is constant, like in the case when Z_t is stationary, then (12.22) holds.

Corollary 12.4 (Periodic Solutions for Positive Characteristic Exponents). *A periodic solution analogous to (12.20) can also be constructed in the case when some of the λ_k, e.g., for $k = m+1, m+2, \ldots, d$, have positive real parts whereas the remaining λ_k have negative real parts. Therefore the matrix function $\exp(R\alpha)$ has to be written in the known form $\sum_k R_k(\alpha)\exp(\lambda_k\alpha)$, cf. paragraph 3.4 in [64], and Y_t^0 has to be set as*

$$\begin{aligned} Y_t^0 &:= \sum_{k=1}^{m} \int_{-\infty}^{t} P(t)\, R_k(t-\tau) \exp\left((t-\tau)\alpha_k\right) P^{-1}(\tau)\, Z_\tau d\tau \\ &\quad - \sum_{m+1}^{d} \int_{t}^{\infty} P(t)\, R_k(t-\tau) \exp\left((t-\tau)\alpha_k\right) P^{-1}(\tau)\, Z_\tau d\tau\,. \end{aligned}$$

Concerning the relationship between periodicity and stationarity, Theorem 12.4 immediately implies the following theorem:

Theorem 12.5 (Stationary Solutions - Part 1). *Let the following conditions be satisfied:*

1. *$A(t) \equiv A$ is a constant matrix and all of its eigenvalues have negative real part.*
2. *Z_t is a path-wise continuous stationary process.*

Then, the path-wise solution Y_t^0 of (12.18) as well as $W_t^0 = (Y_t^0, Z_t)$ are stationary processes.

We will show the analogous result for periodic processes:

Theorem 12.6 (Path-Wise θ-Periodic Solutions - Part II). *Let the following conditions be satisfied:*

1. *$A(t)$ is a continuous θ-periodic matrix process.*
2. *Z_t, $t \in I$, is a path-wise second order θ-periodic stochastic process.*
3. *$\int_{-\infty}^{t} \mathbb{E}\left(\|\exp(-R\tau)\, P^{-1}(\tau) Z_\tau\|\right) d\tau < \infty$, for all $t \in I$.*
4. *$\int_{-\infty}^{t_1}\int_{-\infty}^{t_2} \mathbb{E}\left(\left\|\exp(-R\tau_1)\, P^{-1}(\tau_1) Z_{\tau_1} Z_{\tau_2} \left(\exp(-R\tau_2)\, P^{-1}(\tau_2)\right)^T\right\|\right) d\tau_1 d\tau_2 < \infty$ for all $t_1, t_2 \in I$.*
5. *$\int_{-\infty}^{t_1} \mathbb{E}\left(\|\exp(-R\tau_1)\, P^{-1}(\tau_1) Z_{\tau_1} Z_{t_2}\|\right) d\tau_1 < \infty$, for all $t_1, t_2 \in I$.*

Then, the path-wise solution Y_t^0 of (12.18) as well as $W_t^0 = (Y_t^0, Z_t)$ are θ-periodic processes.

Proof. Following [45], pp. 46, we have that if conditions 1 to 3 of Theorem 12.4 hold, then Y_t^0 is a path-wise solution of (12.18). Due to conditions 3, 4 and 5 of this theorem we have according to Lemma 12.1 for the mean of Y_t^0

$$m_{Y^0}(t) \;=\; \Phi(t)\int_0^{\infty} \Phi^{-1}(t-\alpha)m_Z(t-\alpha)\mathrm{d}\alpha\,, \qquad (12.23)$$

for the common covariance of Y_t^0 and Z_t

$$C_{Y^0Z}(t_1,t_2) \;=\; \Phi(t_1)\int_0^{\infty} \Phi^{-1}(t_1-\alpha_1)C_Z(t_1-\alpha_1,t_2)\mathrm{d}\alpha_1\,, \qquad (12.24)$$

as well as for the auto-covariance of Y_t^0

$$\begin{aligned} C_{Y^0}(t_1,t_2) \;&=\; \Phi(t_1)\int_0^{\infty}\int_0^{\infty} \Phi^{-1}(t_1-\alpha_1)C_Z(t_1-\alpha_1,t_2-\alpha_2) \\ &\quad \cdot\Phi^{-T}(t_2-\alpha_2)\mathrm{d}\alpha_1\mathrm{d}\alpha_2\Phi^T(t_2)\,. \end{aligned} \qquad (12.25)$$

Because of the periodicity of m_Z, C_Z and $\Phi(t)\Phi^{-1}(t-\alpha)$ condition 2 implies

$$m_{W^0}(t+\theta) \;=\; (m_{Y^0}(t), m_Z(t)) \;=\; m_{W^0}(t)\,,$$

as well as

$$C_{W^0}(t_1+\theta,t_2+\theta) \;=\; \begin{pmatrix} C_{Y^0}(t_1,t_2) & C_{Y^0Z}(t_1,t_2) \\ C_{ZY^0}(t_1,t_2) & C_Z(t_1,t_2) \end{pmatrix} \;=\; C_{W^0}(t_1,t_2)\,.$$

Hence, W_t^0 is a second order θ-periodic process. □

As in the case of periodicity and stationarity in the narrow sense, here the following result also immediately follows from the relationship between periodicity and stationarity:

Theorem 12.7 (Stationary Solutions - Part II). *Let the following conditions be satisfied:*

1. $A(t)\equiv A$ *is a constant matrix, and all of its eigenvalues have negative real part.*

2. Z_t *is a second order stationary stochastic process.*

Then, the path-wise solution Y_t^0 *of (12.18) and* $W_t^0 = (Y_t^0, Z_t)$ *are both stationary processes.*

In the next two theorems, we discuss the existence of periodic and stationary mean-square solutions of (12.18).

Theorem 12.8 (Mean-Square θ-Periodic Solutions). *Let the following conditions be satisfied:*

1. *$A(t)$ is a continuous θ-periodic matrix function.*
2. *All characteristic exponents λ_k of the matrix function $A(t)$ have negative real parts:* $\max_k Re(\lambda_k) < -\rho < 0$.
3. *Z_t is a second order mean-square continuous θ-periodic stochastic process.*

Then

$$Y_t^* \;=\; P(t)\int_{-\infty}^{t} \exp\left(t-\tau\right) P^{-1}(\tau) Z_\tau d\tau \tag{12.26}$$

is a mean-square solution of (12.18), and $W_t^* = (Y_t^*, Z_t)$ is θ-periodic.

Proof. Following [45], pp. 47, we first show that the mean-square integral in (12.26) is defined for all $t \in I$. Due to the continuity and periodicity of $m_Z(t)$ it holds that

$$\int_0^{2\theta} \left\| P^{-1}(\tau) m_Z(\tau) \right\| \mathrm{d}\tau \;=\; \gamma \;<\; \infty\,,$$

and thus, analogously to Corollary 12.2 of Theorem 12.4 we have

$$\int_{-\infty}^{0} \left\| \exp\left(-R\tau\right) P^{-1}(\tau) m_Z(\tau) \right\| \mathrm{d}\tau \;=\; c\cdot\gamma\cdot\sum_{k=0}^{\infty} \exp\left(-2k\theta\rho\right) \;<\; \infty\,.$$

Hence, for every $t \in I$ the mean

$$m^*(t) \;=\; \int_{-\infty}^{t} \exp\left(-R\tau\right) P^{-1}(\tau) m_Z(\tau) \mathrm{d}\tau$$

is finite. And analogously one shows that the covariance function

$$C^*(t_1,t_2) = \int_{-\infty}^{t_1}\int_{-\infty}^{t_2} \exp\left(-R\tau_1\right) P^{-1}(\tau_1) C_Z(\tau_1,\tau_2) \left(\exp\left(-R\tau_2\right) P^{-1}(\tau_2)\right)^T \mathrm{d}\tau_1 \mathrm{d}\tau_2\,.$$

is finite, too. Next, let us consider the processes

$$X_t^k \;=\; \int_{-k}^{t} \exp\left(-R\tau\right) P^{-1}(\tau) Z_\tau \mathrm{d}\tau\,, \quad \text{for } k = 1,2,\ldots\,.$$

We have

$$m_{X^k}(t) \;=\; \int_{-k}^{t} \exp\left(-R\tau\right) P^{-1}(\tau) m_Z(\tau) \mathrm{d}\tau$$

and

$$C_{X^k}(t_1,t_2) = \int_{-k}^{t}\int_{-k}^{t} \exp(-R\tau_1)\,P^{-1}(\tau_1)C_Z(\tau_1,\tau_2)\left(\exp(-R\tau_2)\,P^{-1}(\tau_2)\right)^T \mathrm{d}\tau_1\mathrm{d}\tau_2\,.$$

For $U_t^k = X_t^k - m_{X^k}(t)$ it holds that

$$m_{U^k}(t) \;\equiv\; 0\,, \quad \text{and} \quad C_{U^k}(t_1,t_2) \;=\; C_{X^k}(t_1,t_2)\,.$$

Because of $\lim_{k\to\infty} C_{U^k}(t_1,t_2) = C^*(t_1,t_2)$ property 14, of Sec. 4.3 implies the existence of a second order process $U_t^*(t)$ such that

$$\text{m.s.-}\lim_{k\to\infty} U_t^k \;=\; U_t^*\,, \quad m_{U^*}(t) \;\equiv\; 0\,, \quad \text{and} \quad C_{U^*}(t_1,t_2) \;\equiv\; C^*(t_1,t_2)\,.$$

Thus, by (12.26) a second order stochastic process

$$Y_t^* \;=\; P(t)\exp(Rt)\,X_t^*$$

is defined with mean

$$m_{Y^*}(t) \;=\; P(t)\exp(Rt)\,m^*(t)\,,$$

and auto-covariance function

$$C_{Y^*}(t_1,t_2) \;=\; P(t_1)\exp(Rt_1)\,C^*(t_1,t_2)\,(P(t_2)\exp(Rt_2))^T\,.$$

Moreover, $Y_t = \Phi(t)\int_{t_0}^{t}\Phi^{-1}(\tau)Z_\tau\mathrm{d}\tau$ is a mean-square solution of (12.18) according to Theorem 12.1 and $Y_t^1 = \Phi(t)X_{t_0}^*$ has the mean-square derivative $\dot{Y}_t^1 = A(t)Y_t^1$ according to property 11 of Sec. 4.3. Hence, $Y_t^* = Y_t^1 + Y_t$ is a mean-square solution of (12.18).

Substitution of $\alpha := t-\tau$ leads to

$$m_{Y^*}(t) \;=\; \Phi(t)\int_0^\infty \Phi^{-1}(t-\alpha)m_Z(t-\alpha)\mathrm{d}\alpha\,, \tag{12.27}$$

and with $\alpha_1 = t_1-\tau_1$ and $\alpha_2 = t_2-\tau_2$ we get

$$C_{Y^*}(t_1,t_2) \;=\; \Phi(t_1)\int_0^\infty\int_0^\infty \Phi^{-1}(t_1-\alpha_1)C_Z(t_1-\alpha_1,t_2-\alpha_2)$$
$$\cdot\left(\Phi^{-1}(t_2-\alpha_2)\right)^T \mathrm{d}\alpha_1\mathrm{d}\alpha_2\Phi^T(t_2)\,, \tag{12.28}$$

as well as

$$C_{Y^*Z}(t_1,t_2) \;=\; \Phi(t_1)\int_0^\infty \Phi^{-1}(t_1-\alpha)C_Z(t_1-\alpha,t_2)\mathrm{d}\alpha\,. \tag{12.29}$$

Because of the periodicity of $\Phi(t)\Phi^{-1}(t-\alpha)$, $m_Z(t)$ and $C_Z(t_1,t_2)$ the equations (12.27), (12.28), and (12.29) immediately imply the asserted periodicity of the process W_t^*. □

If the conditions of Theorem 12.8 are satisfied and if Z_t is path-wise continuous on I, then the conditions of Theorem 12.6 also hold and the periodic path-wise solution Y_t^0 of (12.18) and the periodic mean-square solution Y_t^* of (12.18) are equivalent stochastic processes.

Due to the relationship between periodicity and stationarity, Theorem 12.8 immediately implies the following theorem:

Theorem 12.9 (Stationary Solutions -- Part III). *Let the following conditions be satisfied:*

1. *$A(t) \equiv A$ is a constant matrix function and all of its eigenvalues have negative real parts.*
2. *Z_t is a second order mean-square continuous stationary process on I.*

Then,

$$Y_t^* = \int_{-\infty}^{t} \exp\left(A(t-\tau)\right) Z_\tau d\tau$$

is a stationary mean-square solution of (12.18), and $W_t^* = (Y_t^*, Z_t)$ is a stationary process, too.

Theorem 12.9 derived the stationary mean-square solution Y_t^* of (12.18). The corresponding spectral distribution function F_{Y^*} is gained directly from the spectral distribution F_Z of the forcing process Z_t:

Theorem 12.10 (Spectral Distribution of the Stationary Mean-Square Solution). *Let the conditions of Theorem 12.9 be satisfied. Then, the spectral distribution*

$$F_{Y^*}(\lambda) = \left(F_{j,k}^{Y^*}(\lambda)\right)_{j,k=1,2,\dots,d}$$

of the stationary mean-square solution Y_t^* of (12.18) is given by

$$F_{j,k}^{Y^*}(\lambda) := \sum_{l,m=1}^{d} \int_{-\infty}^{\lambda} H_{j,l}(i\rho)\overline{H_{k,m}(i\rho)} dF_{l,m}^{Z}(\rho) , \qquad (12.30)$$

where

$$H(i\lambda) := \int_{-\infty}^{0} \exp\left(-\lambda\tau\right) \exp\left(-A\tau\right) .$$

If Z_t has the spectral density Φ_Z, then

$$\Phi_{Y^*}(\lambda) = H(i\lambda)\Phi_Z(\lambda)\overline{H^T(i\lambda)} . \qquad (12.31)$$

holds.

Proof. Following [45], p. 49, we study the correlation function Y_t^* with the aid of the spectral decomposition of C_Z and the fundamental matrix $\Phi(t) = \exp(At)$ of a stationary solution:

$$\begin{aligned} C_{j,k}^{Y^*}(t) &= \sum_{l,m=1}^{d} \mathbb{E}\left(\int_{-\infty}^{0} \Phi_{j,l}(-\tau_1) Z_{\tau_1}^{(l)} \mathrm{d}\tau_1 \int_{-\infty}^{0} \Phi_{k,m}(t-\tau_2) Z_{\tau_1}^{(m)} \mathrm{d}\tau_2\right) \\ &= \sum_{l,m=1}^{d} \int_{-\infty}^{0}\int_{-\infty}^{t} \Phi_{j,l}(-\tau_1)\mathbb{E}\left(Z_{\tau_1}^{(l)} Z_{\tau_2}^{(m)}\right) \Phi_{k,m}(t-\tau_2)\mathrm{d}\tau_1\mathrm{d}\tau_2 \\ &= \sum_{l,m=1}^{d} \int_{-\infty}^{0}\int_{-\infty}^{t} \Phi_{j,l}(-\tau_1)\left(\int_{-\infty}^{\infty} \exp\left(i(\tau_2-\tau_1)\lambda\right) \mathrm{d}F_{l,m}^{Z}(\lambda)\right) \\ &\quad \cdot \Phi_{k,m}(t-\tau_2)\mathrm{d}\tau_1\mathrm{d}\tau_2 \,. \end{aligned}$$

Because of $|\Phi_{l,m}(\tau)| \leq c\exp(-\tau\rho)$, $\tau \in \mathbb{R}$, $l, m = 1, 2, \ldots, d$, and

$$\int_{-\infty}^{0}\int_{-\infty}^{t} \exp\left(\tau_1\rho\right) \exp\left((\tau_2 - t)\rho\right) \mathrm{d}\tau_1\mathrm{d}\tau_2 \,,$$

the sequence of integrations can be exchanged, and utilizing the substitution $\theta_2 := t - \tau_2$ leads to

$$\begin{aligned} C_{j,k}^{Y^*}(t) &= \sum_{l,m}^{d} \int_{-\infty}^{\infty} \exp\left(i\lambda t\right) \left(\int_{-\infty}^{0} \Phi_{j,l}(-\tau_1) \exp\left(-i\tau_1\right) \mathrm{d}\tau_1\right. \\ &\quad \left.\cdot \int_{0}^{\infty} \Phi_{k,m}(\theta_2) \exp\left(-i\theta_2\right) \mathrm{d}\theta_2\right) \mathrm{d}F_{l,m}^{Z}(\lambda) \\ &= \int_{-\infty}^{\infty} \exp\left(i\lambda t\right) \sum_{l,m}^{d} \left(H_{j,l}(i\lambda)\overline{H_{k,m}(i\lambda)}\right) \mathrm{d}F_{l,m}^{Z}(\lambda) \,. \end{aligned}$$

If $F_{j,k}^{Y^*}$ is defined by (12.30), then the above representation implies

$$F_{j,k}^{Y^*}(t) = \int_{\infty}^{\infty} \exp\left(-i\theta_2\right) \mathrm{d}F_{j,k}^{Y^*}(\lambda) \,, \tag{12.32}$$

i.e., F_{Y^*} is the spectral distribution of Y_t^*. □

12.3.2 Convergence Towards Periodic and Stationary Solutions

In this section, we discuss conditions for the asymptotic convergence of pathwise solutions of the inhomogeneous linear random differential equation

$$\dot{X}_t = A(t)X_t + Z_t \,, \tag{12.33}$$

towards a periodic or stationary solution Y_t^0. First note that under the conditions (i) that Z_t is path-wise continuous on I, and (ii) that every solution of the homogeneous (deterministic) differential equation

$$\dot{x}(t) = A(t)x(t) \tag{12.34}$$

converges asymptotically towards zero for two arbitrary path-wise solutions $X_t^{(1)}$ and $X_t^{(2)}$, it then holds that

$$\lim_{t\to\infty} \left\| X_t^{(1)} - X_t^{(2)} \right\| \hat{=} 0\,, \tag{12.35}$$

and for $X_t^{(1)} \in L_d^2$ and $X_t^{(2)} \in L_d^2$ it also holds that

$$\lim_{t\to\infty} \mathbb{E}\left(\left\| X_t^{(1)} - X_t^{(2)} \right\|^2 \right) = 0\,. \tag{12.36}$$

Equations (12.35) and (12.36) immediately follow from

$$\left\| X_t^{(1)} - X_t^{(2)} \right\| \leq \|\Phi(t)\| \left\| X_{t_0}^{(1)} - X_{t_0}^{(2)} \right\|, \quad \text{and} \quad \lim_{t\to\infty} \|\Phi(t)\| = 0\,.$$

Before you continue, make sure to answer the following questions:

Quiz: Section 12.3 – Part I

Let X_t be a solution of $\dot{X}_t = A(t)X_t + Z_t$.

Q1 Under which conditions is X_t a (strictly) path-wise θ-periodic process?

Q2 Give a formula for the (path-wise) solution process X_t under the conditions from Q1.

Q3 Sketch the proof of your assertions from Q1 and Q2.

Q4 Under which conditions is the path-wise solution X_t a stationary process?

Q5 Under which conditions is X_t a (strictly) mean-square θ-periodic process?

Q6 Give a formula for the (mean-square) solution process X_t under the conditions from Q1.

Q7 Sketch the proof of your assertions from Q5 and Q6.

Q8 Under which conditions is the mean-square solution X_t a stationary process?

Theorem 12.11 (Convergence of Path-Wise Solutions -- Part I). *Let the conditions of Theorems a) 12.4, b) 12.5, c) 12.6, and d) 12.7, respectively, hold. Moreover, let all characteristic exponents λ_k of A_t have negative real parts $\lambda_* = \max_k Re(\lambda_k) < \rho < 0$. Then every path-wise solution X_t of (12.33) converges with probability one exponentially towards the a) strictly periodic, b) strictly stationary, c) periodic and, d) stationary solution Y_t^0, respectively:*

$$\lim_{t\to\infty} \|X_t - Y_t^0\| \;\hat{=}\; 0\,. \tag{12.37}$$

Proof. Following [45], p. 50, the assertion results immediately from the estimates

- $\|\exp(Rt)\| \le c\exp(-\rho_* t)$, $\rho_* = 2^{-1}(\lambda_* + \rho)$,
- $\max_{t\in I}\|P(t)\| = \max_{0\le t\le 2\theta}\|P(t)\| = \chi < \infty$,
- $\|X_t - Y_t^0\| = \|P(t)\exp(Rt)X_0 - P(t)\exp(Rt)\int_{-\infty}^{0}\exp(-R\tau)Z_\tau \mathrm{d}\tau\|$
  $\overset{I}{\le} \chi c\exp(-\rho_* t)\left(\|X_0\| + \int_{-\infty}^{0}\|\exp(-R\tau)Z_\tau \mathrm{d}\tau\right)$,

as the integral $\int_{-\infty}^{0}\|\exp(-R\tau)T_\tau\|\mathrm{d}\tau$ is finite for almost all $\omega\in\Omega$. □

An analogous result holds for mean-square solutions:

Theorem 12.12 (Exponential Decay of Mean-Square Solutions). *Let the conditions of Theorem 12.8 (or Theorem 12.9, respectively) be satisfied. Moreover, let all characteristic exponents λ_k of A_t have negative real parts $\lambda_* = \max_k Re(\lambda_k) < \rho < 0$. Then, for every mean-square solution X_t of (12.33) it holds that*

$$\lim_{t\to\infty} \exp(\rho t)\,\mathbb{E}\left(\|X_t - Y_t^0\|^2\right) \;=\; 0\,, \tag{12.38}$$

where Y_t^0 is the mean-square solution given in Theorem 12.8 (or Theorem 12.9, respectively).

Proof. Following [45], p. 51, the assertion results with $\rho_* = 2^{-1}(\lambda_* + \rho)$ from

$$\begin{aligned}\mathbb{E}\left(\|X_t - Y_t^0\|^2\right) \;&\le\; 2\chi c\exp(-\rho_* t)\,\mathbb{E}\left(\|X_0\|^2\right)\\ &\quad + 2\chi c\exp(-\rho_* t)\,\mathbb{E}\left(\left\|\int_{-\infty}^{0}\exp(-R\tau)\,P^{-1}(\tau)\mathrm{d}\tau\right\|^2\right),\end{aligned}$$

and from the existence of the above mean-square integral shown in Theorem 12.8. □

Next we study conditions with respect to which the path-wise solutions of (12.33) are asymptotically periodic or asymptotically stationary, provided the coefficients are asymptotically periodic or asymptotically stationary, respectively.

Theorem 12.13 (Convergence of Path-Wise Solutions -- Part II). *Let the following three conditions be satisfied:*

1. *$A(t)$ is a continuous θ-periodic matrix function on I, and all its characteristic exponents λ_k have negative real parts: $\max_k Re(\lambda_k) < -\rho < 0$.*
2. *$C(t)$ is a continuous matrix function on I that satisfies the estimate*
$$\|C(t)\| \leq c\exp(-\chi t) , \quad t \geq t_0 , \quad \chi > 0 ,$$
for a sufficiently small c (with $c < \gamma k^{-1}$).
3. *Z_t is a continuous θ-periodic stochastic process on I such that*
$$\int_0^\theta \mathbb{E}(\|Z_\tau\|)\, d\tau < \infty .$$

Then every path-wise solution of the inhomogeneous random differential equation
$$\frac{dX_t}{dt} = (A(t) + C(t))\, X_t + Z_t \tag{12.39}$$
converges for increasing t with probability one towards the stochastic process Y_t^0 given in (12.20):
$$\lim_{t\to\infty} \left\|X_t - Y_t^0\right\| \mathrel{\hat{=}} 0 . \tag{12.40}$$

Proof. Following [45], pp. 51, we set $t \geq t_0 > 0$ in the proceeding. Let $\Phi(t)$ denote the fundamental matrix of the homogeneous (deterministic) differential equation
$$\frac{dx}{dt} = (A(t) + C(t))\, x \tag{12.41}$$
such that $\Phi(t_0) = \mathbb{I}$. Let us first note, that every solution of (12.41) satisfies the equation
$$u(t) = Q(t)u(t_0) + \int_{t_0}^t Q(t)Q^{-1}(\tau)C(\tau)u(\tau)d\tau , \quad t \in I ,$$
where $Q(t) = P(t)\exp(Rt)$ is the fundamental matrix of (12.34). Together with
$$\left\|Q(t)Q^{-1}(\tau)\right\| \leq k\exp(-\rho(t-\tau))$$

we obtain the inequality

$$\exp\left(\rho(t-t_0)\right)\|u(t)\| \;\leq\; ku(t_0) + kc\int_{t_0}^{t}\exp\left((\rho-\chi)\tau\right)u(\tau)\mathrm{d}\tau\,,$$

which, due to Gronwall's lemma, implies that

$$\exp\left(\rho(t-t_0)\right)\|u(t)\| \;\leq\; ku(t_0)\exp\left(kc(t-t_0)\right)\,.$$

Thus, for $kc<\rho$ it holds that $\lim_{t\to\infty}\|u(t)\|=0$ for every solution of (12.41), and hence also $\lim_{t\to\infty}\|\Phi(t)\|=0$. This implies for any path-wise solution X_t of (12.39) and for the path-wise solution

$$X_t^0 \;=\; \Phi(t)\int_{t_0}^{t}\Phi^{-1}(\tau)Z_\tau\mathrm{d}\tau \;=\; \int_{t_0}^{t}Q(t)Q^{-1}(\tau)\left(Z_\tau + C(\tau)X^0\right)\mathrm{d}\tau \quad (12.42)$$

of (12.39) the following convergence:

$$\lim_{t\to\infty}\left\|X_t-X_t^0\right\| \;\hat{=}\; \lim_{t\to\infty}\left\|\Phi(t)X_t^0\right\| \;\hat{=}\; 0\,.$$

The assertion of the theorem is shown once the convergence

$$\lim_{t\to\infty}\left\|X_t^0-Y_t^0\right\| \;\hat{=}\; 0$$

is established. First of all, it holds that

$$\begin{aligned}\left\|X_t^0-Y_t^0\right\| \overset{I}{\leq} &\left\|\int_{-\infty}^{t_0}Q(t)Q^{-1}(\tau)Z_\tau\mathrm{d}\tau\right\| + \left\|\int_{t_0}^{t}Q(t)Q^{-1}(\tau)C(\tau)\left(X_\tau^0-Y_\tau^0\right)\mathrm{d}\tau\right\| \\ &+ \left\|\int_{t_0}^{t}Q(t)Q^{-1}(\tau)C(\tau)Y_\tau^0\mathrm{d}\tau\right\|\,. \end{aligned} \quad (12.43)$$

With $\alpha=\frac{1}{2}\chi$ and $\gamma=\min(\rho,\alpha)$ we get

$$\left\|\int_{-\infty}^{t_0}Q(t)Q^{-1}(\tau)Z_\tau\mathrm{d}\tau\right\| \;\overset{I}{\leq}\; k\exp\left(-\gamma t\right)V\,, \quad (12.44)$$

here

$$V \;=\; \int_{-\infty}^{t_0}\exp\left(\rho\tau\right)\|Z_\tau\|\,\mathrm{d}\tau \quad (12.45)$$

is finite for almost all $\omega\in\Omega$ according to condition 3 (cf. Corollary 12.2 to Theorem 12.4).

Moreover, it holds that

$$\begin{aligned}&\left\|\int_{t_0}^{t}Q(t)Q^{-1}(\tau)C(\tau)\left(X_\tau^0-Y_\tau^0\right)\mathrm{d}\tau\right\| \\ \overset{I}{\leq}\;& kc\exp\left(-\rho t\right)\int_{t_0}^{t}\exp\left((\rho-\chi)\tau\right)\left\|X_\tau^0-Y_\tau^0\right\|\mathrm{d}\tau \quad (12.46)\\ \overset{I}{\leq}\;& kc\exp\left(-\gamma t\right)\int_{t_0}^{t}\exp\left(\gamma\tau\right)\left\|X_\tau^0-Y_\tau^0\right\|\mathrm{d}\tau\,.\end{aligned}$$

The last term in (12.43) is approximated as follows:

$$
\begin{aligned}
& \left\| \int_{t_0}^t Q(t)Q^{-1}(\tau)C(\tau)Y_\tau^0 \mathrm{d}\tau \right\| \\
& \overset{I}{\leq} \; kc\exp(-\rho t) \int_{t_0}^t \exp((\rho-\chi)\tau) \left\| \int_{-\infty}^{\tau} Q(\tau)Q^{-1}(\theta) Z_\theta \mathrm{d}\theta \right\| \mathrm{d}\tau \\
& \overset{I}{\leq} \; k^2 c\exp(-\rho t) \left(\int_{t_0}^t \exp(-\chi\tau) \left(\int_{-\infty}^{t_0} \exp(\rho\theta) \|Z_\theta\| \mathrm{d}\theta \right) \mathrm{d}\tau \right. \\
& \qquad \left. + \int_{t_0}^t \exp((\rho-\alpha)\tau) \left(\int_{t_0}^{\tau} \exp(-\alpha\theta) \|Z_\theta\| \mathrm{d}\theta \right) \mathrm{d}\tau \right) \qquad (12.47) \\
& \overset{I}{\leq} \; k^2 c\exp(-\rho t) \left((\exp(-\chi t_0) - \exp(-\chi t))\chi^{-1} V \right. \\
& \qquad \left. + (\alpha-\rho)^{-1} (\exp((\rho-\alpha)t_0) - \exp((\rho-\alpha)t)) W \right) \\
& \overset{I}{\leq} \; k^2 c\exp(-\rho t) \left(\chi^{-1} V + (\alpha-\rho)^{-1} W (1 + \exp((\rho-\alpha)t_0)) \right),
\end{aligned}
$$

where

$$
W \;:=\; \int_{t_0}^{\infty} \exp(-\alpha\theta) \|Z_\theta\| \mathrm{d}\theta \qquad (12.48)
$$

is finite for almost all $\omega \in \Omega$.

With (12.44)-(12.48) we obtain from (12.48) the inequality

$$
\exp(\gamma t) \|X_t^0 - Y_t^0\| \;\overset{I}{\leq}\; K + kc \int_{t_0}^{t} \exp(\gamma t) \|X_\tau^0 - Y_\tau^0\| \mathrm{d}\tau, \quad t \geq t_0,
$$

where

$$
K \;=\; kV + ck^2 \left(\chi^{-1} V + (\alpha-\rho)^{-1} W (1 + \exp((\rho-\alpha)t_0)) \right).
$$

Gronwall's lemma implies that

$$
\exp(\gamma t) \|X_t^0 - Y_t^0\| \;\overset{I}{\leq}\; K\exp(kct).
$$

Finally, if $c < \gamma k^{-1}$ holds, we get

$$
\lim_{t\to\infty} \|X_t^0 - Y_t^0\| \;\hat{\leq}\; \lim_{t\to\infty} K\exp((kc-\gamma)t) \;=\; 0,
$$

which shows the assertion. □

12.4 Higher-Order Linear Random Differential Equations with Constant Coefficients and Stationary Inhomogeneity

Next, we study higher-order linear random differential equations of the form

$$
\frac{\mathrm{d}^d X_t}{\mathrm{d}t^d} + a_1 \frac{\mathrm{d}^{d-1} X_t}{\mathrm{d}t^{d-1}} + \cdots + a_{d-1} X_t + a_d \;=\; Z_t \qquad (12.49)
$$

Before you continue, make sure to answer the following questions:

Quiz: Section 12.3 – Part II

Let X_t be a solution of $\dot{X}_t = A(t)X_t + Z_t$.

Q1 Give the conditions that lead to the convergence of every path-wise solution X_t towards a

- periodic,
- stationary,
- strictly periodic, or
- strictly stationary

path-wise solution of $\dot{Y}_t = A(t)Y_t + Z_t$.

Q2 Give the conditions that lead to the convergence of every mean-square solution X_t towards a

- periodic, or
- stationary,

mean-square solution of $\dot{Y}_t = A(t)Y_t + Z_t$.

Finally, let X_t now be a path-wise solution of $\dot{X}_t = (A(t) + C(t))\, X_t + Z_t$.

Q5 Give the conditions that lead to the convergence of every path-wise solution X_t towards a

- periodic, or
- stationary,

path-wise solution of $\dot{Y}_t = A(t)Y_t + Z_t$.

Q6 Sketch the proof of the assertions you used in Q5.

where Z_t is a mean-square continuous stationary stochastic process. Without loss of generality, we assume that

$$Z_t = BU_t = \sum_{k=0}^{m} b_{m-k}\overline{U}_t^{(k)}$$

holds, where U_t is a m-times mean-square continuously differentiable stationary process, the differential operator B is defined by mean-square derivatives $\overline{U}_t^{(k)}$ of degree k, and $m \geq 0$. We seek a stationary mean-square solution Y_t^* of (12.49), i.e., of

$$AX_t = BU_t \,, \quad \left(AX_t = \sum_{k=1}^{d} a_{d-k} \overline{X}_t^{(k)} \,, \quad a_0 = 1 \right). \tag{12.50}$$

If we are interested in path-wise solutions, then, under the additional condition of Z_t being path-wise continuous, the equivalence of mean-square and path-wise solutions (cf. Theorem 12.1) allows us to carry over all results to this setting.

The polynomials

$$A(\xi) = \sum_{k=1}^{d} a_{d-k}\xi^k \,, \quad \text{and} \quad B(\xi) = \sum_{k=0}^{m} b_{m-k}\xi^k$$

are called *characteristic polynomials* of the differential operators A and B, respectively. We assume that all roots of $A(\xi)$ have negative real parts. If $\varphi(t)$ is the solution of the homogeneous differential equation $A\varphi = 0$ with the initial condition

$$\frac{\mathrm{d}^k \varphi(0)}{\mathrm{d}t^k} = 0 \,, \quad \text{for } k = 0, 1, \dots, d-2, \text{ and} \quad \frac{\mathrm{d}^{d-1}\varphi(0)}{\mathrm{d}t^{d-1}} = 1 \,,$$

then

$$Y_t^* = \int_{-\infty}^{t} \varphi(t-\tau) Z_\tau \mathrm{d}\tau \tag{12.51}$$

is a stationary mean-square solution of (12.49) or (12.50), respectively. We obtain this result directly from Theorem 12.9 by considering the corresponding first order system of differential equations

$$\frac{\mathrm{d}\hat{X}_t}{\mathrm{d}t} = \hat{A}\hat{X}_t + \hat{Z}_t \,, \tag{12.52}$$

where $\hat{X}_t^T = \left(X_t, \overline{X}_t^{(1)}, \dots, \overline{X}_t^{(d-1)} \right)$ and $\hat{Z}_t^T = (0, \dots, 0, Z_t)$. Then, Y_t^* is the first component of the stationary mean-square solution $\hat{Y}_t^* = \int_{\infty}^{t} \exp(\hat{A}(t - \tau))\hat{Z}_\tau \mathrm{d}\tau$ of equation (12.51).

Thus, identities (12.27) and (12.51) imply

$$m_{Y^*}(t) = \int_{-\infty}^{t} \varphi(t-\tau) m_Z(\tau) \mathrm{d}\tau = \sum_{k=0}^{m} b_{m-k} \int_0^{\infty} \varphi(\alpha) \frac{\mathrm{d}^k}{\mathrm{d}t^k} m_U(t-\alpha) \mathrm{d}\alpha \,, \tag{12.53}$$

and according to (12.28)

$$C_{Y^*}(t) = \int_0^\infty \int_0^\infty \varphi(\alpha) C_Z(t-\beta+\alpha)\,\varphi(\beta)\mathrm{d}\alpha\mathrm{d}\beta\,. \tag{12.54}$$

Next, let us seek an easy to evaluate the representation of the spectral distribution $F_{Y^*}(\lambda)$ of $Y^*(t)$. With the notation $H_{j,l}$ of Sec. 12.3 we have

$$\sum_{l,m=1}^{d} \hat{H}_{1,l}(i\rho)\overline{\hat{H}_{1,m}(i\rho)} = \int_0^\infty \exp(i\rho t)\,\varphi(t)\mathrm{d}t = (A(i\rho))^{-1}\,,$$

and we obtain with (12.30), or directly from (12.54)

$$F_{Y^*}(\lambda) = \int_{-\infty}^{\lambda} |A(i\rho)|^{-2}\,\mathrm{d}F_Z(\rho)$$

and finally due to Lemma 10.3

$$F_{Y^*}(\lambda) = \int_{-\infty}^{\lambda} \left|\frac{B(i\rho)}{A(i\rho)}\right|^{-2} \mathrm{d}F_U(\rho) \tag{12.55}$$

as the representation of the spectral distribution of $Y^*(t)$.

If U_t has a discrete spectrum, i.e.,

$$R_U(t) = \sum_{k=1}^{\infty} u_k \exp(i\lambda_k t)\,,$$

then it consequently holds that

$$R_{Y^*}(t) = \sum_{k=1}^{\infty} u_k \left|\frac{B(i\lambda_k)}{A(i\lambda_k)}\right|^{-2} \exp(i\lambda_k t)\,.$$

If U_t has the spectral density Φ_U, then

$$\Phi_{Y^*}(\lambda) = \left|\frac{B(i\lambda)}{A(i\lambda)}\right|^{-2} \Phi_U(\lambda) \tag{12.56}$$

is the spectral density of Y_t^*.

Example 12.5 (The Mathematical Oscillator with Random Forcing, cf. [45], pp. 61). As an application of the theory developed in this chapter let us finally discuss the randomly forced mathematical oscillator equation

$$\frac{\mathrm{d}^2 X_t}{\mathrm{d}t^2} + 2a\frac{\mathrm{d}X_t}{\mathrm{d}t} + bX_t = Z_t\,, \tag{12.57}$$

where $a > 0$, $b \geq 0$ and $Z_t = c\sin(\alpha t) + U_t$. Here, U_t is a stationary second order process on I that is path-wise as well as mean-square continuous with $\mathbb{E}(U_t) = 0$ and spectral density $\Phi(\lambda)$. The characteristic polynomial $A(\rho) = \rho^2 + 2a\rho + b$ has the roots

$$\rho_{1/2} = -a \pm \sqrt{a^2 - b}\,,$$

both with negative real parts. We discuss the three usual special cases

1. $\chi^2 = b - a^2 > 0$,
2. $\chi^2 = a^2 - b > 0$, and
3. $a^2 = b$.

Case 1: Let $\chi^2 = b - a^2 > 0$. Then, $\varphi(t) = \chi^{-1}\sin(\chi t)\exp(-at)$ is the solution of the homogeneous oscillator equation with initial condition $\varphi(0) = 0$ and $\dot{\varphi}(0) = 1$. The stochastic process Z_t is periodic, and thus with (12.51)

$$Y_t^0 = m_{Y^0}(t) + \chi^{-1}\int_0^\infty \exp(-a\tau)\sin(\chi\tau)\,U_{t-\tau}\mathrm{d}\tau \tag{12.58}$$

is a periodic (and for $c = 0$ stationary) path-wise and mean-square solution of (12.57) with the same period as Z_t. Here,

$$m_{Y^0}(t) = \chi^{-1}c\int_0^\infty \exp(-a\tau)\sin(\chi\tau)\sin(\alpha(t-\tau))\,\mathrm{d}\tau = c\beta\sin(\alpha(t-\gamma))\,, \tag{12.59}$$

where $c \in \mathbb{R}$ is a constant, $\beta^{-2} = (b-\alpha)^2 + (2a)^2\alpha^2$, and

$$\gamma = \tfrac{1}{\alpha}\arctan\left(\frac{2a\alpha}{b-\alpha^2}\right).$$

With (10.1) from Chap. 10.2 and (12.55) the covariance function of Y_t^0 reads as

$$\begin{aligned} C_{Y^0}(t,t+\tau) = C_{Y^0}(\tau) &= \int_{-\infty}^\infty \exp(i\lambda\tau)\,|A(i\lambda)|^{-2}\,\Phi(\lambda)\mathrm{d}\lambda \\ &= \int_{-\infty}^\infty \cos(\lambda\tau)\left((\lambda^2-b)^2 + (2a\lambda)^2\right)^{-1}\Phi(\lambda)\mathrm{d}\lambda\,. \end{aligned} \tag{12.60}$$

Because $C_{Y^0}(\tau)$, $|A(i\lambda)|$ and $\Phi(\lambda)$ are real functions, it holds that

$$\int_{-\infty}^\infty \sin(\lambda\tau)\,|A(i\lambda)|^{-2}\,\Phi(\lambda)\mathrm{d}\lambda = 0\,.$$

Due to (12.60) we get for the variance of Y_t^0

$$\sigma_{Y^0}(t) = C_{Y^0}(0) = \int_{-\infty}^\infty \left((\lambda^2-b)^2 + (2a\lambda)^2\right)^{-1}\Phi(\lambda)\mathrm{d}\lambda\,. \tag{12.61}$$

If the spectral density $\Phi(\lambda)$ is given numerically, then the integral in (12.61) can be computed numerically.

Coarse approximations for the covariance and variance functions are obtained for small damping coefficients a if $\Phi(\lambda)$ is sufficiently smooth and does not possess sharp peaks. In this case the function $|A(i\lambda)|^{-2}$ exhibits a peak at $\lambda = b$, and, moreover, the essential contribution to the integral (12.60) comes from a small interval around $\lambda = b$.

Thus, it approximately holds that

$$C_{Y^0}(\tau) \approx \Phi(b) \int_{-\infty}^{\infty} \cos(\lambda\tau) \left((\lambda^2 - b)^2 + (2a\lambda)2\right)^{-1} \mathrm{d}\lambda\,,$$

and

$$\sigma_{Y^0}(t) \approx \Phi(b) \int_{-\infty}^{\infty} \left((\lambda^2 - b)^2 + (2a\lambda)2\right)^{-1} \mathrm{d}\lambda\,.$$

For every path-wise solution X_t of (12.57) we get with a suitably small $\rho < a$ by application of Theorem 12.11 on the system of differential equations (12.52) that:

$$\lim_{t\to\infty} \exp(\rho t) \left|X_t - Y_t^0\right| \mathrel{\hat{=}} \lim_{t\to\infty} \exp(\rho t) \left|\dot{X}_t - \dot{Y}_t^0\right| \mathrel{\hat{=}} 0\,. \tag{12.62}$$

Due to Theorem 12.12 we know that for every mean-square solution and every path-wise solution with initial condition $(X_{t_0}, \dot{X}_{t_0}) \in L_2^2$ that it is also a mean-square solution according to Theorem 12.1, and that:

$$\text{m.s.-lim}_{t\to\infty} \exp(\rho t) \left|X_t - Y_t^0\right| \mathrel{\hat{=}} \text{m.s.-lim}_{t\to\infty} \exp(\rho t) \left|\dot{\overline{X}}_t - \dot{\overline{Y}}{}_t^0\right| \mathrel{\hat{=}} 0\,.$$

Thus, for large times t the mean and the covariance function of every mean-square solution and every path-wise solution of (12.57) with initial conditions in L_2^2 can be approximated by the mean and covariance functions, respectively, of the periodic solution Y_t^0, because it holds due to (12.62) with $V_t = X_t - Y_t^0$ that

$$\lim_{t\to\infty} |m_X(t) - m_{Y^0}(t)| \le \lim_{t\to\infty} \left(\mathbb{E}\left(V_t^2\right)\right)^{1/2} = 0\,, \tag{12.63}$$

and

$$\begin{aligned}
&\lim_{t\to\infty} |C_X(t, t+s) - C_{Y^0}(t, t+s)| = \\
&\lim_{t\to\infty} \left|\mathbb{E}\left(V_t V_{t+s}\right) + \mathbb{E}\left(Y_t^0 V_{t+s}\right) + \mathbb{E}\left(V_t Y_{t+s}^0\right)\right. \\
&\qquad \left. - m_X(t) m_X(t+s) + m_{Y^0}(t) m_{Y^0}(t+s)\right| \le \\
&\lim_{t\to\infty} \left(\left(\mathbb{E}\left(V_t^2\right)\mathbb{E}\left(V_{t+s}^2\right)\right)^{1/2} + \left(C_{Y^0}(0)\mathbb{E}\left(V_{t+s}^2\right)\right)^{1/2} + \left(C_{Y^0}(0)\mathbb{E}\left(V_t^2\right)\right)^{1/2}\right) = 0\,.
\end{aligned} \tag{12.64}$$

If Z_t or U_t are Gaussian processes, then Y_t^0 is a Gaussian process, too, according to Sec. 12.2.2, and with the aid of (12.59) and (12.60), probabilities can be computed for different interesting events. For instance,

$$\mathbb{P}\left(Y_t^0 > y\right) \;=\; 1 - F\left(\sigma_{Y^0}^{-1}\left(y - m_{Y^0}(t)\right)\right) ,$$

where

$$F(x) \;=\; \int_{-\infty}^{x} (2\pi)^{-1/2} \exp(-t^2/2) \mathrm{d}t$$

is the well-known probability distribution function of the standard normal distribution.

Case 2: Let $\chi^2 = a^2 - b > 0$. Then, $\varphi(t) = \chi^{-1}\sinh(\chi t)\exp(-at)$ and we obtain, analogously to the first part, the periodic solution

$$Y_t^0 \;=\; m_{Y^0}(t) + \chi^{-1}\int_0^{\infty} \exp\left(-a\tau\right)\sinh\left(\chi\tau\right) U_{t-\tau}\mathrm{d}\tau\,, \tag{12.65}$$

with the mean value

$$m_{Y^0}(t) \;=\; \chi^{-1} C\int_0^{\infty} \exp\left(-a\tau\right)\sinh\left(\chi\tau\right)\sin\left(\alpha(t-\tau)\right)\mathrm{d}\tau \tag{12.66}$$

(with a constant $C \in \mathbb{R}$), and covariance function (12.60).

As in the first part, the mean value $M_X(t)$ and covariance function $C_X(t, t+s)$, $s > 0$, of a mean-square solution of (12.57) or of a path-wise solution of (12.57), respectively, for the initial value $(X_{t_0}, \dot{X}_{t_0}) \in L_2^2$ can be approximated by (12.66) and (12.60) for large t.

Case 3: Let $a^2 = b$, then $\varphi(t) = t\exp(-at)$ and we obtain the periodic solution

$$Y_t^0 \;=\; m_{Y^0}(t) + \int_0^{\infty} \tau\exp\left(-a\tau\right) U_{t-\tau}\mathrm{d}\tau \tag{12.67}$$

with covariance function (12.60) and mean value

$$m_{Y^0}(t) \;=\; \int_0^{\infty} \tau\exp\left(-a\tau\right)\sin\left(\alpha(t-\tau)\right)\mathrm{d}\tau\,. \tag{12.68}$$

12.5 Chapter's Summary

Based on [45], pp. 38, this chapter focused on the mathematical analysis of linear inhomogeneous random differential equations of the form $\dot{X}_t = A(t)X_t + Z_t$, where the randomness appeared only in the inhomogeneous driving term Z_t. Analogously to the deterministic setting, we derived, with the aid of the fundamental matrix of the deterministic homogeneous system

Before you continue, make sure to answer the following questions:

Quiz: Section 12.4

Q1 Give the form of the stationary mean-square solution Y_t^* of

$$X_t^{(n)} + a_1 X_t^{(n-1)} + \cdots + a_{n-2}\dot{X}_t + a_{n-1}X_t + a_n \;=\; Z_t\,.$$

Q2 Which requirements have to be fulfilled such that your assertions from Q1 holds?

Q3 Give representations for the mean, the covariance and the spectral distribution of Y_t^*.

Q4 Discuss the undamped oscillator

$$\ddot{X}_t + bX_t \;=\; W_t\,,$$

where W_t is a 1-dimensional standard Wiener process.

$\dot{x} = A(t)x$, the general solution formulas (12.4) in the path-wise setting and (12.5) in the mean-square setting, respectively. These allowed us to study stochastic characteristics like the mean m_X of the solution process and especially the common covariance C_{YZ} of a particular solution process Y_t with the driving process Z_t. In particular, we gave conditions for the asymptotic un-correlation of Y_t and Z_t and discussed the special case of Gaussian inhomogeneities.

The majority of this chapter was devoted to the study of path-wise/ mean-square periodic and path-wise/ mean-square stationary solutions of $\dot{X}_t = A(t)X_t + Z_t$. Roughly speaking and under some further clarifying assumptions, a periodic solution exists if $A(t)$ and Z_t are periodic, too. In the same way, a stationary solution exists if A is constant and has eigenvalues with exclusively negative real parts, and Z_t is a stationary process. In particular, we obtained important results on the (exponential) convergence behavior of solutions towards periodic and stationary ones.

We concluded our analysis with the study of higher-order random differential equations and a complete discussion of the mathematical oscillator with random forcing.

Problems

Classification: ☼ easy, ⌚ easy with longer calculations, ☆ a little bit difficult,
⚄ challenging.

Exercise 12.6. [☼] Properties of a Ornstein-Uhlenbeck Excited Oscillator
Consider the second-order random differential equation

$$\ddot{X}_t + 2a\dot{X}_t + bX_t = -Z_t\,, \qquad X_0 = \dot{X}_0 = 0\,, \tag{12.69}$$

where Z_t is a Gaussian process and $a, b > 0$.

1. Give the general formula for the path-wise solution of (12.69). What can you say about the mean-square solution?
2. Give the deterministic ordinary evolution equations for the mean and the common covariance function of the path-wise solution.

Exercise 12.7. [☼] Properties of a Ornstein-Uhlenbeck Excited Oscillator
Consider the second-order random differential equation

$$\ddot{X}_t + 2a\dot{X}_t + bX_t = -O_t\,, \qquad X_0 = \dot{X}_0 = 0\,, \tag{12.70}$$

where O_t is a Ornstein-Uhlenbeck process and $a, b > 0$.

1. Give the general formula for the path-wise solution of (12.70). What can you say about the mean-square solution?
2. Give the deterministic ordinary evolution equation for the mean of the path-wise solution and solve it analytically.
3. Give the deterministic ordinary evolution equation for the common covariance function of the path-wise solution and solve it analytically.

Exercise 12.8. [⌚] Properties of a Brownian Bridge Excited Oscillator
The Brownian bridge B_t is a Gaussian process whose increments are not independent. If $W_t \sim \mathcal{N}(0, t)$ is a standard Wiener process, then the process $B_t := W_t - tW(1)$ is called a Brownian bridge for $t \in [0, 1]$.

Consider the second-order random differential equation

$$\ddot{X}_t + 2a\dot{X}_t + bX_t = -B_t\,, \qquad X_0 = \dot{X}_0 = 0\,, \tag{12.71}$$

where $a, b > 0$ for $t \in [0, 1]$.

1. Give the general formula for the path-wise solution of (12.71). What can you say about the mean-square solution?

2. Give the deterministic ordinary evolution equation for the mean of the path-wise solution and solve it analytically.

3. Give the deterministic ordinary evolution equation for the common covariance function of the path-wise solution and solve it analytically.

Exercise 12.9. [⏲] **Properties of a Fractal Brownian Motion Excited Oscillator – Part 1**

The fractal Brownian motion $B_H(t)$ is a Gaussian process whose covariance function is a generalisation of the Wiener process. It is a continuous-time Gaussian process on $[0,T]$ which starts at zero, has expectation zero for all $t \in [0,T]$, and has the following covariance function:

$$\mathbb{E}\left(B_H(t)B_H(s)\right) = \tfrac{1}{2}\left(|t|^{2H} + |s|^{2H} - |t-s|^{2H}\right),$$

where H is a real number in $(0,1)$, called the Hurst index or Hurst parameter associated with the fractional Brownian motion. The Hurst exponent describes the raggedness of the resultant motion, with a higher value leading to a smoother motion. The value of H determines what kind of process the fractal Brownian motion $B_H(t)$ is:

- if $H = 1/2$ then the process is in fact a Wiener process,
- if $H > 1/2$ then the increments of the process are positively correlated, and
- if $H < 1/2$ then the increments of the process are negatively correlated.

Let us choose $H = 1/4$ and consider the second-order random differential equation

$$\ddot{X}_t + 2a\dot{X}_t + bX_t = -B_{1/4}(t), \qquad X_0 = \dot{X}_0 = 0, \tag{12.72}$$

where $a, b > 0$.

1. Give the general formula for the path-wise solution of (12.72). What can you say about the mean-square solution?

2. Give the deterministic ordinary evolution equation for the mean of the path-wise solution and solve it analytically.

3. Give the deterministic ordinary evolution equation for the common covariance function of the path-wise solution and solve it analytically.

Exercise 12.10. [⏲] **Properties of a Fractal Brownian Motion Excited Oscillator – Part 2**

Let us choose a Hurst coefficient $H = 3/4$ for the fractal Brownian motion as discussed in problem 12.9 and consider the second-order random differential equation

$$\ddot{X}_t + 2a\dot{X}_t + bX_t = -B_{3/4}(t), \qquad X_0 = \dot{X}_0 = 0, \tag{12.73}$$

where $a, b > 0$.

1. Give the general formula for the path-wise solution of (12.73). What can you say about the mean-square solution?
2. Give the deterministic ordinary evolution equation for the mean of the path-wise solution and solve it analytically.
3. Give the deterministic ordinary evolution equation for the common covariance function of the path-wise solution and solve it analytically.

Exercise 12.11. [☼] Examples of θ-Periodic Processes

1. Give an example of a θ-periodic and a strict θ-periodic process.
2. Are there θ-periodic processes that are not strictly θ-periodic? Give an example for such a process.

Exercise 12.12. [☼] Is the Kanai-Tajimi Excitation a θ-Periodic Process?
Consider the stochastic ground motion excitation $\ddot{u}_g(t)$ in the sense of the Kanai-Tajimi model which is given as

$$\ddot{u}_g = \ddot{x}_g + w_t = -2\zeta_g\omega_g\dot{x}_g - \omega_g^2 x_g,$$

where x_g is the solution of a zero-mean Gaussian white noise w_t driven stochastic oscillator

$$\ddot{x}_g + 2\zeta_g\omega_g\dot{x}_g + \omega_g^2 x_g = -w_t, \qquad x_g(0) = \dot{x}_g(0) = 0.$$

1. Give $\ddot{u}_g$ in terms of a random differential equation by applying the Doss/ Sussmann & Imkeller/ Schmalfuss correspondence.
2. Is $\ddot{u}_g$ a (strict) θ-periodic process?

Exercise 12.13. [☼] Convergence Towards Periodic Solutions – Part 1
Let Z_t be a suitable strictly 2-periodic path-wise continuous stochastic process. Let us consider the randomly forced Mathieu equation

$$\ddot{x} + (a - 2b\cos(2t))\,x + Z_t = 0,$$

where $a, b > 0$. What can you say about the existence of θ-periodic solutions in this system and the convergence towards them?

Exercise 12.14. [⌨] **Convergence Towards Periodic Solutions – Part 2**
Let Z_t be a suitable strictly θ-periodic mean-square continuous stochastic process. Let us consider the randomly forced Duffing equation

$$\ddot{x} + a\dot{x} + bx + cx^3 - d\cos(\theta t) + Z_t = 0,$$

where $a, b, c, d > 0$. What can you say about the existence of θ-periodic solutions in this system and the convergence towards them?

Exercise 12.15. [☼] **Stationary Processes**

1. Give an example of a strictly stationary path-wise/ mean-square continuous stochastic process.
2. Let $\{X_n\}_{n\in\mathbb{N}}$ be a set of uncorrelated random variables with vanishing mean and variance 1. Show that $\{X_n\}$ is a strictly stationary process.
3. Let $X_t := A_1 + A_2 t$, where A_1, A_2 are independent random variables with $\mathbb{E}(A_i) = a_i$ and $\mathbb{V}ar(A_i) = \sigma_i^2$, for $i = 1, 2$. Show that $\{X_n\}$ is not stationary.

Exercise 12.16. [☼] **Romeo and Juliet's Love Affair (1)**
Let us discuss the random perturbation of a simple model for love affairs, cf. [241] and [242], pp. 138:

Romeo is in love with Juliet, but in our version of this story, Juliet is a fickle lover. The more Romeo loves her, the more Juliet wants to run away and hide. But when Romeo gets discouraged and backs off, Juliet begins to find him strangely attractive. On the other hand, Romeo tends to echo her: he warms up when she loves him, and grows cold when she hates him.

Let

$$\begin{aligned} R(t) &= \text{Romeo's love/ hate for Juliet at time } t, \\ J(t) &= \text{Juliet's love/ hate for Romeo at time } t. \end{aligned}$$

Positive values of R and J signify love, negative values signify hate. Then a randomly perturbed model for their star-crossed romance is

$$\dot{R} = aJ + O_t, \quad \text{and} \quad \dot{J} = -bR,$$

where the parameters a and b are positive, to be consistent with the story, and O_t is an Ornstein-Uhlenbeck process.

In the deterministic setting (without O_t), the sad outcome of their affair is, of course, a never ending cycle of love and hate; the governing system has a center at $(R, J) = (0, 0)$. At least they manage to achieve simultaneous love one-quarter of the time.

What can you say about their affair when following the randomly perturbed dynamics?

Exercise 12.17. [☼] **Romeo and Juliet's Love Affair (2)**
As a continuation of problem 12.16, let us consider the forecast for lovers governed by the randomly perturbed general linear system

$$\dot{R} = aR + bJ + O_t \,, \quad \text{and} \quad \dot{J} = cR + dJ \,,$$

where the parameters a, b, c, d may have either sign. Especially in the deterministic setting, a choice of signs specifies romantic styles. As coined by Strogatz (cf. [242], p. 139), the choice of $a, b > 0$ means that Romeo is an "eager beaver" – he gets excited by Juliet's love for him, and is further spurred on by his own affectionate feelings for her. It's entertaining to name the other three romantic styles, and to predict the outcomes for the various pairings. For instance, can a "cautious lover" ($a < 0, b > 0$) find true love with an eager beaver?

What can you say about the affairs in the randomly perturbed setting?

Exercise 12.18. [☼] **Romeo and Juliet's Love Affair (3)**
In a study sociologist found that women were blind to the mindset of their opposite-sex friends; because females generally were not attracted to their male friends, they assumed that this lack of attraction was mutual. As a result, men consistently "overestimated" the level of attraction felt by their female friends – and women consistently "underestimated" the level of attraction felt by their male friends.

1. Based on problem 12.17 set-up a deterministic model for this behavior by specifying the right sign combinations of the coefficients involved.
2. Incorporate random effects that take into account that men and women are not good in reading their opposite-sex friends.
3. What can you say about the existence of stationary solutions in this case and the asymptotic convergence towards such solutions (if they exist).

Exercise 12.19. [☼] **Romeo and Juliet's Love Affair (4)**
Continuing problems 12.16 and 12.17 we follow [242], p. 139, and give some further questions related to the love affair dynamics presented in problem 12.17. In each of the following scenarios, predict the course of the love affair, depending on the signs and the relative sizes of a and b. (Of course you are encouraged to use a Wiener process or a fractional Brownian motion as random perturbance, too, and alter the system accordingly).

1. Out of touch with their own feelings: Suppose Romeo and Juliet react to each other, but not to themselves, i.e., $\dot{R} = aJ$ and $\dot{J} = bR$. What happens?

2. Fire and water: Do opposites attract? Analyze $\dot{R} = aR + bJ$ and $\dot{J} = -bR - aJ$.

3. Peas in a pod: Suppose Romeo and Julia are romantic clones, i.e., $\dot{R} = aR + bJ$ and $\dot{J} = bR + aJ$. Do they expect boredom or bliss?

4. Romeo the robot: Nothing could ever change the way Romeo feels about Juliet, i.e., $\dot{R} = 0$ and $\dot{J} = aR+bJ$. Does Juliet end up loving him or hating him?

Exercise 12.20. [☼] Convergence Towards Stationary Solutions in a LRC-Circuit System

Consider the randomly perturbed circuit equation $L\ddot{I} + R\dot{I} + I/C + O_t$, where $L, R, C > 0$ and O_t is an Ornstein-Uhlenbeck process. Show that there is a stationary solution and that all other solutions tend towards it.

Chapter 13

Linear RODEs with Stochastic Coefficients

Here we extend the discussion from the preceding chapter on linear random differential equations with stochastic inhomogeneity to linear equations that have a random effect in their coefficient functions as well. Analogously to the deterministic theory, we give the general solution formulas for these types of equations together with equivalence results for path-wise and mean-square solutions as a special case of our considerations in Chap. 4. On the one hand, we analyse the asymptotic properties of path-wise solutions with a focus on (exponential) decay towards the null-solution as well as on upper bounds for path-wise solutions. On the other hand, we also study the properties of the moments of path-wise solutions with respect to the (exponential) decay as well as the existence of asymptotically θ-periodic solutions. As an excursion, the general solution formula of linear non-commutative path-wise continuous noisy systems is constructed.

13.1 Key Concepts

In this chapter, we discuss the random differential equation

$$\frac{\mathrm{d}X_t}{\mathrm{d}t} = A_t X_t + Z_t \,, \tag{13.1}$$

where Z_t is a d-dimensional stochastic vector process and A_t is a stochastic $d \times d$-matrix with elements that are stochastic processes. Studying equation (13.1) is considerably harder than the random differential equation (12.1) from Chap. 12 with a non-random matrix $A(t)$. In general, it is not possible to give simple, closed form formulas for the statistical characteristics of the solution depending on the statistical characteristics of the coefficients of (13.1). Rather, we aim for qualitative results like the asymptotic behavior of the realisations or the moments of the solution.

A special case of (13.1) is given, when the coefficients are approximately white noise processes. Many works treat this type of differential equations either by utilizing (approximate) Fokker-Planck equations or Itô-/ Stratonovich stochastic differential equations, see [33], [103], [222].

Example 13.1 (Randomized Coefficients at a Monod Kinetics Model, cf. [235]). In porous media, as in other aqueous environments, microbial cells

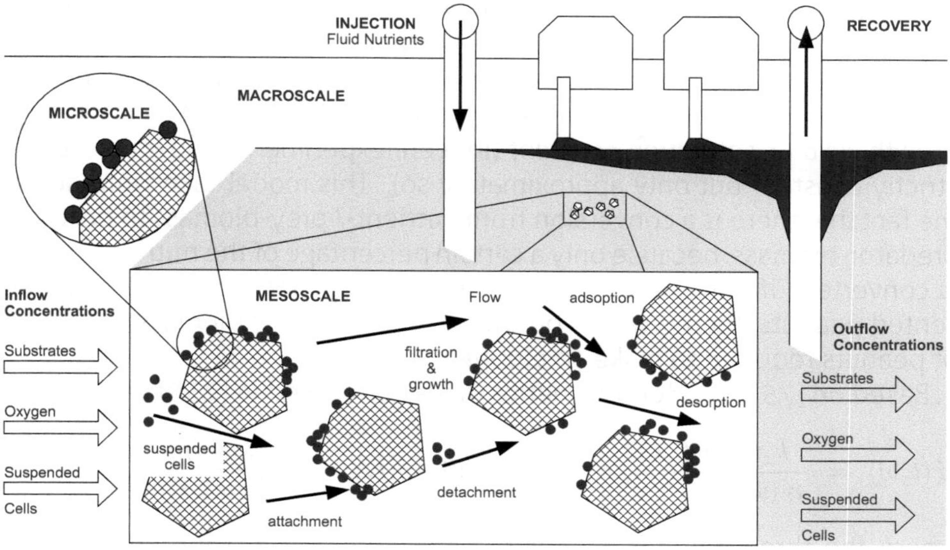

Figure 13.1. Sketch of an injection/ recovery scheme for enhancing in situ bioremediation of contaminated aquifer. Several individual microbial transport processes contribute to biofilm accumulation and activity in porous media, cf [75].

may exist in suspension or adsorbed firmly to solid surfaces comprising the effective pore space, see Fig. 13.1. Following [235], many models for such bacterial growth are given by differential equations where the coefficients are deterministic or random. In particular, random coefficients have been increasingly used in the last few decades to deal with error and uncertainty, and the corresponding stochastic and random differential equations constitute a growing field of great scientific interest.

With respect to the growth of a biofilm, the micro-scale dynamics are well described by a model introduced by the French biologist and Nobel Prize laureate Jacques Lucien Monod (1910 – 1976) who proposed it to relate microbial growth rates in an aqueous environment to the concentration of a limiting nutrient, [191]. Monod performed experiments on bacteria by feeding them a single limiting nutrient (say glucose) in order to see if the logistic equation accurately describes bacterial growth. Let x denote the concentration of the microbes and y that of the nutrients; his experiments suggested that the specific growth rates are given by

$$\dot{x} = \frac{k_x y}{K+y} x\,, \quad \text{and} \quad \dot{x} = -\frac{1}{\gamma}\frac{k_x y}{K+y} x\,,$$

where k_x is called the *maximum growth rate*, despite the fact that it can never be reached, and K is called the *half-saturation constant* because when

$y = K$, the right hand side of the equation for $\dot{x}$ becomes $\frac{1}{2}k_x$, half the maximum growth rate. The key feature of the Monod function is that the specific growth rate increases with nutrient concentration y as expected but it levels out for high nutrient concentrations. The constant $\gamma \in (0, 1)$ is called the *growth yield* or *saturation constant* (indeed, experiments show that it is not strictly constant but only approximately so). This model takes into account the fact that there is a conversion from nutrient-/ prey-biomass to bacterial-/ predator-biomass, because only a certain percentage of the nutrient-biomass is converted. If, for example, x represented elephant biomass and y represented peanuts, then γ would be the ratio of elephant biomass to the mass of peanuts required to make an elephant.

Based on [75], the Monod kinetics model discussed in [235] reads as

$$\dot{x}(t,\omega) = \frac{k_x(\omega)y(t,\omega)}{K(\omega)+y(t,\omega)}x(t,\omega), \quad \text{and} \quad \dot{y}(t,\omega) = \frac{k_y(\omega)y(t,\omega)}{K(\omega)+y(t,\omega)}x(t,\omega),$$

where the three random parameters are uniformly distributed with $K \sim U(0.28, 0.36)$, $k_x \sim U(0.04, 0.06)$ and $k_y \sim U(-0.325, 0.3)$.

Linearisation/ formal first-order Taylor expansion of these equations around the equilibria of the corresponding deterministic model leads to the homogeneous systems of the form $\dot{X}_t = A(\omega)X_t$ (see problem 13.12); a special case of linear random differential equations with stochastic coefficients that we discuss in this chapter.

When reading this chapter note the answers to the following questions

1. How can the moments of the coefficient functions of a linear (inhomogeneous) random differential equation with stochastic coefficients be utilized to show decay towards the null-solution?

2. What is required to ensure decay of solutions of a linear (inhomogeneous) random differential equation with stochastic coefficients towards the null-solution?

3. What does asymptotic θ-periodicity mean and how can the existence of a θ-periodic solution be verified?

as well as the following key concepts

1. Solution formulas for $\dot{X}_t = A_tX_t + Z_t$, in particular if $A_t = A(\omega)$ is a random matrix.

2. Equivalence of path-wise and mean-square solutions of $\dot{X}_t = A_tX_t + Z_t$.

3. Decay of path-wise solutions towards the null-solution for homogeneous systems of the type

- $\dot{X}_t = (A + F_t)\, X_t$, and
- $\dot{X}_t = (A + F_t + C(t))\, X_t$.

4. Upper bounds for the path-wise solution of $\dot{X}_t = (A + F_t + C(t))\, X_t + Z_t$.
5. Decay of the r-th moments $\mathbb{E}\left(\|X_t\|^r\right)$ of path-wise solutions X_t for systems of the type
   - $\dot{X}_t = AX_t$,
   - $\dot{X}_t = AX_t + Z_t$, and
   - $\dot{X}_t = (A + F_t + C(t))\, X_t + Z_t$.
6. Asymptotic θ-periodicity.
7. General representation

$$X_t \;=\; \exp\left(B\Xi_t\right) \cdot \exp\left(At\right) \cdot \sum_{m=0}^{\infty} G_m(t) X_0$$

   of the solution X_t of the linear non-commutative path-wise continuous noise system $\dot{X}_t = (A + B\xi_t)\, X_t$.

This chapter is structured as follows: As in the deterministic setting from Chap. 6, Sec. 13.2 gives the general solution formulas for linear random differential equations with stochastic coefficients and an equivalence result for path-wise and mean-square solutions as a special case of our considerations in Chap. 4. Next, Sec. 13.3 provides us with results on the asymptotic properties of path-wise solutions focusing on (exponential) decay towards the null-solution as well as on upper bounds for path-wise solutions. In Sec. 13.4, we change our point of view from the path-wise solution itself to the properties of its moments by studying the (exponential) decay of moments for homogeneous and inhomogeneous linear random differential equations with stochastic coefficients as well as the existence of asymptotically θ-periodic solutions. As an excursion, the general solution formula of linear non-commutative path-wise continuous noisy systems is constructed in Sec. 13.5. Finally, Sec. 13.6 wraps up the contents of this chapter.

13.2 The General Solution Formula

Let $I = [t_0, \infty)$, $t_0 \geq 0$. We start our study of equation (13.1), i.e., of

$$\frac{\mathrm{d}X_t}{\mathrm{d}t} \;=\; A_t X_t \,+\, Z_t\,,$$

by a recap of the conditions that lead to (unique) solutions and some general solution formulas for considerable easy special cases.

If A_t and Z_t are path-wise continuous stochastic processes on I, then for every initial condition $(X_0, t_0) \in S_d \times I$ there is a unique path-wise solution of (13.1) on I, see Theorem 3.2 for instance.

If Z_t is a second order mean-square continuous stochastic process on I, A_t is path-wise continuous on I and if there is a continuous real function $L(t)$ on I such that $\|A_t\| \overset{I}{\leq} L(t)$, then for every initial condition $(X_0, t_0) \in L_d^2 \times I$ there is a unique mean-square solution of (13.1) on I, see Theorem 4.3 for instance.

If A_t and Z_t are path-wise continuous on I and if there are real continuous functions $L(t)$ and $N(t)$ on I and a square integrable $U \in L_1^2$ on I such that $\|A_t\| \overset{I}{\leq} L(t)$ as well as $\|Z_t\| \overset{I}{\leq} UN(t)$ then for every initial condition $(X_0, t_0) \in L_d^2 \times I$ there are a unique path-wise solution and a mean-square solution of (13.1) on I, and these two are equivalent, see Theorem 4.4 for instance.

If A_t and Z_t are path-wise continuous on I, then analogously to the deterministic setting the path-wise solution of (13.1) with respect to the initial condition (X_0, t_0) has the general form

$$X_t \overset{I}{=} \Phi_t X_0 + \int_{t_0}^{t} \Phi_t \Phi_\tau^{-1} Z_\tau \mathrm{d}\tau \,, \tag{13.2}$$

where for almost all $\omega \in \Omega$ the function Φ_t is the fundamental matrix of the homogeneous differential equation

$$\dot{X}_t(\omega) = A_t(\omega) X_t(\omega) \,,$$

with $\Phi_{t_0} = \mathbb{I}$.

Let $A_t = A + F_t$, where A is a constant real matrix, then the path-wise solution (13.2) reduces to the following form:

$$X_t \overset{I}{=} \exp\left(A(t-t_0)\right) X_0 + \int_{t_0}^{t} \exp\left(A(t-\tau)\right) F_\tau X_\tau \mathrm{d}\tau + \int_{t_0}^{t} \exp\left(A(t-\tau)\right) Z_\tau \mathrm{d}\tau \,. \tag{13.3}$$

It is well known that there are numbers a and b such that $\|\exp(At)\| \leq b\exp(at)$. For (13.3) this implies the following inequality

$$\|X_t\| \overset{I}{\leq} b\exp(at)\left(\|X_0\|\exp(-at_0) + \int_{t_0}^{t} \exp(-a\tau)\|F_\tau\|\,\|X_\tau\|\,\mathrm{d}\tau + \int_{t_0}^{t} \exp(-a\tau)\|Z_\tau\|\,\mathrm{d}\tau\right)$$

which leads to

$$
\begin{aligned}
\|X_t\| &\overset{I}{\leq} b\,\|X_0\| \exp\left(a(t-t_0)\right) \exp\left(b\int_{t_0}^{t} \|F_\tau\|\,\mathrm{d}\tau\right) \\
&\quad + b\exp\left(at\right)\int_{t_0}^{t}\|Z_\tau\|\exp\left(b\int_{\tau}^{t}\|F_\theta\|\,\mathrm{d}\theta - a\tau\right)\mathrm{d}\tau\,. \qquad (13.4)
\end{aligned}
$$

according to Lemma 3.10.

We introduce the notations

$$
\begin{aligned}
U_t &:= \|X_0\|\cdot\exp\left(b\int_{t_0}^{t}\|F_\tau\|\,\mathrm{d}\tau\right),\\
V_t &:= \int_{t_0}^{t}\|Z_\tau\|\exp\left(b\int_{\tau}^{t}\|F_\theta\|\,\mathrm{d}\theta - a\tau\right)\mathrm{d}\tau\,.
\end{aligned}
$$

With the estimate (13.4) we easily gain weaker conditions for the equivalence of path-wise and mean-square solutions of (13.1) as presented in Theorem 4.4:

Theorem 13.1 (Equivalence of Path-Wise and Mean-Square Solutions). *Let the stochastic processes A_t and Z_t be path-wise continuous and let there be a positive function $H(t)$ on I such that Z_t and with the aid of $T_t := (t-H(t), t+H(t))$ also*

$$
\begin{aligned}
U_t^* &= \sup\nolimits_{\tau\in T_t} U_\tau\,, & V_t^* &= \sup\nolimits_{\tau\in T_t} V_\tau\,,\\
S_t &= \sup\nolimits_{\tau\in T_t} \|F_\tau\|\,U_\tau\,, & Q_t &= \sup\nolimits_{\tau\in T_t} \|F_\tau\|\,V_\tau\,,
\end{aligned}
$$

are second order stochastic processes, then the path-wise solution of (13.1) is also a mean-square solution.

Proof. Following [45], pp. 69, let X_t be a path-wise solution of (13.1). With the conditions and the estimate (13.4) it follows that X_t, $t\in I$, and $A_tX_t+Z_t$, $t\in I$, are second order processes.

Next, we have to show that for every $t\in I$ it holds that

$$
\lim_{h\to 0}\mathbb{E}\left(\left(\frac{X_{t+h}-X_t}{h}-\dot{X}_t\right)^2\right) = 0\,. \qquad (13.5)
$$

Because X_t is a path-wise solution of (13.1), it holds that

$$
\left(\frac{X_{t+h}-X_t}{h}-\dot{X}_t\right)^2 \hat{\leq} 2\sup_{\tau\in T_t}\left\|\dot{X}_\tau\right\|^2 + 2\left\|\dot{X}_t\right\|^2, \quad |h| < H(t)\,, \quad t\in I\,. \qquad (13.6)
$$

Substituting for X_t in the right hand side of (13.6), the right hand side of the random differential equation (13.1) and taking the estimate (13.4) as well as the conditions into account, we have that the mean of the right hand side of (13.6) is finite. Thus, the right hand side of (13.6) is an integrable majorant for the left hand side. Hence, with Lebesgue's convergence theorem we can interchange the limit and the mean in the left hand side of (13.5). Since X_t is a path-wise solution this implies equation (13.5). □

Remark 13.2 (Conditions w.r.t. Which the Assumptions of Theorem 13.1 are Easily Fulfilled). For instance, if A_t and Z_t are path-wise continuous on I and if $\|A_t\| \overset{I}{\leq} L(t)$ and $\|Z_t\| \overset{I}{\leq} UN(t)$ holds with continuous real functions $L(t)$ and $N(t)$ and with $U \in L_1^2$ and if $X_0 \in L_d^2$, then the conditions of Theorem 13.1 are satisfied.

Remark 13.3 (The Solutions of $\dot{X}_t = AX_t$). Let $A_t \equiv A$, where A is a random matrix. If $X_0 \in L_d^2$ and A are stochastically independent such that $\exp(\|A\|t)$ and $\|A\| \exp(\|A\|t)$ are second order stochastic processes on I, then according to Theorem 13.1, the function $\exp(A(t-t_0))\, X_0$ is a path-wise as well as a mean-square solution of $\dot{X}_t = AX_t$ on I with respect to the initial condition (X_0, t_0).

For instance, $\exp(\|A\|t)$ and $\|A\| \exp(\|A\|t)$ are second order processes if

$$\sum_{k=0}^{\infty} \frac{\mathbb{E}\left(\|A\|^k\right) t^k}{k!} < \infty \quad \text{and} \quad \sum_{k=0}^{\infty} \frac{\mathbb{E}\left(\|A\|^{k+2}\right) t^k}{k!} < \infty \quad \text{for } t \in I$$

hold.

Before you continue, make sure to answer the following questions:

Quiz: Section 13.2

Q1 Under which conditions do path-wise or mean-square solutions of $\dot{X}_t = AX_t + Z_t$ exist and are unique?

Q2 Under which general conditions are the unique path-wise and the unique mean-square solution of $\dot{X}_t = AX_t + Z_t$ equivalent?

Q3 Prove the statement you used in Q2.

Q4 Give examples for stochastic processes A_t and Z_t such that the conditions of Q_2 are satisfied.

13.3 Asymptotic Properties of Path-Wise Solutions

If certain sufficient conditions on the parameter functions $A(t)$ and $Z(t)$ hold for the asymptotic behavior of the solutions of the (deterministic) ordinary differential equation

$$\dot{x} \;=\; A(t)x \;+\; Z(t)\,, \tag{13.7}$$

and if the realisations of the stochastic processes A_t and Z_t satisfy these conditions with a certain probability, then the realisations of the path-wise solution of

$$\frac{\mathrm{d}X_t}{\mathrm{d}t} \;=\; A_t X_t \;+\; Z_t \tag{13.8}$$

have the same asymptotic behavior as the solutions of (13.7) with the same probability as stated for the stochastic parameters A_t and Z_t.

Unfortunately, the properties of the realisations of A_t and Z_t are not always known sufficiently. Thus, for practical purposes, it is important to determine properties of the path-wise solution of (13.8) from the properties of the moments of the coefficients or the moments of the solution, respectively.

13.3.1 (Exponential) Decay of Path-Wise Solutions

Let

$$\frac{\mathrm{d}X_t}{\mathrm{d}t} \;=\; (A + F_t)\,X_t\,, \tag{13.9}$$

where A is a constant real $d \times d$-matrix and F_t is a path-wise continuous $d \times d$-matrix process. The next theorem follows from inequality (13.4).

Theorem 13.2 (Exponential Decay of Solutions - Version 1). *Let the following conditions be satisfied:*

1. *All eigenvalues of the matrix A have negative real parts such that* $\|\exp(At)\| \le b\exp(-at)$*,* $a, b > 0$.
2. *It holds that* $F_t = F_{1,t} + F_{2,t}$*, where* $F_{1,t}$*,* $F_{2,t}$ *are path-wise continuous and* $\limsup_{t\to\infty} \|F_{1,t}\| \hat{<} (a-\varepsilon)/b$ *and* $\int_{t_0}^{\infty} \mathbb{E}\left(\|F_{2,t}\|\right) d\tau < \infty$ *with* $\varepsilon > 0$.

Then it holds for all path-wise solutions of (13.9) and arbitrary $\rho < \varepsilon$

$$\lim_{t\to\infty} \exp(\rho t)\,\|X_t\| \;\hat{=}\; 0\,.$$

Proof. Following [45], pp. 71, we have according to inequality (13.4)

$$\|X_t\| \;\overset{I}{<}\; b\,\|X_0\|\exp\left(-a(t-t_0) + b\int_{t_0}^{t}\left(\|F_{1,\tau}\| + \|F_{2,\tau}\|\right)\mathrm{d}\tau\right).$$

According to Lemma 12.1, we have with condition 2 that $\gamma_1 = \int_{t_0}^{\infty} \|F_{2,\tau}\| \, d\tau \hat{<} \infty$ holds, and for almost all $\omega \in \Omega$ there is a $t(\omega)$ such that $\|F_{1,t}\| < (a-\varepsilon)/b$ for $t > t(\omega)$. Moreover, $\gamma_2 = \int_{t_0}^{t(\omega)} \|F_{1,\tau}\| \, d\tau \hat{<} \infty$ holds. Then, finally, we have for almost all $\omega \in \Omega$ and $t \geq t(\omega)$

$$\|X_t\| \leq b \|X_0\| \exp\left((\gamma_1 + \gamma_2) b\right) \exp\left(-(t - t(\omega))\varepsilon\right) ,$$

which shows the assertion. □

Next, we give a result which is formulated in [166] for path-wise solutions in the extended sense:

Theorem 13.3 (Decay of Solutions). *Let the following conditions be satisfied:*

1. *Assume that F_t is path-wise continuous such that*

$$\|A + F_t\| \overset{I}{<} K , \quad K \in S_1 .$$

2. *Assume that the path-wise solution X_t of (13.9) obeys*

$$\int_{t_0}^{\infty} \mathbb{E}\left(\|X_t\|\right) dt < \infty .$$

Then it holds that

$$\lim_{t \to \infty} \|X_t\| \hat{=} 0 .$$

Proof. Following [45], pp. 71, we have with condition 2 due to Lemma 12.1

$$\lim_{t \to \infty} \|X_t\| \hat{<} \infty . \tag{13.10}$$

From (13.9), it follows from condition 1 for almost all $\omega \in \Omega$ that

$$\begin{aligned} \|X_{t_2} - X_{t_1}\| &= \left\| \int_{t_1}^{t_2} (A + F_\tau) X_\tau d\tau \right\| \leq \int_{t_1}^{t_2} \|A + F_\tau\| \, \|X_\tau\| \, d\tau \\ &\leq K \int_{t_1}^{t_2} \|X_\tau\| \, d\tau , \quad \text{for } t_1, t_2 \in I \text{ with } t_2 > t_1 . \end{aligned} \tag{13.11}$$

Next, inequality (13.10) implies

$$\lim_{t_1, t_2 \to \infty, \, t_2 > t_1} \int_{t_1}^{t_2} \|X_\tau\| \, d\tau \hat{=} 0 . \tag{13.12}$$

Because of (13.11) and (13.12) the limit $X_\infty = \lim_{t \to \infty} X_t$ exists for almost all $\omega \in \Omega$, and hence $X_\infty \hat{=} 0$ due to (13.10). □

Note, that we obtain the result $\lim_{t\to\infty} \|X_t\| \hat{=} 0$ also under the following conditions:

1. F_t is path-wise continuous such that $\mathbb{E}\left(\|F_t\|^2\right) < \chi < \infty$, $t \in I$.
2. $\displaystyle\int_{t_0}^{\infty} \left(\mathbb{E}\left(\|X_t\|^2\right)\right)^{1/2} \mathrm{d}t < \infty$.

Because of $\mathbb{E}\left(\|X_t\|\right) \leq \left(\mathbb{E}\left(\|X_t\|^2\right)\right)^{1/2}$, condition 2 implies $\displaystyle\int_{t_0}^{\infty} \|X_t\| \mathrm{d}t \hat{<} \infty$, and we have due to Lemma 12.1

$$\lim_{t_1,t_2\to\infty\,,\,t_2>t_1} \|X_{t_2} - X_{t_1}\| \;\hat{\leq}\; \lim_{t_1,t_2\to\infty\,,\,t_2>t_1} \int_{t_1}^{t_2} \|A + F_\tau\|\,\|X_\tau\|\,\mathrm{d}\tau \;\hat{=}\; 0\,,$$

according to

$$\begin{aligned}\int_{t_0}^{\infty} \mathbb{E}\left(\|A + F_\tau\|\,\|X_\tau\|\right)\mathrm{d}\tau \;&\leq\; \int_{t_0}^{\infty} \left(\mathbb{E}\left(\|A + F_\tau\|^2\right)\right)^{1/2} \left(\mathbb{E}\left(\|X_\tau\|^2\right)\right)^{1/2} \mathrm{d}\tau \\ &\leq\; \chi_1 \int_{t_0}^{\infty} \left(\mathbb{E}\left(\|X_\tau\|^2\right)\right)^{1/2} \mathrm{d}\tau \;<\; \infty\,.\end{aligned}$$

The following result was obtained 50 years ago in 1963 by Frank Kozin in [165] by using the norms $\|x\| = \sum_{i=1}^{d} |x_i|$ and $\|A\| = \sum_{i,j=1}^{d} |x_{i,j}|$:

Theorem 13.4 (Exponential Decay of Solutions - Version 2). *Let the following conditions be satisfied:*

1. *All eigenvalues of the matrix A have negative real parts such that $\|\exp(At)\| < b\exp(-at)$, $a, b > 0$, holds.*
2. *The matrix F_t is path-wise continuous, stationary in the narrow sense and ergodic.*
3. *It holds that $\mathbb{E}\left(\|F_t\|\right) \leq c < (a-\varepsilon)/b$, $\varepsilon > 0$ for a constant c.*

Then,

$$\lim_{t\to\infty} \exp(\varepsilon t)\,\|X_t\| \;\hat{=}\; 0$$

holds for any path-wise solution X_t of (13.9).

Proof. Following [45], pp. 72, we have that (13.4) implies

$$\|X_t| \;\overset{I}{\leq}\; b\,\|X_0|\exp\left(-a(t-t_0) + b\int_{t_0}^{t} \|F_\tau\|\,\mathrm{d}\tau\right),$$

and hence $\lim_{t\to\infty} \frac{1}{t}\displaystyle\int_{t_0}^{t} \|F_\tau\|\,\mathrm{d}\tau \hat{=} \mathbb{E}\left(\|F_{t_0}\|\right)$ leads to the assertion. □

Based on Ettore F. Infante's 1968 article [137] we give stronger conditions for convergence towards the null-solution. In particular, let us study the random differential equation

$$\frac{\mathrm{d}X_t}{\mathrm{d}t} = AX_t + F_t X_t + C(t)X_t\,, \tag{13.13}$$

where A is a constant $d\times d$-matrix, F_t a stochastic $d\times d$-matrix process and $C(t)$ a real $d \times d$-matrix function. We denote the largest and smallest eigenvalue of a matrix B by $\lambda_{\max}(B)$ and $\lambda_{\min}(B)$, respectively.

Theorem 13.5 (Exponential Decay of Solutions - Version 3). *Let the following conditions be satisfied:*

1. *The matrix process F_t is strictly stationary, path-wise continuous and ergodic.*
2. *The matrix function $C(t)$ is continuous on I.*
3. *There is a symmetric positive definite matrix B such that*

$$\mathbb{E}\left(\rho_1(t_0)\right) + \lim_{t\to\infty} \frac{1}{t-t_0}\int_{t_0}^{t} \rho_2(\tau)d\tau \;\le\; -\varepsilon\,, \quad \varepsilon > 0\,, \tag{13.14}$$

where

$$\rho_1(t) \;=\; \lambda_{\max}\left(A^T + F_t^T + B\left(A + F_t\right)B^{-1}\right)$$

and

$$\rho_2(t) \;=\; \lambda_{\min}\left(C^T(t) + BC(t)B^{-1}\right)\,.$$

Then,

$$\lim_{t\to\infty} \exp\left(\alpha t\right)\|X_t\| \;\hat{=}\; 0$$

holds for every path-wise solution (13.13) and every $\alpha < \varepsilon$.

Proof. Following [45], pp. 73, we assume that D and B are two real symmetric $d\times d$-matrices and that B is positive definite. Then we have for the minimum and maximum of the quadratic forms

$$\min_{x\in\mathbb{R}^d\setminus\{0\}} \frac{x^T Dx}{x^T Bx} \;=\; \lambda_{\min}\left(DB^{-1}\right)\,, \tag{13.15}$$

and

$$\max_{x\in\mathbb{R}^d\setminus\{0\}} \frac{x^T Dx}{x^T Bx} \;=\; \lambda_{\max}\left(DB^{-1}\right)\,, \tag{13.16}$$

respectively.

If X_t is the null-solution, then the assertion trivially follows. In order to obtain a contradiction, let X_t be some other solution different from the null-solution, and consider the function

$$v(x) \;=\; x^T B x \,. \tag{13.17}$$

With

$$\dot{v}_t(x) \;=\; x^T \left(\left(A^T + F_t^T + C^T(t) \right) B + B \left(A + F_t + C(t) \right) \right) x \tag{13.18}$$

equations (13.16) and (13.17) imply

$$\frac{1}{v(X_t)} \frac{\mathrm{d}v(X_t)}{\mathrm{d}t} \;\stackrel{I}{=}\; \frac{\dot{v}_t(X_t)}{v(X_t)} \;\stackrel{I}{\leq}\; \max_{x \in \mathbb{R}^d \setminus \{0\}} \frac{\dot{v}_t(x)}{v(x)} \;=\; \rho(t)\,, \tag{13.19}$$

where

$$\rho(t) \;:=\; \lambda_{\mathsf{max}} \left(A^T + F_t^T + C^T(t) + B \left(A + F_t + C(t) \right) B^{-1} \right) \,. \tag{13.20}$$

If D_1, D_2 and B are real symmetric matrices and if B is positive definite, then

$$\lambda_{\mathsf{max}} \left(\left(D_1 + D_2 \right) B^{-1} \right) \;\leq\; \lambda_{\mathsf{max}} \left(D_1 B^{-1} \right) + \lambda_{max} \left(D_2 B^{-1} \right) \tag{13.21}$$

holds due to (13.16). Because of (13.21) equation (13.20) implies

$$\rho(t) \;\leq\; \rho_1(t) + \rho_2(t)\,. \tag{13.22}$$

(13.19) and (13.22) lead to

$$v(X_t) \;\stackrel{I}{\leq}\; v(X_0) \exp \left((t - t_0) \left(\frac{1}{t - t_0} \int_{t_0}^{t} \rho_1(\tau) \mathrm{d}\tau + \frac{1}{t - t_0} \int_{t_0}^{t} \rho_2(\tau) \mathrm{d}\tau \right) \right) . \tag{13.23}$$

According to condition 1, it holds that

$$\lim_{t \to \infty} \frac{1}{t - t_0} \int_{t_0}^{t} \rho_1(\tau) \mathrm{d}\tau \mathrel{\hat{=}} \mathbb{E} \left(\rho_1(t_0) \right) \,. \tag{13.24}$$

(13.23), (13.24) and condition 3 imply $\lim_{t \to \infty} \exp(\alpha t)\, v(X_t) \hat{=} 0$ and thus, finally, because of $v(X_t) \stackrel{I}{\geq} \lambda_{\mathsf{min}}(B) \, \|X_t\|^2$, the assertion of this theorem. □

In (13.19) we utilized the quantities $\dot{v}_t(X_t)/v(X_t)$ and $\dot{v}_t(x)/v(x)$. Of course, one has to show that these, and in particular the stochastic quantity $\dot{v}_t(X_t)/v(X_t)$, are well defined. This is the task of problem 13.17.

In the discussion of specific systems one can strive for stability boundaries that are as sharp as possible by using an optimal choice of B.

Remark 13.4 (Upper Bounds for $\mathbb{E}(\rho_2)$). Let $C(t)$ have the form $C(t) = C_1(t) + C_2(t)$, where $C_1(t)$ is a continuous periodic function with period θ. Then

$$\rho_2(t) \;\le\; \lambda_{\max}\left(C_t^T(t) + BC_1(t)B^{-1}\right) + \lambda_{\max}\left(C_2^T(t) + BC_2(t)B^{-1}\right)$$

holds, and for instance

$$\begin{aligned}\lim_{t\to\infty}\frac{1}{t-t_0}\int_{t_0}^{t}\rho_2(\tau)\mathrm{d}\tau \le \frac{1}{\theta}\int_0^{\theta}\lambda_{\max}\left(C_1^T(\tau) + BC_1(\tau)B^{-1}\right)\mathrm{d}\tau \\ + \sup_{\tau\in[t_0,\infty)}\lambda_{\max}\left(C_2^T(\tau) + BC_2(\tau)B^{-1}\right) \qquad (13.25)\end{aligned}$$

is valid. Because of $\lambda_{\max}(D) \le \|D\|$ we gain from (13.25) that

$$\begin{aligned}\lim_{t\to\infty}\frac{1}{t-t_0}\int_{t_0}^{t}\rho_2(\tau)\mathrm{d}\tau \;\le\; \frac{1}{\theta}\int_0^{\theta}\left\|C_1^T(\tau) + BC_1(\tau)B^{-1}\right\|\mathrm{d}\tau \\ + \sup_{\tau\in[t_0,\infty)}\left\|C_2^T(\tau) + BC_2(\tau)B^{-1}\right\| . \qquad (13.26)\end{aligned}$$

For example, if $C(t)$ has the form

$$C(t) \;=\; \sum_{i=1}^{N} c_i(t)C_i\,,$$

where the c_i, $i = 1, 2, \ldots, N$, are continuous periodic functions with period θ_i, and the C_i are constant real matrices, then

$$\begin{aligned}&\lambda_{\max}\left(C^T(t) + BC(t)B^{-1}\right)\\ &\le \sum_{i=1}^{N}\max_{x\in\mathbb{R}^d} c_i(t)\frac{x^T\left(C_i^T B + BC_i\right)c}{x^T Bx}\\ &\le \sum_{i=1}^{N}\left(c_i^+(t)\lambda_{\max}\left(C_i^T + BC_iB^{-1}\right) + c_i^-(t)\lambda_{\min}\left(C_i^T + BC_iB^{-1}\right)\right)\end{aligned} \qquad (13.27)$$

where

$$c_i^+(t) \;=\; \begin{cases} c_i(t) & \text{if } c_i(t)\ge 0\\ 0 & \text{if } c_i(t)\le 0\end{cases} \quad\text{and}\quad c_i^-(t) \;=\; \begin{cases} c_i(t) & \text{if } c_i(t)< 0\\ 0 & \text{if } c_i(t)> 0\end{cases}.$$

Let $\int_0^{\theta i} c_i(\tau)\mathrm{d}\tau \;=\; 0$, then

$$\int_0^{\theta i} c_i^+(\tau)\mathrm{d}\tau \;=\; -\int_0^{\theta i} c_i^-(\tau)\mathrm{d}\tau \;=\; \tfrac{1}{2}\int_0^{\theta i}|c_i(\tau)|\,\mathrm{d}\tau$$

holds, and we gain from (13.28)

$$\begin{aligned}\lim_{t\to\infty} \frac{1}{t-t_0}\int_{t_0}^{t} \rho_2(\tau)\mathrm{d}\tau \\ \leq \sum_{i=1}^{N} \frac{1}{2\theta_i}\int_0^{\theta i} |c_i(\tau)|\,\mathrm{d}\tau\left(\lambda_{\max}\left(C_i^T + BC_iB^{-1}\right) - \lambda_{\min}\left(C_i^T + BC_iB^{-1}\right)\right).\end{aligned} \tag{13.28}$$

If in (13.14) we replace the limit $\lim_{t\to\infty} \frac{1}{t-t_0}\int_{t_0}^{t} \rho_2(\tau)\mathrm{d}\tau$ with its bounds from (13.25), (13.26) and (13.28), respectively, we gain implications form Theorem 13.5 which may be verified more easy.

Further consequences from Theorem 13.5 are obtained by using upper bounds on ρ_1.

Remark 13.5 (Upper Bounds for ρ_1). Inequality (13.21) implies

$$\rho_1(t) \leq \lambda_{\max}\left(A^T + BAB^{-1}\right) + \lambda_{\min}\left(F_t^T + BF_tB^{-1}\right). \tag{13.29}$$

If A is a stable matrix, i.e, all eigenvalues of the matrix A have negative real parts, then the matrix B can, for instance, be chosen as the solution of the matrix equation

$$A^TB + BA = \mathbb{I},$$

see [181]. Because of

$$\lambda_{\max}\left(A^T + BAB^{-1}\right) = \frac{1}{\lambda_{\max}(B)} \tag{13.30}$$

inequality (13.29) leads to

$$\rho_1(t) \leq -\frac{1}{\lambda_{\max}(B)} + \lambda_{\max}\left(F_t^T + BF_tB^{-1}\right) \tag{13.31}$$

or

$$\rho_1(t) \leq -\frac{1}{\lambda_{\max}(B)} + \left\|F_t^T + BF_tB^{-1}\right\|. \tag{13.32}$$

Applying the transformation $x = B^{-1/2}y$ we get the equation

$$\max_{x\in\mathbb{R}^d\setminus\{0\}} \frac{x^T\left(F_t^TB + BF_t\right)x}{x^TBx} = \max_{y\in\mathbb{R}^d\setminus\{0\}} \frac{y^T\left(B^{-1/2}F_t^TB^{1/2} + B^{1/2}F_tB^{-1/2}\right)y}{y^Ty},$$

and thus that

$$\lambda_{\max}\left(F_t^T + BF_tB^{-1}\right) = \lambda_{\max}\left(B^{-1/2}F_t^TB^{1/2} + B^{1/2}F_tB^{-1/2}\right). \tag{13.33}$$

From (13.31) and (13.33) it follows that

$$\rho_1(t) \;\le\; -\frac{1}{\lambda_{\max}(B)} + \left\| B^{-1/2} F_t^T B^{1/2} + B^{1/2} F_t B^{-1/2} \right\| . \tag{13.34}$$

The condition

$$\mathbb{E}\left(\left\| B^{-1/2} F_t^T B^{1/2} + B^{1/2} F_t B^{-1/2} \right\| \right) \;<\; \frac{1}{\lambda_{\max}(B)} - \varepsilon$$

is sufficient for (13.14) in the case $C(t) \equiv 0$.

Remark 13.6 (A Special Case of F_t). Let F_t be of the form $F_t = \sum_{i=1}^{N} Z_{(i)t} F_i$, where the F_i are constant real matrices and the $Z_{(i)t}$ are one-dimensional, path-wise continuous, weakly stationary and ergodic processes. Then

$$\lambda_{\max}\left(F_t^T + B F_t B^{-1}\right) \;\le\; \sum_{i=1}^{N} \left|Z_{(i)t}\right| \left|\lambda_{\max}\left(F_i^T + B F_i B^{-1}\right)\right| \tag{13.35}$$

is valid. Inequality (13.31) and

$$\lambda_{\max}\left(F_i^T + B F_i B^{-1}\right) \;=\; \lambda_{\max}\left(B^{-1/2} F_i^T B^{1/2} + B^{1/2} F_i B^{-1/2}\right) \;=\; \mu_i$$

lead to

$$\mathbb{E}\left(\rho_1(t)\right) \;\le\; -\frac{1}{\lambda_{\max}(B)} + \sum_{i=1}^{N} \mathbb{E}\left(\left|Z_{(i)t}\right|\right) |\mu_i| . \tag{13.36}$$

If additionally $\mathbb{E}\left(Z_{(i)t}\right) = 0$ holds, then the estimate for $\mathbb{E}\left(\rho_1(t)\right)$ can be tightened: Because of (13.15) and (13.16) it holds that

$$\begin{aligned}
&\lambda_{\max}\left(F_t^T + B F_t B^{-1}\right) \\
&= \max_{x\in\mathbb{R}^d} \sum_{i=1}^{N} Z_{(i)t} \frac{x^T\left(F_i^T B + B F_i\right) x}{x^T B x} \\
&\le \sum_{i=1}^{N} \left(Z^+_{(i)t} \lambda_{\max}\left(F_i^T + B F_i B^{-1}\right) + Z^-_{(i)t} \lambda_{\min}\left(F_i^T + B F_i B^{-1}\right)\right) ,
\end{aligned} \tag{13.37}$$

where

$$Z^+_{(i)t} = \begin{cases} Z_{(i)t} & \text{if } Z_{(i)t} \ge 0 \\ 0 & \text{if } Z_{(i)t}) < 0 \end{cases} \quad \text{and} \quad Z^-_{(i)t} = \begin{cases} Z_{(i)t} & \text{if } Z_{(i)t} < 0 \\ 0 & \text{if } Z_{(i)t} \ge 0 \end{cases} .$$

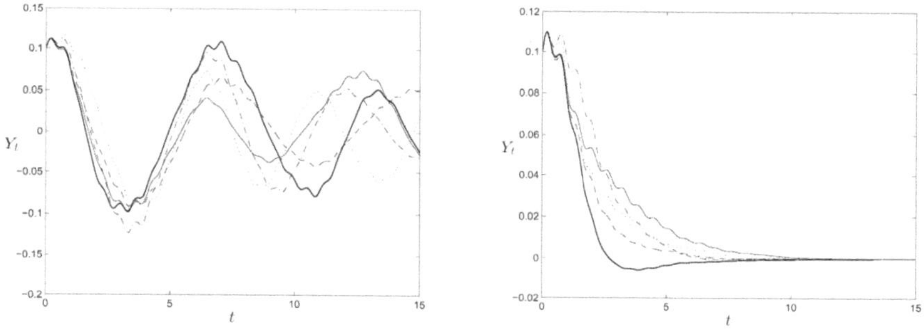

(a) (b)

Figure 13.2. Simulations of the solution process of the damped oscillation equation (13.39) with $Z_t = O_t$, $\omega = 10$, $a = 5$, and $b = 1$ in (a) as well as $b = 0.1$ in (b).

According to $\mathbb{E}\left(Z^+_{(i)t}\right) = -\mathbb{E}\left(Z^-_{(i)t}\right) = \frac{1}{2}\mathbb{E}\left(|Z_{(i)t}|\right)$ we obtain from (13.29) and (13.37) that

$$\begin{aligned}\mathbb{E}(\rho_1(t)) \;\le\; & \lambda_{\max}\left(A^T + BAB^{-1}\right) + \tfrac{1}{2}\sum_{i=1}^{N}\mathbb{E}\left(|Z_{(i)t}|\right)\left(\lambda_{\max}\left(F_i^T + BF_iB^{-1}\right)\right. \\ & \left. - \lambda_{\min}\left(F_i^T + BF_iB^{-1}\right)\right) . \end{aligned} \tag{13.38}$$

For the discussion of the next two examples we utilize methods presented in detail in Ettore F. Infante's article [137].

Example 13.7 (A Second Order Random Differential Equation With Stochasticity of the Position Component, cf. [45], pp. 77). We study the second order random differential equation

$$\ddot{Y}_t + 2b\dot{Y}_t + (1 + Z_t + a\sin(\omega t))\,Y_t = 0\,, \tag{13.39}$$

where $a \ge 0$ and $b > 0$. For $a = 0$ this equation was studied, e.g., in [165], [52] or [137].

Let Z_t be a stationary stochastic process in the narrow sense that is ergodic and path-wise continuous. Fig. 13.2 shows some simulations of this damped oscillation equation (13.39) with $Z_t = O_t$ such that $\mathrm{d}O_t = -O_t\mathrm{d}t + \mathrm{d}W_t$, $\omega = 10$, $a = 5$, and two different values of the parameter b; namely $b = 1$ in Fig. 13.2 (a) and $b = 0.1$ in Fig. 13.2 (b). The simulations were computed with the averaged Euler method that we will discuss in Chap. 14 with a step size $h = 3 \cdot 10^{-3}$.

Equation (13.39) can be rewritten in the form (13.13) with

$$X_t^T = \left(Y_t, \dot{Y}_t\right), \quad A = \begin{pmatrix} 0 & 1 \\ -1 & -2b \end{pmatrix}, \quad F_t = Z_t \begin{pmatrix} 0 & 0 \\ -1 & 0 \end{pmatrix},$$

and

$$C(t) = a \sin(\omega t) \begin{pmatrix} 0 & 0 \\ -1 & 0 \end{pmatrix}.$$

We write the matrix B in the general form

$$B = \begin{pmatrix} \alpha_1^2 + \alpha_2 & \alpha_1 \\ \alpha_1 & 1 \end{pmatrix}, \quad \alpha_2 > 0,$$

and calculate

$$\rho_1(t) = -2b + \sqrt{4(b-\alpha_1)^2 + \alpha_2^{-1}\left(\alpha_2 + \alpha_1^2 - 1 - Z_t + 2\alpha_1(b-\alpha_1)\right)^2} \tag{13.40}$$

and

$$\lambda_{\max}\left(C^T + BCB^{-1}\right) = \frac{1}{\sqrt{\alpha_2}} a\,|\sin(\omega t)|,$$

$$\lambda_{\min}\left(C^T + BCB^{-1}\right) = -\frac{1}{\sqrt{\alpha_2}} a\,|\sin(\omega t)|.$$

From (13.25) we obtain

$$\lim_{t\in\infty} \frac{1}{t-t_0}\int_{t_0}^{t} \rho_2(\tau)\mathrm{d}\tau \le \frac{a\omega}{2\pi\sqrt{\alpha_2}}\int_0^{2\pi/\omega} |\sin(\omega t)|\,\mathrm{d}t = \frac{2a}{\pi\sqrt{\alpha_2}}. \tag{13.41}$$

$\rho_1(t)$ becomes minimal for

$$\begin{aligned} \alpha_1 &= b, & \alpha_2 &= 1-b^2, & &\text{if } b \le \tfrac{1}{2}\sqrt{2}, \\ \alpha_1 &= b, & \alpha_2 &= b^2, & &\text{if } b \ge \tfrac{1}{2}\sqrt{2}. \end{aligned} \tag{13.42}$$

Plugging (13.42) into (13.40) and (13.41) we get with (13.14) the following condition

$$\begin{aligned} \mathbb{E}\left(|Z_t|\right) &\le (2b-\varepsilon)\sqrt{1-b^2} - \frac{2a}{\pi}, & &\text{if } b \le \tfrac{1}{2}\sqrt{2}, \\ \mathbb{E}\left(\left|Z_t + 1 - 2b^2\right|\right) &\le (2b-\varepsilon)b - \frac{2a}{\pi}, & &\text{if } b \ge \tfrac{1}{2}\sqrt{2}. \end{aligned} \tag{13.43}$$

With $\alpha_1 = b$ and $\alpha_2 = b^2 + 1$ we obtain at $\mathbb{E}(Z_t) = 0$ with (13.40) and (13.41) from (13.14) by applying the Cauchy-Schwarz inequality also the condition

$$\mathbb{E}\left(Z_t^2\right) \le \left((2b-\varepsilon)\sqrt{1+b^2} - \frac{2a}{\pi}\right)^2 - 4b^4 \tag{13.44}$$

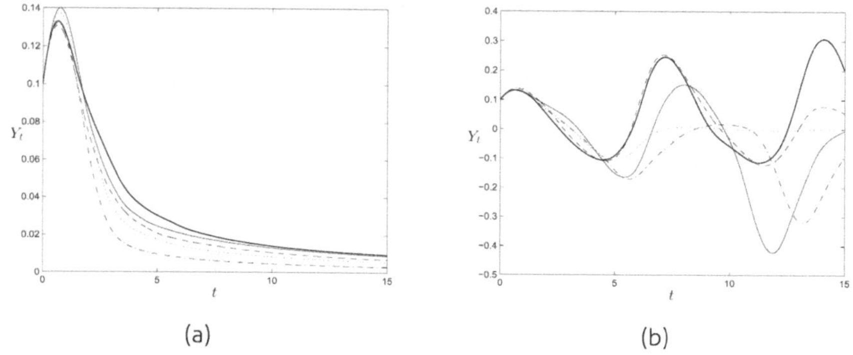

Figure 13.3. Simulations of the solution process of the damped oscillation equation (13.45) with $Z_t = O_t$, $b = 0.1$ and $c(t) = t$ in (a) as well as $c(t) = \sin(t)$ in (b).

and in the case $a = 0$

$$\mathbb{E}\left(Z_t^2\right) \le 4b^2 - \varepsilon^* .$$

respectively.

Example 13.8 (A Second Order Random Differential Equation with Stochasticity at the Velocity Component, cf. [45], pp. 78). Next, let us study the second order random differential equation

$$\ddot{Y}_t + (2b + Z_t + c(t))\,\dot{Y}_t + Y_t = 0\,, \tag{13.45}$$

where Z_t again is a stationary process in the narrow sense that is ergodic and path-wise continuous on I such that $\mathbb{E}(Z_t) = 0$. The function $c(t)$ is continuous and periodic with period θ and such that

$$\frac{1}{\theta}\int_0^\theta c(\tau)\mathrm{d}\tau = 0\,, \quad \text{and} \quad \frac{1}{\theta}\int_0^\theta |c(\tau)|\,\mathrm{d}\tau = \bar{c}\,.$$

Fig. 13.3 shows some simulations of this damped oscillation equation (13.45) with $Z_t = O_t$ such that $\mathrm{d}O_t = -O_t\mathrm{d}t + \mathrm{d}W_t$, $b = 0.1$, and two different values of the function c; namely $c(t) = t$ in Fig. 13.3 (a) and $c(t) = \sin(t)$ in Fig. 13.3 (b). The simulations were computed with the averaged Euler method that we will discuss in Chap. 14 with a step size $h = 3 \cdot 10^{-3}$.

We re-write equation (13.45) in the form (13.14) with $X_t^T = (Y_t, \dot{Y}_t)$, and

$$A = \begin{pmatrix} 0 & 1 \\ -1 & -2b \end{pmatrix}, \quad F_t = Z_t \begin{pmatrix} 0 & 0 \\ 0 & -1 \end{pmatrix},$$

as well as

$$C_t = c(t)\begin{pmatrix} 0 & 0 \\ 0 & -1 \end{pmatrix} =: c(t)C\,.$$

Applying the very same matrix B as in example 13.7 we get

$$\rho_1(t)=-2b-Z_t+\sqrt{(Z_t+2b-2\alpha_1)^2+\alpha_2^{-1}\left(\alpha_2+\alpha_1^2-1+\alpha_1 Z_t+2\alpha_1\left(b-\alpha_1\right)\right)} \tag{13.46}$$

and

$$\lambda_{\max}\left(C^T+BCB^{-1}\right)-\lambda_{\min}\left(C^T+BCB^{-1}\right) \;=\; \sqrt{1+\frac{\alpha_1^2}{\alpha_2}}\,. \tag{13.47}$$

(13.28) and (13.47) imply

$$\lim_{t\to\infty}\frac{1}{t-t_0}\int_{t_0}^{t}\rho_2(\tau)\mathrm{d}\tau \;\leq\; \frac{\overline{c}}{2}\sqrt{1+\frac{\alpha_1^2}{\alpha_2}}\,. \tag{13.48}$$

$\rho_1(t)$ becomes minimal for

$$\begin{aligned}
\alpha_1 &= b\,, & \alpha_2 &= 1-b^2\,, & &\text{if } b\leq \tfrac{1}{2}(\sqrt{5}-1)\,,\\
\alpha_1 &= \frac{1}{\sqrt{b^2+1}}\,, & \alpha_2 &= \frac{b^2}{b^2+1}\,, & &\text{if } b\geq \tfrac{1}{2}(\sqrt{5}-1)\,.
\end{aligned}$$

Thus, with (13.46) and (13.48) we obtain from (13.14) the condition

$$\mathbb{E}\left(|Z_t|\right) \;\leq\; (2b-\varepsilon)\sqrt{1-b^2}-\tfrac{1}{2}\overline{c}\,,$$

if $b\leq \frac{1}{2}(\sqrt{5}-1)$,

$$\mathbb{E}\left(\left|Z_t+2b-2\left(1+b^2\right)^{-1/2}\right|\right) \;\leq\; (2b-\varepsilon)b\left(1+b^2\right)^{-1/2}-\tfrac{1}{2}\overline{c}\,,$$

if $b\geq \frac{1}{2}(\sqrt{5}-1)$. The Cauchy-Schwarz inequality and the values

$$\alpha_1 \;=\; \frac{b}{1+b^2}\,, \quad \text{and} \quad \alpha_2 \;=\; 1-\frac{b^2}{(1+b^2)^2}$$

applied to (13.14) lead with (13.45) and (13.47) also to the condition

$$\mathbb{E}\left(Z_t^2\right) \;\leq\; \left((2b-\varepsilon)\left(1+b^2\right)^{-1}\left(1+b^2+b^4\right)^{1/2}-\tfrac{1}{2}\overline{c}\right)^2-4\left(1+b^2\right)^{-2}b^6\,,$$

or to

$$\mathbb{E}\left(Z_t^2\right) \;\leq\; 4b^4\left(1+b^4\right)\left(1+b^2\right)^{-2}-\varepsilon^*$$

in the case $\overline{c}=0$, respectively.

13.3.2 Boundedness of Path-Wise Solutions

Next, we discuss the random differential equation

$$\frac{\mathrm{d}X_t}{\mathrm{d}t} = AX_t + F_tX_t + C(t)X_t + Z_t\,, \tag{13.49}$$

where A, F_t and $C(t)$ are defined as in equation (13.13) and Z_t is a d-dimensional path-wise continuous stochastic process. We look for conditions under which the realisations of the path-wise solutions of (13.49) are bounded.

Theorem 13.6 (Upper Bounds for the Solution). *Let the conditions 1 and 2 of Theorem 13.5 be satisfied, and let there be a symmetric positive definite matrix B such that $\rho(t) \overset{I}{\leq} -\varepsilon < 0$ holds, where $\rho(t)$ is defined in (13.20). Moreover, let there be a finite function $M(\omega)$ defined in Ω such that $\|Z_t\| \overset{I}{\leq} M$. Then, for every path-wise solution X_t of (13.49) there are positive numbers h and k such that*

$$\|X_t\| \overset{I}{\leq} h\,\|X_0\| + kM\,.$$

Proof. Following [45], p. 79, we first note that according to (13.19) for any path-wise solution Y_t of the homogeneous differential equation

$$\dot{Y}_t = (A + F_t + C(t))\,Y_t$$

the estimate

$$\|Y_t\| \overset{I}{\leq} (\lambda_{\min}(B))^{-1}\,\lambda_{\max}(B)\,\|Y_\tau\|\,\exp\left(-\varepsilon(t-\tau)\right)\,, \quad t \in I\,, \quad \tau \in [t_0, t]\,,$$

Before you continue, make sure to answer the following questions:

Quiz: Section 13.3 – Part I ((Exponential) Decay of Path-Wise Solutions)

Q1 Under which conditions does the unique path-wise solution of $\dot{X}_t = (A + F_t)\,X_t$ decay towards the null-solution?

Q2 Sketch the proof of the theorem you used in Q1.

Q3 Under which conditions does the unique path-wise solution of $\dot{X}_t = (A + F_t)\,X_t + C(t)X_t$ decay towards the null-solution?

Q4 Sketch the proof of the theorem you used in Q3.

holds. Thus, for the fundamental matrix Φ_t of the homogeneous equation (with $\Phi_{t_0} = \mathbb{I}$) estimates of the form

$$\left\|\Phi_t \Phi_\tau^{-1}\right\| \overset{I}{\leq} h \exp\left(-\varepsilon(t-\tau)\right), \quad \text{and} \quad \|\Phi_t\| \overset{I}{\leq} h \exp\left(-\varepsilon(t-\tau)\right)$$

are valid. Every path-wise solution X_t of (13.49) now has the form (13.2), which leads to

$$\begin{aligned} \|X_t\| &\overset{I}{\leq} h\,\|X_0\| \exp\left(-\varepsilon(t-\tau)\right) + Ch\varepsilon^{-1}\left(1 - \exp\left(-\varepsilon(t-\tau)\right)\right) \\ &\overset{I}{\leq} h\,\|X_0\| + Mh\varepsilon\,, \end{aligned}$$

and thus the assertion is shown. □

Sufficient conditions for $\rho(t) \overset{I}{\leq} -\varepsilon$ are obtained with the estimates from Sec. 13.3.1. For instance, if we assume that the real parts of all eigenvalues of A are negative, that $C(t) \equiv 0$ and that B is the solution of the matrix equation $A^T B + BA = -\mathbb{I}$. Then such a sufficient condition is given by

$$\lambda_{\max}\left(F_t^T + BF_tB^{-1}\right) \overset{I}{\leq} \left(\lambda_{\max}(B)\right)^{-1} - \varepsilon\,.$$

On the other hand, for this inequality

$$\left\|F_t^T + BF_tB^{-1}\right\| \overset{I}{\leq} \left(\lambda_{\max}(B)\right)^{-1} - \varepsilon$$

or

$$\left\|B^{-1/2}F_t^T B^{1/2} + B^{1/2}F_tB^{-1/2}\right\| \overset{I}{\leq} \left(\lambda_{\max}(B)\right)^{-1} - \varepsilon$$

are sufficient.

Before you continue, make sure to answer the following questions:

Quiz: Section 13.3 – Part II (Boundedness of Path-Wise Solutions)

Q1 Given the inhomogeneous random differential equation $\dot{X}_t = (A + F_tC(t))\,X_t + Z_t$. What can you say about the boundedness of $\|X_t\|$?

Q2 Sketch the proof of the assertion you used in Q1.

13.4 Asymptotic Properties of the Moments of Path-Wise Solutions

Conversely to the methods of Sec. 13.3.2 one can deduce the asymptotic behavior of the moments from the properties of the realisations of the coefficients, if certain integrability conditions are satisfied.

First, we inspect the differential equation

$$\dot{X}_t = AX_t, \qquad (13.50)$$

where A is a t-independent random $d \times d$-matrix.

Conditions for the asymptotic convergence of the moments of the path-wise solutions of (13.50) towards the null-solution are also discussed, e.g., in [43], [44], or [192].

13.4.1 Exponential Decay of the Moments

Theorem 13.7 (Exponential Decay of the Moments - Homogeneous Case). *Let the following conditions be satisfied by the random matrix A:*

1. *The real parts of the eigenvalues $\lambda_i(A)$ are almost surely negative:* $\max_i Re(\lambda_i) = -\lambda \hat{<} 0$.

2. $\left(\|A\|\lambda^{-1}\right)^{d-1}\|X_0\| \in L_1^r$ *holds for some $X_0 \in S_d$ and for some $r > 0$.*

Then $\lim_{t\to\infty} \mathbb{E}\left(\|X_t\|^r\right) = 0$ holds for the path-wise solution X_t of (13.50) with respect to the initial condition (X_0, t_0).

If additionally $\lambda \hat{\geq} \rho$ holds with a positive real constant ρ, then

$$\lim_{t\to\infty} \exp(r\alpha t)\, \mathbb{E}\left(\|X_t\|^r\right) = 0$$

holds for an arbitrary $\alpha < \rho$.

Proof. Following [45], pp. 80, we have

$$\|\exp(At)\| \leq c_d \left(\|A\|\,\varepsilon^{-1}\right)^{d-1} \exp\left(-(\lambda - \varepsilon)t\right), \quad 0 < \varepsilon < \lambda, \qquad (13.51)$$

and

$$\|\exp(At)\| \leq c'_d \left(\|A\|\,\lambda^{-1}\right)^{d-1} \exp\left(-\lambda t/2\right),$$

respectively, with constants c_d and c'_d depending on d only. Thus, due to condition 2,

$$\mathbb{E}\left(\sup_{t\in I}\|X_t\|^r\right) = \mathbb{E}\left(\sup_{t\in I}\|\exp(At)\,X_0\|^r\right) < \infty, \qquad (13.52)$$

and because of condition 1,

$$\lim_{t\to\infty} \|X_t\| \;\hat{=}\; 0\,. \tag{13.53}$$

(13.52) and (13.53) imply according to Lebesgue's convergence theorem the first assertion. The second assertion follows immediately from the first one because $Y_t = \exp(\alpha(t-t_0))\,X_t$ is a path-wise solution of

$$\dot{Y}_t \;=\; \overline{A}Y_t\,, \quad \text{with} \quad \overline{A} \;=\; A + \alpha\mathbb{I} \tag{13.54}$$

with respect to the initial condition (X_0, t_0) and as the conditions 1 and 2 hold for the differential equation (13.54):

$$\max_i Re\left(\lambda_i(\overline{A})\right) \;=\; \max_i\left(\lambda_i(A)+\alpha\right) \;=\; \alpha - \lambda \,\hat{\leq}\, \alpha - \rho \;<\; 0\,,$$

as well as

$$\begin{aligned}
&\mathbb{E}\left(\left(\left(\|\overline{A}\|\left(\min_i\left|Re\left(\lambda_i(\overline{A})\right)\right|\right)^{-1}\right)^{d-1}\|X_0\|\right)^r\right)\\
&\leq\; \mathbb{E}\left(\left(\left(\left(\|\overline{A}\| + \sqrt{d}\alpha\right)(\lambda-\alpha)^{-1}\right)^{d-1}\|X_0\|\right)^r\right)\\
&\leq\; (1-\rho^{-1}\alpha)^{(1-d)r}\,\mathbb{E}\left(\left(\left(\lambda^{-1}\|\overline{A}\| + \sqrt{d}\right)^{d-1}\|X_0\|\right)^r\right)\\
&\leq\; (1-\rho^{-1}\alpha)^{(1-d)r}\,\chi\left(\mathbb{E}\left(\left((\lambda^{-1}\|\overline{A}\|)^{d-1}\|X_0\|\right)^r\right) + \mathbb{E}\left(\|X_0\|^r\right)\right)\\
&\leq\; (1-\rho^{-1}\alpha)^{(1-d)r}\,2\chi\mathbb{E}\left(\left((\lambda^{-1}\|\overline{A}\|)^{d-1}\|X_0\|\right)^r\right) \;<\; \infty\,.
\end{aligned}$$

This completes the proof. □

Next, we study the inhomogeneous random differential equation

$$\dot{X}_t \;=\; AX_t \,+\, Z_t\,, \tag{13.55}$$

where A again is a random $d\times d$-matrix and Z_t is a path-wise continuous d-dimensional stochastic process.

Theorem 13.8 (Exponential Decay of the Moments -- Inhomogeneous Case). *Let the following conditions be satisfied:*

1. *A fulfills the conditions 1 and 2 of Theorem 13.7.*
2. *It holds that $\|Z_t\| \overset{I}{\leq} \beta\exp(-bt_0)$ with $b,\beta\in S_1$, and $b\hat{\geq}0$.*
3. *$\left(\|A\|\,\lambda^{-1}\right)^{d-1}\beta b^{-1}\exp(-bt_0)\in L_1^r$ for some $r>0$.*

Then, $\lim_{t\to\infty} \mathbb{E}(\|X_t\|^r) = 0$ *holds for the path-wise solution* X_t *of (13.55) with respect to the initial condition* (X_0, t_0).

If additionally $\lambda = -\max_i Re(\lambda(A)) \hat{\geq} \rho$ *and* $b \hat{\geq} \rho$ *with a positive constant* ρ*, then*

$$\lim_{t\to\infty} \exp(\alpha t)\, \mathbb{E}(\|X_t\|^r) = 0$$

holds for an arbitrary $\alpha < \rho$.

Proof. Following [45], pp. 81, we have that the form (13.2) holds and we have

$$\Phi_t = \exp(A(t-t_0)), \quad \text{and} \quad \Phi_t\Phi_\tau^{-1} = \exp(A(t-\tau)).$$

Thus, we get

$$\|X_t\| \overset{I}{\leq} \chi\left(\|A\|\,\lambda^{-1}\right)^{d-1}\left(\|X_0\| + \frac{\beta}{b}\exp(-bt_0)\right),$$

and consequently

$$\mathbb{E}\left(\sup_{t\in I}\|X_t\|^r\right) < \infty. \tag{13.56}$$

Because of conditions 1 and 2 it also holds that

$$\lim_{t\to\infty}\|X_t\| \hat{=} 0. \tag{13.57}$$

Finally, (13.56) and (13.57) imply the first assertion due to Lebesgue's convergence theorem. The second assertion is shown analogously to the second assertion of Theorem 13.7. □

Next, we discuss the random differential equation

$$\dot{X}_t = AX_t + F_tX_t + C(t)X_t, \tag{13.58}$$

where A is a random matrix, F_t is a path-wise continuous matrix function and $C(t)$ is a continuous real matrix function.

Theorem 13.9 (Decay of the r-th Moments). *Let the following conditions be satisfied:*

1. *Let* $\max_i Re(\lambda_i(A)) \hat{<} -\rho$ *and* $\|A\| \hat{<} a$ *with positive constants* a *and* ρ.
2. *Let* $\lim_{t\to\infty}\|C(t)\| = 0$.
3. *Let* $\|F_t\| \overset{I}{\leq} c$*, where* c *is a sufficiently small positive constant.*

Then,

$$\lim_{t\to\infty} \mathbb{E}\left(\|X_t\|^r\right) = 0$$

holds for every path-wise solution X_t of (13.58) with $X_{t_0} \in L_d^r$ and arbitrary $r > 0$.

Proof. Following [45], p. 82, we have that (13.4) implies with (13.51)

$$\|X_t\| \overset{I}{\leq} \|X_0\| \, \chi \exp\left(-\tfrac{1}{2}\rho(t-t_0) + \chi \int_{t_0}^{t} \|F_\tau + C(\tau)\| \, \mathrm{d}\tau\right), \qquad (13.59)$$

where $\chi = c_d' a^{d-1}(\rho^{d-1})^{-1}$.

Let $c < \rho/(2\chi)$. Due to conditions 2 and 3 there are $\varepsilon > 0$ and $t^* > t_0$ such that

$$\|F_t + C(t)\| \overset{[t^*,\infty)}{<} \frac{\rho}{2\chi} - \varepsilon\,,$$

such that

$$\mathbb{E}\left(\sup_{t\geq t_0} \|X_t\|^r\right) < \infty\,, \qquad (13.60)$$

and

$$\lim_{t\to\infty} \|X_t\| \mathrel{\hat{=}} 0 \qquad (13.61)$$

hold.

Again, (13.60) and (13.61) imply the assertion due to Lebesgue's convergence theorem. □

Tighter bounds for $\|F_t\|$ may eventually be obtained with the methods of Theorem 13.5.

13.4.2 Periodic & Stationary Solutions

Let us study the inhomogeneous random differential equation

$$\frac{\mathrm{d}X_t}{\mathrm{d}t} = A_t X_t + Z_t\,, \qquad (13.62)$$

where A_t is a strictly θ-periodic matrix process and Z_t is a second order strictly θ-periodic vector process that is independent of A_t such that

$$\sup_{t\in[t_0,t_0+\theta]} \mathbb{E}\left(\|Z_t\|^2\right) < \chi\,.$$

We call the path-wise solution of (13.62) *asymptotically θ-periodic*, if

$$\lim_{t\to\infty} \left(m_X(t+\theta) - m_X(t)\right) = 0 \qquad (13.63)$$

and if

$$\lim_{t,s\to\infty} \left(C_X(t+\theta, s+\theta) - C_X(t,s)\right) \;=\; 0 \tag{13.64}$$

hold. For comparison reasons we make use of the homogeneous equation

$$\frac{\mathrm{d}Y_t}{\mathrm{d}t} \;=\; A_t Y_t\,. \tag{13.65}$$

Theorem 13.10 (Asymptotic θ-Periodicity). *Let $I = \mathbb{R}$. Assume that*

$$\mathbb{E}\left(\|Y_t\|^2\right) \;\leq\; \alpha \mathbb{E}\left(\|Y_0\|^2\right) \exp\left(-\rho(t-t_0)\right) \tag{13.66}$$

holds for every path-wise solution of the homogeneous equation (13.65), and for every initial condition (Y_0, τ) with positive constants α and ρ. Moreover, let $X_0 \in L_d^2$ be independent from the process (A_t, Z_t). Then every path-wise solution of (13.62) with respect to the initial condition $(X_0, t_0) \in L_d^2 \times I$ is asymptotically θ-periodic.

Proof. Following [45], pp. 84, we first show that there is a positive constant c such that

$$\mathbb{E}\left(\|X_t\|^2\right) \;\leq\; c\,, \quad t \in I\,. \tag{13.67}$$

Before you continue, make sure to answer the following questions:

Quiz: Section 13.4 – Part I (Exponential Decay of the Moments)

Q1 Under which condition does the r-th moment $\mathbb{E}\left(\|X_t\|^r\right)$ of the unique path-wise solution of $\dot{X}_t = AX_t$ decay towards the null-solution for $t \to \infty$?

Q2 Sketch the proof of the theorem you used in Q1.

Q3 Under which condition does the r-th moment $\mathbb{E}\left(\|X_t\|^r\right)$ of the unique path-wise solution of $\dot{X}_t = AX_t + Z_t$ decay towards the null-solution for $t \to \infty$?

Q4 Sketch the proof of the theorem you used in Q3.

Q5 Under which condition does the r-th moment $\mathbb{E}\left(\|X_t\|^r\right)$ of the unique path-wise solution of $\dot{X}_t = \left(A + F_t + C(t)\right) X_t$ decay towards the null-solution for $t \to \infty$?

Q6 Sketch the proof of the theorem you used in Q5.

It holds that

$$X_t \stackrel{I}{=} Y_t^0 + \int_{t_0}^{t} \Phi_{t,\tau} Z_\tau \mathrm{d}\tau\,, \tag{13.68}$$

where Y_t^0 are path-wise solutions of (13.65) with respect to the initial condition (X_0, t_0) and $\Phi_{t,\tau}$ is defined as in (13.2) by $\Phi_{t,\tau} = \Phi_t \Phi_\tau^{-1}$. Because of (13.65) there are constants β and ρ such that

$$\mathbb{E}\left(\|\Phi_{t,\tau}\|^2\right) \le \beta \exp\left(-\rho(t-\tau)\right)\,. \tag{13.69}$$

(13.68) implies together with (13.66) and (13.69) that

$$\begin{aligned}\mathbb{E}\left(\|X_t\|^2\right) &\le 2\alpha\mathbb{E}\left(\|X_0\|^2\right) \exp\left(-\rho(t-\tau)\right) + 2\mathbb{E}\left(\left\|\int_{t_0}^{t} \Phi_{t,\tau} Z_\tau \mathrm{d}\tau\right\|^2\right) \\ &\le 2\alpha\mathbb{E}\left(\|X_0\|^2\right) + 2\chi \int_{t_0}^{t}\int_{t_0}^{t} \mathbb{E}\left(\|\Phi_{t,\tau}\|\,\|\Phi_{t,s}\|\right) \mathrm{d}\tau \mathrm{d}s \\ &\le 2\alpha\mathbb{E}\left(\|X_0\|^2\right) + 8\chi\beta\rho^{-2}\,.\end{aligned}$$

This shows (13.70).

Due to our conditions, the process (X_0, A_t, Z_t) has the same corresponding distribution as the process $(X_0, A_{t+\theta}, Z_{t+\theta})$. Thus, the process X_t has the same corresponding distribution as the path-wise solution U_t of the random differential equation

$$\frac{\mathrm{d}U_t}{\mathrm{d}t} = A_{t+\theta} U_t + Z_{t+\theta} \tag{13.70}$$

with respect to the initial condition $U_{t_0} = X_0$.

With the notation $W_t = U_{t-\theta} - X_t$ we obtain

$$\frac{\mathrm{d}W_t}{\mathrm{d}t} = A_t W_t\,, \quad W_{t_0+\theta} = X_0 - X_{t_0+\theta}\,.$$

Due to (13.66) and (13.67) it holds that

$$\begin{aligned}\|\mathbb{E}\left(X_{t+\theta}\right) - \mathbb{E}\left(X_t\right)\| = \|\mathbb{E}\left(W_{t+\theta}\right)\| &\le \left(\mathbb{E}\left(\|W_{t+\theta}\|^2\right)\right)^{1/2} \\ &\le \left(\alpha\mathbb{E}\left(\|X_{t_0+\theta} - X_0\|^2\right) \exp\left(-\rho(t-t_0-\theta)\right)\right)^{1/2}\end{aligned} \tag{13.71}$$

which implies (13.63).

Similarly it holds that

$$
\begin{aligned}
&\lim_{t,s\to\infty} \|C_X(t+\theta, s+\theta) - C_X(t,s)\| \\
&= \lim_{t,s\to\infty} \left\| \mathbb{E}\left(X_{t+\theta}X_{s+\theta}^T\right) - \mathbb{E}\left(X_t X_s^T\right) + m_X(t)\left(m_X(s)\right)^T \right. \\
&\quad \left. - m_X(t+\theta)\left(m_X(s+\theta)\right)^T \right\| \\
&\le \lim_{t,s\to\infty} \left(\mathbb{E}\left(\|X_{t+\theta}\|\,\|X_{s+\theta} - U_s\|\right) + \mathbb{E}\left(\|X_{t+\theta} - U_t\|\,\|U_s^T\|\right)\right. \\
&\quad \left. + \|m_X(t)\|\,\|m_X(s) - m_X(s+\theta)\| + \|m_X(s+\theta)\|\,\|m_X(t) - m_X(t+\theta)\|\right) .
\end{aligned}
\tag{13.72}
$$

As $\|m_X(t)\| \le c^{1/2}$, $t \in I$, and due to (13.63), the last two terms in (13.72) tend to zero with increasing values of t. An application of the Cauchy-Schwarz inequality to (13.72) under consideration of (13.67) and (13.71) leads to

$$
\begin{aligned}
&\lim_{t,s\to\infty} \|C_X(t+\theta, s+\theta) - C_X(t,s)\| \\
&\le c^{1/2} \lim_{s\to\infty} \left(\mathbb{E}\left(\|W_{s+\theta}\|^2\right)\right)^{1/2} + c^{1/2} \lim_{s\to\infty} \left(\mathbb{E}\left(\|W_{t+\theta}\|^2\right)\right)^{1/2} = 0 .
\end{aligned}
$$

This shows the assertion. □

Before you continue, make sure to answer the following questions:

Quiz: Section 13.4 – Part II (Periodic & Stationary Solutions)

Q1 Under which conditions is the unique path-wise solution of $\dot{X}_t = A_t X_t + Z_t$ asymptotically θ-periodic?

Q2 Sketch the proof of the assertions you applied in Q1.

13.5 Excursion: The Solution Formula for Linear Non-Commutative Colored Noise Systems

Following [157], pp. 134, we discuss a special case of the homogeneous random differential equation $\dot{X}_t = (A + F_t)\, X_t$ and give the recursive solution formula for linear (non-commutative) path-wise continuous colored (i.e. non-white) noise systems of the form

$$\dot{X}_t = (A + B\xi_t)\, X_t\,, \tag{13.73}$$

where $X_t \in \mathbb{R}^d$, A and B are constant real $d \times d$-matrices, and the perturbation process ξ_t is 1-dimensional, real, path-wise continuous non-white, with vanishing mean $\mathbb{E}(\xi_t) = 0$ and autocorrelation function $\mathbb{C}ov(\xi_t, \xi_s) = \mathbb{E}(\xi_t \xi_s)$. Moreover, we assume that the initial conditions $X_{t_0} = X_0$ and the process ξ_t, $t \geq 0$ are independent, which is clear for deterministic initial conditions.

According to the results from Chap. 3, the colored-noise stochastic differential equation (13.73) has a unique path-wise solution X_t for all initial conditions $X_0 = X_{t_0}$ and all continuous realisations of ξ_t. This solution is given by

$$X_t = \Omega(t) \cdot X_0 , \qquad \text{for } t \geq 0 , \tag{13.74}$$

where $\Omega(t)$ is the matrizant of the linear system (13.73) for the realisation of ξ_t. The matrizant can be calculated via its von Neumann/ Peano-Baker series

$$\Omega(t) = \sum_{k=0}^{\infty} P_k(t) \text{ and } \begin{cases} P_0(t) := \mathbb{I} , \\ P_k(t) := \int_0^t (A + B\xi_s) P_{k-1}(s) \mathrm{d}s , \ k = 1, 2, 3, \dots , \end{cases} \tag{13.75}$$

see, e.g., [41], pp. 19, [102], pp. 125, or [81]. This series is absolutely and uniformly convergent, see [41], pp. 20, or [102], pp. 125. (To show this will be the task of problem 13.20.)

Hence, the solution process X_t is the union of all thus obtained realisations, and as ξ_t is almost surely continuous almost all realisations of X_t are given. The remaining null-set of (discontinuous) realisations can be neglected when considering properties of the expectation $\mathbb{E}(X_t)$.

Lemma 13.9 (Properties of the Picard Iterates). *The recursively defined $d \times d$-matrix valued processes*

$$\begin{aligned} P_0(t) &:= \mathbb{I} , \\ P_k(t) &:= \int_0^t (A + B\xi_s) P_{k-1}(s) ds , \ k = 1, 2, 3, \dots , \end{aligned} \tag{13.76}$$

are given by

$$P_k(t) = \sum_{q=0}^{k} \frac{1}{q!} (B\Xi_t)^q \left(\frac{1}{(k-q)!} (At)^{k-q} + f_{k-q}(t) \right) \quad \forall k = 1, 2, 3, \dots \tag{13.77}$$

such that

$$\begin{aligned} f_{l+1}(t) = &\sum_{p=1}^{l} (AB^p - B^p A) A^{l-p} \int_0^t \frac{1}{p!} \Xi_s^p \frac{1}{(l-p)!} s^{l-p} ds \\ &+ \sum_{p=0}^{l} AB^p \int_0^t \frac{1}{p!} \Xi_s^p f_{l-p}(s) ds \end{aligned}$$

$$- \sum_{p=0}^{l} B^{p+1} \int_0^t \frac{1}{(p+1)!} \Xi_s^{p+1} \left(\frac{d}{ds} f_{l-p}(s) \right) ds \,, \qquad (13.78)$$

where $f_0(t) \equiv 0$ *and* $\Xi_t := \int_0^t \xi_s ds, t \geq 0.$

Proof. Following [157], pp. 134, the proof will utilize complete induction. First, we notice that the assertion holds for $k = 1$, as $P_1(t)$ explicitly reads according to (13.76) as

$$P_1(t) = \int_0^t (A + B\xi_s)\, \mathrm{d}s = At + B\Xi_t \,,$$

and, on the other hand, according to (13.77) we get

$$P_1(t) = \mathbb{I} \cdot (At + f_1(t)) + B\Xi_t \cdot (\mathbb{I} + f_0(t)) = At + B\Xi_t \,,$$

as

$$f_1(t) = A \cdot \int_0^t f_0(s) \mathrm{d}s + B \cdot \int_0^t \Xi_s \left(\frac{\mathrm{d}}{\mathrm{d}s} f_0(s) \right) \mathrm{d}s \,.$$

Next, assume the assertion to be valid for an arbitrary but fixed index k. This allows us to show the assertion for $k + 1$: According to (13.76), we have

$$\begin{aligned} P_{k+1}(t) &= \int_0^t (A + B\xi) \sum_{q=0}^{k} \frac{1}{q!} \left((As)^{k-q} + f_{k-q}(s) \right) \mathrm{d}s \\ &= \sum_{q=0}^{k} AB^q A^{k-q} \int_0^t \frac{1}{q!(k-q)!} \Xi_s^q s^{k-q} \mathrm{d}s \\ &\quad + \sum_{q=0}^{k} B^{q+1} A^{k-q} \int_0^t \frac{1}{q!(k-q)!} \Xi_s^q \xi_s s^{k-q} \mathrm{d}s \\ &\quad + \sum_{q=0}^{k} AB^q \int_0^t \frac{1}{q!} \Xi_s^q f_{h-q}(s) \mathrm{d}s + \sum_{q=0}^{k} B^{q+1} \int_0^t \frac{1}{q!} \Xi_s^q \xi_s f_{h-q}(s) \mathrm{d}s \,. \end{aligned}$$

Integration by parts shows

$$\begin{aligned} \int_0^t \frac{1}{q!(k-q)!} \Xi_s^q \xi_s s^{k-q} \mathrm{d}s &= \frac{1}{(q+1)!(k-q)!} \Xi_t^{q+1} t^{k-q} \\ &\quad - \int_0^t \frac{1}{(q+1)!(k-q-1)!} \Xi_s^{q+1} s^{k-q-1} \mathrm{d}s \end{aligned}$$

for $0 \leq q \leq k-1$ and for $q = k$ as

$$\int_0^t \frac{1}{k!} \Xi_s^k \xi_s \mathrm{d}s = \frac{1}{(k+1)!} \Xi_t^{k+1} \,.$$

because of $\Xi_0 = 0$. Moreover, integration by parts at the fourth sum results, for $0 \le q \le k$ in

$$\int_0^t \frac{1}{q!}\Xi_s^q \xi_s f_{k-q}(s)\mathrm{d}s = \frac{1}{(q+1)!}\Xi_t^{q+1} f_{k-q}(t) - \int_0^t \frac{1}{g+1)!} X_s^{q+1}\left(\frac{\mathrm{d}}{\mathrm{d}s} f_{k-q}(s)\right)\mathrm{d}s\,.$$

This immediately leads to

$$\begin{aligned}
P_{k+1}(t) &= \sum_{q=0}^{k} AB^q A^{k-q} \int_0^t \frac{1}{q!(k-q)!}\Xi_s^q s^{k-q}\mathrm{d}s \\
&\quad + \sum_{q=0}^{k} \frac{1}{(q+1)!(k-q)!}\left(B\Xi_t\right)^{q+1}(At)^{k-q} \\
&\quad - \sum_{q=0}^{k-1} B^{q+1}A^{k-q} \int_0^t \frac{1}{(q+1)!(k-q-1)!}\Xi_s^{q+1} s^{k-q-1}\mathrm{d}s \\
&\quad + \sum_{q=0}^{k} AB^q \int_0^t \frac{1}{q!}\Xi_s^q f_{k-q}(s)\mathrm{d}s + \sum_{q=0}^{k} \frac{1}{(q+1)!}\left(B\Xi_t\right)^{q+1} f_{k-q}(t) \\
&\quad - \sum_{q=0}^{k} B^{q+1} \int_0^t \frac{1}{(g+1)!}\Xi_t^{q+1}\left(\frac{\mathrm{d}}{\mathrm{d}s}\right)\mathrm{d}s\,.
\end{aligned}$$

Performing the index shift $q+1 \to q$ at the second, third and fifth sum, we obtain

$$\begin{aligned}
P_{k+1}(t) &= \sum_{q=0}^{k} AB^q A^{k-q} \int_0^t \frac{1}{q!(k-q)!}\Xi_s^q s^{k-q}\mathrm{d}s \\
&\quad + \sum_{q=1}^{k+1} \frac{1}{q!(k+1-q)!}\left(B\Xi_t\right)^{q}(At)^{k+1-q} \\
&\quad - \sum_{q=1}^{k-1} B^{q}A^{k+1-q} \int_0^t \frac{1}{q!(k-q)!}\Xi_s^{q} s^{k-q}\mathrm{d}s \\
&\quad + \sum_{q=0}^{k} AB^q \int_0^t \frac{1}{q!}\Xi_s^q f_{k-q}(s)\mathrm{d}s + \sum_{q=1}^{k+1} \frac{1}{q!}\left(B\Xi_t\right)^{q} f_{k+1-q}(t) \\
&\quad - \sum_{q=0}^{k} B^{q+1} \int_0^t \frac{1}{(g+1)!}\Xi_t^{q+1}\left(\frac{\mathrm{d}}{\mathrm{d}s}\right)\mathrm{d}s\,.
\end{aligned}$$

Here, at the first sum for $q = 0$ we have

$$A^{k+1}\int_0^t \frac{1}{k!}s^k\mathrm{d}s = \frac{1}{(k+1)!}(At)^{k+1}\,.$$

This expression can be added as the $q = 0$ term to the second sum. Furthermore, when combining the remaining summands of the first sum with the third, fourth and sixth sum, we get for

$$
\begin{aligned}
f_{k+1}(t) &= \sum_{p=1}^{k} (AB^p - B^p A)\, A^{k-p} \int_0^t \frac{1}{p!}\, \Xi_s^p \, \frac{1}{(k-p)!}\, s^{k-p}\, \mathrm{d}s \\
&\quad + \sum_{p=0}^{k} AB^p \int_0^t \frac{1}{p!}\, \Xi_s^p\, f_{k-p}(s) \mathrm{d}s \\
&\quad - \sum_{p=0}^{k} B^{p+1} \int_0^t \frac{1}{(p+1)!}\, \Xi_s^{p+1} \left(\frac{\mathrm{d}}{\mathrm{d}s} f_{k-p}(s) \right) \mathrm{d}s\,,
\end{aligned}
$$

which is (13.78) for $l = k$ and together with the enlarged second and the fifth sum we finally have

$$
P_{k+1}(t) = \sum_{q=0}^{k+1} \frac{1}{q!}\, (B\Xi_t)^q \left(\frac{1}{(k+1-q)!}\, (At)^{k+1-q} + f_{k+1-q}(t) \right).
$$

This proves the assertion. □

Applying Cauchy's product formula for series, (13.75) and (13.77) immediately lead to

$$
\begin{aligned}
\Omega(t) &= \sum_{k=0}^{\infty} \sum_{q=0}^{k} \frac{1}{q!}\, (B\Xi_t)^q \left(\frac{1}{(k-q)!}\, (At)^{k-q} + f_{k-q}(t) \right) \\
&= \sum_{j=0}^{\infty} \frac{1}{j!}\, (B\Xi_t)^j \left(\sum_{k=0}^{\infty} \frac{1}{k!}\, (At)^k + \sum_{k=0}^{\infty} f_l(t) \right),
\end{aligned}
$$

i.e.,

$$
\Omega(t) = \exp(B\Xi_t) \left(\exp(At) + \sum_{k=0}^{\infty} f_l(t) \right).
$$

Finally, because of (13.74), the solution process X_t of (13.73) reads as

$$
X_t = \exp(B\Xi_t) \left(\exp(At) + \sum_{k=0}^{\infty} f_l(t) \right) \cdot X_0\,, \qquad t \geq 0\,.
$$

We gain a clearer picture of the actual solution process by collecting the terms in $\sum_{k=0}^{\infty} f_l(t)$ of mth order in Ξ_t and B, respectively. The corresponding new terms $F_m(t)$ obey a certain ordinary differential equation as shown in the following lemma.

Lemma 13.10 (Recursive Formula for $F_m(t)$). *The $d \times d$-matrix valued processes $F_m(t)$ that collect the terms in $\sum_{k=0}^{\infty} f_l(t)$ of mth order in Ξ_t and B, respectively, obey the following recursive system of ordinary differential equations*

$$\dot{F}_m(t) = AF_m + \sum_{q=0}^{m-1} \frac{1}{(m-q)!}\Xi_t^{m-q}\left(AB^{m-q}F_q - B^{m-q}\dot{F}_q\right) \tag{13.79}$$

where $F_m(0) = 0$ for all $m = 1, 2, 3, \ldots$, and $F_0(t) = \exp(At)$.

Proof. First of all, in each process $f_l(t)$, we group all those terms together to a process $f_l^{(m)}(t)$ in which Ξ_t and B respectively occur in the mth order ($l = 1, 2, 3, \ldots, m = 1, 2, 3, \ldots$). To this end, we formally obtain, according to (13.78), after the index shift $j - k \to k$

$$\begin{aligned} f_{j+1}^{(m)}(t) &= (AB^m - B^m A)\, A^{j-m} \int_0^t \frac{1}{m!(j-m)!}\Xi_s^m s^{j-m} \mathrm{d}s \\ &\quad + \sum_{k=0}^{j} AB^{j-k} \int_0^t \frac{1}{(j-k)!}\Xi_s^{j-k} f_k^{(m-j+k)}(s)\mathrm{d}s \\ &\quad - \sum_{k=0}^{j} B^{j-k+1} \int_0^t \frac{1}{(j-k+1)!}\Xi_s^{j-k+1}\left(\frac{\mathrm{d}}{\mathrm{d}s} f_k^{(m-j+k-1)}(s)\right)\mathrm{d}s\,. \end{aligned}$$

Hereby, the first term will not occur as long as $j < m$. Moreover, the middle sum will not occur for $j \le m$ and the last sum not as along as $j \le m-1$ holds. The successive evaluation of (13.78) leads at least to

$$f_{j+1}^{(p)} \equiv 0 \qquad \text{for } p \le 0 \text{ and for } p \ge j+1\,. \tag{13.80}$$

And because of $f_0(t) \equiv 0$ we get

$$F_m(t) = \sum_{l=0}^{\infty} f_l^{(m)}(t) = \sum_{j=0}^{\infty} f_{j+1}^{(m)}(t)\,,$$

and thus

$$\begin{aligned} F_m(t) &= \sum_{j=m}^{\infty} (AB^m - B^m A)\, A^{j-m} \int_0^t \frac{\Xi_s^m}{m!(j-m)!} s^{j-m}\mathrm{d}s \\ &\quad + \sum_{j=m+1}^{\infty} \sum_{k=j-m+1}^{j} AB^{j-k} \int_0^t \frac{\Xi_s^{j-k}}{(j-k)!} f_k^{(m-j+k)}(s)\mathrm{d}s \\ &\quad - \sum_{j=m}^{\infty} \sum_{k=j-m+2}^{j} B^{j-k+1} \int_0^t \frac{\Xi_s^{j-k+1}}{(j-k+1)!}\left(\frac{\mathrm{d}}{\mathrm{d}s} f_k^{(m-j+k-1)}(s)\right)\mathrm{d}s\,, \end{aligned} \tag{13.81}$$

for $m = 1, 2, 3, \ldots$. Whereby the last sum does not occur for $m = 1$.

Due to the absolute and uniform convergence of the matrizant series (13.75), cf. [41], pp. 20, [102], pp. 125, or problem 13.20, the following modifications of (13.81) are justified: First, (13.81) becomes in the case $m = 1$

$$F_1(t) = \sum_{j=1}^{\infty} (AB - BA)\, A^{j-1} \int_0^t \frac{1}{(j-1)!} \Xi_s s^{j-1} \mathrm{d}s + \sum_{j=2}^{\infty} A \int_0^t f_j^{(1)}(s) \mathrm{d}s \,.$$

Because of $f_0(t) \equiv 0$ and $f_1(t) \equiv 0$, this leads after the index shift $j - 1 \to j$ in the first sum to

$$\begin{aligned} F_1(t) &= A \int_0^t \sum_{j=0}^{\infty} f_j^{(1)}(s) \mathrm{d}s + (AB - BA) \int_0^t \Xi_s \sum_{j=0}^{\infty} \frac{1}{j!} (As)^j \, \mathrm{d}s \\ &= A \int_0^t F_1(s) \mathrm{d}s + (AB - BA) \int_0^t \Xi_s \exp(As) \mathrm{d}s \,. \end{aligned}$$

Differentiation immediately leads to

$$\dot{F}_1(t) = AF_1(t) + \Xi_t \,(AB - BA) \exp(At)$$

and we have $F_1(0) = 0$. Together with $F_0(t) = \exp(At)$ the assertion of our lemma holds for $m = 1$.

Next, let $m \geq 2$ be arbitrary but fixed. Then, for $q = j - m$ and $h = j - m - 1$ (13.81) reads as

$$\begin{aligned} F_m(t) &= \sum_{q=0}^{\infty} (AB^m - B^m A) \int_0^t \frac{1}{m!q!} \Xi_s^m \, (As)^q \, \mathrm{d}s \\ &\quad + \sum_{h=0}^{\infty} \sum_{k=h+2}^{h+m+1} AB^{h+m+1-k} \int_0^t \frac{\Xi_s^{h+m+1-k}}{(h+m+1-k)!} f^{(k-h-1)}(s) \mathrm{d}s \\ &\quad - \sum_{q=0}^{\infty} \sum_{k=g+2}^{q+m} B^{q+m+1-k} \int_0^t \frac{\Xi_s^{q+m+1-k}}{(g+m+1-k)!} \left(\frac{\mathrm{d}}{\mathrm{d}s} f_k^{(k-q-1)}(s) \right) \mathrm{d}s \,. \end{aligned}$$

When considering the case $k = h + m + 1$ separately in the second sum, then this can be re-grouped after the index shift $h \to q$ and later on $j = k - q - 1$ to

$$\begin{aligned} F_m(t) &= (AB^m - B^m A) \int_0^t \frac{1}{m!} \Xi_s^m \exp(As) \mathrm{d}s + A \int_0^t \sum_{q=0}^{\infty} f_{q+m+1}^{(m)}(s) \mathrm{d}s \\ &\quad + \sum_{q=0}^{\infty} \sum_{j=1}^{m-1} \int_0^t \frac{\Xi_s^{m-j}}{(m-j)!} \left(AB^{m-1} f_{j+q+1}^{(j)} - B^{m-j} \left(\frac{\mathrm{d}}{\mathrm{d}s} f_{j+q+1}^{(j)} \right) \right) \mathrm{d}s \,. \end{aligned}$$

According to (13.80) and together with $F_0(t) = \exp(At)$ and $f_{l+1}^{(j)} \equiv 0$ for $j \geq l+1$ this leads to

$$
\begin{aligned}
F_m(t) &= (AB^m - B^m A)\int_0^t \frac{1}{m!}\Xi_s^m \exp(As)\mathrm{d}s + A\int_0^t \sum_{l=0}^{\infty} f_l^{(m)}(s)\mathrm{d}s \\
&\quad + \sum_{j=1}^{m-1}\int_0^t \frac{\Xi_s^{m-j}}{(m-j)!}\left(AB^{m-1}\sum_{l=0}^{\infty} f_l^{(j)} - B^{m-j}\left(\frac{\mathrm{d}}{\mathrm{d}s}\sum_{l=0}^{\infty} f_l^{(j)}\right)\right)\mathrm{d}s\,,
\end{aligned}
$$

and hence

$$
\dot{F}_m(t) = AF_m(t) + \sum_{q=0}^{m-1}\frac{1}{(m-q)!}\Xi_t^{m-q}\left(AB^{m-q}F_q(t) - B^{m-q}\dot{F}_q(t)\right),
$$

and $F_m(0) = 0$ holds.

Thus, the assertion of this lemma is true for arbitrary $m \geq 2$. □

The classical deterministic theory of systems of inhomogeneous differential equations, see Chap. 6 or [41], immediately allows us to see that the solutions of (13.79) are given by

$$
F_m(t) = \exp(At)\cdot\Gamma_m(t)\,, \qquad m = 0,1,2,\ldots.
$$

In particular, we can show the following calculation scheme for the processes $\Gamma_m(t)$.

Lemma 13.11 (The General Solution Formula for $F_m(t)$). *The $d\times d$-matrix valued processes $\Gamma_m(t)$ in the expressions*

$$
F_m(t) = \exp(At)\cdot\Gamma_m(t)\,, \qquad m = 0,1,2,\ldots, \tag{13.82}
$$

are recursively given as

$$
\Gamma_m(t) = \sum_{q=0}^{m-1}\int_0^t \frac{1}{(m-q)!}\Xi_s^{m-q}L_{m-q}(s)\Gamma_q(s)ds\,, \tag{13.83}
$$

for $m = 1,2,3,\ldots$ and $G_0(t) = \mathbb{I}$, where

$$
L_k(s) = \exp(-At)K_k\exp(At) \quad \text{and} \quad K_k = K_{k-1}B - BK_{k-1}\,, \tag{13.84}
$$

for $k = 1,2,3,\ldots$ and $K_0 = A$.

The kth order commutators K_k of the matrices A and B are directly given as

$$
K_k = \sum_{q=0}^{k}\binom{k}{q}(-1)^q B^q A B^{k-q}\,, \tag{13.85}
$$

for $k = 1,2,3,\ldots$.

Proof. We start by first applying complete induction to show the second part of the lemma concerning the explicit representation of K_k: For $k = 1$ equation (13.85) holds as it leads to $K_1 = AB - BA$ just like the defining equation (13.84). Next, assume (13.85) to be valid for an arbitrary but fixed $k \in \mathbb{N}$. We will show that it then holds for $k + 1$.

Together with this assumption, the defining equation (13.84) reads for $k+1$ as

$$K_{k+1} = \sum_{q=0}^{k} \binom{k}{q} (-1)^q B^q A B^{k-q+1} - \sum_{q=0}^{k} \binom{k}{q} (-1)^q B^{q+1} A B^{k-q} .$$

The index shift $q + 1 \to q$ at the second sum leads to

$$K_{k+1} = AB^{k+1} + \sum_{q=1}^{k} \left(\binom{k}{q} + \binom{k}{q-1} \right) (-1)^q B^q A B^{k-q} + (-1)^{k+1} B^{k-1} A .$$

As $\binom{k}{q} + \binom{k}{q-1} = \binom{k+1}{q}$ holds for all $1 \leq q \leq k$, we obtain

$$K_k = \sum_{q=0}^{k} \binom{k}{q} (-1)^q B^q A B^{k-q} ,$$

which proves equation (13.85).

Consider now the first part of the lemma: According to Lemma 13.10 we have

$$F_m(t) = \exp(At) \int_0^t \exp(-At) \sum_{q=0}^{m} \frac{\Xi_s^{m-q}}{(m-q)!} \left(AB^{m-q} F_q(s) - B^{m-q} \dot{F}_q(s) \right) \mathrm{d}s ,$$

for $m = 1, 2, 3, \ldots$ because of $F_m(0) = 0$. Plugging (13.82) as well as $\dot{F}_m(t) = A \exp(At) \Gamma_m(t) + \exp(At) \dot{\Gamma}_m(t)$ into this identity results in

$$\begin{cases} \Gamma_m(t) = \int_0^t J_m(s) \mathrm{d}s \\ J_m(s) = \sum_{q=0}^{m-1} \frac{\Xi_s^{m-q}}{(m-q)!} \exp(-As) \big((AB^{m-q} - B^{m-q} A) \exp(As) \Gamma_q(s) \\ \qquad - B^{m-q} \exp(As) \dot{\Gamma}_q(s) \big) \end{cases} \tag{13.86}$$

for $m = 1, 2, 3, \ldots$, whereby $F_0(t) = \exp(At)$ and $\Gamma_0(t) = \mathbb{I}$ correspond according to (13.82).

The remaining parts of the proof will be carried out via complete induction. First, the assertion holds for $m = 1$, because (13.86) leads to the integrand function

$$\begin{aligned} J_1(s) &= \Xi_s \exp(-As)\,(AB - BA)\exp(As)\Gamma_0(s) \\ &= \Xi_s \exp(-As)K_1 \exp(As)\Gamma_0(s)\Xi_s L_1(s) \;=\; \Gamma_0(s)\,. \end{aligned}$$

Hence, (13.86) and (13.83) both result in

$$\Gamma_1(t) \;=\; \int_0^t \Xi_s L_1(s)\Gamma_0(s)\mathrm{d}s\,.$$

Next, let the assertion be true for an arbitrary but fixed index m. We will show that it holds for $m+1$ as well.

The validity of our assertion for m leads immediately to the ordinary differential equation

$$\dot{\Gamma}_q(s) \;=\; \sum_{k=0}^{q-1} \frac{1}{(q-k)!}\Xi_s^{q-k} \exp(-As)K_{q-k}\exp(As)\Gamma_k(s)\,,$$

for $q = 1, 2, 3, \ldots$. Thus, the integrand $J_{m+1}(s)$ for $G_{m+1}(s)$ reads, according to (13.86), as

$$\begin{aligned} J_{m+1}(s) \;&= \sum_{q=0}^{m} \frac{\Xi_s^{m+1-q}}{(m+1-q)!} \exp(-As)\left(AB^{m+1-q} - B^{m+1-q}A\right)\exp(As)\Gamma_q(s) \\ &- \sum_{q=0}^{m}\sum_{k=0}^{q-1} \frac{\Xi_s^{m+1-k}}{(m+1-q)!(q-k)!} \exp(-As)B^{m+1-q}K_{q-k}\exp(As)\Gamma_k(s)\,, \end{aligned}$$

because of $\dot{\Gamma}_0(s) \equiv 0$. When splitting the first sum into

$$\begin{aligned} &\sum_{q=0}^{m} \frac{\Xi_s^{m+1-q}}{(m+1-q)!} \exp(-As)AB^{m+1-q}\exp(As)\Gamma_q(s) \\ &\quad - \sum_{q=0}^{m} \frac{\Xi_s^{m+1-q}}{(m+1-q)!} \exp(-As)B^{m+1-q}A\exp(As)\Gamma_q(s) \end{aligned}$$

the second part completes the double sum for $k = q$ and $q = 0, 1, 2, \ldots, m$, because of $K_0 = A$, such that

$$\begin{aligned} J_{m+1}(s) \;&= \sum_{q=0}^{m} \frac{\Xi_s^{m+1-q}}{(m+1-q)!} \exp(-As)AB^{m+1-q}\exp(As)\Gamma_q(s) \\ &- \sum_{q=0}^{m}\sum_{k=0}^{q} \frac{\Xi_s^{m+1-k}}{(m+1-q)!(q-k)!} \exp(-As)B^{m+1-q}K_{q-k}\exp(As)\Gamma_k(s)\,. \end{aligned} \qquad (13.87)$$

Now, the summation in the double sum can be carried out either by summing k from 0 to q for fixed $q = 0, 1, \ldots, m$ or by summing q from k to m for fixed $k = 0, 1, \ldots, m$. For a general coefficient sequence $\{\rho_{qk}\}$ that obeys the conditions of the re-summation theorem

$$\sum_{q=0}^{m}\sum_{k=0}^{q}\rho_{qk} = \sum_{k=0}^{m}\sum_{q=k}^{m}\rho_{qk}\,, \qquad \text{for } m = 1, 2, 3, \ldots . \tag{13.88}$$

Recall that we are dealing with the matrizant series (13.75), which is absolutely and uniformly convergent, cf. [41], pp. 20, or [102], pp. 125. From (13.87) we gain with the index change $q \leftrightarrow k$ in the double sum the integrand function $J_{m+1}(s)$ as

$$\begin{aligned} J_{m+1}(s) &= \sum_{q=0}^{m}\frac{\Xi_s^{m+1-q}}{(m+1-q)!}\exp(-As)\cdot[AB^{m+1-q} \\ &\quad -\sum_{k=q}^{m}\frac{(m+1-q)!}{(m+1-k)!(k-q)!}B^{m+1-k}K_{k-q}]\cdot\exp(As)\Gamma_q(s)\,. \end{aligned}$$

This difference term within the squared brackets can be re-written as

$$\begin{aligned} &AB^{m+1-q}-\sum_{k=q}^{m}\frac{(m+1-q)!}{(m+1-k)!(k-q)!}B^{m+1-k}K_{k-q} \\ &= AB^{p}+\sum_{k=0}^{p-1}\sum_{j=0}^{k}\frac{p!}{(p-k)!(k-j)!}(-1)^{k+1-j}B^{p-j}AB^{j} \end{aligned}$$

by applying $p := m+1-q$, the index shift $k-g \to k$ and the modification

$$K_k = \sum_{j=0}^{k}\binom{k}{j}(-1)^{k-j}B^{k-j}B^{k-j}AB^{j}\,, \qquad \text{for } k = 1, 2, 3, \ldots \tag{13.89}$$

of the previously proven formula (13.85). Moreover, the interchange of the summation order according to (13.88) leads to

$$\begin{aligned} &AB^{p}+\sum_{k=0}^{p-1}\sum_{j=0}^{k}\frac{p!}{(p-k)!(k-j)!}(-1)^{k+1-j}B^{p-j}AB^{j} \\ &= AB^{p}+\sum_{k=0}^{p-1}\frac{p!}{j!}\left(\sum_{k=j}^{p-1}(-1)^{p+1-k}\frac{1}{(p-k)!(k-j)!}\right)(-1)^{p-j}B^{p-j}AB^{j}\,. \end{aligned}$$

Hereby, the inner k-sum becomes with the index shift $p-k \to k$

$$\begin{aligned}
\sum_{k=j}^{p-1}(-1)^{p+1-k}\frac{1}{(p-k)!(k-j)!} &= \frac{1}{(p-j)!}\sum_{k=1}^{p-j}(-1)^k\frac{(p-j)!}{k!(p-j-k)!} \\
&= \frac{1}{(p-j)}\left(1-\sum_{k=0}^{p-j}(-1)^k\frac{(p-j)!}{k!(p-j-k)!}\right) \\
&= ,\frac{1}{(p-j)!}\,.
\end{aligned}$$

Next, the general binomial formula $(a+b)^n$ reads with $a=1$, $b=-1$ and $n=p-j$ as

$$0 = \sum_{k=0}^{p-j}(-1)^k\frac{(p-j)!}{k!(p-j-k)!}\,,$$

which leads according to (13.89) to

$$\begin{aligned}
&AB^p+\sum_{k=0}^{p-1}\frac{p!}{j!}\left(\sum_{k=j}^{p-1}(-1)^{p+1-k}\frac{1}{(p-k)!(k-j)!}\right)(-1)^{p-j}B^{p-j}AB^j \\
&= AB^p+\sum_{j=0}^{p-1}\frac{p!}{j!(p-j)!}(-1)^{p-j}B^{p-j}AB^j = \sum_{j=0}^{p}\binom{p}{j}(-1)^{p-j}B^{p-j}AB^j \\
&= K_p\,,
\end{aligned}$$

and thus, because of $p=m+1+q$

$$J_{m+1}(s) = \sum_{q=0}^{m}\frac{1}{(m+1-q)!}\Xi_t^{m+1-q}\exp(-As)K_{m+1-q}\exp(As)\Gamma_q(s)\,.$$

According to (13.86) we finally obtain

$$\Gamma_{m+1}(t) = \sum_{q=0}^{m}\int_0^t\frac{1}{(m+1-q)!}\Xi_t^{m+1-q}\exp(-As)K_{m+1-q}\exp(As)\Gamma_q(s)\mathrm{d}s\,,$$

which completes the proof. □

Altogether, the convergence properties of the matrizant series imply that

$$\sum_{l=0}^{\infty}f_l(t) = \sum_{m=1}^{\infty}F_m(t) = \exp(At)\cdot\sum_{m=1}^{\infty}G_m(t)\,.$$

Moreover, $F_0(t)=\exp(At)$ and $G_0(t)\equiv\mathbb{I}$. Thus, we immediately have that the solution process X_t of (13.73) is given as

$$X_t = \exp(B\Xi_t)\cdot\exp(At)\cdot\sum_{m=0}^{\infty}G_m(t)\cdot X_0\,,\qquad t\geq 0\,. \tag{13.90}$$

Before you continue, make sure to answer the following questions:

Quiz: Section 13.5

Q1 Give the general formula for the solution $\dot{X}_t = (A + B\xi_t)\,X_t$, where A and B are constant real quadratic matrices and ξ_t is a Gaussian process with vanishing mean. Moreover, assume that you only use initial values for this equation that are independent from ξ_t.

Q2 Sketch the intermediate steps that are necessary to arrive at the formula given in Q1.

Q3 Consider the random differential equation

$$\dot{X}_t = \begin{pmatrix} 3 - 4W_t & 2 + W_t \\ 1 & -2 + W_t \end{pmatrix} X_t ,$$

where W_t is the 1-dimensional standard Wiener process. Which form does the solution of this equation take by application of the result from Q1?

13.6 Chapter's Summary

Analogously to the deterministic setting from Chap. 6, we gave the general solution formulas for linear random differential equations with stochastic coefficients $\dot{X}_t = A_t X_t + Z_t$, in particular if $A_t = A(\omega)$ is a random matrix, and an equivalence result for path-wise and mean-square solutions of $\dot{X}_t = A_t X_t + Z_t$ as a special case of our considerations in Chap. 4.

We provided results on the asymptotic properties of path-wise solutions with a focus on (exponential) decay towards the null-solution or homogeneous systems of the types $\dot{X}_t = (A + F_t)\,X_t$ and $\dot{X}_t = (A + F_t + C(t))\,X_t$, as well as on upper bounds for path-wise solutions of $\dot{X}_t = (A + F_t + C(t))\,X_t + Z_t$.

Next, instead of staying with the path-wise solutions themselves we switched to the properties of their moments by studying the (exponential) decay of the r-th moments for homogeneous and inhomogeneous linear random differential equations with stochastic coefficients of the types $\dot{X}_t = AX_t$, $\dot{X}_t = AX_t + Z_t$, and $\dot{X}_t = (A + F_t + C(t))\,X_t + Z_t$. Moreover, we analysed the existence of asymptotically θ-periodic solutions.

Finally, the general representation

$$X_t = \exp(B\Xi_t) \cdot \exp(At) \cdot \sum_{m=0}^{\infty} G_m(t) X_0$$

of the solution X_t of the linear non-commutative path-wise continuous noise system $\dot{X}_t = (A + B\xi_t) X_t$ was constructed with the aid of deterministic matrizant methods.

Problems

Classification: ☼ easy, ⌚ easy with longer calculations, ✩ a little bit difficult, ◻ challenging.

Exercise 13.12. [☼] **Linearized Monod Kinetics**
As motivated in example 13.1 let us consider the non-linear random system

$$\dot{x}(t,\omega) = \frac{k_x(\omega)y(t,\omega)}{K(\omega)+y(t,\omega)} x(t,\omega), \quad \text{and} \quad \dot{y}(t,\omega) = \frac{k_y(\omega)y(t,\omega)}{K(\omega)+y(t,\omega)} x(t,\omega),$$

where $K \sim U(0.28, 0.36)$, $k_x \sim U(0.04, 0.06)$ and $k_y \sim U(-0.325, 0.3)$.

1. Determine the linear systems by performing a formal first order Taylor approximation around the equilibria of the corresponding deterministic system.

2. Give the explicit solution of this system and determine the mode of convergence/ divergence towards the null-solution. Can you give good upper bounds for the solutions of this system?

Exercise 13.13. [☼] **Mathematical Theories of War – Part 1, cf. [38], p. 380**
According to Richardson's theory of conflict, cf. [38], pp. 374, let x denote the war potential, or armaments, of the first nation or alliance A and y that of a second nation or alliance B. The rate of change $\dot{x}$ is positive proportional to the war readiness of y and on the grievances g felt towards B, and negative proportional to the cost of armament, represented by a term $-\alpha x$. For A this leads to the differential equation

$$\dot{x} = ky - \alpha x + g.$$

A similar differential equation for y is set-up immediately.

Suppose that what moves a government to arm is not the magnitude of armaments of the other nation, but the difference between its own and theirs. Then, their armament dynamics read as

$$\begin{aligned} \dot{X}_t &= k(\omega)\left(Y_t - X_t\right) - \alpha(\omega)X_t + g(\omega)\,, \\ \dot{Y}_t &= l(\omega)\left(Y_t - X_t\right) - \beta(\omega)X_t + h(\omega)\,, \end{aligned}$$

where $k(\omega)$, $l(\omega)$, $\alpha(\omega)$, $\beta(\omega)$, $g(\omega)$ and $h(\omega)$ are suitable random variables.

1. Interpret the model, i.e., give a meaning to the coefficient functions.
2. Assume that the coefficients are non-random positive numbers. Show that every solution of this system of deterministic equations is stable, if $kl < (\alpha + k)(\beta + l)$, and unstable if $kl > (\alpha + k)(\beta + l)$.
3. Give the general form of the solution of the randomized model and study the boundedness of the solutions and their decay towards the null-solution.

Exercise 13.14. [☼] **Mathematical Theories of War – Part 2**
As a continuation of problem 13.13 assume that the costs of armament depend on economic cycles and are θ-periodic functions:

$$\begin{aligned} \dot{X}_t &= k(\omega)\left(Y_t - X_t\right) - \alpha_t(\omega)X_t + g(\omega)\,, \\ \dot{Y}_t &= l(\omega)\left(Y_t - X_t\right) - \beta_t(\omega)X_t + h(\omega)\,. \end{aligned}$$

What can you say about the boundedness of the solutions, their decay towards the null-solution and their (asymptotic) θ-periodicity?

Exercise 13.15. [☼] **Mathematical Theories of War – Part 3, cf. [38], pp. 380**
Consider the case of three nations, each having the same defense coefficient $k(\omega)$ and the same restraint coefficient $\alpha_t(\omega)$. Then, their armament dynamics read as

$$\begin{aligned} \dot{X}_t &= -\alpha_t(\omega)X_t + k(\omega)\left(Y_t + Z_t\right) + g_1(\omega)\,, \\ \dot{Y}_t &= -\alpha_t(\omega)Y_t + k(\omega)\left(X_t + Z_t\right) + g_2(\omega)\,, \\ \dot{Z}_t &= -\alpha_t(\omega)Z_t + k(\omega)\left(X_t + Y_t\right) + g_3(\omega)\,, \end{aligned}$$

where $\alpha_t(\omega)$ is a suitable stochastic process and $k(\omega)$, $g_1(\omega)$, $g_2(\omega)$, $g_3(\omega)$ are suitable random variables.

1. As a starting point assume that all coefficients are positive non-random constants, and that the armament dynamics are given in the matrix-vector notation $\dot{x} = Ax + g$.

- Show that the characteristic polynomial $\chi_A(\lambda)$ vanishes for $\lambda = -\alpha - k$, and use this information to find its remaining two roots.
- Show that every solution is stable if $2k < \alpha$, and unstable if $2k > \alpha$.

2. Give the general form of the solution of the randomized model and study the boundedness of the solutions and their decay towards the null-solution.

Exercise 13.16. [☼] **Mathematical Theories of War – Part 4, cf. [38], pp. 381**
As a continuation of problem 13.15 assume that the third nation with war potential Z_t is a pacifist nation, while the other two nations are pugnacious. Then, their armament dynamics are given as

$$\begin{aligned}
\dot{X}_t &= -\alpha_t(\omega)X_t + k(\omega)\left(Y_t + Z_t\right) + g_1(\omega)\,,\\
\dot{Y}_t &= -\alpha_t(\omega)Y_t + k(\omega)\left(X_t + Z_t\right) + g_2(\omega)\,,\\
\dot{Z}_t &= -\alpha_t(\omega)Z_t + 0\left(X_t + Y_t\right) + g_3(\omega)\,,
\end{aligned}$$

where $\alpha_t(\omega)$ is a suitable path-wise continuous stochastic process and $k(\omega)$, $g_1(\omega)$, $g_2(\omega)$, $g_3(\omega)$ are suitable random variables.

1. As a starting point assume again that all coefficients are positive non-random constants. Show that every solution is stable if $k < \alpha$, and unstable if $k > \alpha$.

2. Give the general form of the solution of the randomized model and study the boundedness of the solutions and their decay towards the null-solution.

Exercise 13.17. [☼] **Details of the Proof of Theorem 13.5**
In the proof of Theorem 13.5 we made use of the stochastic fraction $\dot{v}_t(X_t)/v(X_t)$. Show that this is well defined.

Exercise 13.18. [☼] **Randomisation of Kendall's Mathematical Marriage Problem, cf. [167], p. 313**
Let F denote the number of unmarried females, M the number of unmarried males and C the number of couples in a community. We consider the following version of Kendall's mathematical marriage dynamics, cf. [154]:

$$\begin{aligned}
\dot{F}_t &= -\left(\delta(\omega) + \rho_t(\omega)\right)F_t + \left(\beta(\omega) + \delta(\omega)\right)C_t\,,\\
\dot{M}_t &= -\delta(\omega)M + \left(\beta(\omega) + \delta(\omega)\right)C_t - \rho_t(\omega)F\,,\\
\dot{C}_t &= \rho_t(\omega)F - 2\delta(\omega)C\,,
\end{aligned}$$

where the randomized death (δ) and birth (β) rates are equal for females and males, and $\rho_t(\omega)$ describes the time-dependent rate of changes from single persons to married couples.

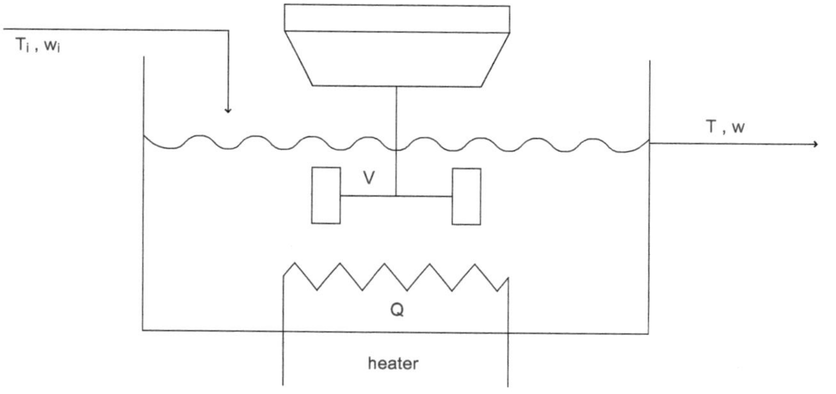

Figure 13.4. Sketch of an electrically heated stirred tank as discussed in problem 13.19.

1. As a starting point assume that all coefficients are positive non-random constants, and discuss the deterministic dynamics.
2. Give the general form of the solution of the randomized model and study the boundedness of the solutions and their decay towards the null-solution.
3. Assume that $\rho_t(\omega)$ is θ-periodic. What can you say about the boundedness of the solutions, their decay towards the null-solution and their (asymptotic) θ-periodicity in this case?

Exercise 13.19. [☼] **Electrically Heated Stirred Tank**
Consider a stirred-bank heating system as shown in Fig 13.4 with a mass inflow rate w_i and an inlet temperature T_i. Let T be the temperature of the tank contents and T_e be the temperature of the heating element. The dynamics of the these temperatures are

$$
\begin{aligned}
mC\dot{T} &= w_i C\,(T_i - T) + H_e\,(T_e - T)\ , \\
m_e C_e \dot{T}_e &= Q - H_e\,(T_e - T)\ ,
\end{aligned}
$$

where $mC, m_e C_e$ are the thermal capacities of the tank contents and the heating element, respectively. Q is an input variable, H_e is the heat transfer coefficient.

Assume that Q and w_i are suitable path-wise continuous stochastic processes. Give the general form of the solution of the randomized model and study the boundedness of the solutions and their decay towards the null-solution.

Exercise 13.20. [☼] Absolute and Uniformly Convergence of the von Neumann/ Peano-Baker Series
Show that the von Neumann/ Peano-Baker series

$$\Omega(t) = \sum_{k=0}^{\infty} P_k(t) \text{ and } \begin{cases} P_0(t) & := \mathbb{I}, \\ P_k(t) & := \int_0^t (A + B\xi_s) P_{k-1}(s) \mathrm{d}s, \; k = 1, 2, 3, \ldots, \end{cases}$$

which was stated in equation (13.75) is absolutely and uniformly convergent.

Exercise 13.21. [🗊] Extending the Results of Sec. 13.5
In Sec. 13.5, the solution formula for linear systems with non-commutative colored noise was presented. It was assumed that the noise is path-wise continuous. Generalize the results and methods of this paragraph to the setting of mean-square continuous noise processes.

Chapter 14

Simulating Path-Wise Solutions

Relevant aspects for simulating random (ordinary) differential equation (RODE) problems are discussed in this chapter. We present lower-order explicit RODE schemes (Euler and Heun) as well as higher-order K-RODE Taylor schemes. Detailed information on the corresponding MATLAB implementation for the wireframe model is given and numerical results show the validity of this approach.

14.1 Key Concepts

As we have seen, random (ordinary) differential equations (RODEs) are path-wise equivalent to deterministic ordinary differential equations (ODEs). Thus, the numerical schemes for deterministic ordinary differential equations can, in principle, be applied for random (ordinary) differential equations. Though, as [149], p. 9, explains, there is a serious drawback:

> [For random (ordinary) differential equations] the driving stochastic process has at most Hölder continuous sample paths, so the solutions are certainly continuously differentiable, but the derivatives of the sample paths are at most Hölder continuous in time. Thus, after insertion of the driving stochastic process, the resulting vector field is at most Hölder continuous in time, no matter how smooth the vector field is in its original variables. Consequently, although classical numerical schemes for ODEs can be used path-wise for RODEs, they rarely attain their traditional order and new forms of higher order schemes are required, cf. [149], p. 9.

The solution to this problems lies in *subcycling*: We need to invest more fine-granular time steps for the integration of the stochastic effects of, for example, the Kanai-Tajimi earthquake model (see Sec. 3.2.2) before performing the coarser deterministic integration of the wireframe building model.

We restrict the considerations to explicit schemes and present two lower-order schemes for the integration of the wireframe model using the Kanai-Tajimi earthquake model: the Euler and the Heun scheme. Special care is taken to keep the global order of convergence of these methods. A family of higher-order methods, the K-RODE Taylor schemes, are presented, and two examples are given to illustrate how such schemes may look for the concrete scenario of wireframe buildings combined with the Kanai-Tajimi model.

When reading this chapter note the answers to the following questions

1. What do averaged schemes look like?
2. What is the basic idea of K-RODE schemes?
3. How are K-RODE schemes formulated and what are the challenging aspects?

as well as the following key concepts

1. (Global) order of convergence,
2. Subcycling for the numerical time stepping,
3. MATLAB implementation details.

This chapter is structured as follows: First, we briefly present a central result on the global discretisation error of one-step methods for the numerical integration of RODEs in Sec. 14.2. Next, in Sec. 14.3, the lower-order Euler and Heun scheme are presented together with MATLAB examples containing all relevant implementation aspects. Corresponding numerical results for different wireframe buildings as well as statistical example data for many more stochastic realisations are also given. We extend the approach to the higher-order K-RODE Taylor schemes in Sec. 14.4. Again, the corresponding MATLAB examples and selected numerical results are indicated for this method. Finally, Section 14.5 wraps up the contents of this chapter.

14.2 Discretisation Error of Explicit One-Step Methods for RODEs

Following [120], the function f in (3.5) is assumed to be infinitely often continuously differentiable in X_t and ω, whereas the resulting right-hand side $F_\omega(x,t)$ is typically only continuous (at most Hölder continuous) and not differentiable in t (cf. [120, 158, 149]). Consider the following RODE

$$\dot{X}_t = -X_t + \sin(W_t(\omega)), \tag{14.1}$$

where $W_t(\omega)$ is a sample path of a Wiener process. Theory of RODEs shows (see [45]) that the solution of (14.1) is given by

$$X_t = e^{-t}X_0 + e^{-t}\int_{t_0}^{t} e^{\tau}\sin\left(W_\tau(\omega)\right)\mathrm{d}\tau. \tag{14.2}$$

The sample paths of $W_t(\omega)$ are Hölder continuous but nowhere differentiable, hence the solution X_t is only once differentiable. This property directly translates to the order of convergence of the standard numerical schemes (such as Euler, Heun, and Runge-Kutta), because the Taylor expansions necessary for the error analyses cannot be carried out. We are now going to take a closer look at this fact which is presented in detail in [120].

Consider the class of explicit one-step numerical schemes for Eq. (3.5) given via

$$x_{n+1} = x_n + h_n \phi(h_n, t_n, x_n), \tag{14.3}$$

at discrete timesteps $t_n \in [t_0, T]$ and time step sizes $h_n = t_{n+1} - t_n \in (0, h]$ for all $n = 0, ..., N-1$. In particular, for a fixed ω, let $F_\omega(x,t) := f(x,t)$ in Eq. (3.5), where f is assumed to be locally Lipschitz continuous in x (with $|x(t)| \leq R$, $R(w) = R > 0$, $\forall t \in [t_0, T]$) with Lipschitz constant L_R. Denote $B[0;R] := \{x \in \mathbb{R}^d : |x| \leq R\}$. Furthermore, let $M_R = \max_{t\in[0,T], x\in B[0;R]} |f(x,t)|$. The discretisation error of one-step schemes (14.3) is given via the moduli of continuity of f and ϕ, defined as

$$\omega_f(h) := \omega_f(h; R, T) = \sup_{\substack{s,t\in[0,T] \\ 0\leq|s-t|\leq h}} \sup_{x\in B[0;R]} |f(t,x) - f(s,x)|, \tag{14.4a}$$

$$\omega_\phi(h) := \omega_\phi(h; R, T) = \sup_{0\leq h_n\leq h} \sup_{\substack{t\in[0,T] \\ x\in B[0;R]}} |\phi(h_n, t, x) - \phi(0, t, x)|\,. \tag{14.4b}$$

Under these assumptions, the following theorem holds.

Theorem 14.1. *The global discretisation error of the numerical scheme 14.3 satisfies the estimate*

$$|x_n - x(t_n; t_0, x_0)| \;\leq\; \frac{\omega_f(h) + \omega_\phi(h) + L_R M_R h}{L_R} \cdot e^{L_R T}\,. \tag{14.5}$$

Proof. The proof is given in [120]. □

14.3 Lower-Order Schemes for Random Differential Equations

14.3.1 The Euler & Heun Schemes for RODEs

The simplest scheme, the explicit Euler scheme (see Sec. 7.3.1), uses $\phi(h,t,x) = f(t,x)$, where (14.4b) gives $\omega_\phi(h) = 0$. The corresponding global discretisation error is given by

$$|x_n - x(t_n; t_0, x_0)| \;\leq\; \frac{\omega_f(h) + L_R M_R h}{L_R} \cdot e^{L_R T} \;=\; \mathcal{O}(\omega_f(h))\,, \tag{14.6}$$

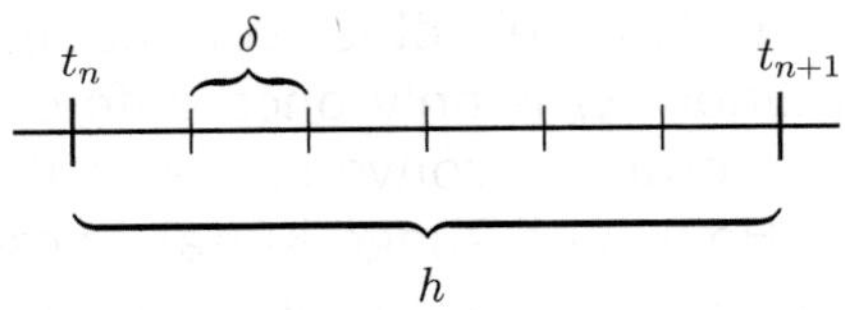

Figure 14.1. Subcycling for the averaging: A subdivision using a (considerably) smaller step size $\delta = h/N$ has to be used in each time step of step size h.

the last step followd from the Hölder continuity of f in t, Eq. (1.1). We now compare this result with the usual error estimate of Euler's method for ODEs, which is of order h (cf. [114]). The error estimate reflects the close relation of the order of the scheme and the order of the Hölder continuity of f. This is illustrated in [149] for the following RODE

$$\frac{\mathrm{d}x}{\mathrm{d}t} = -x + \zeta_t ,$$

where ζ_t is a stochastic process that is path-wisely Hölder continuous of order $\frac{1}{2}$. In this case, the Euler scheme is of path-wise order $\frac{1}{2}$.

It has been shown in [120] how to attain the upper bound $|x_n - x(t_n; t_0, x_0)| = \mathcal{O}(h)$ for a family of RODEs with separable vector field,

$$\frac{\mathrm{d}x}{\mathrm{d}t} = G(t) + g(t)H(x) , \tag{14.7}$$

where $g : [0,T] \to \mathbb{R}$ has a modulus of continuity $\omega_g(\delta)$ on $[0,T]$, $G : [0,T] \to \mathbb{R}^d$ a modulus of continuity $\omega_G(\delta)$ on $[0,T]$, and $H : \mathbb{R}^d \to \mathbb{R}^d$ is at least once continuously differentiable. This method is called the *averaged Euler scheme*, since the basic idea is to substitute the function g by its average

$$\bar{g}^{(1)}_{h,\delta}(t) = \frac{1}{N} \sum_{j=0}^{N-1} g(t + j\delta) \tag{14.8}$$

in every interval $[t, t+h]$ with a subcycling step size $\delta = h/N$ (see Fig. 14.1).

Applying the same averaging procedure for the function $G(t)$, we can rewrite (14.3) as

$$\begin{aligned} x_{n+1} &= \frac{1}{N} \sum_{j=0}^{N-1} \{x_n + hG(t_n + j\delta) + hg(t_n + j\delta)H(x_n)\} \qquad (14.9) \\ &= x_n + \frac{1}{N} \sum_{j=0}^{N-1} hG(t_n + j\delta) + \frac{H(x_n)}{N} \sum_{j=0}^{N-1} hg(t_n + j\delta) . \end{aligned}$$

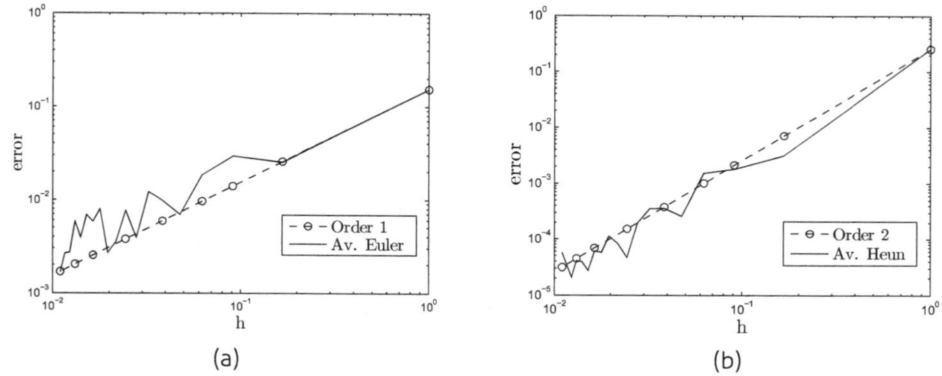

Figure 14.2. Convergence plots for the averaged Euler scheme (a) and the averaged Heun scheme (b) applied to $\dot{X}_t = -X_t + \sin(W_t(\omega))$ (i.e. Eq. (14.1)). One can clearly observe the order of strong convergence 1 for the averaged Euler scheme and of order 2 for the averaged Heun scheme.

The averaged Euler scheme above has a *local* discretisation error $L(h; x, t)$ given by

$$L(h; x, t) = \mathcal{O}(h(\omega_g(\delta) + \omega_G(\delta))),$$

which is of order 2 if δ is chosen such that $\max\{\omega_g(\delta), \omega_G(\delta)\} = \mathcal{O}(h)$. Hence, a global error of order 1 is obtained. Figure 14.2 (a) visualises the convergence properties of the averaged Euler scheme applied to Eq. (14.1).

Heun's method is derived in an analogous manner (see Sec. 7.3.2) by defining

$$\phi(h, t, x) = \frac{1}{2}\{f(x, t) + f(t + h, x + hf(x, t))\}.$$

Since the modulus of continuity $\omega_\phi(h)$ satisfies $\omega_\phi(h) \leq K_1 \omega_f(h)$ for a constant K_1 depending on T and R, the estimate (14.6) for the global discretisation error is still valid.

The *averaged Heun scheme* for RODEs with separable right-hand side f follows the same idea as above, except that the function H has to be twice continuously differentiable. We now need the the double-averaged function

$$\bar{g}^{(2)}_{h,\delta}(t) = \frac{2}{N^2} \sum_{i=0}^{N-1} \sum_{j=0}^{i} g(t + j\delta) = \frac{2}{N^2} \sum_{j=0}^{N-1} (N - j) g(t + j\delta), \tag{14.10}$$

in addition to the single-averaged one above. The resulting scheme is:

$$x_{n+1} = x_n + h\bar{G}^{(1)}_{h,\delta}(t_n) + \frac{h}{2} \bar{g}^{(1)}_{h,\delta}(t_n) H(x_n) \tag{14.11}$$

$$+\frac{h}{2}\bar{g}^{(1)}_{h,\delta}(t_n)H\left(x_n + h\bar{G}^{(2)}_{h,\delta}(t_n) + h\bar{g}^{(2)}_{h,\delta}(t_n)H(x_n)\right).$$

The local discretisation error for the scheme (14.11) is given by

$$L(h;x,t) = \mathcal{O}((h+h^2)(\omega_g(\delta)+\omega_G(\delta))+h^3),$$

which is of order 3 if δ is chosen such that $\max\{\omega_g(\delta),\omega_G(\delta)\} = \mathcal{O}(h^2)$. This ensures that the global discretisation error of the averaged Heun scheme is of order 2.

In Fig. 14.2 (b), the convergence properties of the averaged Heun scheme applied to Eq. (14.1) are visualised. Since the Wiener process is Hölder continuous with exponent $1/2$, we obtain $\omega_G(h) = \mathcal{O}(h^{(1/2)})$ and, hence, $\delta = h^2$ for the averaged Euler scheme and $\delta = h^4$ for the averaged Heun scheme. This fact is discussed in more detail in Sec. 14.5.

14.3.2 MATLAB Examples for Hybrid Deterministic & Averaged Euler & Heun Schemes

For the ground motion excitation of multi-storey buildings via the Kanai-Tajimi earthquake model, the corresponding system of equations (3.7) has a separable right-hand side $f(z,t) = G(t) + g(t)H(z)$, where $z = (z_1, z_2) \in \mathbb{R}^2$, with

$$G(t) = -O_t\begin{pmatrix}1\\ 2\zeta_g\omega_g - 1\end{pmatrix}, \quad H(z) = -\begin{pmatrix}z_2\\ 2\zeta_g\omega_g z_2 - \omega_g^2 z_1\end{pmatrix}, \quad g(t) = 1\,. \tag{14.12}$$

Therefore, the implementation of the averaged Euler and Heun schemes is straightforward for a given realisation of O_t. Recall that the Ornstein-Uhlenbeck process satisfies the equation $\mathrm{d}O_t = -O_t\mathrm{d}t + \mathrm{d}W_t$, for which the solution is obtained in explicit form (cf. [110]) as

$$O_{t+\Delta t} = \mu O_t + \sigma_X n_1\,, \tag{14.13}$$

where $\mu := e^{-\Delta t}$, $\sigma_X^2 := (1-\mu^2)/2$, and n_1 denotes a sample value of a normally distributed random variable $\mathcal{N}(0,1)$. Now, Eq. (14.13) allows for the generation of sample paths for O_t. From the relation of the Ornstein-Uhlenbeck process and the Wiener process, we conclude, as before, that $\omega_G(h) = \mathcal{O}(h^{1/2})$. Hence, the sampling step size is again $\delta = h^2$ for the averaged Euler scheme and $\delta = h^4$ for the averaged Heun scheme in order to keep the full order of convergence of the corresponding methods.

At this stage, we could either perform a complete computation of the multi-storey building subject to the Kanai-Tajimi excitation with the averaged Euler or Heun schemes as just discussed, or aim for a less expensive alternative: We can argue that the excitation

$$\ddot{u} = -2\zeta_g \omega_g z_2 - \omega_g^2 z_1 ,$$

that drives the deterministic dynamics of the multi-storey building, is the sum of two functions. These functions are the path-wise unique solution of

$$\dot{z} = G(t) + H(z) , \quad z = (z_1, z_2)^T ,$$

as stated in (14.12) with a continuous right hand side. Thus, for any realisation of the underlying continuous random driving process as computed in (14.13), z_1 and z_2 are continuously differentiable. Hence, from the numerical point of view, we indeed apply a $\mathcal{C}^1$-function $\ddot{u}$ as the excitation for our multi-storey buildings. This justifies the use of (low-order) deterministic methods for the final simulation aspects.

In particular, these observations naturally lead to a hybrid scheme consisting of the combination of averaged schemes for the computation of the first transmissions of the stochastic effects through the Kanai-Tajimi model and deterministic schemes for the computation of the oscillations that occur in the single storeys of the building.

The necessary implementation to run a 4-storey wireframe building simulation with the Kanai-Tajimi model is contained in the following MATLAB examples. The overall simulation is triggered by calling the method WireFrame4Storey displayed in MATLAB Example 14.1. The user specifies the deterministic time step size h and the final time tf, assuming $t_0 = 0$. Depending on the method and stochasticMethod, the corresponding deterministic and stochastic integration scheme is selected, for the solution of the wireframe system and the Kanai-Tajimi model respectively. In order to formulate the wireframe model of second order in time as a first-order system, the system matrix M is created using the damping and stiffness matrices C and K (see Sec. 3.5.2).

The stochastic accelerations are computed (see MATLAB Example 14.2 below) and substituted into the corresponding deterministic time integration schemes (the explicit Euler and Heun methods in MATLAB Examples 14.3 and 14.4) which use the system matrix M in order to perform their explicit time updates. Of course, as we noted, the explicit Euler is consistent with the $\mathcal{C}^1$ excitation to which we expose the ground floor of our building. To take full advantage of the Heun scheme, on the other hand, a $\mathcal{C}^2$ excitation such as the Clough-Penzien earthquake model would be preferable. Interestingly, when inspecting the final results given in Fig. 14.6 we see a smoothing of the initial irregular stochastic impacts occuring through the coupled oscillator equations (a well known filtering phenomenon). Hence, for any additional oscillator equation or storey of the building the smoothness of the driving excitation

MATLAB Example 14.1 WireFrame4Storey.m: Method for the RODE simulation of a wireframe model for a 4-storey building using the Kanai-Tajimi earthquake model.

```
function [Y F] = WireFrame4Storey(h,tf,method,stochasticMethod,k,c)

Nt = tf/h + 1;
% damping and spring matrices
C = [c(1)+c(2), -c(2), 0, 0; -c(2), c(2)+c(3), -c(3), 0;...
    0, -c(3), c(3)+c(4), -c(4); 0, 0, -c(4), c(4)];
K = [k(1)+k(2), -k(2), 0, 0; -k(2), k(2)+k(3), -k(3), 0;...
    0, -k(3), k(3)+k(4), -k(4); 0, 0, -k(4), k(4)];
% 2nd order system as a 1st order 2D system: system matrix
M = [zeros(4), eye(4); -K, -C];
% Choose RDE method for Kanai-Tajimi model: Euler, Heun, or RODE3
[F] = KT_Stochastic(Nt,h,tf,stochasticMethod);
% Choose the time integration
switch method
    case 'Euler'
        Y = euler(Nt,h,M,F);
    case 'Heun'
        Y = heun(Nt,h,M,F);
    case 'RK4'
        Y = RK4(Nt,h,tf,M,F)
end
```

MATLAB Example 14.2 KT_stochastic.m: Implementation of the Kanai-Tajimi model using the different RODE integration methods.

```
function ddu = KT_Stochastic(Nt,h,t_end,method)

x0 = 0;                     % Initial value of the position
v0 = 0;                     % Initial value of the velocity
zeta_g = 0.64;              % Eq. parameter
omega_g = 15.56;            % Eq. parameter
noSubcyclingTimeSteps = ceil(1/h);

% Solve for x_g and x'_g in the Kanai-Tajimi model
xMat = KT_Solvers(x0, v0, Nt, t_end, h, zeta_g, omega_g, ...
   noSubcyclingTimeSteps, method);

% Output ground acceleration u''_g
ddu = -2*zeta_g*omega_g*xMat(2,:) - omega_g^2*xMat(1,:);
```

for the next storey increases and the potential of higher-order (deterministic) methods begins thus to work in our advantage.

MATLAB Example 14.3 euler.m: Explicit Euler method for the wireframe model.

```
function [Y] = euler(Nt,h,M,F)
%solution: displacements & velocities of 4 floors
Y = zeros(8,Nt);

for n = 1:Nt-1
    Fv = [0;0;0;0;F(n);0;0;0];
    Y(:,n+1) = Y(:,n) + h*(M*Y(:,n) + Fv);
end
```

MATLAB Example 14.4 heun.m: Explicit Heun method for the wireframe model.

```
function [Y] = heun(Nt,h,M,F)
%solution: displacements & velocities of 4 floors
Y = zeros(8,Nt);

for n = 1:Nt-1
    Fv = [0;0;0;0;F(n);0;0;0];
    Fv2 = [0;0;0;0;F(n+1);0;0;0];
    Y(:,n+1) = Y(:,n) + (h*M + h^2*M^2/2)*Y(:,n) + ...
                (h/2)*(eye(8) + h*M)*Fv + (h/2)*Fv2;
end
```

In MATLAB Example 14.2 some preliminary work is done before the averaged schemes are called via KT_Solvers. The resulting data is postprocessed to give the Kanai-Tajimi ground accelerations ddu. The method KT_Solvers in MATLAB Example 14.5 sets the necessary subcycling time step size δ depending on the deterministic time step size h and the method of choice. Then, the desired averaged method is called. In MATLAB Example 14.6, the averaged Euler method is shown which uses the precomputed Ornstein-Uhlenbeck process data to average it over the number of subcycling time steps (noSCTS) within each (coarse) time step of size h (cf. Eq.14.9). The implementation of the averaged Heun method of Eq. 14.11 is given in MATLAB Example 14.7. It is similar to the averaged Euler but uses the double averaged values of the function G in addition.

The only implementation detail still missing is the realisation of the Orenstein-Uhlenbeck process. This is contained in the method OU_Matrix

MATLAB Example 14.5 KT_solvers.m:

```
function xMat = KT_Solvers(x0, v0, Nt, t_end, h, zeta, om, noSCTS, ...
    method)

tau = 1;       % Relaxation time of the OU process
c = 1;         % Diffusion constant of the OU process
% Initial values
z1 = x0;
z2 = -v0-x0;
Z = [z1; z2]; % Solution vector Z

if ( strcmp(method,'stochasticEuler') || ...
    strcmp(method,'stochasticHeun') )
    if strcmp(method,'stochasticHeun')
        noSCTS = ceil(1/h^3); % => h^4
    end
    delta = h/noSCTS;           % h^2 for av.Euler
    % compute sample path of the OU process
    ou = OU_Matrix(0, t_end, delta, tau, c);
    % Coarse grid OU processes
    ou_Coarse = ou(:,1:noSCTS:end);
end

switch method
    case 'stochasticEuler'
        Z = stochasticEuler(Nt, noSCTS, ou, ou_Coarse, Z, h, zeta, ...
            om);
    case 'stochasticHeun'
        Z = stochasticHeun(Nt, noSCTS, ou, ou_Coarse, Z, h, zeta, ...
            om);
    case 'RODE3'
        Z = RODE3(Z);
    otherwise
        error('Method not correctly specified.')
end
% Transform Z vector to x and v coordinates
xMat(1,:) =  Z(1,:);
xMat(2,:) = -Z(2,:) - ou_Coarse;
return
```

shown in MATLAB Example 14.8. Now, we have all necessary implementation parts to run RODE wireframe simulations in MATLAB.

MATLAB Example 14.6 stochasticEuler.m: Averaged Euler method for the Kanai-Tajimi model.

```
function [Z] = stochasticEuler(Nt, noSCTS, ou, ou_Coarse, Z, h, ...
    zeta, om)

% loop over time
for n = 1:Nt-1
    ou_av = sum(ou((n-1)*noSCTS+1 : n*noSCTS));
    ou_av = ou_av/noSCTS;
    G = [-1; 1 - 2*zeta*om]*ou_av;  % First average of G

    H = [-Z(2,n); -2*zeta*om*Z(2,n) + om^2*Z(1,n)];
    Z(:,n+1) = Z(:,n) + h*( G(:) + H );
end
```

MATLAB Example 14.7 stochasticHeun.m: Averaged Heun method for the Kanai-Tajimi model.

```
function [Z] = stochasticHeun(Nt, noSCTS, ou, ou_Coarse, Z, h, ...
    zeta, om)

% loop over time
factor = noSCTS - (0:(noSCTS-1));
for n = 1:Nt-1
    ou_av = sum(ou((n-1)*noSCTS+1 : n*noSCTS));
    ou_av2 = sum(factor.*ou((n-1)*noSCTS+1 : n*noSCTS));
    ou_av = ou_av/noSCTS;
    ou_av2 = 2*ou_av2/noSCTS^2;

    G1 = [-1; 1 - 2*zeta*om]*ou_av;       % First average of G
    G2 = [-1; 1 - 2*zeta*om]*ou_av2;      % Double average of F

    H = [-Z(2,n); -2*zeta*om*Z(2,n) + om^2*Z(1,n)];
    H2 = [ -( Z(2,n) + h*G2(2) + h*H(2) ); ...
           -2*zeta*om*( Z(2,n) + h*G2(2) + h*H(2)) + ...
           om^2*( Z(1,n) + h*G2(1) + h*H(1) )];
    Z(:,n+1) = Z(:,n) + h*( G1(:) + H/2 + H2/2 );
end
```

14.3.3 Numerical Results for Euler & Heun Schemes

We apply the averaged schemes presented in the previous sections to a wire-frame model of two different types of 4-storey buildings: a "normal" building (Fig. 14.3 (a)) and an "L-shaped" 4-storey building (Fig. 14.3 (b)). The general-

MATLAB Example 14.8 OU_Path.m: Computation of the Orenstein-Uhlenbeck process.

```
function x = OU_Path(x0, t_end, dt, tau, c)
% Parameters of the method
mu = exp(-dt/tau);
sigma_x = sqrt(c*tau/2*(1-mu^2));

Ntt = ceil(t_end/dt)+1;
% Computation of the sample path  matrix x
x = zeros(1,Ntt);
% initial value
x(1) = x0;

for i = 2:Ntt
    n1 = randn;
    % Matrix of OU Processes
    x(i) = x(i-1)*mu + sigma_x*n1;
end
```

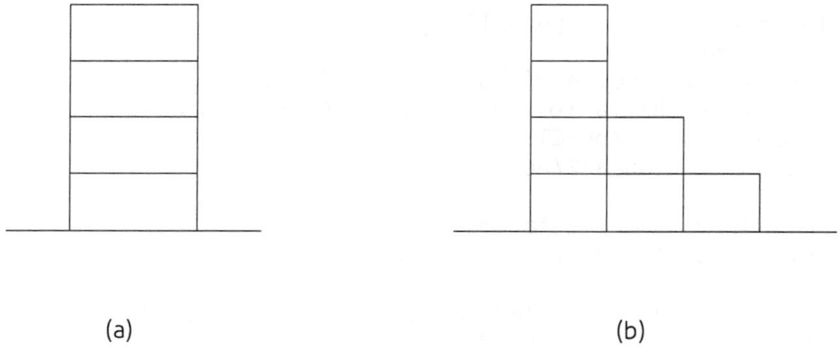

Figure 14.3. Sketches of the two building types considered in this contribution: "normal" 4-storey building (a) and "L-shaped" 4-storey building (b).

isation of the implementation for the "L-shaped" building is straightforward and does not affect the averaged schemes but only the setup of the system matrix M. Figure 14.4 shows one simulation run of different 4-storey buildings under stochastic ground-motion excitation, i.e. approximate solutions of Eq. (3.18) with a right-hand side given by the solution of the Kanai-Tajimi model $\ddot{u}_g = \ddot{x}_g + \xi_t$ with parameters $\zeta_g = 0.64$ and $\omega_g = 15.56$ (in rad/s).

The Kanai-Tajimi equation is solved with the averaged Euler scheme using $\delta = 1/1024^2$, and the system of equations (3.18) is solved using a deterministic Euler scheme with $h = 1/1024$.

A large amount of realisations ($M = 10,000$) of the normal 4-storey building 1 has been computed for an overall time interval of $T = 5$. The explicit Euler method with step size $h = 1/2048$ is used to numerically integrate the wireframe model while a subcycling step size of $\delta = h/2048 \approx 2.4e-7$ is applied for the averaged Euler method to solve the Kanai-Tajimi model (with same parameters as before). In total, about $21,000,000$ values of the OU process are computed and $10,240$ of them have to be stored for the deterministic time steps of size h for each of the M runs.

In Fig. 14.5 (a), the corresponding mean and variance of u_1 (the first floor) is plotted over time. The hitting times are visualised in Fig. 14.5 (b): The histogram shows the amount of realisations that fulfil the criterion $|u_1| \geq 5e-2$ within a given subinterval of length 0.5. This way, one can determine if the stiffness and damping parameters of the building are appropriate to keep the horizontal displacement of a floor within a given tolerance.

14.4 Higher-Order Schemes through Implicit Taylor-like Expansions

Exploiting the smoothness of the function f in (3.5) with respect to its variables X_t and ω results in an alternative approach to obtain explicit schemes of type (14.3). These so-called *K-RODE Taylor schemes* were introduced by Kloeden and Jentzen in 2007 ([158]; see also [149]). We present the most important aspects of these schemes in the following (for details, see [158]).

14.4.1 The K-RODE Taylor Schemes for RODEs

Let us again consider the initial value problem

$$\frac{\mathrm{d}x}{\mathrm{d}t} = F_\omega(t,x) := f(\omega(t),x), \qquad x(t_0) = x_0, \tag{14.14}$$

for $\omega \in \Omega$.

The goal is to carry out the Taylor expansion of the function f with respect to ω and x, for which we consider the *multi-index* $\alpha = (\alpha_1, \alpha_2) \in \mathbb{N}_0^2$. Its magnitude is given by $|\alpha| := \alpha_1 + \alpha_2$, which may take a weight $\gamma \in (0,1]$ such that $|\alpha|_\gamma := \gamma\alpha_1 + \alpha_2$. Similarly, for each $K \in \mathbb{R}_+$ with $K \geq |\alpha|_\gamma$, define $|\alpha|_\lambda^K := K - |\alpha|_\lambda$. Also, let $\alpha! := \alpha_1!\alpha_2!$. Finally, let $f_\alpha := (\partial_1)^{\alpha_1}(\partial_2)^{\alpha_2} f$, with $f_0 = f$.

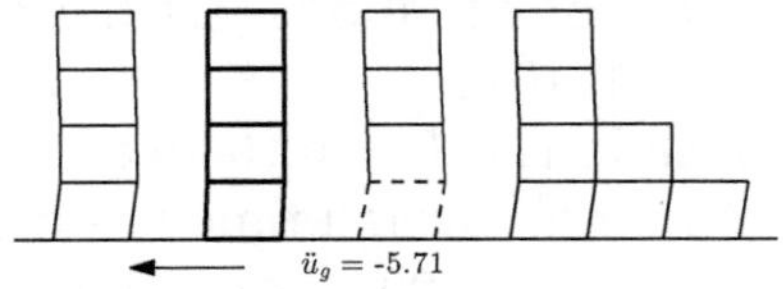

(a)

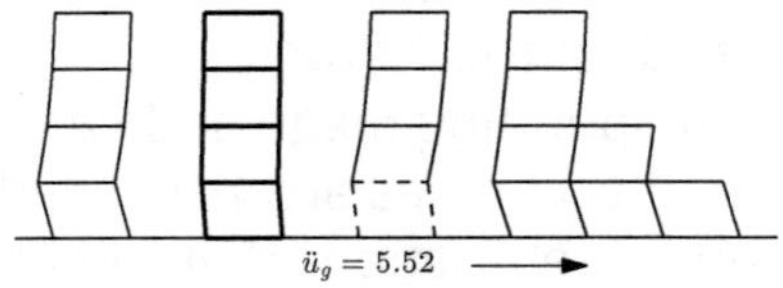

(b)

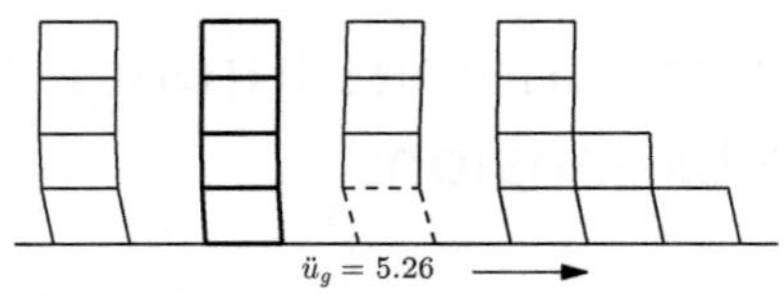

(c)

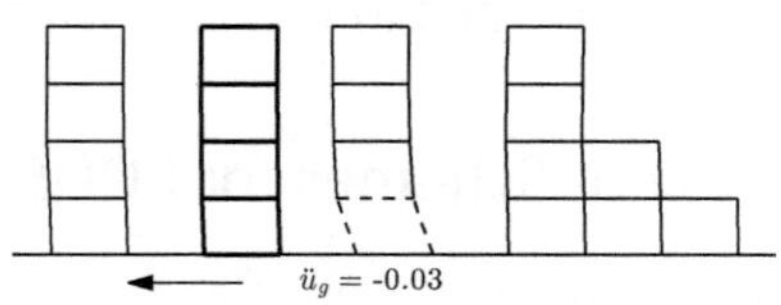

(d)

Figure 14.4. Movement of four different 4-storey buildings under the same stochastic excitation at times $t = 0.5$ (a), $t = 1.5$ (b), $t = 2.5$ (c), and $t = 3.5$ (d). Building 1 (leftmost): Normal 4-storey building with $k_1 = \ldots = k_4 = 25$, $c_1 = \ldots = c_4 = 10$. Building 2: Same as building 1 but with higher damping coefficients $c_1 = \ldots = c_4 = 40$. Building 3: same as building 1 but with smaller $k_1 = 2.5$. Building 4 (rightmost): L-shaped with the same parameters as building 1.

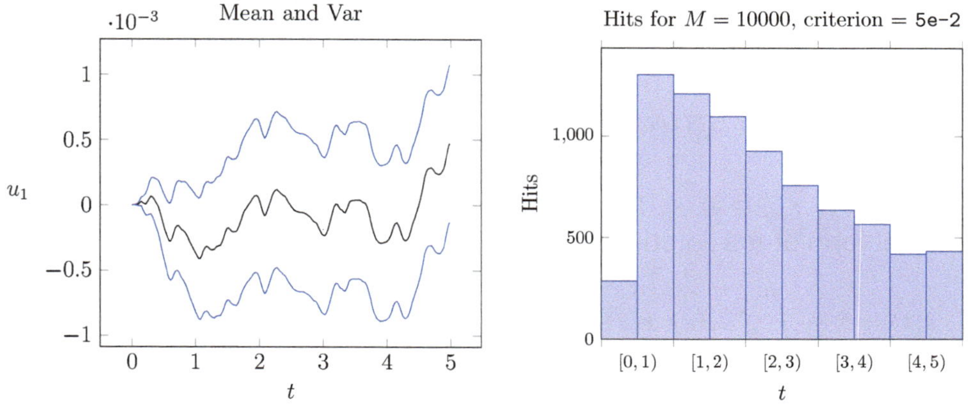

(a) (b)

Figure 14.5. $M = 10,000$ realisations of the 4-storey building model computed with the (averaged) Euler method using $h = 1/2048$: mean and variance of u_1 (a) and histogram of the hitting times (b), i.e. the amount of realisations that show $|u_1| \geq 5e-2$ for corresponding time subintervals within the overall time $T = 5$.

The k-th order Taylor expansion of f then has the following form

$$f(\omega(s), x(s)) = \sum_{|\alpha| \leq k} \frac{f_\alpha(\hat{\omega}, \hat{x})}{\alpha!} (\Delta\omega_s)^{\alpha_1} (\Delta x_s)^{\alpha_2} + R_{k+1}(s) \,, \tag{14.15}$$

Before you continue, make sure to answer the following questions:

Quiz: Section 14.3

Q1 What is the main computational drawback of using the averaged schemes? What are the benefits?

Q2 Suppose $H(x)$ in Eq. (14.7) is expensive to evaluate. Would it be preferable to use an averaged scheme with a large deterministic step size h instead of a deterministic scheme with a smaller step size? Why?

Q3 Derive the averaged Euler scheme and the averaged Heun scheme for Eq. (14.1) and compare the expressions to the exact solution (14.2). What do you observe?

Q4 What is an appropriate measure to compute the error of the averaged schemes? (cf. Fig. 14.2)

with $\Delta\omega_s := \omega(s) - \hat{\omega}$, where $\hat{\omega} := \omega(\hat{t})$, and $\Delta x_s := x(s) - \hat{x}$, where $\hat{x} := x(\hat{t})$ for some $\hat{t} \in [t_0, T)$. The remainder R_{k+1} in (14.15) is then given by

$$R_{k+1}(s) = \sum_{|\alpha|=k+1} \frac{1}{\alpha!} f_\alpha(\hat{\omega} + \xi_{\omega_s}\Delta\omega_s, \hat{x} + \xi_{x_s}\Delta x_s)(\Delta\omega_s)^{\alpha_1}(\Delta x_s)^{\alpha_2}, \quad (14.16)$$

for some $\xi_{\omega_s}, \xi_{x_s} \in [0,1]$. The Taylor expansion for f is now inserted into the integral form of the solution of (14.14):

$$x(t) = \hat{x} + \int_{\hat{t}}^{t} f(\omega(s), x(s))\mathrm{d}s, \quad (14.17a)$$

$$= \hat{x} + \underbrace{\sum_{|\alpha|\le k} \frac{f_\alpha(\hat{\omega},\hat{x})}{\alpha!} \int_{\hat{t}}^{t} (\Delta\omega_s)^{\alpha_1}(\Delta x_s)^{\alpha_2}\mathrm{d}s}_{:=T_\alpha(t;\hat{t})} + \int_{\hat{t}}^{t} R_{k+1}(s)\mathrm{d}s, \quad (14.17b)$$

$$= \hat{x} + \sum_{|\alpha|\le k} T_\alpha(t;\hat{t}) + \int_{\hat{t}}^{t} R_{k+1}(s)\mathrm{d}s\,. \quad (14.17c)$$

Even though this expansion is implicit in $x(t)$ (which appears after computing the integral inside $T_\alpha(t;\hat{t})$), one can construct explicit higher-order schemes in a recursive manner. This is realised by approximating the Δx_s terms inside T_α using a scheme of one order lower. To obtain a K-RODE Taylor scheme, let us truncate Eq. (14.17) by ignoring the remainder and set $t \to \hat{t} + h$. Then,

$$x(\hat{t}+h) \approx x(\hat{t}) + \sum_{|\alpha|\le k} \frac{f_\alpha(\hat{\omega},\hat{x})}{\alpha!} \int_{\hat{t}}^{\hat{t}+h} (\Delta\omega_s)^{\alpha_1}(\Delta x_s)^{\alpha_2}\mathrm{d}s\,. \quad (14.18)$$

We can now derive the corresponding numerical scheme of order $K \in \mathbb{R}_+$, given by the approximated solution $y_n^{K,h}$, defined for the sets of multi-indices

$$\mathcal{A}_K := \{\alpha = (\alpha_1,\alpha_2) \in \mathbb{N}_0^2 \; : \; |\alpha|_\theta = \theta\alpha_1 + \alpha_2 < K\}. \quad (14.19)$$

The *K-RODE-Taylor scheme* is then given by

$$y_{n+1}^{K,h} := y_n^{K,h} + \sum_{\mathcal{A}_K} N_\alpha^{(K)}(t_{n+1}, t_n, y_n^{K,h}), \quad (14.20)$$

where

$$N_\alpha^{(K)}(\hat{t}+h,\hat{t},\hat{y}) := \frac{1}{\alpha!} f_\alpha(\hat{\omega},\hat{y}) \int_{\hat{t}}^{\hat{t}+h} (\Delta\omega_s)^{\alpha_1} \left(\Delta y_{\Delta s}^{(|\alpha|_\theta^K)}(\hat{t},\hat{y})\right)^{\alpha_2} \mathrm{d}s, \quad (14.21a)$$

$$\Delta y_h^{(\ell)}(\hat{t},\hat{y}) := \sum_{|\alpha|_\theta<\ell} N_\alpha^{(\ell)}(\hat{t}+h,\hat{t},\hat{y}), \quad (14.21b)$$

and $\Delta s = s - \hat{t}$. In the notation of Eq. (14.3), this family of schemes possesses the increment function

$$\phi^{(K)}(h,\hat{t},\hat{y}) := \frac{1}{h}\sum_{\mathcal{A}} N_{\alpha}^{(K)}(\hat{t}+h,\hat{t},\hat{y}). \tag{14.22}$$

Eqs. (14.21a)–(14.21b) reflect the recursivity of the scheme (14.20), since the term $\Delta y_{\Delta s}^{(|\alpha|_\theta^K)}$ is of order $|\alpha|_\theta^K = K - |\alpha|_\theta < K$. The weight θ is taken to be the supremum of the Hölder coefficients of the sample paths derived from the noise process of the RODE [149]. Hence, two cases have to be distinguished:

A) The Hölder continuity holds for the supremum θ;

B) The Hölder continuity does not hold for the supremum θ.

This distinction paves the way for an expression for the local discretisation error of the K-RODE-Taylor schemes, defined at the initial step $y_1^{(K,h)}(\hat{t},\hat{y})$ as

$$L_h^{(K)}(\hat{t},\hat{y}) := |x(\hat{t}+h,\hat{t},\hat{y}) - y_1^{(K,h)}(\hat{t},\hat{y})|. \tag{14.23}$$

Now, define $\tilde{R}_0 := 0$ and, for $K > 0$,

$$\tilde{R}_K := \sup_{0<L\le K} \max_{\substack{(h,t,x)\in\\ [0,1]\times[t_0,T]\times[-R,R]}} |\phi^{(L)}(h,t,x)|.$$

Finally, let

$$k = k_K := \left\lfloor \frac{K}{\theta} \right\rfloor, \quad R_K := \max\{\tilde{R}_K, \|f\|_{k+1}\}.$$

Under these considerations, the following theorem holds.

Theorem 14.2. *The local discretisation error for a RODE-Taylor scheme in case A satisfies*

$$\left|L_h^{(K)}(\hat{t},\hat{x})\right| \le C_K h^{K+1}, \tag{14.24}$$

for each $0 \le h \le 1$*, where*

$$C_K := (\exp(\|\omega\|_\theta + 2R_K))^{K+1}.$$

In case B, it satisfies

$$\left|L_h^{(K)}(\hat{t},\hat{x})\right| \le C_K^\epsilon h^{K+1-\epsilon}, \tag{14.25}$$

for $\epsilon > 0$ *arbitrarily small, where*

$$C_K^\epsilon := (\exp(\|\omega\|_{\gamma_\epsilon} + 2R_K))^{K+1}, \quad \gamma_\epsilon := \theta - \frac{\epsilon}{(k+1)^2}.$$

Proof. The proof is given in [158]. □

Furthermore, the RODE-Taylor schemes converge for each $K > 0$, and the global error behaves as in the ODE case, i.e. it is one order lower than the local error (of order K for case A, and $K - \epsilon$ for case B, cf. [148, 158]).

14.4.2 MATLAB Examples for the K-RODE Taylor Scheme

Due to the inherent recursivity of the higher-order K-RODE Taylor schemes their implementation is a demanding task. Nevertheless, it turns out that the particular form of Eq. (3.7) simplifies the problem in such way that, for moderate K, an explicit form of the scheme can be calculated. For $K = 1.0$, e.g., the resulting numerical scheme from Eq. (14.20) is

$$y_{n+1}^{1.0,h} = y_n^{1.0,h} + hf(O_t, z) + f_{(1,0)}(O_t, z) \int_{t_n}^{t_{n+1}} \Delta O_s \mathrm{d}s,$$

which is similar to the Euler method with an additional "correction" term. Using $K = 3.0$ for the right-hand side $f(O_t, z)$, the corresponding RODE Taylor scheme yields

$$y_{n+1}^{(3),h} = y_n^{(3),h} + hf + f_{(0,1)} f \frac{h^2}{2} + \frac{h^3}{6} f_{(0,1)}^2 f + f_{(1,0)} \int_{t_n}^{t_{n+1}} \Delta O_s \mathrm{d}s \qquad (14.26)$$

$$+ f_{(0,1)} f_{(1,0)} \int_{t_n}^{t_{n+1}} \int_{t_n}^{s} \Delta O_v \mathrm{d}v \mathrm{d}s + f_{(0,1)}^2 f_{(1,0)} \int_{t_n}^{t_{n+1}} \int_{t_n}^{s} \int_{t_n}^{v} \Delta O_w \mathrm{d}w \mathrm{d}v \mathrm{d}s,$$

where $f_{(0,1)}(O_t, z)$ corresponds to the Jacobian of f, since $z \in \mathbb{R}^2$.

The derived formulas for the 3-RODE Taylor scheme have been implemented in MATLAB Example 14.9[1]. This function computes the corresponding output (Z) of the Kanai-Tajimi model which is then used to compute the coordinate transformation to obtain the acceleration contributions identically to the averaged methods. As a deterministic time integration, we now use a Runge-Kutta method of fourth order to be consistent in the order of convergence. The corresponding implementation is given in MATLAB Example 14.10. Both MATLAB Examples can directly be used with WireFrame4Storey.m given in MATLAB Example 14.1 which has the corresponding function calls already prepared.

14.4.3 Numerical Results for K-RODE Taylor Schemes

Figures 14.6 (a) and 14.6 (b) show the resulting motion of two independent runs with different stochastic realisations of the earthquake accelerations applied to the normal 4-storey building (visualised in Fig. 14.3(a)) using the 3.0-RODE Taylor scheme for the Kanai-Tajimi model with a step size $h = 1/32$. The eight-dimensional system of equations (3.18) (one position and one velocity for each of the four levels) was solved by the deterministic Runge-Kutta 4 scheme, with stiffness coefficients $k_1 = \ldots = k_4 = 15$ and damping coefficients $c_1 = \ldots = c_4 = 5$.

[1] Note that a more stable alternative for the computation of the multiple time integrals is to interpret them as spatial integrals, see [199].

MATLAB Example 14.9 RODE3.m: 3-RODE scheme.

```
function ZZ = RODE3(Z)
ZZ = Z;
% Number of steps to calculate the integrals
delta = h/ceil(1/h^3);
% OU process and its integral, evaluated with the necessary steps
[ou,oui] = OU(0,t_end,delta,tau,c);
ou_Coarse = ou(1:m:end);
oui_Coarse = oui(1:m:end);
% Define functions and derivatives
f = [-z2-ou_Coarse(1); -2*zeta*om*(z2+ou_Coarse(1)) + om^2*z1 + ...
    ou_Coarse(1)];
f10 = [-1; -2*zeta*om + 1];
f01 = [0, -1; om^2, -2*zeta*om];
f012 = f01*f01;
% Calculate the solution via 3.0-RODE Taylor scheme
for n = 1:N_t-1
    delY1 = oui_Coarse(n+1) - oui_Coarse(n);
    % Integrals using trapezoidal rule
    delY2 = oui(m*(n-1)+2:n*m+1) - oui_Coarse(n);
    I1 = delta*sum(delY2);
    I2 = delta^2*sum((1:m).*delY2);
    % Solution vector using the 3-RODE scheme
    ZZ(:,n+1) = ZZ(:,n) + h*f + f10*(delY1 - h*ou_Coarse(n)) + ...
        f01*f*h^2/2 + f01*f10*(I1 - h^2*ou_Coarse(n)/2) + ...
        h^3*f012*f + f012*f10*(h*I1 - I2 - h^3*ou_Coarse(n)/6);
    % Update f function with new (z1,z2) values
    f = [-ZZ(2,n+1)-ou_Coarse(n+1); -2*zeta*om*(ZZ(2,n+1) + ...
        ou_Coarse(n+1)) + om^2*ZZ(1,n+1) + ou_Coarse(n+1)];
end
```

14.5 Chapter's Summary

We presented two different types of explicit schemes for the numerical solution of RODEs: low-order methods—deterministic Euler and Heun schemes with corresponding averaged versions—and higher-order methods—deterministic Runge-Kutta with K-RODE Taylor schemes. The two families of schemes may appear very different at first sight, but their numerical implementation shows various similarities. Actually, both require a further subdivision of the chosen timestep size h in order to compute the averaging (Eqs.(14.8),(14.10)) or to approximate the integrals (e.g. Eq. (14.26))[2], respec-

[2] Such integrals may be interpreted as the mean of the functions up to a multiplicative constant

MATLAB Example 14.10 RK4.m: Runge-Kutta method of fourth order.

```
function [Y] = RK4(Nt,h,tf,M,F)
%solution: displacements & velocities of 4 floors
Y = zeros(8,Nt);
FF = zeros(8,Nt);
FF(5,:) = F;
A = [0,0,0,0;.5,0,0,0;0,.5,0,0;0,0,1,0];
b = [1/6;1/3;1/3;1/6];  %weight of RK step vectors
t = 0:h:tf;
s = length(b);          %amount of RK step vectors
N = length(t);          %number of full time steps
dim = length(Y);
kk = zeros(dim,s);
y = zeros(dim,N);
% Initial value
y(:,1)=Y;
% expl. Runge Kutta method:
for i = 1:N-1
    for j = 1:s
        kk(:,j)= M*(y(:,i)+h*kk*A(j,:)')+FF(:,i);
    end
    y(:,i+1)=y(:,i)+h*kk*b;
end
Y = y;
```

tively. The choice of the subcycling step size $\delta = h/N$ (and h/m, resp.) is as crucial to the overall order of the scheme as the choice of h itself. In order to achieve an absolute error of order $\mathcal{O}(10^{-4})$ in the examples we have presented, one could use an averaged Euler method with $h = 10^{-4}$ and $\delta = 10^{-8}$, or an averaged Heun scheme with $h = 10^{-2}$ and $\delta = 10^{-8}$, or even more, a regular (non-averaged) Euler scheme with $h = 10^{-8}$ (since it will be of order $\mathcal{O}(h^{1/2})$). Finally, a 3rd order RODE Taylor scheme requires each integral from t_n to t_{n+1} to be calculated with $\delta = h^3 \cdot h$ (order 3 plus performing N such integrals). Hence, all four methods need the same degree of refinement in time. Following [120], the ease of computation of the averages (or Riemann sums) as well as stability properties make higher-order methods preferable.

Problems

Classification: ☼ easy, ⊕ easy with longer calculations, ☆ a little bit difficult,
◫ challenging.

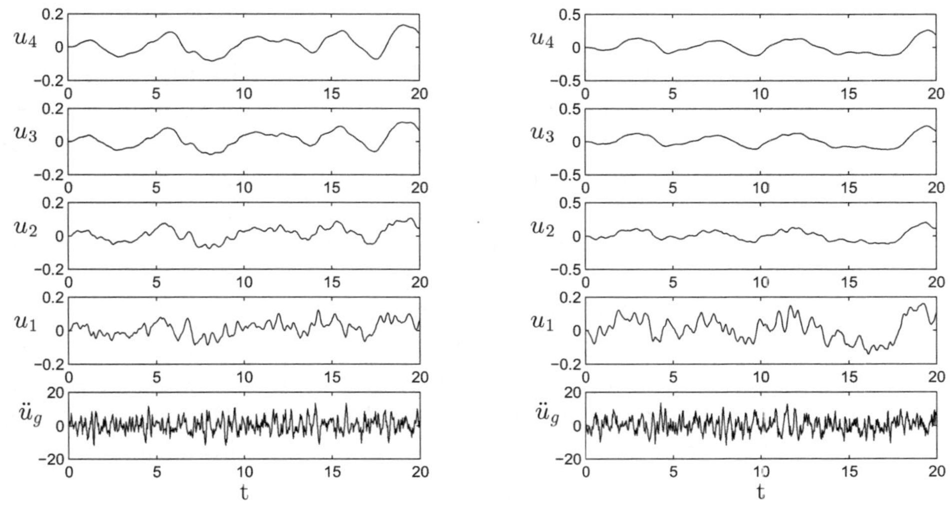

(a) (b)

Figure 14.6. Two different realisations (a) and (b) of the ground-motion excitation applied to a usual 4-storey building. The ground-motion acceleration $\ddot{u}_g$ and the resulting displacement of the four storeys of the building are visualised as functions of time.

Exercise 14.1. [☼] **Averaged Schemes: Hands On**

Implement a MATLAB program to simulate the path-wise solutions of Eq. (14.1) by means of the Averaged Euler and Heun schemes. Compare your results with the exact solution (14.2). How small should h be in order to have a global error of the order 10^{-2}? What does this mean in terms of the subcycling step size δ?

Exercise 14.2. [☼] **More Sample Paths**

Adapt the MATLAB Example 14.8 so that it generates M independent sample paths of the Ornstein-Uhlenbeck process and returns the corresponding $\mathrm{M}\times\mathrm{Ntt}$ matrix of realisations.

Exercise 14.3. [☼] **Simulating a second order RODE**

In Chap. 13, we encountered the equation

$$\ddot{Y}_t + 2b\dot{Y}_t + (1 + Z_t + a\sin(\omega t))Y_t = 0\,,$$

where $a \geq 0$ and $b > 0$. Derive the averaged Euler scheme for this equation. Implement your solution with MATLAB and plot five sample paths of Y_t.

Analyse the behaviour of the solution as a funciton of the different equation parameters.

Exercise 14.4. [◷] **A Higher-Order Scheme for the Kanai-Tajimi Model**
Eq. (14.26) gives the general form of the 3.0-RODE Taylor scheme applied to the Kanai-Tajimi model Eq. (14.12). Discretise the integrals by means of Riemann sums and calculate the necessary partial derivatives. Implement the resulting numerical scheme for the solution $y^{(3),h}$ with MATLAB, taking $h = 1/16$. How small should the integration stepsize be for this choice of h?

Exercise 14.5. [☆] **Deriving a K-RODE Taylor Scheme**
Eq. (14.20) may look innocent, but there is more to it than meets the eye. Our simple Kanai-Tajimi model (14.12) will allow us to understand the inner workings of these schemes. Start by deriving the 1.0-RODE Taylor scheme given at the beginning of Sec. 14.4.2. Begin by defining the appropriate $\mathcal{A}_1$ set and iteratively find the expressions for $N_\alpha^{(1.0)}$
Once you have warmed up with this example, go for the third order scheme (14.26).

Before you continue, make sure to answer the following questions:

Quiz: Section 14.4

Q1 The derivation of many numerical methods starts off by considering Taylor expansions of the differential equation w.r.t. time. What is different about The K-RODE Taylor schemes? Why is it not possible to carry out an expansion in time?

Q2 Observe from Eq. (14.17b) that the solution x appears implicitly inside the first integral of the right-hand side. Does this mean that the resulting K-RODE scheme is implicit? Why?

Q3 Analyse the resulting scheme of Eq. (14.26). Which numerical integration method would you choose for the integrals in the right-hand side? What do you know about the smoothness of the integrand? Would it make sense to use higher-order integration schemes?

Q4 For a given RODE of the form (14.7), what would be the advantages/disadvantages of implementing a 1.0-RODE Taylor scheme compared to an averaged Euler scheme? What about a 2.0-RODE Taylor scheme and the averaged Heun scheme?

Chapter 15

Stability of Path-Wise Solutions

This chapter begins with the study of various notions of stability of the null solution of a random (ordinary) differential equation with a focus on path-wise equi-stability, h-, $\mathbb{P}$-, and W-stability. In particular, the relations/ implications and inter-connections between these concepts are discussed and the results of Chap. 13 on the path-wise stability of linear random differential equations with stochastic coefficients are re-framed in the context of these concepts. Moreover, we extend the deterministic Lyapunov method to random differential equations. Based on suitable Lyapunov-functions, necessary conditions for h-stability and path-wise equi-stability are given. Finally, the stability of deterministic systems subject to different classes of continuously acting random perturbations is analysed.

15.1 Key Concepts

As [149], p. 11, mentions and as we have already seen in Chap. 3, deterministic calculus can be used path-wise for random differential equations, and contrary to stochastic differential equations no new rules of calculus, like Itô's celebrated formula, need to be invented. This greatly facilitates the investigation of dynamical behavior and other qualitative properties of random differential equations.

Example 15.1 (Existence of an Equilibrium Solution, cf. [149], p. 11). As in example 3.9 let the following random differential equation

$$\dot{Z}_t = f(Z_t + O_t) + O_t$$

be given, where O_t is a stationary Ornstein-Uhlenbeck process, and assume that f satisfies the one-sided dissipative Lipschitz-condition

$$\langle x - y \,, f(x) - f(y) \rangle \le -L\|x - y\|^2 \,,$$

for all $x, y \in \mathbb{R}^d$ and some $L > 0$. Then, for any two solutions $Z_1(t)$ and $Z_2(t)$ of the random differential equation we have

$$
\begin{aligned}
\tfrac{\mathrm{d}}{\mathrm{d}t}\|Z_1(t) - Z_2(t)\|^2 &= 2\langle Z_1(t) - Z_2(t)\,,\, \tfrac{\mathrm{d}}{\mathrm{d}t}Z_1(t) - \tfrac{\mathrm{d}}{\mathrm{d}t}Z_2(t)\rangle \\
&= 2\langle Z_1(t) - Z_2(t)\,,\, f(Z_1(t) + O_t) - f(Z_2(t) + O_t)\rangle \\
&= 2\langle (Z_1(t) + O_t) - (Z_2(t) + O_t)\,,\, f(Z_1(t) + O_t) - f(Z_2(t) + O_t)\rangle \\
&\leq -2L\|Z_1(t) - Z_2(t)\|^2\,,
\end{aligned}
$$

from which it follows that

$$\|Z_1(t) - Z_2(t)\|^2 \leq \exp(-2Lt)\|Z_1(0) - Z_2(0)\|^2 \to 0\,, \quad \text{as } t \to \infty\,.$$

Thus, there exists a path-wise asymptotically stable stochastic stationary solution.

In particular, as we have seen in example 3.9, this random differential equation is equivalent to the stochastic differential equation $\mathrm{d}X_t = f(X_t)\mathrm{d}t + \mathrm{d}W_t$. Thus, this stochastic differential equation has a path-wise asymptotically stable stochastic stationary solution, too.

When reading this chapter note the answers to the following questions

1. How do stochastic and deterministic stability concepts differ from each other? How are they connected? Do they imply each other?
2. What is path-wise equi-stability, h-, $\mathbb{P}$-, and W-stability?
3. What is a stochastic Lyapunov-function?
4. How does a stochastic Lyapunov-function imply equi-stability or h-stability?

as well as the following key concepts

1. The various notions of stochastic stability and their implications/ relations amongst each other,
2. Lyapunov-functions for random ordinary differential equations and their implications, and
3. Application of stochastic stability to the deterministic stability of systems subject to continuously acting random perturbations.

Note that we will use the following abbreviations in this chapter: U_t and V_t will be one-dimensional stochastic processes on I, and we will define a point-wise probabilistic size-relation $U_t \overset{(p)}{<} V_t$ as

$$U_t \overset{(p)}{<} V_t \qquad :\Leftrightarrow \qquad \mathbb{P}(U_t < V_t) > p$$

for $t \in I$, and

$$U_t \overset{(I,p)}{<} V_t \quad :\Leftrightarrow \quad \mathbb{P}\left(U_t < V_t : t \in T\right) > p$$

which holds on the whole of I. Clearly, $U_t \overset{(p)}{<} V_t$ for all $t \in I$ implies $U_t \overset{(I,p)}{<} V_t$.

Moreover, during our discussion of stability we will frequently apply the following estimate:

Lemma 15.2 (A Growth Inequality for Continuous Gaussian Processes). *Let $\alpha, \beta, \gamma > 0$ and Z_t, $t \in I := [t_0, \infty)$ be a path-wise as well as $q.m.$-continuous Gaussian vector-valued process such that*

$$\|\mathbb{E}(Z_t)\| \leq \alpha, \quad \|K^Z(t,t)\| = \sigma_Z^2(t) \leq \beta^2, \quad \textit{and} \quad \int_0^\infty \|K^Z(t,\tau)\| d\tau \leq \gamma,$$

hold for all $t \in I$. Then, the following growth inequality holds for arbitrary $k > 0$ and $t \geq t_0$:

$$\mathbb{E}\left(\exp\left(k \int_{t_0}^t \|Z_\tau\| d\tau\right)\right) \leq \exp\left(k\left(\alpha + \beta + \tfrac{1}{2}k\gamma\right)(t - t_0)\right).$$

Proof. See [155]. □

This chapter is structured as follows: Sec. 15.2 studies the various notions of stability of the null solution of a random (ordinary) differential equation with a focus on path-wise equi-stability, h-, $\mathbb{P}$-, and W-stability. The relationship between these concepts are discussed and the results of Chap. 13 on the path-wise stability of linear random differential equations with stochastic coefficients are re-framed in the context of these concepts. Next, Sec. 15.3 extends the deterministic Lyapunov method to random differential equations. In particular, based on suitable Lyapunov-functions, necessary conditions for h-stability and path-wise equi-stability are given here. As an excursion, Sec. 15.4 holds results concerning the stability of deterministic systems subject to different classes of continuously acting random perturbations. Finally, Sec. 15.5 wraps up the contents of this chapter.

15.2 Stability Notations for Path-Wise Solutions of RODEs and Their Connections

As in the deterministic case, the problem of whether a specific solution of a random differential equation is stable or not is decided by studying the null-solution (trivial solution) of a transformed equation if necessary.

Let $f(x,t,\omega) : \mathbb{R}^d \times I \times \Omega \to \mathbb{R}^d$. For almost all $\omega \in \Omega$ let the differential equation $\dot{x} = f(x,t,\omega)$ satisfy the following three properties:

1. **Unique existence of solutions**: There is a unique solution $x(t, x_0, t_0, \omega)$ on I for any initial condition $(x_0, t_0) \in \mathbb{R}^d \times I$.

2. **Continuity of solutions**: For fixed (t, ω) the solution $x(t, x_0, t_0, \omega)$ is a continuous vector-valued function on $\mathbb{R}^d \times I$.

3. **Isolated equilibria at zero**: There is a neighborhood $U_\omega \ni \{0\}$ around $x = 0$ such that $f(x, t, \omega) = 0$, $t \in I$ holds in U_ω if and only if $x = 0$.

We assume that for any initial condition $(X_0, t_0) \in S_d \times I$ there is a unique path-wise solution $X_t(X_0, t_0)$ of the random differential equation

$$\frac{\mathrm{d}X_t}{\mathrm{d}t} = f(X_t, t, \omega) \,. \tag{15.1}$$

Thus, $X_t(X_0, t_0) \stackrel{I}{=} x(t, X_0(\omega), t_0, \omega)$. The path-wise solution $X_t(0, t_0) \stackrel{I}{=} 0$ with respect to the initial condition $(0, t_0)$ is called *null-solution* or *trivial solution* of (15.1).

15.2.1 The Zoo of Stochastic Stability Concepts

Analogously to the deterministic setting, there exist many stability concepts that are tailored specifically for random differential equations. Due to the various convergence notions for stochastic processes, though, the volume of these concepts is increased.

Let us start by defining those stochastic stability concepts that can be played back easily to their deterministic counterparts:

Definition 15.3 (Path-Wise Stability Concepts). The null solution of the random differential equation (15.1) is called

1. *path-wise stable,*
2. *asymptotically path-wise stable,*
3. *asymptotically path-wise stable in the whole,*
4. *uniformly path-wise stable,*
5. *exponentially path-wise stable,* or
6. *path-wise stable under continuously acting disturbances*, respectively,

if for almost all $\omega \in \Omega$ the null-solution of deterministic differential equation $\dot{x} = f(x, t, \omega)$ is

1. stable,

2. asymptotically stable,
3. asymptotically stable in the whole,
4. uniformly stable,
5. exponentially stable, or
6. stable under continuously acting disturbances, respectively.

We get a strong notion of stability if we can choose the constants $\delta_1, \delta_2, \dots$ that occur in these stability definitions independently of ω:

Definition 15.4 (Path-Wise Equi-Stability). The null solution of the random differential equation (15.1) on I is called *path-wise equi-stable*, if for every $\varepsilon > 0$ there is a $\delta > 0$ such that

$$\|X_t(X_0, t_0)\| \stackrel{I}{<} \varepsilon ,$$

for all $X_0 \in S_d$ with $\|X_0\| \hat{<} \delta$.

If, additionally, there is a $\eta > 0$ such that

$$\lim_{t \to 0} \|X_t(X_0, t_0)\| \hat{=} 0$$

for all $\|X_0\| \hat{<} \eta$, then the null-solution of (15.1) is called *path-wise asymptotically equi-stable.*

The null solution of (15.1) is called *path-wise exponentially equi-stable* if there are positive constants a, b, and η such that

$$\|X_t(X_0, t_0)\| \stackrel{I}{\leq} a\|X_0\| \exp(-b(t - t_0))$$

holds for all $t_0 \in I$ and $X_0 \in S_d$ with $\|X_0\| \hat{<} \eta$.

Similarly, one obtains generalisations of the deterministic stability concepts to the random setting by taking the norm of the expected value of the realisations instead of the realisations of a path-wise solution themselves. Alternatively, one gets further notions of stochastic stability by considering a function $h(X_t)$ that depends on the initial value X_0.

Let $h : \mathbb{R}^d \to \mathbb{R}$ be continuous, monotonously increasing in $|x_k|$ ($k = 1, 2, \dots, d$) for otherwise fixed $(x_1, \dots, x_{k-1}, x_{k+1}, \dots, x_d)$, and for which $\lim_{\|x\| \to 0} h(x) = 0$ as well as $\lim_{\|x\| \to \infty} h(x) = \infty$ hold.

Definition 15.5 (h-Stability). The null solution of the random differential equation (15.1) on I is called *h-stable*, if for every $\varepsilon > 0$ there is a $\delta > 0$ such that

$$\mathbb{E}\left(h(X_t(X_0, t_0))\right) < \varepsilon$$

for all $t \in I$ and $X_0 \in S_d$ with $\mathbb{E}(h(X_0)) < \delta$.

If, additionally, there is a $\eta > 0$ such that

$$\lim_{t\to\infty} \mathbb{E}\left(h(X_t(X_0, t_0))\right) = 0$$

for all $X_0 \in S_d$ with $\mathbb{E}(h(X_0)) < \eta$, then the null-solution of (15.1) is called *asymptotically h-stable*.

The null solution of (15.1) is called *exponentially h-stable* if there are positive constants a, b, and η such that

$$\mathbb{E}\left(h(X_t(X_0, t_0))\right) \le a\mathbb{E}(h(X_0)) \exp(-b(t - t_0))$$

holds for all $0 \le t_0 \le t$ and all $X_0 \in S_d$ with $\mathbb{E}(h(X_0)) < \eta$.

If $h(x) = \|x\|$, then h-stability is often called m-stability for *stability in the mean*. Analogously, if $h(x) = \|x\|^2$, then h-stability is often called $q.m.$-stability for *stability in the quadratic mean* or *mean square stability*.

Definition 15.6 ($\mathbb{P}$-Stability). The null solution of the random differential equation (15.1) on I is called $\mathbb{P}$*-stable* (stable in probability), if for every $\varepsilon > 0$ and every $p \in (0,1)$ there are numbers $\delta > 0$ and $\gamma \in (0,1)$ such that

$$\|X_t(X_0, t_0)\| \overset{(p)}{<} \varepsilon$$

for all $t \in I$ and $X_0 \in S_d$ with $\|X_0\| \overset{(\gamma)}{<} \delta$.

Definition 15.7 (W-Stability). The null solution of the random differential equation (15.1) on I is called W*-stable* (stochastically stable), if for every $\varepsilon > 0$ and every $p \in (0,1)$ there are numbers $\delta > 0$ and $\gamma \in (0,1)$ such that

$$\|X_t(X_0, t_0)\| \overset{(I,p)}{<} \varepsilon$$

for all $t \in I$ and $X_0 \in S_d$ with $\|X_0\| \overset{(\gamma)}{<} \delta$.

The null solution of (15.1) is called *asymptotically W-stable*, if additionally, for every $\pi \in (0,1)$ there are positive numbers η and χ such that

$$\mathbb{P}\left(\lim_{t\to\infty} \|X_t(X_0, t_0)\| = 0\right) > \pi$$

for all $X_0 \in S_d$ with $\|X_0\| \overset{(\chi)}{<} \eta$.

Assume that in the definitions 15.6 and 15.7 the numbers δ, γ and χ, η, respectively, can be chosen independently of $t_0 \in I$ such that the required inequalities hold for all $t_0 \in I$. Then, the null solution of (15.1) is called *uniformly $\mathbb{P}$-, W-stable* or *uniformly asymptotically W-stable*.

Remark 15.8 (On Initial Conditions From Subsets of S_d). All definitions of stability were formulated with respect to initial conditions $X_0 \in S_d$. Of course, they can also be given in terms of initial conditions from subsets M of S_d, e.g. for non-random initial conditions $X_0 \in M = \mathbb{R}^d$ or for

$$X_0 \in M = \gamma(K) := \{X_0 \in S_d : \|X_0\| \hat{<} K\} ,$$

where $0 < K < \infty$.

For instance, we say that the null solution of (15.1) is $\mathbb{P}$*-stable with respect to* M if for every $\varepsilon > 0$ and $p \in (0,1)$ there are numbers $\delta > 0$ and $\gamma \in (0,1)$ such that

$$\|X_t(X_0, t_0)\| \overset{(p)}{<} \varepsilon$$

for all $t \in I$ and $X_0 \in M$ with $\|X_0\| \overset{(\gamma)}{<} \delta$.

The notions of (stochastic) stability described above enable us to study the stability properties which the null-solution of a random differential equation exhibits. In the deterministic limit, i.e., if $f(x,t,\omega)$ and $X_0(\omega)$ are independent of ω, each of the stability concepts provided in definitions 15.3 to 15.7 becomes a byword for the stability concepts discussed in Sec. 8. Theses stochastic stability concepts are thus really generalisations of the deterministic ones.

15.2.2 Relations Between the Different Stability Notions

Following [45], pp. 97, we next discuss some important relations between the different stochastic stability notions.

Before you continue, make sure to answer the following questions:

Quiz: Section 15.2 – Part I (The Zoo of Stochastic Stability Concepts)

Q1 Give the definition of uniform path-wise stability.

Q2 Give the definition of path-wise equi-stability/ asymptotic path-wise equi-stability/ exponential path-wise equi-stability.

Q3 Give the definition of h-stability/ asymptotic h-stability/ exponential h-stability.

Q4 Give the definitions of stability in the mean and of mean-square stability.

Q5 Give the definitions of $\mathbb{P}$-stability and W-stability. Explain how these two stability notions differ from each other and give simple examples for which they lead (i) to the same and (ii) to a different result.

Proposition 15.9 (Path-Wise Stability Implies Uniform Path-Wise Stability). *Let $f(x,t,\omega)$ be independent of t for almost all $\omega \in \Omega$ or periodic in t for almost all $\omega \in \Omega$, respectively. Then, (asymptotic) path-wise stability of the null-solution of equation (15.1) implies uniform (asymptotic) path-wise stability.*

If the null-solution of equation (15.1) is uniformly asymptotically path-wise stable, then it is also path-wise stable under continuously acting disturbances.

Proof. The assertions follow immediately from the deterministic results covered in Sec. 8. □

The null-solution of equation (15.1) can be path-wise stable without being h-stable.

Example 15.10 (A Path-Wise Stable Solution That is not h-Stable, cf. [45], p. 97, and [42]). Let Z be a random variable and consider the random differential equation

$$\dot{X}_t = \left(\frac{2}{t} - Z\right) X_t$$

for $t_0 > 0$. Its path-wise solution is given by

$$X_t = \frac{X_0}{t_0^2} t^2 \exp\left(-Z(t-t_0)\right) .$$

If Z is a uniformly distributed random variable on $(0,1)$, then it holds that

$$\mathbb{E}\left(|X_t|\right) = \frac{X_0}{t_0^2} \cdot \frac{t^2}{(t-t_0)} \left(1 - \exp\left(-(t-t_0)\right)\right) .$$

Thus, the null-solution is path-wise stable but not h-stable.

Proposition 15.11 (r-th Mean Stability Implies r'-th Mean Stability). *Let $r' < r$. Then, stability in the r-th mean ($h(x) = \|x\|^r$) with respect to deterministic initial conditions ($M = \mathbb{R}^d$) implies stability in the r'-th mean with respect to deterministic initial conditions.*

Proof. This follows from the fact that $(\mathbb{E}(\|X_t\|^r))^{1/r}$ is a monotonously non-decreasing function in r. □

Proposition 15.12 (Path-Wise Stability Implies W-Stability). *Let the null-solution of equation (15.1) be path-wise stable. Then it is W-stable, too.*

Proof. Following [45], pp. 97, assume the null-solution of (15.1) is path-wise stable and let $\varepsilon > 0$ and $p \in (0,1)$ be given. Then, there is a positive $\delta_\varepsilon \in S_1$

such that $\|X_t\| \overset{I}{<} \varepsilon$ holds, if $\|X_0\| \hat{<} \delta_\varepsilon$. Moreover, there is a positive number $\delta(\varepsilon, p)$ such that $\delta_\varepsilon > \delta(\varepsilon, p)$ holds. Thus, for $\|X_0\| < \delta(\varepsilon, p)$ it follows that

$$p < \mathbb{P}\left((\|X_0\| < \delta(\varepsilon, p)) \cap (\delta_\varepsilon > \delta(\varepsilon, p))\right) \leq \mathbb{P}\left(\|X_t\| < \varepsilon \,:\, t \in I\right),$$

which means $\|X_t\| \overset{(I,p)}{<} \varepsilon$. □

Proposition 15.13 (W-Stability Implies $\mathbb{P}$-Stability). *Let the null-solution of equation (15.1) be W-stable. Then it is $\mathbb{P}$-stable, too.*

Proposition 15.14 (Stability w.r.t. Initial Value Sets is Passed on to Subsets). *Let $M_1 \subset M_2 \subset S_d$. Let the null-solution of equation (15.1) be stable with respect to M_2 in the sense of any of the stability definitions 15.3 to 15.7. Then, it is stable with respect to M_1 in the sense of the same stability definition.*

Proposition 15.15 (h-Stability Implies $\mathbb{P}$-Stability). *Let K be an arbitrary positive constant and $\gamma(K) := \{X_0 \in S_d : \|X_0\| \hat{<} K\}$ as defined above. Let the null-solution of equation (15.1) be h-stable with respect to $\gamma(K)$. Then, it is also $\mathbb{P}$-stable with respect to $\gamma(K)$.*

Proof. Following [45], pp. 98, let

$$\overline{h}(\varepsilon) := \sup_{x \in \mathbb{R}^d,\, \|x\| \leq \varepsilon} h(x) \quad \text{and} \quad \underline{h}(\varepsilon) := \inf_{x \in \mathbb{R}^d,\, \|x\| \geq \varepsilon} h(x).$$

Let the null-solution of (15.1) be h-stable with respect to $\gamma_n(K)$. For given $\varepsilon > 0$ and $p \in (0,1)$ choose $\delta' > 0$ such that

$$\mathbb{E}\left(h\left(X_t(X_0, t_0)\right)\right) < \underline{h}(\varepsilon)(1-p)\,, \text{for all } t \in I\,, \tag{15.2}$$

holds for $\mathbb{E}(h(X_0)) < \delta'$.

Moreover, for this δ' we choose $\delta > 0$ and $\gamma \in (0,1)$ such that

$$(1-\gamma)\overline{h}(K) < \tfrac{1}{2}\delta', \qquad \text{and} \qquad \overline{h}(\delta) < \tfrac{1}{2}\delta'$$

holds. Thus,

$$\mathbb{E}\left(h\left(\|X_0\|\right)\right) \leq \overline{h}(\delta) + (1-\gamma)\overline{h}(K) < \delta' \tag{15.3}$$

for $\|X_0\| \overset{(\gamma)}{<} \delta$.

Next, (15.3) implies (15.2) and consequently $\|X_t(X_0, t_0)\| \overset{(p)}{<} \varepsilon$ for all $t \in I$ because of

$$\underline{h}(\varepsilon)(1-p) > \mathbb{E}\left(h\left(X_t(X_0, t_0)\right)\right) \geq \underline{h}(\varepsilon)\mathbb{P}\left(\|X_t(X_0, t_0) \geq \varepsilon\|\right), \quad \text{for all } t \in I.$$

This shows the assertion. □

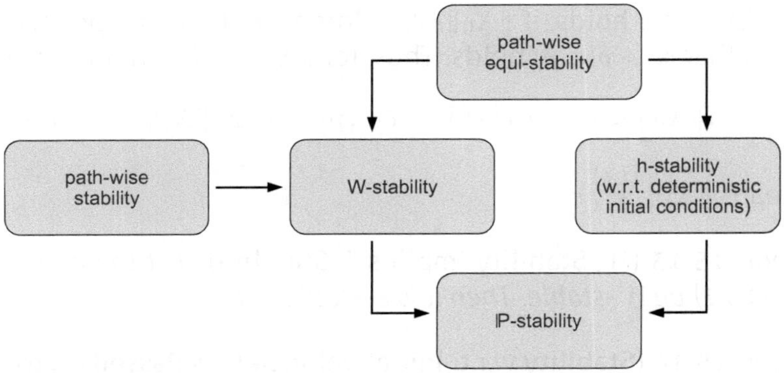

Figure 15.1. Connections between the different notions of the stochastic stability concepts for random differential equations.

Proposition 15.16 (Path-Wise Equi-Stability Implies W- & h-Stability). *Let the null-solution of equation (15.1) be path-wise equi-stable. Then it is W-stable (and thus $\mathbb{P}$-stable) as well as h-stable with respect to deterministic initial conditions ($M \subset \mathbb{R}^d$), too.*

Proof. Following [45], pp. 98, the first part is easy. We continue with the second part: Let the null-solution of (15.1) be path-wise equi-stable. As in the proof of the previous lemma, let $\overline{h}(\varepsilon) := \sup_{x\in\mathbb{R}^d,\, \|x\|\leq\varepsilon} h(x)$.

For given $\varepsilon > 0$ choose $\varepsilon' > 0$ and $\delta > 0$ such that $\overline{h}(\varepsilon') < \varepsilon$ and $\|X_t(X_0, t_0)\| \overset{I}{<} \varepsilon'$ for $\|X_0\| \hat{<} \delta$. Then, the estimate

$$\mathbb{E}\left(h(X_t)\right) \leq \mathbb{E}\left(\sup_{t\geq t_0} h(X_t)\right) \leq \overline{h}(\varepsilon') < \varepsilon$$

follows from $\|X_0\| \hat{<} \delta$. □

As a summary, Fig. 15.1 sketches the connections that we just derivedand implications between the different notions of the stochastic stability concepts for random differential equations.

To have a set of examples ready, let us apply at least the path-wise stability notions from definition 15.3 to linear random differential equations as studied in Chaps. 12 and 13.

15.2.3 Stability of Path-Wise Solutions of Linear RODEs

As a first example, let us discuss the inhomogeneous linear random differential equation

$$\dot{X}_t = A(t)X_t + Z_t \tag{15.4}$$

with a real (time-dependent) coefficient matrix, see Chap. 12. Let X_t^0 be an arbitrary fixed path-wise solution of (15.4), then the difference $U_t := X_t - X_t^0$ satisfies the homogeneous linear deterministic differential equation

$$\dot{U}_t = A(t)U_t\,. \tag{15.5}$$

Let the null-solution of equation (15.5) be stable/ asymptotically stable/ exponentially stable, then any arbitrary path-wise solution of (15.4) is path-wise

Before you continue, make sure to answer the following questions:

Quiz: Section 15.2 – Part II (Relations Between the Different Stability Notions)

Q1 Which conditions are required such that (asymptotic) path-wise stability implies uniform (asymptotic) path-wise stability?

Q2 Show that uniform asymptotic path-wise stability implies path-wise stability w.r.t. continuously acting disturbances.

Q3 Does path-wise stability always imply h-stability? Give an example or a counter-example.

Q4 Let X_t be a mean-square stable solution of $\dot{X} = f(X_t, t, \omega)$ with initial condition $X_{t_0} \in \mathbb{R}$. Is X_t stable in the mean? Why?

Q5 Show that path-wise stability of the null-solution implies its W-stability.

Q6 Show that W-wise stability of the null-solution implies its $\mathbb{P}$-stability.

Q7 Under which conditions does h-stability imply $\mathbb{P}$-stability?

Q8 Sketch the proof of the assertion you used in Q7.

Q9 Under which conditions does path-wise equi-stability imply W-stability?

Q10 Sketch the proof of the assertion you used in Q9.

Q11 Why does the setting of Q9 also imply h-stability?

equi-stable/ path-wise asymptotically equi-stable/ path-wise exponentially equi-stable.

Next, we study linear random differential equation with stochastic coefficient matrices as done in Chap. 13, namely

$$\dot{X}_t = (A + F_t)\, X_t\,, \tag{15.6}$$

and

$$\dot{X}_t = (A + F_t + C(t))\, X_t\,. \tag{15.7}$$

Let the conditions of Theorem 13.2 or 13.4 be satisfied, then the null-solution of (15.6) is path-wise asymptotically stable in the whole. If the conditions of Theorem 13.5 are satisfied, then the null-solution of (15.7) is path-wise asymptotically stable in the whole.

15.3 Lyapunov-Functions and Stability of Solution of RODEs

In this section, sufficient conditions for important types of stochastic stability are studied, cf. [45], pp. 99. As in the deterministic setting, the tools of choice are specifically designed Lyapunov-functions. Together with the form of the random differential equations, these (comparison) functions enable us to gain results on the stability of solutions without any further knowledge of the solutions themselves.

15.3.1 Lyapunov-Functions and h-Stability

Let $h(x)$ be as given in definition 15.5 and let $v(x, t, \omega)$ be a real function defined on $\mathbb{R}^d \times I \times \Omega$ that is $\mathcal{B}^d \times \mathcal{A}$-measurable for fixed $t \in I$. Moreover, let us use the following notations

$$\begin{cases} H_t := h(X_t(X_0, t_0))\,, & h(t) := \mathbb{E}(H_t) \\ V_t := V(X_t(X_0, t_0), t, \omega)\,, & v(t) := \mathbb{E}(V_t) \end{cases}. \tag{15.8}$$

Theorem 15.1 (h-Stability w.r.t. $M \subset S_d$). *Let there be a function $v(x, t, \omega)$ and a set $M \subset S_d$ such that the following three assumptions hold:*

1. *For an arbitrary sequence $\{X_0^{(\nu)}\}_{\nu \in \mathbb{N}} \subset M$ such that $\lim_{\nu\to\infty} \mathbb{E}(h(X_0^{(\nu)}))$ it holds that*
$$\lim_{\nu\to\infty} \mathbb{E}\left(v(X_0^{(\nu)}, t_0, \omega)\right) = 0\,.$$
2. *There is a real constant $a > 0$ such that*
$$v(x, t, \omega) \overset{\mathbb{R}^d \times I}{\geq} a h(x)\,.$$

Before you continue, make sure to answer the following questions:

Quiz: Section 15.2 – Part III (Stability of Path-Wise Solutions of Linear RODEs)

Q1 Let X_t be a path-wise solution of

$$\dot{X}_t = \begin{pmatrix} \cos(t) & 1 \\ \sin(t) & \exp(-t) \end{pmatrix} X_t + \begin{pmatrix} 0 \\ 1 \end{pmatrix} W_t ,$$

where W_t is the 1-dimensional standard Wiener process. What can you say about the stability of X_t?

Q2 Let the following random differential equation be given:

$$\dot{X}_t = \begin{pmatrix} -1 + W_t & 1 \\ O_t & -3 \end{pmatrix} X_t ,$$

where W_t is the 1-dimensional standard Wiener process and O_t an Ornstein-Uhlenbeck process such that W_t and O_t are independent. What can you say about the stability of the null-solution?

Q3 Let the following random differential equation be given:

$$\dot{X}_t = \begin{pmatrix} -\cos(t) + W_t & \sin(2t) \\ O_t & 3\exp(-t) - 2 \end{pmatrix} X_t ,$$

where W_t is the 1-dimensional standard Wiener process and O_t an Ornstein-Uhlenbeck process such that W_t and O_t are independent. What can you say about the stability of the null-solution?

3. *There is a real constant $k > 0$ such that $\sup_{t \in I} v(t) = v(t_0)$ holds for $X_0 \in M$ and $h(t_0) < k$.*

Then the null-solution of (15.1) is h-stable with respect to M.

Moreover, if additionally the following condition holds:

4. *$\lim_{t \to \infty} v(t) = 0$ holds for $X_0 \in M$, and there is a $\eta > 0$ such that $h(t_0) < \eta$.*

Then the null-solution of (15.1) is asymptotically h-stable with respect to M.

Finally, let conditions 1-4 and the following two assumptions be true:

5. *Let $a > 0$ be as in condition 2 and let there be a constant $b > 0$. Then it holds that*

$$a h(x) \overset{\mathbb{R}^d \times I}{\leq} v(x, t, \omega) \overset{\mathbb{R}^d \times I}{\leq} b h(x) .$$

6. *$h(t)$ is continuous and there is a $\eta > 0$ such that*

$$v(t) - v(t_0) \le -c \int_{t_0}^{t} h(\tau) d\tau \,, \quad \textit{for all } t \in I \textit{ and } t \ge t_0 \,,$$

holds for $X_0 \in M$ and $h(t_0) < \eta$, where $c > 0$ is a suitably chosen constant.

Then the null-solution of (15.1) is exponentially h-stable with respect to M.

Proof. Following [45], p. 100, first let conditions 1-3 hold. For given $\varepsilon > 0$ choose a $\delta > 0$ ($\delta < k$) such that for $X_0 \in M$ and $h(t_0) < \delta$ it holds that $v(t_0) < \varepsilon$. Then, we obtain

$$a \sup_{t \in I} h(t) \le \sup_{t \in I} v(t) = v(t_0) < \varepsilon a \,,$$

which implies the first assertion.

Next, let conditions 1-4 hold. We will give an indirect proof of the second assertion: Therefore, let us assume that $\limsup_{t\to\infty} h(t) = \gamma > 0$ holds for a $X_0 \in M$ with $h(t_0) < \min(k, \eta)$. Then, there exists a sequence $t_\nu \to \infty$ such that $h(t_\nu) > \frac{1}{2}\gamma$ holds for all $\nu = 1, 2, \ldots$. Because of condition 2 this implies $v(t_\nu) \ge \frac{1}{2} a\gamma$. This is a contradiction to condition 4.

Finally, let conditions 1-6 hold, then we have

$$ah(t) - bh(t) \le v(t) - v(t_0) \le -c \int_{t_0}^{t} h(\tau) \mathrm{d}\tau \,, \quad \text{for all } t \in I \text{ and } t \ge t_0 \,.$$

Gronwall's Lemma (Lemma 3.15) now implies

$$h(t) \le \frac{b}{a} h(t_0) \exp\left(-\frac{c}{a}(t - t_0)\right) , \quad \text{for all } t \in I \text{ and } t \ge t_0 \,,$$

which shows the third assertion. □

The h-stability of the null-solution can now be shown for a given system by constructing a Lyapunov-function v that satisfies the conditions of Theorem 15.1. Beyond the usual difficulties in the deterministic case, the construction of a stochastic Lyapunov-function according to Theorem 15.1 is complicated by the form of conditions 3, 4 and 6 respectively. These conditions are given in a way that, in general, does not permit one to check their validity solely based on the given random differential equation and without explicit knowledge about the solution. In particular, before being valuable for concrete applications, conditions 3, 4 and 6, respectively, have to be transformed into a suitable form by making certain (additional) assumptions about the given random differential equations (and their solutions).

The Lyapunov-function method can often be applied successfully to linear systems by constructing a quadratic Lyapunov-function v such that the conditions 1 and 2 of Theorem 15.1 are fulfilled with $h(x) = \|x\|^2$ and it remains to study 3 and 4, respectively. The following lemma gives a sufficient requirement for conditions 3 and 4.

Lemma 15.17 (Tangible Requirements for Conditions 3 and 4). *Let $v(x,t,\omega) = v(x,t)$ be defined on $\mathbb{R}^d \times I_0$ and be continuous. Moreover, the partial derivatives of degree one $(\partial_x v)^T = (\partial_{x_1} v, \partial_{x_2} v, \ldots, \partial_{x_d} v)$ and $\partial_t v$ exist and are continuous on $\mathbb{R}^d \times I_0$.*

For the function

$$\dot{v}(x,t,\omega) = \partial_t v + (\partial_x v)^T f(x,t,\omega)$$

assume that an inequality of the form

$$\dot{v}(x,t,\omega) \overset{\mathbb{R}^d \times I_0}{\leq} \eta_t v(x,t) \tag{15.9}$$

holds, where η_t is a stochastic process independent of all $X_0 \in M$ such that almost all of its realisations on any finite sub-interval of I_0 are Lebesgue-integrable, and such that

$$\mathbb{E}\left(\exp\left(\int_{t_0}^{t} \eta_\tau d\tau\right)\right) = y(t_0,t)$$

exists for all $t \in I_0$. Then, condition 3 of Theorem 15.1 at $y(t_0,t) \leq 1, t \in I_0$, is satisfied, and condition 4 of Theorem 15.1 is satisfied at $\lim_{t\to\infty} y(t_0,t) = 0$.

Proof. Following [45], p. 101, we note that (15.9) implies the inequality

$$\frac{\mathrm{d}V_t}{\mathrm{d}t} = \dot{v}(X_t,t,\omega) \overset{I_0}{\leq} \eta_t V_t$$

and thus

$$V_t \overset{I_0}{\leq} V_{t_0} + \int_{t_0}^{t} \eta_\tau V_\tau \mathrm{d}\tau .$$

Applying Gronwall's Lemma (Lemma 3.15) we obtain

$$v(t) \leq v(t_0)\mathbb{E}\left(\exp\left(\int_{t_0}^{t} \eta_\tau \mathrm{d}\tau\right)\right)$$

due to the independence of X_0 and η_t, $t \in I_0$. This implies the assertion. □

Analogously to the above consideration, we can also obtain sufficient conditions for exponential h-stability.

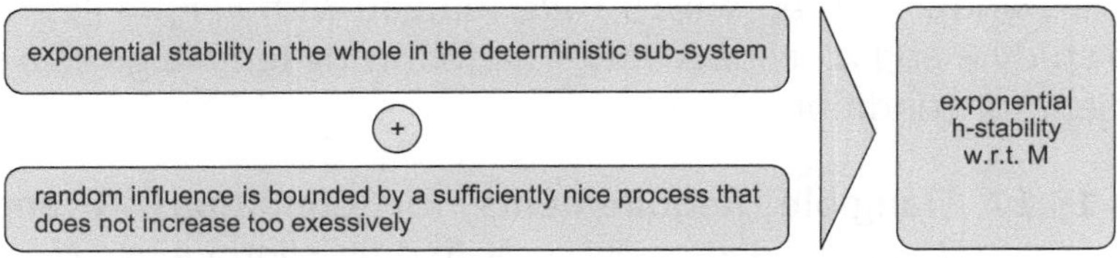

Figure 15.2. Conditions and assertion of Theorem 15.2.

Lemma 15.18 (Sufficient Conditions for Exponential h-Stability). *Let $v(x,t)$ satisfy condition 5 of Theorem 15.1 and the conditions of Lemma 15.17 for all $t \in I_0$. Moreover, let there be positive number c and d such that*

$$y(t_0,t) \;\le\; c\exp(-d(t-t_0))\,, \quad t \in I\,, \quad t_0 \le t\,.$$

Then the null-solution of (15.1) is exponentially h-stable with respect to M.

Next, let us study the random ordinary differential equation

$$\frac{\mathrm{d}X_t}{\mathrm{d}t} \;=\; f(X_t,t) + g(X_t,t,\omega)\,, \tag{15.10}$$

where the same conditions are fulfilled as in equation (15.1), and $f(x,t)$ is defined on $\mathbb{R}^d \times I$ and $g(x,t,\omega)$ is defined on $\mathbb{R}^d \times I \times \Omega$, respectively. Moreover, let $f(0,t) = 0$ for all $t \in I$, and require that the Jacobi matrix $\partial_x f$ exists and is continuous on $\mathbb{R}^d \times I$ and bounded.

Let the deterministic ordinary differential equation

$$\frac{\mathrm{d}x}{\mathrm{d}t} \;=\; f(x,t) \tag{15.11}$$

have a unique solution for every initial condition $(x_0,t_0) \in \mathbb{R}^d \times I$. Then, provided the stochastic influence $g(x,t,\omega)$ in (15.10) is sufficiently bounded, we can deduce exponential h-stability of the null-solution of the random differential equation (15.10) from exponential stability in the whole of the null-solution of the deterministic ordinary differential equation (15.11), see Fig. 15.2:

Theorem 15.2 (Exponential h-Stability w.r.t. M). *Let the following conditions be satisfied:*

1. *The null-solution of the deterministic equation (15.11) is exponentially stable in the whole.*

2. *It holds that*

$$\|g(x,t,\omega)\| \;\overset{\mathbb{R}^d\times I}{\le}\; L_t\|x\|\,,$$

where L_t is a stochastic process such that almost all of its realisations are Lebesgue-integrable on every sub-interval of I.

3. *For sufficiently large $a > 0$ and sufficiently small $\rho > 0$, it holds for a $b > 0$ that*

$$\mathbb{E}\left(\exp\left(a\int_{t_0}^{t} L_\tau d\tau\right)\right) \leq b\exp(\rho(t-t_0)), \quad t_0 \in I, \quad t \geq t_0.$$

Moreover, let M be the set of all random variables from S_d that are independent of the stochastic process L_t, $t \in I$. Then the null-solution of the random equation (15.10) is exponentially h-stable with respect to M, where $h(x) = \|x\|^{2r}$.

Proof. Following [45], p. 102, due to condition 1 and together with Krasovskii's theorem (Theorem 8.1), there is a real function $w(x,t)$ defined on $\mathbb{R}^d \times I$ with continuous partial derivatives $\partial_x w$ and $\partial_t w$ on $\mathbb{R}^d \times I$ such that w has the properties 1 - 3 stated in Theorem 8.1. Because of properties 1 - 3 from Theorem 8.1 and condition 2 we obtain

$$\dot{w}(x,t,\omega) = \partial_t w + (\partial_x w)^T (f(x,t) + g(x,t,\omega)) \overset{\mathbb{R}^d\times I}{\leq} -c_3\|x\|^2 + L_t c_4 \|x\|^2, \tag{15.12}$$

which implies

$$\dot{w}(x,t,\omega) \overset{\mathbb{R}^d\times I}{\leq} \eta_t w(x,t), \tag{15.13}$$

with $\eta_t := c_4 c_1^{-1} L_t - c_3 c_2^{-1}$. Analogously to the proof of Lemma 15.17, this leads to the inequality

$$w(X_t,t) \overset{I_0}{\leq} w(X_0,t)\exp\left(\int_{t_0}^{t} \eta_\tau \mathrm{d}\tau\right). \tag{15.14}$$

From the properties 1 - 3 of Theorem 8.1 and (15.14) together with $h(x) = \|x\|^{2r}$ it follows that

$$h(X_t) \leq c_1^{-r} c_2^r h(X_0)\exp\left(r\int_{t_0}^{t} \eta_\tau \mathrm{d}\tau\right), \quad t_0 \in I, \quad t \geq t_0. \tag{15.15}$$

If condition 3 is satisfied with $\rho < c_3 c_2^{-1} \cdot r$ and $a > bc_1^{-1} \cdot r$, and if $\alpha \in (\rho, c_3 c_2^{-1} \cdot r)$ holds, then (15.15) leads to

$$\begin{aligned} h(t) &\leq c_1^{-r} c_2^r h(t_0)\mathbb{E}\left(\exp\left(a\int_{t_0}^{t} L_\tau \mathrm{d}\tau\right)\right)\exp\left(-\alpha(t-t_0)\right) \\ &\leq c_1^{-r} c_2^r h(t_0)\exp\left(-(\alpha-\rho)(t-t_0)\right), \; t_0 \in I, \; lt \geq t_0. \end{aligned} \tag{15.16}$$

This shows the assertion. □

Finally, we study the linear random differential equation with stochastic coefficients

$$\dot{X}_t = A_t X_t , \tag{15.17}$$

where A_t is a path-wise continuous process on I. We set $v(x,t) = x^T Bx$, with a symmetric positive definite matrix B. Following the same considerations as in the proof of Theorem 13.5 we obtain inequality (15.9) with the process

$$\eta_t = \lambda_{\text{max}}[A_t^T + BA_t B^{-1}] .$$

Let M be the set of all initial conditions $X_0 \in L_d^2$ that are independent of the matrix-process A_t, $t \in I$. Due to

$$\lambda_{\text{min}}[B]\|x\|^2 \le v(x,t)\ \lambda_{\text{max}}[B]\|x\|^2 ,$$

Theorem 15.1, Lemma 15.17, and Lemma 15.18 imply

Theorem 15.3 (Mean-Square Stability w.r.t. M). *Let there be a symmetric positive definite matrix B such that*

$$y(t_0,t) = \mathbb{E}\left(\exp\left(\int_{t_0}^{t} \lambda_{\text{max}}[A_\tau^T + BA_\tau B^{-1}]d\tau\right)\right) \le 1 , \quad t \in I_0 . \tag{15.18}$$

Then, the null-solution of (15.17) is $m.s.$-stable with respect to M.

If additionally $\lim_{t\to\infty} y(t_0,t) = 0$ holds, then the null-solution of (15.17) is asymptotically $m.s.$-stable with respect to M.

Moreover, if

$$y(t_0,t) \le c \cdot \exp\left(-d(t-t_0)\right) , \quad t_0 \in I , \quad t > t_0$$

also holds, then the null-solution of (15.17) is exponentially $m.s.$-stable with respect to M.

Analogously to the corollaries to Theorem 13.5, one may get more practical conditions by replacing η_t with upper bounds. For instance, if $A_t = A + F_t$ holds with a matrix A the eigenvalues of which all have negative real part, and if we choose B as the solution of the matrix equality $A^T B + BA = -\mathbb{I}$, then we get

$$\eta_t \overset{I}{\le} -\lambda_{\text{max}}^{-1}[B] + \left\|F_t^T + BF_t^T B^{-1}\right\| .$$

For equation (15.18) we thus obtain the following sufficient condition:

$$\mathbb{E}\left(\exp\left(\int_{t_0}^{t} \|F_\tau^T + BF_\tau^T B^{-1}\|\, d\tau\right)\right) \le \exp\left(\lambda_{\text{max}}[B](t-t_0)\right) , \quad t \in I_0 .$$

Example 15.19 (A Second Order Random Differential Equation with Stochastic Perturbation at the Position Term, cf. [45], p. 104). Let us discuss the second order random differential equation

$$\ddot{X}_t + 2b\dot{X}_t + (1 + Z_t)\, X_t \;=\; 0\,, \tag{15.19}$$

with $b > 0$ and a path-wise continuous stochastic process Z_t.

Applying the solution methods from Chap. 13 we get

$$\begin{aligned} \eta_t &= \lambda_{\max}[A_t^T + BA_tB^{-1}] \\ &= \begin{cases} -2b + |Z_t|(1-b^2)^{-1/2} & \text{if } 2b^2 \le 1 \\ -2b + |2b^2 - 1 - Z_t|b^{-1} & \text{if } 2b^2 \ge 1 \end{cases} \end{aligned}$$

for a certain choice of B. For the mean-square stability of the null-solution of (15.19)

$$\left(\mathbb{E}\left(|X_t|^2\right) + \mathbb{E}\left(|\dot{X}_t|^2\right) \;<\; \varepsilon \quad \text{at} \quad \mathbb{E}\left(|X_0|^2\right) + \mathbb{E}\left(|\dot{X}_0|^2\right) \;<\; \delta\right)$$

we hence have the following sufficient conditions:

$$u_1(t) \;=\; \mathbb{E}\left(\exp\left((1-b^2)^{-1/2}\int_{t_0}^{t}|Z_\tau|\mathrm{d}\tau\right)\right) \;\le\; \exp\left(2b(t-t_0)\right)\,, \quad t \in I_0 \tag{15.20}$$

at $2b^2 \le 1$ or

$$u_2(t) \;=\; \mathbb{E}\left(\exp\left(b^{-1}\int_{t_0}^{t}|2b^2 - 1 - Z_\tau|\mathrm{d}\tau\right)\right) \;\le\; \exp\left(2b(t-t_0)\right)\,, \quad t \in I_0 \tag{15.21}$$

at $2b^2 \ge 1$.

For the exponential mean-square stability of the null-solution of (15.19), a sufficient condition is the existence of positive numbers c and d such that

$$u_i(t) \;\le\; c \cdot \exp\left((2b - d)(t - t_0)\right)\,, \quad t_0 \in I_0\,, \quad t \ge t_0$$

with $i = 1$ at $2b^2 \le 1$ and $i = 2$ at $2b^2 \ge 1$, respectively.

Next, let us assume that Z_t is a Gaussian process with mean $m_Z(t)$, covariance $C_Z(\tau, t)$ and variance $\sigma_Z^2(t) = C_Z(t,t)$. Assume for all $t \ge 0$

$$|m_Z(t)| \;\le\; \alpha\,, \quad \sigma_Z^2(t) \;\ge\; \beta^2\,, \quad \text{and} \quad \int_0^\infty C_Z(\tau, t)\mathrm{d}\tau \;\le\; \gamma\,.$$

In the case $2b^2 \le 1$ we obtain, from (15.20) together with Lemma 15.2, the sufficient condition

$$\alpha + \beta + \left(2\sqrt{1-b^2}\right)^{-1}\gamma \;\le\; 2b\sqrt{1-b^2} \tag{15.22}$$

for mean-square stability. In particular, if

$$\alpha+\beta+\left(2\sqrt{1-b^2}\right)^{-1}\gamma \;<\; 2b\sqrt{1-b^2} \tag{15.23}$$

holds, then the null-solution of (15.19) is even exponentially mean-square stable.

In the case $2b^2 \geq 1$ we apply Lemma 15.2 to the Gaussian process $Z_t^T = Z_t + 1 - 2b^2$. Then, analogously to (15.23) we obtain the sufficient condition

$$\left|\mathbb{E}\left(Z_t\right)+1-2b^2\right| \;<\; 2b^2-\beta-(2b)^{-1}\gamma-\varepsilon\,, \quad t\in I \tag{15.24}$$

with an arbitrary $\varepsilon > 0$. Let $\mathbb{E}\left(Z_t\right) \geq m$.

Squaring both sides of (15.24), we get a sufficient condition for the exponential mean-square stability of the null-solution of (15.19), namely

$$2b^2 \;\geq\; 1\,, \quad 2b^2 \;>\; Q\,, \quad \alpha^2+2m+1-Q^2 \;<\; 4b^2\left(1+m-Q\right)\,, \tag{15.25}$$

where $Q := \beta + (2b)^{-1}\gamma$.

If Z_t is a stationary Gaussian process such that

$$m_Z(t) \;=\; 0\,, \quad \sigma^2(t) \;=\; \sigma^2\,, \quad \text{and} \quad C_Z(\tau,t) \;=\; \sigma^2\exp\left(-\rho|t-\tau|\right)\,,$$

then, from (15.23) and (15.24), we get the following sufficient conditions for exponential mean-square stability:

$$\sigma+\sigma^2\left(2\rho\sqrt{1-b^2}\right)^{-1} \;<\; 2b\sqrt{1-b^2}\,,$$

if $2b^2 \leq 1$ and

$$q \;<\; 1\,, \quad 4b^2 \;>\; \frac{1-q^2}{1-q}$$

if $2b^2 \geq 1$ with $q := \sigma + \sigma^2(2b\rho)^{-1}$.

15.3.2 Lyapunov-Functions and Path-Wise Equi-Stability

Let K be an arbitrary fixed positive number. We say that a real-valued function $v(x,t,\omega)$ on $\mathbb{R}^d \times I \times \Omega$ belongs to the class $F(K)$ if the following holds:

1. $v(0,t,\omega) \hat{=} 0$, $t \in I_0$.
2. There is a continuous, real-valued and monotonically increasing function φ on $[0,K]$ such that $\varphi(0) = 0$ and

   $$|v(x,t,\omega)| \overset{R_K^d\times I_0}{\geq} \varphi\left(\|x\|\right)\,,$$

   where $R_K^d := \{x \in \mathbb{R}^d : \|x\| < K\}$.

3. There is a continuous real-valued function ψ on R^d_K such that $\psi(0) = 0$ and

$$v(x, t, \omega) \overset{R^d_K}{\leq} \psi(x) .$$

Theorem 15.4 (Path-Wise Equi-Stability). *Let there be a function v that belongs to the class $F(K)$ such that for every $X_0 \in \gamma(K)$ the function $V_t = v(X_t(X_0, t_0), t, \omega)$ has the following property: for almost all $\omega \in \Omega$ it holds that $V_t(\omega)$ is absolutely continuous on I_0 with $\dot{V}_t(\omega) \leq 0$ for almost all $t \in I_0$. Then, the null-solution of (15.1) is path-wise equi-stable.*

If, additionally, it holds for almost all $\omega \in \Omega$ that

$$\dot{V}_t(\omega) \leq -cV_t(\omega) \qquad \text{for almost all } t \in I ,$$

Before you continue, make sure to answer the following questions:

Quiz: Section 15.3 – Part I (Lyapunov-Functions and h-Stability)

Q1 With the aid of a suitably chosen Lyapunov-function give sufficient conditions for h-stability of the null-solution of $\dot{X} = f(X_t, t, \omega)$.

Q2 Sketch the proof the assertion you used in Q1.

Q3 Which conditions additional to those given in Q1 are required for (i) asymptotic and (ii) exponential h-stability?

Q4 Let the following random differential equation be given:

$$\dot{X}_t = \begin{pmatrix} -1 & 0 \\ 0 & -1 \end{pmatrix} X_t + g(X_t, t, \omega) .$$

Which requirements have to be imposed on the random function g and the initial conditions such that the null-solution of this equation is exponentially h-stable in some sense?

Q5 Consider the linear homogeneous random differential equation

$$\dot{X}_t = \begin{pmatrix} -1 & 0 \\ 1 & -1 \cdot \exp(-t) \end{pmatrix} W_t X_t ,$$

where W_t is the 1-dimensional standard Wiener process. According to Theorem 15.3 mean-square stability w.r.t. M of the null-solution can be shown if there is a suitable quadratic matrix B. Which conditions have to be imposed on B? Construct a suitable matrix B for the given equation.

where c is a positive constant, then the null-solution of (15.1) is path-wise asymptotically equi-stable.

Moreover, if for positive constants a and b

$$a\|x\|^r \overset{R_K^d \times I_0}{\geq} v(x,t,\omega) \overset{R_K^d \times I_0}{\geq} b\|x\|^r \qquad (r > 0)$$

holds together with

$$\dot{V}_t(\omega) \leq -cV_t(\omega) \qquad \textit{for almost all } t \in I\,,$$

for almost all $\omega \in \Omega$, then the null-solution of (15.1) is path-wise exponentially equi-stable.

Proof. Following [45], p. 106, let $\varepsilon > 0$ with $\varepsilon < K$ be given. We choose $\delta > 0$ such that $\sup_{\|x\|<\delta} |v(x,t,\omega)| \hat{<} \phi(\varepsilon)$.

According to the conditions, we have at $\|X_0\| \hat{<} \delta$ that

$$V_t \overset{T}{\leq} V_{t_0} \,\hat{<}\, \varphi(\varepsilon) \tag{15.26}$$

holds. Let us next assume that $\|X_0\| \hat{<} \delta$ does not imply $\|X_t\| \hat{<} \varepsilon$, then there exist $X_0 \in S_d$ such that $\|X_0\| \hat{<} \delta$, and $\Omega_0 \in \mathcal{A}$ as well as a mapping $\tau : \Omega_0 \to I_0$ such that $\mathbb{P}(\Omega_0) > 0$ and $\|x(\tau(\omega), X_0(\omega), t_0, \omega\| \geq \varepsilon$ for $\omega \in \Omega_0$. Hence, for $\omega \in \Omega_0$ it would hold that $V_{\tau(\omega)} \geq \varphi(\varepsilon)$ which is a contradiction to (15.26). This shows the path-wise equi-stability.

If $\dot{V}_t \overset{I_0}{\leq} -cV_t$ holds with a positive constant c, then Gronwall's Lemma (Lemma 3.15) implies for $\|X_0\| \hat{\leq} K$

$$\varphi\left(\|X_t\|\right) \overset{I_0}{\leq} V_t \overset{T_0}{\leq} V_0 \exp\left(-c(t-t_0)\right) \overset{I_0}{\leq} \psi_K \exp\left(-c(t-t_0)\right)\,, \tag{15.27}$$

where $\psi_K = \sup_{x \in R_K^d} \psi(x)$.

This implies $\lim_{t\to\infty} \varphi\left(\|X_t\|\right) \hat{=} 0$. Due to the continuity of φ and $\varphi(r) > 0$ at $r > 0$ this leads to the claimed convergence $\lim_{t\to\infty} \|X_t\| \hat{=} 0$.

If the final requirements of the theorem are met, then one can show at $\|X_0\| \leq K$ the following inequality

$$\|X_t\| \overset{I_0}{\leq} a^{-1/r} b^{1/r} \|X_0\| \exp\left(-\tfrac{c}{r}(t-t_0)\right)\,, \qquad t_0 \in I\,. \tag{15.28}$$

analogously to (15.27), see problem 15.28. □

Example 15.20 (Path-Wise Exponential Equi-Stability in the Whole, cf. [45], pp. 106). Let $A_0 = (a_{ij}^0)$ be a constant real $d \times d$-matrix such that all its eigenvalues have negative real part. Then there is a symmetric positive definite matrix B that is the solution of the matrix equation

$$A_0^T B + BA_0 = -\mathbb{I}\,. \tag{15.29}$$

Let $C[A] := A^T B + BA$. The eigenvalues of the symmetric matrix $C[A]$ depend continuously on the matrix A and are all real. Let $\rho(A)$ denote the largest eigenvalue of $C[A]$ and let $-\rho$ be a fixed number from the interval $(-1, 0)$. Then there is a neighborhood $Q := \{A : \|A - A_0\| < \pi\}$ of A_0, such that

$$\rho(A) \leq -\rho < 0, \qquad A \in Q \tag{15.30}$$

holds. Let f be a linear or non-linear vector-valued function such that

$$f(x, t, \omega) = A(x, t, \omega) x, \tag{15.31}$$

where A is a matrix-valued function with the property

$$\|A(x, t, \omega) - A_0\| \overset{\mathbb{R}^d \times I}{<} \pi. \tag{15.32}$$

Then, the null-solution of (15.1) is path-wise exponentially equi-stable in the whole.

To see this, we discuss the Lyapunov-function $v(x) = x^T B x \in F(\infty)$. Due to (15.29) and (15.30) the path-wise derivative of V_t satisfies

$$\begin{aligned} \dot{V}_t &= X_t^T C[A(X_t, t, \omega)] X_t \overset{I}{\leq} \sup_{A \in Q} X_t^T C[A] X_t \overset{I}{\leq} -\rho X_t^T X \\ &\overset{I}{\leq} -\rho \lambda_{max}^{-1}[B] V_t, \end{aligned} \tag{15.33}$$

where $\lambda_{max}[B]$ is the largest eigenvalue of the symmetric positive definite matrix B. The function $v(x)$ satisfies the conditions of Theorem 15.4, which shows the assertion.

15.4 Excursion: Stability of Deterministic Systems Subject to Continuously Acting Random Perturbations

Finally, as in [45], p. 116, we study the stability of the null-solution of a deterministic ordinary differential equation subject to continuously acting random perturbations.

Let $f(x, t) : \mathbb{R}^d \times I \to \mathbb{R}^d$ be continuous and $g(x, t, \omega) : \mathbb{R}^d \times I \times \Omega \to \mathbb{R}^d$ be path-wise continuous. We discuss the differential equations

$$\dot{x} = f(x, t), \tag{15.34}$$

and

$$\dot{X}_t = f(X_t, t) + g(X_t, t, \omega), \tag{15.35}$$

such that for any initial condition $(x_0, t) \in \mathbb{R}^d \times I$ there is a unique solution $x(t)$ of (15.34) on I as well as a unique path-wise solution X_t of (15.35) on I. Assume $x(t) = 0$, $t \in I$, is an isolated singular point of (15.34) on I. Let H denote the class of all functions g for which the corresponding random differential equation (15.35) has an isolated singular point at zero for almost all $\omega \in \Omega$. Moreover, let G denote the class of all path-wise continuous functions g on $\mathbb{R}^d \times I$ such that there are real functions $l(t)$ and $m(t)$ for which

$$\|g(x_1, t, \omega) - g(x_2, t, \omega)\| \overset{\mathbb{R}^d \times \mathbb{R}^d \times I}{\leq} l(t)\, \|x_1 - x_2\| \,,$$

and

$$\|g(x, t, \omega)\| \overset{\mathbb{R}^d \times I}{\leq} m(t)\, (\|x\| + Z(\omega))$$

hold, where $Z(\omega) \in L_1^2$.

Definition 15.21 (Mean-Square Stability w.r.t. Continuously Acting Random Perturbations). The null-solution of the deterministic ordinary differential equation (15.34) is called mean-square stable with respect to continuously acting random perturbations in G, if for arbitrary $\varepsilon > 0$ there exists a $\delta > 0$ such that at

$$\|x_0\| + \sup_{t \in I\,,\, x \in \mathbb{R}^d} \mathbb{E}\left(\|g(x, t, \omega)\|^2\right) \leq \delta \,,$$

$x_0 \in \mathbb{R}^d$ and $g \in G$ the following inequality holds

$$\mathbb{E}\left(\|X_t\|^2\right) < \varepsilon \,, \qquad t \in I \,.$$

Before you continue, make sure to answer the following questions:

Quiz: Section 15.3 – Part II (Lyapunov-Functions and Path-Wise Equi-Stability)

Q1 With the aid of a suitably chosen Lyapunov-function give sufficient conditions for path-wise equi-stability of the null-solution of $\dot{X} = f(X_t, t, \omega)$.

Q2 Sketch the proof of the assertion you used in Q1.

Q3 Which conditions in addition to thBefore you continue, make sure to answer the following questions:

ose given in Q1 are required for (i) asymptotic and (ii) exponential path-wise equi-stability?

Definition 15.22 ($\mathbb{P}$-Stability w.r.t. Continuously Acting Random Perturbations). The null-solution of the deterministic ordinary differential equation (15.34) is called $\mathbb{P}$-stable with respect to continuously acting random perturbations in G if for arbitrary $\varepsilon > 0$ and $p \in (0, 1)$ there exists a $\delta > 0$ such that with $Z_t = \sup_{x \in \mathbb{R}^d} \|g(x, t, \omega)\|^2$ at

$$\|x_0\| + \sup_{t \in I} \mathbb{E}(Z_t) < \delta ,$$

$x_0 \in \mathbb{R}^d$ and $g \in H$ the following inequality holds:

$$\|X_t\|^2 \overset{(p)}{<} \varepsilon , \qquad t \in I .$$

Definition 15.23 (W-Stability w.r.t. Continuously Acting Random Perturbations). The null-solution of the deterministic ordinary differential equation (15.34) is called W-stable with respect to continuously acting random perturbations in G if for arbitrary $\varepsilon > 0$ and $p \in (0, 1)$ there exists a $\delta > 0$ such that with $Z_t = \sup_{x \in \mathbb{R}^d} \|g(x, t, \omega)\|^2$ at

$$\|x_0\| + \mathbb{E}\left(\sup_{t \in I} Z_t\right) < \delta ,$$

$x_0 \in \mathbb{R}^d$ and $g \in H$ the following inequality holds:

$$\|X_t\|^2 \overset{(I,p)}{<} \varepsilon , \qquad t \in I .$$

With Theorem 4.4 and theorem III 6.5 from [22] we get the following result:

Theorem 15.5 (Mean-Square Stability w.r.t. Continuously Acting Random Perturbations in G). *Let he following conditions be fulfilled:*

1. *The function f is continuous on $\mathbb{R}^d \times I$.*
2. *There is a positive number L such that*

$$\|f(x_1, t) - f(x_2, t)\| \leq L \|x_1 - x_2\| , \quad x_1, x_2 \in \mathbb{R}^d , \quad t \in I .$$

3. *The null-solution of the deterministic equation (15.34) is asymptotically equi-stable.*

Then the null-solution of (15.34) is mean-square stable with respect to continuously acting random perturbations in G.

Sufficient conditions for $\mathbb{P}$-stability with respect to continuously acting random perturbations of the null-solution of equation (15.34) are, for instance, given in the seminal monograph [155]. We cite this result in Theorem 15.6.

Particularly in applications, $\mathbb{P}$-stability is not totally satisfying as this stability notion does not imply that the norms of the solution realisations are bounded by ε with a high probability. Thus, we show in Theorem 15.7 that weaker conditions, as stated in Theorem 15.6, already imply W-stability.

Theorem 15.6 ($\mathbb{P}$-Stability w.r.t. Continuously Acting Random Perturbations in H). *Let $v : \mathbb{R}^d \times I \to \mathbb{R}$ be a continuously partially differentiable function such that:*

1. *There is a positive number L with*
$$\left\| \frac{\partial v(x,t)}{\partial x} \right\| \leq L, \quad x \in \mathbb{R}^d, \quad t \in I.$$
2. *On $[0,\infty)$ there are real, continuous, monotonically non-decreasing functions $\varphi(r)$, $\psi(r)$, and $\rho(r)$ with $\varphi(0) = \psi(0) = 0$ and*
$$0 < \varphi(r), \quad 0 < \psi(r), \quad 0 < \rho(r) < \rho_0, \quad \text{for } r > 0,$$
such that
   - *(a) $\varphi(\|x\|) \leq v(x,t) \leq \psi(\|x\|)$, and*
   - *(b) $v_t(x,t) + v'(x,t) \leq -\rho(\|x\|)v(x,t)$ for $x \in \mathbb{R}^d$ and $t \in I$, where v' is the orbital derivative as defined in Sec. 8.3.1.*

Then the null-solution of the deterministic ordinary differential equation (15.34) is $\mathbb{P}$-stable with respect to continuously acting random perturbations in H.

Proof. Following [45], p. 118, let $\delta > 0$ and $\|X_t(\omega)\| > \delta$. Then, due to conditions 1 and 2b together with $V_t(\omega) = V(X_t(\omega), t)$ we get
$$\frac{\mathrm{d}V_t(\omega)}{\mathrm{d}t} \leq -\rho(\delta)V_t(\omega) + LZ_t(\omega).$$
The above inequality and
$$\frac{\mathrm{d}V_t(\omega)}{\mathrm{d}t} \overset{I}{\leq} LZ_t(\omega)$$
lead to
$$\frac{\mathrm{d}V_t(\omega)}{\mathrm{d}t} \overset{I}{\leq} -\rho(\delta)V_t + LZ_t + \rho(\delta)\psi(\delta).$$
Applying Gronwall's Lemma (Lemma 3.15) to this inequality and taking the expectation, we get
$$\mathbb{E}(V_t) \leq v(x_0,t_0)\exp(-\rho(\delta)(t-t_0)) + \frac{L}{\rho(\delta)} \sup_{\tau \in I} \mathbb{E}(Z_\tau) + \psi(\delta), \quad t \in I.$$

Let $\varepsilon > 0$ and $p \in (0,1)$ be given. Due to condition 2 we have for sufficiently small δ, $\|x_0\|$ and $\sup_{t\in I} \mathbb{E}(Z_t)$ on the one hand

$$\mathbb{E}(V_t) < \varphi(\varepsilon)(1-p) ,$$

and on the other hand it holds that

$$\mathbb{E}(V_t) \geq \varphi(\varepsilon)\mathbb{P}(\|X_t\| \geq \varepsilon) .$$

These last two inequalities finally lead to

$$\|X_t\| \overset{(p)}{<} \varepsilon ,$$

showing the assertion. □

Theorem 15.7 (W-Stability w.r.t. Continuously Acting Random Perturbations in H). *Let $v : \mathbb{R}^d \times I \to \mathbb{R}$ be a continuously partially differentiable function such that there are positive numbers L and K with*

$$\left\| \frac{\partial v(x,t)}{\partial x} \right\| \leq L, \quad \|x\| \leq K, \quad t \in I .$$

Moreover, let condition 2 of Theorem 15.6 hold. Then the null-solution of the deterministic ordinary differential equation (15.34) is W-stable with respect to continuously acting random perturbations in H.

Proof. Following [45], p. 118, we assume to obtain a contradiction that the null-solution of (15.34) is not W-stable with respect to continuously acting random perturbations in H. Then there are an $\varepsilon \in (0,K)$, a $p \in (0,1)$ and a sequence of initial conditions $x_0^{(\nu)} \in \mathbb{R}^d$ as well as functions $g^{(\nu)}(x,t,\omega) \in H$, $\nu = 1,2,\ldots$ with the following two properties:

1. It holds that $\lim_{\nu\to\infty} \|x_0^{(\nu)}\| + \mathbb{E}(Z^{(\nu)}) = 0$, where $Z^{(\nu)} = \sup_{x\in\mathbb{R}^d,\, t\in I} \|g^{(\nu)}(x,t,\omega)\|$.

2. Let $X_t^{(\nu)}$ denote the path-wise solution of the random (ordinary) differential

$$\frac{\mathrm{d}X_t}{\mathrm{d}t} = f(X_t,t) + g^{(\nu)}(X_t,t,\omega)$$

   with the initial condition $(x_0^{(\nu)}, t_0)$. Then, it holds that

$$\sup_{t\in I} \left\| X_t^{(\nu)} \right\| \overset{(1-p)}{\geq} \varepsilon .$$

Let $\delta \in (0, \varepsilon)$ be chosen such that

$$\psi(\delta) \; < \; \tfrac{1}{3}\varphi(\varepsilon)(1-p)$$

holds, and N be chosen such that

$$\|x_0^{(\nu)}\| \; < \; \delta \, , \quad \text{and} \quad \mathbb{E}\left(Z^{(\nu)}\right) \; < \; \frac{\rho(\delta)(1-p)\varphi(\varepsilon)}{3L}$$

holds for $\nu > N$.

Moreover, let $\Omega^{(\nu)}$ be the set of all $\omega \in \Omega$ such that $\sup_{t\in I} \|X_t^{(\nu)}(\omega)\| \geq \varepsilon$. Due to the conditions stated we get with $V_t^{(\nu)}(\omega) = v(X_t^{(\nu)}(\omega), t)$

$$\frac{\mathrm{d}V_t^{(\nu)}(\omega)}{\mathrm{d}t} \; \leq \; LZ_t^{(\nu)} \tag{15.36}$$

at $\|X_t^{(\nu)}(\omega)\| < K$ and

$$\frac{\mathrm{d}V_t^{(\nu)}(\omega)}{\mathrm{d}t} \; \leq \; -\rho(\delta) + V_t^{(\nu)} + LZ_t^{(\nu)} \tag{15.37}$$

at $\delta < \|X_t(\omega)\| < K$, respectively.

From (15.36) and (15.37) we obtain for $\|X_t^{(\nu)}(\omega)\| < K$

$$\frac{\mathrm{d}V_t^{(\nu)}(\omega)}{\mathrm{d}t} \; \leq \; -\rho(\delta) + V_t^{(\nu)} + LZ_t^{(\nu)} + \rho(\delta)\psi(\delta) \, . \tag{15.38}$$

For $\omega \in \Omega^{(\nu)}$ let $t^{(\nu)}(\omega) = \inf\{t \geq t_0 : \|X_t^{(\nu)}(\omega)\| \geq \varepsilon\}$. Inequality (15.38) implies for $\omega \in \Omega^{(\nu)}$, $t \in [t_0, t^{(\nu)}(\omega)]$ and $\nu > N$

$$V_t^{(\nu)} \; \leq \; V_0^{(\nu)} - \rho(\delta) \int_{t_0}^t V_\tau^{(\nu)} \mathrm{d}\tau + \int_{t_0}^t \left(LZ_\tau + \rho(\delta)\psi(\delta)\right) \mathrm{d}\tau \, . \tag{15.39}$$

According to Gronwall's Lemma (Lemma 3.15) this inequality (15.39) leads to

$$\begin{aligned} V_{t^{(\nu)}(\omega)}^{(\nu)} \; &\leq \; V_0^{(\nu)} \exp\left(-\rho(\delta)\left(t^{(\nu)}(\omega) - t_0\right)\right) + \psi(\delta) \\ &\quad + L \int_{t_0}^{t^{(\nu)}} Z_\tau^{(\nu)} \exp\left(-\rho(\delta)\left(t^{(\nu)}(\omega) - \tau\right)\right) \mathrm{d}\tau \, , \end{aligned}$$

which implies

$$V_{t^{(\nu)}(\omega)}^{(\nu)} \; \leq \; V_0^{(\nu)} + \psi(\delta) + \frac{LZ^{(\nu)}}{\rho(\delta)} \, . \tag{15.40}$$

Integrating (15.40) over $\Omega^{(\nu)}$ we obtain

$$\int_{\Omega^{(\nu)}} V_{t^{(\nu)}(\omega)}^{(\nu)} \mathrm{d}\mathbb{P} \; \leq \; V_0^{(\nu)} + \psi(\delta) + \frac{L}{\rho(\delta)} \mathbb{E}\left(Z^{(\nu)}\right) \; < \; (1-p)\varphi(\varepsilon) \, . \tag{15.41}$$

for $\nu > N$. But inequality (15.41) contradicts

$$\int_{\Omega^{(\nu)}} V^{(\nu)}_{t^{(\nu)}(\omega)} \, \mathrm{d}\mathbb{P} \;=\; \int_{\Omega^{(\nu)}} v\left(X^{(\nu)}_{t^{(\nu)}(\omega)}, t^{(\nu)}(\omega)\right) \mathrm{d}\mathbb{P} \;\geq\; (1-p)\varphi(\varepsilon)\,,$$

which shows the assertion. □

Remark 15.24 ($\mathbb{P}$- & W-Stability w.r.t. Continuously Acting Random Perturbations in H). If the null-solution of (15.34) is exponentially stable in the whole, and if f_x is continuous and bounded on $\mathbb{R}^d \times I$, then there is a function $v(x,t)$ with the properties (1)-(3) stated in Krasovskii's theorem (Theorem 8.1). The function $v(x,t)$ fulfills the conditions of Theorem 15.7 and its square root $(v(x,t))^{1/2}$ those of Theorem 15.6.

Thus, if the null-solution of (15.34) is exponentially stable in the whole, and if f_x is continuous and bounded on $\mathbb{R}^d \times I$, then the null-solution of (15.34) is $\mathbb{P}$- as well as W-stable with respect to continuously acting random perturbations in H.

15.5 Chapter's Summary

The aim of this chapter was to study the various notions of stability of the null solution of a random (ordinary) differential equation. Besides obvious extensions of the deterministic stability concepts, we introduced the notions of

- path-wise equi-stability,
- h-stability,
- $\mathbb{P}$-stability, and
- W-stability.

We discussed the relations/ implications/ inter-connections between these concepts and re-framed the results of Chap. 13 on the path-wise stability of linear random differential equations with stochastic coefficients in the context of these concepts.

An important aim of this chapter was to extend the deterministic Lyapunov method to random differential equations. Here, based on suitable Lyapunov-functions, necessary conditions for h-stability and path-wise equi-stability were stated and proven.

Finally, we treated the interesting topic of stability of deterministic systems subject to different classes of continuously acting random perturbations.

Before you continue, make sure to answer the following questions:

Quiz: Section 15.4

Q0 State Krasovskii's theorem.

Q1 Give the definition of mean-square stability w.r.t. continuously acting random perturbations.

Q2 Under which conditions is the null-solution of $\dot{x} = f(x,t)$ mean-square stable w.r.t. continuously acting random perturbations from the class G? First of all what does it mean for a random perturbation to be from the class G?

Q3 What does it mean for a random perturbation to be from the class H?

Q4 Under which conditions is the null-solution of $\dot{x} = f(x,t)$ mean-square stable w.r.t. continuously acting random perturbations from the class H?

Q5 Sketch the proof of the assertion you used in Q4.

Q6 Give the definition of $\mathbb{P}$-stability w.r.t. continuously acting random perturbations.

Q7 Under which conditions is the null-solution of $\dot{x} = f(x,t)$ $\mathbb{P}$-stable w.r.t. continuously acting random perturbations from the class H?

Q8 Sketch the proof of the assertion you used in Q7.

Q9 Give the definition of W-stability w.r.t. continuously acting random perturbations.

Q10 Under which conditions is the null-solution of $\dot{x} = f(x,t)$ W-stable w.r.t. continuously acting random perturbations from the class H?

Q11 Sketch the proof of the assertion you used in Q10.

Problems

Classification: ☼ easy, ⊕ easy with longer calculations, ☆ a little bit difficult, ⊞ challenging.

Exercise 15.25. [☼] **Some Easy Proofs**
Prove the Propositions 15.13, 15.14 and the first part of Proposition 15.16:

1. Proposition 15.13: Let the null-solution of equation (15.1) be W-stable. Then it is $\mathbb{P}$-stable, too.

2. Proposition 15.14: Let $M_1 \subset M_2 \subset S_d$. Let the null-solution of equation (15.1) be stable with respect to M_2 in the sense of any of the stability definitions 15.3 to 15.7. Then, it is stable with respect to M_1 in the sense of the same stability definition, too.

3. First part of Proposition 15.16: Let the null-solution of equation (15.1) be path-wise equi-stable. Then it is W-stable and thus $\mathbb{P}$-stable.

Exercise 15.26. [☼] **Stability of Systems driven by Ornstein-Uhlenbeck Noise**
Let X_t be a path-wise solution of

$$\dot{X}_t = \begin{pmatrix} \cos(t) & 1 \\ \sin(t) & \exp(-t) \end{pmatrix} X_t + \begin{pmatrix} 0 \\ 1 \end{pmatrix} O_t ,$$

where O_t is a Ornstein-Uhlenbeck process with mean μ and variance σ^2. What can you say about the stability of X_t?

Exercise 15.27. [☼] **Stability of Solutions driven by Independent Fractional Brownian Motions**
Let the following random differential equation be given:

$$\dot{X}_t = \begin{pmatrix} -1 + B_{1/4}(t) & 1 \\ B_{3/4}(t) & -3 \end{pmatrix} X_t ,$$

where $B_{1/4}(t)$ and $B_{3/4}(t)$ are independent 1-dimensional fractional Brownian motions with Hurst index $1/4$ and $3/4$, respectively. What can you say about the stability of the null-solution?

Exercise 15.28. [☼] **Completion of the Proof of Theorem 15.4**
Complete the proof of Theorem 15.4 by showing the correctness of inequality (15.28).

Chapter 16

Random Dynamical Systems

This short chapter provides a glimpse into the very recent theory of random dynamical systems. We give the fundamental definitions of metric, measurable and random dynamical systems, and some illustrative examples. Moreover, we study the notions of forward and backwards stability and their implications.

16.1 Key Concepts

Traditionally, stochastic processes are considered as objects that evolve with positive time in some infinitely dimensional space of more or less continuous functions (due to the properties of the underlying sample and probability space). On the other hand, deterministic dynamics has achieved marvelous results with trajectory based concepts and the notion of a flow forwards as well as backwards in time. The concept of random dynamical systems unites both these points of view by introducing models equipped with certain flow properties for the noisy perturbations themselves and the system that is disturbed by this noise.

When reading this chapter note the answers to the key concepts of

1. Cocylces and random dynamical systems,
2. Forward and backward stability in random dynamical systems

as well as the following following questions

1. What is a metric/ measurable/ random dynamical system?
2. What is the cocycle property?
3. What are random sets and random attractors (in the sense of forward stability)?
4. What is an equilibrium in a random dynamical system?

This chapter is structured as follows: In Sec. 16.2 ,the fundamental definitions of metric, measurable, and random dynamical systems are provided together with some illustrative examples. Next, in Sec. 16.3, the notions of forward and backwards stability are studied as well as their implications and connections to (stochastic) Lyapunov funtions. Finally, Section 16.4 wraps up the contents of this chapter.

16.2 Definition of a Random Dynamical System

In [11], pp. 5, a dynamical system influenced by stochastic noise is called a *random dynamical system* (see, e.g., [9], [10], [7] as well) and defined in its continuous two-sided real time version on the state space $\mathbb{R}^d$ as follows:

Definition 16.1 (Continuous Random Dynamical System). A *continuous random dynamical system* on the state space $\mathbb{R}^d$ $(d \geq 1)$ consists of two ingredients:

1. A *model of the noise*, namely a metric dynamical system $(\Omega, \mathcal{F}, \mathbb{P}, (\theta_t)_{t\in\mathbb{R}})$, where $(\Omega, \mathcal{F}, \mathbb{P})$ is a probability space and $(t, \omega) \mapsto \theta_t\omega$ is a measurable flow, i.e.,

$$\theta_0\omega \;=\; \omega\,, \quad \text{and} \quad \theta_{t+s}\omega \;=\; \theta_t\left(\theta_s\omega\right)\,, \quad \text{for all } s, t \in \mathbb{R}\,,$$

holds for all $\omega \in \Omega$, which leaves $\mathbb{P}$ invariant, i.e., $\theta_t\mathbb{P} = \mathbb{P}$ for all $t \in \mathbb{R}$. For simplicity we also assume that θ is ergodic under $\mathbb{P}$, meaning that a θ-invariant set has either probability 0 or 1.

2. A *model of the system perturbed by the noise*, namely a measurable dynamical system $(\mathbb{R}^d, \mathbb{R}, (\varphi_t)_{t\in\mathbb{R}})$ on the measurable space $(\mathbb{R}^d, \mathcal{B}^d)$, endowed with the Borel σ-algebra $\mathcal{B}^d$, over the metric dynamical system $(\Omega, \mathcal{F}, \mathbb{P}, (\theta_t)_{t\in\mathbb{R}})$ stated in part (i). The mapping

$$\varphi : \begin{cases} \mathbb{R} \times \Omega \times \mathbb{R}^d \;\to\; \mathbb{R}^d \\ (t, \omega, x) \;\mapsto\; \varphi(t, \omega, x) \;=:\; \varphi(t, \omega)x \end{cases}$$

has the following properties

- *Measurability:* φ is $\mathcal{B} \otimes \mathcal{F} \otimes \mathcal{B}^d, \mathcal{B}^d$-measurable, where $\mathcal{B}$ is the usual topology of Borel sets of $\mathbb{R}$.
- *Continuity:* $(t, x) \mapsto \varphi(t, \omega, x)$ is continuous for all $\omega \in \Omega$.
- *Cocycle Property:* the family $\varphi(t, \omega) := \varphi(t, \omega, \cdot) : \mathbb{R}^d \to \mathbb{R}^d$ of random self-mappings of $\mathbb{R}^d$ form a *cocycle* over θ, i.e., they satisfy

$$\varphi(0, \omega) \;=\; \text{id}\,, \qquad \text{for all } \omega \in \Omega \tag{16.1}$$

$$\varphi(t + s, \omega) \;=\; \varphi(t, \theta_s\omega) \circ \varphi(s, \omega)\,. \tag{16.2}$$

Here, "$\circ$" denotes the usual composition of two functions.

If the identity in (16.2) holds

- for all $s, t \in \mathbb{R}$ and $\omega \in \Omega$, we say that φ is a perfect cocycle. Obviously the concept of a cocycle as the stochastic analogy of a deterministic flow makes most sense in this situation only.
- for fixed $s \in \mathbb{R}$ and all $t \in \mathbb{R}$ $\mathbb{P}$-almost surely, we say that φ is a crude cocycle. In this situation the exceptional set $N_s \subset \Omega$ may depend on s.
- for fixed $s, t \in \mathbb{R}$ $\mathbb{P}$-almost surely, we say that φ is a very crude cocycle. In this situation the exceptional set $N_{s,t} \subset \Omega$ may depend on both s and t.

Finally, we call the continuous random dynamical system a *smooth random dynamical system of class $\mathcal{C}^k$* or a *$\mathcal{C}^k$ random dynamical system*, $1 \leq k \leq \infty$, if additionally the mapping

$$\varphi(t,\omega) \;=\; \varphi(t,\omega,\cdot) \;:\; \mathbb{R}^d \;\rightarrow\; \mathbb{R}^d\,, \qquad x \;\mapsto\; \varphi(t,\omega,x)$$

is $\mathcal{C}^k$ (i.e. k-times differentiable with respect to x), and the derivatives are continuous with respect to (t, x).

Before we give some examples for random dynamical systems, we start with an illustration of metric dynamical systems:

Example 16.2 (Metric Dynamical Systems from ODEs, cf. [60], p. 11). Let us consider the system of ordinary differential equations on $\mathbb{R}^d$:

$$\dot{x}_i \;=\; f_i(x_1, x_2, \ldots, x_d)\,, \qquad i \;=\; 1, 2, \ldots, d\,, \tag{16.3}$$

and assume that the initial value problem for this system is well-posed. We define $\{\theta_t\}_{t\in\mathbb{R}}$ by the identity $\theta_t x_0 = x(t)$, where $x(t)$ is the (unique) solution of (16.3) for the initial condition $x(0) = x_0$.

Next, let us assume that there is a non-negative function $\rho(x_1, x_2, \ldots, x_d)$ which satisfies the stationary Liouville equation

$$\sum_{i=1}^{d} \partial_{x_i}\left(\rho(x_1, x_2, \ldots, x_d) \cdot f_i(x_1, x_2, \ldots, x_d)\right) \;=\; 0\,, \tag{16.4}$$

and possesses the property $\int_{\mathbb{R}^d} \rho(x)\mathrm{d}x = 1$. Then, $\rho(x)$ is a density of a probability measure on $\mathbb{R}^d$. By Liouville's theorem we have

$$\int_{\mathbb{R}^d} f(\theta_t x_0)\rho(x)\mathrm{d}x \;=\; \int_{\mathbb{R}^d} f(x)\rho(x)\mathrm{d}x$$

for any bounded continuous function $f(x)$ on $\mathbb{R}^d$ and thus in this situation a metric dynamical system arises with $\Omega = \mathbb{R}^d$, $\mathcal{F} = \mathcal{B}(\mathbb{R}^d)$ and $\mathbb{P}(\mathrm{d}x) = \rho(x)\mathrm{d}x$. Here, $\mathcal{B}(\mathbb{R}^d)$ is the Borel σ-algebra of the open sets of $\mathbb{R}^d$.

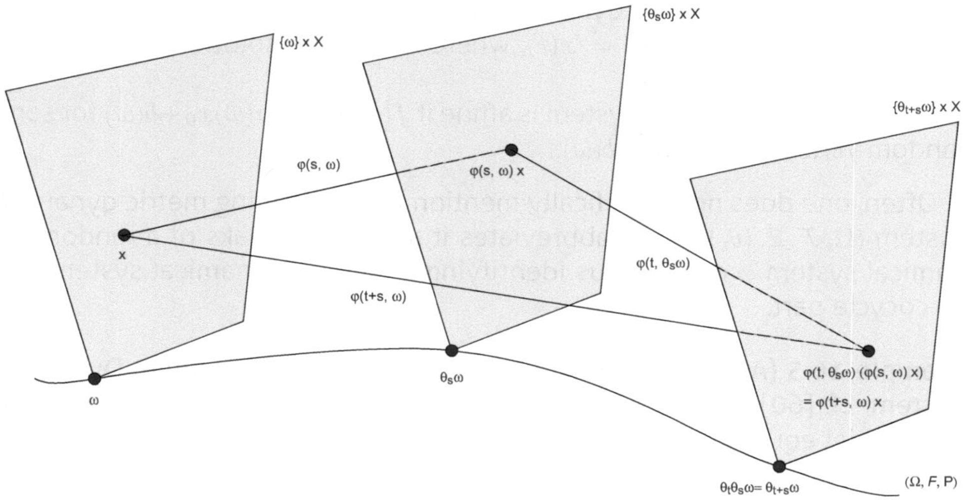

Figure 16.1. A random dynamical system φ over θ on the state space X as an action on the (trivial) bundle $\Omega \times X$, see [11], pp. 6.

Example 16.3 (Stationary Random Processes and Metric Dynamical Systems, cf. [60], p. 11). Let $\xi = \{\xi_t\}_{t\in\mathbb{R}}$ be a stationary random process on a probability space $(\Omega, \mathcal{F}, \mathbb{P})$, where $\mathcal{F}$ is the σ-algebra generated by ξ. Assume that ξ possesses the cadlag property (i.e. all trajectories are right-continuous and have limits form the left). Then the shifts $\xi_t \mapsto (\theta_\tau\xi)(t) = \xi_{t+\tau}$ generate a metric dynamical system.

It is illustrative to imagine a random dynamical system move on the (trivial) bundle $\Omega \times \mathbb{R}^d$, as shown in Fig. 16.1: While ω is shifted by the metric dynamical system θ in time s to the point $\theta_s\omega$ on the base space Ω, the cocycle $\varphi(s,\omega)$ moves the point x in the fiber $\{\omega\}\times\mathbb{R}^d$ over ω to the point $\varphi(s,\omega)x$ in the fiber $\{\theta_s\omega\}\times\mathbb{R}^d$ over $\theta_s\omega$. The cocycle property is clearly visible on this bundle, see [11], pp. 5.

Note, that deterministic dynamical systems in the sense of Chap. 8 are special cases of random dynamical systems, as any (perfect) cocycle reduces to a flow if ω is absent.

Example 16.4 (A 1D Random Differential Equation Inducing a Random Dynamical System, cf. [60], p. 16). Let $\theta(\Omega, \mathcal{F}, \mathbb{P}, \{\theta\}_{t\in\mathbb{R}})$ be a metric dynamical system. Let us consider the path-wise ordinary differential equation

$$\dot{x} = f(\theta_t\omega, x(t)) \,. \tag{16.5}$$

If the right-hand side function f admits a path-wise unique solution of the corresponding random differential equation on the whole of $\mathbb{R}$, then equation

(16.5) generates a random dynamical system with state space $\mathbb{R}$ and with the cocycle given by $\varphi(t,\omega)x_0 = x(t)$, where $x(t)$ is the solution to (16.5) with initial condition $x(0) = x_0$.

This random dynamical system is affine if $f(\omega, x_0) = a(\omega)x_0 + b(\omega)$ for some random variables $a(\omega)$ and $b(\omega)$.

Often, one does not specifically mention the underlying metric dynamical system $(\Omega, \mathcal{F}, \mathbb{P}, (\theta_t)_{t\in\mathbb{R}})$ or abbreviates it as θ_t and speaks of a random dynamical system φ over θ, thus identifying a random dynamical system with its cocycle part.

Example 16.5 (A Binary Biochemical Model Inducing a Random Dynamical System, cf. [60], p. 16). Let us consider the system of path-wise ordinary differential equations

$$\dot{x}_1 = g(x_2) - \alpha_1(\theta_t\omega)x_1 \,, \qquad \text{and} \qquad \dot{x}_2 = x_1 - \alpha_2(\theta_t\omega)x_2 \,, \tag{16.6}$$

over a metric dynamical system θ. If we assume that $g(x)$ is a globally Lipschitz-continuous function and that the coefficients $\alpha_i(\omega)$, $i = 1, 2$, are random variables such that $\alpha_i(\theta_t\omega) \in L^1_{loc}(\mathbb{R})$, $i = 1, 2$, and $\omega \in \Omega$. Then the system (16.6) generates a random dynamical system on $\mathbb{R}^2$ with the cocycle given by $\varphi(t,\omega)x_0 = x(t)$, where $x(t) = (x_1(t), x_2(t))$ is the solution of (16.6) with $x(0) = x_0$.

It is straightforward to verify that from the cocycle property it follows that $\varphi(t,\omega)$ is a homeomorphism of $\mathbb{R}^d$ and $\varphi(t,\omega)^{-1} = \varphi(-t, \theta_t\omega)$, see [11], pp. 7:

Theorem 16.1 (Basic Properties of Random Dynamical Systems). *Let φ be a continuous random dynamical system over θ, then for all $(t,\omega) \in \mathbb{R} \times \Omega$, $\varphi(t,\omega)$ is a bi-measurable bijection of $(\mathbb{R}^d, \mathcal{B}^d)$, and*

$$\varphi(t,\omega)^{-1} = \varphi(-t, \theta_t\omega) \,, \qquad \textit{for all } (t,\omega) \in \mathbb{R} \times \Omega \,, \tag{16.7}$$

or, equivalently,

$$\varphi(t,\omega) = \varphi(-t, \theta_{-t}\omega)^{-1} \,, \qquad \textit{for all } (t,\omega) \in \mathbb{R} \times \Omega \,,$$

Moreover, $(t,x) \mapsto \varphi(t,\omega)^{-1}x$ is continuous for all $\omega \in \Omega$.

If φ is a $\mathcal{C}^k$ random dynamical system, $1 \le k \le \infty$, then φ is a diffeomorphism on $\mathbb{R}^d$ and $(t,x) \mapsto \varphi(t,\omega)^{-1}x$ is $\mathcal{C}^k$ with respect to x for all $\omega \in \Omega$.

Proof. Let us follow [11], pp. 8, for the proof. First, we use the fact that $g = f^{-1}$ holds if and only if $f \circ g = id$ and $g \circ f = id$, i.e., that the right- and the left inverse are identical. Inserting $t = -s$ in (16.2) and applying (16.1) results in

$$id = \varphi(-s, \theta_s\omega) \circ \varphi(s,\omega) \,, \qquad \text{for all } s \in \mathbb{R}, \omega \in \Omega \,,$$

as a statement for the left inverse of $\varphi(s, \omega)$. Next, taking (16.2) for $s = -t$ and $\tilde{\omega} := \theta_t\omega$ together with (16.1) gives

$$id = \varphi(t, \theta_{-t}\tilde{\omega}) \circ \varphi(-t, \tilde{\omega}) = \varphi(t, \omega) \circ \varphi(-t, \theta_t\omega), \qquad \text{for all } s \in \mathbb{R}, \omega \in \Omega,$$

as a statement for the right inverse of $\varphi(t, \omega)$. Together, for $s = t$ this yields equation (16.7).

The mapping $(t, \omega, x) \to \varphi(t, \omega)^{-1}x = \varphi(-t, \theta_t\omega)x$ is measurable, since it is the composition of the measurable maps $(t, \omega, x) \mapsto (-t, \theta_t\omega, x)$ and $(t, \omega, x) \mapsto \varphi(t, \omega)x$. The measurability of the inverse mapping with respect to x follows from that of $(t, \omega, x) \mapsto \varphi(t, \omega)^{-1}x$ by fixing (t, ω).

Equation (16.7) says that the left-hand side is continuous with respect to x as long as the right-hand side is. Hence, φ is a homeomorphism. We observe, that $(t, x) \mapsto (t, \varphi(t, \omega)x)$ is a continuous and bijective mapping of $\mathbb{R} \times \mathbb{R}^d$ onto itself. This mapping is thus a homeomorphism (i.e. its inverse $(t, x) \mapsto (t, \varphi(t, \omega)^{-1}x)$ is continuous and consequently $(t, x) \mapsto \varphi(t, \omega)^{-1}x$ is continuous, too).

The $\mathcal{C}^k$ diffeomorphism property of $\varphi(t, \omega)$ follows from equation (16.7). For the last statement, we use the fact that the derivative of a diffeomorphism is nonsingular and the derivative of the inverse is given by the formula

$$D\varphi(t, \omega)^{-1}x = \left(D\varphi(t, \omega)y_{|y=\varphi(t,\omega)^{-1}x}\right)^{-1}.$$

Hence $(t, x) \mapsto D\varphi(t, \omega)^{-1}x$ is continuous since

- $(t, x) \mapsto D\varphi(t, \omega)x$ is continuous by assumption,
- $(t, x) \mapsto \varphi(t, \omega)^{-1}x$ is continuous (as we have just proven), and
- $g \mapsto g^{-1}$ is continuous in $GL(d\mathbb{R})$.

Analogously for higher order derivatives. □

In the following we will denote by $Hom(\mathbb{R}^d)$ the group of homeomorphisms on $\mathbb{R}^d$ endowed with its compact-open topology; in particular, the evaluation mapping $(f, x) \mapsto f(x)$ is continuous. We will denote by $Diff^k(\mathbb{R}^d)$, $1 \leq k \leq \infty$, the group of $\mathcal{C}^k$-diffeomorphisms on $\mathbb{R}^d$ endowed with its compact-open topology; the evaluation mapping $(f, x) \mapsto f(x)$ is $\mathcal{C}^k$ with respect to x.

Remark 16.6 (Random Dynamical Systems as Skew Products). The mapping

$$\begin{cases} \Theta : \mathbb{R} \times \Omega \times \mathbb{R}^d \to \Omega \times \mathbb{R}^d, \\ \Theta_t(\omega, x) := (\theta_t\omega, \varphi(t, \omega)x) \end{cases}$$

gives the *skew product flow* on $\Omega \times \mathbb{R}^d$. However, continuity of the mappings $(t, \omega) \mapsto \theta_t\omega$ and $(t, \omega, x) \mapsto \varphi(t, \omega, x)$ will exclude this concept in our studies of stochastic differential equations.

Before you continue, make sure to answer the following questions:

Quiz: Section 16.2

Q1 Give the definitions of a metric and a measurable dynamical system.

Q2 Give the definition of a cocycle and explain the difference between cocycles and flows. Illustrate the cocycle property.

Q3 Give an example for a metric and a measurable dynamical system.

Q4 Give the definition of a random dynamical system and construct an example of a random dynamical system based on a random differential equation.

Q5 Give some properties of random dynamical systems and prove your assertions.

Q6 Give the definition of a skew product flow and show the connection between skew products and random dynamical systems.

16.3 Stability and Lyapunov-Functions

As we have already seen, stability is the fundamental concept studied in dynamical systems. For random dynamical systems there are again different notions of stability. Here, the concepts of forward and backward stability are discussed .

16.3.1 Forward Stability

Similar to the definition of deterministic Lyapunov stability in $\mathbb{R}^d$ the stability of a reference solution with respect to a random norm is stated in [11], p. 372. Hereby, a *random norm* $\|x\|_\omega : \Omega \times \mathbb{R}^d \to \mathbb{R}_0^+$ is defined as a measurable function such that $\|\cdot\|_\omega$ is a norm on $\mathbb{R}^d$ for every fixed $\omega \in \Omega$.

Definition 16.7 (Stability of a Reference Solution). Let φ be a continuous $\mathbb{R}^d$-valued random dynamical system over θ. A reference solution $\varphi(\cdot, \omega, x_0)$ is called *stable* (with respect to the random norm $\|\cdot\|_\omega$), if for any $\varepsilon : \Omega \to (0, \infty)$ there exists a $\delta : \Omega \to (0, \infty)$ such that

$$\sup_{0 \le t < \infty} \|\varphi(t, \omega, x) - \varphi(t, \omega, x_0)\|_{\theta_t \omega} < \varepsilon(\omega)$$

whenever $\|x - x_0\|_\omega < \delta(\omega)$.

The reference solution is called *asymptotically stable* if it is stable and, in addition, if

$$\lim_{t\to\infty} \|\varphi(t,\omega,x) - \varphi(t,\omega,x_0)\|_{\theta_t\omega} = 0 .$$

holds for any x with $\|x - x_0\|_\omega < \delta(\omega)$.

Moreover, if the reference solution is stable, and if there exists a $c < 0$ such that for any x with $\|x - x_0\|_\omega < \varepsilon(\omega)$ it holds that

$$\sup_{0\le t<\infty} \|\varphi(t,\omega,x) - \varphi(t,\omega,x_0)\|_{\theta_t\omega} < \exp(ct) , \qquad \text{for } t > T(\omega, x) ,$$

then, the reference solution is called *exponentially stable.*

Let us follow [7], definition 4.1, to extend the concept of stability of a reference solution to that of whole random sets. First note, that a function $\omega \mapsto M(\omega)$ taking values in the non-empty compact subsets of $\mathbb{R}^d$ is called a *random compact set* if $\omega \mapsto d(x, M(\omega))$ is measurable for each $x \in \mathbb{R}^d$, where $d(x, M) := \inf_{y\in M} \|x - y\|$.

Definition 16.8 ((Asymptotic) Stability of Random Compact Sets). Let φ be a random dynamical system over θ and A be a random compact set which is invariant under φ.

1. A is called *stable* under φ if for any $\varepsilon > 0$ there exists a random compact set C which is a neighborhood of A (i.e. $C(\omega)$ is a neighborhood of $A(\omega)$ for all ω) such that
   - $\mathbb{P}(d(C|A) \ge \varepsilon) < \varepsilon$ (i.e. C is ε-close to A in the metric of convergence in probability[1])
   - $\varphi(t,\omega,\cdot)C(\omega) \subset C(\theta_t\omega)$ for all $t \ge 0$ (i.e. C is forward invariant under $\varphi : (\mathbb{R} \times \Omega \times \mathbb{R}^d) \to \mathbb{R}^d)$ with Ω denoting the sample space and θ_t the shift underlying the stochastic dynamical system).
2. A is called a (global) *attractor* of φ if for any random variable X the following holds

$$\lim_{n\to\infty} \mathbb{P}\left(\omega \in \Omega : \|d(\varphi(t,\omega,\cdot)X(\omega), A(\theta_t\omega)) - 0\| > \varepsilon\right) = 0 \qquad \forall \varepsilon > 0 .$$

3. A is called (globally) *asymptotic stable* if it is stable and is an attractor.

Following [7] a stochastic Lyapunov-function for the random dynamical system (RDS) is given as:

[1] Recall: $d(C|A) := \sup_{x\in C} d(x, A) = \sup_{x\in C} \inf_{y\in A} \|x - y\|$.

Definition 16.9 ((Stochastic) Lyapunov-Function --- RDS Version). Let φ be a random dynamical system in $\mathbb{R}^d$ and A be a random compact set which is invariant under φ. A function $V : \Omega \times \mathbb{R}^d \to \mathbb{R}^+$ is called a *(stochastic) Lyapunov-function* for A (under φ) if it has the following properties:

1. $\omega \mapsto V(\omega, x)$ is measurable for each $x \in \mathbb{R}^d$, and $x \mapsto V(\omega, x)$ is continuous for each $\omega \in \Omega$ (this implies that $(\omega, x) \mapsto V(\omega, x)$ is measurable),
2. V is *uniformly unbounded*, i.e. $\lim_{||x||\to\infty} V(\omega, x) = \infty$ for all ω,
3. V is *positive-definite*, i.e. $V(\omega, x) = 0$ for all $x \in A(\omega)$ and $V(\omega, x) > 0$ for all $x \notin A(\omega)$,
4. V is *strictly decreasing along orbits* of φ, i.e.

$$V(\theta_t\omega, \varphi(t, \omega, x)) \ < \ V(\omega, x) \qquad \text{for all } t > 0 \text{ and } x \notin A(\omega)\,. \qquad (16.8)$$

Property (16.8) seems to be "incorrect" in the sense that it compares the value of V in the fiber over t with the one in the fiber over time 0. However writing

$$V(\Theta_t(\omega, x)) \ = \ V(\theta_t\omega, \varphi(t, \omega, x)) \ = \ C(t, \omega, x)V(\omega, x)\,,$$

the flow property of Θ implies that C is a cocycle over Θ with values in the multiplicative group $\mathbb{R}^+_*$ of positive real numbers, and (16.8) "correctly" reads as $C(t, \omega, x) < \mathcal{E}(t, \omega, x)$, where $\mathcal{E}(t, \omega, x) \equiv 1$ is the trivial cocycle in $\mathbb{R}^+_*$ over Θ. The choice $C(t, \omega, x) = e^{-t}$ will be made below.

Theorem 16.2 (Asymptotic Stability and Lyapunov-Function --- RDS Version). *Let φ be a random dynamical system in $\mathbb{R}^d$ and A be a random compact set which is invariant under φ. Then A is asymptotically stable (in the sense of definition 16.8) if and only if there exists a (stochastic) Lyapunov-function for A.*

If A is asymptotically stable, then the (stochastic) Lyapunov-function can be chosen to satisfy

$$V(\theta_t\omega, \varphi(t, \omega, x)) \ = \ e^{-t}V(\omega, x) \qquad \textit{for all } t \in \mathbb{R} \textit{ and } x \in \mathbb{R}^d$$

on a θ-invariant ω set of full measure.

Proof. See [7]. □

16.3.2 Backwards Stability

Following [60], we next study the stability in the backwards sense of expressions of the form $x(t) = \varphi(t, \theta_{-t}\omega)x$ for $t \to \infty$. This might look strange at a first glance, however there are some reasons to discuss the limiting structure of $\varphi(t, \theta_{-t}\omega)x$:

- In many applications random dynamical systems are generated by equations whose coefficients depend on $\theta_t\omega$. These coefficients describe the internal evolution of the environment and $\theta_{-t}\omega$ represents the state of the environment at time $-t$ which transforms to the real state ω at the time of the observation. Furthermore, the two-parameter mapping $U(\tau, s) := \varphi(\tau - s, \theta_s\omega)$ describes the evolution of the system from moment s to time τ, $\tau > s$. Therefore, the limiting structure of $U(0, -t)x = \varphi(t, \theta_{-t}\omega)x$, when $t \to \infty$ can be interpreted as the state of our system which we observe now ($t = 0$) provided it was in state x in the infinitely distant past $t = -\infty$. Thus, the union of all these limits provides us with the real picture of the present state of the system.

- The asymptotic behavior of $\varphi(t, \theta_{-t}\omega)x$ provides us with some information about the long time future. Since, θ_t was assumed to be ergodic, i.e., measure preserving, we have that

$$\mathbb{P}\left(\omega \in \Omega : \varphi(t,\omega)x \in D\right) \;=\; \mathbb{P}\left(\omega \in \Omega : \varphi(t,\theta_{-t}\omega)x \in D\right)$$

  for any element $x \in \mathbb{R}^d$ of the state space and any Borel-set $D \in \mathcal{B}(\mathbb{R}^d)$. Therefore,

$$\lim_{t\to\infty} \mathbb{P}\left(\omega \in \Omega : \varphi(t,\omega)x \in D\right) \;=\; \lim_{t\to\infty} \mathbb{P}\left(\omega \in \Omega : \varphi(t,\theta_{-t}\omega)x \in D\right) ,$$

  if the limit on the right hand side exists. Thus, the limiting behavior of $\varphi(t, \theta_{-t}\omega)x$ for all $\omega \in \Omega$ determines the long time behavior of $\varphi(t,\omega)x$ with respect to the convergence in probability.

- It follows that if on the set of random variables $A(\omega)$ with values in $\mathbb{R}^d$ we define the operators T_t via

$$(T_t(a))(\omega) \;=\; \varphi(t,\theta_{-t}\omega)a(\theta_{-t}\omega)\,, \qquad t \,\in\, \mathbb{R}^+\,,$$

  then the family $\{T_t\}_{t\in\mathbb{R}^+}$ is a one-parameter semi-group. Indeed, by applying the cocycle property, we have that

$$\begin{aligned}(T_s \circ T_t)(a)(\omega) &= \varphi(s,\theta_s\omega)(T_{-t}a)(\theta_{-s}\omega)\\ &= \varphi(s,\theta_s\omega) \circ \varphi(t,\theta_{-t-s}\omega)a(\theta_{-t-s}\omega)\\ &= \varphi(t+s,\theta_{-t-s}\omega)a(\theta_{-t-s}\omega) \;=\; T_{t+s}(a)(\omega)\,.\end{aligned}$$

  Thus, it becomes possible to apply ideas from the theory of deterministic (autonomous) dynamical systems for which the semi-group structure of the evolution operator is crucial (just think of the existence of an attractor).

The three observations above motivate the following

Definition 16.10 (Equilibrium of a Random Dynamical System). Let $(\mathbb{R}^d, \mathbb{R}, \varphi)$ be a random dynamical system over the metric dynamical system $(\Omega, \mathcal{F}, \mathbb{P}, (\theta_t)_{t\in\mathbb{R}})$. A random variable $u : \Omega \to \mathbb{R}^d$ is said to be an equilibrium of the random dynamical system if it is invariant under φ, i.e., if

$$\varphi(t, \omega)u(\omega) \;=\; u(\theta_t\omega)\,, \qquad \text{for all } t \in \mathbb{R} \text{ and all } \omega \in \Omega.$$

Before you continue, make sure to answer the following questions:

Quiz: Section 16.3

Q1 Give the definitions of a random norm and a random compact set.

Q2 Give the notion of a stable/ asymptotically stable/ exponentially stable reference solution in a random dynamical system.

Q3 Give the definitions a stable compact random set and a (global) random attractor.

Q4 Give the definition of a (stochastic) Lyapunov-function. How can you use such a function to determine the stability of a random set?

Q5 Illustrate the concept of backwards stability and give the definition of an equilibrium of a random dynamical system.

16.4 Chapter's Summary

Following [11] and [60], this short chapter provided us with a glimpse into the recent theory of random dynamical systems. First, we gave the fundamental definitions of metric, measurable, and random dynamical systems together with some illustrative examples. Moreover, we briefly studied the notions of forward and backwards stability and their implications.

Problems

Classification: ☼ easy, ◷ easy with longer calculations, ☆ a little bit difficult, ◫ challenging.

Exercise 16.11. [☼] **Linear Random Differential Equations and Random Dynamical Systems**
Let $A : \Omega \to \mathbb{R}^{d\times d}$ be a measurable function such that $A \in L^1(\Omega, \mathcal{F}, \mathbb{P})$, and $f_\omega(t,x) := A(\theta_t\omega)x$ satisfies $f_\omega \in L_{loc}(\mathbb{R}, \mathcal{C}^\infty)$. Show the following assertions:

1. The linear random differential equation $\dot{x}_t = A(\theta_t\omega)x_t$ generates a unique linear $\mathcal{C}^\infty$ random differential equation φ satisfying
$$\varphi(t,\omega) = \mathbb{I} + \int_0^t A(\theta_s\omega)\varphi(s,\omega)\mathrm{d}s$$
and
$$\det(\varphi(t,\omega)) = \exp\left(\int_0^t \mathrm{tr}\left(A(\theta_s\omega)\right)\mathrm{d}s\right).$$
2. Also differentiating $\varphi(t,\omega)\left(\varphi(t,\omega)\right)^{-1} = \mathbb{I}$ yields
$$\left(\varphi(t,\omega)\right)^{-1} = \mathbb{I} + \int_0^t \left(\varphi(s,\omega)\right)^{-1} A(\theta_s\omega)\mathrm{d}s\,.$$

Exercise 16.12. [☼] **Affine Random Differential Equations and Random Dynamical Systems**
Let $A : \Omega \to \mathbb{R}^{d\times d}$ be a measurable function such that $A \in L^1(\Omega, \mathcal{F}, \mathbb{P})$, and $f_\omega(t,x) := A(\theta_t\omega)x$ satisfies $f_\omega \in L_{loc}(\mathbb{R}, \mathcal{C}^\infty)$. Show that the equation
$$\dot{x}_t = A(\theta_t\omega)x_t + b(\theta_t\omega)\,, \qquad A, b \in L^1(\Omega, \mathcal{F}, \mathbb{P})\,,$$
generates a unique $\mathcal{C}^\infty$ random dynamical system. The variation of constants formula yields
$$\begin{aligned}\varphi(t,\omega)x &= \Phi(t,\omega)x + \int_0^t \Phi(t,\omega)\Phi^{-1}(u,\omega)b(\theta_u\omega)\mathrm{d}u \\ &= \Phi(t,\omega)x + \int_0^t \Phi(t-u,\theta_u\omega)b(\theta_u\omega)\mathrm{d}u\,,\end{aligned}$$
where Φ is the

Exercise 16.13. [☼] **Random Differential Equations with Polynomial Right-Hand Side and Random Dynamical Systems**
Let $f(\omega,x) = \sum_{k=0}^N a_k(\omega)x^k$ be a polynomial of degree $N \geq 2$ with random coefficients. If $a_k \in L^1(\Omega, \mathcal{F}, \mathbb{P})$ for all $k = 1, 2, \ldots, N$, then the random differential equation
$$\dot{x}_t = \sum_{k=0}^N a_k(\theta_t\omega)x_t^k$$

uniquely generates a local $\mathcal{C}^\infty$ random dynamical system on $\mathbb{R}^1$.

The case

$$\dot{x}_t = a(\theta_t\omega)x_t + b(\theta_t\omega)x_t^N$$

can be explicitly solved by transforming the equation to an affine equation via $y = x^{1-N}/(1-N)$. Write down explicitly all ingredients of the local random dynamical system.

The cases $N = 2$ and $N = 3$ are, e.g., treated in [262].

Part V

The Workshop Project

Computer programming is tremendous fun. Like music, it is a skill that derives from an unknown blend of innate talent and constant practice. Like drawing, it can be shaped to a variety of ends—commercial, artistic, and pure entertainment. Programmers have a well-deserved reputation for working long hours, but are rarely credited with being driven by creative fevers. Programmers talk about software development on weekends, vacations, and over meals not because they lack imagination, but because their imagination reveals worlds that others cannot see.

LARRY O'BRIEN & BRUCE ECKEL

Chapter 17

The Workshop Idea

17.1 Key Concepts

Whether or not they realize it, students have the freedom to explore and to think about problems in new ways.
LEWIS D. GIROD

In this chapter, we explain the rationale behind the workshop project that we conducted during the summer term 2012 with 10 students at the wonderful location of the Studienzentrum Josefstal.

The integration of the workshop in the overall student-centred course is presented in Sec. 17.2. Afterwards, we present important details on the design of the project in Sec. 17.3. Finally, Section 17.4 summarises the workshop idea(s).

17.2 Integration of the Workshop in the Course

The workshop has been integrated in a larger course "Interdisciplinary Guided Research at the Edge of Dynamical Systems & Scientific Computing" at the Technische Universität München. The course introduces the theory and numerics of randomly disturbed differential equations from the point of view of both Dynamical Systems and Scientific Computing. It is designed as a seminar with a dedicated workshop part supplemented by exercises and some few lectures (see Fig. 17.1). It is open (and meant for) students of mathematics as well as computer science: This interesting combination of backgrounds allows for fruitful interdisciplinary discussions and solution approaches.

The focus of this course is on the student-centred approach to the topics during the seminar. In the three-day workshop, the participants applied their knowledge to a specific problem on earthquake responses in multi-storey buildings (cf. Chap. 18). Most parts of the simulation pipeline are, thus, tackled via a hands-on approach: theory, modelling, numerics & algorithms, implementation & data structures, visualisation, and verification.

The didactical approach of the course covers three aspects with corresponding challenges. First, interdisciplinarity is key of our concept, the seminar topics are taken from the intersection of mathematics and informatics. Furthermore, the participating students—mostly of Master level—come from

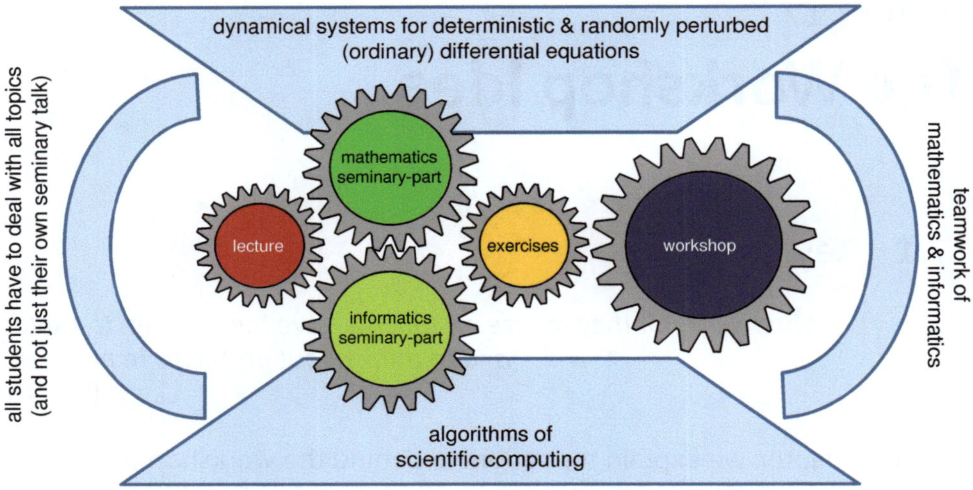

Figure 17.1. Design of the course combining the workshop with seminar contributions, exercises, and few lectures to increase student-centred learning effects.

these two departments. Hence, the various levels of previous knowledge in the corresponding disciplines are a considerable challenge, but a challenge that appears both in academia and industry in "real work life" in particular for interesting and promising projects. Hence, experiencing this interdisciplinarity in the curriculum is good preparation for the students. Second, realising a student-centred learning approach mainly means generating a stimulating learning environment and changing the perception of learning from mere exam preparation to a continuous and exciting process. We tried to tackle the corresponding challenge of leaving the traditional lecture routine—both for students and supervisors—and motivating the students to personally contribute via literally leaving the campus for the workshop. The results were surprisingly positive (see Sec. 18.3 and 18.4). Third, all the tasks culminate in the workshop (i.e. in a fully application-oriented task). Exciting examples and problems including work assignments for the students require massive teamwork. The challenges are to define a new course format and "realistic" scenarios while still being restricted to the time-frame of one term and academic examples.

The major innovative element of this course unifying the three above-mentioned aspects is the workshop[1]. Its design is presented in the following section.

[1] at least with respect to the classical concept of von Humboldt in Germany

17.3 Design of the Workshop

The storyline of the workshop is to realise a small software product in MATLAB as an independent virtual software company. We briefly explain this concept and the chosen environment. The required information had been passed to the participating students at least one week prior to the workshop start in form of documents and a dedicated workshop preparation meeting.

17.3.1 The Concept of a Virtual Software Company

To make the workshop context realistic, the students realise the tasks as a small, independent software company. A customer (a virtual one in this case) asks for the delivery of a specific software product within a given time window: We designed the net working time for the workshop to be 2.5 days. This rather short period has been chosen for two reasons. First, the workshop has to fit into the general schedule of the participants who attend other courses besides this one. Moving the workshop into the holidays after the term does not solve the scheduling problems due to other events (internships, summer schools, etc.). Furthermore, our focus is on the strong connection of the workshop to the rest of the course. Second, we intended to create a scenario with limited resources. The resulting time pressure is one important factor to make the workshop realistic and prepare the students for future tasks in their work life.

Similar to a real-world context, a detailed specification of the desired software has been carefully designed (see Sec. 17.3 for details). In order to estimate the possible results and necessary workload for each task, we decided to pre-code large parts of the specification internally.

The internal team structure of the company is described in Sec. 17.3.3 whereas some details on the workshop environment are presented in the following section.

17.3.2 Choice of the Workshop Environment

Concerning the location for the workshop, a couple of aspects are important. First, a quiet environment outside the usual campus life is necessary to put the team together. If everybody is continuously distracted by meeting fellow students not participating in the workshop, by important facebook messages on how to spend the evening in town, and needs to travel back and forth for passing the nights, then forming the team and realising the workload is more difficult and will take longer (at least considerably longer than the 2.5 days available). Second, a decent infrastructure—concerning accommodation and generally available quiet rooms or space for the individual subgroups of the team as well as for plenary meetings—is of course important but typically not fully available on campus. Therefore, we chose the seminar centre Josefstal (approx. one hour south of Munich by car/train to avoid losing too much time with the transfer).

Two reasons convinced us to choose MATLAB as programming language for the workshop project. Since the programming background of the participants is heterogeneous, we need a common language for both mathematics and informatics students. In addition, struggling with mere programming details such as type declarations or cross-platform compilation aspects does not provide helpful contributions for the learning effects of the workshop we intend to have and simply is out of scope under the time restriction of 2.5 days.

Concerning the computing infrastructure, we rely on the notebooks of the participants. This assures that the students are comfortably working on the platform that they are used to (Windows vs. Linux, vi vs. emacs vs. eclipse, etc.) and that the computers are easily installed in the seminar centre. To assure a smooth collaboration, one of the seminar talks covers various aspects of software engineering (design, repositories, automated testing, continuous integration) with a specific touch on MATLAB. A CVS repository[2] has been provided by the supervisors to exchange, merge, and backup the reports, source code, etc.

In a workshop preparation meeting, all relevant information mentioned above has been passed to the participants to guarantee a smooth and fast startup phase of the workshop. Furthermore, the MATLAB and CVS infrastructure has been prepared at this occasion, too.

17.3.3 Team and Role Descriptions

In the workshop project specification, the students get a survey on possible classical team roles in order to organise their company (see Sec. 17.3). The proposed possible roles are:

- scientific expert(s), i.e. someone who has expertise in the theory or background of a work package,
- programming expert(s), i.e. someone who has expertise in programming MATLAB for realising specific implementation tasks, maybe in combination with a scientific expert,
- infrastructure expert(s), i.e. someone who sets up and maintains a repository (CVS, e.g.), or is responsible for a common design, etc.,

[2] Concurrent Versioning System, see http://savannah.nongnu.org/projects/cvs.

- interface coordinator(s), i.e. someone who organises the teamwork aspects of work packages that are related, and
- a team coordinator who is responsible for the overall picture (global view on all work packages, their realisation in time, and problem resolution between different sub-teams).

The participants have the complete freedom of choice whether to use (some of) these roles or not, whether to create additional ones or work with a complete anarchic structure. There is not going to be any interaction from the supervisors in this assignment (except if they are directly asked by the students). This full freedom is one of the great advantages of having this project in the curriculum instead of real work life: even in the worst case, no company will go bankrupt; but the impression and learning effect for the students will be considerable.

Figure 17.2. A typical classroom situation in the course "Interdisciplinary Guided Research at the Edge of Dynamical Systems & Scientific Computing". (Photo: TUM/ Andreas Heddergott)

17.4 Chapter's Summary

We presented the didactical concept of the overall course and the major role of the workshop project. The story plot of the workshop—a software company delivering a MATLAB product—has been elaborated together with a decent amount of design aspects. What is now still missing is the concrete topic of the project from the content point of view. This will be presented in the next chapter.

Chapter 18

The Workshop Project: Stochastic Excitations of Multi-Storey Buildings

18.1 Key Concepts

The concrete topic of the workshop project is on the stochastic excitation of multi-storey buildings due to earthquakes. Three main arguments led to this choice of the topic. First, research on geophysical effects such as earthquakes is an important scientific field[1]. A considerable amount of damage (both with respect to human lives and to money) is caused every year by earthquake-related events. Knowing the relevance of the context of their work increases the motivation of the workshop participants. Second, effects related to earthquakes and shaking buildings are easy to comprehend; everybody, in particular the participating students, possesses an intuitive understanding of what typically happens. Third, the modelling of multi-storey buildings works nicely on different scales of resolution and complexity: A basic wireframe model related to ODEs can be generalised to a continuum mechanics approach using PDEs, etc. Besides, the stochastic effects fit naturally into the scene. Hence, this topic is well-suited for our interdisciplinary RODE approach.

The workshop project brings together the different fields presented in the previous chapters: The wireframe model represents a typical approach involving deterministic ODEs (Chap. 6), dynamical systems (Chap. 8, and the corresponding numerical solution (Chap. 7). Random excitations of the earthquake coupled with the wireframe building model are a direct example of how to include random effects into ODEs to obtain RODE systems, a topic covered in Chap. 1, 3, and 4. Simulating such RODE systems in MATLAB has been prepared in Chap. 12, 13, 14, 15, and 16. Extending the RODE setup towards PDEs by using a continuum mechanics approach (Chap. 2) enables us to model a very simple but spatially dependent version of a multi-storey building and to discretise it via finite differences (for the sake of simplicity). This spatial discretisation involves various aspects of the choice of algorithms. Space-filling curves (Chap. 11) are used for a clever ordering of the discrete cells of the building combined with analysis of effects on the intermediate level (fill-in of direct solvers etc.). Concerning the solution of the Poisson-

[1] Actually, 2013 is the year of mathematics of planet earth: http://mpe2013.org/

like system, different variants of implementation of the discrete Fourier and trigonometric transforms (Chap. 9) are realised.

The rest of this chapter is organised as follows: In Sec. 18.2, the project specification is listed. This is the original specification the students obtained during the preparation of the workshop; therefore, the team roles appear once more. Analogous to the exercises, the work packages of the workshop project have been classified according to the corresponding workload and level of challenge:

☼ easy, ⌚ easy with longer calculations, ✩ a little bit difficult, 🕮 challenging.

We list results of the project team in Sec. 18.3 before presenting feedback of the participants and impressions of the supervisors in Sec. 18.4. Finally, Section 18.5 summarises the workshop project.

18.2 Project Specification

Your software company "perfect design" is contracted by the public relations officer of a well-recognized university. In 5 days, the faculties of mathematics and computer science have their annual joint information day for prospective students. One of the program topics is an interactive earthquake simulation that will show the effects of ground motion excitation on various models of multi-storey buildings. This topic is the pet project of the public relations officer who already committed a professor to this task six month ago. At a routine call last week, the public relations officer got aware that the professor is on a sailing turn through the Caribbean and the responsible young researcher left to another university as assistant professor. An illustrative description of the project and a couple of notes could be found by the secretary and are now handed to your team by your boss together with some inspiring words that even Faustus did some work in the evening hours. It is now shortly after lunch time, your boss expects the final presentation in about 50 hours.

Figure 18.1 illustrates the aspects which the public relations officer, your client, expects as the minimum requirements: First, the prospective students should have the opportunity to test the effects of specific stochastic models for ground motion accelerations on multi-storey buildings that are modelled as a wireframe coupled oscillator structure. In particular, the mathematics department had stressed the fact that analytic existence and uniqueness results have to be shown at the information day as well as considerations about stability of such randomly perturbed structures. Interesting shapes of the multi-storey buildings are a tall 4-storey building and an L-shaped structure to see the effects of a broader contact surface with the ground. Of course, the stochastic ground motion accelerations will serve as inputs for the following

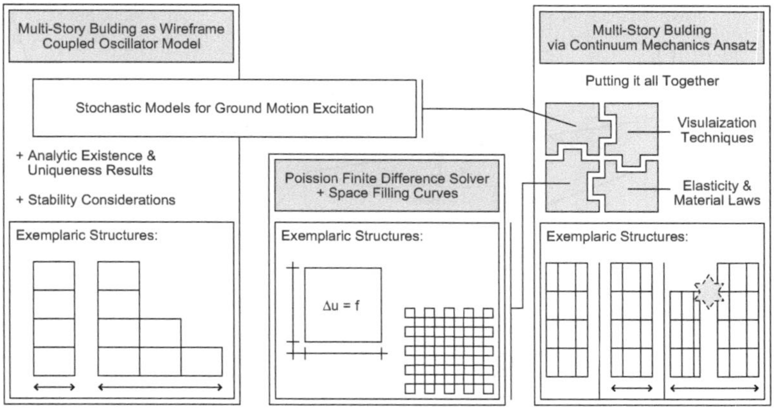

Figure 18.1. The three major parts of the earthquake simulation your team has to execute in the following 2 days: (i) simulations and analytic discussions of multi-storey buildings as a wireframe coupled oscillator model, (ii) a finite-difference solver for the Poisson equation together with effective data storage based on space filling curves, and (iii) simulations of multi-storey buildings as a continuum mechanics model.

tasks, too. Second, a finite-difference solver for the Poisson equation $\Delta u = f$ on the unit square shall be implemented together with an effective data storage based on space filling curves. This task is an essential prerequisite for the following one. Finally, an abstract two-dimensional multi-storey building and its response to ground motion acceleration shall be considered. Hereby, a continuum mechanics ansatz has to be carried out to describe the building as a solid homogeneous block subject to specific material constants. It would be great if your team could model two multi-storey buildings nearby with different material properties such that one may see, under certain conditions, collisions of them. Of course, "perfect design" was chosen because of its particular expertise in visualisation. The public relations officer stresses the fact that he would like to have really tangible and nice graphical user interfaces (GUIs) that inspire prospective students. Note, that explaining posters are common at university information days.

Before she leaves to the opera, your boss tells you and your team that she already took the notes from the university and used them to brake down your designated tasks and wrote detailed specifications for each of these work packages, see Fig. 18.2:

1. visualisation/ graphical user interface
2. stochastic excitations (mathematical models and real data)

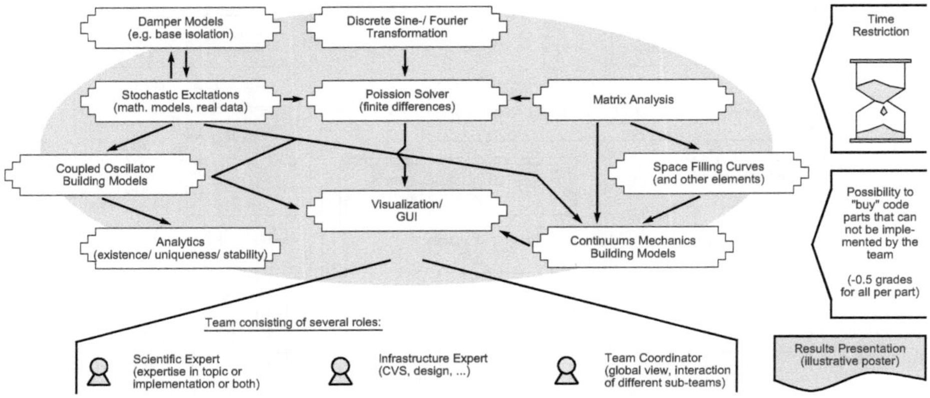

Figure 18.2. Work packages and their connections together with side conditions imposed on the work of "perfect design".

3. modelling of multi-storey buildings as a wireframe coupled oscillator structure
4. analytic discussion (existence and uniqueness of solutions, stability)
5. Poisson solver based on a finite-difference scheme
6. discrete Sine-/ Fourier-Transformation
7. space filling curves
8. matrix analysis
9. modelling of multi-storey buildings as a continuum mechanics structure
10. damper models (e.g. base isolation)

Now, you are left alone with your team. Before you can begin to work on specific work packages you have to make some infrastructure considerations. In particular, it would be wise to assign roles to the team members. These roles could be

- scientific expert(s), i.e. someone who has expertise in the theory or background of a work package,
- programming expert(s), i.e. someone who has expertise in programming MATLAB for realising specific implementation tasks, maybe in combination with a scientific expert,
- infrastructure expert(s), i.e. someone who sets up and maintains a repository (CVS, e.g.), or is responsible for a common design, etc.,

- interface coordinator(s), i.e. someone who organises the teamwork aspects of work packages that are related, and
- a team coordinator who is responsible for the overall picture (global view on all work packages, their realisation in time, and problem resolution between different sub-teams).

As usual in industry you may purchase knowledge by outsourcing. That means you can buy parts of the required code which your team cannot implement within the required time on its own. Note, that this decreases the profit of "perfect design" and this makes your boss rather unhappy, leading to a decrease of your performance evaluation by 0.5 grades (and consequently to less money for you at the end of the year).

Note that your boss forgot a dedicated task for designing an illustrative results presentation poster. The public relations officer will be extremely happy if such a poster is delivered since it is a perfect eye-catcher to attract prospective students to the simulation.

T 1: Visualisation/ Graphical User Interface [⊕]

In this work package, the usage of MATLAB's *graphical user interface design environment* (guide) will be essential.

1. Test the basic guide features with a prototype project. Make sure to use both basic functionality (different types of buttons, etc.) as well as filling the prototype with life in the corresponding MATLAB .m-file.
2. Realize a suitable visualisation/GUI together with the team "oscillator structure".
3. Realize a suitable visualisation/GUI together with the team "Poisson problem".
4. Realize a suitable visualisation/GUI together with the team "continuum mechanics structure".

Except for the first task, you have to be in close contact with the corresponding teams of the respective other work packages. Make sure to communicate! Hint: It may be useful to tackle the 3 major steps in parallel (at least concerning prototype versions).

T 2: Stochastic Excitations [☼]-[⊕]

In order to have a variety of tangible stochastic models of earthquake excitations $y(t)$ that affect the basement of a ground-bound structure, you have to implement (at least) the following three variants:

1. First, and foremost, according to [207], pp. 18, the stochastic differential equation formulation for the ground motion excitation $y(t)$ in the sense of the Kanai-Tajimi model is given as
$$y(t) = \ddot{x}_g + \xi_t = -2\zeta_g\omega_g\dot{x}_g - \omega_g^2 x_g ,$$
where x_g is the solution of a zero-mean Gaussian white noise ξ_t driven stochastic oscillator
$$\ddot{x}_g + 2\zeta_g\omega_g\dot{x}_g + \omega_g^2 x_g = -\xi_t , \qquad x_g(0) = \dot{x}_g(0) = 0 .$$
Here, ζ_g and ω_g are model parameters reflecting the local geological site conditions. For instance, in [133], the use of $\zeta_g = 0.64$ and $\omega_g = 15.56$ (in rad/s) is recommended for firm soil conditions in a frequency range from $f = 2.1 - -21$ (in rad/s).

   To handle this stochastic differential equation, the Doss-Sussmann/Imkeller-Schmalfuss correspondence should be applied to transform it into an equivalent random differential equation.

2. Second, Bogdanoff et al. [31], [144], considered the following model for ground accelerations
$$y(t) = \begin{cases} \sum_{j=1}^{n} ta_j \exp(-\alpha_j t)\cos(\omega_j t + \Theta_j) & \text{for } t \geq 0 \\ 0 & \text{for } t < 0 \end{cases} ,$$
where a_j, α_j and ω_j are given real positive numbers and the parameters Θ_j are independent random variables uniformly distributed over an interval of length 2π.
In particular, we have (all parameters with units rad/s): $\omega_1 = 6$, $\omega_2 = 8$, $\omega_3 = 10$, $\omega_4 = 11.15$, $\omega_5 = 20, 75$, $\omega_6 = \cdots = \omega_{11} = 0$, $\omega_{12} = 21.50$, $\omega_{13} = 23.25$, $\omega_{14} = 25$, $\omega_{15} = 27$, $\omega_{16} = 29$, $\omega_{17} = 30.5$, $\omega_{18} = 32$, $\omega_{19} = 34$, $\omega_{20} = 36$, as well as $\alpha_1 = \cdots = \alpha_{20} = 1/3$, $a_1 = \cdots = a_{20} = 0.5$.

3. Third, "real" data should be used. Based on

   http://peer.berkeley.edu/peer_ground_motion_database/

   and

   http://www.emsc-csem.org/Earthquake/ ,

   a program is required that takes these real data and returns $y(t)$. Be particularly careful in choosing and extrapolating the corresponding time steps as these depend on the requirements of other work packages.

Note, the users should switch between these models easily and be able to view $y(t)$ over a suitable time horizon.

T 3: Modelling of Multi-Storey Buildings as a Wire-Frame Coupled Oscillator Structure [ ]

As a first illustrative model for a d-storey building ($d = 1, 2, \ldots$), a wire-frame structure is considered where the single floors are described by coupled damped oscillators. The displacement u_i of the floor $i \in \{1, \ldots, d\}$ is given by the following dimension-less second order matrix-vector differential equation

$$\ddot{u} + C\dot{u} + Ku = F(t)\,,$$

where $u = (u_1, \ldots, u_d)^T$, $C \in \mathbb{R}^{d \times d}$ is the matrix of the damping coefficients and $K \in \mathbb{R}^{d \times d}$ the matrix of the inelastic deformation forces. $F(t)$ is an external driving force. Its first component corresponds to a ground motion excitation $y(t)$ as discussed in the work package "stochastic excitations".

1. Set up the first order differential equation system that describes, in the above way, a 4-storey building with the four floors on top of each other. This leads to rather simple band matrices C and K. Assume different damping and deformation properties for each floor.

2. Analogously, formulate the first order differential equation system describing an L-shaped 4-storey building consisting of seven blocks with different damping and deformation properties. Here, 3 blocks form the basement (1st floor) and 2 blocks the 2nd floor on top of which the two single-block floors 3 and 4 follow, see Fig. 18.1.

Next, a numerical solver for these random differential equation systems is required.

1. Suggest suitable values for the damping and deformation coefficients used in the two 4-storey building models such that a "soft" structure and a rigid one can be simulated.

2. Implement the explicit (deterministic) Euler and Heun scheme to solve the above problems as well as a suitable Runge-Kutta scheme. Are our systems stiff or non-stiff?

3. Implement the explicit averaged Euler and Heun scheme to solve the above problems as well as a suitable random ordinary differential equation Runge-Kutta scheme, e.g. RODE 3.0.

Finally, test the accuracy of these methods for the 4-storey building with the four floors on top of each other. Recall that the analytic solution in this case is gained by the frequency domain method for response analysis. How well do these methods work for the Kanai-Tajimi excitation, the Bogdanoff excitation and a real earthquake?

T 4: Analytic Considerations [☆]

Of course, before the first line of code can be written, we have to be sure that we get unique solutions and do not run into problems inherent to the system under consideration.

1. Show that the two systems of random differential equations built in the work package "modelling of multi-storey buildings as a wireframe coupled oscillator structure" and equipped with the three stochastic excitations from the work package "stochastic excitations" have a unique path-wise solution.
2. Are these solutions bounded?
3. What is the expected value of these solutions?

In order to have at least some kind of a hint as to whether our two buildings will collapse, the analytic stability of the models needs to be discussed (or simulated) with respect to different combinations of the damping and deformation coefficients. What can we say about the stability of the two buildings and do these mathematical results match our knowledge about earthquakes and their effects on buildings?

T 5: Poisson Solver [☼]-[◷]

This work package comprises 3 tasks: Implement the analytical and discrete (FD) solution of the Poisson equation for 1D and 2D via the standard sparse linear system of equations, applying a Fast Poisson solver in 1D and 2D as an alternative solution of the system, and reformulate the matrix assembly to a stencil-wise approach allowing arbitrary enumerations of the nodes. The following sub-tasks need to be realised in this work package:

1. Write a function solve_poisson_1d_LSE() that computes the sparse matrix and right-hand side of the 1D Poisson problem

$$\begin{aligned} \Delta u(x) &= f(x)\,, & x \in \Omega := [0,1]\,, \\ u(x) &= 0\,, & x \in \Gamma := \partial\Omega\,, \end{aligned}$$

   for a given right-hand side function f and the number of inner points in x direction. Provide the capability of using arbitrary intervals $[0,a]$ of a given size $a > 0$. Implement a method poisson_1d() that takes $p \in \mathbb{N}$ as input parameter, computes the function $f(x) = -\sin(\pi x) \cdot \pi^2$ for $n = 2^p$ grid points inside $[0,1]$ (without boundaries), and calls solve_poisson_1d_LSE() to generate the discrete solution.

2. What does the analytical solution for this problem look like? Make sure to compare the 1D discrete solution with the analytical one using the maximum and the discrete L2 norm. Visualise the discrete and analytical solution.

3. Extend the previous implementation to the two-dimensional case using two methods solve_poisson_2d_LSE() and poisson_2d() for the problem

$$\begin{aligned} \Delta u(x,y) &= -\sin(\pi x)\cdot\sin(\pi y)\cdot 2\pi^2\,, & (x,y)\in\Omega := [0,1]^2\,,\\ u(x,y) &= 0\,, & (x,y)\in\Gamma := \partial\Omega\,, \end{aligned}$$

   Assume a lexicographic numbering of the vertices in the grid.

4. Compare the 2D discrete solution in the maximum and discrete L2 norm and visualise the results using MATLAB's surf() method.
   Hint: The method reshape() may be helpful.

5. Create a table showing the error of the discrete solution for increasing size of the resolution (i.e. increase p to get higher values for $n_x = 2^p$; consider both 1D and 2D). What is the order of the method? How can you experimentally check that using error plots?

6. Once the work package "space-filling curves" is ready, combine it with the 2D Poisson solver. For this purpose, change the matrix setup to a stencil-wise computation with arbitrary enumeration input to compute the actual indices of the vertices under consideration. Cross-check that the system with different vertex enumerations still correctly computes the results of the Poisson problem (use norms and visualisations!).

7. Once the work package "matrix analysis" is ready, use it to compare the 5 variants of enumerations with respect to the different properties. In particular, use the ILU considerations.

8. Realize a GUI allowing to specify the different variants of your solver (1D/2D, norms, enumeration type, etc.).

9. (optional): Use additional right-hand sides for the Poisson equation (eventually also allowing for analytical solutions to compare with). Implement the corresponding functions and their usage in a modular way such that the GUI may directly use it.

T 6: Discrete Sine-/ Fourier-Transform [☼]-[⊕]

For the following tasks, rely on the comparison (visual + norms) of the discrete solution with the analytical solution of task "Poisson solver". In particular, the discrete solution using the Fast Poisson algorithm has to be numerically identical to the direct solution of the sparse linear system of equations.

1. Implement a forward and inverse discrete sine transform (DST) for 1D using two methods $\mathrm{dst\_manual}()$ and $\mathrm{idst\_manual}()$ with a vector as input parameter.
   (optional): How can the program be sped up without using explicit loops?

2. Make sure to verify that your DST implementation is doing the correct transform.

3. Implement the Fast Poisson algorithm of Sec. 9.6.2 using the DST for the Poisson problem (with all-zero boundary conditions) in 1D. Take care to scale the right-hand side of the system with h^2!

4. Extend the algorithm to the two-dimensional case assuming equal mesh sizes $h = h_x = h_y$. In particular, use the 1D (i)DST methods extended to matrix input to realise the 2D transforms.

5. (optional): Write a DST variant using MATLAB's FFT implementation directly together with pre- and post-processing.

6. (optional): How can the idea of a Fast Poisson solver be kept if Neumann boundary conditions are used or if a combination of Dirichlet and Neumann is applied?

T 7: Space-Filling Curves [⊕]

This work package will allow different variants of ordering the 2D nodes in a regular Cartesian mesh besides the standard lexicographic enumeration. In particular, a space-filling curve approach will be used.

1. Implement a turtle grammar for the 2D Hilbert curve on $[0, 1]^2$ in a function $\mathrm{turtleHilbert}()$ which takes the depth $p \in \mathbb{N}$ of the corresponding cell tree as input parameter (and uses it to stop the recursion) and which returns the vectors x, y containing the coordinates of the cell centers of the Hilbert curve. Visualise and cross-check your curve by plotting x vs. y in the mesh.

2. In order to apply the discrete Hilbert curve as an alternative ordering of the vertices of a given Cartesian mesh, it is useful to interpret the vertices as the cell centers x, y of the Hilbert curve. Hence, implement the following additional steps in a function $\mathrm{computeHilbertIndices}()$ which receives the depth of the tree p and the sizes a and b of an arbitrary rectangle as input parameters and which returns a matrix of indices:

- Compute the discrete Hilbert curve coordinates by calling turtle-Hilbert().
- Translate the Hilbert curve coordinates (cell centers) in the appropriate manner to the vertices of the computational grid.
- Scale all coordinates such that discrete rectangles of arbitrary size $[0, a] \times [0, b]$ may be indexed with the curve (under the assumption that the number of vertices in each direction is identical).
- Store the Hilbert curve indices of the corresponding vertices in a 2D array $\text{indices}_{i,j}$ where (i, j) is a global access index indicating the ith vertex in x and the jth vertex in y direction.
- Write a short function enum_hilbert() with input parameters i, j, m, n, and indices representing the index pair (i, j), the number of vertices in y and x direction, and the matrix of Hilbert curve indices (cf. Fig. 18.3 (b) for an example situation in the case of lexicographic numbering). enum_hilbert() shall return the resulting absolute index stored in indices.

3. Now realise four alternative orderings besides the Hilbert curve: the standard lexicographic ordering (row-wisely ordered in x direction, increasing in y direction), a column-wise lexicographic enumeration, a diagonal one and a meander-like variant (see Fig. 18.3 (a)). Implement these enumerations via the four new functions enum_lexicographic(), enum_lexicographic_col(), as well as enum_diagonal() and enum_meander() using the same signature as for enum_hilbert() (ignoring the stored array of indices in the implementation which is only necessary for the Hilbert-curve ordering). Write a short test script to visualise the resulting enumerations (similar to Fig. 18.3 (a), e.g.) for different depths p.
4. Apply the resulting code to the "Poisson solver" as well as the elasticity problem. In order to facilitate this step, think of providing a simple prototype of enumerations (the lexicographic one, e.g.) to the teams of the corresponding tasks 5 and 9.
5. (optional): Realize a second variant for the computation of the Hilbert curve indices using the "inverse" mapping of the Hilbert curve for each pair of vertex coordinates (x, y).
6. (optional): Implement a vertex enumeration using the Peano curve. Cross-check potential different behavior compared to the Hilbert curve mapping when applying the "matrix analysis" to both work packages "Poisson solver" and the elasticity problem.

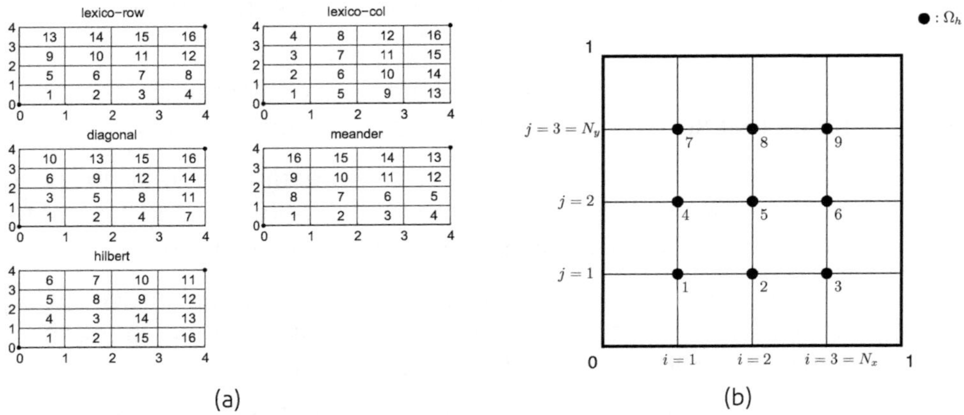

Figure 18.3. The five different enumeration types (a) for $p = 2$ (for the sake of simplicity, the numbers are plotted in cell centers, but represent vertex indices). A visualisation of the index mapping for lexicographic numbering (b).

T 8: Matrix Analysis [☼]

The target of this task is to analyse different properties of (sparse) matrices stemming from PDE discretisations such as the "Poisson solver" or the "continuum mechanics structure". Note: You don't have to wait for the realisation of these packages to start working on the matrix analysis!

1. Implement a function analyseMatrices() that receives five matrices (corresponding to the five different enumeration types used in the work package "space-filling curves") as input parameters. Based on the template code of

   http://www.mathworks.de/products/matlab/demos.html?\\file=
   /products/demos/shipping/matlab/sparsity.html#3,

   implement for each of the matrices:

   - the computation of the corresponding bandwidth;
   - the LU decomposition (instead of the Cholesky decomposition) using MATLABs built-in function lu()[2]. Take care to avoid pivoting;

[2] Additional information on MATLAB's LU decomposition is found, e.g., at:
http://www.mathworks.de/help/techdoc/math/f6-8856.html\#f6-9633

- a figure subplot showing the sparsity pattern and indicating the bandwith;
- a figure subplot showing the sparsity pattern of the matrix L+U and indicating the nonzeros (absolutely and as percentage; this shows the fill-in of the decomposition);
- the measuring of the execution time for the corresponding decomposition.

In addition, provide a summary plot of the amount of relative fill-in for the LU decomposition as well as the execution times.

2. Apply $\mathrm{analyseMatrices}()$ to the matrices of the work package "Poisson solver" and "continuum mechanics structure". What do you observe?

3. Extend $\mathrm{analyseMatrices}()$ by an incomplete LU decomposition (ILU, $\mathrm{ilu}()$) for the case of that an additional input parameter $\mathrm{shallILUBeUsed}$ is true. This decomposition is often used as a preconditioner for sparse systems. Take care to only work on the original non-zero structure of the matrix (no fill-in) as well as to avoid, again, pivoting. Track the runtime for each decomposition and compute and display the condition number of the original matrix A and of the preconditioned matrix $L^{-1} \cdot A \cdot U^{-1}$ where L and U are the results of the ILU decomposition[3]. Apply ILU to the matrices of the work package "Poisson solver". What do you observe?

4. (optional): Extend the decomposition survey by the Cholesky decomposition in case an additional input parameter $\mathrm{shallCholeskyBeUsed}$ is true.

5. (optional): If the Peano curve is realised in the work package SFC, extend the matrix analysis to modularly use either Hilbert or Peano numbering (because the number of vertices is either 2^p or 3^p, disjointly) in the analysis.

T 9: Modelling of Multi-Storey Buildings as a Continuum Mechanics Structure [ ]

For this work package, the documents on material laws are extremely relevant. The basic explicit equations are

$$0 = (2\mu + \lambda)\frac{\partial^2 u_1}{\partial x_1^2} + \mu\frac{\partial^2 u_1}{\partial x_2^2} + (\lambda + \mu)\frac{\partial^2 u_2}{\partial x_1 \partial x_2} \quad \text{in} \quad \Omega \tag{18.1}$$

3 Note: For the sake of simplicity, you may use MATLAB's $\mathrm{inv}()$ method to really compute the inverse of a matrix—which you usually should never do!

$$0 = (2\mu + \lambda)\frac{\partial^2 u_2}{\partial x_2^2} + \mu\frac{\partial^2 u_2}{\partial x_1^2} + (\lambda + \mu)\frac{\partial^2 u_1}{\partial x_1 \partial x_2} \quad \text{in} \quad \Omega \tag{18.2}$$

$$u(x,t) = u_0(x,t) \quad \text{on} \quad \Gamma_0 \tag{18.3}$$

$$\sigma(u(x,t))n(x) = \tau(x,t) \quad \text{on} \quad \Gamma_1 \,. \tag{18.4}$$

It may be advantageous to communicate with the team "Poisson problem" from time to time concerning discretisation issues or other aspects.

1. Implement a steady-state solution for the elasticity problem using a FD discretisation on a rectangular domain ($[0,1]^2$, e.g.). In particular:

    - Derive the necessary finite difference (FD) terms appearing in the explicit equations (18.1) and (18.2).
    - Use Young's modulus E and Poisson's ratio ν as parameters to compute the Lamé constants λ and μ. Make sure to use $\nu = 0$ for all simulations below since only 2D scenarios are considered.
    - Implement the assembly of the system matrix A. Hint: An ordering of the unknown displacements such that the x displacements of all vertices are collected before all y displacements are considered may be advantageous together with MATLAB's "reshape()" method.
    - Use Dirichlet boundary conditions (18.3) for the bottom edge (no ground motion). Since the right-hand side of the system (18.1), (18.2) is supposed to be zero, i.e. no gravity or other volume forces, the right-hand side vector b consists only of such Dirichlet boundary contributions. For the steady-state realisation, apply all-zero Dirichlet conditions, but keep in mind that this will be generalized below.
    - Use Neumann boundary conditions (18.4) on the other edges (free surface of the building). Derive the necessary finite difference approximations for vertices on the Neumann boundary (Hint: first order is sufficient). For the surface forces τ on the boundary, realise first a fixed surface force in a given direction for all nodes of the left edge. Extend it to a more modular approach allowing a prescribed surface force (realised as a function) on an arbitrary edge as well as a point force on a specific surface location.
    - Solve the whole linear system trivially by using MATLAB's backslash operator (i.e. $u = A\backslash b$).
    - Work together with the "visualisation/GUI" team to realise the interactive simulations. You may start with prescribed dummy displacements to prepare a prototype before the overall simulation is ready.

Keep in mind that a basic visualisation is typically helpful for developing the application.

- (optional): Extend the system to consider gravity as an example of volume forces, too.

2. Once the work package "space-filling curves" is ready, combine it with the elasticity solver. For this purpose, change the matrix setup to a stencil-wise computation with arbitrary enumeration input to compute the actual indices of the vertices under consideration. Cross-check that the system with different vertex enumerations still correctly computes the results of the elasticity problem (use norms and visualisations!).

3. Compute a time-dependent solution to the problem. In particular:

   - Implement the Newmark-β method as time integration schemes. Eventually take a scalar test equation of second order whose analytical solution you know to verify and validate your implementation. Make sure to have input parameters for the start and the end time as well as the number and/or size of the time steps.
   - Realize initial conditions for the elasticity problem. You may use the simplification of prescribing a fixed vector-valued function for initial displacements and initial velocities: All displacements and velocities in x and y direction, respectively, get a given initial vector value.
   - Extend your visualisation to represent time-dependent problems ("movie" as a sequence of figures, or similar). Adapt the GUI together with the respective team in order to select the steady-state or time-dependent setup. These features will help already for debugging.
   - Use a prescribed time-depending ground motion as Dirichlet boundary condition (such as a sine function or similar).
   - (optional): Choose another time integration scheme (such as explicit Euler (simple) or other), translate the system into first order, and apply it to the problem. Compare the results with those of the Newmark-β method.

4. Use the results of the work package "stochastic excitation" as the time-dependent Dirichlet boundary condition for the bottom edge of the building (Attention: right-hand side b has to be modified!). Include these excitations together with corresponding parameters and a plot into the GUI. Apply the different Young's moduli from above and identify alterna-

tive values to generate "nice" responses with visible oscillations of the building.

5. (optional): Create two buildings of possibly different height and with different Young's modulus. Hint: The usage of a flag field to distinguish inner (i.e. building) cells from outer (air) cells in an overall domain embedding the two geometries may be advantageous.

T 10: Damper Models (optional) [☼]

In order to minimise the effects of earthquakes, dampers are installed in modern buildings (see Fig. 18.4). One possible way of damping is "base isolation with shock absorbers" where the building is separated from the surrounding earth crust and placed upon shock absorbers. These absorbers can be modelled mathematically in terms of a dimension-less second order scalar oscillator equation, i.e.,

$$\ddot{z} + c\dot{z} + kz = y(t),$$

where $y(t)$ is the input signal and $\ddot{z}(t)$ the output which serves as an excitation of the building.

1. Apply such dampers to the models from the work packages "modelling of multi-storey buildings as a wireframe coupled oscillator structure" and "modelling of multi-storey buildings as a continuum mechanics structure".

2. How does one have to choose the coefficients c and k in order to have an effective protection from earthquakes?

18.3 Project Results

We briefly present the results the participants achieved throughout the workshop. In general, all students showed a very large commitment to realise their final product. All tasks of the specification have been tackled except task 10 (Sec. 18.2) which was optional anyway. The students managed to obtain nice results in the work packages. We only list those for the most prominent tasks in the following.

The wireframe model in different variants has been tackled in T3. In Fig. 18.5, simulation results of four different types of 4-storey wireframe buildings are visualised at three different times $t = 0.5$ (Fig. 18.5 (a)), $t = 1.5$ (Fig. 18.5 (b)), and $t = 2.5$ (Fig. 18.5 (c)): The leftmost and rightmost buildings are a normal and L-shape version using stiffness and damping coefficients of

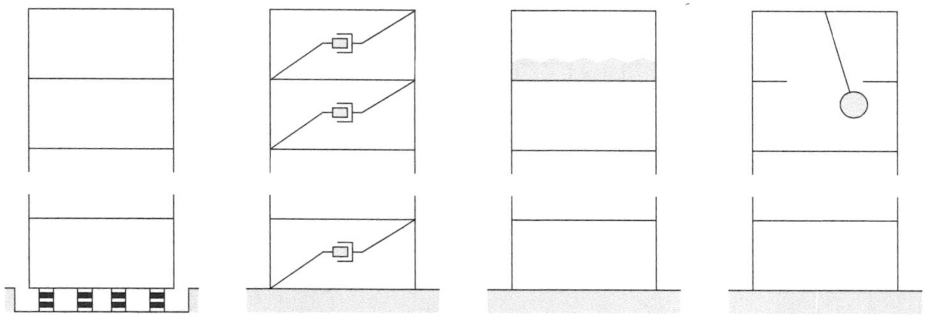

Figure 18.4. Illustration of several possibilities of securing buildings from the effects of earth excitations.

$k_1 = \ldots = k_4 = 25$ and $c_1 = \ldots = c_4 = 10$. Building 2 (second from left) possesses higher damping values $c_1 = \ldots = c_4 = 40$ resulting in less displaced storeys. For Building 3 (second from right), the standard damping ($c_i = 10$) has been used in combination with standard stiffness ($k_i = 25$) except for coefficient $k_1 = 2.5$.

The participants started working on both tasks T1 (GUI) and T9 (continuum mechanics model) relatively late, but then invested high efforts in transferring the GUI, in particular, to the different demanded tasks. A snapshot of the GUI for T9 for two different steady-state scenarios is given in Fig. 18.6. The students included a lot of variants to specify forces on the various boundaries of the building: External forces can be applied in both x and y direction onto the surface of the building. An example of a wind load from left is visible in the top picture of Fig. 18.6. Below, an artificial test case of "squeezing" the building from left and right is shown.

Several time-dependent simulations using the Newmark scheme in combination with the Kanai-Tajimi excitation has been carried out. Figure 18.7 shows the resulting displacements at different time steps: The building moves, due to the Kanai-Tajimi excitation, slightly to the right, back to the centre, and finally to the left for this realisation of the random input. 2047 time steps of size $dt = 0.0049$ have been used to reach the final time $t_e = 10.0$ for a coarse grid with 4×20 degrees of freedom (DoF) resulting in $h_x = 0.05/3, h_y = 0.25/19$ to discretise the building with Young's modulus equal to $E = 1e - 4$.

Furthermore, the participants were eager to make the optional task T9-5 possible: In Fig. 18.8, two buildings (green patches) of different Young's moduli (left: $E_1 = 500$, right: $E_2 = 1500$) react on a surface force from the left.

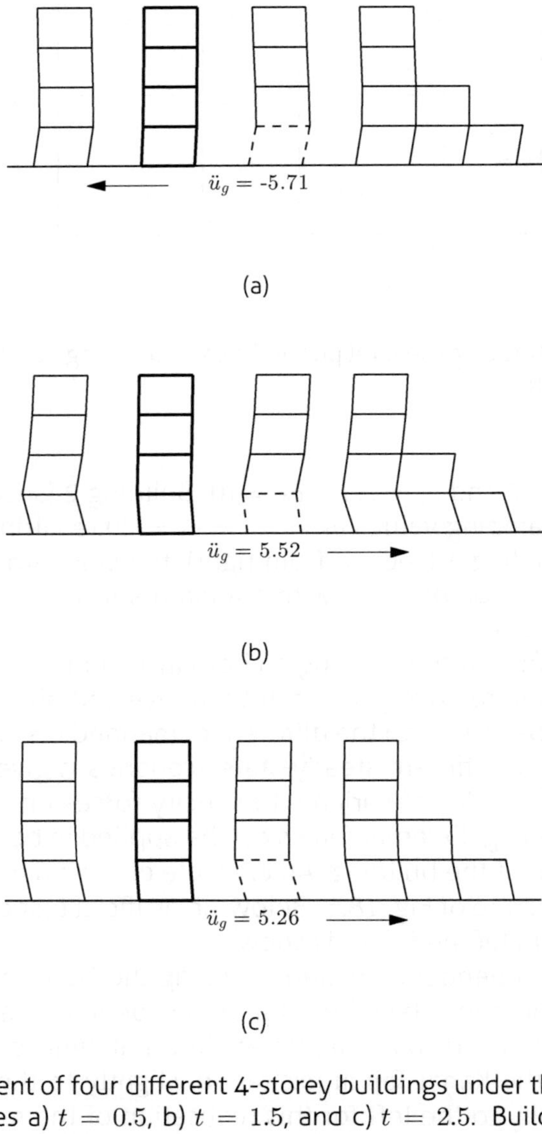

Figure 18.5. Movement of four different 4-storey buildings under the same stochastic excitation at times a) $t = 0.5$, b) $t = 1.5$, and c) $t = 2.5$. Building 1 (leftmost): Normal 4-storey building with $k_1 = \ldots = k_4 = 25$, $c_1 = \ldots = c_4 = 10$. Building 2: Same as building 1 but with $c_1 = \ldots = c_4 = 40$. Building 3: same as building 1 but with smaller $k_1 = 2.5$. Building 4 (rightmost): L-shaped with the same parameters as building 1.

Only those cells that are located completely inside the building patches are used for the building geometry.

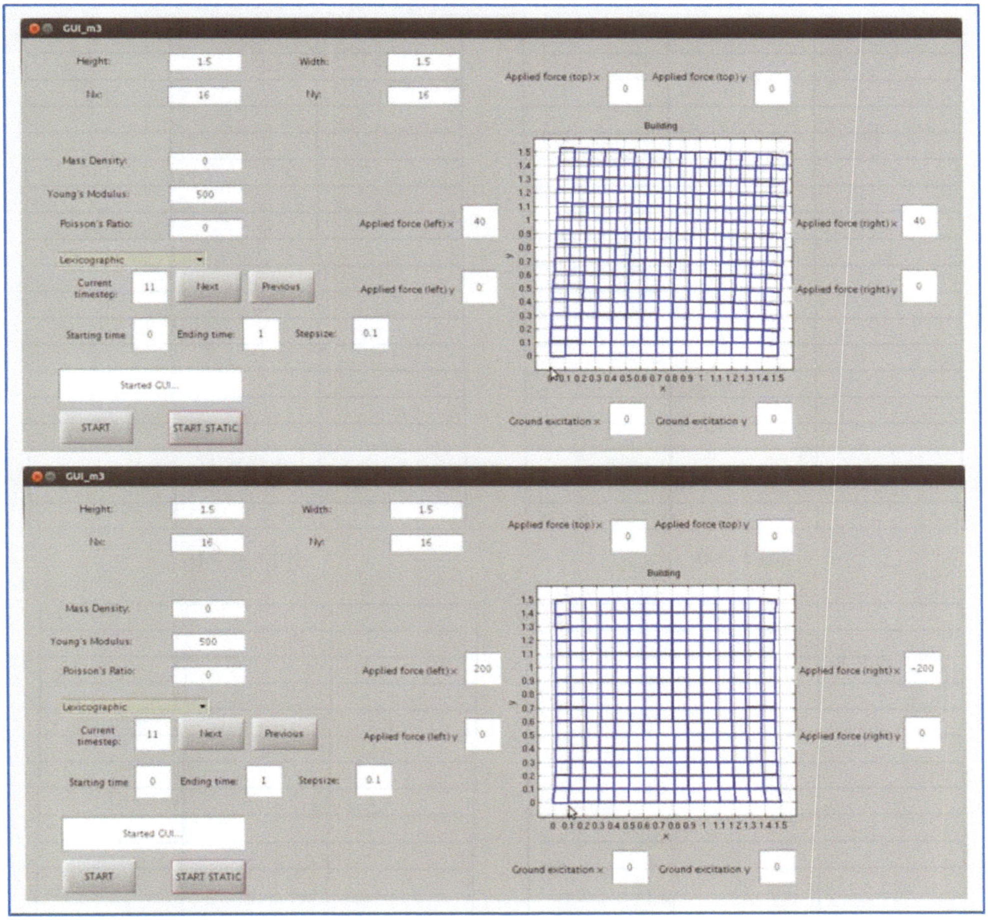

Figure 18.6. Example of the GUI provided for T9: Steady-state solutions of the 2D continuum mechanics model without earthquake incitation for 16×16 nodes under specified forces: top: pushing forces from left (wind load, e.g.); bottom: symmetric "squeezing" of the building from left and right.

For all obtained results and tackled work packages, it really payed off to have the corresponding specifications carefully designed (up to having a possible solution at hands before the workshop started): It payed off from the mere content point of view, but also concerning the ambiance and motivation which will be discussed in the following section.

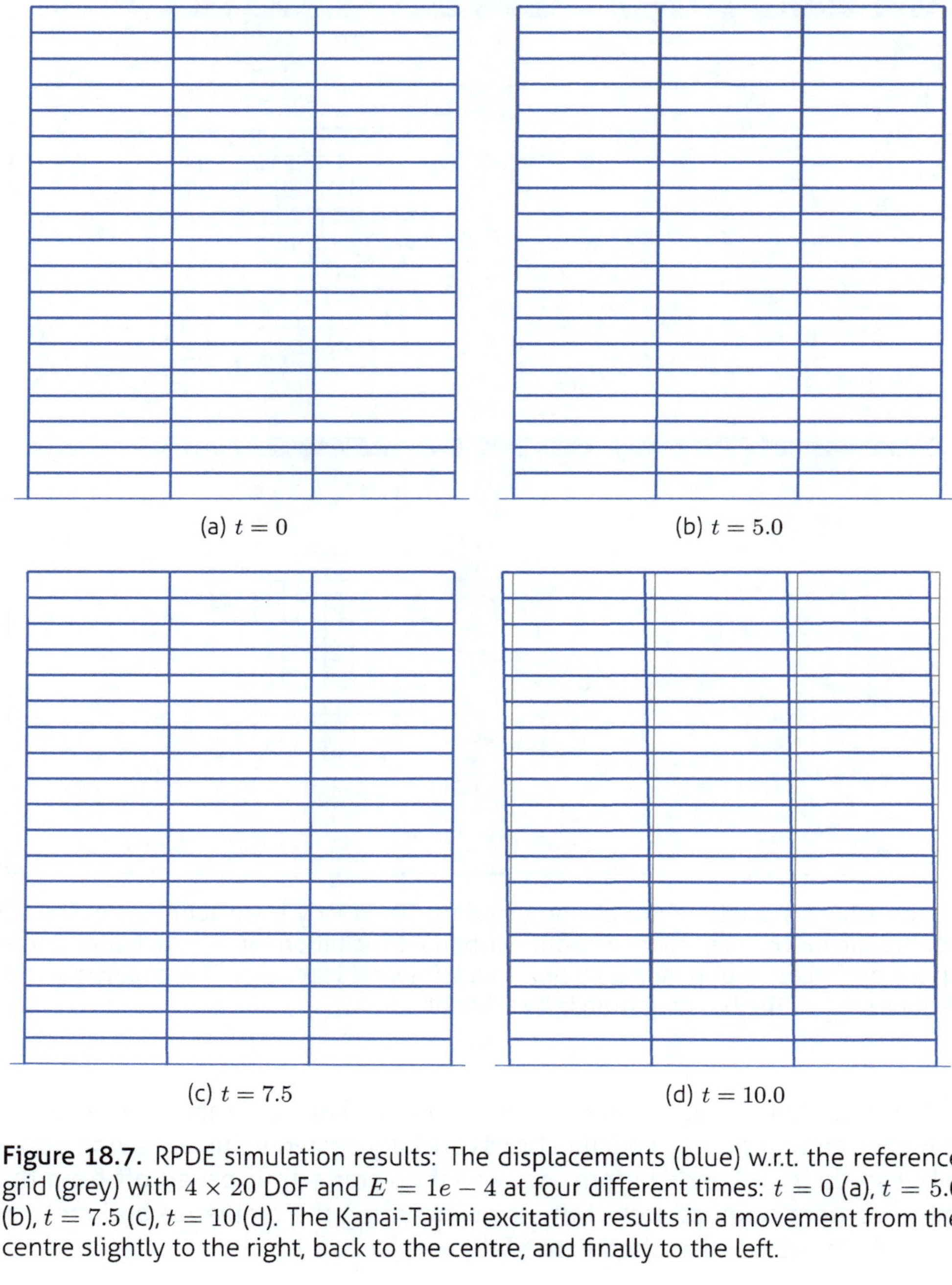

(a) $t = 0$ (b) $t = 5.0$

(c) $t = 7.5$ (d) $t = 10.0$

Figure 18.7. RPDE simulation results: The displacements (blue) w.r.t. the reference grid (grey) with 4×20 DoF and $E = 1e - 4$ at four different times: $t = 0$ (a), $t = 5.0$ (b), $t = 7.5$ (c), $t = 10$ (d). The Kanai-Tajimi excitation results in a movement from the centre slightly to the right, back to the centre, and finally to the left.

Figure 18.8. "Workshop: The Two Towers": Extension of the T9 simulations to two buildings of different Young's moduli (left: $E_1 = 500$, right: $E_2 = 1500$) and different sizes in an overall reference grid (grey) of 20×20 cells.

18.4 Lessons Learnt

Both the participating students and the supervisors had a lot of nice and sometimes surprising experiences throughout the workshop. We are going to list our impressions from the supervisor point of view in the following subsection before presenting the feedback of the students and concluding this section on lessons learnt.

18.4.1 General Impressions

Our role as supervisors throughout the runtime of the workshop was restricted to observing and giving (technical) advice when being asked. Even for such a relatively small group of 10 participants, being two supervisors and not just one was important.

Teamwork

The participants actively used their freedom in the design of their team structure: They decided on having nearly no designated roles. Except for one student acting as the infrastructure expert due to her experience in that field, no other roles had been assigned. The group initially analysed the specification and estimated the time and workload necessary for the tasks. With this information, the participants decided on the size of the subgroups and its members together with milestones.

Already in the first couple of hours, some first results were obtained which considerably helped to create a motivational environment for the team from the very beginning. The team decided not to create or use prototypes and interfaces. For the tasks that involved cross-task collaboration, this turned out to have an overall negative effect: Fixing and smoothing existing work packages took more time than investing in a smooth design in the beginning.

The team showed a surprisingly efficient internal organisation without any interaction from the supervisors. Regular checkpoint meetings were held with a fast survey on the overall status of each subtask. All members contributed constructively and used a decent level of abstraction such that all team mates were able to follow and all necessary details to cross-check dependencies of different work packages, e.g., were provided. In case somebody missed a point or joined in a bit later, a direct "briefing" with a short repetition took place. In certain cases, a specific task was terminated faster than expected; the assigned team immediately tried to tackle another work package in order to not lose time. Sometimes, even extraordinary plenary meetings arose from corresponding individual discussions to better balance the workload.

The atmosphere in the team was very positive. All team members gave honest indications of their status; problems that eventually arose could, thus, be openly discussed. Furthermore, individual subgroups supported each other according to their strengths and weaknesses. A nice indicator for the open policy in the group is the dealing with mistakes:

participant A: "Oh, we did it on your laptop, you didn't commit."
participant B: "Yeah, that's right..."

So, no accusation and no defending took place because it was simply not necessary. Only pure constructive criticism arose. Perhaps this fact is also related to the relatively short duration of the project of 2.5 days where no "historic backpack" of emotions related to work is gathered.

In total, all students participated very constructively. We observed a large will and commitment of all participants. Sometimes, the motivation to finish a specific subtask was so high that we had problems to get people to lunch.

Choice of the Workshop Environment

We briefly list a couple of important impressions concerning the organisational and technical aspects of the workshop environment. The expectations with regard to the seminar centre (see Sec. 17.3.2) were completely fulfilled: Being abroad from campus in a nice environment pushed the motivation and the commitment of the participants considerably. We as the supervisors could focus on our role as advisors while most of the organisational work had been done in advance and the operational level now was realised by the seminar centre. Using MATLAB as programming language worked very well. The environment ran without interruptions, and the students were able to fix all questions using numerous tutorials and documentation available online. With the help of the infrastructure expert, all minor CVS problems could be solved directly within the team. While the participants used the repository only rarely in the beginning (where more independent work packages were tackled), they started to rely more and more on it when it came to the more complex tasks.

18.4.2 Feedback of Participants

In this section, we list the unfiltered feedback of the participants.

Concerning the teamwork, the students were aware of the minor usage of roles. They appreciated the volunteering of certain participants for organising/moderating/leading aspects as well as the surprisingly good cooperation, communication, and active contribution of everybody resulting in a constantly high motivation. In the rare cases that some participants took a short break on working on the project, the students enjoyed that this was always done with respect to some "investments" for the group: baking cookies, preparing the barbecue dinner, offering a MATLAB GUI short tutorial, etc. Besides, the teamwork across different disciplines throughout the workshop project and the whole course was regarded as a really enriching experience.

Concerning the workshop specifications and their realisation, the participants appreciated the technical environment of MATLAB and CVS since it is easy to handle. The work packages as well as the overall topic were considered to be interesting. Some criticism arose concerning the interconnection of task T5 (Sec. 18.2)and T9 (Sec. 18.2): Why and how T5 serves as a prerequisite for T9 could be explained in more detail. Furthermore, the students proposed a tighter formulation of the connection of T4 (Sec. 18.2) and T3 (Sec. 18.2) and vice versa. So obviously (and actually not surprisingly), some smoothing of the specification is still possible and will be applied for the second run of the course in the summer term 2013.

The students mentioned several times that they highly appreciated the freedom that the specification allowed in the implementation. Only in rare cases such as in T7 (Sec. 18.2), tight restrictions appeared in order to assure the success of the task. The large freedom of realisation in the other packages considerably increased motivation and creativity within the team.

18.4.3 Conclusion

From our point of view as the supervisors, nearly all desired effects listed in Sec. 17.2 and 17.3 could be realised throughout the course and the project. The strongest argument that the concept worked is the fact that the students succeeded in tackling an unexpectedly large amount of tasks in the workshop.

It was interesting to see that the students only rarely thought about how to "sell" the their results, i.e. how to present them in a decent way. Perhaps this focus is due to the specific culture dominating the field of mathematics and natural sciences that our students come from: The content of a topic or work is esteemed to be more important than the selling effects surrounding it. The experience and hints that some small and cheap add ons to the work in the visualisation or reports could considerably improve the mood of the client are an important feedback for the participants on their way to a full professional working life.

Already in the very first plenary assembly of the participants, an interesting effect took place that continued throughout the whole workshop: Those students who had the feeling of not being very strong in theory or in one concrete subtopic did not hide away but proactively tried to compensate this technical weakness by contributing more in on the field of organisation (moderating, time keeping, preparing reports, etc.). Everybody really tried to contribute her/his best to make the project a success and, thus, everybody got accepted in the complete group. Knowing each other via the seminar talks and exercises represents a major advantage to correctly judge the strengths and weaknesses and supports a smooth running and nice outcomes of the project. This is an additional argument that the concept of the course connected to the workshop project is a valid alternative to classical existing course formats.

18.5 Outlook: Extension to Future Projects

To terminate this chapter, we use the experiences made so far to give some hints for the preparation of future projects. From both the feedback obtained from the participants and our own observations, it is absolutely necessary to be well-prepared in order to turn such a short-term project into success. In particular, the specifications have to be well designed. Everything that remains non-smooth in the preparatory steps will turn into smaller or larger problems throughout the workshop realisation and has to solved ad hoc. Furthermore, the trick to design some work packages rather compact and keep them separate instead of merging them with others into larger tasks allows for fast first results of the subteams working on them and, thus, leads to a considerable boost in motivation especially in the beginning of the project. Making some tasks optional and stand-alone from the very beginning avoids frustration and offers flexibility when things turn out to run very well (avoiding bored participants). It turned out to be particularly useful to discuss the specifications and prepare the infrastructure with the students in advance. Hence, they were ready to start productive work by the very first hour of the workshop. Being off the campus in a seminar centre or similar is, from our point of view, an indispensable key to success. A clear definition and distribution of team roles is, for sure, advantageous. But, as our example showed, it is no real must to push the participants into that since considerable learning effect will be realised anyway. In general, our experience and the provided feedback show that a decent amount of freedom in the specification and organisation has huge positive effects on the motivation and, thus, also on the amount of realised tasks. Hopefully, the above-mentioned aspects contribute to turn other workshops or similar projects into a successful experience for both students and supervisors.

Index

Bibliography

[1] N. Ahmed, T. Natarajan, and K. R. Rao. Discrete cosine transform. *IEEE Transactions on Computers*, C-23(1):90–93, 1974.

[2] K. T. Alligood, T. D. Sauer, and J. A. Yorke. *Chaos – An Introduction to Dynamical Systems*. Springer-Verlag, Berlin, Heidelberg, New York, 1996.

[3] Hans Wilhelm Alt. *Lineare Funktionalanalysis*. Springer-Verlag, Berlin, Heidelberg, New York, 3. edition, 1999.

[4] J. C. Anderson and V. V. Bertero. Uncertainties in establishing design eaerthquakes. *Journal of Structural Engineering, ASCE*, 113(8):1709–1724, 1978.

[5] A. A. Andronov, A. A. Vitt, and S. E. Khaikin. *Theory of Oscillators*. Dover Publications, New York, 1966.

[6] D. V. Anosov, S. Kh. Aranson, V. I. Arnold, I. U. Bronshtein, V. Z. Grines, and Yu. S. Il'yashenko. *Ordinary Differential Equations and Smooth Dynamical Systems*. Springer-Verlag, Berlin, Heidelberg, New York, 1997.

[7] L. Arnold and B. Schmalfuss. Lyapunov's second method for random dynamical systems. *Journal of Differential Equations*, 177(1):235–265, November 2001.

[8] Ludwig Arnold. *Stochastic Differential Equations*. John Wiley & Sons, New York, London, Sydney, Toronto, 1974.

[9] Ludwig Arnold. The unfolding of dynamics in stochastic analysis. *Computational and Applied Mathematics*, 16:3–27, 1997.

[10] Ludwig Arnold. Trends and open problems in the theory of random dynamical systems. In L. Accardi and C. C. Heyde, editors, *Probability towards 2000*, volume 128 of *Springer Lecture Notes in Statistics*, pages 34–46. Springer, Berlin, Heidelberg, New York, 1998.

[11] Ludwig Arnold. *Random Dynamical Systems*. Springer, Berlin, Heidelberg, New York, 2003.

[12] Vladimir Igorevich Arnold. *Ordinary Differential Equations*. Springer-Verlag, Berlin, Heidelberg, New York, 3. edition, 1992.

[13] Nakhle H. Asmar. *Partial Differential Equations with Fourier Series and Boundary Value Problems*. Pearson Education International, Upper Saddle River, NJ, 2. edition, 2005.

[14] K. B. Athreya and S. N. Lahiri. *Measure Theory and Probability Theory*. Springer-Verlag, New York, Heidelberg, Berlin, 2006.

[15] Bernd Aulbach. *Gewöhnliche Differenzialgleichungen*. Spektrum Akademischer Verlag, München, 2. edition, 2004.

[16] I. Babuška and T. Strouboulis. *The Finite Element Method and its Reliability*. Numerical Mathematics and Scientific Computations. Oxford University Press, 2001.

[17] L. Bachelier. Theorie de la speculation. *Annales Scientifiques de l'Ecole Normale Superieure*, 3(17):21–86, 1900.

[18] L. Bachelier. *Theorie de la Speculation*. Gauthier-Villars, Paris, 1900.

[19] M. Bader. *Space-Filling Curves - An Introduction with Applications in Scientific Computing*, volume 9 of *Texts in Computational Science and Engineering*. Springer-Verlag, 2013.

[20] Michael Bader, Robert Franz, Stephan Guenther, and Alexander Heinecke. Hardware-oriented implementation of cache oblivious matrix operations based on space-filling curves. In Roman Wyrzykowski, Jack Dongarra, Konrad Karczewski, and Jerzy Wasniewski, editors, *Parallel Processing and Applied Mathematics, 7th International Conference, PPAM 2007*, volume 4967 of *Lecture Notes in Computer Science*, pages 628–638. Springer, May 2008.

[21] C. A. L. Bailer-Jones, T. J. Sabin, D. J. C. MacKay, and P. J. Withers. Prediction of deformed and annealed microstructures using bayesian neural networks and gaussian processes. In T. Chandra, S. R. Leclair, J. A. Meech, B. Verma, M. Smith, and B. Balachandran, editors, *Proceedings of the Australasia Pacific Forum on Intelligent Processing and Manufacturing of Materials*, pages 913–919. Watson Ferguson & Co., Brisbane, 1997.

[22] E. A. Barbasin. *Einführung in die Theorie der Stabilität [russ.]*. Nauka, Moscow, 1967.

[23] June Barrow-Green. *Poincaré and the Three Body Problem*. American Mathematical Society, Providence, Rhode Island, 1997.

[24] K. J. Bathe. *Finite Element Procedures*. Prentice Hall, 1996.

[25] Heinz Bauer. *Wahrscheinlichkeitstheorie und Grundzüge der Maßtheorie*. Walter de Gruyter, Berlin, New York, 2. edition, 1974.

[26] J. Behrens. Multilevel optimization by space-filling curves in adaptive atmospheric modeling. In Frank Hülsemann, Markus Kowarschik, and Ulrich Rüde, editors, *Proceedings of the 18th symposium simulationtechnique ASIM 2005*, volume 15 of *Fortschritte in der Simulationstechnik – Frontiers in Simulation*, pages 186–196, Erlangen, Germany, 2005. SCS Publishing House e.V.

[27] Nam Parshad Bhatia and Giorgio Philip Szegö. *Stability Theory of Dynamical Systems*. Springer-Verlag, Berlin, Heidelberg, New York, 1970. reprinted in 2002.

[28] F. Black and M. Scholes. The pricing of options and corporate liabilities. *Journal of Political Economy*, 81(3):637–654, 1973.

[29] D. J. Block, K. J. Aström, and M. W. Spong. *The Reaction Wheel Pendulum*. Morgan & Claypool Publishers, San Rafael, 2007.

[30] Dobieslaw Bobrowski. *Wstep do Losowych Rownan Rozniczkowych Zwyczajnych [Introduction to Random Ordinary Differential Equations]*. Panstwowe Wydawnictwo Naukowe (PWN), Warsaw, 1987.

[31] J. L. Bogdanoff, J. E. Goldberg, and M. C. Bernard. Response of a simple structure to a random earthquake-type disturbance. *Bulletin of the Seismic Society America*, 51:293–310, 1961.

[32] D. M. Boore. Stochastic simulation of high-freequency ground motions based on seismological models of radiated spectra. *Bulletin of the Seismological Society of America*, 76(6):1865–1894, 1983.

[33] Vivek S. Borkar. *Stochastic Approximation: A Dynamical Systems Viewpoint*. Cambridge University Press, Cambridge, 2008.

[34] A. A. Borovkov. *Probability Theory*. Gordon and Breach Science Publishers, Amsterdam, 1998.

[35] P. Boyle and M. Frean. Dependent gaussian processes. In L. K. Saul, Y. Weiss, and L. Bottou, editors, *Advances in Neural Information Processing Systems 17*, pages 217–224. The MIT Press, Cambridge, 2005.

[36] R. N. Bracewell. The fourier transform. *Scientific American*, pages 62–69, June 1989.

[37] D. Braess. *Finite Elements—Theory, Fast Solvers, and Applications in Solid Mechanics*. Cambridge University Press, 3rd edition, 2007.

[38] Martin Braun. *Differential Equations and Their Applications*. Springer-Verlag, New York, Heidelberg, Berlin, 2. edition, 1978.

[39] H. Bremer. *Dynamik und Regelung mechanischer Systeme*. Teubner Verlag, Stuttgart, 1988.

[40] D. Brigo and F. Mercurio. *Interest Rate Models – Theory and Practice with Smile, Inflation and Credit*. Springer-Verlag, New York, Heidelberg, Berlin, 2. edition, 2006.

[41] R. W. Brocket. *Finite Dimensional Linear Systems*. John Wiley & Sons, New York, London, Sydney, Toronto, 1970.

[42] H. Bunke. Stabilität bei stochastischen differentialgleichungssystemen. *Zeitschrift fr Angewandte Mathematik und Mechanik*, 43:63–70, 1963.

[43] H. Bunke. Über das asymptotische verhalten von lösungen linearer stochastischen differentialgleichungssysteme. *Zeitschrift für Angewandte Mathematik und Mechanik*, 45:1–9, 1965.

[44] H. Bunke. Zum asymptotischen verhalten von lösungen linearer stochastischen differentialgleichungssysteme. *Zeitschrift für Angewandte Mathematik und Mechanik*, 47:64, 1967.

[45] H. Bunke. *Gewöhnliche Differentialgleichungen mit zufälligen Parametern*. Akademie-Verlag, Berlin, 1972.

[46] J. C. Butcher. *The numerical analysis of ordinary differential equations: Runge-Kutta and general linear methods.* Wiley-Interscience, 1987.

[47] Eugene I. Butikov. The rigid pendulum - an antique but evergreen physical model. *European Journal of Physics*, 20(6):429–441, 1999.

[48] Eugene I. Butikov. On the dynamic stabilization of an inverted pendulum. *American Journal of Physics*, 69(7):755–768, 2001.

[49] Eugene I. Butikov. Regular and chaotic motions of the parametrically forced pendulum: Theory and simulations. In Peter M.A. Sloot, C.J. Kenneth Tan, Jack J. Dongarra, and Alfons G. Hoekstra, editors, *Computational Science – ICCS 2002, Part III*, Lecture Notes in Computer Science, pages 1154–1169. Springer-Verlag, Berlin, Heidelberg, New York, 2002.

[50] M. I. Caiado and A. V. Sarychev. Remarks on stability of inverted pendula. *Rendiconti del Seminaro Matematico, Universit e Politecnico di Torino*, 63(4):333–347, 2005.

[51] C. G. Canuto, Y. Hussaini, A. Quarteroni, and T. A. Zang. *Spectral Methods: Evolution to Complex Geometries and Applications to Fluid Dynamics.* Scientific Computation. Springer-Verlag Berlin Heidelberg, 2007.

[52] T. K. Caughey and A. H. Gray Jr. On the almost sure stability of linear dynamic systems with stochastic coefficients. *Journal of Applied Mechanics*, 32:365–372, 1965.

[53] P. Du Chateau. *The Cauchy-Goursat Problem.* American Mathematical Society, Providence, Rhode Island, 1972.

[54] Kui Fu Chen. Standing human – an inverted pendulum. *Latin-American Journal of Physics Education*, 2(3):197–200, 2008.

[55] Ward Cheney and David Kincaid. *Numerical Mathematics and Computing.* Brooks/ Cole, Cengange Learning, Stamford, CT, U.S.A., 7. (international) edition, 2013.

[56] Carmen Chicone. *Ordinary Differential Equations with Applications.* Springer-Verlag, New York, Berlin, Heidelberg, 1999.

[57] Anil K. Chopra. *Dynamics of Structures: Theory and Applications in Earthquake Engineering.* Prentice Hall, Upper Saddle River, New Jersey, 3 edition, 2006.

[58] A. J. Chorin. Numerical solution of the Navier-Stokes equations. *Math. Comp.*, 22:745–762, 1968.

[59] W. Chu, Z. Ghahramani, F. Falciani, and D. L. Wild. Biomarker discovery in microarray gene expression data with gaussian processes. *Bioinformatics*, 21(16):3385–3393, 2005.

[60] Igor Chueshov. *Monotone Random Systems – Theory and Applications.* Springer-Verlag, New York, Heidelberg, Berlin, 2002.

[61] K. L. Chung. *Elementary Probability Theory with Stochastic Processes.* Springer-Verlag, New York, Heidelberg, Berlin, 1979.

[62] A. B. Clarke and R. L. Disney. *Probability and Random Processes for Engineers and Scientists.* John Wiley & Sons, New York, London, Sydney, 1970.

[63] R. W. Clough and J. Penzien. *Dynamics of Structures.* McGraw-Hill, New York, 1975.

[64] E. A. Coddington and N. Levinson. *Theory of Ordinary Differential Equations.* McGraw-Hill, New York, 1955.

[65] J. W. Cooley and J. W. Tukey. An algorithm for the machine calculation of complex fourier series. *Mathematics of Computation,* 19:297–301, 1965.

[66] A. W. Coppel. *Stability and Asymptotic Behavior of Differential Equations.* Heath, Boston, 1965.

[67] R. Courant, K. O. Friedrichs, and H. Lewy. Über die partiellen Differenzengleichungen der mathematischen Physik. *Math. Ann.,* 100:32–74, 1928.

[68] R. Courant and D. Hilbert. *Methoden Der Mathematischen Physik I.* Number Bd. 1 in Heidelberger Taschenbücher: Mathematik. Springer-Verlag, 1968.

[69] J.-M. Courtault, Y. Kabanov, B. Bru, P. Crepel, I. Lebon, and A. le Marchand. Louis bachelier: On the centenary of theorie de la speculation. *Mathematical Finance,* 10:341–353, 2000.

[70] Harald Cramer and M. R. Leadbetter. *Stationary and Related Stochastic Processes: Sample Function Properties and Their Applications.* Dover Publications, New York, 2005.

[71] Noel A. C. Cressie. *Statistics for Spatial Data.* John Wiley & Sons, New York, revisited edition, 1993.

[72] L. Csato. *Gaussian Processes — Iterative Sparse Approximations.* Dissertation, Aston University, 2002.

[73] K. M. Cuddington and P. Yodzis. Black noise and population persistence. *Proceedings of the Royal Society, London B,* 266:969–973, 1999.

[74] R. Cuerno, A. F. Ranada, and J. J. Ruiz-Lorenzo. Deterministic chaos in the elastic pendulum: A simple laboratory for nonlinear dynamics. *American Journal of Physics,* 60(1):73–79, 1992.

[75] A. B. Cunningham, W. G. Characklis, F. Abedeen, and D. Crawford. Influence of biofilm aaccumulation on porous media hydrodynamics. *Environmental Science and Technology,* 25(7):1305–1311, 1991.

[76] E. Cuthill and J. McKee. Reducing the bandwidth of sparse symmetric matrices. In *Proceedings of the 1969 24th national conference,* ACM '69, pages 157–172, New York, NY, USA, 1969. ACM.

[77] S. Cyganowski, L. Grüne, and P.E. Kloeden. Maple for stochastic differential equations. In J.F. Blowey, J.P. Coleman, and A.W. Craig, editors, *Theory and Numerics of Differential Equations,* pages 127–178. Springer-Verlag, Berlin, Heidelberg, New York, 2001.

[78] S. Cyganowski and P. E. Kloeden. Stochastic stability examined through maple. In A. Sydow, editor, *Proc. 15th IMACS World Congress, Volume 1: Computational Mathematics*, pages 432–437. Wissenschaft & Technik Verlag, Berlin, 1997.

[79] S. Cyganowski, P. E. Kloeden, and J. Ombach. *From Elementary Probability to Stochastic Differential Equations with MAPLE.* Springer-Verlag, Berlin, Heidelberg, New York, 2001.

[80] H. Cyr and I. Cyr. Temporal scaling of temperature variability from land and oceans. *Evolutionary Ecology Research*, 5:1183–1197, 2003.

[81] J. J. DaCunha. Transition matrix and generalized matrix exponential via the peano-baker series. *Journal of Difference Equations and Applications*, 11(15):1245–1264, 2005.

[82] K. Deimling. *Ordinary Differential Equations in Banach Spaces.* Springer-Verlag, Berlin, Heidelberg, New York, 1977.

[83] J. P. den Hartog. *Mechanics.* Dover Publications, New York, 1961.

[84] P. Deuflhard and F. Bornemann. *Scientific Computing with Ordinary Differential Equations.* Texts in Applied Mathematics 42. Springer, 2002.

[85] J. Dieudonne. *Foundations Of Modern Analysis.* Academic Press, New York, London, 1969.

[86] J. Donea and A. Huerta. *Finite Element Methods for Flow Problems.* John Wiley & Sons Ltd, 2003.

[87] J. L. Doob. *Stochastic Processes.* John Wiley & Sons, New York, 1953.

[88] J. L. Doob. Wiener's work in probability theory. *Bulletin of the American Mathematical Society*, 72(1):69–72, 1966.

[89] H. Doss. Liens entre equations differentielles stochastiques et ordinaires. *Annales de l'Institut Henri Poincare (B)*, 13(2):99–125, 1977.

[90] V. Kaitala E. Ranta, P. Lundberg and J. Laakso. Visibility of the environmental noise modulating population dynamics. *Proceedings of the Royal Society, London B*, 263:1851–1856, 2000.

[91] M. Eichler. Spectral Analysis of Time Series, 2003. Lecture Notes of a course given at the University of Chicago.

[92] A. Einstein. über die von der molekularkinetischen theorie der wärme geforderte bewegung von in ruhenden flüssigkeiten suspendierten teilchen. *Annalen der Physik*, 17:549–560, 1905.

[93] J. Elstrodt. *Maß- und Integrationstheorie.* Springer-Verlag, Berlin, Heidelberg, New York, 2 edition, 1999.

[94] M. Emery. *Stochastic Calculus in Manifolds — with an Appendix by P.A. Meyer.* Springer-Verlag, Berlin, Heidelberg, New York, 1989.

[95] Y. Engel, P. Szabo, and D. Volkinshtein. Learning to control an octopus arm with gaussian process temporal difference methods. In Y. Weiss, B. Schölkopf, and J. C. Platt, editors, *Advances in Neural Information Processing Systems*, pages 347–354. The MIT Press, Cambridge, 18 edition, 2006.

[96] J. A. Evans. *Divergence-free B-spline Discretizations for Viscous Flows.* PhD thesis, The University of Texas at Austin, 2011.

[97] L. C. Evans. An Introduction to Stochastic Differential Equations, 2002. Lecture Notes (version 1.2), UC Berkeley.

[98] R. Eymard, T. Gallouët, and R. Herbin. Finite volume methods. *Handbook of Numerical Analysis*, 7:713–1018, 2000.

[99] H. Föllmer. On kiyosi itô's work and its impact — gauss lecture at the icm 2006. *Proceedings of the International Congress of Mathematicians, Madrid 2006*, 1:109–124, 2006.

[100] A. Friedman. *Stochastic Differential Equations and Applications.* Dover Publications, Mineola, 2006.

[101] M. Frigo and S. G. Johnson. The design and implementation of FFTW3. *Proceedings of the IEEE*, 93(2):216–231, 2005. Special issue on "Program Generation, Optimization, and Platform Adaptation".

[102] F. R. Gantmacher. *The Theory of Matrices – Volume 2.* Chelsea Publishing Company, New York, 1974.

[103] C. Gardiner. *Stochastic Methods: A Handbook for the Natural and Social Sciences.* Springer-Verlag, Berlin, 4. edition, 2009.

[104] M. Garrido, J. Grajal, and O. Gustafsson. Optimum circuits for bit reversal. *IEEE Trans. on Circuits and Systems*, 58-II(10):657–661, 2011.

[105] C. Gasquet and P. Witomski. *Fourier Analysis and Applications: Filtering, Numerical Computation, Wavelets.* Springer-Verlag, Berlin, 1999.

[106] Peter Giesl. Nonlinear Dynamics, 2002. lecture notes at the TU München, winter term 2001 / 2002.

[107] Peter Giesl. Einzugsbereiche dynamischer Systeme und ihre algorithmische Bestimmung, 2004. lecture notes at the TU München, winter term 2003 / 2004.

[108] Peter Giesl. *Construction of Global Lyapunov Functions Using Radial Basis Functions.* Springer Verlag, Berlin, Heidelberg, 2007.

[109] I. I. Gikhman and A. V. Skorokhod. *The Theory of Stochastic Processes I.* Springer-Verlag, Berlin, Heidelberg, New York, 2004.

[110] D.T. Gillespie. Exact numerical simulation of the ornstein-uhlenbeck process and its integral. *Physical Review E*, 54(2):2084–2091, 1996.

[111] M. Gitterman. *The Noisy Oscillator – The First Hundred Years, From Einstein until Now.* World Scientific, New Jersey, London, Singapore, 2005.

[112] M. Gitterman. *The Noisy Pendulum.* World Scientific, New Jersey, London, Singapore, 2008.

[113] P. W. Goldberg, C. K. I. Williams, and C. M. Bishop. Regression with input-dependent noise: A gaussian process treatment. In M. I. Jordan, M. J. Kearns, and S. A. Solla, editors, *Advances in Neural Information Processing Systems 10*, pages 493–499. The MIT Press, Cambridge, 1998.

[114] G.H. Golub and J.M. Ortega. *Scientific Computing and Differential Equations.* Academic Press, 1992.

[115] Günter Gramlich. *Anwendungen in der Linearen Algebra mit MATLAB.* Fachbuchverlag Leipzig im Carl Hanser Verlag, München, Wien, 2004.

[116] W. Greiner. *Mechanik – Teil 1.* Verlag Harri Deutsch, Thun, Frankfurt am Main, 5. edition, 1989.

[117] P. M. Gresho and R. L. Sani. Resume and remarks on the open boundary condition minisymposium. *Int. J. Num. Meth. Fluids,* 18:983–1008, 1994.

[118] P. M. Gresho and R. L. Sani. *Incompressible Flow and the Finite Element Method.* John Wiley & Sons, 1998.

[119] M. Griebel, Th. Dornseifer, and T. Neunhoeffer. *Numerical Simulation in Fluid Dynamics, a Practical Introduction.* SIAM, 1998.

[120] L. Grüne and P. E. Kloeden. Pathwise approximation of random ordinary differential equations. *BIT Numerical Mathematics,* 41(4):711–721, 2001.

[121] F. Günther, M. Mehl, M. Pögl, and C. Zenger. A cache-aware algorithm for PDEs on hierarchical data structures based on space-filling curves. *SIAM Journal on Scientific Computing,* 28(5):1634–1650, 2006.

[122] Allan Gut. *Probability: A Graduate Course.* Springer, New York, 2005.

[123] E. Hairer, C. Lubich, and G. Wanner. *Geometric Numerical Integration.* Springer-Verlag, Berlin, 2002.

[124] E. Hairer, S. P. Nørset, and G. Wanner. *Solving Ordinary Differential Equations I.Nonstiff Problems.,* volume 8 of *Springer Series in Comput. Mathematics.* Springer-Verlag, second edition, 1993.

[125] E. Hairer, S. P. Nørset, and G. Wanner. *Solving Ordinary Differential Equations II.Stiff and Differential-Algebraic Problems.,* volume 14 of *Springer Series in Comput. Mathematics.* Springer-Verlag, second edition, 1996.

[126] A. Hald. T. n. thiele's contributions to statistics. *International Statistic Review,* 49:1–20, 1981.

[127] J. M. Halley and P. Inchausti. The increasing importance of 1/f-noises as models of ecological variability. *Fluctuation and Noise Letters,* 4(2):R1–R26, 2004.

[128] Harro Heuser. *Gewöhnliche Differentialgleichungen: Einführung in Lehre und Gebrauch.* Vieweg+Teubner Verlag, Wiesbaden, 6 edition, 2009.

[129] Desmond J. Higham. An algorithmic introduction to numerical simulation of stochastic differential equations. *SIAM Review,* 43(3):525–546, 2001.

[130] A. Hindy and M. Novak. Pipeline response to random ground motion. *Journal of the Engineering Mechanics Division, ASCE,* 106(EM 2):339–360, 1980.

[131] Frank C. Hoppensteadt. *Analysis and Simulation of Chaotic Systems.* Springer-Verlag, Berlin, Heidelberg, New York, 1993.

[132] G. W. Housner. The behavior of inverted pendulum structures during earthquakes. *Bulletin of the Seismological Society of America,* 53(2):403–417, 1963.

[133] G. W. Housner and P. C. Jenning. Generation of artificial earthquakes. *Journal of the Engineering Mechanical Division, ASCE*, 90(EM 1):113–150, 1964.

[134] P. Howard. Solving ODE in MATLAB, 2007. Lecture Notes, Texas A & M University, Fall 2007.

[135] T. Huckle and S. Schneider. *Numerik für Informatiker.* Springer, Berlin, September 2002.

[136] P. Imkeller and B. Schmalfuss. The conjugacy of stochastic and random differential equations and the existence of global attractors. *Journal of Dynamics and Differential Equations*, 13(2):215–249, 2001.

[137] E. F. Infante. On the stability of some linear nonautonomous random systems. *Journal of Applied Mechanics*, 35(1):7–12, 1968.

[138] L. Isserlis. On a formula for the product-moment coefficient of any order of a normal frequency distribution in any number of variables. *Biometrika*, 12(1/2):134–139, 1918.

[139] K. Itô. Stochastic integral. *Proceedings of the Imperial Academy, Tokyo*, 20:519–524, 1944.

[140] K. Itô. Multiple wiener integral. *Journal of the Mathematical Society of Japan*, 3(1):157–169, 1951.

[141] K. Itô. Complex multiple wiener integral. *Japanese Journal of Mathematics*, 22:63–86, 1952.

[142] Wilfred D. Iwan and Leonidas G. Paparizos. The stochastic response of strongly yielding systems. *Probabilistic Engineering Mechanics*, 3(2):75–82, 1988.

[143] R. N. Iyengar and C. S. Manohar. Nonstationary random critical seismic excitations. *Journal of Engineering Mechanics*, 113(4):529–541, 1987.

[144] J. L. Bogdanoff J. E. Goldberg and D. R. Sharpe. The response of simple nonlinear systems to random disturbance of the earthquake type. *Bulletin of the Seismic Society America*, 54:263–274, 1964.

[145] P. Ludberg J. Ripa and V. Kaitala. A general theory of environmental noise in ecological food webs. *American Naturalist*, 151:256–263, 1998.

[146] Richard C. Jaffe. *Random Signals for Engineers Using MATLAB and Mathcad.* Springer, Berlin, Heidelberg, New York, 1999.

[147] Robert Jarrow and Philip Protter. A short history of stochastic integration and mathematical finance the early years, 1880-1970. *IMS Lecture Notes Monograph*, 45:1–17, 2004.

[148] A. Jentzen. Numerische Verfahren hoher Ordnung für zufällige Differentialgleichungen, 2007. Diploma Thesis, J.W. Goethe Universität, Frankfurt am Main.

[149] A. Jentzen and P.E. Kloeden. *Taylor Approximations for Stochastic Partial Differential Equations.* SIAM Press, Philadelphia, 2011.

[150] R. Kan. From moments of sum to moments of product. *Journal of Multivariate Analysis*, 99:542–554, 2008.

[151] K. Kanai. Semi-empirical formula for the characteristics of the ground. *Bulletin of the Earthquake Research Institute*, 35, 1957.

[152] K. Kanai. An empirical formula for the spectrum of strong earthquake motions. *Bulletin of the Earthquake Research Institute*, 39, 1961.

[153] Steven Kay. *Intuitive Probability and Random Processes Using MATLAB*. Springer, Berlin, Heidelberg, New York, 2006.

[154] D. G. Kendall. Stochastic processes and population growth. *Journal of the Royal Statistic Society, Series B*, 11:230–264, 1948.

[155] R. Z. Khas'minskii. *Stochastic stability of differential equations*. Sijthoff and Noordhoff, Allphen aan den Rijn (NL), Rockville (MD, USA), 1980. Translation of the Russian edition, Nauka, Moscow, 1969.

[156] R. Z. Khas'minskii. *Stochastic stability of differential equations*. Springer, Berlin, Heidelberg, New York, 2 edition, 1980.

[157] A. Kistner. Strenge Aussagen Über Lösung, Momente und Stabilität Linearer Systeme mit Parameteranregung Durch Farbiges Rauschen, 1978. PhD-thesis, Universität Stuttgart.

[158] P. E. Kloeden and A. Jentzen. Pathwise convergent higher order numerical schemes for random ordinary differential equations. *Proc. R. Soc. A*, 463:2929–2944, 2007.

[159] Peter E. Kloeden and Eckhard Platen. *Numerical Solution of Stochastic Differential Equations*. Springer-Verlag, Berlin, Heidelberg, New York, corrected 3. edition, 1999.

[160] P. Knabner and L. Angerman. *Numerical Methods for Elliptic and Parabolic Partial Differential Equations*. Texts in Applied Mathematics. Springer, 2003.

[161] P. Knabner and L. Angermann. *Numerik partieller Differentialgleichungen*. Springer-Verlag, 2000.

[162] K. Königsberger. *Analysis 2*. Springer-Verlag, Berlin, Heidelberg, New York, 2. edition, 1997.

[163] T. W. Körner. *Fourier Analysis*. Cambridge University Press, Cambridge, 1989.

[164] T. W. Körner. *Exercises for Fourier Analysis*. Cambridge University Press, Cambridge, 1993.

[165] F. Kozin. On almost sure stability of linear systems with random coefficients. *Journal of Mathematics and Physics*, 42:59–66, 1963.

[166] F. Kozin. On realtions between moment properties and almost sure lyapunov stability for linear stochastic systems. *Journal of Mathematical Analysis and Applications*, 10:342–353, 1965.

[167] Anil G. Ladde and Gangaram S. Ladde. *An Introduction to Differential Equations: Deterministic Modeling, Methods and Analysis, Volume 1*. World Scientific, New Jersey, London, Singapore, 2012.

[168] L. D. Landau and E. M. Lifschitz. *Lehrbuch der Theoretischen Physik – Hydrodynamik*, volume VI. Akademie, 1991.

[169] J. H. Lawton. More time means more variation. *Nature*, page 563, 1988.

[170] Joan-Andreu Lazaro-Cami and Juan-Pablo Ortega. Stochastic hamiltonian dynamical systems, 2007. Preprint: arXiv:math/0702787v3.

[171] R. LeVeque. *Finite Difference Methods for Ordinary and Partial Differential Equations: Steady-State and Time-Dependent Problems (Classics in Applied Mathematics Classics in Applied Mathemat)*. SIAM, Philadelphia, PA, USA, 2007.

[172] R. J. LeVeque. *Finite Volume Methods for Hyperbolic Problems*. Cambridge Texts in Applied Mathematics. Cambridge University Press, 2002.

[173] M. Levi. Stability of the inverted pendulum – a topological explanation. *Siam Review*, 30(4):639–644, 1988.

[174] M. Y. Li and L. Wang. A criterion for stability of matrices. *journal of Mathematical Analysis and Applications*, 225:249–264, 1998.

[175] Y. K. Lin and G. Q. Cai. *Probabilistic Structural Dynamics*. McGraw-Hill, New York, Chicago, San Francisco, 2004.

[176] J. Lindenstrauss and L. TZafriri. *Classical Banach Spaces*. Springer-Verlag, Berlin, Heidelberg, New York, 1973.

[177] G. R. Lindfeld and J. E. T. Penny. *Numerical Methods using Matlab*. Academic Press, Amsterdam, Boston, Heidelberg, 3. edition, 2012.

[178] M. A. Little, P. E. McSharry, S. J. Roberts, D. A. E. Costello, and I. M. Moroz. Exploiting nonlinear recurrence and fractal scaling properties for voice disorder detection. *BioMedical Engineering OnLine*, 6(1):23, 2007.

[179] Michael Loeve. *Probability Theory I*. Springer-Verlag, Berlin, Heidelberg, New York, 4. edition, 1977.

[180] Michael Loeve. *Probability Theory II*. Springer-Verlag, Berlin, Heidelberg, New York, 4. edition, 1977.

[181] J. G. Malkin. *Theorie der Stabilität einer Bewegung*. Akademie Verlag, Berlin, 1959.

[182] B. B. Mandelbrot. *The Fractal Geometry of Nature*. W. H. Freedman, San Francisco, 1983.

[183] H. J. Marquez. *Nonlinear Control Systems: Analysis & Design*. John Wiley & Sons, New York, 2003.

[184] S. A. Martucci. Symmetric convolution and the discrete sine and cosine transforms. *IEEE Transactions on Signal Processing*, 42(5):1038–1051, 1994.

[185] Gustavo J. Mata and E. Pestana. Effective hamiltonian and dynamic stability of the inverted pendulum. *European Journal of Physics*, 25(6):717–721, 2004.

[186] E. J. McShane. Toward a stochastic calculus, i. *Proceedings of the National Academy of Sciences*, 63(2):275–280, 1969.

[187] E. J. McShane. Toward a stochastic calculus, ii. *Proceedings of the National Academy of Sciences*, 63(4):1084–1087, 1969.

[188] Kurt Meyberg and Peter Vachenauer. *Höhere Mathematik 2*. Springer-Verlag, New York, Berlin, Heidelberg, 4 edition, 2003.

[189] Kenneth Ray Meyer and Glen R. Hall. *Introduction to Hamiltonian Dynamical Systems and the N-Body Problem*. Springer-Verlag, New York, Berlin, Heidelberg, 1992.

[190] E. Milotti. 1/f Noise: A Pedagogical Review, 2002. arXiv:physics/0204033v1 [physics.class-ph].

[191] Jacques Monod. The growth of bacterial cultures. *Annual Review of Microbiology*, 3:371–394, 1949.

[192] T. Morozan. Remarques sur un travail de h. bunke. *C. R. Acad. Sci., Paris Ser. A-B*, 261:2049–2052, 1965.

[193] Eva-Marie Müller-Stüler. *Mathematische Analyse des Dopplersignals zur quantitativen Bestimmung des Blutflusses*. Shaker, Berlin, 2012.

[194] James Murdock. *Normal Forms and Unfoldings for Local Dynamical Systems*. Springer-Verlag, Berlin, Heidelberg, New York, 2003.

[195] R. Murga, T. S. Forster, E. Brown, J. M. Pruckler, B. S. Fields, and R. M. Donlan. Role of biofilms in the survival of legionella pneumophila in a model potable-water system. *Microbiology*, 147(11):3121–3126, 2001.

[196] N. H. Naqvi and M. D. Blaufox. *Blood Preasure Measurement – An Illustrated History*. Parthenon Publishing, New York, 1998.

[197] T. Neckel. *The PDE Framework Peano: An Environment for Efficient Flow Simulations*. Verlag Dr. Hut, 2009.

[198] T. Neckel, M. Mehl, and C. Zenger. Enhanced divergence-free elements for efficient incompressible flow simulations in the PDE framework Peano. In *Proceedings of the Fifth European Conference on Computational Fluid Dynamics, ECCOMAS CFD 2010, 14th-17th June 2010*, Lissabon, 2010.

[199] Tobias Neckel, Alfredo Parra, and Florian Rupp. Path-wise algorithms for random & stochastic odes with efficient quadrature rules for multiple ou- & wiener-integrals. 2013.

[200] E. Nelson. *Dynamical Theories of Brownian Motion*. Princeton University Press, Princeton, 1967.

[201] N. M. Newmark. A method of computation for structural dynamics. *Journal of the Engineering Mechanics Division, ASCE*, 85:67–94, 1959.

[202] D. Nikolic-Despotovic. One type of random diffusion equation. *Review of Research Faculty of Science, University of Novi Sad*, 9:19–22, 1979.

[203] Alan V. Oppenheim and George C. Verghese. Signals, Systems, and Inference, 2010. Class Notes for 6.011: Introduction to Communication, Control and Signal Processing Spring 2010, at the Massachusetts Institute of Technology.

[204] F. Otto. The geometry of dissipative evolution equations: The porous medium equation. *Communications in Partial Differential Equations*, 26(1-2):101–174, 2001. DOI: 10.1081/PDE-100002243.

[205] W. J. Padgett, G. Schultz, and C. P. Tsoks. A random differential equation approach to the probability distribution of bod and do in streams. *SIAM Journal of Applied Mathematics*, 32(2):467–483, 1977.

[206] Antonis Papachristodoulo and Stephen Prajna. On the construction of Lyapunov functions using the sum of squares decomposition. *Proceedings of the 41st IEEE Conference on Decision and Control, Las Vegas, Nevada USA*, pages 3472–3487, December 2002.

[207] Leonidas G. Paparizos. Some Observations on the Random Response of Hysteretic Systems, 1986. Earthquake Engineering Research Laboratory, Report No. EERL 86-02, Pasadena, California.

[208] Pablo A. Parrilo. Structured Semidefinite Programs and Semialgebraic Geometry Methods in Robustness and Optimization, 2000. California Institute of Technology, Pasadena, California USA.

[209] F. Pfeiffer. *Einführung in die Dynamik*. Teubner Verlag, Stuttgart, 1989.

[210] Umberto Picchini. SDE Toolbox User's Guide for Version 1.4.1 — Simulation and Estimation of Stochastic Differential Equations with MATLAB, 2007. http://sdetoolbox.sourceforge.net/.

[211] Mark A. Pinsky. Stochastic stability and the dirichlet problem. *Communications on Pure and Applied Mathematics*, 27(3):311–350, May 1974.

[212] Philip E. Protter. *Stochastic Integration and Differential Equations*. Springer-Verlag, Berlin, Heidelberg, New York, 2 edition, 2005.

[213] Philip E. Protter. The work of kyosi itô. *Notices of the AMS*, 54(6):744–745, 2007.

[214] Carl Edward Rasmussen. *Evaluation of Gaussian Processes and other Methods for Non-linear Regression*. Dissertation, University of Toronto, 1996.

[215] Carl Edward Rasmussen and Chris Williams. *Gaussian Processes for Machine Learning*. MIT Press, Cambridge, 2006.

[216] C. Reinsch. Numerische Mathematik 1 und 2, 1995. Mathematisches Institut der TU München.

[217] Sanaz Rezaeian and Armen der Kiureghian. Stochastic Modeling and Simulation of Ground Motions for Performance-Based Earthquake Engineering, 2010. PEER Report 2010/02 Pacific Earthquake Engineering Research Center, College of Engineering, University of California, Berkely.

[218] Dennis Ward Ricker. *Echo Signal Processing*. Springer-Verlag, Berlin, Heidelberg, New York, 2003.

[219] J. Ripa and P. Ludberg. Noise color and the risk of population extinction. *Proceedings of the Royal Society of London, Series B*, 263:1751–1753, 1996.

[220] Joseph P. Romano and Andrew F. Siegel. *Counterexamples in Probability and Statistics.* Wadsworth & Brooks/ Cole Advanced Books & Software, Monterey (California), 1986.

[221] F. Rupp. Waves, spectra and sounds: Fourier analysis. *Homo Oeconomicus,* 25(3/4):1–10, 2008.

[222] F. Rupp. Stochastic Dynamics, 2011. Extended Lecture Notes, Summer Term 2011 (Felix-Klein Prize), Technische Universität München.

[223] H. Sagan. *Space-filling curves.* Springer-Verlag, New York, 1994.

[224] Peter Sagirow. *Stochastic Methods in the Dynamics of Satellites.* Courses and Lectures (No. 57), Springer-Verlag, Undine, 1970.

[225] R. Sainudiin and D. Lee. Computational Statistical Experiments in MATLAB, 2011. Lecture Notes, Department of Mathematics and Statistics, University of Canterbury, Christchurch, New Zealand.

[226] J. Salençon. *Mécanique des milieux continus.* Editions de l'Ecole Polytechnique, 2002.

[227] Zeev Schuss. *Theory and Applications of Stochastic Processes.* Springer-Verlag, Berlin, Heidelberg, New York, 2010.

[228] C. U. Schwermer, G. Lavik, R. Abed, B. Dunsmore, T. G. Ferdelman, P. Stoodley, A. Gieseke, and D. de Beer. Impact of nitrate on the structure and function of bacterial biofilm communities in pipelines used for injection of seawater into oil fields. *Applied and Environmental Microbiology,* 74(9):2841–2851, 2008.

[229] R. Seydel. *Tools for Computational Finance.* Springer-Verlag, New York, 5th edition, 2012.

[230] Young-Kyun Shin, Mohammad A. Fard, Hikaru Inooka, and Il Hwan Kim. Human postural dynamics in response to the horizontal vibration. *International Journal of Control, Automation, and Systems,* 4(3):325–332, 2006.

[231] G. Simmons. *Differential Equations with Applications and Historical Notes.* McGrow-Hill Book Company, New York, San Francisco, St. Louis, 1972.

[232] E. Snelson. *Flexible and efficient Gaussian process models for machine learning.* Dissertation, University College London, 2007.

[233] T. T. Soong. *Random Differential Equations in Science and Engineering.* Academic Press, New York, 1973.

[234] P. J. C. Spreij. Stochastic Integration, 2008. Lecture Notes, Korteweg-de Vries Institute for Mathematics, Universiteit van Amsterdam.

[235] Dan Stanescu and Benito Chen-Charpentier. Random coefficient differential equation models for monod kinetics. *Discrete and Continuous Dynamical Systems,* Supplement 2009:719–728, 2009.

[236] J. H. Steele. A comparison of terrestrial and marine ecologcal systems. *Nature,* 313:355–358, 1985.

[237] J. H. Steele and E. W. Henderson. Modeling long-term fluctuations in fish stocks. *Science*, 224:985–987, 1984.

[238] Michael L. Stein. *Statistical Interpolation of Spatial Data: Some Theory for Kriging*. Springer-Verlag, Berlin, Heidelberg, 1999.

[239] J. Stoer and R. Bulirsch. *Numerische Mathematik 2*. Springer-Verlag, 4th edition, 2000.

[240] R. L. Stratonovich. A new representation for stochastic integrals and equations. *SIAM Journal on Control*, 4(2):362–371, 1966.

[241] Steven H. Strogatz. Love afffairs and differential equations. *Mathematics Magazine*, 61(35):35, 1988.

[242] Steven H. Strogatz. *Nonlinear Dynamics and Chaos*. Addison-Wesely Publishing Company, Reading, Melon Parl, NEw York, 1994.

[243] D. W. Strook and S. R. S. Varadhan. *Multidimensional Diffusion Processes*. Springer-Verlag, Berlin, Heidelberg, New York, reprint of the 1997 edition, 2006.

[244] A. M. Stuart and A. R. Humphries. *Dynamical Systems and Numerical Analysis*. Cambridge University Press, Cambridge, 1998.

[245] H. J. Sussmann. On the gap between dterministic and stochastic differential equations. *Annals of Probability*, 6(1):19–41, 1978.

[246] W. Szemplinska-Stupnicka and E. Tyrkiel. The oscillation rotation attractors in the forced pendulum and their peculiar properties. *International Journal of Bifurcation and Chaos*, 12(1):159–168, 2002.

[247] S. R. Taghizadeh. Digital Signal Processing: Part 3 – Discrete-Time Signals & Systems Case Studies, 2000. Lecture Notes at the University of North London, School of Communications Technology and Mathematical Sciences.

[248] H. Tajimi. A Statistical Method of Determining the Maximum Response of a Building During an Earthquake, 1960. Proceedings of the Second World Conference on Earthquake Engineering, Tokyo and Kyoto, Japan, Vol. II.

[249] R. Témam. Sur l'approximation de la solution des équations de Navier-Stokes par la méthode des pas fractionnaires (I). *Archive for Rational Mechanics and Analysis*, 32(2):135–153, 1969.

[250] T. N. Thiele. Sur la compensation de quelques erreurs quasisystematiques par la methode des moindres carres, 1880. Copenhagen, C. A. Reitzel.

[251] S. Turek and M. Schäfer. Benchmark computations of laminar flow around a cylinder. In E. H. Hirschel, editor, *Flow Simulation with High-Performance Computers II*, number 52 in NNFM. Vieweg, 1996.

[252] David A. Vasseur and Peter Yodzis. The color of environmental noise. *Ecology*, 85(4):1146–1152, April 2004.

[253] Ferdinand Verhulst. *Nonlinear Differential Equations and Dynamical Systems*. Springer-Verlag, Berlin, Heidelberg, New York, 1980.

[254] T. Weinzierl. *A Framework for Parallel PDE Solvers on Multiscale Adaptive Cartesian Grids.* Verlag Dr. Hut, 2009.

[255] Y. K. Wen. Inelastic structures under nonstationary random excitation. In W. Kliemann and N. S. Namachchivaya, editors, *Nonlinear Dynamics and Stochastic Mechanics*, CRC Mathematical Modelling Series, pages 411–435. CRC Press, Boca Raton, Florida, 1995.

[256] Dirk Werner. *Funktionalanalysis.* Springer, Berlin, Heidelberg, New York, 3. edition, 2000.

[257] Edmund Taylor Whittaker. *A Treatise on the Analytical Dynamics of Particles and Rigid Bodies.* Cambridge University Press, Cambridge, 4. edition, 1947.

[258] N. Wiener. Differential space. *Journal of Mathematical Physics*, 2:131–174, 1923.

[259] David Williams. *Probability with Martingales.* Cambridge University Press, Cambridge, 2000.

[260] Eugene Wong and Moshe Zakai. Riemann-stieltjes approximations of stochastic integrals. *Zeitschrift für Wahrscheinlichkeitstheorie und verwandte Gebiete*, 12:87–97, 1969.

[261] R. Wüchner. *Mechanik und Numerik der Formfindung und Fluid-Struktur-Interaktion von Membrantragwerken.* PhD thesis, Technische Universität München, 2006.

[262] Kedai Xu. Bifurcations of random differential equations in dimension one. *Random & Computational Dynamics*, 1:277–305, 1993.

[263] A. M. Yaglom. *An Introduction to the Theory of Stationary Random Functions.* Dover Publications, Mineola, 2004.

[264] C. H. Yeh and Y. K. Wen. Modeling of Nonstationary Earthquake Ground Motion and Biaxial and Torsional Response of Inelastic Structures, 1989. Civil Engineering Studies, Structural Research Series, No. 546, University of Illinois.

[265] O. C. Zienkiewicz and R. L. Taylor. *The Finite Element Method for Solid and Structural Mechanics.* Elsevier Butterworth Heinemann, 6th edition, 2005.

Bei Fragen zur Produktsicherheit wenden Sie sich bitte an:
If you have any questions regarding product safety,
please contact:

Walter de Gruyter GmbH
Genthiner Straße 13
10785 Berlin
productsafety@degruyterbrill.com